Ecological Intensification of Natural Resources for Sustainable Agriculture

Manoj Kumar Jhariya •
Ram Swaroop Meena • Arnab Banerjee
Editors

Ecological Intensification of Natural Resources for Sustainable Agriculture

Springer

Editors
Manoj Kumar Jhariya
Department of Farm Forestry
Sant Gahira Guru Vishwavidyalaya
Ambikapur, Chhattisgarh, India

Ram Swaroop Meena
Department of Agronomy
Banaras Hindu University
Varanasi, Uttar Pradesh, India

Arnab Banerjee
Department of Environmental Science
Sant Gahira Guru Vishwavidyalaya
Ambikapur, Chhattisgarh, India

ISBN 978-981-33-4202-6 ISBN 978-981-33-4203-3 (eBook)
https://doi.org/10.1007/978-981-33-4203-3

This Springer imprint is published by the registered company Springer Nature Singapore Pte Ltd.
The registered company address is: 152 Beach Road, #21-01/04 Gateway East, Singapore 189721, Singapore

Preface

The population explosion has taken place at an unprecedented rate which is expected to reach more than 9 billion by 2050. Thus, it was observed that 70% higher production in the agricultural sector is required in 2050 when compared to the last two decades (FAO 2018). These indicate a higher level of agricultural intensification is required through ecological intensification. It is also questionable whether the earth's carrying capacity would sustain such an unprecedented rate of intensification which is totally unsustainable. Under this context, the concept of eco-intensification is the need of the hour which aims to reduce the pressure on earth resources along with maintaining the balance and harmony in production sectors.

Ecological intensification comprises genetic intensification and socio-economic intensification to give an all-round eco-friendly development. Policies under ecological intensification should be synergistic in the approach to keep the balance between the production sector and consumer sector. The development of new farming systems of intensive to semi-intensive in nature may promote natural resources conservation. Ecological intensification is such an issue which has not been explored properly till date. It encompasses better food production at a low environmental cost, broader perspectives in environmental conservation, and maintaining the integrity of the earth ecosystem. Under these circumstances, new research and development need to be done to exploit the possibility and opportunity for sustainable eco-intensification, hence the target to develop new principles and management policies towards sustainable development. Ecological intensification tends to improve the productivity of various production systems as well as reduce the ecological footprint. It also helps to conserve the diverse ecosystem services such as maintaining soil quality, inhibition of soil degradation, reducing GHGs emission, establishing proper source–sink relationship of carbon to maintain carbon balance, soil and water conservation, maintaining bio-resource, ecosystem resistance and resilience to autochthonous and allochthonous changes along with overall sustainability of the ecosystem.

The present book discussed the critical issue of ecological intensification to fulfill the current demand for food as well as address the issue of sustainability in relation to natural resources and sustainable agriculture. Natural resource is the central point of all social, economic, and environmental development. Therefore, proper management requires proper priority. The present title is an attempt to understand the concept of ecological intensification, its role towards natural resource management and its approach towards sustainability of the agroecosystem. In the introduction, various aspects of ecological intensification have been clarified for resource management and sustainable productive perspective. Further, specific issues such as food security, biodiversity conservation, climate change, sustainable agriculture, soil contaminant, eco-modeling, eco-designing, and animal breeding in relation to ecological intensification were addressed. The book also covered some allied aspects of mulching, vertical greenhouses, pollination, ecosystem services, and soil carbon stock and sequestration in a holistic way to provide a pathway of sustainable agricultural practices for the learned society of the globe. The book concluded the proper management strategies with various issues related to natural resource, environment, ecology, sustainable agriculture, and allied fields with new updated knowledge that would enrich and create a platform of discussion on ecological intensification at the global level.

From a global perspective, multidisciplinary approach is required to address the issue of sustainability and conservation. It includes wide disciplines such as forestry, agriculture, environmental science, and ecology. Reference textbook and separate edited volumes are not available addressing specific issues of "Ecological Intensification of Natural Resources for Sustainable Agriculture." However, most of the books are focused on natural resources and their conservation. The integration of the concept of ecological intensification with natural resource is the biggest challenge of twenty-first century. It is also a limiting factor in terms of knowledge for academicians, scientists, research scholars, and policymakers of the present time. This edited book would act as a basic to update knowledge base for the scientists and academicians for the future goal. The objectives of this book are: (1) to address the issue of ecological intensification for natural resources, (2) to generate awareness and proper understanding of the concept and its associated issues and challenges, and (3) to educate the learned society about the recent trend and development to formulate strategies for future research and development.

The present attempt is for the national and international audience to clearly understand the concept of ecological intensification and its applicability in the field of natural resource management and sustainable agriculture. Highly professional and internationally renowned researchers are invited to contribute, authoritative and cutting-edge scientific information on a broad range of topics covering agroecology, environment, ecological footprints and sustainability. All the chapters are well illustrated with appropriately placed data, tables, figures, and photographs and supported with extensive and most recent references. The submitted chapters are reviewed by the members of the Editorial Committee in the relevant field for further improvement and authentication of the information provided. The editors also provided a roadmap for ecological intensification for natural resources aiming towards sustainable agricultural development.

Ambikapur, India — Manoj Kumar Jhariya
Varanasi, India — Ram Swaroop Meena
Ambikapur, India — Arnab Banerjee

Contents

About the Editors

Manoj Kumar Jhariya is Assistant Professor at the Department of Farm Forestry, Sant Gahira Guru Vishwavidyalaya (formerly, Sarguja University), Ambikapur (Chhattisgarh), India, and the author or co-author of more than 68 scientific papers, 9 books, 33 book chapters, and several extension articles. He acquired his B.Sc. (Agriculture) and M.Sc. and Ph.D. (both in Forestry) degrees from Indira Gandhi Krishi Vishwavidyalaya, Raipur, Chhattisgarh, India. He was awarded a UGC-RGNF Fellowship, New Delhi, India, and received a Chhattisgarh Young Scientist Award from Chhattisgarh Council of Science and Technology in 2013. He is an editorial board member for several journals and a life member of numerous societies.

Ram Swaroop Meena is working as an Assistant Professor in the Department of Agronomy, I.Ag. Scs., BHU, Varanasi (UP). He has been awarded Raman Research Fellowship by the MHRD, GOI. He has completed his postdoctoral research on soil carbon sequestration under Padma Shri Prof. Rattan Lal World Food Prize 2020, Laureate 2020, Director, CMASC, Columbus, USA. He has supervised 23 PG and 6 PhD students and has 10 years of research and teaching experience. He is working on three externally funded projects (DST, MHRD, and ICAR) with one patent. He has published more than 110 research and review papers and has an H-index ~41, as well 4 published books at the national and 14 books at the international levels, and contributed to the books with 15 chapters at national and 50 at the international levels. He has worked as an expert for the NCERT, MHRD, and GOI. He has contributed to several agricultural extension activities, trainings, meetings, workshops, etc.

Arnab Banerjee is an Assistant Professor, Department of Environmental Science, Sant Gahira Guru Vishwavidyalaya, Ambikapur, Chhattisgarh, India. He completed his M.Sc. and Ph.D. (Environmental Science) from Burdwan University and M.Phil. in Environmental Science from Kalyani University, West Bengal. He won the University Gold Medal for securing first class first position in M.Sc. examination. He has been awarded Young Scientist Award for the best oral presentation at the international conference held at the University of Burdwan. He was a project fellow under UGC-sponsored major research project. To his credit, he has published 67 research papers in reputed national and international journals, 8 books, and 27 book chapters. He is a life member of the Academy of Environmental Biology. He has supervised 23 postgraduate students and engaged in postgraduate teaching and research.

Ecological Intensification of Natural Resources Towards Sustainable Productive System

1

Manoj Kumar Jhariya, Ram Swaroop Meena, and Arnab Banerjee

Abstract

As per the estimates of FAO (Food and Agriculture Organization) world's population would be requiring 60% more food in comparison to present times till 2050. The situation is worser due to limitation in terms of availability of arable lands. In this context, intensification in the agricultural sector is the basic requirement for both developed and developing nations. Intensification towards sustainability is an important aspect for sustainable utilization of resource and its management. Policy formulation, strategies and technological growth should take place at the scientific level and executed at the farmers level in order to reduce inputs and maximize the yield and productivity. This would also help in maintaining agro-biodiversity along with ecosystem services followed by livelihood sustenance. Therefore, innovation in the field of agroecology through incentive-based practices may give fruitful results. Ecological intensification (EI) has an integrated approach by improving production along with maintenance of environmental quality. EI addresses various issues such as food security as well as technological intervention in the form of organic farming, conservation agriculture, climate smart practices, etc. Above all it addresses the issues of environmental sustainability through proper strategy formulations, good

M. K. Jhariya (✉)
Department of Farm Forestry, Sant Gahira Guru Vishwavidyalaya, Sarguja, Ambikapur, Chhattisgarh, India

R. S. Meena
Department of Agronomy, Institute of Agricultural Sciences, Banaras Hindu University, Varanasi, Uttar Pradesh, India

A. Banerjee
Department of Environmental Science, Sant Gahira Guru Vishwavidyalaya, Sarguja, Ambikapur, Chhattisgarh, India

M. K. Jhariya et al. (eds.), *Ecological Intensification of Natural Resources for Sustainable Agriculture*, https://doi.org/10.1007/978-981-33-4203-3_1

governance and generation of awareness for adoption of EI for economic and ecological gain.

Keywords

Agriculture · Climate change · Ecological intensification · Environment · Forestry · Sustainability

Abbreviations

EI	Ecological Intensification
EU	European Union
FAO	Food and Agricultural Organization
GHGs	Greenhouse Gases
NR	Natural Resource
R&D	Research and Development
SD	Sustainable Development
SI	Sustainable Intensification

1.1 Introduction

In the present era modernized technique and process used in agriculture is creating environmental degradation along with the loss of biodiversity. The impact is severe as it reduces the agricultural productivity as well as total destruction of agroecosystem on long term. Various factors play effective role towards agriculture and economy. Another big issue includes feeding of growing population of human beings across the world. In this context, intensification in agricultural practices is the need of the hour. But, one must take care about the process should be eco-friendly. Maintaining the harmony of natural resource (NR) is also required to maintain the integrity of agroecosystem (Kumar et al. 2020; Jhariya et al. 2019a). Therefore, intensification can be achieved in the sector of farm inputs, mechanized inputs, man power inputs as well as proper functioning of the ecosystem.

Such approaches should promote efficient and sustainable NR utilization. However, such approaches are sometimes criticized for lesser efficiency and less protection of biodiversity (Raj et al. 2018). In this context, ecological intensification (EI) focuses on sustainable functioning of ecosystem, ecological processes as well as ecological interactions (Shaver et al. 2015). To achieve sustainability in agriculture sector eco-intensive farming practices should be approached using natural assets and services within the carrying capacity of the habitat. One major problem on this aspect is practicing eco-farming technologies does not fulfil the growing demand of food worldwide (Meena et al. 2018; Harvey et al. 2014). In Indian perspective as there is gradual increase in the footprint of agroecosystem, it is creating a crisis situation for other NRs as well as human civilization (Banerjee et al. 2020; Raj et al. 2020). Intensification of agricultural activities should therefore be properly investigated by the global scientific community.

Ecosystem is an integrated unit of biotic and abiotic factors. Therefore, efficient functioning of ecosystem is very much essential to maintain the ecosystem homeostasis. It has been observed that high level of biodiversity promotes EI in agricultural productivity and economically effective. Multifaceted of benefits lies within agricultural biodiversity such as maintenance of soil quality and nutrient flow in agroecosystem, protection and conservation of soil and water as well as the integrity of ecosystem. Loss of biodiversity in agroecosystem has revealed loss of genetic diversity as well as more susceptibility of ecosystem towards various stresses (Jhariya et al. 2019a).

To achieve sustainability, one has to work in a combined way to promote conservation of agro-diversity as well as move towards increased production. Intensification in the negative sense leads to the use of modernized technology causing agro-pollution and therefore, world is looking for suitable alternatives through eco-friendly practices. This way of deleterious agricultural practices has caused a drastic reduction in world's biodiversity at macro level. Intensive use of agrochemicals has converted the complex nature of ecosystem and species interaction into simpler forms of reduced number of food chains and food webs operating in a particular ecosystem (Meena et al. 2020). When one considers intensification in terms of agricultural productivity more than half of flora and fauna in UK has depleted due to increased cultivation practices (UNEPWCMC 2011).

The problem of biodiversity loss is a century old problem which is aggravated through modernization of agro-technology causing a crisis situation in the entire Europe along with loss of ecosystem services (Storkey et al. 2012). As a consequence of that agro-biodiversity becomes the key issue in the policy matter of European Union (EU) promoting R&D (research and development) in various schemes of agriculture and environment (Sutcliffe et al. 2015). As a consequence of that vision of EU Biodiversity Strategy-2020, focused on strengthening ecosystem function and promote sustainable development (SD) in the sectors of forestry and farming (Mace et al. 2010). It has been observed that by preserving biological diversity one can achieve sustainability in the field of agricultural production (Tryjanowski et al. 2011).

The present chapter deals with various forms of EI practices leading towards SD in agroecosystem and NR conservation.

1.2 Problems Associated with the Resources

India is an agriculture based country in which 75% of the people are dependent upon agricultural activities for maintaining their livelihood. Rest of the people is also dependent upon the natural assets in terms of consumption and habitat. The prosperity and well-being of human civilization is dependent upon the reserve base of NR along with environmental quality. In Indian perspective the major issue is that resource depletion appears to be a common aspect reducing the quantity and quality of NR. Each of the components of the ecosystem is under threat from various forms of environmental degradation reducing the agricultural productivity to a considerable extent. For instance in the hydrosphere the ground water extraction is taking place in an unprecedented rate, water pollution in an uncontrolled way and as a consequence fresh and safe water is a scare resource (Meena and Lal 2018). Pollution in every sphere of environment hampers the productivity and economic growth of a country. However, in developed nations conservation approaches seem to be a luxury to maintain the aesthetics. On the other hand developing countries are under stress to promote such approaches for their existence (Singh 2009).

Intensification to boost up agricultural productivity has huge negative impact upon biodiversity and other associated ecosystem services. Such losses promote ecological invasion, loss of indigenous crop diversity, making more species critically endangered or rare and overall decline in agricultural productivity (Kennedy et al. 2002; Jhariya and Yadav 2017). Researches have revealed better productive agroecosystem nurtures higher level of biodiversity of flora and fauna (Tscharntke et al. 2005). Decline in avifaunal species have been reported by various researchers due to reduction in agricultural productivity (Soderstrom et al. 2001). Considering the facts the concept of sustainable agriculture emerged to maintain both the quantity and quality of food as well as promote EI process to operate within the agroecosystem in the form of organic farming, green farming, etc.

With the unprecedented growth of human population, problems of food security and crisis came into our forefront. As per the report of Food and Agriculture Organization (FAO) (2012a) more than 800 million people are suffering from the problem of food crisis along with improper diet and nutrition globally. The situation is much worser in the developing countries. It is a biggest challenge in the area of intensification and as a consequence to cope up with such problems, EI is the solution. Further, wastage of food materials is also putting pressure on the agroecosystem by increasing the demand for human civilization (Alexandratos and Bruinsma 2012; Meena et al. 2020a, b). Therefore, implementing EI is a hard task to perform for betterment of quality of life (FAO 2012b).

Agriculture comprises of diverse form of activities including crop cultivation and management of animal husbandry which provides multifaceted of economic benefits for people both in developed and developing countries. Therefore, eco-intensified agriculture can be considered as a combating measure towards the problem of food security, crisis and poverty. The major problem in terms of global economy includes growth and development of non-agricultural sectors in comparison to agricultural sectors which is prevalent in developing countries (FAO 2012a). Therefore, the objective of agriculture is not only to provide food and employment opportunities but also act subsequently to combat environmental challenges as well as with other non-agricultural sectors. Thus, a paradigm shift is required towards sustainable agriculture through capacity building and comprehensive policy framework.

Green revolution is a mega event in the area of intensification of agricultural productivity. Under this event adoption of hybrid seeds, chemical inputs and modernized technologies have intensified the agricultural productivity to a considerable extend (Stevenson et al. 2011). Such agricultural intensification activities have also addressed the issues of poverty up to a certain level. Higher productivity would lead to decline in economy of agricultural products which indirectly contributes to socio-economic development of rural livelihoods. On the other hand agricultural intensification has promoted environmental degradation in various forms and thus has become a bane for the modern technological world. Under such situation intensification practices should be modified to EI practices.

The EI process in agriculture seems to focus on zero wastage and more production strategy. On the other hand it should encompass for socio-economic upliftment and economy of rural stakeholders. Another aspect of EI of agriculture includes developing the agroecosystem as a shock absorber under the face of various biotic and abiotic stresses. The agroecosystem should be capable of reducing greenhouse gases (GHGs) emission, providing ecosystem services such as pest and disease control along with maintenance of fertile land. The modern agriculture should also be energy intensive which will consume less energy and eco-friendly sources of energy. The major challenge would be to focus on development of such an agricultural system which reduces the negative impact on the ecosystem. Sustainable agriculture in this perspective requires an integrated approach for sustainable use of NR and efficient management of ecosystem services. In this way the negative impacts on agroecosystem can be reduced and one can move towards climate resilient agroecosystem in future. Under this purview, the focus point should be on conservation of crop varieties, climate resilient agriculture practice, maintenance of germplasm stock and proper utilization of genetic resources in terms of ecosystem services they provide.

Level of awareness and willingness of farming community towards EI activities is also required from unsustainable to sustainable practices. Further, effective implementation of policies should also be required for their better accountability. In these aspects recognizing traditional knowledge and experience of farming communities may be the better option for adopting EI practices. Economic incentives, rights over land would help the rural poor farmers to adopt EI practices in their agricultural system. Capacity building is also an important aspect in order to achieve sustainability through EI practices.

1.3 Agricultural Intensification and Environmental Sustainability

Use of NR is associated with the agricultural activities to a maximum extent. As per the reports maintenance of livestock tends to be the largest user of land on the earth surface, using 3/4th of the geographical area of cultivable land. Also, in agriculture sector >65% of water resource is usually consumed (Kabat 2013). More than half of the geographical habitat and assets are consumed unsustainably through agricultural activities (MEA 2005). As per FAO (2011b) approximately 1/3rd of food materials (>1.2 billion t/yr) are gradually wasted across the world which is severe in front of food crisis problem. By comparing the economic conditions of developed and developing nations, it was observed that wastage of food occurs at consumer level for high income countries and loss during agricultural activities in low income countries due to the lack of proper infrastructure. Reducing the wastage of food one can minimize the food demand as well as associated cost in cultivation practices.

In the present context, sustainable agriculture is the need of the hour. It involves approaches of wide dimension such as conservation agriculture, integrated nutrient and pests management, various forms of intensification practices and technologies (e.g. system of rice intensification), region based tested model, proper management of livestock along with conservation of NRs such as water and soil (Table 1.1).

Sustainable production system has various components which include eco-friendly practices, application of agroecological principles supported by good legal framework along with proper planning, execution and monitoring process (Fig. 1.1).

EI is such a process or approach which acts upon the community level through eco-governance leading to ecosystem quality improvement followed by increase in ecological value. In this process there is growth in technological innovation leading to equilibrium between ecosystem quality improvement and ecological values (Fig. 1.2).

1.4 Challenges for Ecological Intensification towards Sustainability

Increasing productivity in an unscientific manner is creating the problem of agricultural pollution and land degradation. Intensive use of pesticides and fertilizers is increasing the energy footprint, economic cost, loss of soil health and agro-biodiversity and many more irreplaceable problems (Meena et al. 2020; Jhariya et al. 2018a, 2018b). On the other hand we have to increase the food production to feed the growing human population of the globe. Therefore, the concept of EI becomes very handy to address these problems. Further, it would also lead to sustainability.

Implementing EI at the grassroot level is very challenging as because it requires an integrated approach as well as proper scientific planning and suitable strategies. In the agroecosystem the soil and water environment is affected mostly at the cost of more production. For example, problem such as soil salinization, desertification, soil erosion are the result of faulty land-use practices. In the water component it was observed that non-judicious use of chemical fertilizer and pesticides pollute the water in the form of growth of algal blooms known as eutrophication. Another major issue is the pest and disease outbreaks within the agroecosystem which reduces the crop yield in significant level. Therefore, an approach of integration between agriculture and ecology is very much essential to overcome these challenges.

Two major challenges associated with agroecosystem includes the problem of hunger and malnutrition followed by too much of anthropogenic influence causing ecological overshoot and crossing carrying capacity of the earth. Secondary

Table 1.1 Various schemes of ecological intensification across the globe

Practices/schemes/ methods/models	Region	Outcomes	References
The comprehensive assessment of water management in agriculture (2007)	Burkina Faso alone	Rehabilitation of more than 2,50,000 hectares of land and production of more than 75,000 tons of food material annually.	Reij et al. (2009)
	Southern Niger	Farming community are actively engaged in regeneration and multiplication of trees of higher economy which has improved the land quality of more than 4.5 million hectares with extra production of 500,000 tons of food annually. This has been contributed significantly to ensure food security for more than 2.5 million people. Further economic earning increases up to >200 $ due to baobab production on household basis annually.	
	Ethiopia	Farming community is capturing the agricultural runoff from various natural structures by creating temporary water reservoir and then utilizing it for irrigation purpose. In this process it was found that >60,000 hectare land area has come under irrigation followed by benefits to over >3 lakh people due to sorghum production. It has also benefited the agricultural extension of various horticultural productions up to 3/4th times.	Binyam and Desale (2015)
Conservation agriculture (CA)	Brazil	Conservation based agriculture is practiced for >20 million hectares accounting for >20% cultivable land combating the events of drought and erosion. As per report in 2008–2009 the yield loss	Altieri et al. (2012)

(continued)

Table 1.1 (continued)

Practices/schemes/ methods/models	Region	Outcomes	References
		in production of maize almost reached half of the production. However, farmers who practiced CA reflected approximately 20% loss in maize production. This approach leads to ecosystem resilience of CA.	
	Developing countries	An average increase in yield of 79% was reflected in CA projects from pilot study of >250 projects across >50 countries of the globe. It helps to increase the water use efficiency of crops, improves C sequestration potential and reduces the dependency on pesticides for eradication of pests.	Pretty et al. (2006)
Sustainable intensification (the foresight project)	African countries	Sustainable intensification approach were adopted by >15 African countries in >35 projects with an economic benefit to >ten million farming community along with improvement in the environment up to 12 million hectare land area	Pretty et al. (2011)
System of rice intensification (SRI)	Various region of the world	SRI has been widely adopted for staple food crops along with other vegetable crops for sustainable yield. The benefit of SRI has been reflected for >45 countries with an yield increase up to 100% with 9/10th reduction in seed requirement as well as half reduction of water requirements	SRI International Network and Resources Center (2014)
	India	From Indian perspective in the past five decades development of small holds farmer were found	Vidal (2013)

(continued)

Table 1.1 (continued)

Practices/schemes/ methods/models	Region	Outcomes	References
		among the global population of 500 million. SRI such a technology which requires lesser inputs and thus become economically feasible. Bihar government is actively promoting this programme	
Participatory plant breeding (PPB)	China	The south West China, maize based PPB programme was initiated and increase in yields was recorded up to 30%. Organic supplements have increased the production of maize by 30% in comparison to villages not adopting the PPB programme and therefore promoted the economic flow towards the villages adopting PPB models. Such approaches help in regulation of pest population, use of organic amendments. Area under risks may be adapted with monocropping system to reduce the risk of crop failure, and adaptation to local condition, and was found to be more efficient in increasing the quality and quantity of yield in comparison to the hybrid variety	Song and Li (2011)
One acre fund	Western Kenya	In western Kenya till 2012 investment of *One Acre Fund* has promoted three times yield increment of raw material after harvesting with per acre of plantation. Further, the economic gain has increased twice.	Pretty et al. (2011); Royal Society (2009)
Microdosing	Niger, Mali and Burkina Faso	The microdose concept adopted in various African countries have reduced the	ICRISAT (2009)

(continued)

Table 1.1 (continued)

Practices/schemes/ methods/models	Region	Outcomes	References
		chemical fertilizer use with higher yield of millet crops along with other crops having better adoptability of water	
The zai system	Burkina Faso	The Zai approach includes seed showing through organic amendments such as leaves and compost manure along with rain water is done during summer season which enriches the soil biota. Further, sowing of sorghum and millet crop tend to increase the yield up to 120% giving additional yield of >75,000 tons grain annually	CGIAR (2011); Sawadogo (2011)
Agroforestry with *Faidherbia Albida*	African countries	Leguminous crop tend to add nitrogen through biological nitrogen fixation as well as decomposition of plant materials. Plantation of crops under leguminous tree is a suitable alternative which can give better yield without the application of fertilizers. Additionally the leguminous trees add >2 tons per hectare basis carbon into the soil and it has been reported that mature trees can add carbon up to 30 tons per hectare basis	World agroforestry Centre (n.d.) (http://www.worldagroforestry.org/sites/default/files/F.a_keystone_of_Ev_Ag.pdf)
IPM with FFS model (farmer field schools)	African countries	Under the leadership of FAO and the partnership with civil society IPM and FFS models was launched in Ghana in which training and extension programme were given to the farmers and as an outcome 23% yield increase was recorded with a decline in 75% pesticide use	FAO (2001), (http://www.fao.org/fileadmin/templates/agphome/documents/IPM/IPPM_West_Africa.pdf)

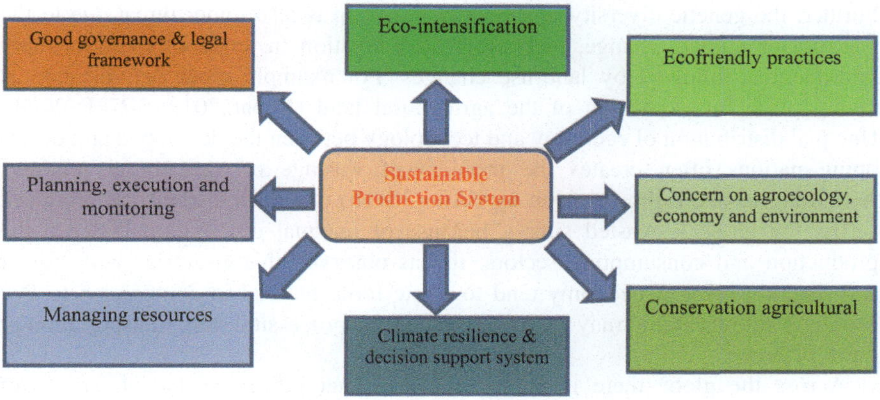

Fig. 1.1 Components of sustainable production system

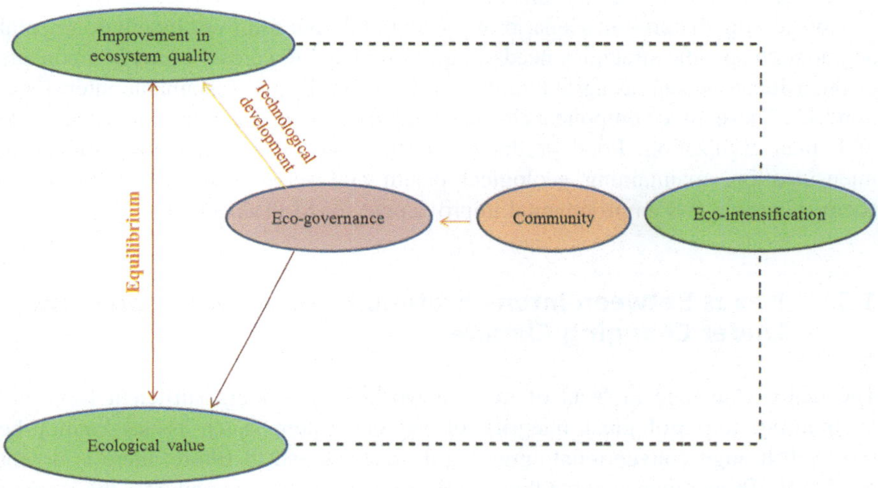

Fig. 1.2 Pathway of ecological intensification

challenge includes loss of agro-biodiversity, higher level of GHGs emission followed by rising temperature of the earth surface leading to climate change. Further, scarcity of fresh water resources, higher rate of deforestation, high concentration of nutrients are adding more challenges in the soil environment of the agroecosystem. In developing countries poverty is another biggest challenge for the agricultural sector to feed the population (FAO 2012a). Therefore, the multifaceted of challenges in agriculture is to provide nutritious and adequate food to people on one hand and mitigate environmental issues on the other (FAO 2012a).

Under the objective of increasing productivity it was observed that 60% of the ecosystem across the globe is under severe threat due to unsustainable utilization.

Further, the genetic diversity of the agroecosystem is also under threat due to the pressure of climate change, technological innovation in terms of hybrid variety introduction, followed by land-use changes. For example rapid deforestation is taking place for expansion of the agricultural land (Kabat 2013; UNEP 2010). Unequal distribution of economy and technology between the developed and developing nations often creates the problem of wastage of agricultural produce. According to one estimate given by FAO (2011b) one third of the food produced across the globe is wasted simply because of unequal distribution between the production and consumption sectors. It was observed that countries with higher and medium level of economy tend to waste food material in comparison to low economy country. Food may be wasted during the processing step, transportation or during storage.

Across the globe there is range of factors that influences the effectivity of intensification practices. Among them the most important one includes economic incentive, lack of appropriate technology, food crisis, ever increasing human population, soil and climatic conditions (Table 1.2).

Considering the present scenario of population boom followed by environmental degradation specific strategies needs to be formulated towards various components of intensification such as agricultural intensification, EI and sustainable intensification (SI). These three components have a complex interaction in order to achieve SD of human civilization. Food production is the biggest challenge for agricultural intensification, maintaining ecological health and services for EI followed by socio-economic and environmental improvement for SI (Table 1.3).

1.5 Nexus Between Intensification, Food Security and Crisis Under Changing Climate

The major challenge in front of modern world is producing sufficient food and maintaining the ecological integrity of agroecosystem. Such issues cannot be resolved through conventional agricultural practices and therefore, there is urgent need of EI. Proper management through EI is the requirement to address the issue of sustainable agriculture. To implement such strategies proper scientific knowledge, technical skills, adequate infrastructure is required to focus EI and sustainable agriculture. It was observed that the issue of food security at local level can be resolved through adoption of low input agricultural practices leading to socio-economic upliftment of local community stakeholders. This leads to development of concepts such as organic farming, biofertilizer based farming along with introduction of advance molecular techniques towards climate resilient agriculture practice (Halberg et al. 2015).

As per FAO a massive increase in agricultural output is required within 40 years span of which significant contribution should come from the under developed world, where the production of biomass needs to be increased for the said period (Alexandratos and Bruinsma 2012). Another challenge is in the form of competition to be maintained between productivity and energy production in the form of biomass

Table 1.2 Challenges of ecological intensification across the globe

Location/region	Study level	Factors influencing	References
Amazon region of Brazil	Field/farm	Accessibility and policy for marketing, commodity price, facilities for ranching, etc.	Cortner et al. (2019)
Malawi (central and southern part)	Farming community	Lack of infrastructure, high population strength, low productivity of soil, farmers response, etc.	David et al. (2016)
Malawi central region	Field/farm	Male female ratio, climatic feature, commodity price, availability of land, population strength, etc.	Snapp et al. (2018)
Andes central region	Regional	Climatic perturbations, mining and grazing activities, improper plantation, etc.	Willy et al. (2019)
Africa (eastern and southern region)	Field/plot	Extension activities, accessibility of markets, lack of information technology, technological upgradation, etc.	Kassie et al. (2015)
Kenya (eastern and western region)	Farming community	Male female ratio, lack of technology, quality of land, extension activities and accessibility of markets, etc.	Ndiritu et al. (2014)
Kenya, Uganda and Ethiopia (eastern region of Africa)	Nation	Insufficient incentives for economic growth, extension activities, improper infrastructure	Yami and Van (2017)
Uganda (eastern part)	Farming community	Lack of information technology, price variability, climatic perturbations, insufficient yield, livelihood of farming community, population strength, etc.	Rahn et al. (2018)
Kenya (eastern part)	Field/farm	Population strength, soil fertility and productivity, lack of information technology, farmer's attitude, local climatology, etc.	Rolando et al. (2017)
Germany	Field/plot	Soil edaphic features, food requirement, population strength, etc.	Schiefer et al. (2015)
Sub-Saharan Africa	Farming community	Soil quality, population strength, improper production, productivity, livelihood, accessibility to market, etc.	Vanlauwe et al. (2014)
Kenya (southern part)	Rural setup /village level	Population strength, commodity price variability, precipitation, etc.	Zaal and Oostendorp (2002)
Tropical Reunion	Field/farm	Population strength, rising food requirement, biotic and abiotic stress, nature of land, etc.	Jonathan et al. (2011)
India	–	Population rise, small land holding, subsidy for agricultural inputs, poverty level, information technology, infrastructure, soil and climatic conditions, etc.	Vidal (2013), Nath et al. (2016)

Table 1.3 Strategies for various components of intensification (Source: Tittonell 2014; Tittonell and Giller 2013; Wezel et al. 2015; Clay 2018; Xie et al. 2019)

Agricultural intensification	Ecological intensification	Sustainable intensification
✓ Intensify agricultural productivity. ✓ Greater sales and marketing. ✓ Increase productivity on region basis. ✓ Increase market share for socio-economic upliftment. ✓Environment intensive farming practices	✓ Minimizing the negative impact of modern agriculture. ✓ Maintain the soil health and biodiversity. ✓ Optimum use of resources. ✓ Ecology based processes and services. ✓ Minimizing external inputs and focusing in on-farm inputs. ✓ Maintenance of ecosystem services. ✓ Recognizing traditional knowledge for resource conservation	✓ Reduce energy subsidy. ✓ Maintain ecosystem resilience. ✓ Agricultural diversion for proper land-use. ✓ Improve economic potential and quality. ✓ Adaptation and mitigation through capacity building. ✓ Managing landscape for multidimensional services. ✓ Resource efficient technologies. Harmonization of input-output agricultural practices. ✓ Restoration of degraded ecosystem

till 2050. Higher production of biomass would lead to development of alternate energy source by replacing fossil fuels which would help to mitigate the mega event of climate change in agricultural sector. On the other hand to feed the growing human population one needs to improve the productivity at highest level without damaging the soil and land environment (Harvey and Pilgrim 2011). Another problem is the availability of land for cultivation for increasing the agricultural productivity (Raj et al. 2019a, 2019b). As per the research data, six persons to be fed from per hectare area of agriculture land globally since 2000 onwards (Cassidy et al. 2013).

As per Halberg et al. (2009) the unsustainable consumption pattern of agroecosystem and their subsequent ecosystem services creates the issue of food security. The ecosystem service and benefits are undermined through non-judicious use of chemical fertilizer and other agrochemicals altering the soil quality to a considerable extent (Meena et al. 2020). Changing climate also poses significant challenge in terms of reduction in agricultural productivity and therefore raising the issue of food security (Porter and Xie 2014). The integrated system of food security, NR depletion and climate change is creating a huge problem for prosperity and well-being of the people (Halberg et al. 2009). Thus, to address these three issues all together, EI is the need of the hour in the agricultural sector (Fig. 1.3) (Tittonell 2014).

The nature of EI depends very much upon sustainability which can be integrated as SI. SI is such a process that aims towards conservative approach to improve soil health, fertility and productivity of agroecosystem. It is also an integrated approach that includes the principles of organic farming, use of biofertilizer and bio-pesticides policies, judicious use of agrochemicals to reduce agricultural pollution and protection from pests and diseases (FAO 2011a). Such approaches also work on case to

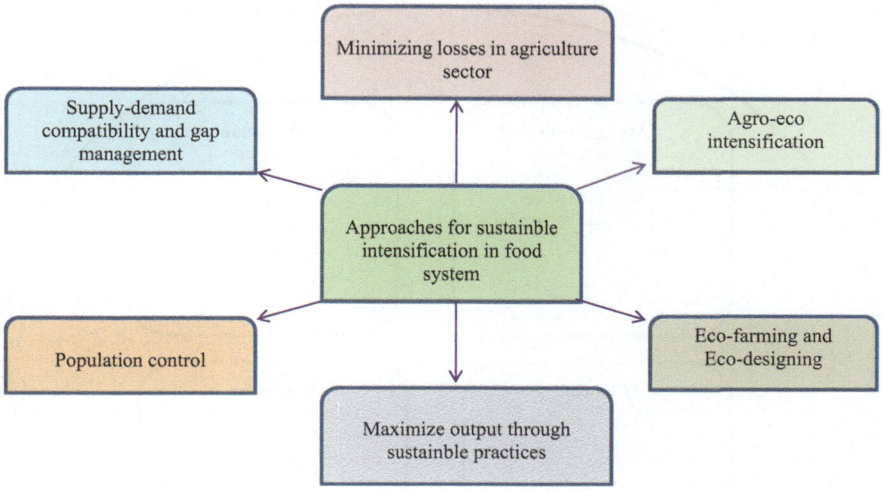

Fig. 1.3 Food production and sustainable intensification

case basis. The policy underlying the mechanism of SI includes reducing wastage of food and demand for products coming from livestock population. However, the aim of improving food production should be continued in an eco-friendly way in order to maintain the balance of the environment. In Europe the major policy behind such factor includes not only to restrict the use of agrochemicals in order to check the pollution but also towards the optimum use through intensive management for proper caring of ecosystem services provided by the agroecosystem to human civilization. In order to do that proper knowledge for specific areas needs to be maintained.

EI is a concept which emphasizes the maintenance of crop biodiversity in order to achieve sustainable yield and reduce the impact on environment (Bommarco et al. 2013). As per the various works and reports, the principle of EI is very much similar with agroecological perspective and organic mode of farming (de Abreu and Bellon 2013). Agroecology is a concept which integrates the application of ecological principles in agroecosystem for increased income for the rural stakeholder as well as maintains the ecosystem health (de Abreu and Bellon 2013). Organic agriculture is a form of cultivation practice that emphasizes more on organic inputs in order to maintained agroecosystem health and diversity. Further, it also aims towards low input agriculture practice in terms of different agrochemicals to maintain the soil health and fertility. Such system also emphasizes the use of traditional knowledge to get sustainable yield by application of modern scientific principles.

Thus, the two concepts of agroecology and organic farming emphasizes on proper maintenance of ecosystem services as well as sustainable NR utilization. Therefore, the practice of organic farming should be intensified in order to build the soil nutrient pool and organic matter. In Europe, the concept of EI was modified to eco-functional intensification by recognizing the traditional knowledge based on biological

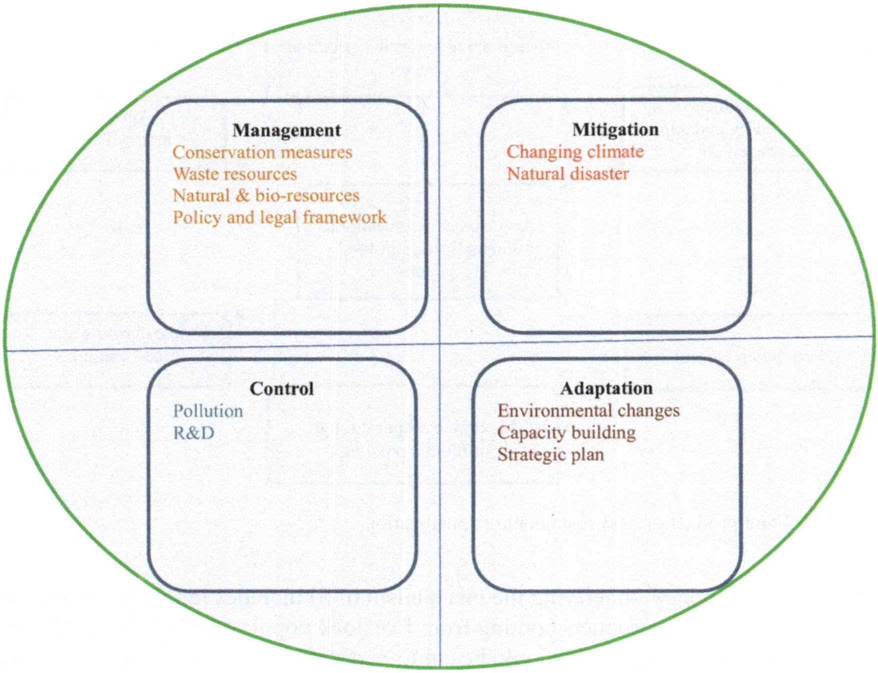

Fig. 1.4 Strategic component of ecological intensification

principles and maintaining the harmony between food production unit and agroecosystem. In this way through low input agriculture practice helps to implement eco-functional intensification principles at the field level (Lipper et al. 2014).

Strategy for implementing intensification in agricultural system comprises of management, mitigation, adaptation and control (Fig. 1.4). Management component comprises of development of policy for proper management of resources followed by effective legal infrastructure through good governance. Mitigation should be aimed towards reducing the impact of natural hazards and adopting towards climate change (Khan et al. 2020a, 2020b). In the adaptation component one needs to cope up with environmental changes followed by framing strategic plan for capacity building of stakeholders. Control components should be aimed towards pollution regulation and future R&D.

1.6 Organic Farming Towards Ecological Intensification

Field based studies were conducted by various researchers by using various combination of organic and conventional farming system to achieve the goal of EI. According to de Ponti et al. (2012) organic farming gives lesser output (1/3rd) in comparison to traditional agricultural practices across various centuries of the

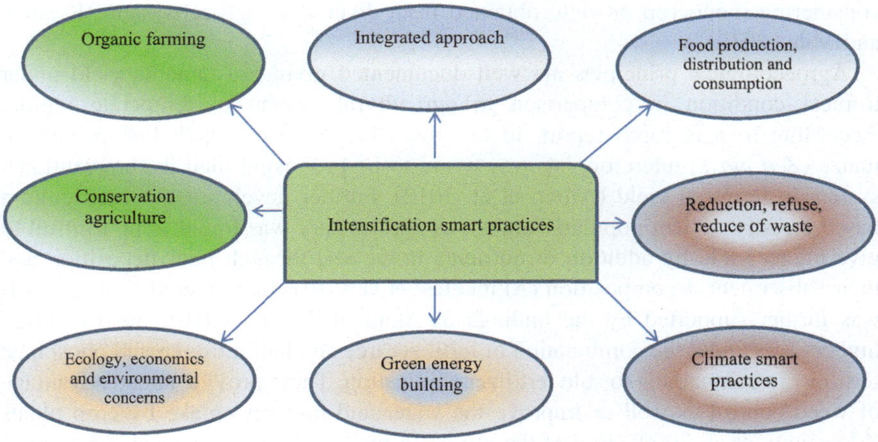

Fig. 1.5 Practices for moving towards smart intensification

globe. However, the result of the yield varies on the basis of agroecological region. For instance in Asian subcontinent yield ranges up to >80% and in European continent it is <80%. It also varies according to various crop species. But some results shown by Te Pas and Rees (2014) revealed higher output in organic farming in comparison to traditional agricultural practices under tropical and sub-tropical conditions.

Based on the type of crop the result in terms of yield also varies significantly. For example, yield gap between conventional and organic farming practices is not very significant in case of leguminous crops but the case is different for categories of non-leguminous crops (Ponisio et al. 2015). Various treatments of chemical fertilizer did not show any significant impact over yield variations. Considering these facts various diversifying practices are required in which the concept of agroecology and organic farming should be included in order to move towards EI and sustainability in agricultural system (Fig. 1.5).

According to Kirchmann et al. (2008) the level of organic farming should be upgraded in such an extent that the yield becomes comparable to conventional system of agriculture practice. The organic farming should be integrated with proper crop rotation as well as use of green manuring. This would provide continuous nutrient flow into the agroecosystem an aid in EI process. One of the major problems in relation to reduction in yield includes pest and disease outbreak which can be regulated through biological diversification of agricultural crops. This would help to maintain the soil fertility as well as soil physical properties. It also helps to build up ecosystem resilience against climate and human perturbations. However, the results seem to be little bit unclear as most of these approaches were conducted on experimental basis and the practices such as intercropping, crop diversification were not included in the field trials. There are also lacunae in terms of yield calculation considering monocrop or yield obtained from diversified agroecosystem (Kremen and Miles 2012).

Agroecological principles are well documented giving sustainable yield under tropical condition in comparison to agricultural system of temperate region. According to a research report, in the area of east Africa *Push Pull System* of maize (*Zea mays*) intercropping was found to be promising than the conventional system in terms of yield (Altieri et al. 2012). Further development of agriculture based forestry system popularly known as agroforestry was found to be fruitfull to give higher yield by addition of nutrients in the soil through litter deposition and their subsequent decomposition (Akinnefesi et al. 2010; Jhariya et al. 2019b). This was further supported by the findings of Akinnefesi et al. (2010), who reported higher yield under the combination of fertilizer tree and half nitrogen and phosphate fertilizer in comparison to sole fertilizer application. These provide added advantage of weed control as well as improve the water and nutrient uptake by crop plants (Malezieux et al. 2009). Beside the aforesaid matter of EI approaches one need to take care about fulfilling the biomass requirement, maintaining ecosystem services as well as mitigating and adopting climatic perturbations.

1.7 Research and Development

It is no doubt that EI appears to be need of the hour considering deteriorating conditions of agroecosystem under the light of environmental pollution and climatic change. The major objective of EI includes sustainable yields along with reducing anthropogenic inputs through organic materials. Such approach would also help to regulate and maintained the ecosystem services of agroecosystem. Future research should be aimed to mitigate the gaps in sustainable yield and incorporate ecosystem services to crop production system (Bommarco et al. 2013).

A lot of work has been done in the area of food security with the help of agroecological process and eco-functional intensification. But, major problems associated with this include recognizing and evaluating the actual potential of food production to meet the present and future food demand and their proper prediction. World research has revealed the promotion of agroecological practices that would help to satisfy the growing food demand up to 2050. The proper outcome of EI at regional and global level is yet to be explored properly and therefore, demand future R&D followed by approaches and awareness (Fig. 1.6). The factor of diet preference should also be considered in relation to global food security issues (Halberg et al. 2015).

The integration of agroecology, organic farming and EI has revealed positive output in terms of agricultural output and providing ecosystem services. Therefore, the principles of agroecology and organic farming are the pillars of EI process which relies on zero chemical use for crop production. Future research needs to be done on the area of combination treatment development including organic and agroecological principles to achieve sustainability and develop ecosystem resilience. To achieve this, such practices should be followed in a wider landscape in order to maintained

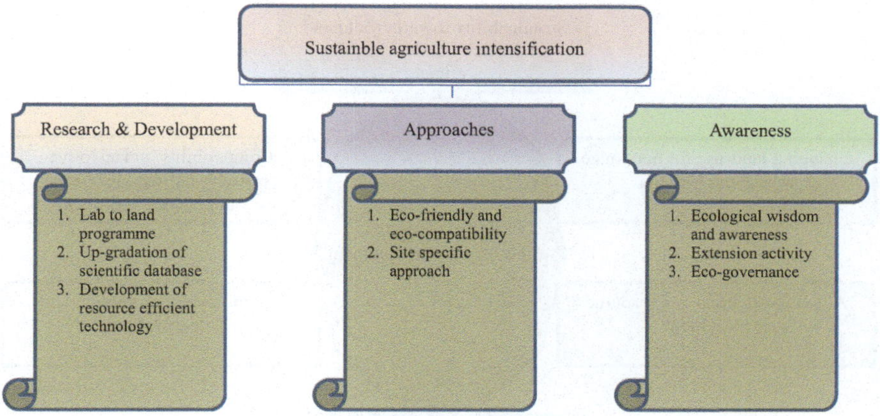

Fig. 1.6 Agroecology—a way towards ecological intensification

agro-diversity and proper land-use for moving towards sustainable agriculture (Halberg et al. 2015).

Systematic and scientific exploration needs to be done in the area of organic farming, agroecological practices and other strategies that increase crop production in a coordinated system to check or arrest the environmental degradation. These would lead to precise information for the farming community in developing nation regarding the benefits of eco-functional intensification as well as ecological sustainability in agroecosystem (Halberg et al. 2015).

1.8 Policy Intervention

Technological intervention through proper policy formulation is the requirement for successful implement of EI practices. Under such circumstances the issue of sustainability in the area of productivity and yield should be addressed in a proper way for mitigating changing climate. In this direction, one should give due consideration about the compatibility and adaptive capacity of the ecosystem in order to get maximum benefits. Further, such approaches can be promoted through upgradation of existing agro-technology and its successful adoption of local farming community. Various exogenous inputs should be minimized in order to check environmental degradation and promote EI. Such approach would lead to proper management of NR and its conservation. Another major aspect in implementing EI includes development of proper land-use policy for maximizing production in one hand and prevents deterioration of soil resource on the other (Fig. 1.7).

As the time progresses the problems or challenges are gradually becoming more complex. Demand for food, fodder, fibre and other agriproducts is increasing day by day and therefore, the agroecosystem needs to be more productive with minimum wastage. Further, to address the issue of poverty, hunger and malnutrition policies

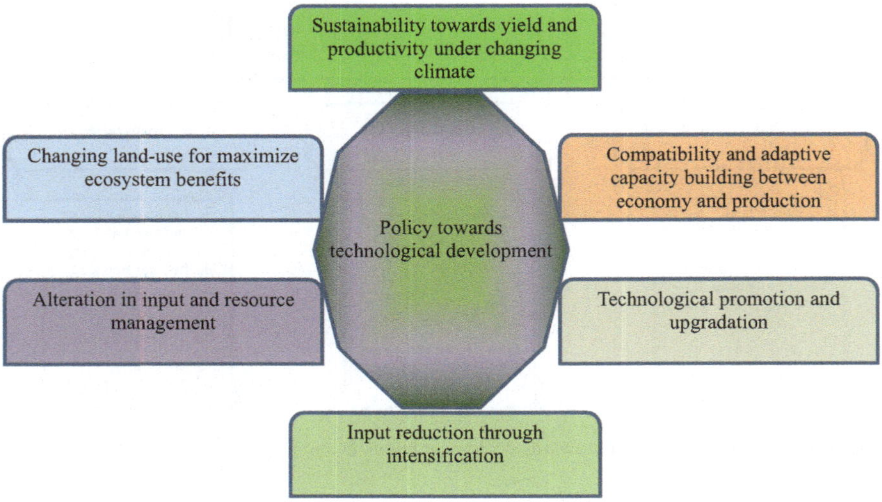

Fig. 1.7 Policy intervention towards intensification

needs to be designed to provide more income opportunities for the farming community, focus on landless farmers and labour and socio-economic upliftment from rural perspective. Another key policy should be sustainable utilization of NRs as well as strategies for mitigating climate change. From the pollution perspective reducing the GHGs emission and maintaining other ecosystem services is the essential requirement. One needs to formulate policy to design eco-friendly technologies with less energy requirement along with proper economic incentive.

1.9 Prospects of Ecological Intensification

The interaction between food production, environmental setup and agricultural practices has originated the concept of SI. The concept of SI is adopting the traditional practices such as conservation agriculture, organic farming, etc. while having major focus on reducing their deficiencies to maximize the output in a sustainable way (Blumenstein et al. 2018). The main motto behind such approaches is to develop scope for maintaining ecological integrity and ecological services. This includes a transfer of technology of normal cultivation practices to intense cultivation practices but in a sustainable way. The success and effective implementation of intensification governs the future prospects of EI. In order to do that one needs to go for model construction and its further analysis. Repeated monitoring of intensification practices across various countries needs to be done regularly on case study basis. However, there are some factors which regulate the technology and practice in the crop production system and effective implementation of intensification. This includes socio-economic conditions followed by farmer's inherent nature to adopt intensification and play of nature.

The concept of SI includes various methods and technologies which need to be explored properly in order to increase the food production in a sustainable way. This is a big problem for effective implementation of SI (Petersen and Snapp 2015). In European countries major stress has been given on SI including the diverse disciplines and multidisciplinary approach. Further, it was observed that inclusion of economics and social sciences was not included properly in the intensification process making it questionable in terms of its cost effectiveness (Weltin et al. 2018). In many areas SI is a simply a concept of discussion but there is no proper guidelines or principles to implement them on case to case basis (Petersen and Snapp 2015). Therefore, to reveal the concept at global level there should be some suitable standard and principles on the basis of which measurement of the effectivity of the SI system can be measured. However, more robust method and principles along with scientific exploration is required in this field to develop a sound base in SI.

1.10 Future Roadmap

Green revolution has helped to increase the productivity to a considerable extent of Indian agriculture. Increase yield and productivity in Europe is associated with higher amount of environmental and economic consequences and on the other hand it has little to do with the issue of food security globally. In this context, Africa show the path of EI in cereals production associated with lesser pollution and economic loss. Another roadway includes optimum production of food depending upon the demand and in case of surplus food it would be economic revenue for farming stakeholders.

In this approach we need to formulate to identify and recognize traditional knowledge and customs for food production which indicates towards lesser input based agriculture. In relation to technology based solution on market economy, private sectors tend to opt for newer technologies and products to get more economic benefits which are their business principles. Such approach should be adopted by the public sector to have more focus on new processed technology as well as give due consideration of eco-intensifying the agricultural outputs.

Future roadmap for EI relies on various factors such as R&D activities in proper direction, designing and formulation of suitable strategies to boost the agro-economy, critical analysis of policy, strategies and technologies formulated along with subsequent evaluation and timely monitoring. In the R&D sector proper exploration should be done in the areas of agriculture intensification, EI and overall SD. Policy designing and formulation includes eco-friendly approach development through eco and green designing. Critical analysis of these strategies must include social, economic, environmental and ecological dimensions. The scale of evaluation should be broader considering the agroecosystem component, followed by legal framework and e-governance (Fig. 1.8).

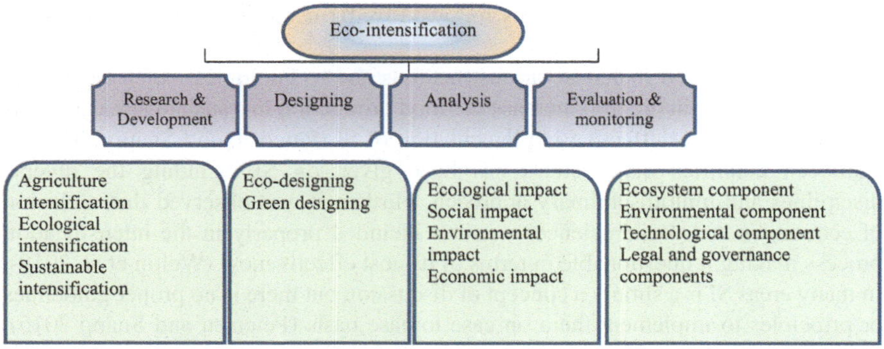

Fig. 1.8 Future roadmap for ecological intensification in sustainable agriculture

The various challenges associated with effective implementation of EI includes maximizing productivity, reduce yield gap, optimum utilization of resource and reducing environmental degradation. On the secondary part it would boost up the economy of rural stakeholders and would promote employment opportunities. Issues such as maintaining ecosystem services, arresting biodiversity loss, reducing various footprints in agroecosystem should be the future pathway for EI approaches. Overall, mitigating food security can be addressed through maintaining nutritional value and giving more emphasis on sustainable production of food. Developing the resiliency in agroecosystem is the need of the hour which would automatically address the issue of food crisis and security as well as developed mitigation attitude among the agroecosystem towards various biotic and abiotic stresses. For effectivity of eco-intensive mechanism public participation in the direction of sustainable agriculture should be promoted as well as economic incentives should be given to the rural livelihoods for their better responsiveness in the EI approaches. Use of advance technologies such as remote sensing and geographical information system could be effectively utilized to stimulate crop productivity and yield, planning and management of various resources. Promote collaboration at various levels to address the issue of sustainability in resource use and capacity building among the community is the pathway to promote EI.

Future perspective of EI comprises of appropriate steps which includes exploration of new technology, newer site inventorization, environmental quality improvement, management of NRs development of energy efficiency and efficient designing and infrastructure development (Fig. 1.9).

1.11 Conclusion

In the modern world, maintaining ecological integrity is a bigger challenge due to population explosion, climate change, increase food production, maintenance of agro-biodiversity and above all environmental degradation. Resource depletion is

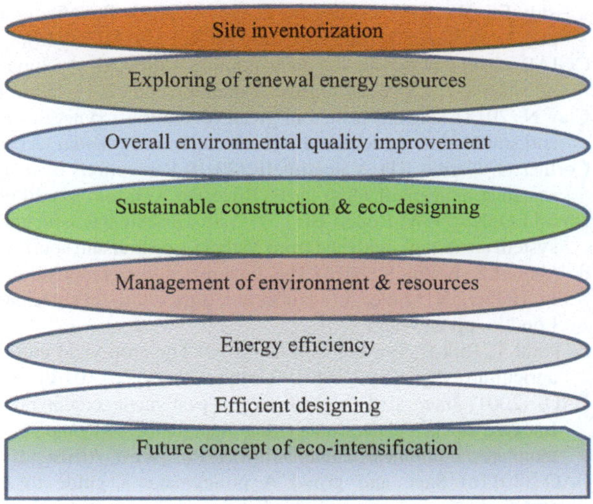

Fig. 1.9 Future perspective of ecological intensification

a significant problem in this aspect. Agricultural intensification should be kept in harmony with environmental sustainability. It, therefore, faces some severe challenges which needs to be monitored and must address the inter-relationship between food security and crisis under changing climate. Technological intervention in terms of organic farming and other such practices needs to be implemented to intensify the agricultural production. Proper research and development is the requirement of the hour for proper policy framing to move towards a greener future.

References

Akinnefesi FK, Ajayi OC, Sileshi G, Chirwa PW, Chianu J (2010) Fertiliser trees for sustainable food security in the maize-based production systems of east and southern Africa. A review. Agron Sustain Dev 30:615–629. https://doi.org/10.1051/agro/2009058

Alexandratos N, Bruinsma J (2012) ESA Working Paper No. 12–03, June. In: FAO. World agriculture towards 2030/2050: the 2012 revision, Rome. http://typo3.fao.org/fileadmin/templates/esa/Global_persepctives/world_ag_2030_50_2012_rev.pdf

Altieri MA, Funes-Monzote FR, Petersen P (2012) Agroecologically efficient agricultural systems for smallholder farmers: contributions to food sovereignty. Agron Sustain Dev 32:1–13. https://doi.org/10.1007/s13593-011-0065-6

Banerjee A, Jhariya MK, Yadav DK, Raj A (2020) Environmental and sustainable development through forestry and other resources. CRC Press, Boca Raton, FL, pp 1–400. https://doi.org/10.1201/9780429276026

Binyam AY, Desale KA (2015) Rain water harvesting: an option for dry land agriculture in arid and semi-arid Ethiopia. Int J Water Resour Environ Eng 7:17–28

Blumenstein B, Siegmeier T, Selsam F (2018) A case of sustainable intensification: stochastic farm budget optimization considering internal economic benefits of biogas production in organic agriculture. Agric Syst 159:78–92

Bommarco R, Kleijn D, Potts SG (2013) Ecological intensification: harnessing ecosystem services for food security. Trends Ecol Evol 28(4):230–238. https://doi.org/10.1016/j.tree.2012.10.012

Cassidy ES, West PC, Gerber JS, Foley JA (2013) Redefining agricultural yields: from tonnes to people nourished per hectare. Environ Res Lett 8(034015):8
CGIAR (2011) Water for dry land farming with ZAI: A climate smart response in arid regions. https://www.iwmi.cgiar.org/wp-content/uploads/2011/11/zai.pdf
Clay N (2018) Seeking justice in green revolutions: synergies and trade-offs between large-scale and smallholder agricultural intensification in Rwanda. Geoforum 97:352–362
Cortner O, Garrett RD, Valentim JF (2019) Perceptions of integrated crop-livestock systems for sustainable intensification in the Brazilian Amazon. Land Policy 82:841–853
David LO, Kurt BW, Robert BR (2016) Sustainable intensification and farmer preferences for crop system attributes: evidence from Malawi's central and southern regions. World Dev 87:139–151
de Abreu LS, Bellon S (2013) The dynamics and recomposition of agroecology in Latin America. In: Halberg N, Müller A (eds) Organic agriculture for sustainable livelihoods, vol 10. Routledge, London, pp 223–245
de Ponti T, Rijk B, van Ittersum M (2012) The crop yield gap between organic and conventional agriculture. Agric Syst 108:1–9. https://doi.org/10.1016/j.agsy.2011.12.004
FAO. (2001) Integrated production and pest management programme in West Africa. Gestion Intégrée de la Production et des Déprédateurs. Rome: FAO. http://www.fao.org/fileadmin/templates/agphome/documents/IPM/IPPM_West_Africa.pdf
FAO (2011a) Save and grow. A policymaker's guide to the sustainable intensification of smallholder crop production. Food and Agriculture Organization of the United Nations, Rome, p 2011
FAO (2011b) Global food losses and food waste: extent, causes and prevention. http://www.fao.org/docrep/014/mb060e/mb060e00.pdf
FAO (2012a) The state of food insecurity in the world, economic growth in necessary but not sufficient to accelerate reduction of hunger and malnutrition. FAO, Rome
FAO (2012b) The state of world fisheries and aquaculture 2012. FAO, Rome
Halberg N, Peramaiyan P, Walaga C (2009) Is organic farming an unjustified luxury in a world with too many hungry people? In: Willer H, Klicher L (eds) The world of organic agriculture, Statistics and Emerging Trends 2009. FiBL and IFOAM, Frick, Switzerland, pp 95–101
Halberg N, Panneerselvam P, Sébastien Treyer S (2015) Eco-functional intensification and food security: synergy or compromise? Sustain Agric Res 4(3):126–139. https://doi.org/10.5539/sar.v4n3p126
Harvey M, Pilgrim S (2011) The new competition for land: food, energy, and climate change. Food Policy 36:S40–S51. https://doi.org/10.1016/j.foodpol.2010.11.009
Harvey CA, Rakotobe ZL, Rao NS (2014) Extreme vulnerability of smallholder farmers to agricultural risks and climate change in Madagascar. Philos Trans R Soc London 369:20130089
ICRISAT (2009) Fertilizer microdosing. Boosting production in unproductive lands. http://www.icrisat.org/impacts/impact-stories/icrisat-is-fertilizer-microdosing.pdf
Jhariya MK, Yadav DK (2017) Invasive alien species: challenges, threats and management. In: Rawat SK, Narain S (eds) Agriculture technology for sustaining rural growth. Biotech Books, New Delhi, India, pp 263–285
Jhariya MK, Yadav DK, Banerjee A (2018a) Plant mediated transformation and habitat restoration: phytoremediation an eco-friendly approach. In: Gautam A, Pathak C (eds) Metallic contamination and its toxicity. Daya Publishing House, New Delhi, India, pp 231–247
Jhariya MK, Banerjee A, Yadav DK, Raj A (2018b) Leguminous trees an innovative tool for soil sustainability. In: Meena RS, Das A, Yadav GS, Lal R (eds) Legumes for soil health and sustainable management. Springer, Cham, pp 315–345. https://doi.org/10.1007/978-981-13-0253-4_10
Jhariya MK, Banerjee A, Meena RS, Yadav DK (2019a) Sustainable agriculture, Forest and environmental management. Springer, Singapore, p 606. https://doi.org/10.1007/978-981-13-6830-1
Jhariya MK, Yadav DK, Banerjee A (2019b) Agroforestry and climate change: issues and challenges. CRC Press, Boca Raton, FL, p 335. https://doi.org/10.1201/9780429057274

Jonathan V, Vigne M, Véronique A (2011) Integrated participatory modelling of actual farms to support policy making on sustainable intensification. Agric Syst 104:146–161

Kabat P (2013) Water at a crossroads. Nat Clim Chang 3:11–12

Kassie M, Teklewold H, Jaleta M (2015) Understanding the adoption of a portfolio of sustainable intensification practices in eastern and southern Africa. Land Use Policy 42:400–411

Kennedy TA, Naeem S, Howe KM, Knops JMH, Tilman D, Reich P (2002) Biodiversity as a barrier to ecological invasion. Nature 417:636–638

Khan N, Jhariya MK, Yadav DK, Banerjee A (2020a) Herbaceous dynamics and CO_2 mitigation in an urban setup- a case study from Chhattisgarh, India. Environ Sci Pollut Res 27(3):2881–2897. https://doi.org/10.1007/s11356-019-07182-8

Khan N, Jhariya MK, Yadav DK, Banerjee A (2020b) Structure, diversity and ecological function of shrub species in an urban setup of Sarguja, Chhattisgarh, India. Environ Sci Pollut Res 27 (5):5418–5432. https://doi.org/10.1007/s11356-019-07172-w

Kirchmann H, Bergström L, Kätterer T, Andrén O, Andersson R (2008) Chapter 3: can organic crop production feed the world. In: Kirchman H and L Bergström: Organic crop production – ambitions and limitations. (pp. 39–72). Springer, Cham

Kremen C, Miles A (2012) Ecosystem services in biologically diversified versus conventional farming systems: benefits, externalities, and trade-offs. Ecol Soc 17(4):40. https://doi.org/10.5751/ES-05035-17440

Kumar S, Meena RS, Jhariya MK (2020) Resources use efficiency in agriculture. Springer, Singapore. https://doi.org/10.1007/978-981-15-6953-1

Lipper L, Thornton P, Campbell BM, Baedeker T, Braimoh A, Bwalya M, Caron P, Cattaneo A, Garrity D, Henry K, Hottle R, Jackson L, Jarvis A, Kossam F, Mann W, McCarthy N, Meybeck A, Neufeldt H, Remington T, Thi Sen P, Sessa R, Shula R, Tibu A, Torquebiau EF (2014) Climate-smart agriculture for food security. Nat Clim Chang 4:1068–1072. https://doi.org/10.1038/nclimate2437

Mace GM, Cramer W, Diaz S, Faith DP, Larigauderie A, Le Prestre P, Palmer M, Perrings C, Scholes RJ, Walpole M, Walther BA, Watson JEM, Mooney HA (2010) Biodiversity targets after 2010. Curr Opin Environ Sustain 2:3–8. https://doi.org/10.1016/j.cosust.2010.03.003

Malezieux E, Crozat Y, Dupraz C, Laurans M, Makowski D, Ozier-Lafontaine H, Rapidel B, de Tourdonnet S, Valantin-Morison M (2009) Mixing plant species in cropping systems: concepts, tools and models. A review. Agron Sustain Dev 29(1):43–62. https://doi.org/10.1051/agro:2007057

MEA (2005) Ecosystems and human well-being: synthesis. http://www.millenniumassessment.org/documents/document.356.aspx.pdf

Meena RS, Lal R (2018) Legumes for soil health and sustainable management. Springer, Singapore, p 541. https://doi.org/10.1007/978-981-13-0253-4_10

Meena RS, Kumar V, Yadav GS, Mitran T (2018) Response and interaction of *Bradyrhizobium japonicum* and Arbuscular mycorrhizal fungi in the soybean rhizosphere: a review. Plant Growth Regul 84:207–223

Meena RS, Kumar S, Datta R, Lal R, Vijaykumar V, Brtnicky M, Sharma MP, Yadav GS, Jhariya MK, Jangir CK, Pathan SI, Dokulilova T, Pecina V, Marfo TD (2020) Impact of agrochemicals on soil microbiota and management: a review. Land (MDPI) 9(2):34. https://doi.org/10.3390/land9020034

Meena RS, Lal R, Yadav GS (2020a) Long term impacts of topsoil depth and amendments on soil physical and hydrological properties of an Alfisol in Central Ohio, USA. Geoderma 363:1141164

Meena RS, Lal R, Yadav GS (2020b) Long-term impact of topsoil depth and amendments on carbon and nitrogen budgets in the surface layer of an Alfisol in Central Ohio. Catena 194:104752

Nath TK, Jashimuddin M, Kamrul HM (2016) The sustainable intensification of agroforestry in shifting cultivation areas of Bangladesh. Agrofor Syst 90:405–416

Ndiritu SW, Kassie M, Shiferaw B (2014) Are there systematic gender differences in the adoption of sustainable agricultural intensification practices? evidence from Kenya. Food Policy 49:117–127

Petersen B, Snapp S (2015) What is sustainable intensification? Views from experts. Land Use Policy 46:1–10

Ponisio L, M'Gonigle LK, Mace KC, Palomino J, de Vaopine P, Kremen C (2015) Diversification practices reduce organic to conventional yield gap. Proc Roy Soc B 282:20141396. https://doi.org/10.1098/rspb.2014.1396

Porter JR, Xie L (2014) Food security and food production systems. In: Field CB et al (eds) Climate change 2014: impacts adaptation and vulnerability. IPCC, Cambridge University Press, Cambridge, UK, pp 458–533

Pretty J, Noble AD, Bossio D, Dixon J, Hine RE, Penning de Vries FWT, Morison JIL (2006) Resource-conserving agriculture increases yields in developing countries. Environ Sci Technol 40(4):1114–1119. https://doi.org/10.1021/es051670d

Pretty J, Toulmin C, Williams S (2011) Sustainable intensification in African agriculture. Int J Agric Sustain 9:5–24

Rahn E, Liebig T, Ghazoul J (2018) Opportunities for sustainable intensification of coffee agro-ecosystems along an altitudinal gradient on Mt. Elgon, Uganda. Agric Ecosyst Environ 263:31–40

Raj A, Jhariya MK, Harne SS (2018) Threats to biodiversity and conservation strategies. In: Sood KK, Mahajan V (eds) Forests, climate change and biodiversity. Kalyani Publisher, India, pp 304–320

Raj A, Jhariya MK, Yadav DK, Banerjee A, Meena RS (2019a) Agroforestry: a holistic approach for agricultural sustainability. In: Jhariya MK, Banerjee A, Meena RS, Yadav DK (eds) Sustainable agriculture, forest and environmental management. Springer, Singapore, pp 101–131. https://doi.org/10.1007/978-981-13-6830-1

Raj A, Jhariya MK, Banerjee A, Yadav DK, Meena RS (2019b) Soil for sustainable environment and ecosystems management. In: Jhariya MK, Banerjee A, Meena RS, Yadav DK (eds) Sustainable agriculture, forest and environmental management. Springer, Singapore, pp 189–221. https://doi.org/10.1007/978-981-13-6830-1

Raj A, Jhariya MK, Yadav DK, Banerjee A (2020) Climate change and agroforestry systems: adaptation and mitigation strategies. CRC Press, Boca Raton, FL, pp 1–383. https://doi.org/10.1201/9780429286759

Reij C, Tappan G, Smale M (2009) Agroenvironmental transformation in the Sahel: another kind of "Green Revolution". International food Policy research institute (IFPRI). IFPRI discussion paper 00914, Washington, DC, p 43

Rolando JL, Turin C, Ramírez DA (2017) Key ecosystem services and ecological intensification of agriculture in the tropical high-Andean Puna as affected by landuse and climate changes. Agric Ecosyst Environ 236:221–233

Royal Society (2009) Reaping the benefits: science and the sustainable intensification of global agriculture. Royal Society, London

Sawadogo H (2011) Using soil and water conservation techniques to rehabilitate degraded lands in northwestern Burkina Faso. Int J Agric Sustain 9:120–128

Schiefer J, Lair GJ, Blum WEH (2015) Indicators for the definition of land quality as a basis for the sustainable intensification of agricultural production. Int Soil Water Conserv Res 3:42–49

Shaver I, Chain-Guadarrama A, Cleary KA (2015) Coupled social and ecological outcomes of agricultural intensification in Costa Rica and the future of biodiversity conservation in tropical agricultural regions. Glob Environ Chang 32:74–86

Singh K (2009) Environmental degradation and measures for its mitigation with special reference to India's agricultural sector. Ind J Agri Econ 64(1):40–61

Snapp SS, Grabowski P, Chikowo R (2018) Maize yield and profitability tradeoffs with social, human and environmental performance: is sustainable intensification feasible? Agric Syst 162:77–88

Soderstrom B, Svensson B, Vessby K, Glimskar A (2001) Plants, insects and birds in semi-natural pastures in relation to local habitat and landscape factors. Biodivers Conserv 10:1839–1863

Song Y, Li J (2011) The role of biodiversity, traditional knowledge and participatory plant breeding in climate change adaptation in karst mountain areas in SW China. Chinese Centre for Agricultural Policy, Beijing

SRI International Network and Resources Center (2014) Improving and scaling up the system of rice intensification in West Africa. Technical Manual for SRI in West Africa, August 2014, Version 2, (SRI-Rice), Ithaca, New York, pp. 1–57

Stevenson J, Byerlee D, Villoria N, Kelley T, Maredia M (2011) Agricultural technology, global land use and deforestation: a review. CGIAR. http://impact.cgiar.org/sites/default/files/images/SPIAlandJune2011.pdf

Storkey J, Meyer S, Still KS, Leuschner C (2012) The impact of agricultural intensification and land-use change on the European arable flora. Proc Royal Soc B 279:1421–1429. https://doi.org/10.1098/rspb.2011.1686

Sutcliffe LME, Batary P, Kormann U, Baldi A, Dicks LV, Herzon I, Kleijn D, Tryjanowski P, Apostolova I, Arlettaz R, Aunins A, Aviron S, Balezentien L, Fischer C, Halada L, Hartel T, Helm A, Hristov I, Jelaska SD, Kaligaric M, Kamp J, Klimek S, Koorberg P, Kostiukova J, Kovacs-Hostyanszki A, Kuemmerle T, Leuschner C, Lindborg R, Loos J, Maccherini S, Marja R, Mathe O, Paulini I, Proenca V, Rey-Benayas J, Sans FX, Seifert C, Stalenga J, Timaeus J, Torok P, van Swaay C, Viik E, Tscharntke T (2015) Harnessing the biodiversity value of central and eastern European farmland. Diversity Distrib 21:722–730. https://doi.org/10.1111/ddi.12288

Te Pas CM, Rees RM (2014) Analyses of differences in productivity, profitability and soil fertility between organic and conventional cropping Systems in the Tropics and sub-tropics. J Integrative Agric 13(10):2299–2310

Tittonell P (2014) Ecological intensification of agriculture – sustainable by nature. Curr Opin Environ Sustain 8:53–61. https://doi.org/10.1016/j.cosust.2014.08.006

Tittonell P, Giller KE (2013) When yield gaps are poverty traps: the paradigm of ecological intensification in African smallholder agriculture. Field Crops Res 143:76–90

Tryjanowski P, Dajok Z, Kujawa K, Kałuski T, Mrówczyński M (2011) Threats to biodiversity in farmland: are results from Western Europe good solution for Poland? Polish J Agron 7:113–119

Tscharntke T, Klein AM, Kruess A, Steffan-Dewenter I, Thies C (2005) Landscape perspectives on agricultural intensification and biodiversity—ecosystem service management. Ecol Lett 8:857–874. https://doi.org/10.1111/j.1461-0248.2005.00782.x

UNEP (2010) A report of the working group on the environmental impacts of products and materials to the international panel for sustainable resource management. In: Hertwich E, van der Voet E, Suh S, Tukker A, Huijbregts M, Kazmierczyk P, Lenzen M, McNeely J, Moriguchi Y (eds) Assessing the environmental impacts of consumption and production: priority product and materials. UNEP, Nairobi

UNEPWCMC (2011) The UK National Ecosystem Assessment: synthesis of the key findings, Cambridge, UK, p 85

Vanlauwe B, Coyne D, Gockowski J (2014) Sustainable intensification and the African smallholder farmer. Curr Opin Environ Sustain 8:15–22

Vidal J (2013) India's rice revolution. The Guardian. http://www.guardian.co.uk/globaldevelopment/2013/feb/16/india-rice-farmers-revolution

Weltin M, Zasada I, Piorr A (2018) Conceptualising fields of action for sustainable intensification-a systematic literature review and application to regional case studies. Agric Ecosyst Environ 257:68–80

Wezel A, Soboksa G, Mcclelland S (2015) The blurred boundaries of ecological, sustainable, and agroecological intensification: a review. Agron Sustain Dev 35:1283–1295

Willy KD, Muyanga M, Jayne T (2019) Can economic and environmental benefits associated with agricultural intensification be sustained at high population densities? A farm level empirical analysis. Land Use Policy 81:100–110

World Agroforestry Centre (n.d.) Faidherbia Albida. Keystone of evergreen agriculture in Africa. http://www.worldagroforestry.org/sites/default/files/F.a_keystone_of_Ev_Ag.pdf
Xie H, Huang Y, Chen Q, Zhang Y, Wu Q (2019) Prospects for agricultural sustainable intensification: a review of research. Land 8:157. https://doi.org/10.3390/land8110157
Yami M, Van AP (2017) Policy support for sustainable crop intensification in eastern Africa. J Rural Stud 55:216–226
Zaal F, Oostendorp RH (2002) Explaining a miracle: intensification and the transition towards sustainable small-scale agriculture in Dryland Machakos and Kitui districts, Kenya. World Dev 30:1271–1287

Ecologically Harmonized Agricultural Management for Global Food Security

2

Yevhen Mishenin, Inessa Yarova, and Inna Koblianska

Abstract

About 11% of global population is undernourished today and the society is expected to run into the grave concerning the 2030 zero hunger goal achievement. The environmental factors are among the keys threat in this case, specifically, climate changes and shocks, deterioration of land and soils, ecosystems' destruction influencing agriculture's capacity to provide enough food of certain quality. These environmental problems are caused by agriculture itself in a large measure. In view of this there is the need to come to grips with socio-ecological and economic aspects of agricultural greening on the way towards the global food security.

The chapter provides the systemic overview of environmental aspects of agriculture and food provision, outlining the main nature-sector interrelations, the most urgent environmental problems associated with feeding the world, as well as reveals social and environmental peculiarities of industrial model of agriculture using the Ukrainian case. It should be outlined, that along with positioning the Ukraine as a "Food Basket of Europe", the National food security is failed as the amount of basic food in the diet of average Ukrainian is lacking. Moreover, every dollar of agricultural output is becoming more expensive for Ukrainians considering all types of environmental impact (air emissions, wastes, including pesticides wastes, sown area with pesticides, freshwater consumption

Y. Mishenin
Department of Economics, Entrepreneurship and Business-Administration, Sumy State University, Sumy, Ukraine

I. Yarova
Department of International Economic Relations, Sumy State University, Sumy, Ukraine

I. Koblianska (✉)
Department of Economics and Entrepreneurship, Sumy National Agrarian University, Sumy, Ukraine

M. K. Jhariya et al. (eds.), *Ecological Intensification of Natural Resources for Sustainable Agriculture*, https://doi.org/10.1007/978-981-33-4203-3_2

and withdrawal). This case highlights the importance of sound policy towards the agriculture sustainability. Under these circumstances, modern challenges of the world were identified and are facing on the way to food and environmental security, namely: the production of sufficient food for the own consumption, as well as for the import; the creation of strategic and insurance food stocks, as well as the food export possibility; the provision of optimal and rational structure of foodstuffs consumed; ensuring the ecological quality of the consumed food within the existing food structure; the socio-economic accessibility of food; the environmental component associated with the agricultural production.

It is stated that the greening of agribusiness and agro-food sphere embraces the transformation of existing technological agro-production methods towards maximizing the output of high-quality ecological agricultural products along with preserving environment. In this context, the chapter investigates and classifies the possible existing innovative solutions within the framework of eco-intensification, climate smart agriculture and sustainable agriculture concepts.

All the same, institutional transformations are the key for the movement towards the environmentally friendly agricultural practices. It embraces the conceptualization of sustainable agriculture and its basic principles such as partnership, integration, ecosystem and environmental management, equity for all generations and civilized competitiveness. The appropriate organizational and economic mechanism needs to be put in place to promote the sustainable agriculture. It is a set of subsystems of supporting, organizing, regulating and controlling agriculture resource use nature, and requires the implementation of environmentally adjusted prices for agriculture resources and food, as well as changes of agriculture producers' behaviour, i.e. more environmentally and socially responsible. Specifically, in order to fully compensate the economic damage from environmental pollution through the pricing system, it is suggested to calculate and use a price increase index considering the ecological component of the production cost.

Finally, the chapter also disclosures the role of agriculture in local communities' development, searching for the best model of agriculture organization and agrarian policy consistent with sustainable rural development goals. The local food concept implementation is seen as a main strategy for the elaboration of a policy addressing the issues of industry and community sustainable development.

Keywords

Agriculture · Agricultural sustainability · Eco-intensification · Environment · Food security · Sustainable development

Abbreviations

FAO	Food Agriculture Organization
IFPRI	International Food Policy Research Institute
R&D	Research and Development
GDP	Gross domestic product
EU	European Union
CSA	Climate smart agriculture
SDG	Sustainable development goals

2.1 Introduction

Modern agri-production methods, which have increased its efficiency and volume, deplete agro-ecosystems at different scales (from local to global) (Mishenin and Koblianska 2016; Spiess 2016; Müller et al. 2016; FAO 2017; Yatsuk 2018; Pingali et al. 2019; Koblianska and Kalachevska 2019; World Bank 2020). This leads to a search for improved methods of agricultural management (Bartolini and Brunori 2014; Pangaribowo and Gerber 2016; Delzeit et al. 2017; Zilberman et al. 2018; Asfaw and Branca 2018; Ickowitz et al. 2019).

The environmental deterioration due to a significant increase in anthropogenic and technogenic load on the environment requires the dominant achievement of resource-ecological safety of nature and agricultural management (Delzeit et al. 2017; Ickowitz et al. 2019; Raj et al. 2020; Banerjee et al. 2020; Kumar et al. 2020). Thus, among the priorities of sustainable socio-economic development is the necessity of environmentally balanced agro-economy (Ullah et al. 2020), which is impossible without reorientation of the agricultural organizational-economic mechanism to the rational use and conservation of natural and land-resource potential (Mishenin et al. 2015; Mishenin and Yarova 2019). That is the desirable way to solve the problem of food security ensuring (Gaffney et al. 2019; Nicholls et al. 2020), which is multifaceted (Breeman et al. 2015; Devaux et al. 2020) and covers issues on providing the enough food supply, the availability, stability and quality of the latter (Delzeit et al. 2017; Ickowitz et al. 2019), and appears both at local (Strochenko et al. 2017; GRFC 2020) and global scale (FAO 2018, 2019b; GRFC 2020; World Bank 2020).

Feeding about 9.7 billion people in the next 30 years will require an increase of food supply over 50% of current volume (Diaz-Ambrona and Maletta 2014; Konuma 2018; World Bank 2020). This poses a significant risk of environmental pressure aggravation (Gowdy 2020), concerning the nature driven character of agriculture (Gaffney et al. 2019; Nkonya et al. 2016; Chakravorty et al. 2007; Andrade et al. 2019; Lipper and Zilberman 2018; Tonitto et al. 2018; Adenuga et al. 2019). It is clear now that disregarding the ecological and economic foundations of agricultural land use will continue the acceleration of the eco-degradation of land-resource potential (Diaz-Ambrona and Maletta 2014), reduce the ecological and economic efficiency of agricultural management (FAO 2017), deepen the socio-environmental problems of food security (GRFC 2020; Meena et al. 2020) and even threaten the achievement of the development goal of zero hunger (FAO 2017, 2018, 2019b; Gowdy 2020). Despite that, many socio-ecological and economic issues in the field of agro-economy (in particular, sustainable land-potential use) are still remaining unresolved concerning development of strategic guidelines and mechanisms for greening agriculture to ensure global food security (Spiess 2016; Ickowitz et al. 2019; Gaffney et al. 2019). Under these circumstances, we have investigated the socio-ecological and economic aspects of agricultural greening on the way towards the global food security.

2.2 Agriculture and Nature: Interrelation, Influence and Issues to Be Solved

Agriculture is a vital industry as it provides humanity with food at any form of social organization (from times of gathering and hunting until modern with genetic and nanotechnologies of food production). At the same time, the industry is very nature-dependent and environmentally driven (Gaffney et al. 2019; Andrade et al. 2019; Tonitto et al. 2018; Adenuga et al. 2019; El Bilali 2019). Agricultural practices carried out in all parts of the world affect a single natural space, resulting in a change in both local and global conditions for agricultural activities, and, in particular, the production of the required amount of food. Therefore, natural constraints for humans in meeting their basic needs are on full display in agriculture (Meena et al. 2018; Meena and Lal 2018). That covers available land areas, water resources, favourable weather and climate, etc. However, the whole range of complex agriculture and nature interrelations is not fully investigated and recognized. Moreover, a fairly large number of agricultural producers neglect scientifically sound principles of rational land use because of lag time in cause-effects and for the sake of need to provide a certain level of income (Mishenin and Koblianska 2016; Fatemi and Rezaei-Moghaddam 2019). This leads to the sweeping and irreversible adverse environmental effects, which are already palpable.

Basically, agricultural activities commit land, water, space (infrastructure) and ecosystem resources under certain climate conditions, consuming nutrients, energy, human-made inputs and thus resulting in water and air pollution and emissions, land and soils deterioration, ecosystem degradation, climatic change, etc. (Fig. 2.1) (Gaffney et al. 2019; Nkonya et al. 2016; Chakravorty et al. 2007; Andrade et al. 2019; Lipper and Zilberman 2018; Tonitto et al. 2018; Adenuga et al. 2019). In particular, agriculture uses 70% of water (World Bank 2020) and consumes about 30% of global energy consumption (FAO 2017). The sector also accounts for 18% of globe carbon dioxide emissions (with animal husbandry accounting for about 64% and 21% for growing rice) (Pingali et al. 2019).

Along with that, there is a need to investigate the whole food chain for the assessment of total environmental impact caused by agriculture food production.

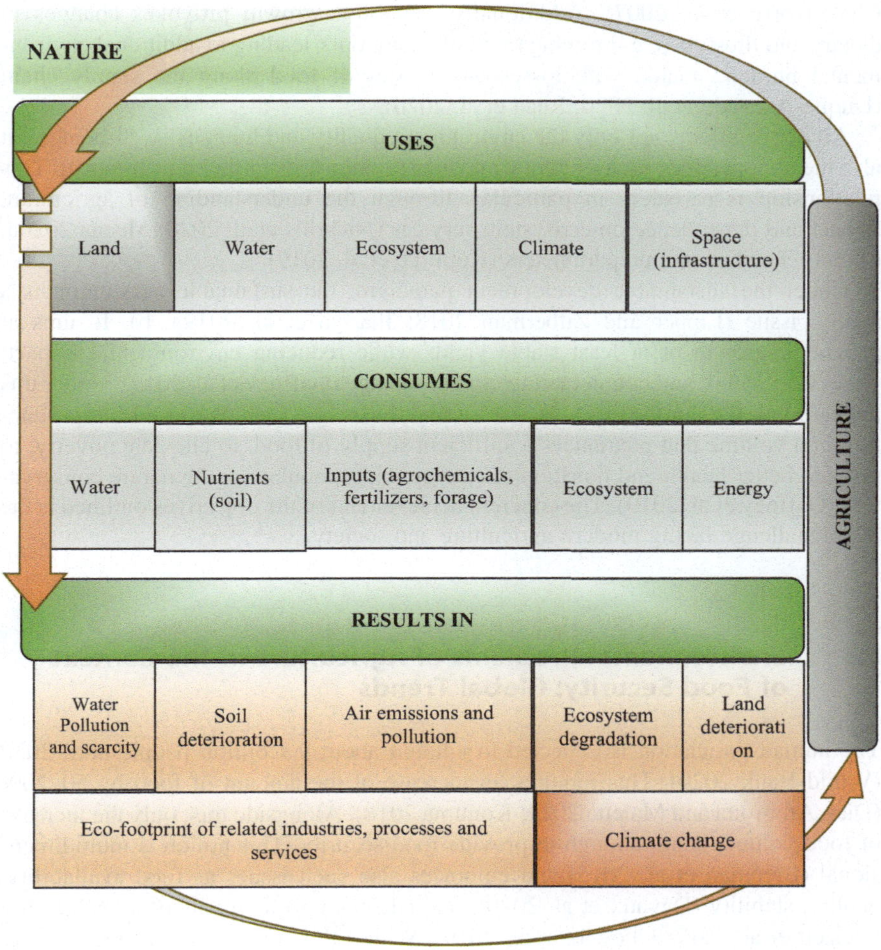

Fig. 2.1 Agriculture and Nature: the main interrelations (developed by Koblianska on the base of Gaffney et al. 2019; Nkonya et al. 2016; Chakravorty et al. 2007; Andrade et al. 2019; Lipper and Zilberman 2018; Tonitto et al. 2018; Adenuga et al. 2019; Pingali et al. 2019)

In this regard, one should emphasize on the significant ecological footprint of agriculture related industries both upper and downstream, e.g. the so-called virtual water footprint of animal husbandry (Spiess 2016). In this context, the production of human-made inputs (agrochemicals, mineral fertilizers, antimicrobials, industrial feed) is gaining attention (Andrade et al. 2019; Lipper and Zilberman 2018; Meena et al. 2020), as well as appropriate wastes generation (Spiess 2016). The application of human-made materials leads to the irreversible change in all components of natural environment (soil and water pollution, pest and weed resistance, losses of biodiversity, etc.) (Gaffney et al. 2019; Nkonya et al. 2016;

Chakravorty et al. 2007). Additionally, economic growth provokes changes in dietary and lifestyle (e.g. growing trend of eating out), leading to additional environmental burden, related with losses and wastes of food along the supply chain (Duque-Acevedo et al. 2020; Read et al. 2020).

All above affects not only the environment quality and humans' well-being, but also the opportunities for agriculture and related industries further development. This relationship is revealed, in particular, through the understanding of agriculture impact and dependence on ecosystem services (Nicholls et al. 2020; Meena 2020a, b) and related economic parameters (Kopittke et al. 2019).

Under the sustainable development paradigm, the sustainable agriculture is a raising issue (Lipper and Zilberman 2018; Jhariya et al. 2019a, b). It aims at providing growth or at least stable yields while reducing environmental impact, preserving nature and counteracting climate change (Gaffney et al. 2019). Under this background, the competing objectives come to the fore, namely: to ensure production in a volume that guarantees a sufficient supply of food, to alleviate poverty, to provide better health and nutrition for the growing population, the nature conservation (Gaffney et al. 2019). The coherent achievement of the objectives outlined is the main challenge facing modern agriculture and society.

2.3 Environmental Problems of Agriculture in the Context of Food Security: Global Trends

The human population is expected to amount about 9.7 billion people until 2050 (World Bank 2020). This requires an increase of the amount of food by 50–70% (Diaz-Ambrona and Maletta 2014; Konuma 2018). Alongside this, only the increase of food volume is not enough to provide food security. This notion is multidimensional (Breeman et al. 2015) and embraces also such issues as food availability, quality, stability (Devaux et al. 2020), and adequacy to the goals of a healthy life (Delzeit et al. 2017; Ickowitz et al. 2019). Within this framework, a food security assurance requires lesser agriculture production growth, than as sound economic, social and technological policy and measures. However, the increase of agriculture productivity and output remains the important target in this context, especially for agriculture dependent people, social groups and even countries (Funk and Brown 2009). Under these circumstances, environmental problems of agriculture are getting new sound, because increase of production results in corresponding increase of input resources and waste (Diaz-Ambrona and Maletta 2014). Thus, ensuring environmental security towards achieving the food security goals at the global scale is one of the main challenges for further agriculture development.

According to FAO's estimates, the scarcity of resources available for food production will increase until 2050 significantly, leading to aggravation of competition, unsustainable and destructive use of resources, thereby endangering the welfare of millions of farmers, foresters, fishermen and other agriculture dependent groups. About 33% of agricultural lands are already medium- and highly degraded and further expansion of agricultural land threatens deforestation, especially in countries of Asia, South and Central America (FAO 2017).

Despite the sufficiency of existing resources to provide food for up to 10 billion people, the major concern is the allocation of available resources to fully meet food security goals (Spiess 2016). This results in a fair amount of undernourished people and even hunger.

According to the FAO, over 820 million people were hungry and undernourished in 2018 (FAO 2019b), that is every ninth (World Bank 2020). Among them, over 113 million people felt severe hunger, being unable to provide the necessary food and nutrition (Table 2.1). That was observed in 53 countries (GRFC 2020). Upward of 100 million people annually was suffering from severe hunger across the world, despite the gradual decline in their number for 2016–2018. Additionally, about 143 million people in 42 countries in 2018 were living on the edge of starvation (GRFC 2020). Alongside this, the existence of millions of people liable to obesity and overweight reveals another dimension of food security concept, i.e. food quality (World Bank 2020).

The problem of hunger and undernourishment is deepened under the background of continued climatic changes and existing climate shocks. Particularly, climatic factors caused 25.7% of severe hunger cases (29 million people) in 2018 (GRFC 2020). Climate changes lead to further deterioration of natural capital, degradation of ecosystems, water scarcity, climate shocks (drought, floods, storms), etc. (Khan et al. 2020a, b). That influences the agricultural production substantially, especially on crop production (49% of the total climate-related losses in agriculture) and animal husbandry (36% of industry's losses) (FAO 2018). This leads to a decline of incomes of agriculture dependent people (FAO 2018), challenging their capacities to provide enough food of a needed quality, as well as national and global food security (Spiess 2016).

Climate changes are expected to continue significantly threatening the food security through droughts, winds, flooding, affecting total food production across the world and small farmer's activities mainly (GRFC 2020). In this context, the maintenance of the achieved level of agricultural productivity needs substantial investments, as well as clear policy aimed at responding to climate change (World Bank 2020; FAO 2018). At the same time, the last data show a decline of agricultural research funding in different regions of the world (Fig. 2.2). Moreover, according to the IFPRI IFPRI (2020), government and donor funding for agro-research in Africa declined by 5% over the 2014–2016 period, and a share of R&D spending in 2016 have been slided to only 0.39% of GDP, which is critical. In particular, it was expected that agricultural productivity in the Sub Sahara African countries could increase by 62% until 2050 compared to the current level with the agricultural R&D funding as of 1% GDP (IFPRI 2020). It should be emphasized that only high-income countries finance agro-research at the level of more than 1% of GDP, whereas for other countries this figure did not exceed 0.6% since the century (Fig. 2.2). Therefore, the significant potential of the industry remains untapped in a global scale.

Table 2.1 Number of undernourished, 1990/92–2030 (FAO 2017)

Region	1990–1992		2000–2002		2005–2007		2014–2016		2030	
	%	million	%	million	%	million	%	Million	%	million
High-income countries	<5.0	32	<5.0	36	2.2	29	1.6	23	1.1	16
Low- and middle-income countries	29.7	978	24.5	894	17.6	920	13.2	775	9.3	637
East Asia	28.2	432	20.3	339	15.9	311	11.1	233	7.8	175
South Asia	25.1	284	19.0	258	20.5	311	14.9	257	9.3	188
Sub-Saharan Africa	45.9	173	40.4	201	29.0	212	23.3	213	17.4	216
Latin America	22.1	66	18.3	60	8.4	47	6.1	37	4.0	27
Near East	14.5	20	24.8	33	8.3	36	6.5	33	4.7	29
World	18.6	1011	14.9	930	14.4	949	11.0	797	7.9	653

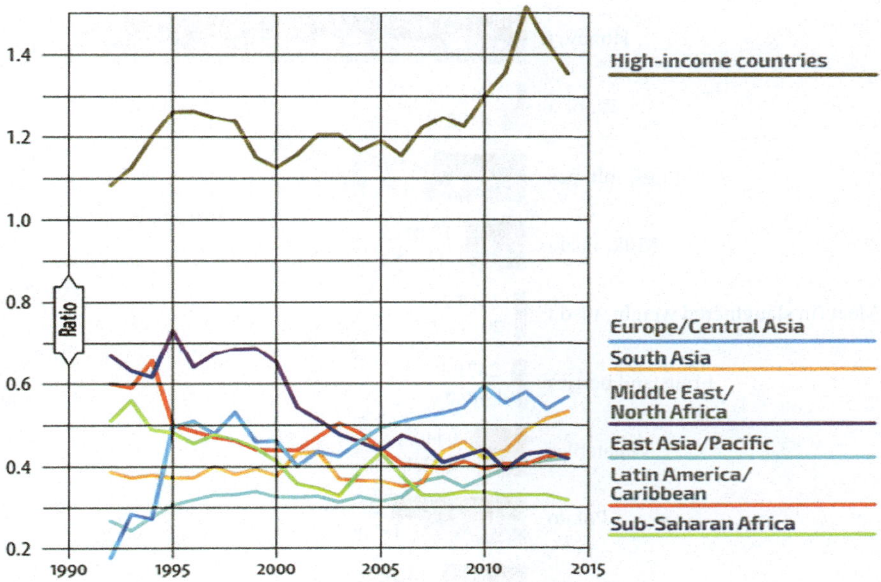

Fig. 2.2 Agricultural investment orientation ratio, 1990–2015 (FAO 2017)

Unfortunately, nowadays it becomes obvious that humans are not able to ensure food security until 2030 (FAO 2018, 2019b) and environmental factors are one of the key affecting that (FAO 2017, 2018).

Following the above, it is desirable to investigate the trends of agriculture development in countries with substantial potential in terms of ensuring the global food security goals and the Ukraine is one of them.

2.4 Food Provision and Environmental Impact: National Peculiarities for Ukraine

According to 2019 data, Ukraine reached the third position in the rankings of food exporters for the EU with exports of EUR 6.3 billion. In 2019 agricultural exports counted for 42.9% of the country's total export with the grain as the main export product. The agricultural products of Ukrainian origin were exported to China (8.9%), India (8.3%), Egypt (8.2%), Turkey (7.6%), the Netherlands (7.1%) and other countries including the EU (BUM 2019). Thus, the Ukraine regained the title "Food Basket of Europe", however, social and environmental consequences of rapid agriculture growth and development remained behind the scenes (Koblianska and Kalachevska 2019).

First of all, it should be noted that Ukrainian agriculture is a bipolar with large scale export targeted agro-holdings and small-scale (mainly of a semi-subsistence nature) producers competing for resources (Strochenko et al. 2017; Koblianska and

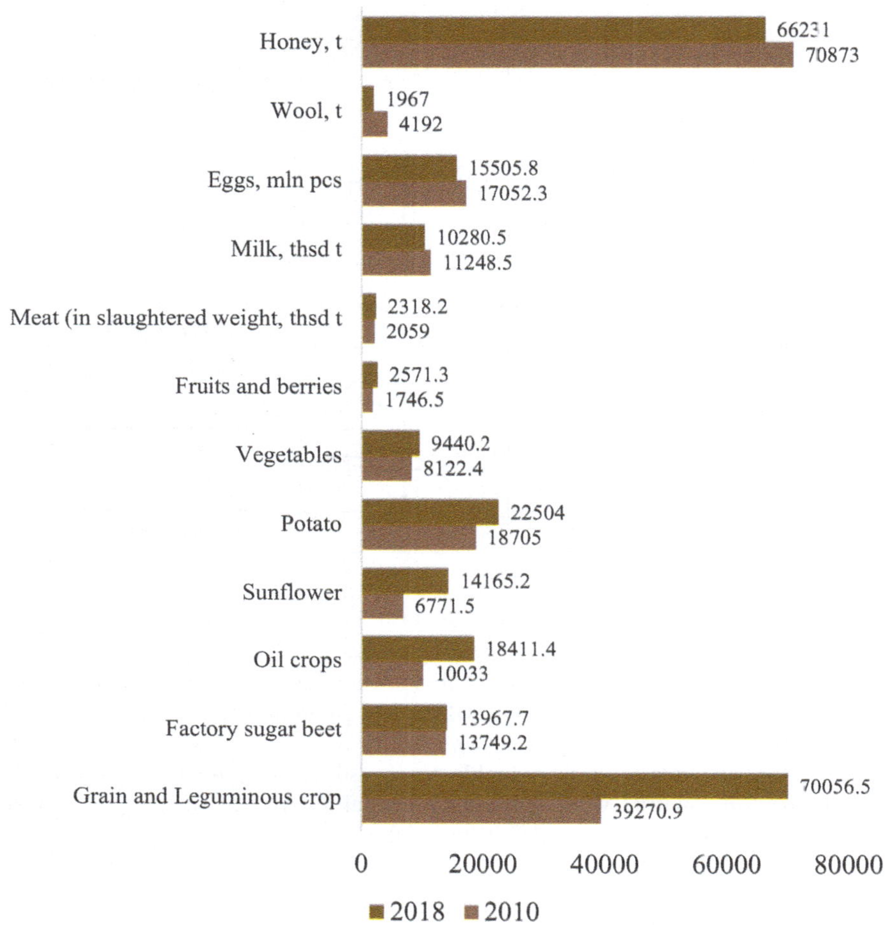

Fig. 2.3 The agricultural output in Ukraine, thsd.ton (http://ukrstat.gov.ua)

Kalachevska 2019). The small agri-producers provide the main share of agricultural products for final consumption, i.e. vegetables, potatoes, milk, meat, fruits, etc.

The offensive development of the industry, specifically, the expansion of a large agribusiness has led to the significant changes of agriculture gross output for the last years (Fig. 2.3). It is notable that the production of grain and leguminous, oil crops has increased by 1.8 times, sunflower—more than 2 times, but livestock output and honey—have decreased in 2018 compared to 2010 (Verner 2019). The significant increase of cereals production is also notable with a view to the gross agricultural output per capita (Fig. 2.4).

Having regard to the above, it is necessary to investigate whether such an increase in agriculture output provides income growth (Fig. 2.5).

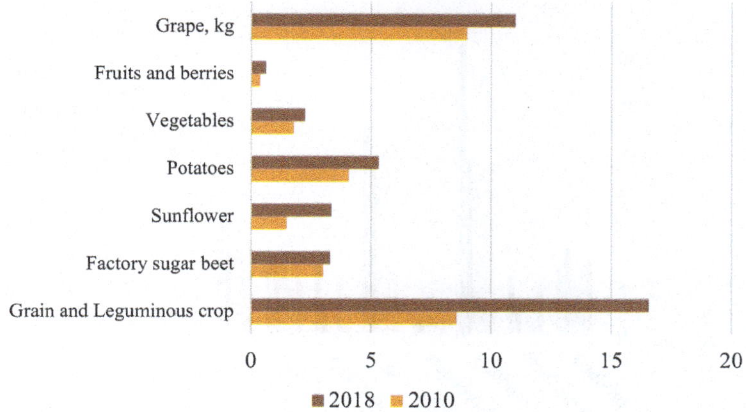

Fig. 2.4 Production of agricultural crops in Ukraine per capita (http://ukrstat.gov.ua)

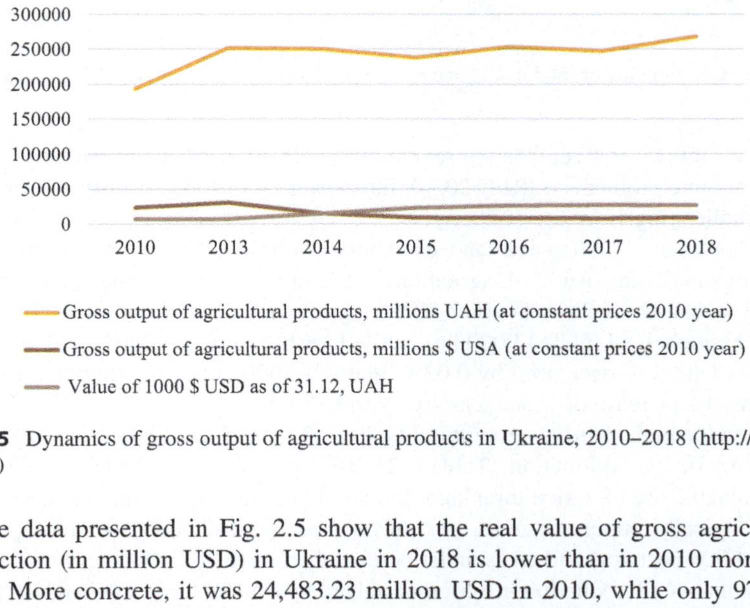

Fig. 2.5 Dynamics of gross output of agricultural products in Ukraine, 2010–2018 (http://ukrstat.gov.ua)

The data presented in Fig. 2.5 show that the real value of gross agricultural production (in million USD) in Ukraine in 2018 is lower than in 2010 more than twice. More concrete, it was 24,483.23 million USD in 2010, while only 9730.34 million USD in 2018. Under these conditions the availability, accessibility and quality of food (i.e. variety of diet) for Ukrainians require in-depth study (Fig. 2.6).

As it is shown in Fig. 2.6, the amounts of food consumed by the Ukrainians are not sufficient almost in respect of all food groups. It is true for both rural and urban residents. The diet of the average Ukrainian resident is not healthy, considering the excess consumption of bread and bakery products and the lack of other necessary products. Moreover, rural residents experience the lack of basic food more than urban habitants, revealing the adverse social and economic effects of agriculture industrialization. Taking into consideration that about 1.1 million of people in the

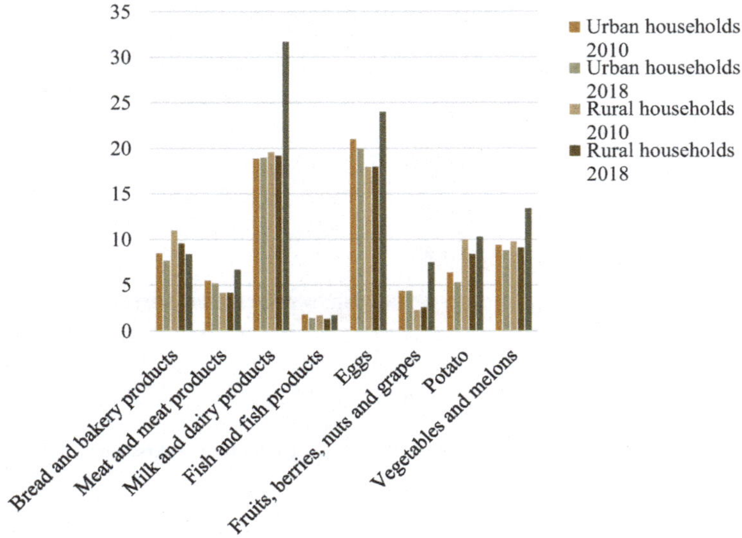

Fig. 2.6 Consumption of food in Ukraine by one person per month, kg (http://ukrstat.gov.ua)

Luhansk and Donetsk regions felt severe hunger in 2018 caused by military conflict and economic problems (GRFC 2020), thus ensuring of food security in Ukraine is quite challenging.

The problem of extensive land use should be pointed the first concerning the environmental dimension of agriculture development in Ukraine. According to official statistics for 2018 (Verner 2019), agricultural land occupies 68.7% of the total Ukraine's land (60354.9000 ha), while forests count only for 17.7% and water 4%. The latter has decreased by 0.02%, that is 1.5000 hectares, compared to 2010, showing the increase of water scarcity (Verner 2019).

Given the high share of agricultural land in Ukraine, it is necessary to investigate the forms of its exploitation (Table 2.2). The data presented show the extensive unsustainable use of agricultural land. In particular, there is a significant decrease of an ecologically important areas, i.e. conversions (by 192.3000 hectares), pastures (by 99.8000 hectares) and areas under perennial crops (by 37.1000 hectares). On the back of an overall reduction in agricultural land by 0.81%, the arable land decreased by only 0.06%. So, as of early 2018, arable land occupies 54% of the country.

Figure. 2.7 shows the allocation of land resources for different crops. It is notable that there is an increase of areas under export targeted crops (wheat, maize, sunflower). The share of area under grain and leguminous crops reached 53.57% of sown areas, under industrial crops—33.45%, fodder—6.39% in 2018. So, it is obvious that export orientation of agriculture leads to the problem of monoculture, resulting in over-exploitation, degradation and depletion of land and soils. In particular, the results of the 10th round of agrochemical survey of Ukrainian soils (2010–2015) indicate that the soils have lost a considerable part of humus, and the

Table 2.2 Dynamics and structure of agricultural land in Ukraine (http://ukrstat.gov.ua)

Land type	31 December 2000		31 December 2017		Change for 2000–2017	
	thsd. ha	Share (%)	thsd. ha	Share (%)	thsd. ha	Share (%)
Agricultural land, total	41,827	100.00	41,489.3	100.00	−337.7	−0.81
Incl. arable land	32,563.6	77.85	32,544.3	78.44	−19.3	−0.06
Perennial crops	931.9	2.23	894.8	2.16	−37.1	−3.98
Conversions	421.6	1.01	229.3	0.55	−192.3	−45.61
Hayfields	2388.6	5.71	2399.4	5.78	10.8	0.45
Pastures	5521.3	13.20	5421.5	13.07	−99.8	−1.81

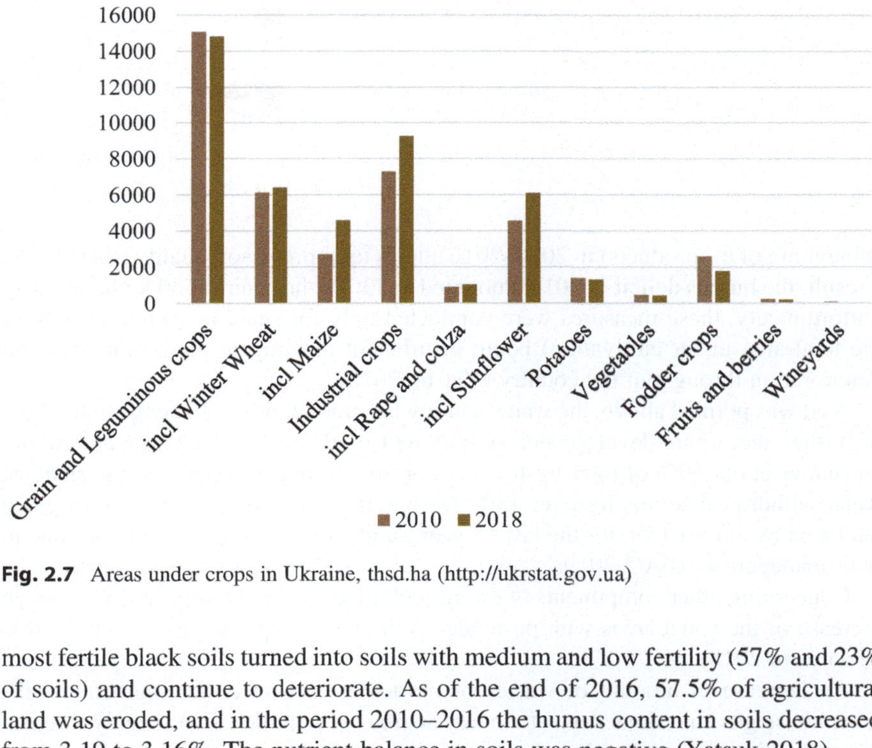

Fig. 2.7 Areas under crops in Ukraine, thsd.ha (http://ukrstat.gov.ua)

most fertile black soils turned into soils with medium and low fertility (57% and 23% of soils) and continue to deteriorate. As of the end of 2016, 57.5% of agricultural land was eroded, and in the period 2010–2016 the humus content in soils decreased from 3.19 to 3.16%. The nutrient balance in soils was negative (Yatsuk 2018).

Farmers have increased the application of mineral fertilizers trying to compensate the losses of natural soil's fertility. However, as of end of 2018, 9% of the area under cultivation remained untreated. The application of organic fertilizers is also insufficient, covering only 4.4% of the sown areas with the amount of 0.6 tons/ha, while the minimally required quantity to support soil fertility is 8 tons/ha. All this leads to a dampening of the soil formation process and further dehumidification (Yatsuk 2018). The implementation of certain measures of agriculture biologization

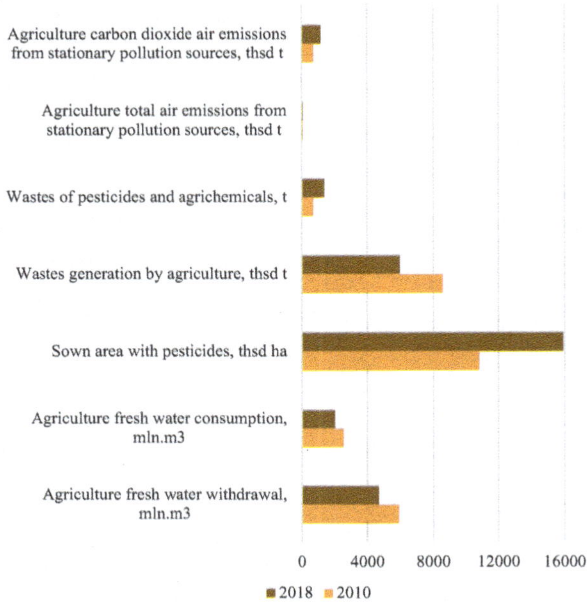

Fig. 2.8 Environmental impact of agriculture in Ukraine (http://ukrstat.gov.ua)

(ploughing of by-products) in 2014–2016 allows to improve soil quality partially. As a result, the humus deficit in 2015 amounted to 130 kg/ha against 530 kg/ha in 2010. Unfortunately, these measures were conducted only for small areas (about 35% of the total area under cultivation) being insufficient to mitigate the problem of soil deterioration throughout the country (Yatsuk 2018).

As it was pointed above, the water scarcity becomes another challenge in the light of further agriculture development. According to 2018 official statistics, agriculture consumes about 30% of total fresh water consumption in Ukraine and the share of water withdrawal counts for over 43%. Despite this, there is a decrease of irrigated land area by almost 15% for the last 15 years, and an increase of water losses due to poor management (FAO 2019a).

Concerning other components of the agriculture ecological footprint (Fig. 2.8), an increase of the sown areas with pesticides is the most palpable (up to 15,908.8000 hectares in 2018).

An increase of amount of wastes of pesticides and unsuitable agrochemicals, greenhouse gas emissions from stationary sources of pollution significantly threaten the environmental quality, but the amount of water consumed and withdrawal, as well as wastes generated demonstrates the positive trends in absolute terms (Fig. 2.8). Alongside this, our estimates of the environmental burden in reference to the value of gross output indicate that every dollar of agricultural output is becoming more expensive for Ukrainians by all types of environmental impact (Fig. 2.9).

Ukraine also does not keep out of climate change processes although the negative impact of these processes on the domestic industry will not have catastrophic

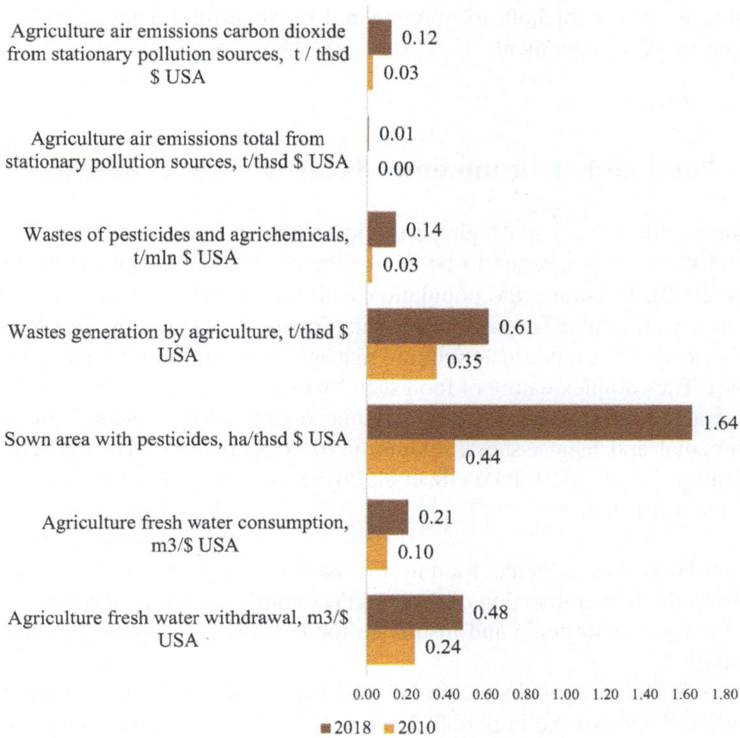

Fig. 2.9 Environmental impact of agriculture in Ukraine per unit of gross output (http://ukrstat.gov.ua)

consequences in the nearest future (Müller et al. 2016; FAO 2015). Specifically, it is expected that wheat yields in a southern Steppe zone will decrease with concurrent modest increase at the North of the Ukraine under the higher emissions scenario, which is seen the most probable (Müller et al. 2016). The similar results were outlined by FAO (2015). The forecast made for 2020 shows that governments' expectations for an increase of grain production have been overestimated. Based on the historical trends of climate changes it is forecasted the reduction of yields in traditional zones due to drought with slightly increase of yields in northern parts of the country (FAO 2015).

The researchers called for the elaboration of the regional specific sound policy and measures dealing with climate change (Müller et al. 2016; FAO 2015). However, government action towards responding the climate changes is relatively slow. In particular, the "Strategy for Prevention and Adaptation to Climate Change of Agriculture, Forestry, Hunting and Fisheries of Ukraine by 2030" is under approval by the ministries as of March 2020 although its approval was planned for 2019–2020 (CMU 2016). As for now, there are no any progressive approaches aimed at promoting measures for both to mitigate and to adapt to climate changes, as it is envisaged by Paris agreement.

2.5 Food and Environmental Security: New Challenges

The food problem has been high on the international agenda for the past several years and therefore it appears to be one of the overarching problems of our times (Gowdy 2020). Providing the population with high-quality, ecologically safe and economically affordable foods as well as the formations of necessary insurance stock are at the heart of the modern agriculture management at different scales (from local to global). The complex nature of food security problem requires a systematic vision and integrated solutions towards the economic, organizational, technological, social, environmental and legal issues (Shkuratov 2016; Kupinets 2010; Mishenin et al. 2015; Gaffney et al. 2019; Ickowitz et al. 2019). Among them the following issues are of great importance:

1. The production of sufficient food for the own consumption, as well as for import. It shows the interconnection of food and economic (national) security.
2. The creation of strategic and insurance food stocks, as well as the food export possibility.
3. The optimal and rational structure of foodstuffs (assortment) is consumed by the population. One of the important indicators of providing the country's residents with food products is the observance of scientifically based norms of rational nutrition. An integral indicator of the rational nutrition is the calorie content of the daily set of food products per capita. An almost complete correspondence between the norms of nutrition and the actual provision of the population with food products in Ukraine was achieved in 1990. However, in 1995 the calorie content of the daily set of food products was only more than 70% compared to the base year (Kupinets 2010). At present, the situation has gone worser. It is important and necessary to evaluate the diversity (assortment, structure) of the actual caloric content for daily food consumption, which can significantly determine the level of public health.
4. The ecological quality of the consumed food is within the existing food structure. At the same time, the quality of agricultural products can largely determine its competitiveness. The ecological quality of food significantly affects the elements of economic and national security and the level of life quality.
   Studies indicate that almost all food products are contaminated with a complex of hazardous substances at a level higher than sanitary and hygienic standards (Kupinets 2010). It leads to large losses, which until recently have not been sufficiently estimated. At least half of cases of morbidity, disability and mortality are due to consumption of contaminated food. These losses as a whole account for more than half of all damage from environmental pollution, which in Ukraine exceeds $12 billion/year (Tsarenko 2001).

5. Socio-economic accessibility of food, taking into account quantitative and qualitative parameters of consumption for harmonious human development.
6. The environmental component is associated with the agricultural production. It is especially worthwhile to environmental safety of the agricultural land use (Kupinets and Zhavnerchik 2016).

The main goal of greening agribusiness and agro-food sphere is to solve ecological and economic contradictions between society and nature by transformation of existing technological agro-production methods in the direction of maximizing output of high-quality ecological agricultural products while the environment preserving. Greening the food security is an objective process, aimed at the more rational agro-natural resources use by reducing the negative environmental effects of agricultural production and avoiding disturbances of ecological equilibrium on the basis of reproductive ecological processes. Therefore, greening of the agro-production cannot be considered as an isolated area of activity, but it should be a harmonious component at all levels of sustainable spatial development. Here it should be noted that agriculture plays a dual role: firstly, it produces food and secondly, it creates jobs for households. As agriculture is the largest employer in the world, at the same time, productivity gains can create additional purchasing power for the rural population, which in turn will use this extra income to purchase more food and other basic consumer goods (Mishenin et al. 2011). Large scale agricultural production will also help expand agrarian-based food industries, which will also stimulate new businesses and jobs.

Improving the agricultural land productivity through the use of safe innovation technologies will stimulate the real incomes and savings increase; job creation and diversification of agricultural production; increasing land value and investments; creation of new agricultural markets; increase of the public purchasing power in the services sphere; increase of public social security. The sustainable agriculture is closely related with the food security (Fig. 2.10).

However, the sustainable agricultural production is not sufficient to achieve food security goals. Even in case of the adequate food supply the lack of employment opportunities can lead to malnutrition. Sustainable agricultural development must be considered in a broader political context: strengthening the role of other employments will help to reduce eco-destructive pressure on lands.

Thus, achieving food security depends on the key prerequisites as follows:

1. The volumes and quality of agricultural production are determined by the following components: production, human capital; greening reproduction processes (agricultural management); natural (land) capital. All these components should be formed on an innovative basis, which implies an entrepreneurial approach to their effective implementation.
2. Food consumption is characterized by the main parameters as follows: socio-economic availability of food, consumption structure and food quality (general, technological and ecological).

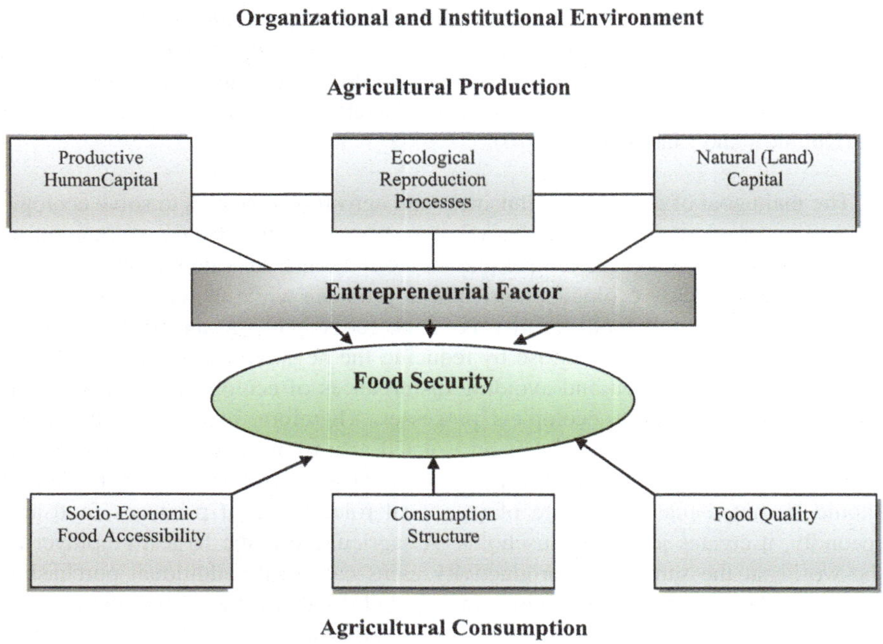

Fig. 2.10 The links between sustainable agriculture and food security (Author's development on the base of Mishenin et al. 2015)

We define *environmentally oriented food security* as a state of development of competitive, eco-balanced, ecologically safe agribusiness, which provides an optimal level of quantity and quality of food production and consumption in accordance with formed socio-environmental parameters of life quality on the basis of legal, technological, innovative, economic, informational and social mechanisms. The wide range of modern technological as well as management solutions is available to support environmentally oriented food security. Among them the eco-intensification measures are of great importance.

2.6 Eco-Intensification in Agro-ecosystem: Possible Ways and their Outcomes

The need to intensify agricultural production becomes evident addressing the problem of feeding the global growing population (Diaz-Ambrona and Maletta 2014). This not only provides enough food supply, but also making food prices lower, thereby ensuring the food security (Delzeit et al. 2017). However, one should emphasize that traditional agriculture intensive practices lead to soil degradation, water pollution, ecosystems' destruction, etc., as a rule (Ickowitz et al. 2019). The regard on environmental perspective has led to the concept of sustainable

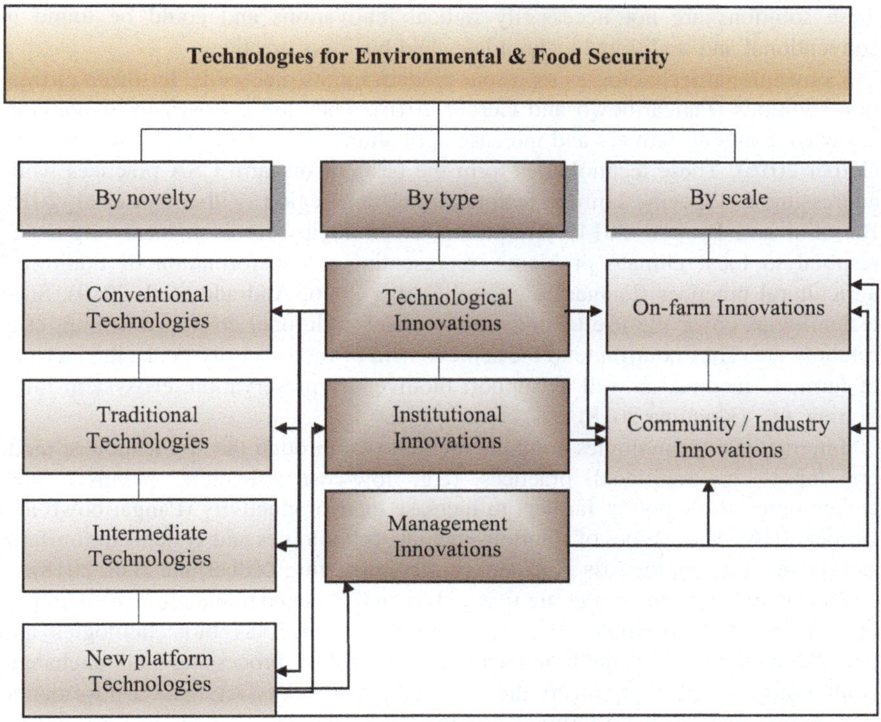

Fig. 2.11 Classification of eco-targeted technologies in agriculture (generalized on the base of Zilberman et al. (2018) and Pangaribowo and Gerber (2016)

intensification mitigating the environmental impact of agriculture industrialization (Ickowitz et al. 2019), ecological intensification concept applying ecosystem services to replace external inputs (Kleijn et al. 2019). Environmentally friendly intensification appears as win-win strategy leading to both an increase of crop yields and decrease of environmental impact (in particular, carbon emissions and nitrogen losses (Ullah et al. 2020), compared to traditional or industrialized agriculture practices. The eco-intensification promotion and implementation require a solid knowledge and technological changes, as well as institutional transformations favourable for innovations' spread and application (Delzeit et al. 2017; Kleijn et al. 2019; Ickowitz et al. 2019; Ullah et al. 2020).

Modern innovative agricultural practices make it possible to achieve the food security goals and improve the environmental quality. In particular, a wide range of solutions was released under the framework of CSA concept, addressing climate changes mitigation and adaptation issues (Zilberman et al. 2018; Asfaw and Branca 2018), and sustainable agri-practices (Pangaribowo and Gerber 2016). It is desirable to classify the prominent eco-intensification solutions for agriculture (Fig. 2.11).

Commenting on Fig. 2.11 data one should indicate that technological innovations are fundamental in terms of responding to climate change in agriculture. However, these solutions are not necessarily radical innovations and could be found in conventional and traditional technologies and business practices.

Conventional technologies represent modern inputs, i.e. seeds, fertilizers, irrigation solutions (Pangaribowo and Gerber 2016). They are exposed to disseminate knowledge among farmers and increase agriculture productivity (Pangaribowo and Gerber 2016). These technologies form the basis of on-farm CSA practices while addressing the specific climate features of certain region (Zilberman et al. 2018; Bartolini and Brunori 2014). Traditional technologies are of a local origin and respond to local climate problems, representing a transformation of traditional agricultural practices (Pangaribowo and Gerber 2016; Andrade et al. 2019). Such technologies cover the use of underutilized and traditional crops, gardening, crop rotation, etc. They contribute to the achievement of food security goals, the increase of farmers' income, as well as support biodiversity preservation, ecosystem functioning, etc. (Konuma 2018).

Intermediate technologies combine the first two through the application of modern inputs in traditional practices (e.g. low-cost irrigation, pumps). Such technologies allow poorer farmers to increase their productivity (Pangaribowo and Gerber 2016). These types of solutions include technologies and systems for on-farm storage that prevent the loss of products after harvesting (Zilberman et al. 2018).

New platform technologies are first and foremost related to the implementation of information and communication technologies, as well as biotechnologies and nanotechnologies. The modern technologies of data processing and exchange, communication gives producers the knowledge and necessary market information access, enhance the local farmers' organizations capacity, and facilitating farmers market entrance (Pangaribowo and Gerber 2016). These technologies are applicable both at the farm level (mobile farm management applications) and at the community/industry level. The latter include, but are not limited to, weather information dissemination technologies, which reduce the production uncertainty and prevent losses (such information should be available to poorer farmers as well) (Zilberman et al. 2018). Successful implementation of such technologies is closely linked to the transformation of management systems both on-farm (automation of processes) and throughout the local community (proper infrastructure, interaction and cooperation, coordination and support from all stakeholders). Therefore, new platform technologies are directly related with management and institutional innovations.

Actually, management innovations are realized by use of data processing and communication technologies use. The improved farm management through implementation of information systems for processes' monitoring, and precision agriculture technologies could serve as an example. The increased productivity and prevention of over-spending are the main outcomes of such innovative decisions (Zilberman et al. 2018).

At the community level information-based innovations may include the following: collective actions to improve the use and management of inputs (above all, sharing of new knowledge, collective action concerning externalities, regional institutions for collaboration and support for public services); insurance products; improved supply chain management (providing market access for SME farmers)

(Zilberman et al. 2018). This implies a change in the management paradigm with the formation of multilateral platforms bringing together different stakeholders and their activities towards the food security and sustainability goals. Such a scheme should provide reflexivity, resilience, response, recovery and generation of required outputs at different levels, from local to global (Breeman et al. 2015). At the same time, an imperfect system of innovation infrastructure, the lack of knowledge and skills, a weakness of the intellectual property rights system, limited funding and weak government form the obstacles to the implementation of the above technologies. In this context, conservative views (counteraction to genetic research and development) are also quite threatening. All listed problem issues are of an institutional nature and require appropriate institutional transformations (Zilberman et al. 2018).

Institutional innovations involve the transformation of values, knowledge, culture and practices of management and governance (Zilberman et al. 2018; Pangaribowo and Gerber 2016). This type of innovations encompasses social and political processes that enhance farmers' ability to act in a coordinated and collective way, combining their interests and technologies (Pangaribowo and Gerber 2016). Institutional innovations embrace an implementation of ecosystem thinking at on-farm and industry scale, the development of advisory services and knowledge dissemination, an enhanced interaction and cooperation, trade assistance and regulation, aid and its distribution (subsidies, reduction of transaction costs), conflict resolution mechanisms, insurance, development of cooperative actions (Zilberman et al. 2018), field schools, regulation of land relations, financial market development, overall market transformation (Pangaribowo and Gerber 2016).

A clear policy to support environmental risks mitigation is an important direction of an institutional transformations, as the adaptation is a restrictive measure.

2.7 Conceptual Basis for the Ecologically Harmonized Multi-Scale Agricultural Management

In the period of globalization of agricultural markets and distribution systems, the problem of ensuring world food security is becoming of great relevance. Therefore, the food security can be defined as an ability of various agricultural systems to satisfy the basic needs of a growing population and to solve environmental problems. Increasing the food production is considered to be the only prerequisite for improving food security. From these perspectives, it is worth analysing the way sustainable agriculture contributes to the food security (Spiess 2016; Horton et al. 2017; Ickowitz et al. 2019; Nicholls et al. 2020).

It is important to characterize major science schools concerning the food security (Mishenin et al. 2011; Pretty 1995; Pretty and Thompson 1996; Thompson 1996; Hazell 1995; McCalla 1994; CGIAR 1994).

1. *Environmental pessimists*: They argue that the population is growing too fast compared to the rate of increase in the yield of basic crops. With the current knowledge level, new technological breakthroughs are unlikely to take place, and

some agro-ecological systems are already so degraded that they are no longer reproducible.
2. "*Business-as-usual*" *optimists*: Proponents of this approach believe that supply will always meet growing demand. Biotechnological innovations will boost food production. The area of arable lands is also expected to increase significantly.
3. *Proponents of the industrialized approach* (*industrialized world to the rescue*): It is argued that developing countries never feed themselves due to a wide range of economic, institutional, political and environmental factors. Increasing the volume of production with the help of innovative technologies is advisable to carry out by creating large scale agrarian industrial complexes.
4. *New modernists*: It is believed that the growth of agricultural production is possible only through the involvement of a large number of external resources. New modernists believe that agricultural producers are using insufficient mineral fertilizers, pesticides, heavy yielders and other modern inputs to increase the agricultural output with simultaneous environmental impact reduction. High-resource agriculture is seen more environmentally friendly compared with the low-resource one, because an intensive use of local resources can lead to their degradation.
5. *Sustainable intensification*: This group of scientists argues for the steady intensification of agricultural production, since sustainable development contributes to the protection or even regeneration of agrarian natural resources. Low-resource agriculture can be highly productive, because land use productivity is primarily a function of human capital, and only then of biological processes.

The better utilization of available resources, in particular, biophysical and human is the main objectives of sustainable agriculture. This requires the minimization of the external resources involved, an optimization of the internal resources use, or a combination of these methods. In light of this, sustainable agriculture aims to integrate a wide range of pest management technologies, nutrients, forest plantations, soils and water resources. The by-products or waste from one element of the agro-ecosystem must be resources for its other component. As external resources are replaced by natural processes, the environmental impact will be reduced.

Consequently, the sustainable agriculture represents an agricultural food production system aimed at achieving the following objectives:

- an involvement, recurrence and restoration of natural processes (nutritive cycle, nitrogen fixation, trophic webs);
- a minimization of the non-replenishable resources use, as well as external ones;
- a participation of agricultural producers in the processes of analysis of problems, technology development, adaptation, monitoring and evaluation;
- equal opportunities and fair access to resources required for agricultural production;
- an effective and fruitful use of local resources and knowledge among others;
- raising self-sufficiency among rural communities.

2.8 Strategies for Sustainable Agriculture Addressing Food Security

Today, the search for sustainable agricultural strategies in the context of food security is necessary and undeniable. For example, it is argued that innovative farming practices that have increased its efficiency and agricultural output are depleting the agro-ecosystem, which has necessitated the search for environmental agricultural land use methods. Concerns about pesticide use, biotechnology and other issues have focused public attention on ecological quality and food security (Horton et al. 2017; Ickowitz et al. 2019), drawing interest in alternative environmentally friendly and balanced methods for its production (Mishenin et al. 2011; Lipper and Zilberman 2018; Diaz-Ambrona and Maletta 2014).

In our opinion, environmental sustainability of agriculture should mean that the used agricultural resources must be renewed by the same process of their use. In order for a system of agrarian nature management to be sustainable, it must be based on natural processes of the local ecosystem, regardless of external resources or chemical systems of agriculture. Environmentally sustainable agriculture should function indefinitely without depleting its land and resource potential. The implementation of the concept of environmentally safe and balanced agriculture development requires a fundamental conceptual departure from the economic perspectives which have guided agrarian science for the last hundred years.

The environmentally safe prospect of agribusiness is characterized by the complexity of the factors that are included in the system, as well as the long-term nature of their analysis and control. In the greening system of agrarian nature management, the complexity of natural ecosystems is the subject of value, and the traditional economic approach tries to simplify them.

It is worth noting that it will not be possible to improve the long-term efficiency of agricultural land use without the application of an ecosystem approach to the agrarian nature management. So, if the agrarian development institutions are unable to ensure the environmental sustainability of the various farming methods, then they are actually damaging to society, households, citizens and individual industries. It is important to note that land use productivity should be improved according to the population growth through increased crop yields.

Most scientists believe that land productivity can be improved, as we have already noted, only through the implementation of innovative technologies based mainly on the use of agrochemicals. According to such an industrial model, the main criteria for success are technical and economic efficiency. Proponents of the ecological model of agrarian environmental management support the development of more efficient low-resource agro-ecosystems based on the biological cycle of energy and chemical elements. The effectiveness criteria for this model include indicators of ecological and economic efficiency of agriculture, socio-environmental sustainability and energy efficiency of agrarian nature management (Targetti et al. 2019).

Thus, the ever-increasing need for productive and sustainable agriculture prompts the introduction of a new vision for the development of agrarian environmental management, and in particular for the land use and its risk reduction.

Agro-ecology should ensure efficient energy and materials circulation within agro-ecosystems. In this regard, the need for a holistic approach that would include agriculture research at enterprise or ecosystem level has been raised. This approach makes it possible to implement complex ecological and economic relations into agriculture. For example, instead of improving one variety at a time, a holistic ecological perspective involves searching for a set of plants and animals that, together, produce high environmental, economic and social outcomes (Mishenin et al. 2011).

As agricultural management extends its environmental approach from conservation of natural agro-resources to the impact of its functioning on larger ecosystems, new problems would arise as a result of concern for human health and external environmental effects. Other issues will include social and environmental responsibility, compliance with regulatory requirements, and monitoring of potential socio-environmental and economic risks associated with agro-resources. The environmental future of agricultural land use will be shaped primarily by socio-economic factors, in particular, the global demand for food, its prices, government programs, international trade agreements, technology and new knowledge of agricultural research.

The basic principles of sustainable development strategy for agrarian nature management should include the aspects as follows:

- *partnership*: active interaction between different groups of stakeholders in order to ensure sustainable agricultural production;
- *integration*: promoting the integration of environmental and social thinking into management decision-making processes and innovative ways of doing eco-business;
- *ecosystem and environmental management*: the focus is on preventing, not eliminating, negative environmental impacts;
- *equity for all generations*: equitable sharing of costs and benefits (effects) between generations to encourage the use of socially environmentally responsible methods to minimize the ecological responsibility of future generations;
- *civilized competitiveness*: supporting the effective market mechanisms that ensure the use of innovative ecological management methods, identifying links between environmental sustainability and economic productivity (Mishenin et al. 2015).

It is possible to implement the above-mentioned principles through the following solutions of a strategic nature:

- improving the essence of sustainable agriculture based on the knowledge economy;
- improving the ability of decision-makers to integrate ecological factors into this process;

- ensuring environmental, ecosystem and resource management;
- providing the management and sustainable resource use in the agricultural food sector;
- development of innovative solutions;
- research focusing on the environmental issues to ensure the agricultural sustainability; identifying market opportunities;
- stimulating the agricultural food marketing and trade that have a positive impact on ecological quality and sustainable development.

2.9 Organizational and Economic Mechanisms for Ecologically Safe Agriculture Land Use

The growing needs for productive and sustainable environmentally safe and balanced agriculture are leading to the need for a new vision for the development of agrarian nature management and, in particular, for resource-saving land use (Ullah et al. 2020). This position requires understanding of the ecological principles of agriculture, as well as the putting in place an organizational and economic framework needed to support the agrarian ecosystem management wide implementation.

We define the *organizational and economic mechanism ensuring the environmentally safe and balanced agricultural land use* as a complex system of forms, methods and tools of organizational, economic and social influence on the environmental behaviour of agricultural entities in the direction of increasing socio-ecological and economic efficiency of use, reproduction, protection and conservation of land-resource potential, as well as the effectiveness of the functioning of agro-natural and land capital.

The general purpose of the organizational and economic mechanism is the effective organization of reproductive processes in the use, reproduction, protection and conservation of land and implementation of land management regulation based on an ecosystem approach.

An integrated function of the organizational and economic mechanism of ensuring environmentally safe and balanced land use is the harmonization of socio-ecological and economic needs and interests of economic entities, society as a whole in the process of practical implementation of the principles of environmentally safe and balanced organization of sustainable land and resource potential use, the functioning of land capital, as well as the resolution of contradictions and certain ecological-economic conflicts.

The formation of an organizational and economic mechanism for ensuring environmentally safe and balanced agriculture involves the interaction of the regulatory subsystems of the external organizational and institutional mechanism and the internal mechanism of agricultural enterprise management, using the principles and tools of ecosystem management that provides motivation for environmental behaviour (Mishenin et al. 2002).

Therefore, the economic assessment of agricultural land use consequences through ecological and economic losses is important to form an innovative market-based mechanism of environmentally safe and balanced agricultural land use.

The determining component of the organizational and economic mechanism for ensuring environmentally safe and balanced agricultural land use is *the result-oriented subsystem*, which can be an integral result of the interaction of the external mechanism with the internal one and determines the economic, environmental and social results of management.

External organizational and institutional mechanism of environmentally safe and balanced agricultural land use includes providing institutional and resource subsystem (sub-mechanism), subsystem (sub-mechanism) of organization and regulation of environmentally safe and balanced agricultural land use, and controlling subsystem.

Supporting institutional-resource subsystem (sub-mechanism) includes regulatory, resource (financial, information, human resources), infrastructure (in particular, it concerns the activity of credit institutions, innovation-investment funds, environmental insurance companies and consulting agencies) security.

The subsystem (sub-mechanism) of the organization and regulation of *environmentally safe and balanced agricultural land use* are aimed at the implementation of mechanisms of state regulation of land relations, as well as programming and planning of protection and conservation of lands at national and regional levels. Regional forecasts and programs for the use and protection of land potential are important for ensuring ecological security and balanced farming. These measures should be preventive and should include scientific analysis of the ecological destructive status of land use, tendencies of negative processes in agro-landscape formations (erosion, pollution by heavy metals, soil fertility decreasing).

The subsystem of control within the external organizational and institutional mechanism of regulation of environmentally safe and balanced *agricultural land use* should have a *program-oriented focus* on ecological and economic indicators of agrarian land management on an ecosystem basis. For example, it requires monitoring the eco-destructive state of the land-resource potential, control over the ecological quality of agricultural products on a logistical basis, etc.

It is important to emphasize that the practical reproduction of the prerequisites for environmentally safe and balanced agricultural land use requires the formation of a favourable economic environment capable to support the ecosystem-based agricultural management expansion. Harmonization of economic interests of agricultural business structures with ecological and economic regional and national goals requires the development of not only administrative and regulatory mechanisms, but also the formation of effective motivational and incentive systems. The administrative and regulatory subsystem is aimed mainly at creating a system of restriction of eco-destructive economic activity in the process of land management, in particular, through the application of ecological expertise, external eco-audit, as well as environmental certification of agricultural enterprises, etc.

Available conceptual and methodological approaches (Kupinets and Zhavnerchik 2016; Medvedev 2010; Stepchin 2006; Shkuratov 2016) to the creation of incentive factors, mechanisms, and levers of ecological and economic regulation of rational nature management are divided into the following types of instrumental support (Stupin 2017):

1. Focused on compliance with rules, requirements, norms of rational nature management and the implementation of the obligatory system of environment protection measures (in particular, normative regulation, penalties, payments for the environmentally destructive state of natural objects).
2. Promoting the implementation of environmental activities (in particular, the environmental tax system and payment system for the use of natural resources, and financial incentives).
3. Incentive, aimed at supporting economic entities to implement environment protective (environmental) measures (subsidies, preferential crediting and taxation, special funding).

In this context, the compensation mechanism for the afforestation of agricultural lands is a prime focus. This includes economic incentive tools important for encouraging environmentally balanced use and protection of agricultural lands based on the creation of protective forest plantings. These tools relate to the measures of economic impact aimed at changing the financial and property status of entities of agrarian land use in order to equalize the imbalance between ecologically balanced, environmentally safe agribusiness within a certain agricultural landscape.

Compensatory and stimulating mechanism of agrarian natural management with an emphasis on the issues of agroforestry production (Stepchin 2006) may include the following components:

7. Partial reimbursement of the lost revenue, in particular, in the form of rent payments in the case of conservation of land, depending on their intended purpose and degree of degradation.
8. Payments for the increase of soil fertility and reduction of their pollution due to agroforestry improvement of agricultural lands.
9. Subventions (grants) for the production of environment-friendly agricultural products under the conditions of the land arrangement on the agricultural forest reclamation basis.
10. Some compensation (reimbursement) of expenses for carrying out works on the conversion of the intended use of land within the limits of expansion of the agroforestry reclamation complex.
11. Compensation (reimbursement) of capital and current expenses for the implementation of investment agroforestry projects.

The presented components are cost-compensatory and can be implemented in the form of rent payments, as well as provide for the transfer of payments by environmentally responsible entrepreneurs. Of course, the system of financial and economic incentives should provide for a variety of tax and credit benefits.

It is important to focus on the environmental taxation systems and environmental policies and payments for nature conservation from the perspectives of agroforestry spatial development (Lindgren et al. 2018; Gaffney et al. 2019). At the conceptual plan, from the point of view of preventing environmental pollution and creating ecological and economic prerequisites for the transition to conservation forms of agriculture, the maximum use of the system of environmental taxes and payments for natural resources that already exist in Ukraine is needed (Mishenin and Yarova 2019). This system provides payments for a land use, environment pollution, a violation of the rules and regulations of environmental legislation, and payments for compensation for the harm caused by environmental offenses. It is worth noting that the adopted system of environmental taxes is almost non-functional in relation to agrarian environmental management (Mishenin and Yarova 2019). This is due to a number of reasons. First, the share of the agricultural sector in environmental pollution is generally thought to be minimal. Second, in agroforestry areas, there is no environmental service to control the quality of the natural agricultural environment, which explains the misconceptions about the extent of the agrarian sector's influence on natural objects. Third, the economic crisis in the agrarian sphere has more severe consequences than in the manufacturing, specifically concerning negative social and economic outcomes for households and agricultural enterprises. However, this is not a reason for not taking measures to prevent pollution and environmental degradation within the agriculture activities. It is obvious that the environmental system function should taking into account the real economic condition of agricultural enterprises (Ovsyannikov 2000).

In a number of European countries, to increase motivation in the transition to conservation forms of agriculture, taxes (payments) are imposed on the use of chemicals (Ovsyannikov 2000; Chakravorty et al. 2007), which can be used to form incentives in the afforestation of agricultural land.

Effective ways of influencing users in the agricultural sector can be:

- payments for the pollution caused by mineral fertilizers, plant protection products and cattle-breeding drains of soil and water bodies;
- payments for the pollution of water bodies with the soil washed off from the fields;
- payments for the soil destruction resulted from erosion.

To increase the arable land protective forest cover, the creation of protective forest plantings on erosion-free agricultural lands, as well as the introduction of environmental methods of cultivation of crops should be used:

- preferential payments (taxes) for the use of land resources;
- preferential taxation of land ownership.

Thus, the development of an organizational and economic mechanism for ensuring environmentally safe and balanced agrarian farming should be carried out on the basis of enhancing the environmental behaviour of business-entrepreneurial structures under the influence of an external organizational and institutional mechanism, which should exert a dynamic regulatory influence on the internal mechanism of the entity.

The theoretical and methodological orientations for the construction of organizational and economic mechanism of agrarian nature of the economy allow to form the systematic management processes for the greening land use at different spatial levels of management (Mishenin et al. 2015).

2.10 Economic Reforms Towards Agricultural Sustainability and Development

Scientific and technological progress may serve the interests of human to a certain extent, but having passed the peak of usefulness, it begins to play the opposite role, although it continues to promote economic growth (Mishenin et al. 2011; Diaz-Ambrona and Maletta 2014). Society is gradually becoming aware that fact. Entrepreneurs are gradually beginning to understand that consumers have increasingly begun to demand a higher life quality, the availability of ecological food. The desire of companies for continuous expansion, the achievement of a monopoly position in the market, maximizing profits by reducing costs and raising sales prices was until recently a commonly accepted motivational basis for entrepreneurship. However, the increasing interdependence of economic, social and environmental interests ultimately led to including the environmental protection, healthy lifestyles, humanization of working conditions, rational nutrition and agro-industrial products safety in the marketing conception (Mishenin et al. 2011).

A necessary prerequisite for sustainable development is to consider environmental, economic and social factors in the pricing mechanism (Lindgren et al. 2018; Gaffney et al. 2019). In general, in a market environment, the role of prices for goods and services is difficult to overestimate because they reflect a shortage of production, natural resources (factors) and consumer goods. The price of a particular product is determined not only by the amount of costs required for its production, but also by the benefits it can bring (which ensures the efficiency of scarce resources). From these perspectives, Hoffmann (1991) noted that the market mechanism broadens the boundaries for profitable investments in the natural resources conservation with increasing their scarcity (Hoffmann 1991). So, this fact explains the achievements of developed countries in the field of reducing environmental intensity of public production.

Market pricing mechanism is based on the assumption that all the benefits and costs are associated with the production and consumption of each environmentally friendly product and it is fully reflected in the market demand and supply curves. In other words, it is considered that there are no externalities in the production and consumption of goods and services (Mishenin et al. 2011; Gaffney et al. 2019).

The value of prices for the normal reproduction processes of environmental quality is determined by their main functions. First of all, prices serve as a measurement and information function. With their help, it is possible to express various natural indicators of natural resource potential (in particular, ecological potential, the total economic value of agricultural lands, etc.), costs and results of environmental activities in a single monetary form, as well as other different environmental and economic indicators. Ultimately, price is an important criterion for choice, and also a benchmark for making optimal eco-management decisions, providing the necessary information about the needs of environmentally friendly goods, the cost of their production and the ability to take into account the negative nature externalities. The environmental factor also affects the distribution function of prices in the pricing mechanism. Price mechanisms help to compensate the negative externalities of environmental management based on the system of ecological taxation and payments for the natural resources use (Mishenin and Yarova 2019).

For a more comprehensive consideration of the environmental factor in pricing, the incentive function of prices is of particular importance. The price incentive effect on the products of enterprises—polluters can be manifested in different directions. Increasing or decreasing price due to the environmental component can stimulate, or vice versa, impede the purchase of products (Chakravorty et al. 2007; Lindgren et al. 2018; Gaffney et al. 2019; Targetti et al. 2019). It should be noted that in developed countries, along with market prices, there remains a place for regulated prices for certain products (Mishenin and Yarova 2019).

State regulation of prices can be carried out using the following methods: setting fixed prices for specific goods (this method is characterized, mainly for the administrative-command economy), direct impact on prices through certain restrictions, or individual components of the price (this method is applied to the functioning of contractual prices system and its liberalization), and indirect impact on prices through the use of state economic levers (for example, ecological taxation). State price regulation, taking into account the environmental factor, can be presented as a system of organizational and institutional measures designed to fully reflect the environmental costs of production based on a corresponding change in price levels to overcome the economic and socio-environmental contradictions related to pollution (degradation) of the environment and rational resources use. Price regulation, taking into consideration environmental factors, can also be defined as a system of influence on prices, which creates new legal conditions in a particular economic and environmental situation (Mishenin and Yarova 2019). Pricing ecologization involves regulating prices for nature-intensive and eco-friendly products in order to reflect the real socio-ecological value of natural resources and their scarcity, as well as the environmental production cost (Gaffney et al. 2019; Targetti et al. 2019).

Involving full environmental costs into enterprise costs is often called *absolute cost accounting* (Schmidheni 1994). Today, this is just a theory, but it also undergoes dynamic changes depending on different conditions, time and place (Gaffney et al. 2019; Targetti et al. 2019). The inaccuracy in determining the actual and future costs of pollution cannot be used to justify a difficulty to determine the cost of environmental disruption. In industry, total cost includes the cost of production plus the cost expression of environmental damage associated with production. It is often claimed that not only the polluter but the consumer pays. However, this is the main task of this principle. The inclusion of environmental costs in the production cost, of course, has the changing effect for the price of goods, because high prices for

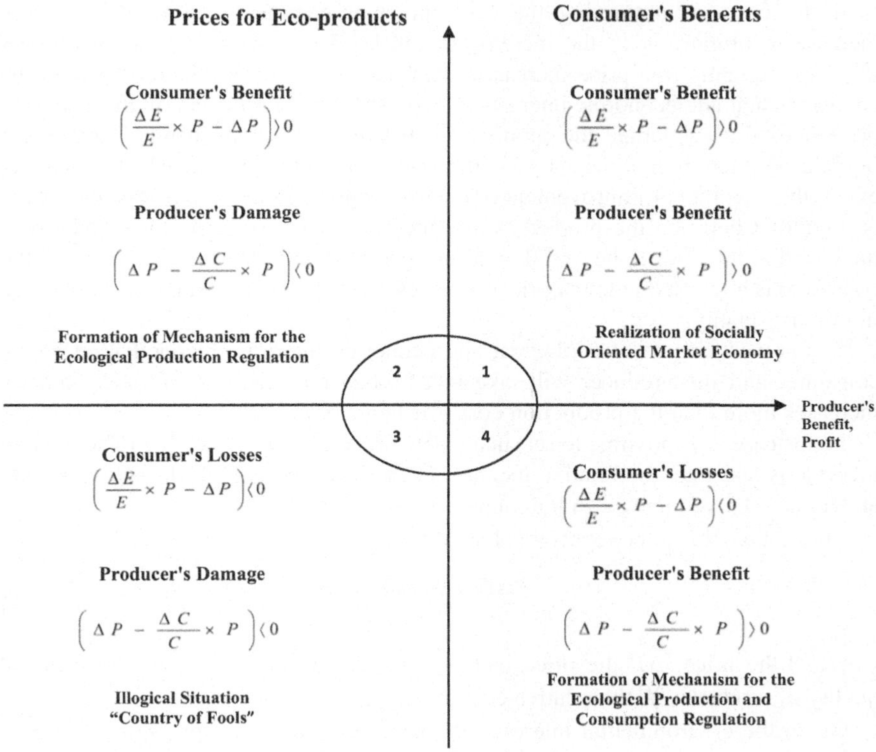

Fig. 2.12 The "benefits—losses" matrix of producers and consumers of ecological agricultural products (formed by Yarova, on the base of Mishenin et al. 2011)

products of environmentally harmful production can be a kind of signal to the consumer to purchase more "environmentally friendly" goods. Possible socio-economic and environmental-economic consequences of different priorities in the approaches to production and consumption of organic products, as well as the need for environmental regulation of production can be illustrated by the "benefits—losses" matrix (Mishenin et al. 2011) (Fig. 2.12).

The market price for products should theoretically meet the normative level of their ecological quality and socially necessary production costs (Chakravorty et al. 2007; Lindgren et al. 2018). The consumer is interested in the fact that with increasing quality the price increases in proportion to the ecological compatibility of the products. But, the manufacturer, of course, is interested in the fact that the new price will offset the costs and brings additional profit to the enterprise. The state should be interested both in meeting the needs of the population with minimum negative socio-environmental and economic consequences, and in obtaining part of the income through an ecological tax system (Mishenin and Yarova 2019).

If the costs of improving the environmental parameters of product quality increase in proportion to the increase in the level of environmental friendliness and, consequently, the price increases, then there is no contradiction between the environmental and economic interests of consumers and producers. In this case, the price does not encourage the producer to the quality improvement. Demand for organic products is holding back. If the cost of production increases to a greater extent than the rate of improvement of environmental quality parameters, then there is a conflict between the producer's and the consumer's interests: the producer is interested in the price to be raised, at least in proportion to the increase in costs; the consumer is interested in lower price increases—at least in proportion to the increase in product quality.

The highest degree of ecological and economic interests' reconciliation of the consumer and the producer will take place when the level of ecological quality increases more than the production cost (Mishenin et al. 2011).

In this case, it is possible to set such a new price ($P + \Delta P$) when the relative price increase is less than the relative increase in the level of ecological quality, $\frac{\Delta E}{E}$ and above the relative increase in the total cost, $\frac{\Delta C}{C}$.

This approach can be represented as follows

$$\left(\frac{\Delta E}{E} \rangle \frac{\Delta P}{P} \rangle \frac{\Delta C}{C}\right), \tag{2.1}$$

where P the price; ΔP the price increase; $\frac{\Delta E}{E}$ the relative increase of ecological quality of products; $\frac{\Delta C}{C}$ the relative cost increase.

Given the environmental interests of society as a whole, expression (2.1) may look like:

$$\left(\frac{\Delta PC}{PC} \geq \frac{\Delta E}{E} \rangle \frac{\Delta P}{P} \rangle \frac{\Delta C}{C}\right), \tag{2.2}$$

where ΔPC reducing external environmental costs of production and consumption; $\frac{\Delta PC}{PC}$ the relative reduction of production external environmental costs.

This is the best balance between the quality (ecological) of products, prices and costs from the perspectives of socio-environmental interests of society as a whole. In this case, utility of the new product (its ecological quality) for the individual consumer has increased to a greater extent than its cost of purchasing this product. For the producer, the price has increased to a greater extent than its cost. So, the consumer and the manufacturer have benefited. The price in this case stimulates the increase of ecological quality and demand for these products. With the increase in the environmental friendliness of new products, external environmental costs (ecological and economic damage) for both the individual consumer and society as a whole are decreasing (Mishenin et al. 2011).

The terms of harmonization of ecological and economic interests of the producer and the consumer can be considered as follows: if the price of products of higher ecological quality for the individual consumer was set in proportion to the quality level:

$$\frac{\Delta PC_E}{PC} = \frac{\Delta E}{E}, \tag{2.3}$$

then

$$\Delta P_E = \frac{\Delta E}{E} \times P, \tag{2.4}$$

where $\boldsymbol{\Delta P_E}$ the price increase due to the ecological quality increasing by $\boldsymbol{\Delta E}$.

If the price for these products was set in proportion to the cost,

$$\frac{\Delta Pc}{\Delta P} = \frac{\Delta C}{C}, \tag{2.5}$$

then

$$\Delta Pc = \frac{\Delta C}{C} \times P, \tag{2.6}$$

where ΔPc an increase in the cost due to the increase in cost by ΔC.

The condition of reconciliation of the consumer's and the producer's interests can be considered as follows:

$$\Delta P_E \rangle \Delta P \rangle \Delta P_C, \tag{2.7}$$

The producer's benefits can be presented as follows:

$$\Delta P - \Delta P_C = \Delta P - \frac{\Delta C}{C} \times P, \tag{2.8}$$

The consumer's benefits can be presented as follows:

$$\Delta P_E - \Delta P = \frac{\Delta E}{E} \times P - \Delta P, \tag{2.9}$$

The total benefit for the producer and the consumer is determined as follows:

$$\Delta P_E - \Delta P_C = \left(\frac{\Delta E}{E} - \frac{\Delta C}{C}\right) \times P, \tag{2.10}$$

This total benefit can be divided into three parts: the benefit of the individual consumer, the benefit of the producer (enterprise) and the society's benefit as a whole.

Consider the content basis of the quadrants of the above-mentioned matrix (Fig. 2.12).

Quadrant 1. "*Consumer's benefit, producer's benefit*" follows the principles of socially oriented market economy and sustainable development, and reflects the complex of long-term mutually beneficial "buyer-producer" relations, as it provides both return on investment and satisfaction of social and environmental needs within acceptable prices.

Quadrant 2. "*Consumer's benefit, producer's damage*" corresponds to the situation when external environmental costs of production, lack of the positive effect assessment of production and ecological products consumption are subject to environmental production regulation in terms of stimulating the "greening" of the enterprise economy.

Quadrant 3. "*Consumer's losses, producer's damage*" corresponds to the situation that, in the terminology of I. Ansoff, is called "the country of fools".

Quadrant 4. "*Consumer's losses, producer's benefit*" reflects the case where the manufacturer makes a profit from sales, but does not provide the consumer with goods of the certain ecological quality. This situation often occurs in industries with the low technological development. However, such a situation may occur in the process of manufacturing new products in highly sophisticated industries based on "fashionable technology", when the buyer's interests are not always taken into account.

Let us now consider some methodological possibilities for a more complete account of environmental costs in the pricing mechanism for enterprises–polluters. Our calculations have shown that about 19–60% of the damage from environmental pollution is compensated through the ecological payment system (Mishenin and Yarova 2019).

Thus, in order to fully compensate the economic damage from environmental pollution through the pricing system, it is possible to calculate *a price increase index*, taking into account the ecological component in the production cost. But it is very difficult to implement this proposal, especially in conditions of unstable economic development, monopoly power and distortions in the pricing system.

The price increase index for environmentally misbalanced industries, but in terms of providing the full economic damage compensation caused by environmental pollution and the relative equal supply and demand $\left(I_P^E\right)$, can be calculated as follows:

$$I_P^E = \frac{D + T_E + P}{T_E + P}, \quad (2.11)$$

where D the economic damage from environmental pollution, which is not compensated through the ecological tax system (payments); T_E ecological tax (payments) within the limits of normative indicators for environmental pollution; P the price of products.

Indicative price indices for the products of enterprises of some industries are presented in Table 2.3.

The calculated price indices, taking into account the environmental factor, implement the principle of *absolute cost accounting*. The presented indices can be used to regulate the external economic activity of enterprises based on the indicative price system. But there is a certain deviation of contract (foreign trade) prices from indicative in crisis market conditions of economy. That is why the proposed and

Table 2.3 Indicative price indices for products in terms of full compensation for economic damage from environmental pollution (Mishenin and Yarova 2019)

Industry	Indices
Chemical	1.142
Manufacture of agricultural machinery	1.143
Manufacture of equipment for various industries	1.153
Food	1,146

calculated *price increase indices*, with absolute accounting for economic loss may be some benchmarks to regulate and balance contract price deviations for enterprises–polluters (Schmidheni 1994).

It should be noted that world prices for natural resources, reflecting the degree of their natural scarcity and demand, also do not take fully into account their external effects to the process of their extraction. For example, timber prices do not reflect the socio-ecological value of forest resources, and therefore deviation of contract prices for timber products from indicative prices may be subject to environmental regulation. Methodical approaches to the cost estimation of social consequences of economic activity (including those that are not in kind and are not valued in money) have been scientifically considered (Bohm 1979; Boulding 1970; Chakravorty et al. 2007; Lindgren et al. 2018; Gaffney et al. 2019; Targetti et al. 2019).

The assessment of social and environmental efficiency in terms of the "cost-benefit" method includes its natural determination, and then a monetary estimate is made (Mishenin et al. 2011). The economic assessment of social and environmental goods and services has nothing to do with market prices because the market is unable to account for them. This aspect stipulates the necessity of environmentally oriented state regulation of prices for nature-intensive and eco-intensive products (Mishenin et al. 2011).

2.11 Social and Environmental Responsibility as Systemic Element for Agricultural Sustainability

Social and ecological responsibility makes a significant contribution to the national food security, as the problem of providing quality of food is critical to the population. From time to time, scandals with newly discovered dangerous substances occur in the world and the emergence of new viral infections types threatens the society's sustainable development in a whole. This situation may lead to the fact that increasing consumption of agricultural products can reduce the life quality in general (Bengtsson et al. 2018). This decline in life quality is primarily defined as a decrease in the level of public health, which leads to the loss of human capital. From these perspectives, we believe that an important idea of the economic paradigm of greening agriculture as a whole is the socio-ecological and economic responsibility for the food quality and the environment in all chains of agribusiness.

Therefore, the creation of an effective mechanism of socio-ecological and economic responsibility is a logical development of the market agricultural sector, which requires effective ecological and economic regulation. The need to increase the level of responsibility for the eco-destructive effects of agricultural management is largely determined by its complex impact on the agricultural sustainability, eco-balanced production results. That is why the formation and further development of organizational, economic and legal mechanisms of social and environmental responsibility should become an integral part of the agrarian economy (Mishenin and Yarova 2015).

Socially and environmentally oriented responsibility in the field of agriculture often acts as a responsibility for the consequences of the irrational use of agrarian natural resources that affect the environmental, economic and social interests of society, economic entities and individuals. Social responsibility directly has a complex socio-ecological and economic character and implies management responsibility that goes beyond the specific (real) mechanisms for generating profit. Also it should be taken into account protecting and enhancing social well-being under various parameters of sustainable development. Therefore, it should be emphasized that the most important structural elements concerning the social responsibility are social commitment and responsiveness in the long-term socially beneficial goals of agribusiness (Robbins and Coutler 2007). Of course, the processes of social responsibility implementation require the formation of appropriate mechanisms of environmental management. At the same time, the social activity of the enterprise is a set of measures for the effective realization of the entrepreneurial social responsibility, which should have both internal and external orientation.

Analysis of theory and methodological background of social responsibility in view of environmental management concept indicate certain versatility concerning essential and meaningful basis of social and ecological responsibility (Belousov 2016; Makarova and Stepanova 2014; Mishenin and Yarova 2015; Pakhomova and Malyshkov 2008). The structural components of social responsibility are also ambiguously identified. In particular, according to Pakhomova and Malyshkov (2008) the social and ecological responsibility is conscious and motivated business participation in various preventive measures concerning environmental damage and irrational nature management; also in providing social and ecological benefits, measures for labour protection, environmental quality improvement and sustainable nature management (Pakhomova and Malyshkov 2008).

Social and ecological responsibility of agricultural enterprises of different forms of ownership and organizational forms of agribusiness must be determined by their responsible attitude regarding a rational use and reproduction of natural resources, as well as towards workers, society in general and individual citizens, as well as concerning negative changes in the ecological and economic parameters of land and resource potential (capital) (Mishenin and Yarova 2015).

The social and ecological responsibility in agribusiness is determined by the certain factors as follows (Mishenin and Yarova 2015):

1. Voluntary and initiative ecological and economic measures of enterprises go, especially at the initial stage, beyond the limits of profit, the legislative regulation of environmental agriculture.

2. Ecological and economic measures to improve agrarian nature and resource potential are of social importance for the local population to contribute its employment.
3. "Greening" of the agricultural production has undoubtedly social effects, both in terms of improving the level of labour safety and increasing incomes for workers. It, no doubt, positively affects the environmental dimension of food security.
4. The relations of enterprises with the public are social in the system of environmentally responsible agricultural management.

We define *social and ecological responsibility of enterprises* as an initiative-voluntary internal and external activity aimed at responding and forming commitments concerning economic, social and environmental aspects of sustainable, environmentally balanced development of rural community under the background of the established system of environmentally oriented regional agricultural management (Mishenin and Yarova 2015).

The peculiarity of this definition is that it reflects the basic signs of social responsibility—*responsiveness and commitment*, as well as its external and internal orientation (Mishenin and Yarova 2015).

Socially responsible enterprise management on an ecological and economic basis is defined as a process of implementation and integration of social and environmental measures into agro-economic activities that go beyond the formation of profit and the legally established principles, rules, norms, standards of rational use and restoration of agricultural resources of nature's origin to ensure the sustainable agricultural development (Mishenin and Yarova 2015).

Social and ecological responsibility in nature management within the framework of the enterprise's activity is formed and determined by the main factors as follows: social and environmental initiative; ecological and economic knowledge management system; ecological culture; ecological and economic technologies of socially responsible agricultural management (Targetti et al. 2019).

Generally, it is necessary and appropriate to talk about a comprehensive organizational and management mechanism of socio-ecological and economic responsibility, which is necessary to support the agricultural land use greening and environmentally safe nature economy in the context of food security. Such a mechanism should be defined as a set of forms and instruments in the system of social, ecological and economic regulation of agricultural development on the basis of simultaneous application of administrative, economic and social management methods (Mishenin and Yarova 2015).

Forms and methods of economic and at the same time legal responsibility should also be optimally combined with instruments not only of purely economic stimulation, but also motives for realization of environmental, social interests of society and individual economic entities (legal entities and individuals), as well as with other functional links of the mechanism of agricultural management (Mishenin and Yarova 2015).

Agricultural economic, social and ecological responsibility should be based on the following principles: ensuring the economic parity, environmental and social values of the agro-economic activity results; achieving the optimal combination of vertical and horizontal responsibilities; the most complete compensation for socio-economic and environmental damage; inevitability of economic and legal sanctions; ensuring a balance between economic sanctions and economic incentives (socially and environmentally responsible behaviour of agribusiness entities should be encouraged through subsidies, tax breaks, preferential lending) (Mishenin and Yarova 2015; Gaffney et al. 2019; Targetti et al. 2019).

The construction of a comprehensive mechanism of social, ecological and economic responsibility in the field of agriculture takes into account some specifics of ecological and economic, social and legal relations within the system of greening agro-production, the need for optimal maintenance of ecological safety of agricultural products (Pingali et al. 2019). The implementation of the functions of economic and legal environmental responsibility affects the behaviour of subjects of agrarian relations, focuses on providing motivation to comply with the rules, requirements and norms of rational agricultural production and land use.

Economic responsibility, administrative and legal sanctions for irrational agro-production, violations of environmental legislation can fulfil the main functions as follows (Mishenin and Yarova 2015):

1. *Incentive function:* This function of responsibility is fundamental, since it is expedient to prevent the negative impact of irrational agricultural land use, eco-destructive production on the level of quality (ecological) of products, and then eliminate them. The implementation of the incentive function requires a wide variety of motivational tools for the greening of agricultural production, rational land use. This function acts as counterparty to the compensation function.
2. *Compensation function:* Full compensation for the loss is a prerequisite and, at the same time, a demand for the development of market relations, one of the factors for ensuring socio-economic sustainable development. Compensatory nature is aimed at the rational use of financial, material and labour resources to eliminate, neutralize and prevent the negative consequences of environmental damages.
3. *The preventive function:* Social, ecological and economic responsibility requires the awareness of agribusiness entities of the extent of material liability for violations of environmental quality, in particular, land and capital. It can be reflected, for example, in the system of contractual relations. The implementation of the preventive function is ensured by the inevitability of economic sanctions, assessment of their impact on the final financial and economic results, which causes the conduct of socio-environmental analysis.
4. *The control-information (communication) function:* Precedes the compensation function, facilitates the detection of environmental violations in the agricultural sector, and provides the information base of natural indicators of loss for their further transformation into cost indicators. The implementation of the control and information function involves the creation and operation of various social and environmental monitoring systems of agricultural use.

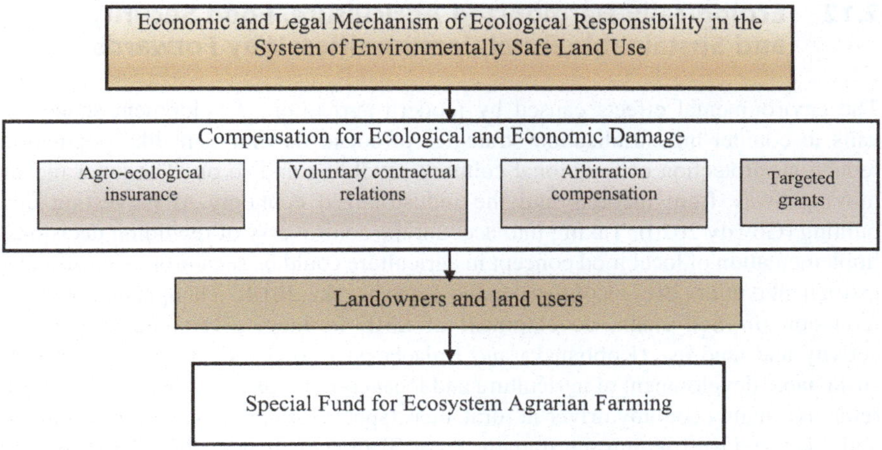

Fig. 2.13 Organizational and economic bases of ecological and economic losses compensation to the system of environmentally safe agrarian farming (developed by Yarova, on the base of Mishenin et al. 2017)

5. *The evaluation function:* Creates opportunities for measuring the level of socio-ecological security of the economic behaviour of agricultural entities and forming relevant conclusions. This function is a process of more accurate determination of the full cost of social work, social and ecological losses due to the environmental agricultural use.
6. *The regulatory function:* Ensures the application of organizational, economic and social instruments to the environmental behaviour of business entities, which largely depends on the application of sanctions and the threat of their use.

Thus, the ecological responsibility's economic and legal mechanism should provide for compensation of ecological and economic damage caused by external and internal ecological destructive factors, to promote the environmentally safe and balanced agrarian farming system (Fig. 2.13).

It is advisable to accumulate a share of the compensation payments within the framework of the special ecosystem agrarian farming fund, which needs to be formed at the regional level in order to solve common regional problems of agri-management.

Thus, it should be noted that the basis of providing environmentally oriented food security is the formation of complex socio-ecological and economic mechanisms that would contribute to the development of agriculture and a whole society in a sustainable manner, as well as guarantee meeting economic and socio-environmental requirements of individual citizens.

2.12 Ecologically Harmonized Agriculture, Food Security and Sustainable Rural Communities: Way Forwards

The environmental effects caused by modern agriculture development shape the calls to counter-industrialization strategies of social development like population reduction, protection of traditional cultures, rewilding and so on. The latter means moving away from markets and the industrialized economy to harvesting and hunting (Gowdy 2020). Taking into account the controversy of rewilding decisions, implementation of local food concept in agriculture could be seen promising strategy (Strochenko et al. 2017; Koblianska and Kalachevska 2019). This presupposes the reduction of food chain, development of SME farming, diversified agricultural activity and land use (Koblianska and Kalachevska 2019). Such strategy supports sustainable development of agriculture and local communities, as the agri-sector still remains a main economy driver in rural areas (specifically, in developing countries) and a key industry within the framework of SDG achievement (Fig. 2.14). Briefly commenting on Fig. 2.14, one should emphasize a substantial role of agriculture in poverty alleviation, economic and income's growth, conflict mitigation, supporting

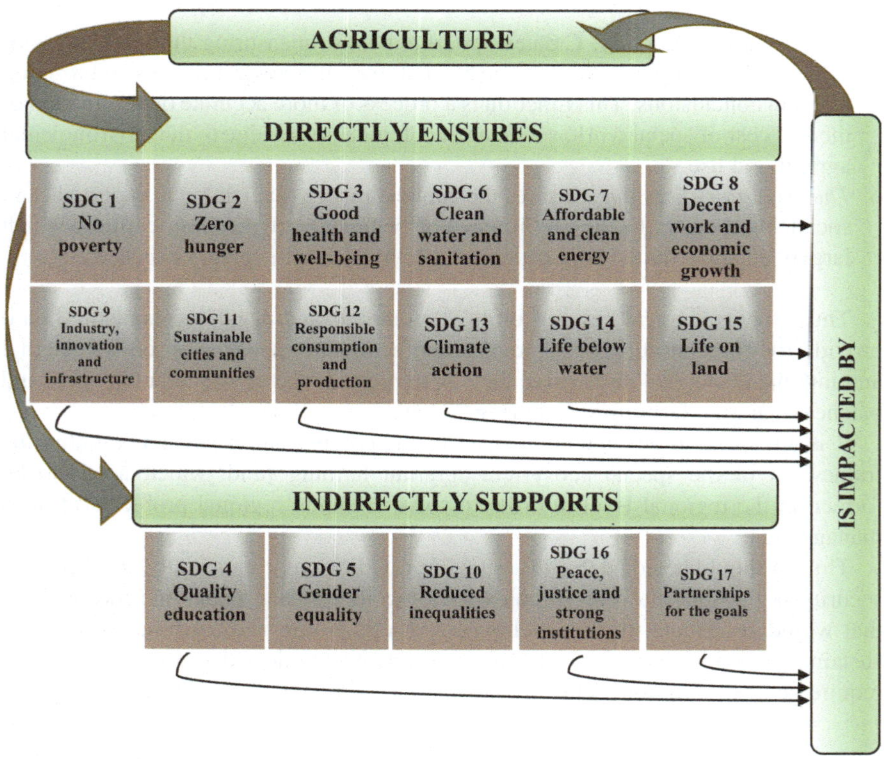

Fig. 2.14 Agriculture within the sustainable development goals' framework (developed by Koblianska with the use of (UNO 2016)

employment. The sector also substantially influences the health through the food quality, especially in developing countries. World Bank's estimates show that agriculture growth is double and even fourfold more effective with regard to the growing incomes of the poorest people than measures in other industries. Actually, agriculture provides the income for more than 65% of poor working adults worldwide (World Bank 2020).

Concerning the above, implementation of local food concept is economically, socially and environmentally beneficial and thus ensures the local community's sustainable development. The development of small farms is contrary to the industrialization of agriculture (Nicholls et al. 2020) and due to traditional farming practices supports ecosystem services (Ickowitz et al. 2019). This strategy not only provides environmental benefits but is also a way to diversify nutrition (especially for the poorest rural residents) and to support institutional shifts and innovative solutions. The key aspects of its implementation are the development of an appropriate set of measures to stimulate the development of local diversified business in rural areas representing the multifunctional and eco-friendly agriculture. In this context the sound market-based economic incentives are of great importance, in particular, to counter the commercialization of production activities (Ickowitz et al. 2019), that is an evident tendency in Ukraine. These incentives should provide, namely:

- The internalization of public benefits generated by diversified agricultural activities (Targetti et al. 2019; Gaffney et al. 2019; Funk and Brown 2009; Adenuga et al. 2019), for example, through the ecosystem service payment mechanism.
- The internalization of negative externalities of farming (Adenuga et al. 2019), for example, through the introduction of compensation for marginal costs associated with livestock contamination and use of chemicals (Hediger and Lehmann 2007).

Monetary incentives have traditionally been regarded as basic to stimulate environmentally friendly agricultural practices, but recent research (van Westen et al. 2019; Targetti et al. 2019) has shown that locally and regionally an institutional context is of great importance, and therefore a development of sustainable agriculture is a result of the spread of knowledge and awareness of public goods and their status (among consumers, local community, government) and possible technologies (among farmers) (Targetti et al. 2019). This requires the improvement of appropriate policy and legal framework.

2.13 Policy and Legal Framework

The policy improvement is an urgent task to be solved for the achievement of coherent goals of food and environmental security, and sustainable development of agri-sector worldwide. Actually, this is an institutional transformation that gives the ground for further technological improvements.

The new policy should be of an inclusive, integrated and multi-sectoral nature, flexible and adaptive with respect to the whole complex of agriculture and nature interrelations (Devaux et al. 2020; El Bilali 2019). This touches both the developed and developing countries.

For developing countries, the major concern is to integrate environmental aspects into a whole agrarian and economic policy context (Fatemi and Rezaei-Moghaddam 2019). At the same time, developed countries need to ensure high flexibility and adaptability of political measures to the new challenges, including climate mitigation efforts (Candel and Biesbroek 2018). The closer communication and interaction between different stakeholders are needed for both.

The change of consumer preferences is another important area of policy reform towards the food and environmental security. The promotion of a healthy and sustainable diet, as well as raising public awareness on the need to prevent food losses and wastes are the important areas of government activities in this area (Bahn et al. 2019).

The policy measures to support small-scale agri-producers are needed to ensure diversified multifunctional agriculture with all related benefits such as food security for the poorest, healthy diet, income growth, ecosystem services, etc. (Ickowitz et al. 2019). The recent studies show that the model of local food is the win-win strategy in this case, providing the reduction of environmental footprint, biodiversity losses without a decrease of food supply per capita. Such a model is the most beneficial for the simultaneous achievement of food security and environmental goals and is opposite to land polarization and specialization that are characteristics of a neoliberal policy (Rega et al. 2019).

However, the deficit of funding is a major modern challenge for food security provision and sustainable agriculture development, especially for developing countries cases. It is possible to stimulate economic growth and poverty alleviation through the investments in public goods like agro-research, consulting, road infrastructure, etc. (Funk and Brown 2009). In this context, the public goods concept (proper identification and assessment of public goods provided by sustainable agriculture practices) could serve as a baseline for elaboration of a new economic policy towards the food and environmentally secure future (Gaffney et al. 2019; Targetti et al. 2019).

2.14 Conclusions

The agriculture appears as a key sector under the food security agenda that is overcoming the hunger in all its forms and manifestations, providing the adequate income and ensuring the quality of food. At the same time, agriculture is the main industry where interrelations with natural conditions are the key factors to be taken into account for successful sector's development. However, the full set of relationships between the sector and the nature is not understood, recognized, reflected and responded to enough, especially in the agriculture dependent countries. Due to this, currently agriculture is one of the main industries influencing natural environment, causing irreversible climate changes, depleting the land and ecosystems, etc. In light of that, the existing and future ability of society to overcome hunger is extremely questionable, and agriculture environmental security is a growing concern with the food one.

Against this background the attention of scientists, international organizations and politicians is riveted on the search for an optimal model of agriculture, i.e. an industrial system, which provides food supply and income, or an organic farming with environmental benefits, or diversified local food systems providing an income to the poorest and necessary ecosystem services, etc. At present, there is no single view concerning these approaches, however, the experience of Ukraine convincingly shows that the industrialized model is an environmentally loosing strategy if proper economic, institutional and technological transformations are not in place. At the same time, it is not possible to provide at least the achieved level of food supply (per capita) without increasing the food production. In this context the eco-friendly intensification appears as a desirable way of agricultural development in a globalized world and requires systemic transformations of society, economic models, policy tools and solutions. In particular, the most important areas for improvements cover environmentally adjustable pricing for food and other agri-products, implementation of the social and environmental responsibility of agribusiness concept, as well as recognition of industry's role regarding the sustainable community development. In this case, the policy of localism (support for diversified local producers) appears to be quite promising as it addresses the whole range of food security dimensions.

2.15 Future Research Roadmap

The ecological and economic regulation of balanced agricultural development in the food security context must ultimately be defined from the point of view of natural capital itself. So, further research should be aimed at shaping the organizational, institutional and economic conditions of capitalization and "securitization" of territorial agrarian natural resources. The elaboration of a fair system of externalities internalization is of great importance in this regard.

References

Adenuga AH, Davis J, Hutchinson G, Donnellan T, Patton M (2019) Environmental efficiency and pollution costs of nitrogen surplus in dairy farms: a parametric hyperbolic technology distance function approach. Environ Resour Econ 74(3):1273–1298. https://doi.org/10.1007/s10640-019-00367-2

Andrade JF, Rattalino Edreira JI, Farrow A, van Loon MP, Craufurd PQ, Rurinda J, Zingore S, Chamberlin J, Claessens L, Adewopo J, van Ittersum MK, Cassman KG, Grassini P (2019) A spatial framework for ex-ante impact assessment of agricultural technologies. Glob Food Sec 20:72–81. https://doi.org/10.1016/j.gfs.2018.12.006

Asfaw S, Branca G (2018) Introduction and overview. In: Lipper L, McCarthy N, Zilberman D, Asfaw S, Branca G (eds) Climate smart agriculture, vol 52. Springer, Cham, pp 3–12. https://doi.org/10.1007/978-3-319-61194-5_1

Bahn R, EL Labban S, Hwalla N (2019) Impacts of shifting to healthier food consumption patterns on environmental sustainability in MENA countries. Sustain Sci 14(4):1131–1146. https://doi.org/10.1007/s11625-018-0600-3

Banerjee A, Jhariya MK, Yadav DK, Raj A (2020) Environmental and sustainable development through forestry and other resources. Apple Academic Press Inc., New York, p 400. https://doi.org/10.1201/9780429276026

Bartolini F, Brunori G (2014) Understanding linkages between common agricultural policy and high nature value (HNV) farmland provision: an empirical analysis in Tuscany Region. Agric Food Econ 2(1):13. https://doi.org/10.1186/s40100-014-0013-2

Belousov KY (2016) The modern stage of evolution of the concept of social responsibility. Theory Pract Publ Dev 3:32–34

Bengtsson M, Alfredsson E, Cohen M, Lorek S, Schroeder P (2018) Transforming systems of consumption and production for achieving the sustainable development goals: moving beyond efficiency. Sustain Sci 13(6):1533–1547. https://doi.org/10.1007/s11625-018-0582-1

Bohm P (1979) Social efficiency. A concise introduction to welfare economics. Macmillan, London

Boulding K (1970) Economics as a science. McGraw-Hill Inc., New York, p 232

Breeman G, Dijkman J, Termeer C (2015) Enhancing food security through a multi-stakeholder process: the global agenda for sustainable livestock. Food Secur 7(2):425–435. https://doi.org/10.1007/s12571-015-0430-4

BUM (2019) Ukraine is emerging as the world's next agricultural superpower. Business Ukraine Magazine, 16 July 2019. http://bunews.com.ua/economy/item/ukraine-feeds-the-world. Accessed 15 April 2020

Candel JJL, Biesbroek R (2018) Policy integration in the EU governance of global food security. Food Secur 10(1):195–209. https://doi.org/10.1007/s12571-017-0752-5

CGIAR (1994) Sustainable agriculture for a food secure world: a vision for international agricultural research. Expert Panel of the CGIAR, Washington DC

Chakravorty U, Fisher DK, Umetsu C (2007) Environmental effects of intensification of agriculture: livestock production and regulation. Environ Econ Policy Stud 8(4):315–336. https://doi.org/10.1007/BF03353963

CMU (2016) Concept of state policy implementation concerning the climate change until 2030. Order of Cabinet of Ministers of Ukraine # 932-p from 7 December 2016. https://zakon.rada.gov.ua/laws/show/932-2016-p

Delzeit R, Zabel F, Meyer C, Václavík T (2017) Addressing future trade-offs between biodiversity and cropland expansion to improve food security. Reg Environ Chang 17(5):1429–1441. https://doi.org/10.1007/s10113-016-0927-1

Devaux A, Goffart JP, Petsakos A, Kromann P, Gatto M, Okello J, Suarez V, Hareau G (2020) Global food security, contributions from sustainable potato agri-food systems. In: Campos H, Ortiz O (eds) The potato crop. Springer, Cham, pp 3–35. https://doi.org/10.1007/978-3-030-28683-5_1

Diaz-Ambrona CGH, Maletta E (2014) Achieving global food security through sustainable development of agriculture and food systems with regard to nutrients, soil, land, and waste management. Curr Sustain Energy Rep 1(2):57–65. https://doi.org/10.1007/s40518-014-0009-2

Duque-Acevedo M, Belmonte-Ureña LJ, Cortés-García FJ, Camacho-Ferre F (2020) Agricultural waste: review of the evolution, approaches and perspectives on alternative uses. Global Ecol Conserv 22:e00902. https://doi.org/10.1016/j.gecco.2020.e00902

El Bilali H (2019) Research on agro-food sustainability transitions: where are food security and nutrition? Food Secur 11(3):559–577. https://doi.org/10.1007/s12571-019-00922-1

FAO (2015) Climate change and food systems: global assessments and implications for food security and trade. Food Agriculture Organization of the United Nations (FAO), Rome, p 356

FAO (2017) The future of food and agriculture: Trends and challenges. Food and Agriculture Organization of the United Nations, Rome

FAO (2018) Building climate resilience for food security and nutrition. Food and Agriculture Organization of the United Nations, Rome

FAO (2019a) Healthy Soils in Ukraine: 2019. Integrated natural resources management in degraded landscapes in the forest-steppe and steppe zones of Ukraine. Food and Agriculture Organization of the United Nations, Rome, p 6

FAO (2019b) The state of food security and nutrition in the world: safeguarding against economic slowdowns and downturns. Food and Agriculture Organization of the United Nations, Rome

Fatemi M, Rezaei-Moghaddam K (2019) Multi-criteria evaluation in paradigmatic perspectives of agricultural environmental management. Heliyon 5(2):e01229. https://doi.org/10.1016/j.heliyon.2019.e01229

Funk CC, Brown ME (2009) Declining global per capita agricultural production and warming oceans threaten food security. Food Secur 1(3):271–289. https://doi.org/10.1007/s12571-009-0026-y

Gaffney J, Bing J, Byrne PF, Cassman KG, Ciampitti I, Delmer D, Habben J, Lafitte HR, Lidstrom UE, Porter DO, Sawyer JE, Schussler J, Setter T, Sharp RE, Vyn TJ, Warner D (2019) Science-based intensive agriculture: sustainability, food security, and the role of technology. Glob Food Sec 23:236–244. https://doi.org/10.1016/j.gfs.2019.08.003

Gowdy J (2020) Our hunter-gatherer future: climate change, agriculture and uncivilization. Futures 115:102488. https://doi.org/10.1016/j.futures.2019.102488

GRFC (2020) Global report on food crises-2019. Joint analysis for better decisions. Food Security Information Network, Rome, p 202

Hazell P (1995) Managing agricultural intensification. IFPRI 2020 brief, 11. International Food Policy Institute, Washington, DC

Hediger W, Lehmann B (2007) Multifunctional agriculture and the preservation of environmental benefits. Swiss J Econ Stat 143(4):449–470. https://doi.org/10.1007/BF03399246

Hoffmann KG (1991) The economic mechanism of nature management in the transition to a market economy. Econ Math Methods 27(2):315–321

Horton P, Banwart SA, Brockington D, Brown GW, Bruce R, Cameron D, Holdsworth M, Lenny Koh SC, Ton J, Jackson P (2017) An agenda for integrated system-wide interdisciplinary agri-food research. Food Secur 9(2):195–210. https://doi.org/10.1007/s12571-017-0648-4

Ickowitz A, Powell B, Rowland D, Jones A, Sunderland T (2019) Agricultural intensification, dietary diversity, and markets in the global food security narrative. Glob Food Sec 20:9–16. https://doi.org/10.1016/j.gfs.2018.11.002

IFPRI (2020) Global food policy report-2020: building inclusive food systems (ed.) International Food Policy Research Institute. https://doi.org/10.2499/9780896293670

Jhariya MK, Banerjee A, Meena RS, Yadav DK (2019a) Sustainable agriculture, forest and environmental management. Springer, Singapore, p 606

Jhariya MK, Yadav DK, Banerjee A (2019b) Agroforestry and climate change: issues and challenges. Apple Academic Press Inc., New York, p 335. https://doi.org/10.1201/9780429057274

Khan N, Jhariya MK, Yadav DK, Banerjee A (2020a) Herbaceous dynamics and CO_2 mitigation in an urban setup- a case study from Chhattisgarh, India. Environ Sci Pollut Res 27(3):2881–2897. https://doi.org/10.1007/s11356-019-07182-8

Khan N, Jhariya MK, Yadav DK, Banerjee A (2020b) Structure, diversity and ecological function of shrub species in an urban setup of Sarguja, Chhattisgarh, India. Environ Sci Pollut Res 27 (5):5418–5432. https://doi.org/10.1007/s11356-019-07172-w

Kleijn D, Bommarco R, Fijen Thijs PM, Garibaldi LA, Potts SG, van der Putten WH (2019) Ecological intensification: bridging the gap between science and practice. Trends Ecol Evol 34 (2):154–166. https://doi.org/10.1016/j.tree.2018.11.002

Koblianska I, Kalachevska L (2019) Implementation of local food concept for social-economic revitalization in rural areas: the case of Ukraine. Indian J Econ Dev 7(10):148032

Konuma H (2018) Status and outlook of global food security and the role of underutilized food resources: sago palm. In: Ehara H, Toyoda Y, Johnson DV (eds) Sago palm. Springer, Singapore, pp 3–16. https://doi.org/10.1007/978-981-10-5269-9_1

Kopittke PM, Menzies NW, Wang P, McKenna BA, Lombi E (2019) Soil and the intensification of agriculture for global food security. Environ Int 132:105078. https://doi.org/10.1016/j.envint.2019.105078

Kumar S, Meena RS, Jhariya MK (2020) Resources use efficiency in agriculture. Springer, Singapore, p 760

Kupinets LE (2010) Greening of the food complex: theory, methodology, mechanisms. IPREEI NAS of Ukraine, Odessa, p 71

Kupinets LE, Zhavnerchik OV (2016) Environmental security of agricultural land use: theory and mechanisms of provision. National Academy of Sciences of Ukraine, Institute of Market Problems and Economic and Ecological Research, p 316

Lindgren E, Harris F, Dangour AD, Gasparatos A, Hiramatsu M, Javadi F, Loken B, Murakami T, Scheelbeek P, Haines A (2018) Sustainable food systems—a health perspective. Sustain Sci 13 (6):1505–1517. https://doi.org/10.1007/s11625-018-0586-x

Lipper L, Zilberman D (2018) A short history of the evolution of the climate smart agriculture approach and its links to climate change and sustainable agriculture debates. In: Lipper L, McCarthy N, Zilberman D, Sfaw SA, Branca G (eds) Climate smart agriculture, vol 52. Springer, Cham, pp 13–30. https://doi.org/10.1007/978-3-319-61194-5_2

Makarova SV, Stepanova NR (2014) Social responsibility - the most important factor of sustainability of the development of the organization and society as a whole. Fundam Res 5 (5):1075–1079

McCalla A (1994) Agriculture and food needs to 2025: why we should be concerned. Sir John Crawford Memorial Lecture, October 27. CGIAR Secretariat. The World Bank, Washington, DC

Medvedev VV (2010) Measures to promote the implementation of soil protection technologies in European countries. Bull Agrar Sci 6:15–17

Meena RS, Lal R (2018) Legumes for soil health and sustainable management. Springer, Singapore, p 541

Meena RS, Kumar V, Yadav GS, Mitran T (2018) Response and interaction of Bradyrhizobium japonicum and Arbuscular mycorrhizal fungi in the soybean rhizosphere: a review. Plant Growth Regul 84:207–223

Meena RS, Kumar S, Datta R, Lal R, Vijaykumar V, Brtnicky M, Sharma MP, Yadav GS, Jhariya MK, Jangir CK, Pathan SI, Dokulilova T, Pecina V, Marfo TD (2020) Impact of agrochemicals on soil microbiota and management: a review. Land 9(2):34. https://doi.org/10.3390/land9020034

Mishenin YV, Koblianska II (2016) Socio-economical aspects of restriction on implementation of the agricultural land ownership right in Ukraine. Balanced Nat Manage 1:112–120

Mishenin YV, Yarova IY (2015) Mechanisms for ensuring socially environmentally responsible agricultural land use. Balanced Nat Manage 2:90–94

Mishenin YV, Yarova IY (2019) Systematic assessment of the effectiveness of environmental taxation in the context of socio-ecological and economic security of spatial development. Balanced Nat Manage 1:38–47. https://doi.org/10.33730/2310-4678.1.2019.170589

Mishenin YV, Rishnyak IN, Tarkhov PV (2002) Organizational-economic mechanism of ecologization of agrarian sphere. Bull Sumy Natl Agrar Univ Ser 1-2:77–81

Mishenin YV, Kosodiy RP, Butenko VM (2011) Socio-economic and financial problems of sustainable rural development. "Papyrus TD" LLC. Sumy. p 334

Mishenin YV, Dutchenko OM, Yarova IY (2015) Sustainable land use in the context of food security: national and global aspects. Bull Sumy Natl Agrar Univ 4(63):4–8

Mishenin YV, Yarova IY, Dutchenko OM (2017) Ecological and economic security of agrarian farming: conceptual guidelines and organizational mechanisms. Balanced Nat Manage 2:41–45

Müller D, Jungandreas A, Koch F, Schierhorn F (2016) Impact of climate change on wheat production in Ukraine. German-Ukrainian Agricultural Policy Dialogue. Kyiv, p 41

Nicholls E, Ely A, Birkin L, Basu P, Goulson D (2020) The contribution of small-scale food production in urban areas to the sustainable development goals: A review and case study. Sustain Sci. https://doi.org/10.1007/s11625-020-00792-z

Nkonya E, von Braun J, Mirzabaev A, Le QB, Kwon HY, Kirui O (2016) Concepts and methods of global assessment of the economics of land degradation and improvement. In: Nkonya E, Mirzabaev A, von Braun J (eds) Economics of land degradation and improvement – a global assessment for sustainable development. Springer, Cham, pp 15–32. https://doi.org/10.1007/978-3-319-19168-3_2

Ovsyannikov YA (2000) Theoretical foundations of ecological-biosphere agriculture. Ural University, Ekaterinburg, p 264

Pakhomova NV, Malyshkov GB (2008) Social and environmental responsibility and business competitiveness: is a synergistic effect possible? Prob Mod Econ 2:310–318

Pangaribowo E, Gerber N (2016) Innovations for food and nutrition security: impacts and trends. In: Gatzweiler FW, von Braun J (eds) Technological and institutional innovations for marginalized smallholders in agricultural development. Springer, New York, pp 41–64. https://doi.org/10.1007/978-3-319-25718-1_3

Pingali P, Aiyar A, Abraham M, Rahman A (2019) Managing climate change risks in food systems. In: Pingali P, Aiyar A, Abraham M, Rahman A (eds) Transforming food systems for a rising India. Springer, New York, pp 241–275. https://doi.org/10.1007/978-3-030-14409-8_10

Pretty JN (1995) Regenerating agriculture: policies and practice for sustainability and self-reliance. Earthscan Publications Ltd., London

Pretty JN, Thompson J (1996) Sustainable agriculture and the overseas development administration. Report for natural resources policy advisory department, ODA, London

Raj A, Jhariya MK, Yadav DK, Banerjee A (2020) Climate change and agroforestry systems: adaptation and mitigation strategies. Apple Academic Press Inc., New York, p 383

Read QD, Brown S, Cuéllar AD, Finn SM, Gephart JA, Marston LT, Meyer E, Weitz KA, Muth MK (2020) Assessing the environmental impacts of halving food loss and waste along the food supply chain. Sci Total Environ 712:136255. https://doi.org/10.1016/j.scitotenv.2019.136255

Rega C, Helming J, Paracchini ML (2019) Environmentalism and localism in agricultural and land-use policies can maintain food production while supporting biodiversity. Findings from simulations of contrasting scenarios in the EU. Land Use Policy 87:5. https://doi.org/10.1016/j.landusepol.2019.05.005

Robbins SP, Coutler M (2007) Management, 8th edn. South Missouri State University, Joplin, p 1056

Schmidheni S (1994) Change of course. Perspectives and environmental problems: an entrepreneurial approach. Helikon, Moscow, p 348

Shkuratov OI (2016) Organizational and economic bases of ecological safety in the agrarian sector of Ukraine: theory, methodology, practice. DKS – Center. Kyiv, p 356

Spiess WEL (2016) Challenges to food security in a changing world. In: Jackson P, Spiess WEL, Sultana F (eds) Eating, drinking: surviving. Springer, Cham, pp 57–73. https://doi.org/10.1007/978-3-319-42468-2_7

Stepchin MV (2006) Ecological and economic problems of rational land use (on the example of water and forest land lands. Author's dissertation for the degree of candidate of economic sciences: specialty 08.00.06. Kyiv, p 18

Strochenko N, Koblianska I, Markova O (2017) Structural transformations in agriculture as necessary condition for sustainable rural development in Ukraine. Spring 1(23):237–249. https://doi.org/10.14505/jarle.v8.1(23).27

Stupin NR (2017) Ecological and economic mechanism of agricultural land reproduction. DKS-Center, Kyiv, p 176

Targetti S, Schaller LL, Kantelhardt J (2019) A fuzzy cognitive mapping approach for the assessment of public-goods governance in agricultural landscapes. Land Use Policy. https://doi.org/10.1016/j.landusepol.2019.04.033

Thompson J (1996) Sustainable agriculture and rural development: challenges for EU Aid. EC Aid and Sustainable Development Briefing Paper, No. 8. International Institute for Environment and Development, London

Tonitto C, Woodbury PB, McLellan EL (2018) Defining a best practice methodology for modeling the environmental performance of agriculture. Environ Sci Pol 87:64–73. https://doi.org/10.1016/j.envsci.2018.04.009

Tsarenko OM (2001) Theoretical substantiation of formation of ecologically safe economic policy in agroindustrial complex. Bull Sumy State Agrar Univ 2:20–24

Ullah S, Ai C, Huang S, Song D, Abbas T, Zhang J, Zhou W, He P (2020) Substituting ecological intensification of agriculture for conventional agricultural practices increased yield and decreased nitrogen losses in North China. Appl Soil Ecol 147:103395. https://doi.org/10.1016/j.apsoil.2019.103395

UNO (2016) About the Sustainable Development Goals. https://www.un.org/. Accessed 22 January 2020

van Westen ACM, Mangnus E, Wangu J, Worku SG (2019) Inclusive agribusiness models in the Global South: the impact on local food security. Curr Opin Environ Sustain 41:64–68. https://doi.org/10.1016/j.cosust.2019.11.003

Verner IY (2019) The statistical yearbook of Ukraine for 2018. BUK-Druk LLC. Zhytomyr. p 481. http://www.ukrstat.gov.ua/druk/publicat/kat_u/2019/zb/11/zb_yearbook_2018.pdf

World Bank (2020) Agriculture and food. Overview 2020. https://www.worldbank.org/en/topic/agriculture/overview#1. Accessed 15 April 2020

Yatsuk IP (ed) (2018) Scientific research on monitoring and investigation of agricultural land of Ukraine (according to results of X round 2011-2015). Institute of Soils Protection of Ukraine, Kyiv, p 66. http://www.iogu.gov.ua

Zilberman D, Lipper L, McCarthy N, Gordon B (2018) Innovation in response to climate change. In: Lipper L, McCarthy N, Zilberman D, Asfaw S, Branca G (eds) Climate smart agriculture, vol 52. Springer, New York, pp 49–74. https://doi.org/10.1007/978-3-319-61194-5_4

Ecological Intensification: A Step Towards Biodiversity Conservation and Management of Terrestrial Landscape

3

Donald Mlambo

Abstract

The current model of agricultural intensification has increased crop yields and profits for farmers. However, this increase takes place by significant loss of biodiversity as well as ecosystem services, which has become a global concern. In agricultural landscapes, biodiversity loss impairs the functionality of ecosystem in the form of pollination, natural pest control, habitat provision and water purification. In order to restore biodiversity along with maintaining agricultural production, there is need for farmers to switch to a novel farming approach that can optimize ecosystem functions and enhance crop yields. Reports reveal that ecological intensification has potential to ameliorate environmental externalities while preserving crop yields and profitability. To intensify ecological processes in agricultural landscapes, a potential strategy is to employ management practices that reduce or substitute synthetic agrochemical use, maintain or enhance landscape heterogeneity and connectivity. Intensification of eco-friendly nature may be achieved by wildlife-friendly approaches in the form of organic farming, conservation farming, agroforestry, integrated pest management and intercropping. However, lack of comprehensive information on the net benefits of ecological intensification farming practices is currently preventing widespread adoption by farmers. To increase uptake, it is critical that scientists address not only the ecological facets of biodiversity-friendly farming practices but also the economic and social facets.

Keywords

Agricultural intensification · Agroforestry · Biodiversity · Conservation agriculture · Ecological intensification · Intercropping · Organic agriculture

D. Mlambo (✉)
Department Forest Resources & Wildlife Management, Faculty of Applied Science, National University of Science &Technology, Bulawayo, Zimbabwe

M. K. Jhariya et al. (eds.), *Ecological Intensification of Natural Resources for Sustainable Agriculture*, https://doi.org/10.1007/978-981-33-4203-3_3

Abbreviations

IPM	Integrated Pest Management
GMOs	Genetically Modified Organisms
NGO	Nongovernmental Organization
SOM	Soil Organic Matter

3.1 Introduction

The conservation of biodiversity in terrestrial landscapes is a key linchpin for combatting the triple Anthropocene challenge of declining biodiversity, climate change and land degradation. Biodiversity is the variety of all living organisms or species within ecosystems and is important for ecosystem functioning. The current best estimate is that there are roughly 8.7 million living species on the planet (Mora et al. 2011). Biodiversity provides many valuable services to the ecosystem and human well-being such as food, income, natural pest control, pollination and water quality (Banerjee et al. 2020; Raj et al. 2020). Together these services have been estimated to be worth over US$125–145 trillion per annum (Costanza et al. 2014).

Globally, the issue of biodiversity loss is of concern given that it will negatively affect ecosystem services that are important for the survival of mankind (Wagg et al. 2014; Jhariya et al. 2019a, 2019b). Across the globe, humankind has intervened tremendously into the natural succession of ecosystems. A large portion (about 40%) of terrestrial landscape has been converted into agriculture globally (Fahrig et al. 2011), making farmers the largest group of ecosystem managers on earth. Currently, farmers are faced with a huge challenge of producing enough food to global population of 9–12 billion till 2050 while reducing the adverse consequences of cultivation towards environment (HLPE 2017; Meena et al. 2018). To meet the ever increasing demand for agricultural products, farmers may be forced to convert more forestland into agricultural production or employ agricultural intensification farming methods on the existing agricultural land, which is likely to increase damage to the environment. There are divergent opinions on how farmers can safeguard loss of biodiversity caused by agriculture. Some have advocated for the adoption of an intensive farming system that should be practiced on small areas, sparing forestland from conversion to farmlands. Other believes that farming and biodiversity conservation can coexist on the same piece of land (Bommarco et al. 2013; Tscharntke et al. 2012; Vongvisouk et al. 2016).

Crop field size enlargement, increased use of agrochemicals, monoculture and reduction in landscape heterogeneity have caused a significant loss of biodiversity (Emmerson et al. 2016; Landis 2017; Pretty 2018; Meena and Lal 2018). Many arable landscapes face pollution by agrochemicals which negatively affect biodiversity and associated ecosystem services on which agriculture depends. Research reports suggest that effective conservation of biodiversity in agricultural sector can be stimulated through reducing fertilizers and pesticides inputs as well as increasing crop diversity and landscape heterogeneity (Gonthier et al. 2014; Sanchez-Bayo and Wyckhuys 2019). A farming system that is ecologically based creates an opportunity for the establishment of a resilient agro-ecosystem, which allows farmers to increase crop yields without exerting loss to biodiversity (Bommarco et al. 2013; Kleijn et al. 2018; Kumar et al. 2020).

3.2 Agricultural Intensification and Biodiversity Conservation

Agricultural intensification considers cultivation practices that rely on high inputs to obtain maximum yield per hectare of crops or livestock. It is often pointed as a strategy for reducing agricultural encroachment into forests. Before the industrialization of agriculture in the 1960s, the most common strategy to increase agricultural production was to clear natural vegetation to expand agricultural land (Tilman et al. 2001, 2002). As human populations increased and arable land dwindled, the novel approach was to maximize the agricultural output on the existing agricultural setup supported by higher inputs. The continued expansion of agricultural land is causing shrinkage of natural habitats (Benton et al. 2003; Meena et al. 2020a, b). Where remnants of natural or semi-natural habitats exist in agro-ecosystems, they support less biodiversity because of increased habitat fragmentation and isolation as well as agrochemical drifts in crop field edges (Egan et al. 2014; Landis 2017). Heavy uses of pesticides and pesticide drift have a risk of limiting organisms that are beneficial to the farmer. Empirical evidence shows that pollinators are reduced in agricultural landscapes with heavy use of agrochemicals (Potts et al. 2010; Kennedy et al. 2013; Shackelford et al. 2013). In agro-ecosystems, pollinators provide important ecosystem services to flowering crops and wild plants.

Agricultural intensification is featured by high usage of agrochemicals per unit area, increased mechanization and soil tillage, reduced landscape heterogeneity, and low crop diversity (Fig. 3.1). There are some reports indicating that in the past 50 years, irrigated land area doubled, the use of fertilizers increased sevenfold and global food production more than tripled, reducing world hunger (Foley et al. 2011; Tilman et al. 2011). However, the increase in global food production was at the expense of biodiversity loss (Tscharntke et al. 2005). Agricultural intensification is one of the factors responsible for loss of biodiversity, including birds (Donald et al. 2006; Mineau and Whiteside 2013; Traba and Morales 2019), vascular plants (Storkey et al. 2012), invertebrates (Medan et al. 2011) and soil organisms (Wagg et al. 2014).

In high-input farming systems, the heavy dependence and overuse of agrochemicals may cause damage the various beneficial organisms (Catarino et al. 2015). The leakage of fertilizer from intensively managed conventional farms pollutes surface waters and causes damage to aquatic organisms (Geiger et al. 2010; Beketov et al. 2013). As a result of pollution from fertilizers, algal blooms proliferate in nutrient-loaded water bodies, causing damage to freshwater biodiversity (Kibria et al. 2013). Many agrochemicals drift far from the point of application,

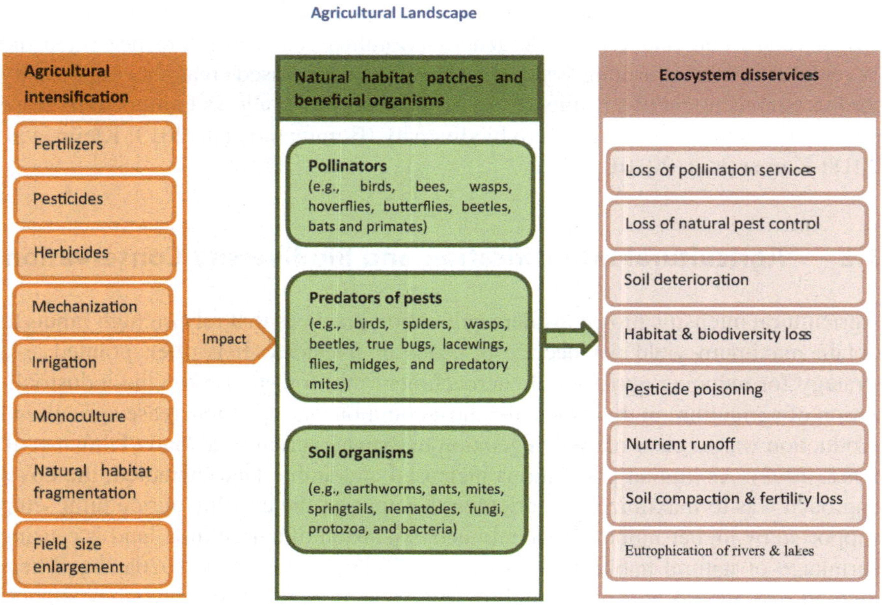

Fig. 3.1 The impact of agricultural intensification on organisms beneficial to farmlands and the associated ecosystem disservices

causing damage to non-target organisms (Martín-López et al. 2011; Egan et al. 2014; Chagnon et al. 2015). Herbicides such as atrazine are endocrine disruptors that can cause reproductive problems in mammals, amphibians, and fish. Wildlife poisoning by highly toxic pesticides can cause substantial decline in species populations including rare ones. Many bird species are directly affected by poisoning from broad-spectrum pesticides such as organophosphates, carbamates and anticoagulant rodenticides (BLI 2008; Mitra et al. 2011). Birds get their exposure by consuming seeds contaminated with pesticides. Broad-spectrum herbicides tend to work for reduction of weeds and insect population which is actually the food material for birds. Furthermore, beneficial insects such as bees, spiders and beetles are negatively affected by broad-spectrum insecticides. In agro-ecosystems, predatory mammals and raptors are often indirectly poisoned by anticoagulant rodenticides. Wildlife habitats can be altered by the application of herbicides, which in turn threaten the survival of predatory mammals. Pesticides can reduce the abundances and activities of earthworms, symbiotic mycorrhizae, and other soil-dwelling organisms (Meena et al. 2020).

Global concern of agro-ecosystem services is the matter of great concern as it may reflect reduced functioning as biodiversity continues to decline due to continued use of biodiversity-unfriendly farming practices (Power 2010). The most affected ecosystem services are biological pest control (Bengtsson 2015), crop pollination (Deguines et al. 2014), biogeochemical cycling and health of soil (Matson et al.

1997). Monoculture is another common feature of intensive agriculture which provides a uniform range of habitats that are colonized by a limited range of species. To prevent loss of ecosystem services due to agricultural intensification, there is need to search for a novel farming system that is ecologically based.

3.3 Ecological Intensification and Biodiversity Conservation

The conventional farming system remains primarily driven by the 'intensification' of external inputs to increase yield and sustain food requirements of the world's growing population (Norton et al. 2009). There are varying ideas on how to increase crop yield and farmers' profits without encroaching further into natural areas or causing loss to biodiversity. Some believe that agricultural intensification can make farming more efficient and productive on limited land area while others believe that it must be replaced with a system that is environmentally friendly. Since the agricultural intensification model is not compatible with biodiversity conservation goals, scientists are advocating for a transition to agro-ecological intensification (Bommarco et al. 2013; de Molina and Casado 2017; Cui et al. 2018; Garibaldi et al. 2019). Ecological intensification refers to a farming system that relies on local rather than external inputs (e.g., agrochemicals) to increase yield while maintaining or enhancing biodiversity and ecological functions (Bommarco et al. 2013; Cassman 1999; Garibaldi et al. 2019). Various bioresource-enhancing farming practices already exist (Fig. 3.2) and, if widely adopted by farmers, they have potential to reverse the damage to the environment caused by conventional farming.

A principal question is whether the ecological intensification approach can 'erase' the ecological footprints of agricultural intensification, while meeting food production needs. As already highlighted by Kovács-Hostyánszki et al. (2017), this approach is surrounded by uncertainties, largely stemming from inadequate information on how to implement it, whether an individual farmer can realize positive net economic benefits and the unpredictability of natural systems. Ecological intensification relies on natural systems such as the use of organic fertilizers (e.g., livestock manure and compost), pollinators, natural pesticides and natural enemies of crop pests. Currently, there are differences between scientists and farmers in the way they perceive potential benefits of ecological intensification. Scientists believe that ecological intensification can replace the expensive external farm inputs with ecosystem services but there is lack of evidence to prove that it can increase yield and farmer's profits. In the absence of motivating factors such as increases in yield and profitability or receiving financial support to conserve biodiversity, it is unlikely that farmers will radically change their way of farming (Pannell 2003; Schoonhoven and Runhaar 2018). Some governments and international conservation NGOs have schemes designed to incentivize farmers to implement biodiversity-friendly farming techniques. For example, schemes such as paying farmers for environmental services (PES) provide incentives to farmers for adopting land management techniques that can reduce negative impacts on biodiversity.

Fig. 3.2 Farming techniques that can intensify ecological processes in agricultural landscapes

Although farmers are slow in adopting the idea of ecological intensification, many scientists believe that maintaining high biodiversity in agricultural landscapes makes farming more sustainable. Biodiversity-friendly farming practices enhance the provision of ecosystem services (Bommarco et al. 2013; Singh and Jhariya 2016; Jhariya et al. 2015, 2018). There are several farming practices that can support farmland biodiversity including various eco-friendly practices (Fig. 3.2).

3.4 The Role of Biodiversity in Agricultural Landscapes

It has been suggested that wildlife-friendly farming practices make ecosystem service delivery more stable (Yachi and Loreau 1999; Isbell et al. 2017). In agro-ecosystems, biodiversity conservation is important in terms of ecological function and process like biogeochemical cycling, soil fertility, erosion and pest control, water quality, carbon sequestration, habitat provision for wildlife, wood and recreation (Table 3.1). There is a large pool of organisms hidden below-ground in soils (Wagg et al. 2014) whose diversity and abundance are altered by farming activities. Soil microorganisms help to maintain soil health and crucial to farming. The arthropod community is another group of wildlife that is beneficial to farmers for

Table 3.1 Examples of ecosystem services provided by biodiversity in agricultural landscapes (Source: Garibaldi et al. 2011; Garbach et al. 2014)

Ecosystem services	Definition	Examples
Regulating services	Ecosystem process regulation	•Erosion control •Flood control •Pollination •Pest control •Wildlife habitat •Water purification •Carbon sequestration •Habitat connectivity •Wind breaks •Microclimate
Provisioning services	Provisioning services relate to products obtained from the ecosystem	•Food •Fibre •Fuel •Forage •Medicines •Ornamental products •Genetic resources
Supporting	Supporting services mean aiding activity for smooth functioning of ecosystem	•Nitrogen fixation •Nutrient cycling •Soil formation and retention •Water cycling •Oxygen production
Cultural	Ecosystem's non-tangible benefits	•Aesthetic landscapes •Recreation •Shade •Sport (game hunting) •Sounds from birds •Scent from plants •Spiritual experience

their importance in nutrient cycling, seed dispersal, pest control, pollination and maintaining soil structure and fertility (Thorbek and Bilde 2004; Scudder 2009). Additionally, they are an important food source for other taxa including farmland birds.

Crop production is supported by ecological functions operating the ecosystem (Winfree et al. 2015; Kovács-Hostyánszki et al. 2017). Pollination services are special to farmers and a wide range of crops grown for human consumption are pollinator-dependent (Klein et al. 2007; Garibaldi et al. 2011; Deguines et al. 2014).

There are more than 100,000 species of pollinators which include bees, butterflies, beetles, birds, flies and bats. Garibaldi et al. (2011) conducted meta-analysis and found a decreasing in pollination services due to isolation from non-cropped habitats which clearly reveals the importance on farmlands.

It is estimated that globally pests damage 1/third of crop productivity yearly (Oerke 2006). This makes regulation of pest population as a significant ecological function in agro-ecosystem since it has the potential to reduce pesticides consumption. Enemies of natural pests are many and include birds, insects (e.g., ladybugs, parasitic wasps and flies), spiders, fungi, pathogens and many other types of organisms. Such biological controls help to reduce the costs of protecting crops and the need for pesticide use.

3.5 The Impact of Integrated Pest Management on Biodiversity and Ecosystem Services

Integrated Pest Management (IPM) aims to control pests, weeds and diseases by relying upon natural enemies rather than the extensive use of synthetic pesticides. Biological pest control is one of the IPM tactics that involves using natural enemies of crop pests such as insect predators, parasitoids or pathogens to suppress pest populations (Hokkanen 2015; Xu et al. 2017). IPM is designed to reduce the negative ecological impacts from over and improper use of synthetic pesticides which in turn helps to maintain or enhance biodiversity (Table 3.2). Although IPM emerged more than seven decades ago, adoption by farmers is still low (Parsa et al. 2014; Pretty and Bharucha 2015).

One interesting example of IPM is the 'push-pull' system which involves growing multiple crops on the piece of land (polyculture) for purposes of protecting crops from pest infestation (Xu et al. 2017; Hassanali et al. 2008). The said approach uses plants that repel insect pests ('push') and plants that attract the pests away from the crop ('pull'). As illustrated in Fig. 3.3, Napier grass (*Pennisetum purpureum*) is often planted by smallholder farmers in East Africa around the edges of crop fields to attract the moths, pulling them away from the main crop (Khan et al. 2011; Pickett

Table 3.2 Examples of IPM strategies that can intensify ecological processes on farmlands

IPM strategy	References
1. Substituting chemical pesticides with bio-pesticides	Pickett et al. (2014)
2. Controlling pesticide applications (e.g., targeted spraying)	Pretty and Bharucha (2015)
3. Breeding crops or livestock with traits that can resist attack by pests, parasites and pathogens	Pretty and Bharucha (2015)
4. Releasing natural enemies of pests (parasitoids and predators) to control pest populations	Hassanali et al. (2008), Khan et al. (2011)
5. Deploying sticky and pheromone compounds to trap pests	Xu et al. (2017)
6. Using crop rotations and crop diversity to limit pests and diseases	Pretty and Bharucha (2015)
7. Using the push-pull system to drive-off pests from crops	Hassanali et al. (2008)

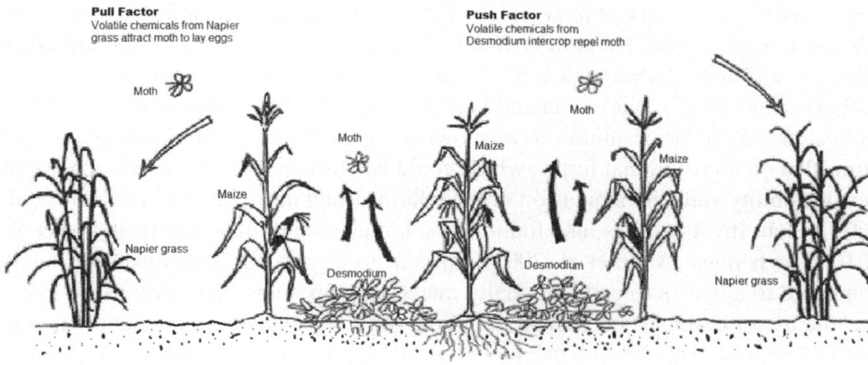

Fig. 3.3 The push-pull system (Adapted from Pickett et al. 2014)

et al. 2014). The farmers also intercrop legumes such as *Desmodium* with cereal crops (maize, millet and sorghum) to control pests as well as fixing nitrogen, which can be up to 100 kg N ha^{-1} $yr.^{-1}$ (Khan et al. 2011). *Desmodium* releases volatile chemicals that repel stem borer moths (push) and attract natural enemies of moths and parasitic wasps (pull). The push-pull system allows farmers to control pests and erosion, enhance soil fertility and reduce synthetic pesticide use. *Desmodium* can be used as fodder for livestock or can be sold to gain income (Hassanali et al. 2008).

3.6 The Impact of Organic Agriculture on Biodiversity and Ecosystem Services

Organic farming aims to increase soil health by nutrient use efficiency and other resources along with application of synthetic pesticides (IFOAM 2008; FAO 2015). Synthetic pesticides are substituted with bio-pesticides which include the use of microbes, entomophagous nematodes, plant-derived pesticides, antibiotics, insect pheromones, and fungal and viral attacks (Copping and Menn 2000). On organic farms, soil fertility is enhanced through the use of environmentally friendly farming systems such as crop rotation, intercropping, polyculture, covering crops and mulching. Weeds are controlled by employing a variety of techniques such as appropriate rotations, timing of seeding, mechanic cultivation, mulching, transplanting and flaming. Soil C concentrations are reported to be more elevated on organic than non-organic crop fields due to greater accumulation of soil organic matter from crop residues, cover crops, manure and compost (Gattinger et al. 2012). Organic farming is wildlife-friendly due to the non-use of chemical pesticides and fertilizers (Norton et al. 2009) and its wide uptake by farmers can benefit a range of taxa (Hole et al. 2005; Bengtsson et al. 2005; Gattinger et al. 2012; Tuck et al. 2014).

A meta-analysis conducted by Tuck et al. (2014) indicates that there is on average 30% higher species richness on organic farms than on conventional farms. Empirical evidence suggests that organic farming generally has a positive influence on richness and diversity of plants (Fuller et al. 2005; Gabriel et al. 2006), invertebrates (Holzschuh et al. 2008; Rundlöf et al. 2010), birds (Smith et al. 2010), soil microbes (Oehl et al. 2004; Verbruggen et al. 2010) and activity density of small mammals such as wood mouse, bank vole and common shrew (Wickramasinghe et al. 2003; Hole et al. 2005). Some studies have shown a higher abundance of birds on organic farms that on conventional farms, which could be attributed largely to an increase in the availability suitable habitats on organic farms (Santangeli et al. 2019; Hole et al. 2005). Plant diversity was also found to be higher in field margins (Rundlöf et al. 2010) and hedges (Aude et al. 2003) adjacent to organically managed crop fields compared to adjacent to conventionally managed crop fields. In conventional farming, the use of herbicides is known to reduce non-crop plant diversity in both arable lands (Winqvist et al. 2011; Schneider et al. 2014) and adjacent habitats (Aude et al. 2003; Rundlöf et al. 2010). Some studies have demonstrated that pollinating arthropods (e.g., bees, hoverflies and butterflies) and true bug communities have higher species richness on organic farms than on conventional farms (Holzschuh et al. 2008; Andersson et al. 2013; Birkhofer et al. 2015; Rundlöf et al. 2016). Furthermore, the presence of farmscape plantings (e.g., hedgerows, windbreaks and filter strips) and the absence of drifting pesticides and herbicides on organically managed farms attract new or re-colonizing species.

Although organic farming is beneficial to farmland biodiversity (Tuck et al. 2014), the major criticism is that it has lower productivity than conventional agriculture (Hodgson et al. 2010; Pickett 2013; Ponisio et al. 2015; Pittelkow et al. 2015; Röös et al. 2018). Studies by Seufert et al. (2012) and Ponisio et al. (2015) revealed that organic farms have on average 19.2% less yield compared to conventional farms. Recently, Knapp and van der Heijden (2018) found relative yield stability (temporal variation per unit yield produced) on organic farms to be significantly lower than on conventional farms. Opponents of organic farming have argued that it requires traditional practices for food production, and may threaten the world's natural habitats if adopted on large scales (Trewavas 2001, 2002; Avery 2006; Pickett 2013). Organic farming can result in reduced crop yields due to crop-weed competition, herbivory and diseases (Pittelkow et al. 2015; Knapp and van der Heijden 2018). Pest and weed control is a challenge in organic farming where agro-chemicals are not used. It is also a huge challenge in the absence of selection for traits such as resistance to diseases, enhanced interaction with plant symbionts and superior weed suppressing abilities (Knapp and van der Heijden 2018; Ponisio et al. 2015). On organic farms, pest control is achieved by using appropriate cropping techniques such as polyculture, natural pesticides and biological control. Weed control is managed by a wide range of techniques including rotations, timing for seeding, mechanic cultivation, mulching, transplanting and flaming (Gomiero et al. 2011).

3.7 The Impact of Agroforestry on Biodiversity and Ecosystem Services

Agroforestry, broadly defined as the integration of woody plants into farming systems is important for food insecurity, income generation and the conservation of biological diversity (Tscharntke et al. 2011; Kuyah et al. 2016; Barrios et al. 2018; Raj et al. 2019a, 2019b). Many smallholder farmers maintain or integrate trees into the farming system for the provision of fuelwood, food, fodder and shade (McNeely and Schroth 2006). The presence of trees on farmlands has been reported to decrease pest abundance (Guenat et al. 2019). Trees provide habitat for a variety of species including pollinators such as bees that would not be able to survive in a landscape with only annual crops (Ricketts et al. 2008; Pumariño et al. 2015). For peasant farmers who depend largely on the natural environment for inputs, more pollinators can increase mean yields (Garibaldi et al. 2016). Trees also provide favourable habitats for soil biota beneath their canopies through microclimate buffering and continuous supply of litterfall inputs which positively influence the activity of soil organisms (Kamau et al. 2017; Barrios et al. 2018).

Agroforestry practices maintain or increase landscape connectivity and heterogeneity, which is essential for the conservation of biological diversity. For example, smallholder farmers in Africa deliberately select and protect valuable indigenous trees located in their crop fields or set aside non-cropped areas that are managed as biodiversity conservation areas (Boffa 1999). Some authors have reported high levels of woody species diversity in agroforestry systems in Africa and Central America (Michon and de Foresta 1995; Khan and Arunachalam 2003; Schroth et al. 2004). Home gardens as a form of agroforestry are considered to be the richest in plant species diversity per unit area (Galluzzi et al. 2010; Kumar and Nair 2004). A home garden is a parcel of land set aside around a homestead where individual households grow a variety of plants including woody species for family consumption. These gardens are a source of many products including food, fodder, fuelwood, medicines, herbs, flowers, construction materials and income (Kumar and Nair 2004; Gbedomon et al. 2017). Due to their abundant litter and good plant cover, multistorey home gardens help to reduce soil erosion especially on sloppy areas (Kumar and Nair 2004; Agbogidi and Adolor 2013).

Estimates by Kim et al. (2016) show that trees in an agroforestry system have potential to sequester annually about 27.2 tons CO_2 eq per ha during the first decade of establishment. Trees in agricultural landscapes help to control soil erosion, floods and pests (Jose 2012; Mbow et al. 2014). The application of leguminous trees in agroforestry can increase crop yields (Branca et al. 2013) but if tree density is too high, crop yields can be affected negatively as crops compete with trees for nutrients, light and water. Empirical evidence shows that isolated trees in agroforestry systems can increase soil organic carbon and nitrogen beneath canopies (Pardon et al. 2017; Bayala et al. 2018), largely attributed to the ability of trees to retrieve nutrients from deeper soil horizons and deposit them on soil surface layers via litterfall inputs (Mlambo et al. 2005; Acharya and Kafle 2009). Trees in agroforestry systems have a positive impact on soil microbes that are responsible for litter decomposition and mineralization of organic matter. Agroforestry practices can support beekeeping (Hill and Webster 1995) which in turn can enhance pollination service that is required to increase crop yield and profitability (Alam et al. 2014).

Farmscape plantings (e.g., hedgerows, windbreaks and woodlots) are common in many agricultural landscapes. Since these plantings are less disturbed than cropped areas, they provide favourable habitats for a variety of wildlife, including natural enemies of crop pests and pollinators (Hannon and Sisk 2009; Morandin and Kremen 2013; Sardiñas and Kremen 2015). Hedgerows planted along field edges protect insects that are sensitive to pesticide drift from neighbouring crop fields (Kjaer et al. 2014). Farmscape plantings create natural places for birds and beneficial insects, and can provide valuable corridors for a wide variety of wildlife. Some studies have reported that birds can be effective on farms in controlling pest insects and in eating significant amounts of weed seeds (Railsback and Johnson 2014). Farmscape plantings also control erosion and water run-off from agricultural fields, which help to reduce the amount of nutrients, pesticides and sediments washed from agricultural land to waterways (Philips et al. 2014). Some farmers plant beetle banks (perennial bunchgrass plantings within fields or on field edges) to provide shelter for crop damaging enemies (Macleod et al. 2004). Beetle banks are important refugia for predacious ground beetles that prey upon multiple crop pests including caterpillars, slugs and snails.

3.8 The Impact of Conservation Agriculture on Biodiversity and Ecosystem Services

Conservation agriculture aims to increase crop yield and profits while preserving the environment (Corsi et al. 2012). Conservation tillage, permanent covering of the soil surface by crop residues or cover crops and crop rotations are the key components of conservation agriculture (Pittelkow et al. 2015; Busari et al. 2015; Knapp and van der Heijden 2018). Cover crops are crops planted to enrich the soil and capture inorganic N. Conservation tillage leaves at least 30% of the soil surface covered with crop residues after planting to reduce soil erosion (Busari et al. 2015). The protection of soil from erosion is important because soil contains a wealth of biodiversity which can exceed 1000 species m^{-2} of soil surface. There are several conservation tillage practices which include no-tillage, reduced tillage and mulch tillage. No-tillage means cultivating land with little or no soil disturbance, the only disturbance being during planting. In reduced tillage, ploughing is done by primary implements to reduce the level of soil manipulation. Mulch tillage involves leaving plant residues permanently in the field during land preparation or planting to cover soil surface.

A number of studies indicate that conservation farming enhances soil quality, natural pest control, soil carbon sequestration and supports a range of soil biodiversity (Hobbs et al. 2008; Pittelkow et al. 2015; Briones and Schmidt 2017). Conservation tillage systems minimize crop residue disturbances and soil loss by wind and water. Crop residues remaining on the soil surface in zero or reduced tillage systems provide a protective environment and food to a wide range of wildlife (Morris et al.

2010). Reduced tillage provides better wildlife habitats because fewer nests are destroyed by farm machinery. Soil tillage can cause disturbances to organisms by injuring, killing, forcing them to migrate, or exposing them to predation (Roger-Estrade et al. 2010). Generally, earthworms are more abundant in reduced tillage systems than under deep inversion tillage (Boatman et al. 2007) because the later buries earthworm food sources and destroys burrows. Jiang et al. (2011) mentioned that deep inversion tillage generally reduces total soil microbial biomass including fungal biomass which is affected by the destruction of fungal hyphal networks. Least tillage enhances organic materials into the soil and the activity of soil organisms in the uppermost soil layer (Gattinger et al. 2012; Mäder and Berner 2012; Cooper et al. 2016; Puerta et al. 2018). This may increase water infiltration rates, aggregate soil stability and soil nutrient cycling (Bender and Van Der Heijden 2014). In a study conducted by Wang et al. (2016) in the dryland of northern China, conservation tillage increased the abundance and diversity of soil bacteria of the genus *Bacillus*. Natural pest control is reported to be higher in reduced tillage than in conventional tillage systems due to higher abundance of pest predators in the former than in the later (Tamburini et al. 2016).

A major drawback of zero or reduced tillage practices is a potential reduction in crop productivity. Cooper et al. (2016) showed decreased crop productivity by an average of 7.6% in reduced till organic systems compared to conventional tillage systems. Zikeli and Gruber (2017) reported 67% reduction of winter wheat yields in organic reduced tillage system compared to organic inversion tillage system due to reduced mineralization and increased weed pressure. Although reduced tillage farming is biodiversity-friendly, uptake by farmers has been low due to concerns about increased weed pressure and low crop productivity (Mäder and Berner 2012; Zikeli and Gruber 2017). While zero or reduced tillage can save farmers' tilling costs, it requires more herbicide use (Soane et al. 2012). Empirical evidence shows that weed pressure is greater in reduced than in deep inversion tillage system (Légère et al. 2011). Cover crop application in no-till or reduced tillage systems provides cover, nesting areas and food to wildlife. Cover crops can suppress weeds through competition for light and soil resources. In the American Upper Midwest, Silva (2014) reported a significant reduction in weed pressure following the use of rye cover crop in organic no-till crop production system. Although cover crops can reduce annual weed growth in the subsequent crop, the control of perennial weeds remains a challenge. Some studies have reported lack of positive effect as shown by reduction in weed infestation (Blanco-Canqui et al. 2015; Batary et al. 2015; Venter et al. 2016). Although cover crops can improve soil structure and reduce soil erosion, they have a disadvantage of competing with the primary crop for resources.

On organic farms, where use of synthetic herbicides is forbidden, diversified crop rotations may be used in combination with reduced tillage to suppress weeds (Cooper et al. 2016). Crop rotation is the sequential planting of individual crops through time, reducing the reliance on chemical fertilizers, pesticides and herbicides. Diversified crop rotations and appropriate timing of management interventions can disrupt life cycles of many weed species and pests (Bàrberi 2002; Smith et al. 2008; Tiemann et al. 2015). Growing different kinds of crops in recurrent succession on the same land can increase soil fertility particularly when nitrogen fixing plants are used in rotations (Smith et al. 2008; Tiemann et al. 2015). Diversified crop rotations produce a diversity of plant litter, which in turn can support a greater diversity of decomposers. In a meta-analysis of 122 studies conducted by McDaniel et al. (2014), crop rotations increased microbial biomass by an average of 21%, which may be attributed to increased plant diversity, ground cover and organic materials in soil (Zak et al. 2003; Venter et al. 2016). Microbial decomposers play an important role in the formation of soil organic matter, which is a source of nutrients for plants. Soil organic matter binds the soil into clumps, preventing soil erosion and improving soil structure and water storage capacity.

3.9 The Impact of Intercropping on Biodiversity and Ecosystem Services

Intercropping involves the growing of two or more crop species on the same piece of land, and the species have to coexist for some time. Although intercropping is often placed on the fringes of conventional agriculture, it has potential to increase yield stability, reduce pests and weed pressure as well as maintain or increase soil fertility (Letourneau et al. 2011; Beizhou et al. 2012; Boudreau 2013). Previous studies have reported that the frequency of disease occurrences in intercrop systems is reduced by 30–40% compared with crop monocultures (Finch and Collier 2012; Brooker et al. 2015; Bybee-Finley and Ryan 2018). A review of more than 200 studies by Boudreau (2013) found that disease incidences were reduced on average by 73% in intercropped systems compared to their respective monocultures. Another review study by Letourneau et al. (2011) revealed that intercrop systems can support greater abundance of natural enemies of pests compared to monocultures, which may help to reduce crop damage in the former. In a recent meta-analysis, weed biomass was found to be 56% higher in non-weeded monoculture fields than in non-weeded intercropped fields (Verret et al. 2017).

3.10 Ecological Intensification and Its Role towards Management of Terrestrial Landscape

Nearly one-third of terrestrial land is under cropland and planted pastures. Field size enlargement and the removal of natural or semi-natural habitats as well as the use of agrochemical inputs in farming landscapes caused significant reduction of bio-resources and ecological functions (Redlich et al. 2018). On farmlands, amount of non-cropped habitats (e.g., hedgerows, woodlands, and permanent grasslands and pasture) is often used as a surrogate measure of landscape heterogeneity or complexity (Tscharntke et al. 2005). The major role of ecological intensification in the terrestrial landscape is to maintain or restore natural or semi-natural habitats, enhance habitat quantity and quality, species diversity and soil health as well as protect water resources (Table 3.3). Empirical evidence shows that in

Table 3.3 Ecological roles of non-cropped habitats on farmlands

Ecological roles	References
Maintain terrestrial landscape diversity	Tscharntke et al. (2005)
Reduce pesticide use: Exploit pest predators and parasitoids	Holland et al. (2017)
Reduce soil erosion	Zuazo and Pleguezuelo (2008)
Buffer agrochemical drift	Kjaer et al. (2014); Earnshaw (2018)
Reduce fertilizer and other pollutant movement, especially in surface run-off	Philips et al. (2014); Garibaldi et al. (2019)
Act as a refuge or corridor for wildlife	Jose (2012)

Table 3.4 Examples of ecosystem services that can be enhanced through ecological intensification and associated benefits at landscape level (Adapted from Garbach et al. 2014)

Ecosystem service	Description	Benefits at the landscape level
Erosion protection	Reducing soil loss caused by wind or water	Reduction of sedimentation in downstream water bodies
Water flow regulation	Reducing water loss (e.g., increasing water infiltration into soils and aquifers, reducing run-off)	Mitigation of flooding to downstream areas and groundwater recharging
Water purification	Mechanical removal of physical impurities in water (e.g., filtration, sedimentation and precipitation)	Clean water available to downstream users
Pollination	Transferring pollen to the stigma of a flower to allow fertilization	Necessary for outcrossing in non-cultivated flowering plants
Pest control	Controlling pests through the use of natural enemies of pests	Reduced use of chemical pesticides minimizes damage to non-target species, contamination of water bodies and risk to human health.
Weed control	Suppressing weeds through intercropping and the use of cover crops	Reduced use of herbicides, enhance biodiversity
Carbon sequestration	Removal of CO_2 from the atmosphere (e.g., by green plants) and storing it as carbon in biomass and soils	Climate change mitigation

agro-ecosystems, landscape heterogeneity supports a greater number of species than crop diversity (Redlich et al. 2018). Wildlife species that rely on forests for nesting and foraging (e.g., birds and bees) can benefit from the presence of non-cropped habitats due to improved resources or habitat availability. Vegetation in non-cropped habitats intercepts nutrients leached out of the crop fields, thus reducing the negative impact of diffuse pollution on aquatic life and potentially improving the quality of drinking water (Table 3.3). For example, Borin et al. (2010) found that vegetation buffer strips can decrease phosphorus and nitrogen in polluted run-off by 60–98% and 70–95%, respectively. Table 3.4 shows that there are many benefits associated with transitioning to ecological intensification that can be realized at landscape level.

Ecological intensification approaches such as organic farming limit the use of chemically synthesized inputs which help reduce nutrient pollution, eutrophication and pesticide residue contamination in water bodies. Non-cropped habitats and farmscape plantings support air and water purification, carbon sequestration and storage (Khan et al. 2020a, 2020b).

3.11 Future Perspectives

Despite the potentiality of ecological intensification in boosting crop productivity while maintaining or enhancing biodiversity and associated ecosystem services, policies promote its adoption among farming community (Kleijn et al. 2018; Garibaldi et al. 2019). A widespread adoption would require a robust evidence base that demonstrates the net agronomic or economic benefits associated with a transition to ecological intensification of mainstream farming. This would require scientists to focus on the costs and benefits of ecological intensification. Such studies should include costs of establishing and maintaining farmscape plantings as well as loss of production on land used for landscape plantings. The benefits should extend beyond crop yields to incorporate the values of ecosystem services. Future studies should also distinguish between benefits delivered by ecological intensification to individual farmers and the public such as increased carbon sequestration, improved human health due to reduced pesticide use and wildlife conservation.

3.12 Conclusion

Conventional agriculture is one of the major causes of biodiversity loss due to unsustainable farming practices such as natural habitat fragmentation, field size enlargement and the use of agrochemicals. Such practices have multiple detrimental impacts on biodiversity, quality of the environment and threaten the sustainability of food production. The environmental damage associated with conventional agriculture implies the existence of huge external or social costs which can be internalized by governments through a variety of mechanisms such as compulsory practices that support biodiversity or introducing charges for the production and use of agrochemicals. Policies that promote nature-based farm management and make external inputs more expensive have potential to make ecological intensification more attractive to farmers, financially. A growing body of literature shows that practices that constitute ecological intensification have potential to reduce or replace the use of external inputs without compromising on crop yield and profitability if properly planned.

References

Acharya AK, Kafle N (2009) Land degradation issues in Nepal and its management through agroforestry. J Agric Environ 10:133–143. https://doi.org/10.3126/aej.v10i0.2138

Agbogidi OM, Adolor EB (2013) Home gardens in the maintenance of biological diversity. Appl Sci Reports 1:19–25

Alam M, Olivier A, Paquette A, Dupras J, Revéret J, Messier C (2014) A general framework for the quantification and valuation of ecosystem services of tree-based intercropping systems. Agrofor Syst 88:679–691

Andersson GKS, Birkhofer K, Rundlöf M, Smith HG (2013) Landscape heterogeneity and farming practice alter the species composition and taxonomic breadth of pollinator communities. Basic Appl Ecol 14:540–546

Aude E, Tybirk K, Pedersen MB (2003) Vegetation diversity of conventional and organic hedgerows in Denmark. Agric Ecosyst Environ 99:135–147

Avery A (2006) The truth about organic foods. Henderson Communications LLC, Chesterfield, MO

Banerjee A, Jhariya MK, Yadav DK, Raj A (2020) Environmental and sustainable development through forestry and other resources. CRC press, Boca Raton, FL, p 400. https://doi.org/10.1201/9780429276026

Bàrberi P (2002) Weed management in organic agriculture: are we addressing the right issues? Weed Res 42:176–193

Barrios E, Valencia V, Jonsson M, Brauman A, Hairiah K, Mortimer PE, Okubo S (2018) Contribution of trees to the conservation of biodiversity and ecosystem services in agricultural landscapes. Int J Biodiv Sci Ecosys Services Mgt 14:1–16. https://doi.org/10.1080/21513732.2017.1399167

Batary P, Dicks LV, Kleijn D, Sutherland WJ (2015) The role of Agri-environment schemes in conservation and environmental management. Conserv Biol 29:1006–1016

Bayala J, Kalinganire A, Sileshi GW, Tondoh JE (2018) Soil organic carbon and nitrogen in agroforestry systems in sub-Saharan Africa: a review. In: Bationo A, Ngaradoum D, Youl S, Lompo F, Fening J (eds) Improving the profitability, sustainability and efficiency of nutrients through site specific fertilizer recommendations in West Africa agro-ecosystems. Springer, Cham. https://doi.org/10.1007/978-3-319-58789-9_4

Beizhou S, Jie Z, Wiggins NL, Yuncong Y, Guangbo T, Xusheng S (2012) Intercropping with aromatic plants decreases herbivore abundance, species richness, and shifts arthropod community trophic structure. Environ Entomol 41:872–879

Beketov MA, Kefford BJ, Schäfer RB, Liess M (2013) Pesticides reduce regional biodiversity of stream invertebrates. Proc Natl Acad Sci 110:11039–11043. https://doi.org/10.1073/pnas.1305618110

Bender SF, Van Der Heijden MGA (2014) Soil biota enhance agricultural sustainability by improving crop yield, nutrient uptake and reducing nitrogen leaching losses. J Appl Ecol 52:228–239

Bengtsson J (2015) Biological control as an ecosystem service: partitioning contributions of nature and human inputs to yield. Ecol Entomol 40:45–55

Bengtsson J, Ahnstrom J, Weibull AC (2005) The effects of organic agriculture on biodiversity and abundance: a meta-analysis. J Appl Ecol 42:261–269

Benton TG, Vickery JA, Wilson JD (2003) Farmland biodiversity: is habitat heterogeneity the key? Trends Ecol Evol 18:182–187

Birkhofer K, Smith HG, Weisser WW, Wolters V, Gossner MM (2015) Land-use effects on the functional distinctness of arthropod communities. Ecography 38:889–900. https://doi.org/10.1111/ecog.01141

Blanco-Canqui H, Shaver TM, Lindquist JL, Shapiro CA, Elmore RW, Francis CA, Hergert GW (2015) Cover crops and ecosystem services: insights from studies in temperate soils. Agron J 107:2449–2474. https://doi.org/10.2134/agronj15.0086

BLI (2008) State of the World's birds: indicators for our changing world. Bird Life International, Cambridge, UK

Boatman ND, Parry HR, Bishop JD, Cuthbertson AGS (2007) Impacts of agricultural change on farmland biodiversity in the UK. Issues in environmental science and technology, No. 25: biodiversity under threat. The Royal Society of Chemistry

Boffa JM (1999) Agroforestry parklands in sub-Saharan Africa. FAO conservation guide 34. Food and agriculture Organization of the United Nations, Rome, Italy

Bommarco R, Kleijn D, Potts SG (2013) Ecological intensification: harnessing ecosystem services for food security. Trends Ecol Evol 28:230–238

Borin M, Passoni M, Thiene M, Tempesta T (2010) Multiple functions of buffer strips in farming areas. Eur J Agron 32:103–111. https://doi.org/10.1016/j.eja.2009.05.003

Boudreau MA (2013) Diseases in intercropping systems. Annu Rev Phytopathol 5:499–519

Branca G, Lipper L, McCarthy N, Jolejole MC (2013) Food security, climate change, and sustainable land management. A review. Agron Sustain Dev 33:635–650

Briones MJI, Schmidt O (2017) Conventional tillage decreases the abundance and biomass of earthworms and alters their community structure in a global meta-analysis. Glob Change Biol 23:4396–4419

Brooker RW, Bennett AE, Cong WF, Daniell TJ, George TS, Hallett PD, Hawes C, Iannetta PPM, Jones HG, Karley AJ, Li L, McKenzie BM, Pakeman RJ, Paterson E, Schob C, Shen J, Squire G, Watson CA, Zhang C, Zhang F, Zhang J, White PJ (2015) Hallett improving intercropping: a synthesis of research in agronomy, plant physiology and ecology. New Phytol 206:107–117. https://doi.org/10.1111/nph.13132

Busari MA, Kukal SS, Kaur A, Bhatt R, Dulazi AA (2015) Conservation tillage impacts on soil, crop and the environment. Int Soil Water Conserv Res 3:119–129. https://doi.org/10.1016/j.iswcr.2015.05.002

Bybee-Finley KA, Ryan MR (2018) Advancing intercropping research and practices in industrialized agricultural landscapes. Agriculture 8:80

Cassman KG (1999) Ecological intensification of cereal production systems: yield potential, soil quality, and precision agriculture. Proc Natl Acad Sci U S A 96:5952–5959

Catarino R, Ceddia G, Areal FJ, Park J (2015) The impact of secondary pests on *Bacillus thuringiensis* (Bt) crops. Plant Biotec J 13:601–612. https://doi.org/10.1111/pbi.12363

Chagnon M, Kreutzweiser D, Mitchell EAD, Morrissey CA, Noome DA, Van der Sluijs JP (2015) Risks of large-scale use of systemic insecticides to ecosystem functioning and services. Environ Sci Pollut Res 22:119. https://doi.org/10.1007/s11356-014-3277-x

Cooper J, Baranski M, Stewart G, Nobel-de Lange M, Bàrberi P, Andreas Fließbach A, Peigne J, Berner A, Brock C, Casagrande M, Crowley O, David C, De Vliegher A, Doring TF, Dupont A, Entz M, Grosse M, Haase T, Halde C, Hammerl V, Huiting H, Leithold G, Messmer M, Schloter M, Sukkel W, van der Heijden MGA, Willekens K, Wittwer R, Mader P (2016) Shallow non-inversion tillage in organic farming maintains crop yields and increases soil C stocks: a meta-analysis. Agron Sustain Dev 36:22. https://doi.org/10.1007/s13593-016-0354-1

Copping LG, Menn JJ (2000) Biopesticides: a review of their action, applications and efficacy. Pest Mgt Sci 56:651–676

Corsi S, Friedrich T, Kassam A, Pisante M, de Moraes Sà JC (2012) Soil organic carbon accumulation and greenhouse gas emission reductions from conservation agriculture: a literature review, vol 16. AGP/FAO, Rome

Costanza R, de Groot R, Sutton P, van der Ploeg S, Anderson SJ, Kubiszewski I, Farber S, Turner RK (2014) Changes in the global value of ecosystem services. Glob Environ Chang 26:152–158. https://doi.org/10.1016/j.gloenvcha.2014.04.002

Cui Z, Zhang H, Chen X, Zhang C, Wenqi M, Huang C, Ma W, Huang C, Zhang W, Mi G, Miao Y, Li X, Gao Q, Yang J, Wang Z, Ye Y, Guo S, Lu J, Huang J, Lv S, Sun Y, Liu Y, Peng X, Ren J, Li S, Deng X, Shi X, Zhang Q, Yang Z, Tang L, Wei C, Jia L, Zhang J, He M, Tong Y, Tang Q, Zhong X, Liu Z, Cao N, Kou C, Ying H, Yin Y, Jiao X, Zhang Q, Fan M, Jiang R, Zhang F, Dou

Z (2018) Pursuing sustainable productivity with millions of smallholder farmers. Nature 555:363–366. https://doi.org/10.1038/nature25785

de Molina MG, Casado GIG (2017) Agroecology and ecological intensification. A discussion from a metabolic point of view. Sustainability 9:86. https://doi.org/10.3390/su9010086

Deguines N, Jono C, Baude M, Henry M, Julliard R, Fontaine C (2014) Large-scale trade-off between agricultural intensification and crop pollination services. Front Ecol Environ 12:212–217. https://doi.org/10.1890/130054

Donald PF, Sanderson FJ, Burfield IJ, van Bommel FPJ (2006) Further evidence of continent-wide impacts of agricultural intensification on European farmland birds, 1990–2000. Agric Ecosyst Environ 116:189–196

Earnshaw S (2018) Hedgerows and farmscaping for California agriculture, a resource guide for farmers. Community alliance with family farmers. www.caff.org. Accessed 10 Mar 2020

Egan JF, Bohnenblust E, Goslee S, Mortensen D, Tooker J (2014) Herbicide drift can affect plant and arthropod communities. Agric Ecosyst Environ 185:77–87

Emmerson M, Morales MB, Onate JJ, Batáry P, Berendse F, Liira J, Aavik T, Guerrero I, Bommarco R, Eggers S, Part T, Tscharntke T, Weisser W, Clement L, Bengtsson J (2016) How agricultural intensification affects biodiversity and ecosystem services. Adv Ecol Res 55:43–97. https://doi.org/10.1016/bs.aecr.2016.08.005

Fahrig L, Baudry J, Brotons L, Burel FG, Crist TO, Fuller RJ, Sirami C, Siriwardena GM, Martin JL (2011) Functional landscape heterogeneity and animal biodiversity in agricultural landscapes. Ecol Lett 14(2):101–112. https://doi.org/10.1111/j.1461-0248.2010.01559.x

FAO (2015) World fertilizer trends and outlooks to 2018. Food and Agriculture Organization of the United Nations, Rome

Finch S, Collier RH (2012) The influence of host and non-host companion plants on the behaviour of pest insects in field crops. Entomol Exp Appl 142:87–96

Foley JA, Ramankutty N, Brauman KA, Cassidy ES, Gerber JS, Johnston M, Mueller ND, O'Connell C, Ray DK, West PC, Balzer C, Bennett EM, Carpenter SR, Hill J, Monfreda C, Polasky S, Rockstrom J, Sheehan J, Siebert S, Tilman D, Zaks DPM (2011) Solutions for a cultivated planet. Nature 478:337–342. https://doi.org/10.1038/nature10452

Fuller RJ, Norton LR, Feber RE, Johnson PJ, Chamberlain DE, Joys AC, Mathews F, Stuart RC, Townsend MC, Manley WJ, Wolfe MS, Macdonald DW, Firbank LG (2005) Benefits of organic farming to biodiversity vary among taxa. Biol Lett 1:431–434. https://doi.org/10.1098/rsbl.2005.0357

Gabriel D, Roschewitz I, Tscharntke T, Thies C (2006) Beta diversity at different spatial scales: plant communities in organic and conventional agriculture. Ecol Appl 16:2011–2021

Galluzzi G, Eyzaguirre P, Negri V (2010) Home gardens: neglected hotspots of agro-biodiversity and cultural diversity. Biodivers Conserv 19:3635–3654. https://doi.org/10.1007/s10531-010-9919-5

Garbach K, Milder JC, Montenegro M, Karp DS, DeClerck FAJ (2014) Biodiversity and ecosystem services in agroecosystems. Reference module in food science. Encycl Agric Food Syst 2:21–40. https://doi.org/10.1016/B978-0-444-52512-3.00013-9

Garibaldi LA, Steffan-Dewenter I, Kremen C, Morales JM, Bommarco R, Cunningham SA, Carvalheiro LG, Chacoff NP, Dudenhöffer JH, Greenleaf SS, Holzschuh A, Isaacs R, Krewenka K, Mandelik Y, Mayfield MM, Morandin LA, Potts SG, Ricketts TH, Szentgyörgyi H, Viana BF, Westphal C, Winfree R, Klein AM (2011) Stability of pollination services decreases with isolation from natural areas despite honey bee visits. Ecol Lett 14:1062–1072. https://doi.org/10.1111/j.1461-0248.2011.01669.x

Garibaldi LA, Carvalheiro LG, Vaissière B, Gemmill-Herren B, Hipolito J, Freitas BM, Ngo HT, Azzu N, Saez A, Astrom J, An J, Blochtein B, Buchori D, Garcia FJC, da Silva FO, Devkota K, de Fatima RM, Freita L, Gaglianone MC, Goss M, Irshad M, Kasina M, Filho AJSP, Kiill LHP, Kwapong P, Parra GN, Pires C, Pires V, Rawal RS, Rizali A, Saraiva AM, Veldtman R, Viana BF, Witter S, Zhang H (2016) Mutually beneficial pollinator diversity and crop yield outcomes in small and large farms. Science 351:388–391. https://doi.org/10.1126/science.aac7287

Garibaldi LA, Mendez NP, Garratt MPD, Gemmill-Heren B, Miguez FE, Dicks LV (2019) Policies for ecological intensification of crop production. Tends Ecol Evol 34(4):282–286

Gattinger A, Muller A, Haeni M, Skinner C, Fliessbach A, Buchmann N, Mader P, Stolze M, Smith P, El-Hage Scialabba N, Niggli U (2012) Enhanced top soil carbon stocks under organic farming. Proc Natl Acad Sci U S A 109(44):18226–18231. https://doi.org/10.1073/pnas.1209429109

Gbedomon RC, Salako VK, Fandohan AB, Idohou AFR, Kakaï RG, Assogbadjo AE (2017) Functional diversity of home gardens and their agrobiodiversity conservation benefits in Benin, West Africa. J Ethnobiol Ethnomed 13:66. https://doi.org/10.1186/s13002-017-0192-5

Geiger F, Geiger F, Bengtsson J, Berendse F, Weisser WW, Emmerson M (2010) Persistent negative effects of pesticides on biodiversity and biological control potential on European farmland. Basic Appl Ecol 11:97–105. https://doi.org/10.1016/j.baae.2009.12.001

Gomiero T, Pimentel D, Paoletti MG (2011) Environmental impact of different agricultural management practices: conventional vs. organic agriculture. Critical Rev Plant Sci 30:95–124. https://doi.org/10.1080/07352689.2011.554355

Gonthier DJ, Ennis KK, Farinas S, Hsieh HY, Iverson AL, Batáry P, Rudolphi J, Tscharntke T, Cardinale BJ, Perfecto I (2014) Biodiversity conservation in agriculture requires a multi-scale approach. Proc R Soc B 281(1791):1358. https://doi.org/10.1098/rspb.2014.1358

Guenat S, Kaartinen R, Jonsson M (2019) Shade trees decrease pest abundances on brassica crops in Kenya. Agrofor Syst 93:641–652. https://doi.org/10.1007/s10457-017-0159-5

Hannon LE, Sisk TD (2009) Hedgerows in an Agri-natural landscape: potential habitat value for native bees. Biol Conserv 142:2140–2154. https://doi.org/10.1016/j.biocon.2009.04.014

Hassanali A, Herren H, Khan ZR, Pickett JA, Woodcock CM (2008) Integrated pest management: the push-pull approach for controlling insect pests and weeds of cereals, and its potential for other agricultural systems including animal husbandry. Philos Trans Royal Soc London 363:611–621

Hill D, Webster T (1995) Apiculture and forestry (bees and trees). Agrofor Syst 29:313–320

HLPE (High Level Panel of Experts on Food Security and Nutrition) (2017) Sustainable forestry for food security and nutrition. A report by the High Level Panel of Experts on Food Security and Nutrition of the Committee on World Food Security. Food and Agriculture Organization on the United Nations (FAO), Rome

Hobbs PR, Sayre K, Gupta R (2008) The role of conservation agriculture in sustainable agriculture. Philos Trans R Soc Lond Ser B Biol Sci 363:543–555

Hodgson JA, Kunin WE, Thomas CD, Benton TG, Gabriel D (2010) Comparing organic farming and land sparing: optimizing yield and butterfly populations at a landscape scale. Ecol Lett 13:1358–1367. https://doi.org/10.1111/j.1461-0248.2010.01528.x

Hokkanen H (2015) Integrated pest management at the crossroads: science, politics, or business (as usual)? Arthropod Plant Interact 9:543–545

Hole DG, Perkins AJ, Wilson JD, Alexander IH, Grice PV, Evans AD (2005) Does organic farming benefit biodiversity? Biol Conserv 122:113–120

Holland JM, Douma JC, Crowley L, James L, Kor L, Stevenson DRW, Smith BM (2017) Semi-natural habitats support biological control, pollination and soil conservation in Europe. A review. Agron Sustain Dev 37:31. https://doi.org/10.1007/s13593-017-0434-x

Holzschuh A, Steffan-Dewenter I, Tscharntke T (2008) Agricultural landscapes with organic crops support higher pollinator diversity. Oikos 117:354–361

IFOAM (International Movement of Organic Agriculture Movements) (2008) The world of organic agriculture - statistics and emerging trends 2008. International Federation of Organic Agriculture Movements-IFOAM, Bonn, Germany

Isbell F, Adler PR, Eisenhauer N, Fornara D, Kimmel K, Kremen C (2017) Benefits of increasing plant diversity in sustainable agroecosystems. J Ecol 105:871–879. https://doi.org/10.1111/1365-2745.12789

Jhariya MK, Bargali SS, Raj A (2015) Possibilities and perspectives of agroforestry in Chhattisgarh. In: Zlatic M (ed) Precious forests-precious earth. InTech, Croatia, UK, pp 237–257. https://doi.org/10.5772/60841

Jhariya MK, Banerjee A, Yadav DK, Raj A (2018) Leguminous trees an innovative tool for soil sustainability. In: Meena RS, Das A, Yadav GS, Lal R (eds) Legumes for soil health and sustainable management. Springer, Cham, pp 315–345. https://doi.org/10.1007/978-981-13-0253-4_10

Jhariya MK, Banerjee A, Meena RS, Yadav DK (2019a) Sustainable agriculture, forest and environmental management. Springer, Singapore, p 606. https://doi.org/10.1007/978-981-13-6830-1

Jhariya MK, Yadav DK, Banerjee A (2019b) Agroforestry and climate change: issues and challenges. CRC press, Boca Raton, FL, p 335. https://doi.org/10.1201/9780429057274

Jiang X, Wright AL, Wang X, Liang F (2011) Tillage-induced changes in fungal and bacterial biomass associated with soil aggregates: a long-term field study in a subtropical rice soil in China. Appl Soil Ecol 48:168–173. https://doi.org/10.1016/j.apsoil.2011.03.009

Jose S (2012) Agroforestry for conserving and enhancing biodiversity. Agrofor Syst 85:1–8

Kamau S, Barrios E, Karanja N, Ayuke F, Lehmann J (2017) Soil macrofauna abundance under dominant tree species increases along a soil degradation gradient. Soil Biol Biochem 112:35–46

Kennedy CM, Lonsdorf E, Neel MC, Williams NM, Ricketts TH, Winfree R, Bommarco R, Brittain C, Burley AL, Cariveau D, Carvalheiro LG, Chacoff NP, Cunningham SA, Danforth BN, Dudenhoffer JH, Elle E, Gaines HR, Garibaldi LA, Gratton C, Holzschuh A, Isaacs R, Javorek SK, Jha S, Klein AM, Krewenka K, Mandelik Y, Mayfield MM, Morandin L, Neame LA, Otieno M, Park M, Potts SG, Rundlof M, Saez A, Steffan-Dewenter I, Taki H, Viana BF, Westphal C, Wilson JK, Greenleaf SS, Kremen C (2013) A global quantitative synthesis of local and landscape effects on wild bee pollinators in agroecosystems. Ecol Lett 16:584–599. https://doi.org/10.1111/ele.12082

Khan ML, Arunachalam A (2003) Traditional agroforestry as a viable choice to conserve agro-biodiversity in the Northeast India. In: Pathak PS, Newaj R (eds) Agroforestry: potentials and opportunities. Agrobios (India) and Indian Society of Agroforestry, Jodhpur, India, pp 95–105

Khan Z, Midega C, Pittchar J, Pickett J, Bruce T (2011) Push-pull technology: a conservation agriculture approach for integrated management of insect pests, weeds and soil health in Africa. Int J Agric Sustain 9:162–170

Khan N, Jhariya MK, Yadav DK, Banerjee A (2020a) Herbaceous dynamics and CO_2 mitigation in an urban setup- a case study from Chhattisgarh, India. Environ Sci Pollut Res 27(3):2881–2897. https://doi.org/10.1007/s11356-019-07182-8

Khan N, Jhariya MK, Yadav DK, Banerjee A (2020b) Structure, diversity and ecological function of shrub species in an urban setup of Sarguja, Chhattisgarh, India. Environ Sci Pollut Res 27 (5):5418–5432. https://doi.org/10.1007/s11356-019-07172-w

Kibria G, Haroon AKY, Nugegoda D (2013) Climate change and agricultural food production: impacts, vulnerabilities and remedies. New India Publishing Agency, New Delhi. https://doi.org/10.13140/2.1.3245.4081

Kim DG, Kirschbaum MUF, Beedy TL (2016) Carbon sequestration and net emissions of CH_4 and N_2O under agroforestry: synthesizing available data and suggestions for future studies. Agric Ecosyst Environ 226:65–78. https://doi.org/10.1016/j.agee.2016.04.011

Kjaer C, Bruus M, Bossi R, Løfstrøm P, Andersen HV, Nuyttens D, Larsen SE (2014) Pesticide drift deposition in hedgerows from multiple spray swaths. J Pesticide Sci 39:14–21. https://doi.org/10.1584/jpestics.D12-045

Kleijn D, Bommarco R, Fijen TPM, Garibaldi LA, Potts SG, van der Putten WH (2018) Ecological intensification: bridging the gap between science and practice. Trends Ecol Evol 34:154–166. https://doi.org/10.1016/j.tree.2018.11.002

Klein AM, Vaissière BE, Cane JH, Steffan-Dewenter I, Cunningham SA, Kremen C, Tscharntke T (2007) Importance of pollinators in changing landscapes for world crops. Proc Royal Soc B: Biol Sci 274:303–313

Knapp S, van der Heijden MGA (2018) A global meta-analysis of yield stability in organic and conservation agriculture. Nat Commun 9:3632. https://doi.org/10.1038/s41467-018-05956-1

Kovács-Hostyánszki A, Espíndola A, Vanbergen AJ, Settele J, Kremen C, Dicks LV (2017) Ecological intensification to mitigate impacts of conventional intensive land use on pollinators and pollination. Ecol Lett 20:673–689. https://doi.org/10.1111/ele.12762

Kumar BM, Nair PKR (2004) The enigma of tropical homegardens. Agrofor Syst 61:135–152

Kumar S, Meena RS, Jhariya MK (2020) Resources use efficiency in agriculture. Springer, Singapore, p 760. https://doi.org/10.1007/978-981-15-6953-1

Kuyah S, Oborn I, Jonsson M, Dahlin AS, Barrios E, Muthuri C, Malmer A, Nyaga J, Magaju C, Namirembe S, Nyberg Y, Sinclair FL (2016) Trees in agricultural landscapes enhance provision of ecosystem services in sub-Saharan Africa. Int J Biodivers Sci Ecosys Services Mgt 12:255–273. https://doi.org/10.1080/21513732.2016.1214178

Landis DA (2017) Designing agricultural landscapes for biodiversity-based ecosystem services. Basic and Appl Ecol 18:1–12

Légère A, Stevenson F, Benoit D (2011) The selective memory of weed seedbanks after 18 years of conservation tillage. Weed Sci 59:98–106

Letourneau DK, Armbrecht I, Rivera BS, Lerma MJ, Carmona EJ, Daza MC, Escobar S, Galindo V, Gutierrez C, Lopez SD, Mejia JL, Rangel AMA, Rangel JH, Rivera L, Saavedra CA, Torres AM, Trujillo AR (2011) Does plant diversity benefit agroecosystems? A synthetic review. Ecol Appl 21(1):9–21. https://doi.org/10.1890/09-2026.1

Macleod A, Wratten SD, Sotherton N, Thomas MB (2004) Beetle banks' as refuges for beneficial arthropods in farmland: long-term changes in predator communities and habitat. Agric Entomol 6:147–154. https://doi.org/10.1111/j.1461-9563.2004.00215.x

Mäder P, Berner A (2012) Development of reduced tillage systems in organic farming in Europe. Renewable Agric Food Syst 27:7–11. https://doi.org/10.1017/s1742170511000470

Martín-López B, Llorente MG, Palomo I, Montes C (2011) The conservation against development paradigm in protected areas: valuation of ecosystem services in the Doñana social-ecological system (southwestern Spain). Ecol Econ 70:1481–1491. https://doi.org/10.1016/j.ecolecon.2011.03.009

Matson PA, Parton WIJ, Power AG, Swift MJ (1997) Agricultural intensification and ecosystem properties. Science 277:504–509. https://doi.org/10.1126/science.277.5325.504

Mbow C, Smith P, Skole D, Duguma L, Bustamante M (2014) Achieving mitigation and adaptation to climate change through sustainable agroforestry practices in Africa. Curr Opinion Environ Sustain 6:8–14

McDaniel MD, Tiemann LK, Grandy AS (2014) Does agricultural crop diversity enhance soil microbial biomass and organic matter dynamics? A meta-analysis. Ecol Appl 24:560–570

McNeely J, Schroth G (2006) Agroforestry and biodiversity conservation –traditional practices, present dynamics, and lessons for the future. Biodivers Conserv 15:549–554. https://doi.org/10.1007/s10531-005-2087-3

Medan D, Torretta JP, Hodara K, de la Fuente EB, Montaldo NH (2011) Effects of agriculture expansion and intensification on the vertebrate and invertebrate diversity in the pampas of Argentina. Biodivers Conserv 20:3077–3100. https://doi.org/10.1007/s10531-011-0118-9

Meena RS, Lal R (2018) Legumes for soil health and sustainable management. Springer, Singapore, p 541. https://doi.org/10.1007/978-981-13-0253-4_10

Meena RS, Kumar V, Yadav GS, Mitran T (2018) Response and interaction of *Bradyrhizobium japonicum* and Arbuscular mycorrhizal fungi in the soybean rhizosphere: a review. Plant Growth Regul 84:207–223

Meena RS, Kumar S, Datta R, Lal R, Vijaykumar V, Brtnicky M, Sharma MP, Yadav GS, Jhariya MK, Jangir CK, Pathan SI, Dokulilova T, Pecina V, Marfo TD (2020) Impact of agrochemicals on soil microbiota and management: a review. Land (MDPI) 9(2):34. https://doi.org/10.3390/land9020034

Meena RS, Lal R, Yadav GS (2020a) Long term impacts of topsoil depth and amendments on soil physical and hydrological properties of an Alfisol in Central Ohio, USA. Geoderma 363:1141164

Meena RS, Lal R, Yadav GS (2020b) Long-term impact of topsoil depth and amendments on carbon and nitrogen budgets in the surface layer of an Alfisol in Central Ohio. Catena 194:104752

Michon G, de Foresta H (1995) The Indonesian agroforest model: Forest resource management and biodiversity conservation. In: Halliday P, Gilmour DA (eds) Conserving biodiversity outside protected areas: the role of traditional agro-ecosystems. IUCN, Gland, Switzerland

Mineau P, Whiteside M (2013) Pesticide acute toxicity is a better correlate of U.S. grassland bird declines than agricultural intensification. PLoS one 8:e57457. https://doi.org/10.1371/journal.pone.0057457

Mitra A, Chatterjee C, Mandal FB (2011) Synthetic chemical pesticides and their effects on birds. Res J Environ Toxicol 5:81–96

Mlambo D, Nyathi P, Mapaure I (2005) Influence of *Colophospermum mopane* on surface soil properties and understorey vegetation in southern African savanna. For Ecol Mgt 212:394–404

Mora C, Tittensor DP, Adl S, Simpson AGB, Worm B (2011) How many species are there on earth and in the ocean? PLoS Biol 9:e1001127. https://doi.org/10.1371/journal.pbio.1001127

Morandin LA, Kremen C (2013) Hedgerow restoration promotes pollinator populations and exports native bees to adjacent fields. Ecol Appl 23:829–839

Morris NL, Miller PCH, Orson JH, Froud-Williams RJ (2010) The adoption of non-inversion tillage systems in the United Kingdom and the agronomic impact on soil, crops and the environment – a review. Soil Tillage Res 108:1–15

Norton L, Johnson P, Joys A, Stuart R, Chamberlain D, Feber R (2009) Consequences of organic and non-organic farming practices for field: farm and landscape complexity. Agric Ecosyst Environ 129:221–227

Oehl F, Sieverding E, Mader P, Dubois D, Ineichen K, Boller T, Wiemken A (2004) Impact of long-term conventional and organic farming on the diversity of arbuscular mycorrhizal fungi. Oecologia 138:574–583. https://doi.org/10.1007/s00442-003-1458-2

Oerke EC (2006) Crop losses to pests. J Agric Sci 144:31–43. https://doi.org/10.1017/S0021859605005708

Pannell DJ (2003) Uncertainty and adoption of sustainable farming systems. In: Babcock BA, Fraser RW, Lekakis JN (eds) Risk management and the environment: agriculture in perspective. pp 67–81. Springer, Dordrecht. https://doi.org/10.1007/978-94-017-2915-4_5

Pardon P, Reubens B, Reheul D, Mertens J, DeFrenne P, Coussement T, Janssens P, Verheyen K (2017) Trees increase soil organic carbon and nutrient availability in temperate agroforestry systems. Agric Ecosyst Environ 247:98–111. https://doi.org/10.1016/j.agee.2017.06.018

Parsa S, Morse S, Bonifacio A, Chancellor TCB, Condori B, Crespo-Perez V, Hobbs SLA, Kroschel J, Ba MN, Rebaudo F, Sherwood SG, Vanek SJ, Faye E, Herrera MA, Dangles O (2014) Obstacles to integrated pest management adoption in developing countries. Proc Natl Acad Sci U S A 111(10):3889–3894. https://doi.org/10.1073/pnas.1312693111

Philips CR, Rogers MA, Kuhar TP (2014) Understanding farmscapes and their potential for improving IPM programs. J Integ Pest Mgt 5(3):C1–C9. https://doi.org/10.1603/IPM13018

Pickett JA (2013) Food security: intensification of agriculture is essential, for which current tools must be defended and new sustainable technologies invented. Food Energy Secur 2:167–173. https://doi.org/10.1002/fes3.32

Pickett JA, Woodcock CM, Midega CAO, Khan ZR (2014) Push-pull farming system. Curr Opion Biotech 26:125–132. https://doi.org/10.1016/j.copbio.2013.12.006

Pittelkow CM, Liang X, Linquist BA, Groenigen KJV, Lee J, Lundy ME, Gestel NV, Six J, Venterea RT, Kessel CV (2015) Productivity limits and potentials of the principles of conservation agriculture. Nature 517:365–368. https://doi.org/10.1038/nature13809

Ponisio L, M'Gonigle L, Mace K, Palomino J, de Valpine P, Kremen C (2015) Diversification practices reduce organic to conventional yield gap. Proc Roy Soc B Biol Sci 282:20141396. https://doi.org/10.1098/rspb.2014.1396

Potts SG, Roberts SPM, Dean R, Marris G, Brown MA, Jones R, Neumann P, Settele J (2010) Declines of managed honey bees and beekeepers in Europe. J Apicultural Res 49:15–22. https://doi.org/10.3896/IBRA.1.49.1.02

Power AG (2010) Ecosystem services and agriculture: tradeoffs and synergies. Philos Trans R Soc Lond Ser B Biol Sci 365:2959–2971

Pretty J (2018) Intensification for redesigned and sustainable agricultural systems. Science 362:6417. https://doi.org/10.1126/science.aav0294

Pretty J, Bharucha ZP (2015) Integrated pest management for sustainable intensification of agriculture in Asia and Africa. Insects 6:152–182. https://doi.org/10.3390/insects6010152

Puerta VL, Pereira EIP, Wittwer R, van der Heijden M, Six J (2018) Improvement of soil structure through organic crop management, conservation tillage and grass-clover ley. Soil Tillage Res 180:1–9. https://doi.org/10.1016/j.still.2018.02.007

Pumariño L, Sileshi GW, Gripenberg S, Kaartinen R, Barrios E, Muchane MN, Midega C, Jonsson M (2015) Effects of agroforestry on pest, disease and weed control: a meta-analysis. Basic Appl Ecol 16(7):573–582. https://doi.org/10.1016/j.baae.2015.08.006

Railsback SF, Johnson MD (2014) Effects of land use on bird populations and pest control services on coffee farms. PNAS 111:6109–6114. https://doi.org/10.1073/pnas.1320957111

Raj A, Jhariya MK, Yadav DK, Banerjee A, Meena RS (2019a) Agroforestry: a holistic approach for agricultural sustainability. In: Jhariya MK, Banerjee A, Meena RS, Yadav DK (eds) Sustainable agriculture, forest and environmental management. Springer, Singapore, pp 101–131. https://doi.org/10.1007/978-981-13-6830-1

Raj A, Jhariya MK, Banerjee A, Yadav DK, Meena RS (2019b) Soil for sustainable environment and ecosystems management. In: Jhariya MK, Banerjee A, Meena RS, Yadav DK (eds) Sustainable agriculture, forest and environmental management. Springer, Singapore, pp 189–221. https://doi.org/10.1007/978-981-13-6830-1

Raj A, Jhariya MK, Yadav DK, Banerjee A (2020) Climate change and agroforestry systems: adaptation and mitigation strategies. CRC Press, Boca Raton, FL, p 383. https://doi.org/10.1201/9780429286759

Redlich S, Martin EA, Wende B, Steffan-Dewenter I (2018) Landscape heterogeneity rather than crop diversity mediates bird diversity in agricultural landscapes. PLoS One 13(8):e0200438. https://doi.org/10.1371/journal.pone.0200438

Ricketts TH, Regetz J, Steffan-Dewenter I, Cunningham SA, Kremen C, Bogdanski A, Gemmill-Herren B, Greenleaf SS, Klein AM, Mayfield MM, Morandin LA, Ochieng A, Potts SG, Viana BF (2008) Landscape effects on crop pollination services: are there general patterns? Ecol Lett 11:499–515. https://doi.org/10.1111/j.1461-0248.2008.01157.x

Roger-Estrade J, Anger C, Bertrand M, Richard G (2010) Tillage and soil ecology: partners for sustainable agriculture. Soil Tillage Res 111:33–40. https://doi.org/10.1016/j.still.2010.08.010

Röös E, Mie A, Wivstad M, Salomon E, Johansson B, Gunnarsson S, Wallenbeck A, Hoffmann R, Nilsson U, Sundberg C, Watson CA (2018) Opportunities of increasing yields in organic farming. A review. Agron Sustain Dev 38:14. https://doi.org/10.1007/s13593-018-0489-3

Rundlöf M, Edlund M, Smith HG (2010) Organic farming at local and landscape scales benefits plant diversity. Ecography 33:514–522

Rundlöf M, Smith HG, Birkhofer K (2016) Effects of organic farming on biodiversity. In: eLS. Wiley, Chichester. https://doi.org/10.1002/9780470015902.a0026342

Sanchez-Bayo F, Wyckhuys KAG (2019) Worldwide decline of the entomofauna: a review of its drivers. Biol Conserv 232:8–27. https://doi.org/10.1016/j.biocon.2019.01.020

Santangeli A, Lehikoinen A, Lindholm T, Herzon I (2019) Organic animal farms increase farmland bird abundance in the boreal region. PLoS One 14(5):e0216009. https://doi.org/10.1371/journal.pone.0216009

Sardiñas HS, Kremen C (2015) Pollination services from field-scale agricultural diversification may be context-dependent. Agric Ecosyst Environ 207:17–25

Schneider MK, Luscher G, Jeanneret P, Arndorfer M, Ammari Y, Bailey D, Balazs K, Baldi A, Choisis JP, Dennis P, Eiter S, Fjellstad W, Fraser MD, Frank T, Friedel JK, Garchi S, Geijzendorffer IR, Gomiero T, Gonzalez-Bornay G, Hector A, Jerkovich G, Jongman RHG, Kakudidi E, Kainz M, Kovacs-Hostyanszki A, Moreno G, Nkwiine C, Opio J, Oschatz ML, Paoletti MG, Pointereau P, Pulido FJ, Sarthou JP, Siebrecht N, Sommaggio D, Turnbull LA, Wolfrum S, Herzog F (2014) Gains to species diversity in organically farmed fields are not propagated to the farm level. Nat Commun 5:4151. https://doi.org/10.1038/ncomms5151

Schoonhoven Y, Runhaar H (2018) Conditions for the adoption of agro-ecological farming practices: a holistic framework illustrated with the case of almond farming in Andalusia. Int J Agric Sustain 16:442–454. https://doi.org/10.1080/14735903.20181537664

Schroth G, Harvey CA, Vincent G (2004) Complex agroforests: their structure, diversity and potential role in landscape conservation. Chapter 10. In: Schroth G, da Fonseca GAB, Harvey CA, Gascon C, Vasconcelos HL, Izac AMN (eds) Agroforestry and biodiversity conservation in tropical landscapes. Island Press, Washington, DC, pp 227–260

Scudder GGE (2009) The importance of insects. In: Insect biodiversity: science and society. Blackwell Publishing, Hoboken, NJ

Seufert V, Ramankutty N, Foley JA (2012) Comparing the yields of organic and conventional agriculture. Nature 485:229–232

Shackelford G, Steward PR, Benton TG, Kunin WE, Potts SG, Biesmeijer JC, Sait SM (2013) Comparison of pollinators and natural enemies: a meta-analysis of landscape and local effects on abundance and richness in crops. Biol Rev 88:1002–1021. https://doi.org/10.1111/brv.12040

Silva EM (2014) Screening five fall-sown cover crops for use in organic no-till crop production in the upper Midwest. Agroecol Sustain Food Sys 38:748–763

Singh NR, Jhariya MK (2016) Agroforestry and Agrihorticulture for higher income and resource conservation. In: Narain S, Rawat SK (eds) Innovative technology for sustainable agriculture development. Biotech Books, New Delhi, India, pp 125–145

Smith RG, Gross KL, Robertson GP (2008) Effects of crop diversity on agroecosystem function: crop yield response. Ecosystems 11:355–366

Smith HG, Dänhardt J, Lindström A, Rundlöf M (2010) Consequences of organic farming and landscape heterogeneity for species richness and abundance of farmland birds. Oecologia 162:1071–1079

Soane BD, Ball BC, Arvidsson J, Basch G, Moreno F, Roger-Estrade J (2012) No-till in northern, western and South-Western Europe: a review of problems and opportunities for crop production and the environment. Soil Tillage Res 118:66–87. https://doi.org/10.1016/j.still.2011.10.015

Storkey J, Meyer S, Still KS, Leuschner C (2012) The impact of agricultural intensification and land-use change on the European arable flora. Proc Biol Sci 279:1421–1429. https://doi.org/10.1098/rspb.2011.1686

Tamburini G, De Simone S, Sigura M, Boscutti F, Marini L (2016) Conservation tillage mitigates the negative effect of landscape simplification on biological control. J Appl Ecol 53:233e241

Thorbek P, Bilde T (2004) Reduced numbers of generalist arthropod predators after crop management. J Appl Ecol 41:526–538

Tiemann LK, Grandy AS, Atkinson EE, Marin-Spiotta E, McDaniel MD (2015) Crop rotational diversity enhances belowground communities and functions in an agroecosystem. Ecol Lett 18:761–771. https://doi.org/10.1111/ele.12453

Tilman D, Fargione J, Wolff B, D'Antonio C, Dobson A, Howarth R, Schindler D, Schlesinger WH, Simberloff D, Swackhamer D (2001) Forecasting agriculturally driven global environmental change. Science 292:281–284

Tilman D, Cassman KG, Matson PA, Naylor R, Polasky S (2002) Agricultural sustainability and intensive production practices. Nature 418:671–677. https://doi.org/10.1038/nature01014

Tilman D, Balzer C, Hill J, Befort BL (2011) Global food demand and the sustainable intensification of agriculture. Proc Natl Acad Sci U S A 108:20260–20264

Traba J, Morales MB (2019) The decline of farmland birds in Spain is strongly associated to the loss of fallowland. Sci Rep 9:9473. https://doi.org/10.1038/s41598-019-45854-0
Trewavas A (2001) Urban myths of organic farming. Nature 410:409–410. https://doi.org/10.1038/35068639
Trewavas A (2002) Malthus foiled again and again. Nature 418:668–670
Tscharntke T, Klein AM, Kruess A, Steffan-Dewenter I, Thies C (2005) Landscape perspectives on agricultural intensification and biodiversity – ecosystem service management. Ecol Lett 8:857–874. https://doi.org/10.1111/j.1461-0248.2005.00782.x
Tscharntke T, Clough Y, Bhagwat SA, Buchori D, Faust H, Hertel D, Holscher D, Juhrbandt J, Kessler M, Perfecto I, Scherber C, Schroth G, Veldkamp E, Wanger TC (2011) Multifunctional shade-tree management in tropical agroforestry landscapes-a review. J Appl Ecol 48 (3):619–629
Tscharntke T, Clough Y, Wanger TC, Jackson L, Motzke I, Perfecto I, Vandermeer J, Whitbread A (2012) Global food security, biodiversity conservation and the future of agricultural intensification. Biol Conserv 151:53–59
Tuck SL, Winqvist C, Mota F, Ahnstreom J, Turnbull LA, Bengtsson J (2014) Land-use intensity and the effects of organic farming on biodiversity: a hierarchical meta-analysis. J Appl Ecol 51:746–755. https://doi.org/10.1111/1365-2664.12219
Venter ZS, Jacobs K, Hawkins HJ (2016) The impact of crop rotation on soil microbial diversity: a meta-analysis. Pedobiologia 59:215–223. https://doi.org/10.1016/j.pedobi.2016.04.001
Verbruggen E, Roling WFM, Gamper HA, Kowalchuk GA, Verhoef HA, van der Heijden MGA (2010) Positive effects of organic farming on below-ground mutualists: large-scale comparison of mycorrhizal fungal communities in agricultural soils. New Phytol 186:968–979
Verret V, Gardarin A, Pelzer E, Médiène S, Makowski D, Valantin-Morison M (2017) Can legume companion plants control weeds without decreasing crop yield? a meta-analysis. Field Crops Res 204:158–168
Vongvisouk T, Broegaard RB, Mertz O, Thongmanivong S (2016) Rush for cash crops and forest protection: neither land sparing nor land sharing. Landuse Policy 55:82–192. https://doi.org/10.1016/j.landusepol.2016.04.001
Wagg C, Bender SF, Widmer F, van der Heijdena MGA (2014) Soil biodiversity and soil community composition determine ecosystem multifunctionality. PNAS 14:5266–5270
Wang Z, Liu L, Chen Q, Wen X, Liao Y (2016) Conservation tillage increases soil bacterial diversity in the dryland of northern China. Agron Sustain Dev 36:28. https://doi.org/10.1007/s13593-016-0366-x
Wickramasinghe LP, Harris S, Jones G, Vaughan N (2003) Bat activity and species richness on organic and conventional farms: impact of agricultural intensification. J Appl Ecol 40:984–993
Winfree R, Fox JW, Williams N, Reilly J, Cariveau D (2015) Abundance of common species, not species richness, drives delivery of a real-world ecosystem service. Ecol Lett 18:626–635
Winqvist C, Bengtsson J, Aavik T, Berendse F, Clement LW, Eggers S (2011) Mixed effects of organic farming and landscape complexity on farmland biodiversity and biological control potential across Europe. J Appl Ecol 48:570–579. https://doi.org/10.1111/j.1365-2664.2010.01950.x
Xu Q, Hatt S, Lopes T, Zhang Y, Bodson B, Chen J, Francis F (2017) A push–pull strategy to control aphids combines intercropping with semiochemical releases. J Pest Sci 91:93–103. https://doi.org/10.1007/s10340-017-0888-2
Yachi S, Loreau M (1999) Biodiversity and ecosystem productivity in a fluctuating environment: the insurance hypothesis. Proc Natl Acad Sci U S A 96:1463–1468
Zak DR, Holmes WE, White DC, Peacock AD, Tilman D (2003) Plant diversity, soil microbial communities, and ecosystem function: are there any links? Ecology 84:2042–2050
Zikeli S, Gruber S (2017) Reduced tillage and no-till in organic farming systems, Germany-status quo, potentials and challenges. Agriculture 7:35. https://doi.org/10.3390/agriculture7040035
Zuazo VHD, Pleguezuelo CR (2008) Soil-Erosion and runoff prevention by plant covers a review. Agronomy Sustain Dev 28(1):65–86. https://doi.org/10.1051/agro:2007062

Climate Change and Agricultural Sustainable Intensification in the Arid Lands

4

Zied Haj-Amor, Latifa Dhaouadi, Abdulrasoul Al-Omran, and Salem Bouri

Abstract

Arid lands face a variety of environmental challenges, principally water scarcity (with rainfall typically <200 mm year^{-1}) and land degradation (it is estimated that around 30% arid lands are degraded globally). Sustainable agricultural intensification is a fundamental strategy to respond to many challenges in arid lands, where the impacts of climate change are more pronounced. This chapter summarizes the manifestations of climate change and their effects on agricultural productivity in the arid lands, and analyzes possible agricultural sustainable intensification practices (e.g. soil and water conservation, agroforestry, etc.) that can help us to respond towards climate change in these contexts. In addition, the chapter describes the key policies and measures required for ensuring sustainable agricultural intensification. All these issues (i.e. climate change manifestations, climate change effects, sustainable intensification practices, and key policies) are clearly discussed with good analysis of positive consequences of agriculture sustainability in the arid lands. Finally, the chapter describes the sustainable intensification practices that have a high potential to conserve natural resources and enhance plant growth in arid areas. Based on the findings reported here, it is concluded that the agricultural productivity in arid lands could be enhanced by 21–33% through actions to address productivity and sustainability, for example,

Z. Haj-Amor (✉) · S. Bouri
Sfax University, Sfax, Tunisia

L. Dhaouadi
Regional Research Center for Oases Agriculture, Tozeur, Tunisia

A. Al-Omran
Soil Science Department, College of Food and Agricultural Sciences, King Saud University, Riyadh, Saudi Arabia
e-mail: rasoul@ksu.edu.sa

M. K. Jhariya et al. (eds.), *Ecological Intensification of Natural Resources for Sustainable Agriculture*, https://doi.org/10.1007/978-981-33-4203-3_4

enhancing crop genetics, better management of natural resources, climate smart agriculture, etc.

Keywords

Agriculture · Arid lands · Climate change · Sustainable intensification

Abbreviations

AI	Aridity Index
C	Carbon
CO_2	Carbon dioxide
CSA	Climate Smart Agriculture
ET	Evapotranspiration
INM	Integrated Nutrient Management
IPM	Integrated Pest Management
P/ET	Precipitation/Evapotranspiration ratio
P	Precipitation
WUE	Water use efficiency

4.1 Introduction

The history of agriculture refers to the development and dissemination of various techniques (especially intensification technique) resulting in the increased productivity of plants, crops, and animals (FAO 2004). The major objective of these techniques today is to meet growing global food demand. Increased global food demand, coupled with population growth and socio-economic development, has exerted significant pressure on water, energy, land, etc. (Xie et al. 2019). Negative consequences of this pressure can include increase in global food prices (Senker 2011) and hunger in several regions (Suhardiman et al. 2016; Godfray 2015; IPCC 2019).

Globally, agricultural intensification began in the 1940s and gradually expanded from1960s (FAO 2004). During this period, significant modifications in plant types and land management options occurred in many countries including in Europe, the USA, and Asia (Hazell 2008). This intensification was successfully ensured through the adoption of suitable irrigation methods, fertilizers, and pesticides (Senker 2011; Meena et al. 2020). This rapid agricultural productivity enhances plant yields and ensures food security worldwide. However, these initial positive results were accompanied by negative impacts including the loss of biodiversity, decline in soil fertility, increase in soil salinity, shallow groundwater rising, etc. (IPCC 2019).

The United Nations forecasts that the global population will reach 9 billion in 2050, many of whom will live in arid lands. This population growth, together with continuous urbanization, means that many countries will face formidable challenges in protecting available natural resources (Tilman et al. 2002; Wezel et al. 2015). Additionally, climate change impacts (e.g. sea level rise, drought, etc.) will adversely affect water, soil, and other resources (Burney et al. 2010; Khan et al. 2020a, b). Given these factors and the importance of agricultural areas, adequate adaptation strategies are essential to mitigate the adverse impacts of agricultural intensification on water and soil resources, and to avoid the future loss of large agricultural areas (Tilman et al. 2011; Meena et al. 2020a, b).

Recently, sustainable agricultural intensification practices have been adopted in a number of countries to enhance agricultural productivity (Xie et al. 2019). These practices have an important role to play in ensuring the sustainability of cultivated areas in the face of climate change (Senker 2011). Furthermore, the viability of these practices depends on many factors, including local environmental conditions and land management regimes (FAO 2004). Accordingly, the nature of sustainable agricultural intensification varies in different regions, and sustainable agriculture in arid lands needs to draw on a wide variety of specific actions and practices (IPCC 2019).

The purpose of this chapter is to identify the main sustainable agricultural intensification practices that can help to address climate change in arid lands, defined as areas in which the annual evapotranspiration (ET) is more than the annual precipitation (P) (Huang et al. 2017).

4.2 Climate Change in the Arid Lands

Climate change can be caused by both natural and human processes, with the former dominating on geological time scales and the latter being the principal drivers of climate change, in the form of global warming since the beginning of the industrial era. As the global climate warms, it results in changes in atmospheric and oceanic circulation, seasonal and interannual climate variability, rainfall patterns, and the behavior of climate and weather extremes such as heat waves, storms, intense rainfall, and droughts. Thermal expansion of the oceans, combined with snow and ice melt, also results in increase in sea levels. In this section, we discuss the following manifestations of climate change: increasing air temperature, sea level rise, changes in rainfall patterns and droughts. These phenomena are particularly relevant for arid lands, where the existing harsh environment amplifies sensitivity and vulnerability to climate change (Earman and Dettinger 2011; Meena et al. 2018).

4.2.1 Increasing Air Temperature

The global average surface temperature record for the period 1901–2001 indicates that the earth is warming (Fig. 4.1). This tracking revealed that the average air temperature of the world has increased in the range between 0.8 and 1.13 °C along the investigated 100 years (IPCC 2013). Furthermore, the simulation results (Fig. 4.1)

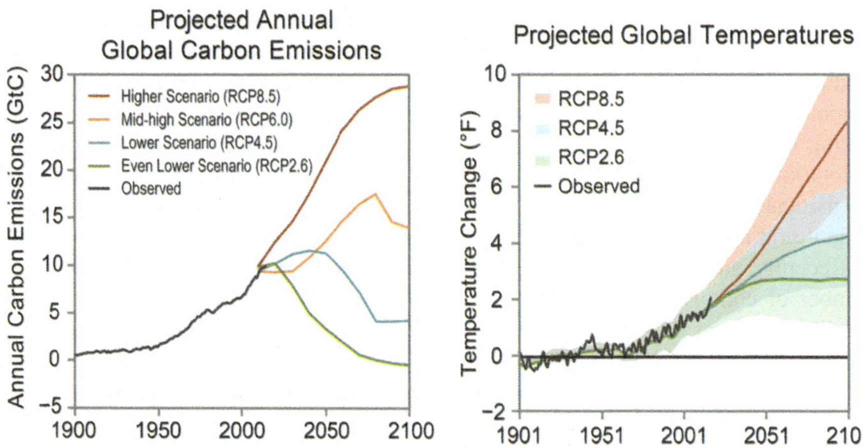

Fig. 4.1 Observed and projected average air temperature of the world from 1901 to 2100 (Adapted from Clarke and Smethurst 2010)

Table 4.1 Contribution of different sectors to greenhouse gas emissions (IPCC 2013)

Sector	Contribution (%)
Electricity	35
Transport	29
Agriculture	27
Others	<10

showed that the temperature would continue to increase progressively to rise as much as 2 °F in 2100 (Clarke and Smethurst 2010). The main contributor of this significant increasing trend is the increase for greenhouse gases (for example: CO_2).

Multiple sectors have resulted in increased greenhouse gases emissions. The major sectors include electricity, transport, and agriculture (Table 4.1).

Temperature records indicate that warming has been more pronounced in low-latitude arid regions than in the mid-latitude humid areas, and climate projections indicate that this trend is likely to continue (IPCC 2019). For example, it is expected that a rise of 2.6 °C in average air temperatures and a 35% decrease in rainfall in 2030 will take place in Iran. The high greenhouse gases emissions (about 617 million tons of CO_2) are a major contributor of climate change in this country (Mansouri Daneshvar et al. 2019).

4.2.2 Changes in Rainfall Patterns

As reported in multiple studies (Putnam and Broecker 2017; Ge et al. 2010; Jain and Kumar 2012; Al Charaabi and Al-Yahyai 2013), climate change is altering rainfall patterns in many arid countries, as wet areas receive more rainfall and dry areas become disproportionately warmer and drier (Held and Soden 2006). Increased air

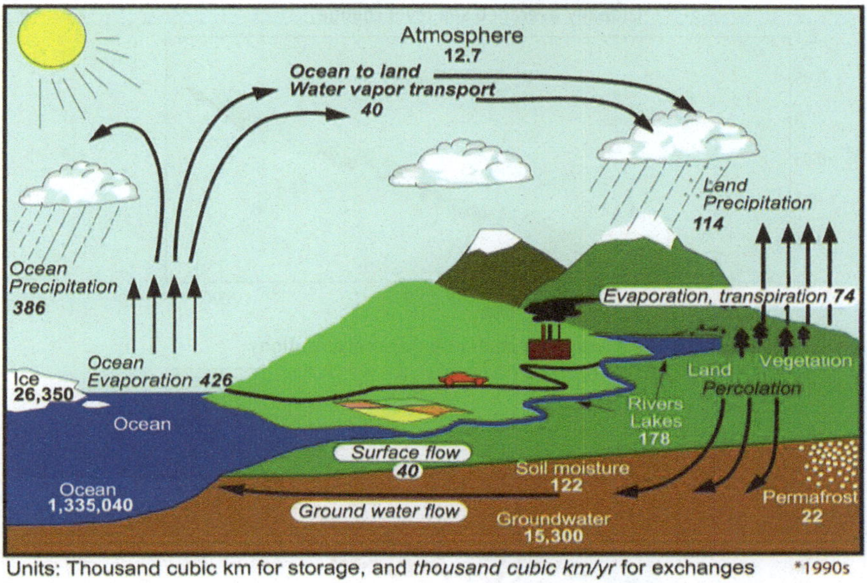

Fig. 4.2 Global water budget based on 2002–2008 flow observations (Adapted from Trenberth et al. 2011)

temperatures have resulted in greater evaporation from land surfaces and water bodies (Fig. 4.2). For each 1 °F increase in air temperature, the atmosphere can hold about 4% more water vapor. Globally, this increase in atmospheric moisture results in an increase in rainfall and heavy downpours (Meena and Lal 2018). However, this extra rainfall is not evenly spread around the globe, and some regions, including most of the world's arid lands, will receive less rainfall as a result of other changes in atmospheric and ocean circulation driven by climate. Consequently, the impact of changing rainfall patterns on water resources will be more serious in arid lands (Leung et al. 2004). In the arid lands, the implications of climate change for the water sector represent a major risk for initiatives to enhance uninterrupted water supplies for multiple uses, and represent a particular risk to agriculture.

4.2.3 Sea Level Rise

Sea level rise represents a significant risk in coastal arid lands. Satellite radar measurements from 1990 to 2010 indicate an increase in global average sea level of about 20 cm. Therefore, the rate of rise is about 1 cm per year. This increase in sea level is the result of increased global air temperature that causes thermal expansion of seawater, combined with melting from glacier and ice sheets (Vermeer and Rahmstorf 2009) (Fig. 4.3).

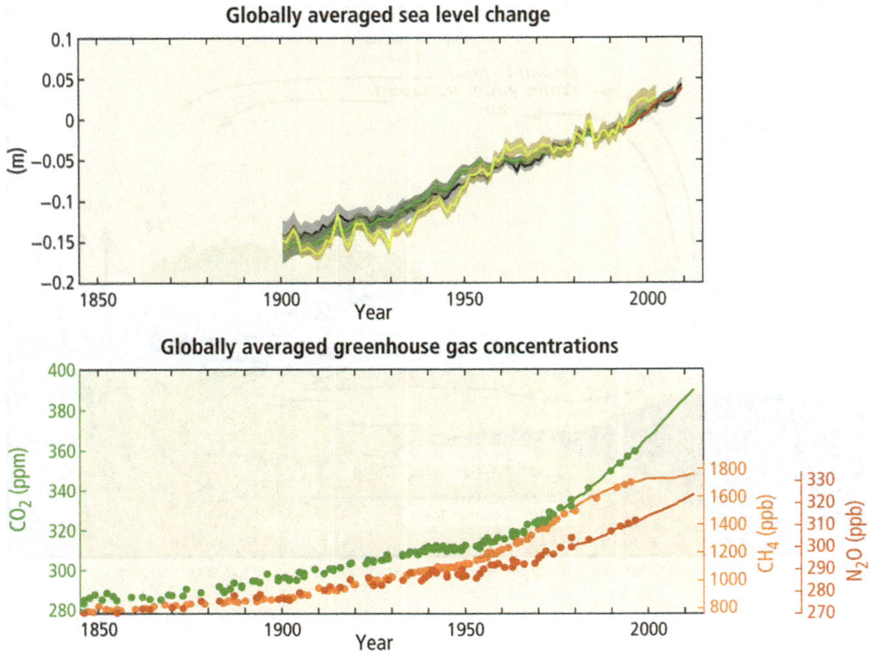

Fig. 4.3 Global average sea level from 1990 to 2010 (Adapted from IPCC 2013)

In addition, climate projections indicate that the annual rate of sea level rise will continue to increase during the next years (Rahmstorf 2012). Precise projections of sea level are difficult because of the complex properties of the climate system and uncertainties relating to ice sheet dynamics. Estimates of sea level rise to 2100 are generally around 1 m, but upper estimates are above 2 m. Worst-case estimates for 2200 are around 7.5 m. Past warming of around 2 °C reflects sea levels several m above those of today and past warming of 3 °C reflects them around 20 m higher (Rahmstorf 2012).

4.2.4 Drought

Drought is a natural feature of arid and semi-arid areas, but changes in drought behavior can result from changes in the hydrological cycle resulting from climate change. In particular, when droughts occur they increasingly do so during period of higher temperature, which increases evapotranspiration and makes the droughts worse. A number of areas are seeing much more severe and protracted drought although it remains difficult to say much about changes in drought frequency. More fundamentally, climate change might result in shifts in climatic and agro-ecological zones, for example, shifts from semi-arid or sub-humid to more arid conditions, i.e. permanent drought. This also occurred in the past, before about 5000 years ago

Fig. 4.4 Impact of drought on maize in Mixteca, Mexico (Adapted from Altieri et al. 2015)

the Sahara was semi-arid, before it transitioned to hyper-aridity. This type of transition could occur in some dry lands in the near future (IPCC 2019). In many arid lands, droughts have frequent negative impacts on the growth of dry-farmed crops (Altieri et al. 2015) such as maize (*Zea mays* L.) (Fig. 4.4). Furthermore, drought has different negative impacts on hydroclimatic variables such as decrease in rainfall, increase in evapotranspiration, and soil moisture reduction (Mukherjee et al. 2018). The drought propagation has occurred due to climate change, especially due to the variability in anomalies to sea-surface temperatures. Due to the continuous change in weather conditions, the effects of drought under these conditions are expected to accelerate in the future (Salvadori and De Michele 2004).

High evapotranspiration rates coupled with higher temperatures resulting from the continuous emission of greenhouse gases have intensified droughts in the arid lands. Climate projections suggest that, globally, about 29% of available agricultural lands will be affected by drought by 2050 (IPCC 2013). In the arid lands, dry conditions contribute to significant decrease in water resources availability (Van Loon and Van Lanen 2013). Accordingly, appropriate strategies are needed to make adequate and sustainable use of increasingly scarce water in the arid lands.

4.2.5 Pest and Disease Outbreaks

The prevalence of insect pests and diseases affecting crops is strongly correlated with climatic variables (especially air temperature and rainfall). Accordingly, climate change has the potential to have significant effects on insect pests and crop diseases (Bhagat et al. 2016). The severity of these effects is related to many factors; however, the major factors include the crop type and the growth status of the cultivated crop (Khan et al. 2009). Usually, the crops that have good potential to resist the stress of salinization, temperature, and show efficient CO_2 use are better than many other crop types (Bergot et al. 2004). Based on the data reported in IPCC (2013), it is expected that the yield of many crops cultivated in the arid lands will be increased by 28% by 2050 as a direct consequence of doubling CO_2 level. The observations of agricultural productivity during the recent years have confirmed that increasing production could be offset in the arid agricultural areas by the insect pests, weeds, and pathogens (Bhagat et al. 2016).

Air temperature and humidity affect the development of pests, and changes in these variables have significantly potential to affect crop yields. Changes in temperature and humidity also have the potential to the emergence of new pests in many arid agricultural areas (Khan et al. 2009). Additionally, a warmer climate may result in changes in the relative abundance of pathogens (Chakraborty and Datta 2003); usually, pathogens with wide host ranges have a high potential for survival. Furthermore, increasing atmospheric CO_2 concentrations have the potential to change the developmental conditions of pathogens (Matros et al. 2006). Higher CO_2 concentrations have been associated with the accelerated growth of several fungal pathogens in some studies (Matros et al. 2006; Khan et al. 2009).

It is important to highlight that the manifestations of climate change, particularly increases in air temperature and changes in rainfall patterns, have exacerbated problems associated with invertebrate pests such as insects and mites (Rao et al. 2014), through changes in populations and dispersal (Fig. 4.5). The research performed by Lastuvka (2009) revealed that a warmer climate coupled with a significant increase in CO_2 concentration and changes in rainfall patterns have a high potential to affect the survival of these pests. Examples of these effects include shifts in species distribution, modifications in phenology, enhanced growth rates, modifications to migratory behavior, disruption to enemy–pest interaction, etc.

Even though there are many manifestations and impacts of climate change in the arid lands, there have been few field-based studies (Holzkämper et al. 2015) on the effects of climate change on crop diseases and insect pests in these regions. Most currently available crop protection technologies have been designed based on laboratory conditions which do not reflect real climate conditions in arid lands. The integration of observational climate data and data relating to the properties of real farming systems into the development of these technologies would enhance the ability of decision makers to develop and promote more relevant crop protection technologies.

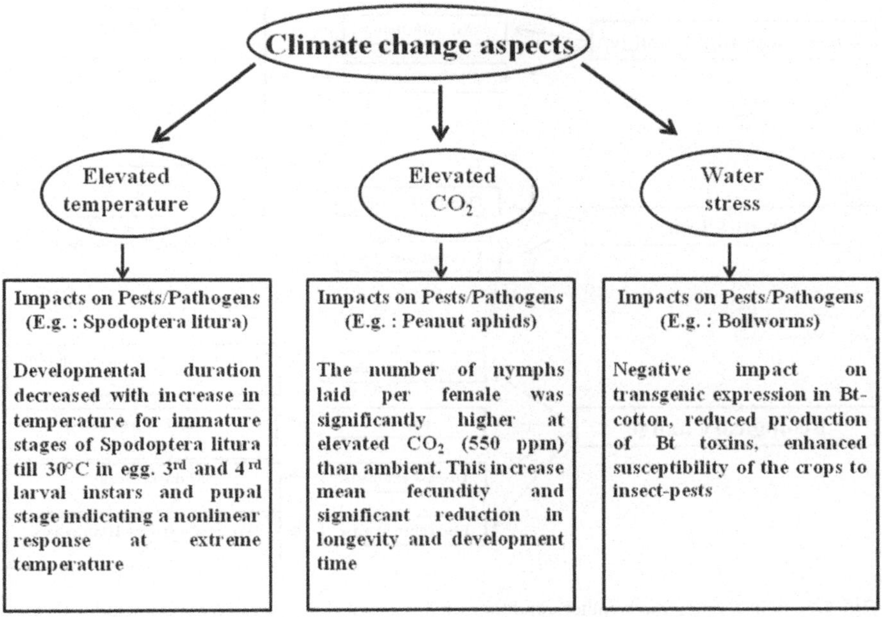

Fig. 4.5 Examples of climate change impacts on pests and pathogens (Adapted from Bhagat et al. 2016)

4.2.6 Eco-environmental Aspects

As confirmed by several research works (Skelly and Weinstein 2003; Diaz 2007; Costello et al. 2009), it is expected that climate change will affect the health of arid eco-environmental systems. As reported by Diaz (2007), the eco-environmental health of a given system can be defined in terms of the interdependencies between climate (measured in terms of specific climatic variables) and community health. Based on this definition, it is expected that any change in climate variables (i.e. as a result of climate change) will have significant effect on eco-environmental health. Usually, the health of ecological systems, community health, and the health of people are closely related (Diaz 2007; Costello et al. 2009). For example, in Fig. 4.6, we show the impacts of some modifications in weather conditions and ecosystem on child health (Stanley and Farrant 2015).

In the arid lands, a warmer climate (i.e. continuous increasing in air temperature) results in multiple negative environmental impacts. The principal environmental problems include increases in heat waves, sea level rise, ocean acidity, droughts, water salinization, etc. These environmental problems have several direct and indirect effects on already scare water resources. A number of negative effects on population health have been noted in relation to these problems (Epstein 2005; Pirard et al. 2005; Sharkey 2007; Cheong et al. 2013). For example, the heat wave that occurred in Melbourne (Australia) in 2009 resulted in increased human

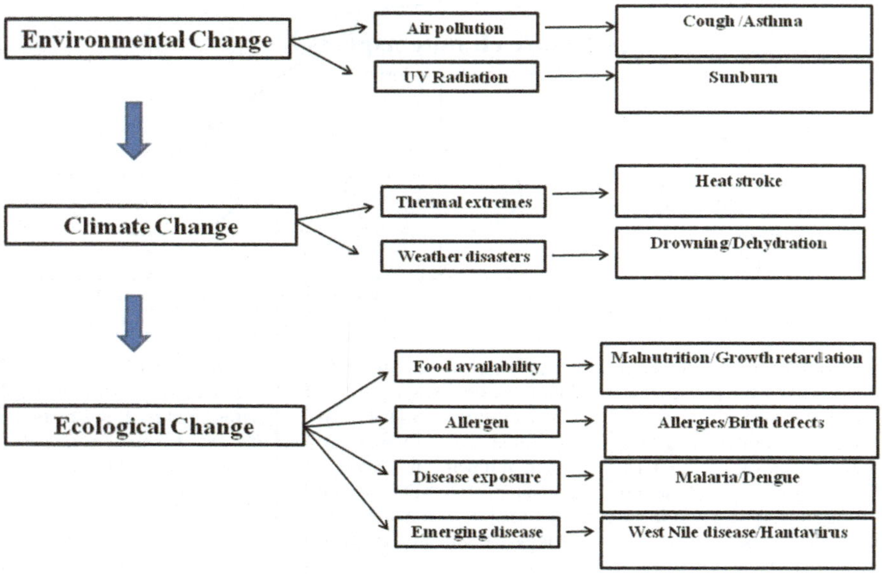

Fig. 4.6 Impacts of some weather conditions and ecological properties on child health (Adapted from Stanley and Farrant 2015)

mortality (Strand et al. 2010), which can be considered as a direct impact of climate change. In Fig. 4.7, we illustrate potential actions and policies that can help us to reduce the critical impacts of climate change on human health (WHO 2009).

Climate change affects human health through a number of pathways and processes. In Table 4.2, we summarize the climatic and environmental processes that can have significant impacts on human health. Based on the information reported in Table 4.2, it can be concluded that there is an urgent need to adapt to the potential effects of climate change and changes in climate and weather extremes on human health. As suggested by McMichael (2013), the development of optimal adaption strategies requires a comprehensive evaluation of the current and potential effects of weather conditions and their consequences on health and effective implementation of a variety of solutions. These solutions include continuous control of environmental changes, respond better to existing and emerging diseases, and health flood protection.

Even though the above environmental impacts and their consequences on human health have been considered by decision makers in order to develop the optimal adaption strategies, there is still a lot of work (especially research work) to be performed to ensure that all aspects of climate change in the arid lands are considered. Furthermore, it is observed that human health may be affected by indirect factors (e.g. soil management) that have not been taken into account in recent years. In future research, including these factors would be helpful for developing effective policies for both human and environmental protection in the arid lands.

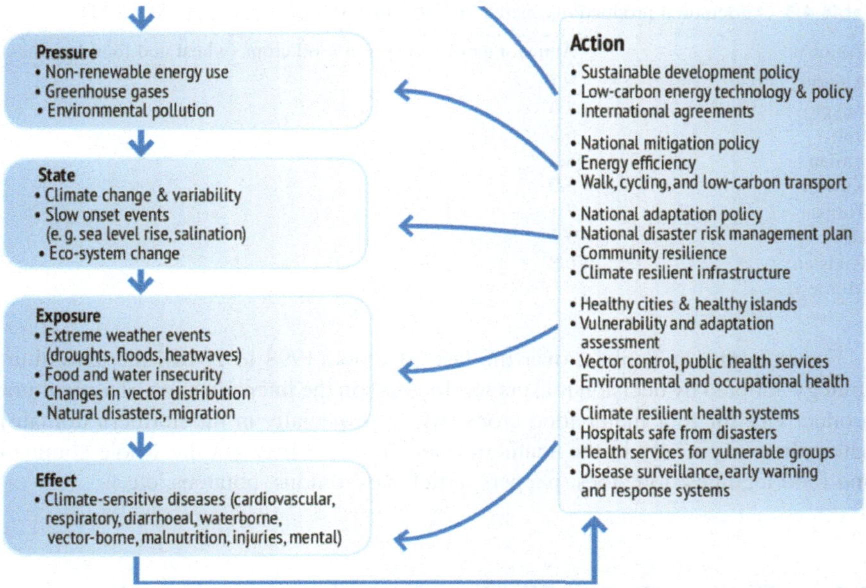

Fig. 4.7 Potential actions and policies for reducing the critical impacts of climate change on human health (WHO 2009)

Table 4.2 Climate change effects on human health (McMichael 2013)

Climate change	Environmental impact	Heath impact
Change in average climate conditions	Glacier loss Sea level rise Altered surface water	Deaths Injuries Infections disease risk Nutrition and child development Mental health Cardio-respiratory diseases
Increased climate variability	Reduced crop yields Ecosystem damage Microbial ecology change	
Change in extreme climate events	Infrastructure damage Conflicts and displacement	

4.3 Scenarios of Agricultural Productivity in Arid Lands

Based on the agricultural productivity data estimated by the FAO (2011) from 1998 to 2008 (Table 4.3), it was revealed that significant agricultural productivity in the arid countries has been ensured through two major strategies. The first involves a doubling of agricultural productivity of major (fundamental) food crops such as wheat, maize, rice, food legumes, etc. The second scenario involves the culture of certain crops in small fields. The latter has been adopted by few arid countries.

Table 4.3 Agricultural productivity increases (1998 to 2008) in arid lands (FAO 2011)

Country	Annual rate of increase in food crops (wheat and food legumes)
Algeria	5.7
Egypt	8.6
Iran	6.3
Jordan	8.4
Kuwait	5.9
Tunisia	7.1
USA	3.3
Pakistan	6.8
Morocco	

For example, in Tunisia, over the last 20 years (1998 to 2018), the agriculture strategy adopted by decision makers has focused on the intensification of agricultural productivity for four major food crops: wheat (especially in the northern domain), date palms (especially in the southern domain), olive trees (in the whole country), and food legumes (tomatoes, peppers, artichokes, onions, potatoes, etc.).

4.4 Climate Change Effect on Agricultural Productivity in the Arid Lands

In the arid lands, climate variability and weather conditions, both of which are affected by climate change, have direct effects on agricultural productivity (Kumar and Gautam 2014). Increases in air temperature and CO_2 concentrations, coupled with a potential decrease in annual rainfall, have the potential to affect the crop growth significantly. Generally, the impact of climate change on food production may be manageable if adequate irrigation water management strategies are adopted (IPCC 2013), at least in the short- to medium-term. In the longer term, changes in temperature, rainfall, evaporation, and groundwater could be so great that irrigation may not be feasible. Worldwide, it is expected that agricultural productivity will be enhanced as a direct result of the carbon dioxide fertilization. This fertilization is defined as the increased rate of photosynthesis in crops because of the continuous increase in the concentration of carbon dioxide in the air (Ellsworth et al. 2004). The carbon dioxide fertilization effect depends on several factors (crop species, nutrients availability, etc.) (Zhu et al. 2010). In Fig. 4.8, a photosynthetic response to increasing CO_2 concentration is illustrated (Medlyn et al. 2002).

Furthermore, in Table 4.4, growth amelioration due to elevated CO_2 is mentioned (Wang 2007). Usually, enhanced rates of photosynthesis in plants due to carbon dioxide fertilization are only partially transferred to enhanced plant growth and any hypothesized carbon dioxide fertilization response is unlikely to reduce the human-made increases in atmospheric CO_2 concentration over the next century (IPCC 2013).

In the arid lands (e.g. South America, Central America, Australia, North Africa, India, etc.), a warmer climate can have several critical effects on the hydrologic

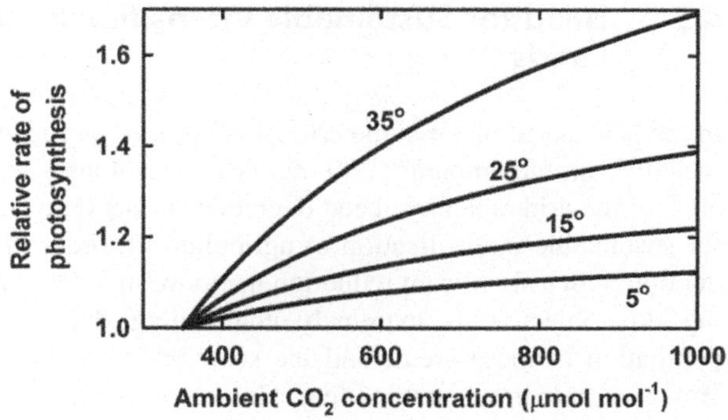

Fig. 4.8 Photosynthetic responses to continuous increase in concentration of carbon dioxide (Adapted from Medlyn et al. 2002)

Table 4.4 Growth amelioration in response to elevated CO_2 (Data sources: Wang 2007)

Observation	CO_2 amelioration (%)
All herbaceous plants	+45
Woody plants	+48
Low-nutrient-grown plants	+25
Low-temperature-grown plants	+27
Dry matter production	+20
Grassland biomass	+12
Forest growth	+23

cycle. The major critical effects include rainfall alteration, increase in intensity of run-off, significant modification of soil moisture, changes in crop evapotranspiration, etc. For example, in Rajasthan state (India), it is expected that a rise of 14% in total crop evapotranspiration requirement will occur in 2030 as a result of a 1 °C increase in air temperature (Gautam and Sharma 2012). Furthermore, a warmer climate can affect groundwater recharge and water resources quality (for example, through saline intrusion into aquifers and rivers) (Xu et al. 2007). All these conditions may result in reduced agricultural productivity.

Several recent analyses (Anand and Khetarpal 2015; Roco et al. 2017) have revealed that over the next years (up to 2030), the warmer climate will significantly affect agricultural productivity in many arid countries where much of the world's population lives. As an option to reduce the negative effects of harsh weather conditions on crop production, these recent analyses have revealed the urgent need for the development and adoption of suitable water resources management strategies (e.g. water saving technologies) that can enhance crops yields. For example, in 2007, the water saving technologies installed in many Indian agricultural areas have contributed positively to agricultural production (Gautam and Sharma 2012). These analyses encouraged farmers to adapt to new harsh environments through several practices such as higher water use efficiency (WUE) and the selection of water salinity tolerant crops, etc. (Malash et al. 2008; Kumar et al. 2020).

4.5 Need for Sustainable Intensification of Agriculture in Arid Lands

In the arid lands, under some critical conditions as such as water resources shortage (usually, rainfall amount <200 mm year^{-1}) and land degradation (worldwide about 30% of the arid area have been degraded so far) (FAO 2011). There is a huge need for sustainable intensification of agriculture in order to ensure adequate food production with a significant reduction in environmental impacts (IPCC 2013). Achieving this objective is extremely challenging due to the continuous growth in population in these areas, and the scarcity of available water and soil resources. However, numerous studies (e.g. Halvorson et al. 2002; Ahlam et al. 2015) have indicated that suitable strategies involving the adequate exploitation of land, water, and energy resources could be helpful in avoiding damage to the environment while increasing food production. Furthermore, it is important to indicate that sustainable intensification of agriculture in the arid lands involves more than respecting the environment, and must extend to delivering a combination of environmental, economic, human, and social benefits (IPCC 2013). Desirable environmental benefits include positive impacts on biodiversity, and soil and water resources. Economic benefits include profitability and managing the variability of profits and labor requirements. Human benefits are focused on ensuring the availability of sufficient quantities of crop foods. Finally, the social benefits involve equity, social cohesion, and collective action.

4.6 Characterization of Sustainable Agricultural Intensification in the Arid Lands

Sustainable agricultural intensification may be represented as increased production in a specific area coupled with the conservation of used resources (energy, water, and soil) and the reduction of negative effects on the agricultural system (FAO 2011; IPCC 2019), while enhancing natural capital and ecosystem services flows (Garnett et al. 2013; Pretty and Bharucha 2014). In the arid lands, due to limited natural resources such as water resources, the main objective of sustainable intensification is to improve management of crop, land, soil, water, etc. (Pretty 1997; Matson et al. 1997). In Table 4.5, key characteristics of sustainable agricultural intensification are presented:

Table 4.5 Key characteristics of agricultural sustainable intensification

Characteristic	Source
Increase agricultural production with environment protection	IPCC (2019)
Balance between inputs and outputs(of the agricultural system)	IPCC (2019)
Increase land productivity to enhance farmer income	Ruerd and Lee (2000)
Ensure nutrient equilibrium within the cultivated soil	Ruerd and Lee (2000)
Improve social and human aspects of production	Pretty (2008)

Several international studies (Schiefer et al. 2016; Scherer et al. 2018) have shown that natural factors such as climatic conditions (especially temperature and rainfall), land slope, soil management options, etc. have the potential to affect the adoption of key characteristics of agricultural sustainable intensification. The manifestations of climate change in the arid lands (increasing air temperature, sea level rise, rainfall patterns change, and droughts) could result in a number of negative effects on sustainable agriculture. Indeed, these manifestations have a high potential to work against sustainable agriculture (through some actions and strategies). Accordingly, there is a need to support and encourage farmers to adapt to the impacts of climate change. For example, the farmers of Uganda are currently encouraged and supported to ensure sustainable coffee (Arabica coffee) production within the agricultural fields at high altitudes (Rahn et al. 2018).

4.7 Agricultural Sustainable Intensification Practices in the Arid Lands

In the arid lands, the manifestations of climate change (e.g. increasing air temperature, sea level rise, rainfall patterns modification, and droughts) can have multiple negative effects on crop yields and agriculture sustainability. Accordingly, it is useful to identify those agricultural practices that have a high potential to increase crop yields without adverse environmental effects (IPCC 2019). In this section, we discuss agricultural practices relevant to sustainable intensification.

4.7.1 Conservation Tillage and Crop Rotation

Conservation tillage practices (e.g. decrease tillage frequencies, decreased intensity of tillage events, etc.) coupled with cropping system intensification can contribute to significant increases in organic carbon levels within the cultivated soil (Halvorson et al. 2002). This increasing depends principally on the quantity of residue input to the soil (Peterson et al. 1998). Furthermore, carbon (C) sequestration in cultivated soil (i.e. the movement of CO_2 from the air to the soil) can be enhanced through better tillage practices. The adoption of suitable tillage practices by farmers on cropland is required to ensure the sustainability of agricultural systems. Crop rotation (i.e. growing different crop species in recurrent succession on the same field) has several benefits in arid agricultural areas. These benefits include enhanced soil fertility, soil erosion reduction, weed control, crop diversity, enhancement of soil health, and improvement of water use efficiency.

A study conducted by Al-Rumikhani (2002) in Saudi Arabia has confirmed that rotations of cereals and alfalfa (*Medicago sativa*) crops contributed to significant amelioration in soil hydrological characteristics and improved yield of the cultivated crops. In addition, an investigation performed by Jones and Singh (2000) in Syria confirmed that a barley (*Hordeum vulgare*)-legume rotation resulted in a substantial increase in crop yield. Furthermore, in several Egyptian agricultural areas, Ahlam et al. (2015) have revealed the advantages of crop rotation in enhancing the growth and development of roots within the cultivated soil.

4.7.2 Integrated Pest Management

Pathogens, weeds, and invertebrates result in large decrease in the area available for growth for many plants. As reported in Flood (2010), in the arid lands, the losses were estimated up to 30% (for several crops) (Birch et al. 2011). Integrated pest management (IPM) is defined as all the biological, cultural, and chemical practices that are applied to control insect pests in agricultural production (Pretty et al. 2006). In Table 4.6, we summarize some examples of these practices, especially for arid agricultural areas. A good pest management program could be ensured if all these practices (i.e. biological, cultural, and chemical practices) are applied (Birch et al. 2011). IPM encourages and supports natural pest control mechanisms (Bebber et al. 2013). Its major objective is the best use of all available technologies to manage pest problems safely. A balanced pest management program could be ensured if the following three major components are ensured: firstly, avoid pest build-up within the agricultural system; secondly, continuous monitoring of pest levels, and finally, adequate and timely intervention (Pretty et al. 2006).

4.7.3 Soil Resources Conservation

The major objective of soil resources conservation is to enhance the health of soils. Healthy soils are essential to ensure the productivity of agricultural fields and for conserving the key functions of ecosystems such as biological activities, enhancing water quality, providing micro- and macronutrients for crops, and ensuring C sequestration (Freidman et al. 2001; Jhariya et al. 2019a, b). Several characteristics of the cultivated soil (such as soil hydraulic characteristics) have significant effects on these key functions. Many of these soil properties are dynamic and have a high

Table 4.6 Integrated pest management practices (Modified, Pretty et al. 2006)

Practices	Actions
Cultural practices	**Plant appropriate crops in the field** (**e.g.** crops with low moisture needs for arid areas). **Plant at appropriate times** allows crop roots to develop before summer heat arrives. **Choose disease- and pest-resistant plant varieties.**
Biological practices	All pests, from weeds to insects, have natural enemies. Biological practices conserve, encourage, and support organisms that predate on and limit the growth of pests.
Chemical practices	Choosing the right chemicals in pest control. Boric acid is a powerful weapon against household pests. Hydramethylnon can also eliminate ants, cockroaches, crickets, and silverfish.

potential for alteration. In contrast, a few soil properties are inherent and more resistant to alteration. Several soil management actions can help us to improve soil health and to decrease the critical impacts of harsh weather conditions (Raj et al. 2019a, b, 2020). For example, the negative effects of dry and wet rainfall extremes can be reduced by increasing soil organic matter, which can enhance water infiltration and decrease nutrient losses during intense rainfall (Anwar et al. 2013). In addition, green infrastructure such as pipes and drains can reduce soil erosion and may enhance inherent soil properties. The major soil management options that can decrease the critical consequences of harsh weather conditions on soil health include:

- Yearly increase of vegetation cover (by adding plant residues) can significantly help to minimize the negative consequences of soil erosion (Derner et al. 2015);
- Continuous enhancing of soil organic matter (by adding organic amendments such as animal manure, compost, mulch, biochar, etc.) can significantly help to enhance several properties of soil (e.g. hydraulic properties). This also helps to decrease soil exposure to water and wind erosion (Shea 2014);
- Land leveling, subsurface drains, and perennial cropland use systems are useful to conserve the physical, chemical, and biological properties of soils. These practices help also to reduce crop damage from water ponding after a heavy rainfall event, and to control run-off without causing soil erosion. This leads to improved crop growth and more sustainable agricultural productivity (Ritzema 1994);
- Good management of soil before crop planting and good selection of planting dates can significantly help to avoid field works under extreme weather conditions such as wet conditions (Wolfe et al. 2011);
- Preventing soil compaction by equipment by avoiding work under wet conditions, reducing tillage, and use of appropriate implements contributes positively to conserving the physical properties of soil (DeJong-Hughes et al. 2001). Soil compaction can be reduced through better soil management (e.g. add adequate organic matter);
- Judicious use of grazing animals enhances soil biological characteristics (Bardgett et al. 1997). Grazing animals also have significant potential for enhancing soil structure, and increasing the capacity of soil organic C storage that could contribute to important increase in soil fertility;
- Installation of windbreaks in the costal arid agricultural areas, especially in those near the coastline, can act to significantly decrease wind erosion. Furthermore, colonization of coastal soil by vegetation whose roots bind sediment can enhance resistance to wind erosion (Lebbe et al. 2008);
- Management rates and timing of harvesting hay can be useful to enhance soil health (Anwar et al. 2013);
- Good and suitable application of soil amendments has a high potential to increase the productivity of several crops, especially under dry and saline conditions (Courtney and Harrington 2012; Al-Omran 1987; Hueso-González et al. 2014;

Alkhasha et al. 2019). Such increases in productivity are closely related to good soil water availability conditions within the cultivated soil;
- Suitable application of biochar enhances the conductive capacity of cultivated soils, especially under climate conditions of the arid agricultural lands (Villagra-Mendoza et al. 2017).

4.7.4 Water Resources Conservation

Water resources such as coastal aquifers and groundwater are vital to all agricultural activities in the arid lands. Agricultural activities under climate change can have critical effects on water quality (Howden et al. 2007; Ames and Dufour 2014). Therefore, it is urgent to identify specific agricultural practices that can help to maintain water quality and respond to extreme climate events. Several actions can help us to conserve water quality, including:

- Manage nutrients in agricultural fields: good nutrient management plans are required to ensure improved use of all sources of nutrients, especially under conditions of climate change. For example, the application of pesticides should be avoided unless there is an identified need. This leads to optimum protection of water resources (Howden et al. 2007);
- Install controlled drainage systems: installation of these controlled drainage systems is useful to avoid pollution of aquifers, especially pollution of shallow groundwaters (Haj-Amor et al. 2018);
- Strict manure management plans: for example, apply appropriate type of manure (cow manure) in appropriate quantities;
- Limit seawater intrusion in coastal arid areas: several practices can help to limit seawater intrusion. The major practices include continuous measurement of coastal groundwater levels, decreasing abstraction from shallow groundwaters, installing barriers to fluvial saltwater intrusion, and increasing sustainable aquifer recharge;
- Minimize various impacts of agricultural waste on surface and groundwater resources. For example, it is essential to respect the appropriate distance from any watercourse (e.g. 20 m) when applying fertilizers and organic wastes (Howden et al. 2007).

4.7.5 Irrigation Water Management

In the arid lands, adequate management of irrigation water (i.e. maintaining appropriate water salinity, irrigation frequency, and irrigation amounts) is critical for improving land and water productivity (Haj-Amor et al. 2018). Several actions can help to achieve successful management of irrigation water, including:

- Increased irrigation efficiency (especially efficiency of water application within irrigated fields) is important for ensuring sustainability of irrigation water resource (Haj-Amor et al. 2018);
- Under dry conditions and where soils have low infiltration rates, increased irrigation capacity, especially for high-value crops, is essential for carrying salts from soils to drainage networks (Haj-Amor et al. 2018);
- Saving water for further use during drought periods is important for ensuring the sustainability of irrigation water resource (Derner et al. 2015);
- Increasing the efficiency of irrigation methods, especially surface irrigation (Howden et al. 2007);
- Using new technology for subsurface irrigation (Derner et al. 2015).

4.7.6 Integrated Nutrient Management (INM)

The major objective of INM is to maximize the productivity of cultivated crops (FAO 1995). The components and the resources of INM coupled with the potential impacts on soil productivity are illustrated in Fig. 4.9.

There are several specific objectives of the INM, including enhancing soil characteristics and resistance to future climate change; enhancing soil health through improved conservation of soil characteristics; ensuring a positive nutrient balance in cultivated soils; increasing efficiency of nutrient use; recycling of organic waste, etc. (Gruhn et al. 2000). In the arid agricultural lands, INM has several benefits. For

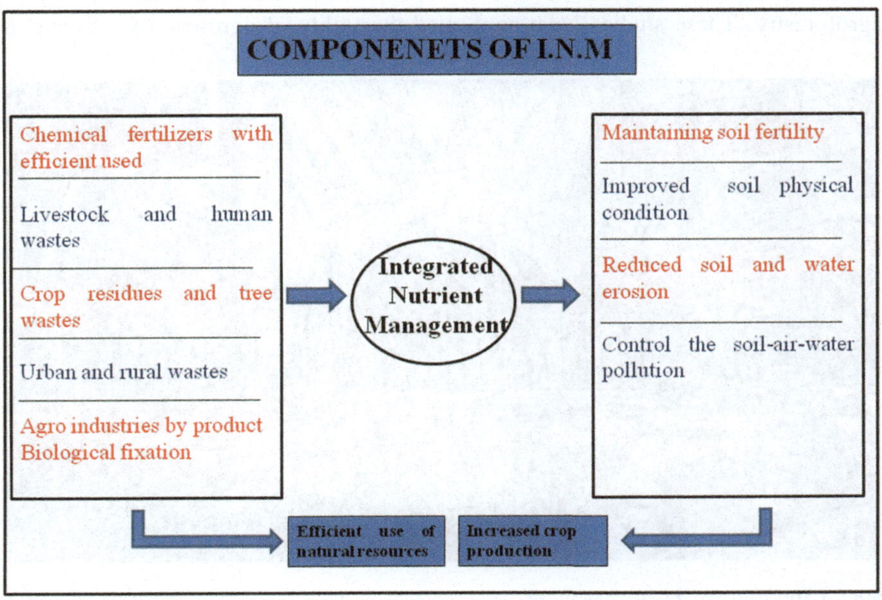

Fig. 4.9 Resources of INM and their role in soil productivity (Adapted, Gruhn et al. 2000)

example, in some Indian arid areas, Suresh (2005) has revealed the ability of INM (especially integrated use of organics and biofertilizers) to increase yields of several field crops (cotton, jowar, bajra, sunflower, etc.). This benefit was mainly attributed to the positive impacts of organics and biofertilizers on key soil properties (for examples: soil moisture, organic C content, and macronutrients). Furthermore, Chander et al. (2013) have revealed that organics and biofertilizers increased Indian crop yields by 6–40% through enhanced water use efficiency.

4.7.7 Agroforestry

Agroforestry is defined as the integration of trees, crops, and/or animal production on one field (Jones 2001). As the population of the arid countries increases, the need for more productive and sustainable land use becomes more urgent. Due to its multifunctional characteristics, agroforestry could be helpful to achieve this sustainable use (Singh and Jhariya 2016; Jhariya et al. 2015; Banerjee et al. 2020). The arid Tunisian oases are perfect examples of agroforestry systems where date palm trees and several other crops such as wheat (*Triticum aestivum* L.) are produced together (Fig. 4.10).

Agroforestry systems are natural resource management systems that diversify and sustain agricultural production and increase social, economic, and environmental benefits for farmers. They are flexible systems that could be applied at the scale of both small and large land holdings (Jones 2001). In recent years, many international studies (Alao and Shuaibu 2011; Nair 2017) have focused on the viability of agroforestry. These studies have confirmed the utility of agroforestry systems as

Fig. 4.10 Tunisian Oasis, South of Tunisia (source: https://lapresse.tn/36836/ecosystemes-oasiens-pour-une-gestion-durable/)

appropriate mechanisms for sustainable land management. Benefits of agroforestry systems include:

- Ensuring more favorable soil physical characteristics than monoculture systems as a positive impact of tree roots;
- Increasing soil water availability in arid agricultural lands;
- Reducing soil acidification and salinization processes due to the positive impacts of tree roots;
- Conserving soil fertility through continuous conservation of soil organic matter;
- Reducing insect pests and associated diseases;
- Reducing negative effects of soil erosion.

Overall, the major objective of agroforestry systems is not to maximize short-term profit, but to ensure long-term sustainable resource use (due to their easy adaptability). They have an important potential to improve food security.

4.7.8 Climate Smart Agricultural Practices

Agricultural performance is closely related to climate, and climate variability has impacts on agricultural outcomes (IPCC 2013). In the arid countries, a warmer climate has a high potential to reduce the yields of major cultivated crops such as wheat, barley, food legumes, corn, soybeans, etc. In 2005, 48% of economically active population in the arid countries depends on agricultural production for their living (World Bank 2008). A warmer climate will contribute to significant increase in prices for major cultivated crops (IFPRI 2009). Generally, we can summarize the adverse impacts of climate variability and climate change on agriculture as follows: significant decrease in crop yields; significant change in outcomes (especially prices and production); and important change in consumption.

In the context of these critical impacts, climate smart agriculture (CSA) is considered as a means to help us to achieve sustainable agriculture under changing climatic conditions. Based on the definition proposed by FAO (2012), CSA is a plan for developing agricultural strategies to ensure sustainable food security under climate change. CSA refers to all the agricultural strategies that can help farmers to develop their agricultural systems, with focus on agriculture productivity. CSA has three major objectives: (1) enhancing agricultural productivity while ensuring the sustainability of natural resources; (2) develop suitable strategies to enhance the climate resilience of agricultural systems; (3) decrease greenhouse gas emissions. The potential practices of CSA that can help to achieve these three objectives are illustrated in Fig. 4.11.

Finally, the major objective of CSA is not to maximize short-term profit, but rather to ensure genuinely sustainable agriculture, especially in the arid lands where natural resources (e.g. water resources) are scarce. CSA has an important role to play in improving food security in these lands.

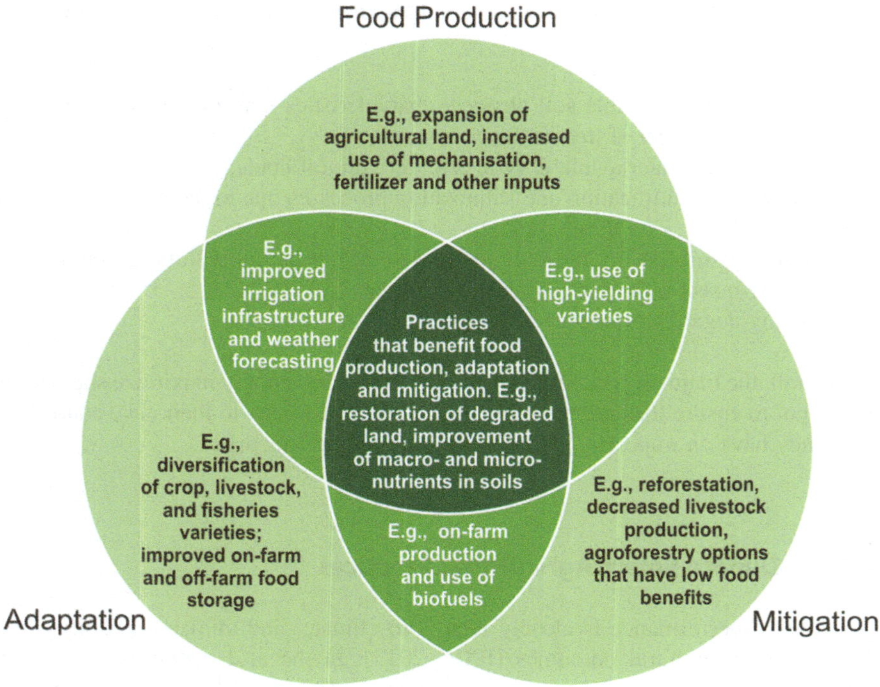

Fig. 4.11 Climate smart agricultural practices (FAO 2012)

4.8 Productivity Potential and Economy of Sustainable Intensification

Under arid climatic conditions, various technologies of sustainable intensification (e.g. smart irrigation technologies, information technologies, automatic weather stations, controlled drainage systems, etc.) are working towards improving productivity and boosting the economy (FAO 2011). Accordingly, over the past 20 years (1998–2018), significant economic growth is noted in many arid countries including India, Australia, the countries of North Africa, etc., as a consequence of agricultural productivity increasing (Jacobsen et al. 2012). As reported in IPCC (2019), it is noted that agricultural modernization, CSA, and perfect commercial investment allowed to a positive trend in agricultural policy and livelihoods. Accordantly, a continuous increase in economic and agricultural revenue is ensured in many arid countries such as Malawi, Tanzania, and Zambia that are experiencing large-scale land acquisition and increasing differentiation of rural wealth (data for the year 2019). Finally, it is so important to indicate that further private finance should be highly attributed to the small farmers and out-grower schemes to ensure sustainability of economic revenue achieved over the recent years.

4.9 Research and Development Towards Agricultural Productivity Under Arid Climate

Some manifestations of climate change in the arid lands (e.g. increasing air temperatures and moisture-deficits) will certainly have negative impacts on agricultural systems (IPCC 2013). Many of these systems are located in the arid and dry lands of the developing world, home to 50% of the world's populations, where significant declines in rainfall and severe droughts and floods continuously disrupt agricultural productivity (IPCC 2019).

Here, in the context of these critical impacts, we review a range of studies whose purpose has been to evaluate agricultural productivity under arid climate conditions. The objective of this evaluation is to point out the advanced practices that could be helpful to ensure sustainable agricultural productivity in the arid and dry lands. Many research works (Bahrun et al. 2002; Siddique et al. 2003; Debaeke and Aboudrare 2004; Liu et al. 2005; Qadir et al. 2007; Hariadi et al. 2011; IPCC 2019) have focused on the factors that can decrease agricultural productivity in arid agricultural systems. Based on the findings of these work, we can summarize the major factors as follows: (1) in appropriate management of natural resources such as soil and water resources; (2) unsuitable control of crop properties such as crop growth; (3) natural factors and climate change impacts that result insignificant decreases in soil fertility; (4) absence of irrigation facilities and water saving technologies in some agricultural fields; and (5) inappropriate and inadequate agricultural management techniques and technologies.

Furthermore, IPCC (2019) summarized all the current conditions that may have negative effects on agricultural productivity. These conditions include unpredictable onset of rainy seasons, significant declines in soil fertility and other soil properties due to unsuitable soil management within agricultural fields, delayed planting, crop pests and declines in arable areas resulting from a number of factors (soil pollution, soil salinity, etc.). Jacobsen et al. (2012) have proposed an integrated crop and resource management approach to address these critical environmental problems (Fig. 4.12). For example, integrated water resources management includes efficient irrigation, use of treated waste water, and suitable irrigation practices when saline water is used for irrigation. Integrated crop management includes enhancing crop genetics, selection of appropriate crop types for specific contexts (e.g. select suitable crops under saline conditions), and continuous control of crop growth, especially under certain harsh climatic conditions (Barba de la Rosa et al. 2009; Ben Hassine and Lutts 2010). All these management options can be critical in ensuring sustainable agricultural productivity in the arid and dry lands. Over recent years, significant research efforts have focused on identifying the minor crops that have a high potential for production, use, and market sale (e.g. Andean lupin). Some of these crops are listed in Jacobsen and Mujica (2009) and Jacobsen et al. (2012).

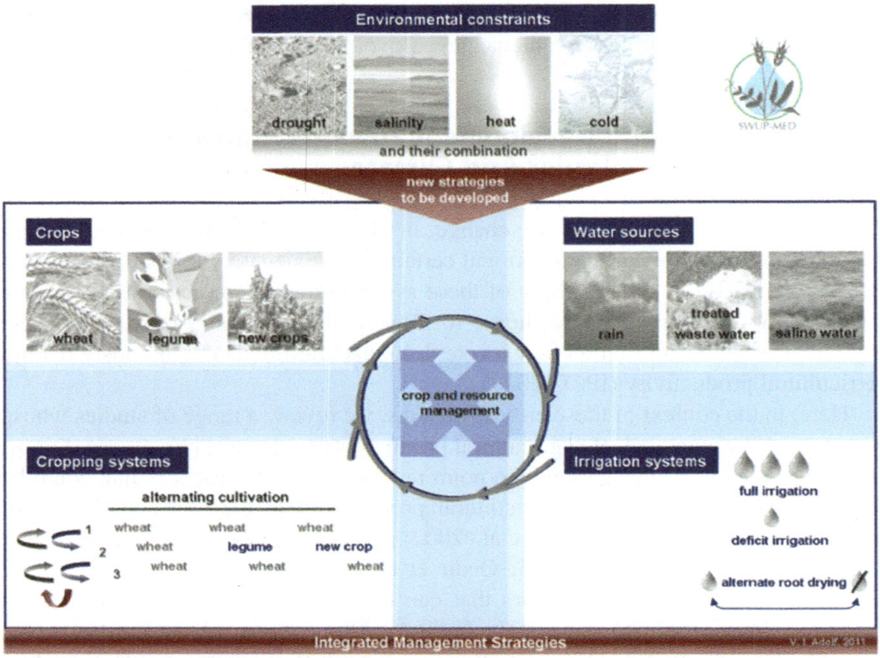

Fig. 4.12 A schematic representation of integrated crop and resource management (Adapted from Jacobsen et al. 2012)

Based on the findings of several research works (Jacobsen and Mujica 2009; Ben Hassine and Lutts 2010; Jacobsen et al. 2012), it can be concluded that many advanced practices could be deployed to ensure sustainable agricultural productivity in the arid and dry lands. These works have revealed that the agricultural productivity could be enhanced by 21 to 33% if the following major productive and sustainable actions are practiced:

- Enhancing crop genetics, including biotechnology and new gene-editing biotechnologies;
- Suitable management of natural resources, including efficient water use (rainfall and irrigation water);
- Transitions to CSA as an optimal strategy for achieving sustainable agriculture under climate change.

4.10 Policy Framework for Agricultural Sustainable Intensification in the Arid Lands

The development of adequate policy frameworks will be instrumental in ensuring sustainable agriculture that delivers continuous increases in food production while sustaining the natural properties of the ecosystems in which agricultural systems are embedded (Firbank et al. 2011). The FAO (2013) reported that sustainable agricultural intensification policies need to focus on integrated production systems in which there is close collaboration with smallholders. In addition, there is an urgent need to spread awareness of CSA measures that enhance crop yields, decrease fertilizer use, and conserve soil and water resources (FAO 2011). Over the past few years, some improvements in environmental services resulting from sustainable agricultural intensification policies have been noted in many arid lands (Pretty et al. 2007). However, further policy options should be developed in order to respond to the host of environmental and agricultural challenges faced in the arid lands. Sustainable agricultural intensification can only be assured through the adoption of a suite of key policy options and technologies. These include:

- Maximizing sustainability and productivity of agricultural systems, with a focus on how this can be achieved under the climate conditions existing in arid lands (Pretty et al. 2007);
- Supporting techniques and technologies that encourage farmers to practice appropriate management of crops, animals, and land (i.e. cultivated land). This management should be pursued alongside agronomic practices that can ensure the sustainability of ecological systems (FAO 2013);
- Enhancing the resilience of production systems under changing climatic conditions. This can deliver higher agricultural productivity and ensure future food security (FAO 2011);
- Offering opportunities and practices to encourage farmers' groups to develop and propose interventions and options for increasing agricultural productivity (Pretty et al. 2007);
- Ensuring the sustainability of ecosystem services through multiple and diverse options (Bulte et al. 2008).

Despite the identification and ongoing implementation of the above policy options, with some demonstrable positive outcomes, there is still much work to be done (especially R&D) to be performed to ensure that agricultural systems in the arid lands increase productivity without or with minimal negative impacts on ecosystems. Furthermore, it is observed that sustainable agricultural intensification may be affected by some factors (e.g. consequences of climate change such as the emergence of new pests and diseases in specific locations) that have not been taken into account by research conducted in recent years. Future research will need to address these factors in order to develop appropriate policies for sustainable agricultural intensification in the arid lands.

4.11 Conclusions

Climate change and sustainable agricultural intensification are closely interlinked concepts. In recent years, in order to ensure higher agricultural productivity while decreasing the critical impacts of climate change and weather extremes on soil and water resources, sustainable agricultural intensification practices have been adopted in a number of countries. Under changing climatic conditions, the present chapter has systematically summarized the main sustainable agricultural intensification practices in the arid lands. Practices taken to improve agricultural productivity and help farmers adapt to climate change have multiple ecological, environmental, and economic benefits. Identifying key ways to formulate appropriate policies for sustainable agriculture is a key priority in the arid lands. Research works such as the current chapter have a fundamental role to play in developing the possible agricultural practices that can allow sustainable agriculture under climate change. There is an urgent need to support farmers in the arid lands to enable them to adopt measures to enhance both productivity and sustainability (e.g. suitable management of natural resources, CSA, etc.).

4.12 Future Perspectives

This chapter provides us with the following perspectives on the future:

- The achievement of food security and sustainable agricultural intensification in the arid lands requires urgent action to reduce greenhouse gases emissions, especially from electricity, transport, and agriculture. Accordingly, more future research on how to achieve such reductions is required. This requires urgent global-scale mitigation. All countries (worldwide) have a role to pay here;
- In order to enhance food security in the arid lands under changing climatic conditions, the following practices are mandatory: sustainable use of natural resources, more efficient use of resources generally, etc. More productive and sustainable agriculture requires significant changes in the management of land, water, soil nutrients, and genetic resources, based on CSA techniques;
- More efficient use of natural resources in the arid lands can be achieved through formulation of resource use efficiency policies and adherence to these policies by all producers. These policies need to address the critical challenges of resource scarcity, environmental impacts, and climate change. Consequently, more research on the development of various policies that can help to achieve efficient use of natural resources is urgently needed;
- Even though significant research into various aspects of climate change in the arid lands has been carried out, few field-based works on the effects of climate change on crop diseases and insect pests have been performed. Accordingly, more fieldwork based research needs to be carried out in this area.

Acknowledgments The authors sincerely thank Dr. Nick Brooks (Director, Garama 3C Ltd., UK) for his helpful comments and language assistance.

References

Ahlam M, El-Nagd WM, Youssef M (2015) Seasonal variations of population density of root knot nematode, Meloidogyne incognita as affected by different cropping sequences. Scientia 10 (1):35–37

Al Charaabi Y, Al-Yahyai S (2013) Projection of future changes in rainfall and temperature patterns in Oman. J Earth Sci Climate Change 4:154

Alao JS, Shuaibu RB (2011) Agroforestry practices and preferential agroforestry trees among farmers in Lafia Local Government Area, Nasarawa State, Nigeria. Waste Manage Biores Technol 1(2):12–20

Alkhasha A, Alomran A, Louki II (2019) Impact of deficit irrigation and addition of biochar and polymer on soil salinity and tomato productivity. Can J Soil Sci 99:380–394

Al-Omran AM (1987) Evaluation of some irrigation water in central region of Saudi Arabia. J CollAgric King Saud Univ 9(2):363–369

Al-Rumikhani YA (2002) Effect of crop sequence, soil sample location and depth on soil water holding capacity under center pivot irrigation. Agric Water Manag 55:93–104

Altieri MA, Nicholls CI, Henao A (2015) Agroecology and the design of climate change-resilient farming systems. Agron Sustain Dev 35:869–890

Ames GK, Dufour R (2014) Climate change and perennial fruit and nut production: investing in resilience in uncertain times. In: ATTRA sustainable agriculture. National Center for Appropriate Technology ATTRA, Butte

Anand A, Khetarpal S (2015) Impact of climate change on agricultural productivity. In: Bahadur B, Venkat Rajam M, Sahijram L, Krishnamurthy K (eds) Plant biology and biotechnology. Springer, New Delhi

Anwar M, Liu D, Macadam I, Kelly G (2013) Adapting agriculture to climate change: a review. Theor Appl Climtol 113(1–2):225–245

Bahrun A, Jensen CR, Asch F, Mogensen VO (2002) Drought-induced changes in xylem pH, ionic composition, and ABA concentration act as early signals in field-grown maize (*Zea mays* L.). J Exp Bot 53:251–263

Banerjee A, Jhariya MK, Yadav DK, Raj A (2020) Environmental and sustainable development through forestry and other resources. Apple Academic Press Inc., CRC Press- a Taylor and Francis Group, US & Canada. ISBN: 9781771888110. p 400. https://doi.org/10.1201/9780429276026

Barba de la Rosa AP, Fomsgaard IS, Laursen B, Mortensen AG, Olvera-Martinez L, Silva-Sanchez C, Mendoza-Herrera A, Gonzalez-Castaneda J, De Leon-Rodnguez A (2009) Amaranth (*Amaranthus hypochondriacus*) as an alternative crop for sustainable food production. Phenolic acids and flavonoids with potential impact on its nutraceutical quality. J Cereal Sci 49:117–121

Bardgett RD, Leemans DK, Cook R, Hobbs PJ (1997) Seasonality of the soil biota of grazed and ungrazed hill grasslands. Soil Biol Biochem 29:1285–1294

Bebber DP, Ramotowski MAT, Gurr SJ (2013) Crop pests and pathogens move polewards in a warming world. Nat Clim Change 3:985–988

Ben Hassine A, Lutts S (2010) Differential responses of saltbush *Atriplex halimus* L. exposed to salinity and water stress in relation to senescing hormones abscisic acid and ethylene. J Plant Physiol 167:1448–1456

Bergot M, Cloppet E, Pérarnaud V, Déqué M, Marçais B, Desprez-Loustau ML (2004) Simulation of potential range expansion of oak disease caused by *Phytophthora cinnamomi* under climate change. Glob Change Biol 10:1–14

Bhagat S, Birah A, Chattopadhyay N, Chattopadhyay C (2016) Climate change: pest incidence in agricultural crops. In: Chattopadhyay N, Chattopadhyay C (eds) Dynamics of crop protection and climate change. Studera Press

Birch ANE, Begg GS, Squire GR (2011) How agro-ecological research helps to address food security issues under new IPM and pesticide reduction policies for global crop production systems. J Exp Biol 62:3251–3261

Bulte E, Boone RB, Stringer R, Thornton PK (2008) Elephants or onions? Paying for nature in Amboseli, Kenya. Environ Dev Econ 13:395–414

Burney JA, Davis SJ, Lobell DB (2010) Greenhouse gas mitigation by agricultural intensification. Proc Natl Acad Sci U S A 107:12052–12057

Chakraborty S, Datta S (2003) How will plant pathogens adapt to host plant resistance at elevated CO2 under a changing climate? New Pathologist 159:733–742

Chander G, Wani SP, Sahrawat KL, Kamdi PJ, Pal CK, Pal DK, Mathur TP (2013) Balanced and integrated nutrient management for enhanced and economic food production: case study from rainfed semi-arid tropics in India. Arch Agron Soil Sci 59(12):1643–1658

Cheong YL, Burkart K, Leitão PJ, Lakes T (2013) Assessing weather effects on dengue disease in Malaysia. Int J Environ Res Publ Health 10:6319–6334

Clarke D, Smethurst J (2010) Effects of climate change on cycles of wetting and drying in engineered clay slopes in England. Q J Eng Geol Hydrogeol 43(4):473–486

Costello A, Abbas M, Allen A, Bell S, Bellamy R, Friel S, Groce N, Johnson A, Kett M, Lee M, Levy C, Maslin M, McCoy D, McGuire B, Montgomery H, Napier D, Pagel C, Patel J, Oliveira JAP, Redclift N, Rees H, Rogger D, Scott J, Stephenson J, Twigg J, Wolff J, Patterson C (2009) Managing the health effects of climate change. Lancet 373:1693–1734

Courtney R, Harrington T (2012) Growth and nutrition of *Holcus lanatus* in bauxite residue amended with combinations of spent mushroom compost and gypsum. Land Degrad Dev 23:144–149

Debaeke P, Aboudrare A (2004) Adaptation of crop management to water-limited environments. Eur J Agron 21:433–446

DeJong-Hughes JM, Swan JB, Moncrief JF, Voorhee WB (2001) Soil compaction: causes, effects and control (Revision). University of Minnesota Extension Service BU-3115-E

Derner J, Joyce L, Guerrero R, Steele R (2015) Northern plains regional climate hub assessment of climate change vulnerability and adaptation and mitigation strategies. In: Anderson T (ed) United States Department of Agriculture. 57 p

Diaz JH (2007) The influence of global warming on natural disasters and their public health outcomes. Am J Disaster Med 2(1):33–42

Earman E, Dettinger M (2011) Potential impacts of climate change on groundwater resources – a global review. J Water Clim Change 2(4):213–229

Ellsworth DS, Reich PB, Naumburg ES, Koch GW, Kubiske ME, Smith SD (2004) Photosynthesis, carboxylation and leaf nitrogen responses of 16 species to elevated pCO2 across four free-air CO_2 enrichment experiments in forest, grassland and desert. Glob Change Biol 10:2121–2138

Epstein PR (2005) Climate change and human health. N Engl J Med 353(14):1433–1436

FAO (1995) Integrated plant nutrition system. FAO Fertilizer and Plant Nutrition Bulletin N°. 12. Food and Agriculture Organization of the United Nations, Rome, Italy, 426 pp

FAO (2004) The ethics of sustainable agricultural intensification. Ethics series. Food and Agriculture Organization of the United Nations, Rome, Italy, pp 3–5

FAO (2011) Save and grow. A policymaker's guide to the sustainable intensification of smallholder crop production. Food and Agriculture Organization, Rome, Italy, 102p

FAO (2012) The Sahel crisis: executive brief. March. Food and Agriculture Organization, Rome, Italy

FAO (2013) Climate-smart agriculture sourcebook. Food and Agriculture Organization, Rome, Italy

Firbank L, Bradbury RB, McCracken DI, Stoate C (2011) Delivering multiple ecosystem services from enclosed farmland in the UK. Agric Ecosyst Environ 166:65–75

Flood J (2010) The importance of plant health to food security. Food Secur 2:215–231
Freidman D, Hubbs M, Tugel A, Seybold C, Sucik M (2001) Guidelines for soil quality assessment in conservation planning. US Government Printing Office, Washington, DC
Garnett T, Appleby MC, Balmford A (2013) Sustainable intensification in agriculture: premises and policies. Science 341:33–34
Gautam HR, Sharma HL (2012) Environmental degradation, climate change and effect on agriculture. J Kurukshetra 60:3–5
Ge X, Li T, Zhang S, Peng M (2010) What causes the extremely heavy rainfall in Taiwan during Typhoon Morakot (2009)? Atmospheric Sci Lett 11(1):46–50
Godfray HCJ (2015) The debate over sustainable intensification. Food Secur 7:199–208
Gruhn P, Goletti F, Yudelman M (2000) Integrated nutrient management, soil fertility and sustainable agriculture: current issues and future challenges, IFRPI 2020 Vision Brief
Haj-Amor Z, Ritzema H, Hashemi H, Bouri S (2018) Surface irrigation performance of date palms under water scarcity in arid irrigated lands. Arab J Geosci 11:27
Halvorson AD, Wienhold BJ, Al B (2002) Tillage, nitrogen, and cropping system effects on soil carbon sequestration. Soil Sci Soc Am J 66:906–912
Hariadi Y, Marandon K, Tian Y, Jacobsen SE, Shabala S (2011) Ionic and osmotic relations in quinoa (Chenopodium quinoa Willd.) plants grown at various salinity levels. J Exp Bot 62:185–193
Hazell PBR (2008) An assessment of the impact of agricultural research in South Asia since the green revolution. Science Council Secretariat, Rome, Italy
Held IM, Soden BJ (2006) Robust responses of the hydrological cycle to global warming. J Clim 19 (21):5686–5699
Holzkämper A, Klein T, Seppelt R, Fuhrer J (2015) Assessing the propagation of uncertainties in multi-objective optimization for agro-ecosystem adaptation to climate change. Environ Model Softw 66:27–35
Howden SM, Soussana JF, Tubiello FN, Chhetri N, Dunlop M, Meinke H (2007) Adapting agriculture to climate change. PNAS 104(50):19691–19696
Huang J, Li Y, Fu C, Chen F, Fu Q, Dai A, Shinoda M, Ma Z, Guo W, Li Z, Zhang L, Liu Y, Yu H, He Y, Xie Y, Guan X, Ji M, Lin L, Wang S, Yan H, Wang G (2017) Dryland climate change: recent progress and challenges. Rev Geophys 55:719–778
Hueso-González P, Martínez-Murillo JF, Ruiz Sinoga JD (2014) The impact of organic amendments on forest soil properties under Mediterranean climatic conditions. Land Degrad Dev 25:604–612
IFPRI (2009) Climate change: impact on agriculture and costs of adaptation. International Food Policy Research Institute, Washington
IPCC (2013) Summary for policymakers. In: Stocker TF, Qin D, Plattner GK, Tignor M, Allen SK, Boschung J, Nauels A, Xia Y, Bex V, Midgley PM (eds) Climate Change 2013: the physical science basis. Contribution of Working Group to the Fifth Assessment Report of the Intergovernmental Panel on Climate Change. Cambridge University Press, Cambridge
IPCC (2019) Climate Change and Land: an IPCC special report on climate change, desertification, land degradation, sustainable land management, food security, and greenhouse gas fluxes in terrestrial ecosystems - Summary for Policymakers
Jacobsen SE, Mujica A (2009) Geographical distribution of the Andean lupin (*Lupinus mutabilis* Sweet). Plant Genetic Res Newslett 155:1–8
Jacobsen SE, Jensen CR, Liu F (2012) Improving crop production in the arid Mediterranean climate. Field Crops Res 128:34–47
Jain SK, Kumar V (2012) Trend analysis of rainfall and temperature data for India. Curr Sci 102 (1):37–49
Jhariya MK, Bargali SS, Raj A (2015) Possibilities and perspectives of agroforestry in Chhattisgarh. pp 237–257. In: Precious forests-precious earth, Edited by Miodrag Zlatic (Ed.). ISBN: 978-953-51-2175-6, 286 pages, InTech, Croatia, Europe, https://doi.org/10.5772/60841

Jhariya MK, Banerjee A, Meena RS, Yadav DK (2019a) Sustainable agriculture, forest and environmental management. Springer Nature Singapore Pte Ltd., 152 Beach Road, #21-01/04 Gateway East, Singapore 189721, Singapore. eISBN: 978-981-13-6830-1, Hardcover ISBN: 978-981-13-6829-5. p 606. https://doi.org/10.1007/978-981-13-6830-1

Jhariya MK, Yadav DK, Banerjee A (2019b) Agroforestry and climate change: issues and challenges. Apple Academic Press Inc., CRC Press- a Taylor and Francis Group, US & Canada. ISBN: 978-1-77188-790-8 (Hardcover), 978-0-42957-274-8 (E-book). p 335. https://doi.org/10.1201/9780429057274

Jones CM (2001) Analog forestry as a conservation and development approach: lessons learned from the International Analogue Forestry Network

Jones MJ, Singh M (2000) Long-term yield patterns in barley-based cropping systems in Northern Syria. 1. Comparison of rotations. J Agric Sci 135:223–236

Khan N, Jhariya MK, Yadav DK, Banerjee A (2020a) Herbaceous dynamics and CO_2 mitigation in an urban setup- A case study from Chhattisgarh, India. Environ Sci Pollut Res 27 (3):2881–2897. https://doi.org/10.1007/s11356-019-07182-8

Khan N, Jhariya MK, Yadav DK, Banerjee A (2020b) Structure, diversity and ecological function of shrub species in an urban setup of Sarguja, Chhattisgarh, India. Environ Sci Pollut Res 27 (5):5418–5432. https://doi.org/10.1007/s11356-019-07172-w

Khan SA, Kumar S, Hussain MJ, Kalra N (2009) Climate change, climate variability and Indian agriculture: impacts vulnerability and adaptation strategies. In: Singh SN (ed) Climate change and crops, environmental science and engineering. C-Springer, Berlin

Kumar R, Gautam HR (2014) Climate change and its impact on agricultural productivity in India. J Climatol Weather Forecast 2:109

Kumar S, Meena RS, Jhariya MK (2020) Resources Use Efficiency in Agriculture. Springer Nature Singapore Pte Ltd., 152 Beach Road, #21-01/04 Gateway East, Singapore 189721, Singapore. eBook ISBN: 978-981-15-6953-1, Hardcover: 978-981-15-6952-4. p 760. https://doi.org/10.1007/978-981-15-6953-1

Lastuvka Z (2009) Climate change and its possible influence on the occurrence and importance of insect Pests. Plant Protect Sci 45:S53–S62

Lebbe L, Van Meir N, Viaene P (2008) Potential implications of sea-level rise for Belgium. J Coastal Res 24:358–366

Leung YK, Yeung KH, Ginn EWL, Leung WM (2004) Global climate change: cause of climate change by human activities. Hong Kong Observatory N° 107

Liu F, Jensen CR, Shahanzari A, Andersen MN, Jacobsen SE (2005) ABA regulated stomatal control and photosynthetic water use efficiency of potato (*Solanum tuberosum* L.) during progressive soil drying. Plant Sci 168:831–836

Malash NM, Flowers TJ, Ragab R (2008) Effect of irrigation methods, management and salinity of irrigation water on tomato yield, soil moisture and salinity distribution. Irrig Sci 26:313–323

Mansouri Daneshvar MR, Ebrahimi M, Nejadsoleymani H (2019) An overview of climate change in Iran: facts and statistics. Environ Syst Res 8:7

Matros A, Amme S, Kettig B, Buck-Sorlin GH, Sonnewald U, Mock HP (2006) Growth at elevated CO2 concentrations leads to modified profiles of secondary metabolites in Tobacco cv. Samsun NN and to increased resistance against Infection with potato virus Y. Plant Cell Environ 29:126–137

Matson PA, Parton WJ, Power AG (1997) Agricultural intensification and ecosystem properties. Science 277:504–509

McMichael AJ (2013) Globalization, climate change, and human health. New England J Med 368:1335–1343

Medlyn BE, Dreyer E, Ellsworth DE, Forstreuter M, Harley PC, Kirschbaum MUF, LeRoux X, Loustau D, Montpied P, Strassemeyer J, Walcroft A, Wang K, Loustau D (2002) Temperature response of parameters of a biochemically based model of photosynthesis. II. A review of experimental data. Plant Cell Environ 25:1167–1179

Meena RS, Lal R (2018) Legumes for Soil Health and Sustainable Management. Springer Singapore, Singapore. pp. 541. ISBN 978-981-13-0253-4 (eBook), ISBN: 978-981-13-0252-7 (Hardcover). https://doi.org/10.1007/978-981-13-0253-4_10

Meena RS, Kumar V, Yadav GS, Mitran T (2018) Response and interaction of *Bradyrhizobium japonicum* and Arbuscular mycorrhizal fungi in the soybean rhizosphere: a review. Plant Growth Regul 84:207–223

Meena RS, Kumar S, Datta R, Lal R, Vijaykumar V, Brtnicky M, Sharma MP, Yadav GS, Jhariya MK, Jangir CK, Pathan SI, Dokulilova T, Pecina V, Marfo TD (2020) Impact of agrochemicals on soil microbiota and management: a review. Land (MDPI) 9(2):34. https://doi.org/10.3390/land9020034

Meena RS, Lal R, Yadav GS (2020a) Long term impacts of topsoil depth and amendments on soil physical and hydrological properties of an Alfisol in Central Ohio, USA. Geoderma 363:1141164

Meena RS, Lal R, Yadav GS (2020b) Long-term impact of topsoil depth and amendments on carbon and nitrogen budgets in the surface layer of an Alfisol in Central Ohio. Catena 194:104752

Mukherjee S, Mishra A, Trenberth KE (2018) Climate change and drought: a perspective on drought indices. Current Climate Change Reports. https://doi.org/10.1007/s40641-018-0098-x

Nair PKR (2017) Managed multi-strata tree + crop systems: An agroecological marvel. Front Environ Sci 5:88

Peterson GA, Halvorson AD, Havlin JL, Jones OR, Lyon DG, Tanaka DL (1998) Reduced tillage and increasing cropping intensity in the Great Plains conserves soil C. Soil Tillage Res 47:207–218

Pirard P, Vandentorren S, Pascal M, Laaidi K, Tertre AL, Cassadou S, Ledrans M (2005) Summary of the mortality impact assessment of the 2003 heat wave in France. Euro Surveill 10 (7):153–156

Pretty J (2008) Agricultural sustainability: concepts, principles and evidence. Philos Trans Biol Sci 363:447–465

Pretty J, Bharucha ZP (2014) Sustainable intensification in agricultural systems. Ann Bot 114:1571

Pretty J, Noble AD, Bossio D, Dixon J, Hine RE, Penning de Vries FWT, Morison JIL (2006) Resource-conserving agriculture increases yields in developing countries. Environ Sci Technol 3:24–43

Pretty J, Noble AD, Bossio D, Dixon J, Hine RE, Penning de Vries FWT, Morison JIL (2007) Resource-conserving agriculture works, despite the scepticism. Environ Sci Technol 40:1114–1119

Pretty JN (1997) The sustainable intensification of agriculture. In: Natural resources forum. Blackwell Publishing Ltd., Oxford

Putnam AE, Broecker WS (2017) Human-induced changes in the distribution of rainfall. Sci Adv 3: e1600871

Qadir M, Sharma BR, Bruggeman A, Choukr-Allah R, Karajeh F (2007) Nonconventional water resources and opportunities for water augmentation to achieve food security in water scarce countries. Agric Water Manage 87:2–22

Rahmstorf S (2012) Modeling sea level rise. Nat Educ Knowl 3(10):4

Rahn E, Liebig T, Ghazoul J (2018) Opportunities for sustainable intensification of coffee agro-ecosystems along an altitudinal gradient on Mt. Elgon, Uganda. Agric Ecosyst Environ 263:31–40

Raj A, Jhariya MK, Yadav DK, Banerjee A, Meena RS (2019a) Agroforestry: a holistic approach for agricultural sustainability. In: MK Jhariya, A Banerjee, RS Meena, DK Yadav (eds) Sustainable agriculture, forest and environmental management. Springer Nature Singapore Pte Ltd, pp 101–131. 152 Beach Road, #21-01/04 Gateway East, Singapore 189721, Singapore. eISBN: 978-981-13-6830-1, Hardcover ISBN: 978-981-13-6829-5. pp 606. https://doi.org/10.1007/978-981-13-6830-1

Raj A, Jhariya MK, Banerjee A, Yadav DK, Meena RS (2019b) Soil for sustainable environment and ecosystems management. In: MK Jhariya, A Banerjee, RS Meena, DK Yadav (eds) Sustainable agriculture, forest and environmental management. Springer Nature Singapore Pte Ltd., pp 189–221. 152 Beach Road, #21-01/04 Gateway East, Singapore 189721, Singapore. eISBN: 978-981-13-6830-1, Hardcover ISBN: 978-981-13-6829-5. pp 606 https://doi.org/10.1007/978-981-13-6830-1

Raj A, Jhariya MK, Yadav DK, Banerjee A (2020) Climate change and agroforestry systems: adaptation and mitigation strategies. Apple Academic Press Inc., CRC Press- a Taylor and Francis Group, US & Canada. ISBN: 9781771888226. pp 383. https://doi.org/10.1201/9780429286759

Rao MS, Manimanjari D, Aruna D, Desai S, Rao VUM, Maheswari M (2014) Prediction of Number of Generations of Helicoverpa armigera (Hub.) on Pigeonpea in India during Climate Change Period, in Compendium-cum-Abstract, International Symposium on New-Dimensions in Agrometeorology for Sustainable Agriculture, GBPUAT, Pantnagar (UA), Association Agrometeorologists

Ritzema HP (1994) Subsurface flow to drains. In: Ritzema HP (ed) Drainage principles and applications, 2nd edn. ILRI Publication 16, Wageningen, pp 263–304

Roco L, Bravo-Ureta B, Engler A, Jara-Rojas R (2017) The impact of climatic change adaptation on agricultural productivity in central Chile: a stochastic production frontier approach. Sustainability 9:1648

Ruerd R, Lee D (2000) Combining internal and external inputs for sustainable intensification. International Food Policy Research Institute (IFPRI), Washington, DC, pp 1–2

Salvadori G, De Michele C (2004) Frequency analysis via copulas: theoretical aspects and applications to hydrological events. Water Resour Res 40

Scherer LA, Verburg PH, Schulp CJE (2018) Opportunities for sustainable intensification in European agriculture. Glob Environ Chang 48:43–55

Schiefer J, Lair GJ, Blum WEH (2016) Potential and limits of land and soil for sustainable intensification of European agriculture. Agric Ecosyst Environ 230:283–293

Senker P (2011) Foresight: the future of food and farming, final project report. Prometheus 29:309–313

Sharkey P (2007) Survival and death in New Orleans: an empirical look at the human impact of Katrina. J Black Stud 37(4):482–501

Shea EC (2014) Adaptive management: the cornerstone of climate-smart agriculture. J Soil Water Conserv 69(6):198A–199A

Siddique KHM, Loss SP, Thomson BD (2003) Cool season grain legumes in dryland Mediterranean environments of Western Australia: significance of early flowering. Management of Agricultural Drought, Agronomic and Genetic Options. ICRISAT, FAO, pp 151–162

Singh NR, Jhariya MK (2016) Agroforestry and agrihorticulture for higher income and resource conservation. In: Narain S, Rawat SK (eds) Innovative technology for sustainable agriculture development. Biotech Books, New Delhi, India, pp. 125–145, ISBN: 978-81-7622-375-1.

Skelly C, Weinstein P (2003) Pathogen survival trajectories: an eco-environmental approach to the modelling of human campylobacteriosis ecology. Environ Health Perspect 111(1):19–28

Stanley F, Farrant B (2015) Climate change and children's health: a commentary. Children 2:412–423

Strand LB, Tong S, Aird R, McRae D (2010) Vulnerability of eco-environmental health to climate change: the views of government stakeholders and other specialists in Queensland, Australia. BMC Public Health 10:441

Suhardiman D, Giordano M, Leebouapao L (2016) Farmers' strategies as building block for rethinking sustainable intensification. Agric Hum Values 33:563–574

Suresh S (2005) Integrated nutrient management in vertisols of Southern zone of Tamil Nadu - a review. Agric Rev 26:67–72

Tilman D, Cassman KG, Matson PA (2002) Agricultural sustainability and intensive production practices. Nature 418:671–677

Tilman D, Balzer C, Hill J (2011) Global food demand and the sustainable intensification of agriculture. Proc Natl Acad Sci U S A 108:20260–20264

Trenberth KE, Fasullo J, Mackaro J (2011) Atmospheric Moisture transports from Ocean to land and global energy flows in reanalyses. J Clim 24:4907–4924

Van Loon AF, Van Lanen HAJ (2013) Making the distinction between water scarcity and drought using an observation-modeling framework. Water Resour Res 49:1483–1502

Vermeer M, Rahmstorf S (2009) Global sea level linked to global temperature. Proc Natl Acad Sci U S A 106:21527–21532

Villagra-Mendoza K, Ortiz-Malavassi E, Otterpohl R (2017) Role of charcoal addition on infiltration processes and soil water content characteristics of a sandy loam soil. Agric Eng Int CIGR J 19:9–15

Wang X (2007) Effects of species richness and elevated carbon dioxide on biomass accumulation: a synthesis using meta-analysis. Oecologia 152:595–605

Wezel A, Soboksa G, Mcclelland S (2015) The blurred boundaries of ecological, sustainable, and agroecological intensification: a review. Agron Sustain Dev 35:1283–1295

WHO (2009) Protecting health from climate change: global research priorities. World Health Organization

Wolfe D, Beem-Miller J, Chambliss L, Chatrchyan A, Menninger H (2011) In: Extension CC (ed) Farming success in an uncertain climate. Cornell University, Ithaca

World Bank (2008) World Development Report 2008: agriculture for development. The World Bank, Washington, DC

Xie H, Huang Y, Chen Q, Zhang Y, Wu Q (2019) Prospects for agricultural sustainable intensification: a review of research. Land 8:157

Xu J, Shrestha AB, Vaidya R, Eriksson M, Hewitt K (2007) The Melting Himalayas-Regional Challenges and Local Impacts of Climate Change on Mountain Ecosystems and Livelihoods. ICIMOD Technical Paper. International Centre for Integrated Mountain Development (ICIMOD), Kathmandu, Nepal

Zhu XG, Long SP, Ort DR (2010) Improving photosynthetic efficiency for greater yield. Annu Rev Plant Biol 61:235–261

Ecological Intensification for Sustainable Development

5

Abhishek Raj, Manoj Kumar Jhariya, Nahid Khan, Arnab Banerjee, and Ram Swaroop Meena

Abstract

The extent and quantity of natural resources (NRs) are going to degrade day by day due to overexploitation, misuse, unscientific management and some other anthropogenic deleterious activity in addition to climate change. NRs are nature's properties that not only sustain life but also maintain ecosystem structure and its services to humankinds. Resources like agriculture, forest, animals, soils and water are global treasure and their extent of utilization must be in optimum, i.e. without overlooking the environment. Agriculture, forestry, animals are integrated unit and linked with each other that deliver various multifarious tangible and intangible products which can be modified by varying level of resources like soil, water and other environmental factors that affect the performance of agriculture and forestry at global level. Today, due to huge application of fertilizers in farm, intensive agricultural practices, illicit felling, deforestation, intensive grazing are affecting the soil health, water purity and its availability that leads to depletion of other NRs which are directly and indirectly linked with food and nutrition security, human and animal health, soil and environmental security. Therefore, the terms ecological intensification (EI) and sustainable intensification (SI) have proven to be a good strategy and play a significant role in conserving and managing these resources without affecting our environment health. FAO has

A. Raj
School of Agriculture, Lovely Professional University, Phagwara, Punjab, India

M. K. Jhariya (✉) · N. Khan
Department of Farm Forestry, Sant Gahira Guru Vishwavidyalaya, Ambikapur, C.G, India

A. Banerjee
Department of Environment Science, Sant Gahira Guru Vishwavidyalaya, Ambikapur, C.G, India

R. S. Meena
Department of Agronomy, Institute of Agricultural Sciences, Banaras Hindu University, Varanasi, UP, India

M. K. Jhariya et al. (eds.), *Ecological Intensification of Natural Resources for Sustainable Agriculture*, https://doi.org/10.1007/978-981-33-4203-3_5

defined the term EI and according to them, EI requires a knowledge intensive process that intensifies the ecosystem services (ES) of NRs by enhancing biodiversity which resulted in higher tree–crop–soil productivity through less use of synthetic inputs. This helps in maintaining food, health and climate security at global scale. However, intensification in agriculture and forestry must be promoted in lieu of maintaining food and nutritional security (FNS) of burgeoning 9.8 billion population along with minimizing global hunger and health issues of people. Therefore, EI in agricultural and forestry not only make sustainable production but also promote other ES, enhance other resource use efficiency (RUE), promote efficient nutrient cycling, maintain soil fertility along with ecological sustainability. However, there is lack of farmer's knowledge regarding EI and SI in agriculture and forestry, effective policies should be framed at government level in relation to knowledge communication among peoples. Lack of scientific oriented research and development (R&D), etc. becomes constraints behind adoption, promotion and application of a better EI in these NRs without affecting our environment. In this context, this chapter gives a framework and outlines the concept and prospects of EI, its utility in various NRs (agriculture and forestry, etc.), its role in ES, RUE, climate change mitigation along with discussions on ongoing trends of hurdles and constraint behind its adoption, related R&D and future roadmap for better applicability of EI in NRs for better environment with sustainable production systems at global scale.

Keywords

Agriculture · Climate change · Ecological intensification · Ecological sustainability · Natural resources · Resource use efficiency

Abbreviations

AI	Agricultural intensification
EI	Ecological intensification
ES	Ecosystem services
FNS	Food and nutritional security
GHGs	Greenhouse gases
NRs	Natural resources
R&D	Research and development
RUE	Resource use efficiency
SI	Sustainable intensification

5.1 Introduction

Expanding agricultural land through deforestation and other anthropogenic activity and intensification of farming systems by higher synthetic inputs for maximizing production have confirmed negative impacts in terms of losing biodiversity, emission of GHGs (greenhouse gases) into the atmosphere, increasing global warming, declining tree–crops–soil productivity, affecting resource use efficiency (RUE) and also disturb other natural resources (NRs) by affecting ecosystem services (ES) for ecological sustainability (Foley et al. 2005; Phalan et al. 2011; Meena et al. 2020; Kumar et al. 2020; Raj et al. 2020; Banerjee et al. 2020). However, intensifying agriculture, i.e. heavy synthetic inputs helps in producing more foods which satisfy billions of people by reducing hunger and malnutrition but at the cost of ecosystem and environment health due to land degradation, depletion of NRs, RUE, declining biodiversity and affecting socio-economic status of peoples (Foley et al. 2011; Godfray et al. 2010).

In this context, both ecological intensification (EI) and sustainable intensification (SI) play emerging role in management and development of agriculture and forestry by minimizing negative impacts of agricultural intensification (AI), having more crop diversification resulting in higher production through intensifying ES. Minimizing nutrient loss, soil erosion, eutrophication reduces GHGs and pollution. Further it helps in building soil fertility, crop productivity and food nutritional security along with ecological sustainability. These are possible through promotion of better EI methodology, i.e. intensifying agro-ecosystem and varying farming systems (e.g. agriculture, agroforestry and other farming systems). Various processes such as changing biological interaction, modifying tree–crop interaction, minimizing pesticide and insecticide application, enhancing beneficial microbial, fungal and plant interaction (e.g. Mycorrhizae, etc.), minimizing the application of inorganic, promoting healthy livestock's population and organic based farming systems would help in building sustainability in farming systems in various agro-ecological region of the world (Fig. 5.1) (Gaba et al. 2014; Xie et al. 2019). However, many farmers and stakeholders are advocating AI and giving solutions which vary from drastic change of food system to smaller field based improvement (Clay 2011; Foley et al. 2005; Royal Society London 2009). Many terms have emerged which focuses on these issues along with solving strategies and these are EI, SI and agro-ecological intensification. There is a blurred boundary in between EI and SI and very less information are available globally (Petersen and Snapp 2015). The concept of sustainability is totally based on ecological sustainability and related intensification. This promotes the practices of SI which depends on the principle of sustainability of production without ignoring environmental health. But the biggest hurdle is that various controversies arise from significant effects of EI and SI in management and conservation of NRs. The main question is *"Whether the principle of EI is applicable for any farming system and is it viable?"* *"How these varying forms of intensification will be operational for NRs management for higher significant benefits?"* and *"Can EI improve RUE?"*

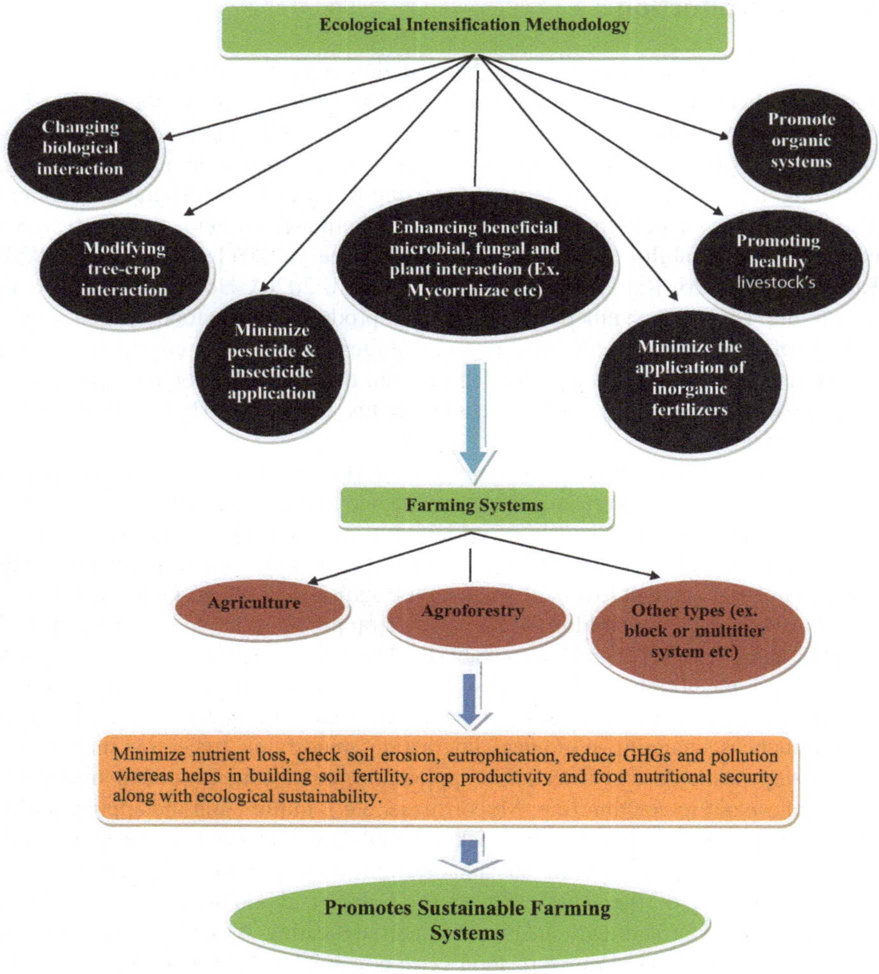

Fig. 5.1 Ecological intensification for sustainable farming systems (Gaba et al. 2014; Xie et al. 2019)

In this context, this chapter covers all relevant concepts regarding EI and SI, its role in NRs management, ES and climate change mitigation by reducing GHGs emission due to intensive farming systems. This chapter also highlights the role of EI in maintaining food and nutritional security (FNS), soil and environmental security. Effective policies, research and development (R&D) and future roadmap for adoption and operation of EI and SI are also discussed. In this chapter we also produce a conceptual framework and models for EI and SI which is quite linked with tree–crop–soil productivity by enhancing biodiversity through intensifying ES.

5.2 Ecological Intensification: Principle and Concept

The term EI itself represents intensification that is based on ecology oriented principle and applied for management and conservation of NRs such as agriculture, forestry, animals, etc. Similarly, the principle relies on the practices and management for higher tree–crop–soil productivity, better soil fertility, efficient RUE, biodiversity management and interaction among resources like plants, animals and soil inhabiting organisms (Agropolis 2013; CIRAD 2008; FAO 2009). The main aim of these intensifications is to make more deep understanding and knowledge for efficient use of NRs and related ecological processes (Doltra and Olesen 2013). As per CIRAD (2008) this intensification gives a better knowledge about tree–crop–soil interactions and linking concepts between biotic and abiotic factors through efficient bio-geochemical and water cycles and also intensifies the interactions among plants and animals.

The principle and practices of EI are based on achieving multiple goals/dimensions such as enhancing biodiversity and its conservation, improvement of tree–crop–soil productivity, maintaining soil fertility with balance flows of nutrient (Meena et al. 2018; Meena and Lal 2018). Further, it helps in efficient cycling in the systems, reducing insect pest infestation in the whole systems through better understanding about plant–insect and insect–insect interactions. This helps in balancing numbers of predators and parasites in the ecosystems and development of climate resilient farming systems. Such types of system have diversified forms of plant breeding technologies which are adapted and operationalized for reducing environmental constraints such as climate change mitigation. Further, EI is based on the principle of simplifying relations between food systems and human factors that initiated the less use of energy which helps in reducing the emissions of GHGs by controlling unstoppable uses of fossil fuels which are non-renewable resources (Dore et al. 2011; Meena et al. 2020a, b). Reducing food wastage, its proper distribution among peoples, recycling of its derived by-products, minimizing negative health among peoples and varying environment externalities are also taken into account for further studies of EI principle and practices at global scale (Tittonell and Giller 2013). Moreover, stakeholder participatory involvement, enhancing local expertise's and understanding about new species introduction along with making of collective form of decision's are also factors on which EI relies (Caron et al. 2014; Tittonell 2014).

5.3 Ecological Intensification: Origin and Historical Perspective

The historical invention of EI is crystal clear and well known among scientists, researchers, stakeholders, policymakers and farmers. Many authors have defined and elaborated the definition of EI along with its origin and historical backgrounds. For example, Egger (1986) proposed this term firstly and he described double approaches such as all practices for enhancing soil fertility on the one hand and establishing a great link among woody perennial trees, crops and animals in same piece of land under agrisilvopastoral systems. One decade later, Cassman and Pingali (1995) have emphasized the role of intensification in farming system. Cassman (1999) has described the goal and objective of EI in agriculture and according to him further AI is needed for satisfying the food requirement of humans without affecting the environmental quality.

After one decade in 2008, the sense of EI totally relies on making a conceptual framework and model which is designed in accordance to control and manage biological invasive species through proper utilization of NRs and its use efficiency with better ES (CIRAD 2008). Similarly, FAO (2009) has emphasized the role of both EI and SI in enhancing production per unit area without compromising any productive capacity of the systems. However, the studies have increased in the form of publications from 2010 afterwards.

Different authors are having their own perceptions to define ecological sustainability for example, focusing on food production with minimizing harmful impacts on environment by some authors such as Doltra and Olesen (2013), Griffon (2013) and Hochman et al. (2013). Dore et al. (2011) and Tittonell and Giller (2013) have emphasized on minimizing synthetic inputs and in contrary enhancing RUE. Thereafter, various authors came as per historical hierarchy and gave proper explanation of EI in successional forms. For example, Dore et al. (2011) emphasized the EI in terms of providing and intensifying ES, whereas Bommarco et al. (2012) made a great link between ES and production system which is managed through EI. After one year, two scientists viz., Dore et al. (2011) and Tittonell and Giller (2013) have reported a great integration of social aspects into EI. Similarly, Tittonell (2014) proposed EI into the landscape approach which provides better ES by enhancing biodiversity of ecosystem. This concept is also supported by Gaba et al. (2014). Thus, we can see the successional evolution of concept and definition of EI which is coming into recent definitions by taking account of ES and landscapes approaches at global scale.

5.4 Sustainable Intensification: Principle and Concept

SI is gaining wide importance in both scientific and development reports (Pretty et al. 2011). The concept of SI is crystal clear which mainly emphasizes on the principle of better environment health. As per Pretty (1997), SI can enhance yield potential in degraded areas along with protecting NRs. Gibon et al. (1999) have defined this term with special reference to livestock production and according to him, subtle changes in input and output in livestock production systems are aimed to maintain health and productivity along with product quality that can meet present and future demand of humans. Similarly, Pretty (2008) used this practices concerning three capital assets viz., natural, social and humans during practices of intensification. This is to be combined with various other technologies along with certain inputs such as recommended plant genotypes and effective ecological management that helps in minimizing negative impacts on our ecosystem and environment.

A very simple concept and principle behind adoption of SI adopted by Royal Society London (2009) reveals that there must be a balance between maintaining environmental quality along with increasing yield and productivity. This should not include more land areas of other land-use type. This is further rectified by FAO (2011) and according to them, enhancing more yield from same piece of land must be followed in parallel to resource conservation. Further, efficient utilization, improving environmental health, intensifying ES along with maintaining natural materials flow in the ecosystem for ecological stability are the need of the hour. This concept is also supported by Firbank et al. (2013). However, in the last decade, SI has gained wide recognition due to its popularity among farmers, scientists, researchers, policymakers and stakeholders due to crystal clear understanding of its concepts and principles. It involves management and conservation of NRs through supports from various national and international organizations. Various organizations such as Consultative Group on International Agricultural Research by (CGIAR 2011), United States Agency for International Development (USAID 2013) and International Fertilizer Industry Association (IFA 2013) are working towards this dimension.

In nutshell, the principles of SI depend on various practices such as conservation tillage (McCune et al. 2011), soil mulching along with better crop rotation practices (FAO 2011), integration of nitrogen fixing leguminous crops along with various other cash and cover crops in the farming systems (Tilman et al. 2011), hedgerow cropping system (Pretty 1997), practices of integrated pest management and soil–water conservation practices (Pretty et al. 2011; FAO 2011; McCune et al. 2011).

5.5 Linking Concept Among Intensification, Ecointensification and Sustainable Intensification

Inter-relationship between intensification, EI and SI reveals that EI and SI have blurred boundaries due to subtle difference in between them. However, link exists in all these three terms which is based on their principles and their applicability in the field and contributions in management and conservation of the NRs. Intensification represents intensive use of all inputs to intensify final products. If this practice is according to the ecological based approach, then it represents EI, whereas SI relies on higher yield and productivity without disturbing our environmental health. Therefore, higher production, RUE, ES, environmental health and ecological sustainability are various indicators/key that makes the difference among intensification, EI and SI at global scale. However, these terms overlap with each other due to similar appearance of their use as key terms. For example, the indicator "higher production" is valid for all these three terms, whereas environmental health and its sustainability are covered by only SI. Therefore, these three terms are linked concept upon which all principle and practices depend. However, some authors have integrated and correlated social dimensions into EI (Dore et al. 2011; Bommarco et al. 2012; Tittonell and Giller 2013), whereas others have integrated into SI (Garnett et al. 2013). Similarly, Kuyper and Struik (2014) reported link and similarity between EI and SI that shared same language worldwide.

Many documents are available on the EI and SI which are based on its concept, principles and significant role in conservation and management of NRs with reference to agriculture system. Xie et al. (2019) and his team did hectic works to review literatures on SI and explore database from various authentic sources and collected around 962 papers between 1980 and 2019. However, documents of research and studies (1956 numbers) were more for SI as compared to EI having 1706 numbers till 2018 which is depicted in Table 5.1. In this table, we can see that after 2010 the number of documents is increasing steeply without any interruptions which represent the work, study efficiency and scientific concern of these two intensifications due to its positive impacts. Also, it leaves various questions and research topics on concept and principles of EI and SI due to already published documental footprint in the world which states that more research studies need to be done in this aspect to explore the inter-relationship of these intensifications with various other fields. More research and topic need to be explored for proper understanding of the characteristics, principles and practices of these two AI and other NRs despite of already existing pools of data represented in Table 5.2. This table summarizes a review on SI in agricultural and other resources based on its varying characteristics, principles and adopted practices. Also these data will be helpful while applying the SI in any farming systems comprising agriculture, forestry and other NRs on the earth.

5.6 Ecointensification in Natural Resources

Conservation and managing NRs are global concern for smoothing of various ecological processes and proper ecosystem structure and its function along with various better ES to maintain the biodiversity (Jhariya et al. 2019a, b). In this context, both EI approaches and SI approaches would be helpful in promoting NRs management and their efficient utilization in ecosystems. Natural RUE will be high along with better ES and various ecological processes in EI approaches, whereas SI approach represents a balance exchange of NRs, i.e. input–output resources along with better environmental services. However, integrated NRs management approach also helps in this context by combining both SI and EI approaches. However, a conceptual model has been developed in this context that is depicted in Fig. 5.2 (Wezel et al. 2015; Lema et al. 2016).

The NRs are important treasure on the earth due to its multifarious benefits and role in overall ecosystem structure and its function that deliver uncountable services to mankind (Khan et al. 2020a, b). Although overexploitation of these resources (forest, agriculture, agroforestry, animals, soils, etc.) are becoming global concern for today and a major challenge for researchers and policymakers. In this context, ecology oriented intensification approach plays an important role in conserving and managing these valuable resources that help in enhancing agricultural productivity

Table 5.1 Literature mining and documents available on ecological and sustainable intensification in the field of agriculture during1990–2018 (Xie et al. 2019)

Ecological intensification (EI)		Sustainable intensification (SI)	
Year	Number of papers	Year	Number of papers
1990	1	1990	1
1991	1	1991	1
1992	8	1992	4
1993	5	1993	2
1994	3	1994	4
1995	9	1995	5
1996	5	1996	6
1997	12	1997	11
1998	10	1998	13
1999	15	1999	13
2000	9	2000	7
2001	15	2001	20
2002	18	2002	17
2003	16	2003	17
2004	26	2004	16
2005	26	2005	18
2006	40	2006	33
2007	49	2007	35
2008	48	2008	36
2009	45	2009	33
2010	61	2010	48
2011	81	2011	60
2012	102	2012	66
2013	107	2013	98
2014	142	2014	163
2015	188	2015	241
2016	182	2016	254
2017	214	2017	322
2018	268	2018	412
	1706		**1956**

for long term basis, maximize forest health and productivity, improve livestock's health, soil health and quality, diversity and management of natural habitat, enhance water resource availability for long term, diversifying food and fruits availability, improve both tangible and intangible services through forestry, maintain soil fertility and population of micro- and macrofloral population and organism and improve overall agro-ecosystem health and productivity under the era of climate change which is depicted in Fig. 5.3 (Mao et al. 2015; Al-Kaisi and Lowery 2017).

Table 5.2 A review on sustainable intensification in agricultural and other resources based on its varying characteristics, principles and adopted practices

Particulars	Sustainable intensification in agriculture and other resources	References
Characteristics	Intensify production along with conservation and protection of other natural resources	Pretty (1997)
	Maintain soil fertility by nutrient availability and its proper balance that signifies the return from land and labour in farming systems	Ruerd and Lee (2000)
	Helps in enhancing tree–crop productivity from unit land without affecting environment and land expansion	Baulcombe et al. (2009), Pretty and Bharucha (2014)
	Enhancing resource use efficiency and promotes the utilization of best technologies with less synthetic inputs that minimize negative impacts on environment	Pretty (2008)
	Intensify productivity and balancing the inputs and outputs in livestock animals production system while taking account of environmental stability	Gibon et al. (1999)
Principles	Based on the principle of less uses of land and utilization of various renewable resources such as light, water and labour to signify the production at farm level	Godfray et al. (2010), Firbank et al. (2013)
	Better use of tree–crop varieties and important cattle breeds	Ruerd and Lee (2000), Pretty (2008)
	Optimization of outside inputs, better resource use efficiency, improves food production systems and reduces its impact on our environment	Pretty (1997), Matson et al. (1997)
	Minimize the wastage of food with enhancing productivity	Garnett et al. (2013)
Practices	Application of mulching to cover soil and minimize losses along with conservation tillage practices in farming systems	Wezel et al. (2015)
	Better practices of integrated pest management	Pretty (1997)
	Integrating cover crops, use cash crop, beans and proper harvesting of crops is going on in crop rotation system	Tilman et al. (2011)
	Cultivation practices of improved varieties of tree–crops–livestocks along with protection of plant genetic resources	FAO (2004)
	Practices are done in favour of soil health and promotion of soil and water conservation	FAO (2004), Wezel et al. (2015)
	Practices that take account of water management in agricultural field and focus on irrigation management and fertigation, etc.	FAO (2004)

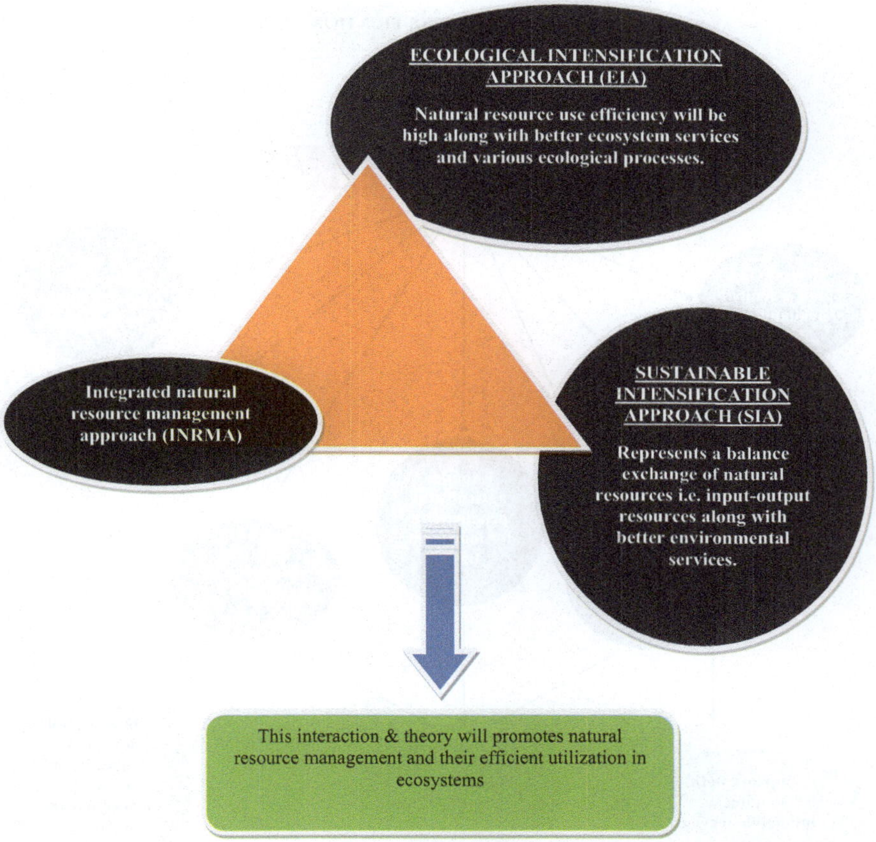

Fig. 5.2 Ecological and sustainable intensification approaches for natural resource management (Wezel et al. 2015; Lema et al. 2016)

5.6.1 Agriculture

The ongoing trends of agricultural land expansion and practices of AI are becoming major hurdles today due to various negative outcomes and deleterious impacts on our ecosystem and environment. Increasing population necessitates the food requirement that leads to expansion of agricultural lands through conversion of various existing natural forest, pastureland and other land-use that causes imbalance among various NRs. Within AI, intensifying synthetic inputs will surely help in enhancing crop productivity but at the cost of environmental health due to emission GHGs leading to global warming and climate change (Kumar et al. 2020). In this context, one question always remains in the mind of scientific community regarding the role of AI towards global warming and climate change. Agriculture itself contributes in climate change through GHGs emission through overuse of synthetic inputs into the

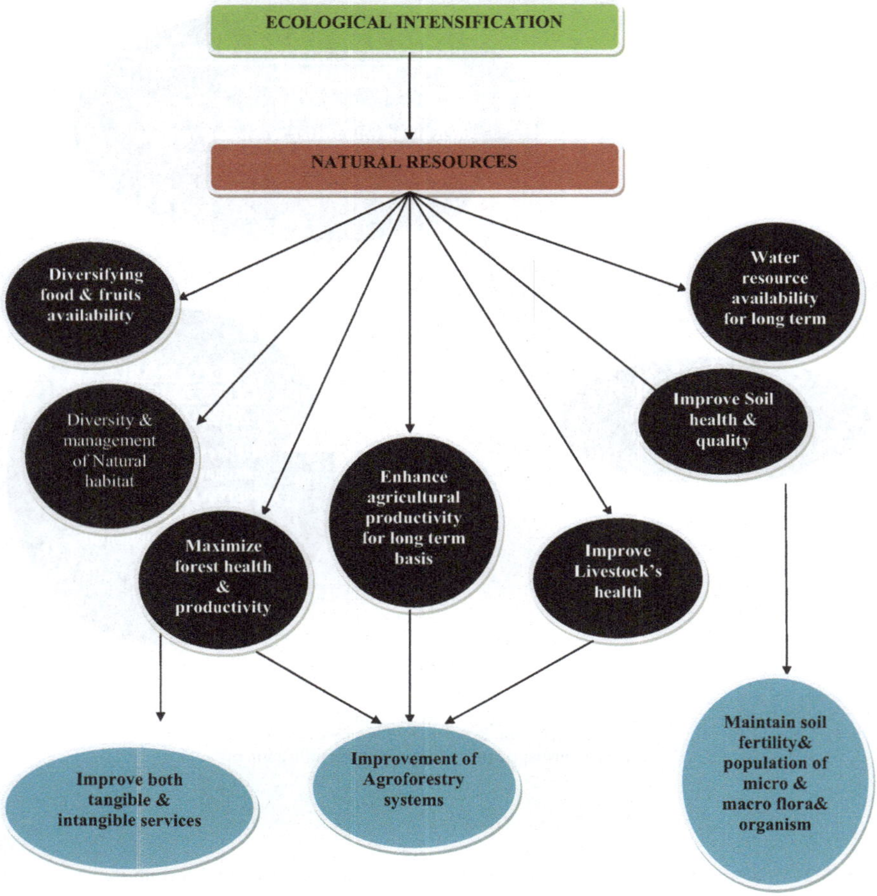

Fig. 5.3 Ecological intensification for natural resource management (Mao et al. 2015; Al-Kaisi and Lowery 2017)

land under the AI and animal intensification that also affects overall tree–crop–soil health and productivity.

In this context, both EI and SI prove themselves to intensify ES by enhancing tree–crop–soil productivity and biodiversity along with reducing GHGs by the practices of ecology based farming systems which are highly ecologically sustainable. Thus, a conceptual model was framed representing the role of EI and SI in agriculture practices for minimizing climate change impacts (Fig. 5.4, Burney et al. 2010). Therefore, various authors have proposed the significance of EI and SI into the agricultural systems, i.e. emphasizing on agro-ecological intensification (based on ecological principles) in terms of enhancing productivity and performance without disturbing environmental health that would lead to food and climate security at global scale. Further, they enhance biodiversity, improve soil fertility, maintain

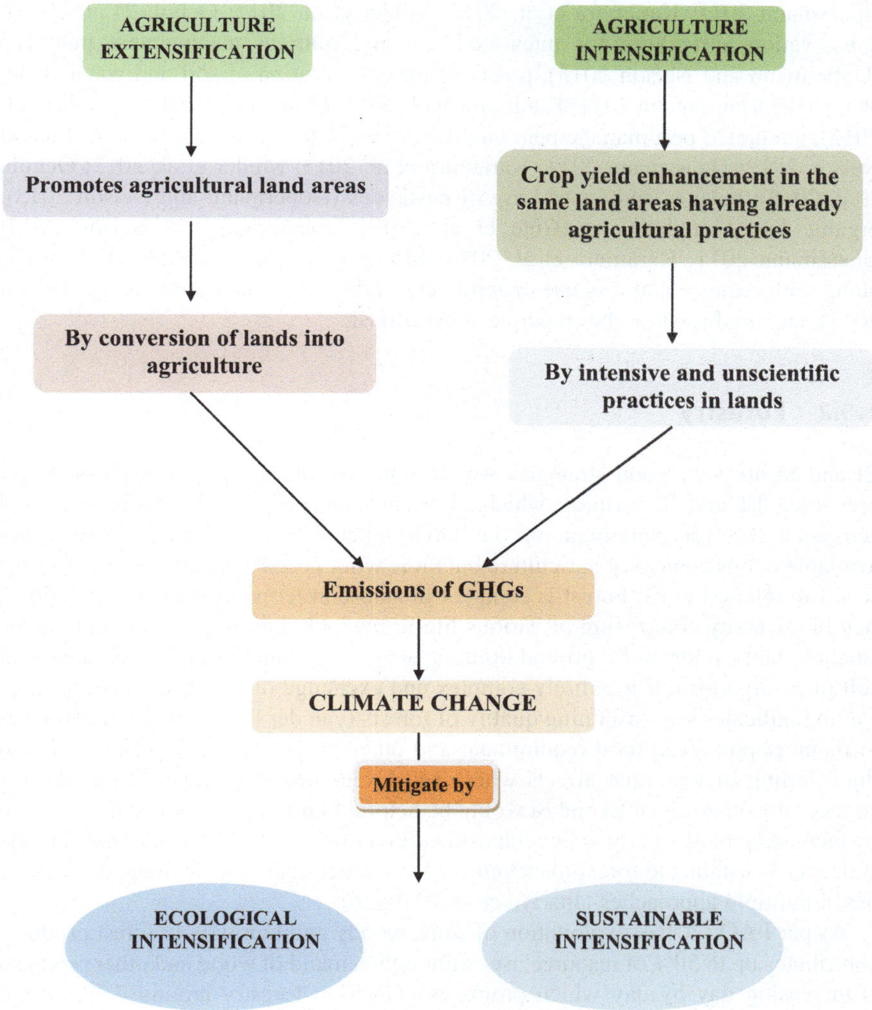

Fig. 5.4 Agriculture practices for climate change and its mitigation strategies (Burney et al. 2010)

soil heath and quality, promote material and nutrients cycling, minimize the leaching losses and soil erosion and RUE. It further increases carbon sink in vegetation and soils through carbon sequestration, optimizes water and nutrient use efficiency, improves socio-economic status of farmers, maintaining food and climate security for ecological sustainability (Milder et al. 2012; CCRP 2013; Dobermann and Nelson 2013).

Different types of practices adopted under EI and SI are described by different authors at various times. For example, the practices of crop rotations, proper soil mulching and better intercropping (Côte et al. 2010; Dobermann and Nelson 2013;

Haussmann 2011; Karamura et al. 2013; Milder et al. 2012; Ochola et al. 2013), conservation agriculture for integrated soil and nutrient management practices (Dobermann and Nelson 2013), practices for conservation of soil and water (Côte et al. 2010; Haussmann 2011; Karamura et al. 2013; Milder et al. 2012; Ochola et al. 2013), integrated pest management into the farms (Côte et al. 2010; Dobermann and Nelson 2013; Haussmann 2011; Karamura et al. 2013; Milder et al. 2012; Ochola et al. 2013), balance and control use of pesticides (Dobermann and Nelson 2013), organic based applications (Côte et al. 2010; Dobermann and Nelson 2013; Haussmann 2011; Karamura et al. 2013; Milder et al. 2012; Ochola et al. 2013) along with balance and less use of fertilizers in the farms (Dobermann and Nelson 2013), etc. are based on the principle of EI and SI.

5.6.2 Forestry

EI and SI are very good strategies which minimize the negative impact of AI by practicing EI and SI farming which helps in enhancing yield, productivity and ecosystem ES for betterment of our environment. However, many studies are available in this context of agriculture but more work needs to be done in the forestry sector in relation to EI. Forest is complex in nature in terms of structure, functions, rich biodiversity comprising of various life forms including woody perennial trees, smaller plants, understory, ground flora, lichens, fungi, animals and other beneficial soil microorganism. It is entirely complex and exchange of biological materials, its cycling indicates self-sustaining quality of forests (van der Plas et al. 2016). But due to rising populations, food requirement and other industrial development, leads to illicit felling of trees that affects whole ecosystem structure and function. In this context, the practices of EI and SI would be helpful in minimizing forest degradation by increasing biodiversity which intensifies ES. However, SI helps in promoting the concept of sustainable forest management by practices and management of ecology based multiple approaches (Jhariya et al. 2019a, b).

As per FAO (2016), exploitation of some woody and non-woody forest products contributes up to 50% of resource use. Although demand of wood and other products is increasing day by day which promotes plantation forestry around 7.3% of the forest globally (FAO 2016). In contrast, the biodiversity and delivery of ES from these plantation forests are very low due to dominancy and characterization of monoculture and sole tree plantation of some exotic species which significantly reduce the ES by less biodiversity and higher susceptibility of insect pest outbreaks (Dwyer et al. 2004; van der Plas et al. 2016). Similarly, the practices of less intensification in forest ecosystem (low-intensified forest management) have maximized biodiversity which delivers prominent ES along with economical and environmental benefits (Tscharntke et al. 2005). Therefore, intensification at certain level is of prime concern for healthy and diverse forest. Obviously, diversified forest promotes occurrence of variety of predators (small mammals, spiders, birds, etc.) and its populations which feed on harmful insect and ensure pest outbreaks prominently (Thompson et al. 2009). Similarly, diversified forest promotes the diversity of soil microbes, actinomycetes and fungal population in the soil which plays an important role in nitrogen fixation and efficient nutrient cycling. Hence, forest diversification enhances the productivity with low inputs (Hiiesalu et al. 2017). Therefore, intensification in forest must be framed and relies on the principles of ecological based approach which promotes biodiversity along with diversified products, improvement in tree–crop–soil health and productivity, reduces the chances of insect outbreaks and enhances ES for making ecological stability (Becerra et al. 2018).

5.6.3 Agroforestry

Agroforestry has various components (tree, crop and pasture/livestocks), structure and different location specific models which varies depending upon the biophysical status, topography and climates in the tropics. Agroforestry is a complex and sustainable farming system. Presently, intensive and unscientific practices and management along with improper understanding of tree–crop interaction affect the overall structure and function of different models of agroforestry which affects ES (Jhariya et al. 2015; Singh and Jhariya 2016). Therefore, EI and SI play a key role in maintaining structure and function (production and protection of tree, crop and soil) of agroforestry systems without affecting the environment and ecological sustainability. However, application of least fertilizers, incorporation of high vigour plant's variety, integration of multipurpose trees, effective soil management practices, etc. intensify the productivity and protection of model that not only enhance biodiversity (both vegetation and soil inhabiting microbes) but also promote ES, maintain FNS, reduce GHGs emission by better carbon sequestration potential for climate security. It also increases socio-economic status of poor farmers at global scale. However, a very little information was available in this context.

Studies of Egger (1986) help in understanding and exploring conservation and management of soils in pasture based agroforestry systems in the tropics. Similarly, Noponen et al. (2013) have conducted a research to evaluate and explore the trade-offs among EI, GHGs emission and profitability of agroforestry systems in the region of Costa Rica. According to them, the application of effective EI along with better management practices would help in enhancing carbon sequestration potential that mitigate the issue of changing climate. Also, it helps in bumper production of agriculture crops and reduces pressure of land conversion. However, there is a clear difference between agriculture and agroforestry intensification in which AI only helps in reducing emissions of GHGs. Intensification in agroforestry will not only mitigate climate change by GHGs emission but also build up higher crop productivity (Burney et al. 2010; Palm et al. 2010). Similarly, SI promotes both productivity and protection of agroforestry without compromising health and security of environment.

5.6.4 Soil

Soil is one of the key resources which hold and sustain various other NRs such as forest, agriculture and wildlife. Health and productivity of both plants and soils are maintained in two way direction such as tree and crops shed their leaves which decompose and add nutrients to the soils that improve soil fertility (better soil health and quality) (Raj et al. 2019a, b). In turn soils release these essential nutrients again to plants, i.e. plants absorb and fix into their body parts for metabolic activity that helps in maintaining proper growth and development of plants (better plants health and quality). These dual profits are proven to be a great link and synergy between them which is represented in Fig. 5.5 (Lal 2008; Pinho et al. 2012; Singh et al. 2017). But due to AI, unsustainable land-use systems, unscientific management of farming

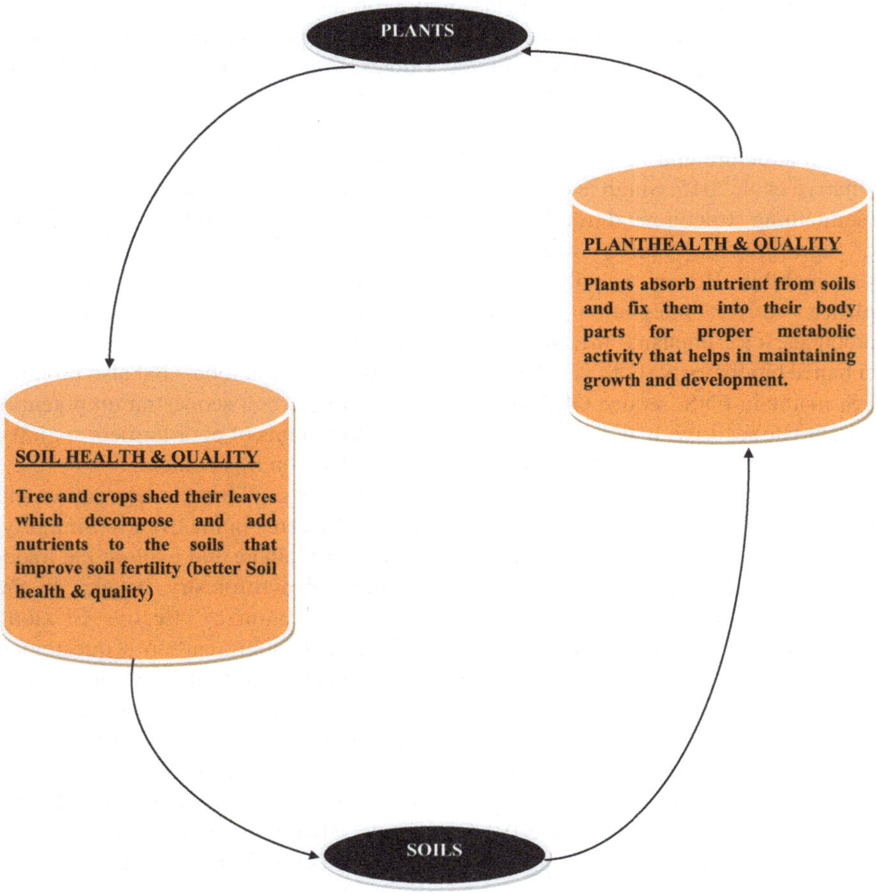

Fig. 5.5 Link between plants and soils in farming systems for better performance (Lal 2008; Pinho et al. 2012; Singh et al. 2017)

technology, use of heavy machine on farms, unstoppable use of inorganic fertilizers on farms, etc. affects health and quality of soils and disturbs related ES.

In this context, EI enhances the tree–crop and soil productivity without adding huge amount of fertilizers into the soil resulting in higher soil organic matter which improves soil microbial biomass and rhizosphere biology. Therefore, EI is used for maintaining soil organic matter which is a critical indicator of soil health (Bommarco et al. 2013). Good and effective management practices in farming systems are the basis for implanting EI. This would help in enhancing soil organic matter in the soils. Conversely decrease in organic matter would lead to loss of important microorganism in the soil that directly or indirectly affects sustainability of farming systems (Tsiafouli et al. 2015). Although enhancing soil ES is controlled by soil biodiversity that relies upon practices of effective EI which helps in controlling and maintaining decomposition and cycling of nutrients in farming ecosystems (Barrios 2007). Thus, strategies for increasing tree–crop diversification, incorporation of leguminous plants in rotation, less use of inorganic fertilizers along with minimum soil disturbance are covered under ecological intensive practices. It helps in building above and below ground biomass, enhances carbon values in both plants and soils and builds physico-chemical properties of soils without affecting overall productivity of the farming systems and degrading our environment (Kremen and Miles 2012; Brady et al. 2015; Jhariya et al. 2018a, b).

5.6.5 Livestock

Livestock maintains social, culture and economic values and plays major role in farming systems. It provides various products such as milk, meat, eggs, feather and other tangible food products. They maintain health and economics of peoples while integrating with farming systems. Integration of animals in agroforestry systems also helps in enhancing biodiversity of the systems but their management practices without affecting animal's health, crop productivity and livestock's potential to produce valuable products through the application of livestock intensification are less properly studied (Fahrig 2017). However, changing biodiversity of any farming systems relies on change in livestock's number too which overall affects the structure and services of the farming system. Therefore, biodiversity conservation is linked with occurrence of animal species and their interactive response to altered farming systems (Phalan et al. 2011; Paul and Knoke 2015). Thus, intensified livestock practices and its management are the important aspects of EI that not only enhance biodiversity but also increase productivity (tree–crop–soil) and profitability of farmers. In this context, Gomes et al. (2014) have studied the impact of EI approach in goat farming systems and made a design for sustainable livestock systems (Dumont et al. 2013) which is based on the five principles of agroecology. The principles include adoption of animal's health perspective management, less inputs for higher productions, minimizing pollution by optimizing different components of farming systems, promoting animal's diversity in the system and conservation of biodiversity in agro-ecosystems by adoption of scientific based management practices.

5.7 Constraint and Limitation in Intensification

EI and SI prove to be a good strategy in every aspects of use efficiency of NRs and intensify the ES for betterment of environment and ecological stability. But these have certain limitations and having hurdles while promoting at ground level to global scale. For example, burgeoning population demands for more foods, timber, fuelwood and other non-wood forest products that promote high input practices and illicit felling of trees for timber either directly or indirectly affects various natural ecosystems (forest, soils, water, etc.). In this context, using the principle of EI and SI would be the best option but they will affect the overall production systems. For example, we stress upon organic agriculture but "*is organic system of practices would satisfy the food requirements of people*?" This is a very conceptual question of today because high quantity of food and other products has to be intensified in agriculture and other farming systems which promote higher use of inputs in the farms. Therefore, adoption level of EI and SI by farmers in their land is questionable, although these strategies fulfil the needs at certain level. Secondly, the practices of EI having certain limitations due to carrying capacity of NRs, type of land, topography, soil types, tree–crop interaction, species natures, social, economics, farmers' attitudes for adoption and political aspects. Therefore, these measures play a major role in practices of EI and SI for NRs. Similarly, farmers and people awareness about significance of EI, farmer to farmer communications, institutional role in strengthening EI, effective policies for promotion and adoption of SI are the key points on which we have to focus while adopting and promoting EI and SI from ground level to large scale.

5.8 Ecointensification for Ecosystem Services

The EI approach in NRs is proven to be a good strategy for enhancing ES from forest, agriculture, soil and animal resources. They provide water and air regulation, soil fertility enhancement, biodiversity conservation and storage and sequestration of carbon, etc. Also, tangible (timber and non-wood products) and intangible services through forestry and agriculture food grains production are very important services that maintain FNS and environmental stability. Therefore, in this context a conceptual model has been developed which is depicted in Fig. 5.6 (Bommarco et al. 2013).

However, the practices of EI in both agriculture and forestry will promote the ES by enhancing biodiversity through mixed plantation, mixed crops and proper crop rotation. Monocropping and sole tree plantations/orchards having less diversified plants that is poor in delivery of various important ES in both direct and indirect ways and highly susceptible to insect pest attacks. For its manipulation, making more diversified form of forestry and agriculture by incorporating middle and

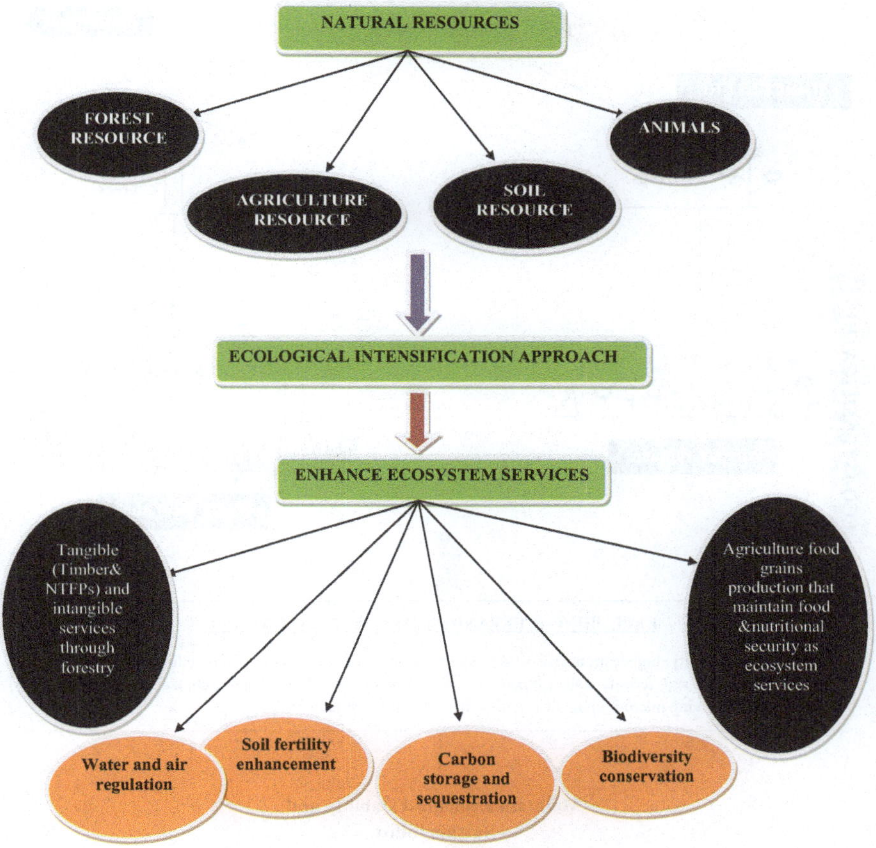

Fig. 5.6 Ecological intensification in natural resources for ecosystem services (Bommarco et al. 2013)

understory plants as mixed plantation and cropping systems would be more significant in delivery of ES. This would promote increasing biodiversity, higher productivity, soil water conservation (understory plants reduce the soil and water erosion problems), soil fertility enhancement, less catastrophic disturbance and stabilization of ecological systems and improving micro-climate of whole systems in particular areas. In this context, a model has been developed for diversified agriculture and forestry plantation and its diversified ES through application of mixed cropping and mixed plantation concepts which is depicted in Fig. 5.7 (Montesinos 2019).

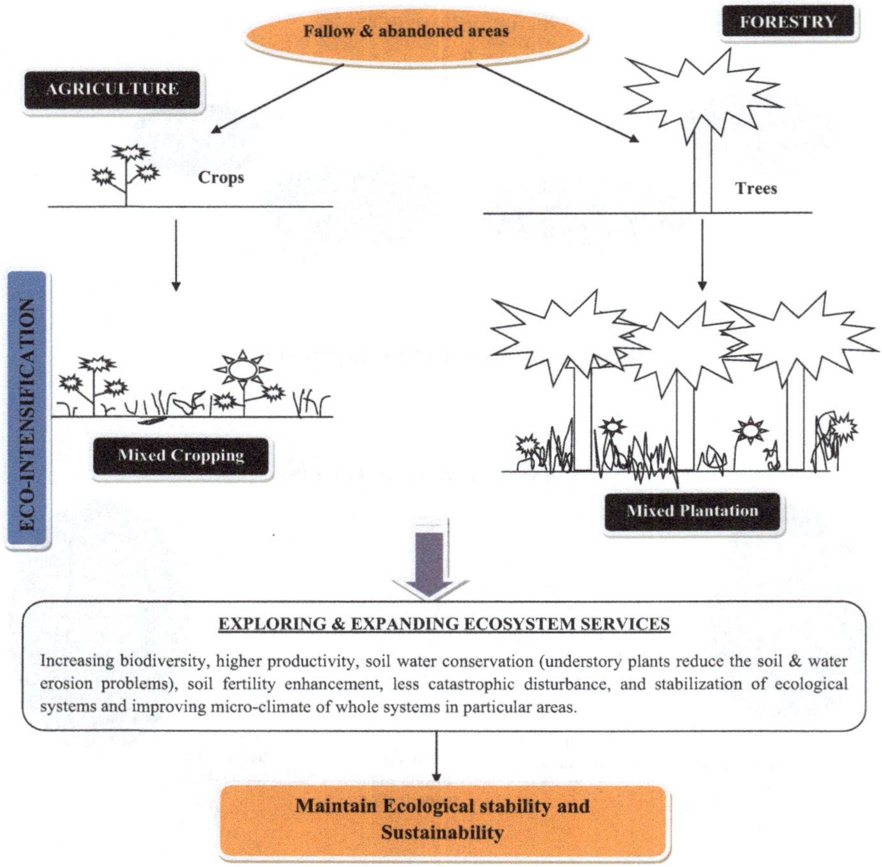

Fig. 5.7 Ecological intensification in agriculture and forestry for ecosystem services (Montesinos 2019)

5.9 Ecointensification for Food Security

Food security and its sustainability are becoming global issues in national and international scientific platforms due to speedy population growth causing hunger and malnutrition problems in these days. Hunger and malnutrition are the major challenge today and it will affect global FNS. In this context, the adoption of some ecological and social approaches is proven to be good strategies for minimizing global hunger and malnutrition problems and makes the availability of quality and nutritious food to the society. Ecological approach comprises both EI and SI. Application of better ecological approach for healthy crops and food grains is achieved by EI. Conversely SI promotes balance production with proper input–

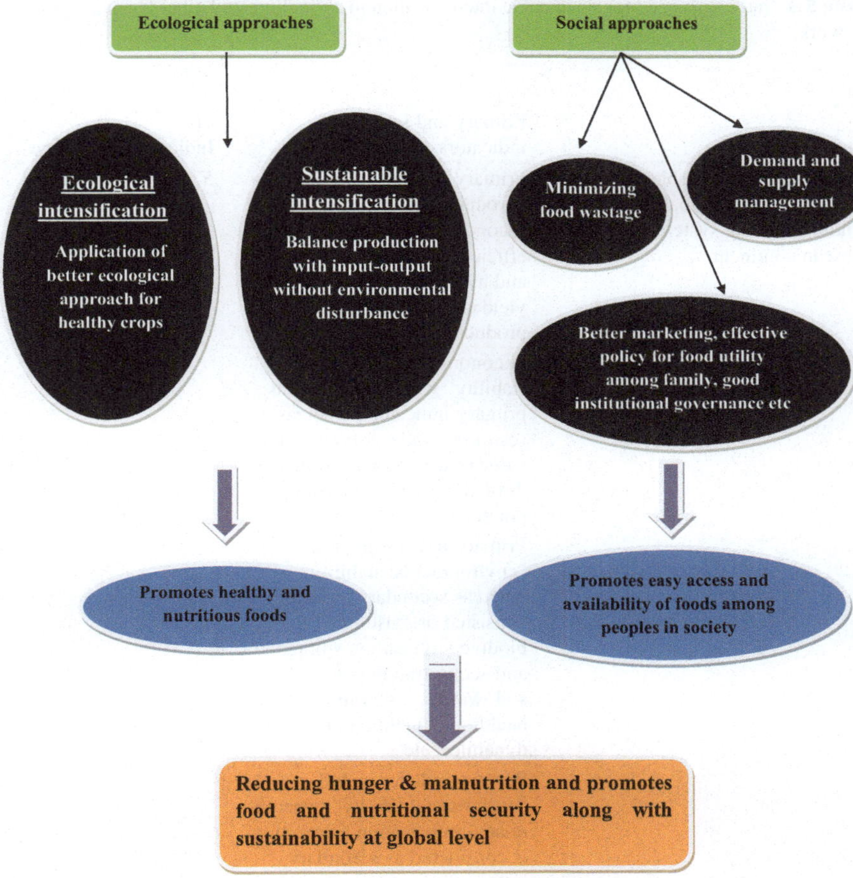

Fig. 5.8 Ecological and social approaches for food security (Garnett and Godfray 2012; Bilali et al. 2019)

output mechanism without environmental disturbance. However, better marketing, effective policy for food utility among family, good institutional governance, minimizing food wastage and demand and supply management are the other strategies that follows social approaches to promote easy access and availability of quality and nutritious foods among peoples in society. Thus, it will be helpful in reducing hunger and malnutrition and promotes FNS along with sustainability at global level. In this context, a model is developed which is depicted in Fig. 5.8 (Garnett and Godfray 2012; Bilali et al. 2019) (Table 5.3).

SI in farming systems produces sustainable production which results in sustainable diet and promotes food system transformation which maintains FNS at global scale. However, FAO (2012) has stressed upon considerable extent of intensification are required for better production that would help in meeting global food demands

Table 5.3 Indicators used for sustainable intensification in agriculture and other farming system in the world

Area of study	Primary and secondary indicators	Total number of Indicators	Source
Indicators of sustainable intensification for small land holding farming systems in the African continent	Primary indictor includes "productivity" for this secondary indicators are efficiency of external inputs and available water along with yield and livestock's health and productivity	57	Smith et al. (2017)
	"Economic balance and stability" were considered as primary indicators, whereas secondary indicators include value of tree–crops and income through agriculture/farming practices		
	Primary indictor includes "environmental stability", whereas secondary indicators consisted of existing biodiversity, carbon storage and sequestration potential, soil–water conservation, soil health and quality, nutrient dynamics, etc.		
	Primary indictor includes "social sustainability", whereas related information acquisition is considered as secondary indicator		
	Human well-being is considered as primary indicator and secondary indicator included food and nutritional security		
Indicators of sustainable intensification for agricultural systems practices in United Kingdom	"Resource unit" is represented as primary indicator for this tree–crop diversity, water table, livestock's population, soil types and biodiversity are considered as secondary indicators	110	Mahon et al. (2018)
	"Resource systems" is represented as primary indicator, whereas total farm size, land holding areas, tree–crop productivity, etc. are		

(continued)

Table 5.3 (continued)

Area of study	Primary and secondary indicators	Total number of Indicators	Source
	considered as secondary indicators		
	"Resource users" is treated as primary indicator, whereas secondary indicators included farmers housing, their age and social networks, employment status, etc.		
	Primary indicator is designated to the term "interaction" for that status of tree–crop–animal interaction, farming quantity, type of mechanization, livestock's rearing, land characteristics and level of farming technologies are considered as secondary indicators		
	"Outcomes" is a very important primary indicator for that yield potential and gaining income from tree–crop systems, GHGs emissions, pollution from agricultural practices, resource use efficiency, farmer welfare, land characteristics, etc. are treated as secondary indicators		
	"Environment" is considered as primary indicator, whereas secondary indicator included occurrence of extreme weather, price of products, competition in varying farming systems, credits, characteristics of consumers and fund amounts, etc.		
Indicators of ecological intensification for coconut based farming system in Brazil	Primary indicator is "landscape ecology", whereas ecological stability, natural locality/ habitat, environmental quality, risks and production status with its diversity are considered as secondary indicators	62	Stachetti and Roberto (2018)
	"Social and cultural status" are considered as primary indicators for this gender		

(continued)

Table 5.3 (continued)

Area of study	Primary and secondary indicators	Total number of Indicators	Source
	quality, educational and employment status, public services, standard of varying consumers, natural heritage and health, etc. are considered secondary indicators		
	"Environmental quality" is considered as primary indicator, whereas secondary indicator includes soil and water quality along with level of GHGs emission, etc.		
	Primary indicator is "economic value", for this secondary indicators are land value in money, source and distribution of income, net income value, debt and housing value, etc.		

up to 2050. FAO (2017) believes on the practices of intensification that enhances diversification of agriculture productions meeting the food demands and maintains FNS at global scale. According to this organization, the practices of SI would help in improving both productivity and ecological sustainability, i.e. food and environmental security by enhancing crop diversity and ES. Although FNS is linked by a broad spectrum of soil quality, climatic situations, socio-economic and political aspects that guarantee SI at large scale (CIRAD 2016). However, food wastage reduction and its proper management are also good strategies covered by SI which enhance the availability of food to people at door steps and helps in achieving FNS by improving food chain efficiency and ecological sustainability (FAO 2011; Waste and Resources Action Programme 2011).

5.10 Ecointensification for Climate Change Mitigation

Today, intensification of agricultural for higher yield and conversion of lands into agriculture (agricultural land expansion) lead to emission of several harmful GHGs that causes global warming and climate changes. Both AI and expansion of agricultural area are the major hurdles towards environmental security and ecological stability. However, both extensification (promotes agricultural land areas) and intensification (crop yield enhancement in the same land areas having already agricultural practices) enhance agricultural productivity but at the cost of our environment due to GHGs emissions and in turn these harmful gases affect all plants, animals, soil and other NRs. In this context, EI and SI will be good strategies which not only help in reducing GHGs emission but also enhance the tree–crops–soil productivity by better ecology oriented and scientific based farming practices with better management. This would help in mitigating climate change and global warming problems at global scale (Burney et al. 2010). The practices of climate-smart agriculture, conservation agriculture, no-tillage practices, organic farming system, mulching practices and integrated farming practices, etc. would help in minimizing deleterious impacts by reducing GHGs emission without affecting the overall yield and productivity of plants.

5.11 Ecointensification for Resource Use Efficiency

Resource and its sustainable uses are having prime importance as they show significant promise towards ecological stability that maintains ecosystem structure and functions along with better delivery of ES. However, unsustainable way of production, deforestation, intensive farming and animal intensifications in farms will affect various resources and their potential of RUE. For example, AI affects the status and availability of nutrients and water in the soils which in turn influence the plant potentials of nutrient and water use efficiency. In this context, EI and SI are gaining wide recognition by making great emphasis to increase soil organic matter, promote nutrient availability, water efficacy, enhance microbial populations along with its plants capacity to utilize all these resources for their proper metabolic activity, growth and developments (Struik and Kuyper 2017). However, intensification promotes unstoppable use of resources, i.e. resource mining in depth that affects overall resource use and its efficiency which is studied at various aspects such as agronomy, socio-economic and environmental aspects. As per Foley et al. (2011), RUE will increase on decreasing NRs that would necessitate targeting more production on even similar amount of inputs. Therefore, many researchers related to this field are having a great conception on EI and according to them, EI and SI are win-a-win strategies which help in increasing tree–crop–soil productivity along with improving and promoting RUE and avoiding from expansion of farming land. Similarly, we can minimize the impact of intensive agricultural practices on our environment by reducing the overuse of inorganic nutrient fertilizers (Mueller et al. 2012).

5.12 Research and Developmental Activity

The intensification in NRs such as agriculture and forestry are not recent practices, it was taken into account from the past when population growth caused food, timbers, fuelwood and other resource depletion. This necessitated intensifying the farming systems by promoting higher synthetic inputs. In past, we have focused only on crop intensification in terms of productivity rather than focusing on other resources such as soil, animals and environment. Research was conducted only in unidirectional approach rather than multidirectional approaches. For example, intensifying crop productivity will always decline the health and fertility status of soil. Further, while approaching economic target we overlooked the ecological sustainability. Therefore, R&D must be framed to balance between economic, ecology and sustainability.

Although in recent past, various R&D were conducted in minimizing the deleterious and negative impacts on our environment and ecosystem from AI but results were not satisfactory due to population rise resulting in higher food demands which were controlled by only unscientific way of AI. Therefore, we intensify only crop productivity rather than development and management of other resources. But in this context, the practices of EI and SI make a harmony with nature by minimizing higher use of synthetic inputs, enhance biodiversity, improve tree–crop–soil productivity and overall intensify ES along with making ecological sustainability. Now a good research has been approved in this context of understanding a difference between EI and SI, exploration of multifarious significant benefits of both EI and SI in NRs management, resource utilization and related RUE.

R&D must emphasize on varying models of intensification according to farming systems prevailing in varying climate, soil types, water availability, socio-economic and political situations in any regions. Moreover, public investments, effective policies, research institute, governmental institution, non-governmental organization and public–private partnership play important role in knowing, understanding and raising awareness among people and farmers for adopting EI and SI. This would lead to betterment and development of our environment and maintenance of ecological stability (Tittonell 2014). Thus, a conceptual model must be developed through better R&D which reflects its significance and multifarious benefits in NRs management and its efficient uses in the ecosystem for maintaining ecological sustainability.

5.13 Policy Framework

Indeed, the EI and SI have proven itself as win-a-win strategy for reducing negative impacts on tree–crop–soil productivity, enhance RUE and intensify ES by enhancing biodiversity. But certain existing policies are not appropriate for promotion and awareness of EI among farmers and people. They have less knowledge and awareness about significant effects of these two types of intensifications in NRs management and related positive impacts on our environment which maintain ecological stability. In this context, many policymakers, academicians, scientists, researchers, consumers and farmers have emphasized on promotion from conventional intensification to ecological and sustainable intensification (Cui et al. 2018). However, policy must be enacted and framed for promoting various multiple indicators/dimensions which are typically used for application of EI in sustainable tree–crop–soil production systems. These various multiple indicators/dimensions are (i) biomass, carbon and microbial diversity on above and underground, (ii) long term delivery of ES, (iii) availability of water resources and use efficiency, (iv) minimizing synthetic inputs, (v) agricultural productivity for long term in sustainable basis, (vi) diversification of natural habitat, (vii) training for farmer's

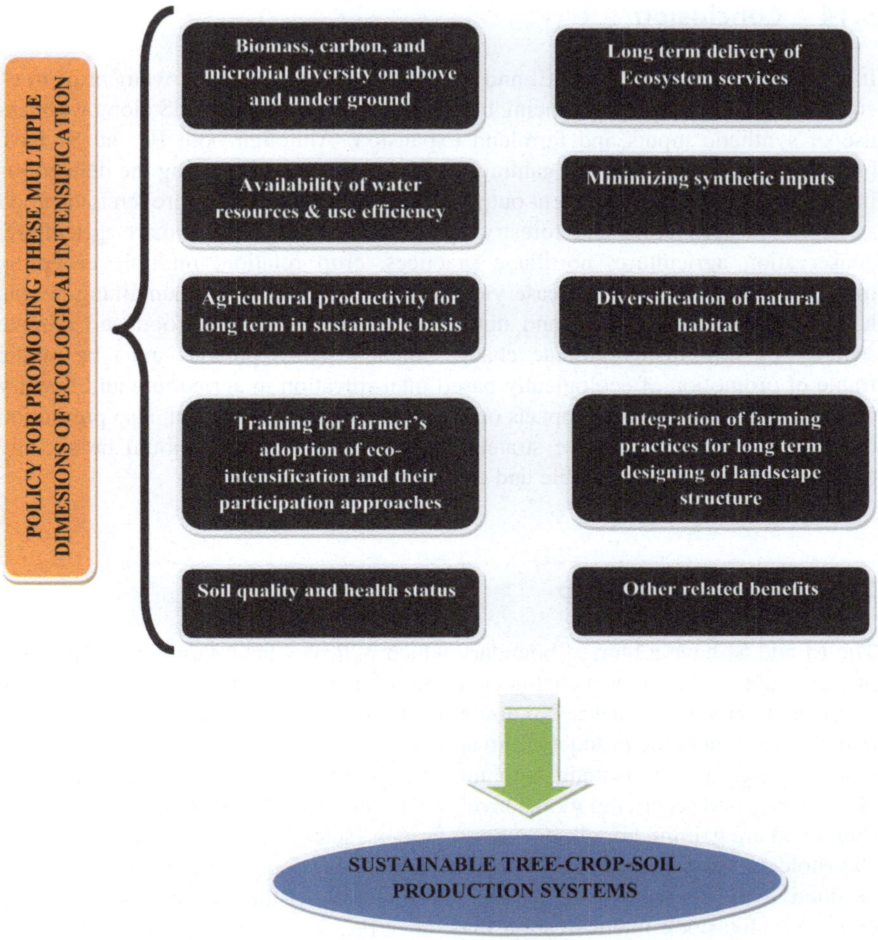

Fig. 5.9 Policy for sustainable tree–crop–soil production systems through ecointensification by considering multiple dimensions (Gemmill-Herren et al. 2019)

adoption of EI and their participation approaches, (viii) integration of farming practices for long term designing of landscape structure, (ix) soil quality and health status and (x) other related benefits. Therefore, policy for sustainable tree–crop–soil production systems through ecointensification by considering multiple dimensions is required and mentioned in Fig. 5.9 (Gemmill-Herren et al. 2019).

5.14 Conclusion

It is clearly understood about EI and its multifarious role in improvement of tree–crop–soil productivity by enhancing biodiversity and intensifying ES along with less use of synthetic inputs and farmland expansion. Although both EI and SI have blurred boundary but they are gaining popularity due to suppressing the deleterious impact of AI and having efficient output. Therefore, EI and SI are proven to be good strategies in agriculture and forestry by application of climate-smart agriculture, conservation agriculture, no-tillage practices, crop rotation, multiple cropping, mixed cropping, etc. that increase yield, improve soil fertility, maintaining people health through quality food and nutrient rich fruits, maintain food and climate security by minimizing climate change impacts. Also, policies must be in the frame of promotion of ecologically based intensification in agriculture and forestry that should buffer negative impacts on both plants and environment, also promotion of farmers for adopting these strategies in their farms which should be socially acceptable, economically viable and ecologically sustainable.

5.15 Future Roadmap

The EI and SI have a blurred boundary which indicates their significance, positive impacts and multifarious benefits in terms of plants productivity by enhancing biodiversity at various scales. No doubt, EI has intensified ES and provides various tangible and intangible products from agriculture and forestry and related other NRs (soil and animals, etc.) along with minimizing emission of GHGs and maintain climate and food security at global level. Therefore, EI and SI both are having bright future and are gaining popularity among farmers, scientists, policymakers and other stakeholders due to significant effects on maintaining food security, tree–crop–soil productivity, water security, better RUE, climate security through minimizing GHGs emissions, livestock intensification and other ES. Thus, we cannot overlook the significance of EI and for the further development a roadmap must be synthesized for adoption and prevalence of intensification at local to global scale.

References

Agropolis (2013) DMC—an ecological intensification engineering tool. www.agropolis.org/agronomy/research.php?id=10

Al-Kaisi MM, Lowery B (2017) Soil health and intensification of agroecosystems. Elsevier Acad Press, p 418

Banerjee A, Jhariya MK, Yadav DK, Raj A (2020) Environmental and sustainable development through forestry and other resources. Apple Academic Press Inc., CRC Press- A Taylor and Francis Group, US & Canada. ISBN: 9781771888110. p 400. https://doi.org/10.1201/9780429276026

Barrios E (2007) Soil biota, ecosystem services and land productivity. Ecol Econ 64

Baulcombe D, Crute I, Davies B, Green N (2009) Reaping the benefits: science and the sustainable intensification of global agriculture. Royal Society, London
Becerra PI, Catford JA, Luce McLeod M, Andonian K, Aschehoug ET, Montesinos D, Callaway RM (2018) Inhibitory effects of Eucalyptus globulus on understorey plant growth and species richness are greater in non-native regions. Glob Ecol Biogeogr 27:68–76
Bilali H, Callenius C, Strassner C, Probst L (2019) Food and nutrition security and sustainability transitions in food systems. Food Energy Secur 8(2):1–20. https://doi.org/10.1002/fes3.154
Bommarco R, Kleijn D, Potts SG (2012) Ecological intensification: harnessing ecosystem services for food security. Trends Ecol Evol 28(4):230–238. https://doi.org/10.1016/j.tree.2012.10.012
Bommarco R, Kleijn D, Potts SG (2013) Ecological intensification: harnessing ecosystem services for food security. Trends Ecol Evol 28:230–238
Brady MV, Hedlund K, Cong RG, Hemerik L, Hotes S, Machado S, Mattsson L, Schulz E, Thomsen IK (2015) Valuing supporting soil ecosystem services in agriculture: a natural capital approach. Agron J 107:1809–1821
Burney JA, Davis SJ, Lobell DB (2010) Greenhouse gas mitigation by agricultural intensification. Proc Natl Acad Sci U S A 107:12052–12057
Caron P, Biénabe E, Hainzelin E (2014) Making transition towards ecological intensification of agriculture a reality: the gaps in and the role of scientific knowledge. Curr Opin Environ Sustain 8:44–52. https://doi.org/10.1016/j.cosust.2014.08.004
Cassman KG (1999) Ecological intensification of cereal production systems: yield potential, soil quality, and precision agriculture. Proc Natl Acad Sci 96(11):5952–5959
Cassman KG, Pingali PL (1995) Intensification of irrigated rice systems: learning from the past to meet future challenges. Geo J 35(3):299–305. https://doi.org/10.1007/BF00989137
CCRP (Collaborative Crop Research Program) (2013) Agroecological intensification. McKnight Foundation, www.ccrp.org/programessentials/how-we-work. Accessed 2 Oct 2013
CGIAR (2011) CGIAR research program on integrated systems for the humid tropics. http://www.cgiar.org/our-research/cgiar-researchprograms/cgiar-research-program-on-integrated-systems-humidtropics/
CIRAD (2016) A literature review about experiences, research and innovation results obtained with a large spectrum of intensification pathways. Deliverable D2.1 of PROIntensAfrica project. Retrieved from http://www.intensafrica.org/documents/#
CIRAD (Centre de coopérationinternationale en rechercheagronomique pour le développement) (2008) CIRAD Strategic Vision 2008–2012. www.cirad.fr/en/content/download/970/31343/version/6/file/CIRAD_Strategie_GB_web.pdf
Clay J (2011) Freeze the footprint of food. Nature 475:287–289
Côte FX, Tomekpe K, Staver C, Depigny S, Lescot T, Markham R (2010) Agro-ecological intensification in banana and plantain (Musa spp.): an approach to develop more sustainable cropping systems for both smallholder farmers and large-scale commercial producers. Acta Horticult (ISHS) 879:457–463
Cui Z, Zhang H, Chen X, Zhang C, Ma W, Huang C, Zhang W, Mi G, Miao Y, Li X, Gao Q, Yang J, Wang Z, Ye Y, Guo S, Lu J, Huang J, Lv S, Sun Y, Liu Y, Peng X, Ren J, Li S, Deng X, Shi X, Zhang Q, Yang Z, Tang L, Wei C, Jia L, Zhang J, He M, Tong Y, Tang Q, Zhong X, Liu Z, Cao N, Kou C, Ying H, Yin Y, Jiao X, Zhang Q, Fan M, Jiang R, Zhang F, Dou Z (2018) Pursuing sustainable productivity with millions of smallholder farmers. Nature 555 (7696):363–366
Dobermann A, Nelson R (2013) Opportunities and solutions for sustainable food production, background research paper. High-Level Panel of Eminent Persons on the Post-2015 Development Agenda. www.post2015hlp.org/wp-content/uploads/2013/05/Doberman-Nelson_Solutions-for-Sustainable-Food-Production.pdf. Accessed 4 Nov 2013
Doltra J, Olesen JE (2013) The role of catch crops in the ecological intensification of spring cereals in organic farming under Nordic climate. Eur J Agron 44:98–108. https://doi.org/10.1016/j.eja.2012.03.006

Dore T, Makowski D, Malézieux E, Munier-Jolain N, Tchamitchian M, Tittonell P (2011) Facing up to the paradigm of ecological intensification in agronomy: revisiting methods, concepts and knowledge. Eur J Agron 34:197–210. https://doi.org/10.1016/j.eja.2011.02.006

Dumont B, Fortun-Lamothe L, Jouven M, Thomas M, Tichit M (2013) Prospects from agroecology and industrial ecology for animal production in the 21st century. Animal 7(6):1028–1043

Dwyer G, Dushohh J, Yee SH (2004) The combined effects of pathogens and predators on insect outbreaks. Nature 430:341–345

Egger K (1986) Ecological intensification. Soil conservation and improvement of tropical soils by pastoral agroforestry systems. Collect Doc Systèmes Agraires 6:129–135

Fahrig L (2017) Ecological responses to habitat fragmentation per se. Ann Rev Ecol Evol Syst 48:1–45

FAO (2004) The ethics of sustainable agricultural intensification; ethics series. Food and Agriculture Organization of the United Nations, Rome, Italy, pp 3–5

FAO (2009) Organic agriculture: glossary on organic agriculture. FAO, Rome, pp 1–173

FAO (2011) Save and grow. A policymaker's guide to the sustainable intensification of smallholder crop production. FAO, Rome, 102 p

FAO (2012) World agriculture towards 2030/2050: the 2012 revision. FAO, Rome, Italy. Retrieved from http://www.fao.org/economic/esa/esag/en

FAO (2016) Global forest resources assessment 2015: how have the World's forests changed? FAO, Rome

FAO (2017) Nutrition-sensitive agriculture and food systems in practice - options for intervention. FAO, Rome, Italy. Retrieved from http://www.fao.org/3/a-i7848e.pdf

Firbank LG, Elliott J, Drake B, Cao Y, Gooday R (2013) Evidence of sustainable intensification among British farms. Agric Ecosyst Environ 173:58–65

Foley JA, DeFries R, Asner GP, Barford C, Bonan G, Carpenter SR, Chapin FT, Coe MT, Daily GC, Gibbs HK, Helkowski JH, Holloway T, Howard EA, Kucharik CJ, Monfreda C, Patz JA, Prentice IC, Ramankutty N, Snyder PK (2005) Global consequences of land use. Science 309 (5734):570–574

Foley JA, Ramankutty N, Brauman KA, Cassidy ES, Gerber JS, Johnston M, Mueller ND, O'Connell C, Ray DK, West PC, Balzer C, Bennett EM, Carpenter SR, Hill J, Monfreda C, Polasky S, Rockström J, Sheehan J, Siebert S, Tilman D, Zaks DP (2011) Solutions for a cultivated planet. Nature 478:337–342. https://doi.org/10.1038/nature10452

Gaba S, Bretagnolle RT, Philippot L (2014) Managing biotic interactions for ecological intensification of agroecosystems. Front Ecol Evol 2:29. https://doi.org/10.3389/fevo.2014.00029

Garnett T, Godfray HCJ (2012) Sustainable intensification in agriculture. Navigating a course through competing food system priorities. (Food Climate Research Network and the Oxford Martin Programme on the Future of Food, University of Oxford

Garnett T, Appleby MC, Balmford A, Bateman IJ, Benton TG, Bloomer P, Burlingame B, Dawkins M, Dolan L, Fraser D, Herrero M, Hoffmann I, Smith P, Thornton PK, Toulmin C, Vermeulen SJ, Godfray HCJ (2013) Sustainable intensification in agriculture: premises and policies. Science 341(6141):33–34

Gemmill-Herren B, Miguez FE, Dicks LV (2019) Policies for ecological intensification of crop production. Trends Ecol Evol 34(4):282–286

Gibon A, Sibbald AR, Flamant JC, Lhoste P, Revilla R, Rubino R, Sørensen JT (1999) Livestock farming systems research in Europe and its potential contribution for managing towards sustainability in livestock farming. Livest Prod Sci 61(2–3):121–137

Godfray HCJ, Beddington JR, Crute IR, Haddad L, Lawrence D, Muir JF, Pretty J, Robinson S, Thomas SM, Toulmin C (2010) Food security: the challenge of feeding 9 billion people. Science 327(5967):812–818

Gomes L, Pailleux J, Dedieu B, Claudete R, Cournut S (2014) An approach for assessing the ecological intensification of livestock systems. International Farming Systems Association, Berlin, p 13, hal-02083429

Griffon M (2013) Qu'est-cequel'agricultureécologiquement intensive? Edition Quae, Versailles, France, 220 p

Haussmann B (2011) How CCRP could support the agro-ecological intensification (AEI) of the Sahel. http://ccrpaeix.drupalgardens.com/sites/ccrpaeix.drupalgardens.com/files/hausssman_ccrp_aei.pdf. Accessed 26 Oct 2013

Hiiesalu I, Bahram M, Tedersoo L (2017) Plant species richness and productivity determine the diversity of soil fungal guilds in temperate coniferous forest and bog habitats. Mol Ecol 26:4846–4858

Hochman Z, Carberry PS, Robertson MJ, Gaydon DS, Bell LW, McIntosh PC (2013) Prospects for ecological intensification of Australian agriculture. Eur J Agron 44:109–123

IFA (2013) Annual Report-2012. Report of the council of the 85th year to be presented to the annual general meeting on 22 May 2013

Jhariya MK, Bargali SS, Raj A (2015) Possibilities and perspectives of agroforestry in Chhattisgarh. In: Zlatic M (ed) Precious forests-precious earth. InTech, Croatia, Europe, pp 237–257, 286 p https://doi.org/10.5772/60841 ISBN: 978-953-51-2175-6

Jhariya MK, Yadav DK, Banerjee A (2018a) Plant mediated transformation and habitat restoration: phytoremediation an eco-friendly approach. . In: Gautam A, Pathak C (eds) Metallic contamination and its toxicity. Daya Publishing House, A Division of Astral International Pvt. Ltd, New Delhi, pp 231–247. ISBN: 9789351248880

Jhariya MK, Banerjee A, Yadav DK, Raj A (2018b) Leguminous trees an innovative tool for soil sustainability. In: Meena RS, Das A, Yadav GS, Lal R (eds) Legumes for soil health and sustainable management. Springer, pp 315–345. ISBN 978-981-13-0253-4 (eBook), ISBN: 978-981-13-0252-7 (Hardcover). https://doi.org/10.1007/978-981-13-0253-4_10

Jhariya MK, Banerjee A, Meena RS, Yadav DK (2019a) Sustainable Agriculture, Forest and Environmental Management. Springer Nature Singapore Pte Ltd., 152 Beach Road, #21-01/04 Gateway East, Singapore 189721, Singapore. eISBN: 978-981-13-6830-1, Hardcover ISBN: 978-981-13-6829-5. https://doi.org/10.1007/978-981-13-6830-1. p 606

Jhariya MK, Yadav DK, Banerjee A (2019b) Agroforestry and climate change: issues and challenges. Apple Academic Press Inc., CRC Press- A Taylor and Francis Group, US & Canada. ISBN: 978-1-77188-790-8 (Hardcover), 978-0-42957-274-8 (E-book). p 335. https://doi.org/10.1201/9780429057274

Karamura EB, Jogo W, Rietveld A, Ochola D, Staver C, Tinzaara W, Karamura DA, Kubiriba J, Weise S (2013) Effectiveness of agroecological intensification practices in managing pests in smallholder banana systems in east and Central Africa. Acta Horticult (ISHS) 986:109–126

Khan N, Jhariya MK, Yadav DK, Banerjee A (2020a) Herbaceous dynamics and CO_2 mitigation in an urban setup- a case study from Chhattisgarh, India. Environ Sci Pollut Res 27(3):2881–2897. https://doi.org/10.1007/s11356-019-07182-8

Khan N, Jhariya MK, Yadav DK, Banerjee A (2020b) Structure, diversity and ecological function of shrub species in an urban setup of Sarguja, Chhattisgarh, India. Environ Sci Pollut Res 27 (5):5418–5432. https://doi.org/10.1007/s11356-019-07172-w

Kremen C, Miles A (2012) Ecosystem services in biologically diversified versus conventional farming systems: benefits, externalities, and trade-offs. Ecol Soc 17

Kumar S, Meena RS, Jhariya MK (2020) Resources Use Efficiency in Agriculture. Springer Nature Singapore Pte Ltd., 152 Beach Road, #21-01/04 Gateway East, Singapore 189721, Singapore. eBook ISBN: 978-981-15-6953-1, Hardcover: 978-981-15-6952-4. pp 760. https://doi.org/10.1007/978-981-15-6953-1

Kuyper TW, Struik PC (2014) Epilogue: global food security, rhetoric, and the sustainable intensification debate. Curr Opin Environ Sustain 8:71–79

Lal R (2008) Carbon sequestration. Philos Trans Royal Soc B Biol Sci 363(1492):815–830. https://doi.org/10.1098/rstb.2007.2185

Lema Z, Mulema AA, Le Borgne E, Duncan AJ (2016) Innovation platforms for improved natural resource management and sustainable intensification in the Ethiopian highlands. In: Dror I,

Cadilhon JJ, Schut M, Misiko M, Maheshwari S (eds) Innovation platforms for agricultural development: evaluating the mature innovation platforms landscape. Routledge
Mahon N, Crute I, Di BM (2018) Towards a broad-based and holistic framework of sustainable intensification indicators. Land Use Policy 77:576–597
Mao LL, Zhang LZ, Zhang SP, Evers JB, van der Werf W, Wang JJ, Spiertz H (2015) Resource use efficiency, ecological intensification and sustainability of intercropping systems. J Integr Agric 14(8):1542–1550
Matson PA, Parton WJ, Power AG (1997) Agricultural intensification and ecosystem properties. Science 277:504–509
McCune NM, González YR, Alcántara EA, Martínez OF, Fundora CO, Arzola NC, Cairo PC, D'Haesse M, DeNeve S, Hernández FG (2011) Global questions, local answers: soil management and sustainable intensification in diverse socioeconomic contexts of Cuba. J Sustain Agric 35(6):650–670
Meena RS, Lal R (2018) Legumes for soil health and sustainable management. Springer, Singapore, p 541. ISBN 978-981-13-0253-4 (eBook), ISBN: 978-981-13-0252-7(Hardcover). https://doi.org/10.1007/978-981-13-0253-4_10
Meena RS, Kumar V, Yadav GS, Mitran T (2018) Response and interaction of *Bradyrhizobium japonicum* and Arbuscular mycorrhizal fungi in the soybean rhizosphere: a review. Plant Growth Regul 84:207–223
Meena RS, Kumar S, Datta R, Lal R, Vijaykumar V, Brtnicky M, Sharma MP, Yadav GS, Jhariya MK, Jangir CK, Pathan SI, Dokulilova T, Pecina V, Marfo TD (2020) Impact of agrochemicals on soil microbiota and management: a review. Land (MDPI) 9(2):34. https://doi.org/10.3390/land9020034
Meena RS, Lal R, Yadav GS (2020a) Long term impacts of topsoil depth and amendments on soil physical and hydrological properties of an Alfisol in Central Ohio, USA. Geoderma 363:1141164
Meena RS, Lal R, Yadav GS (2020b) Long-term impact of topsoil depth and amendments on carbon and nitrogen budgets in the surface layer of an Alfisol in Central Ohio. Catena 194:104752
Milder JC, Garbach K, DeClerck FAJ, Driscoll L, Montenegro M (2012) An assessment of the multifunctionality of agroecological intensification. A report prepared for the Bill and Melinda Gates Foundation, September 2012
Montesinos D (2019) Forest ecological intensification. Trends Plant Sci 1799:2. https://doi.org/10.1016/j.tplants.2019.03.009
Mueller ND, Gerber JS, Johnston M, Ray DK, Ramankutty N, Foley JA (2012) Closing yield gaps through nutrient and water management. Nature 490:254–257. https://doi.org/10.1038/nature11420
Noponen MRA, Haggar JP, Edwards-Jones G, Healey JR (2013) Intensification of coffee systems can increase the effectiveness of REDD mechanisms. Agric Syst 119:1–9
Ochola D, Jogo W, Ocimati W, Rietveld A, Tinzaara W, Karamura DA, Karamura EB (2013) Farmers' awareness and perceived benefits of agro-ecological intensification practices in banana systems in Uganda. Afr J Biotechnol 12(29):4603–4613
Palm CA, Smukler SM, Sullivan CC, Mutuo PK, Nyadzi GI, Walsh MG (2010) Identifying potential synergies and trade-offs for meeting food security and climate change objectives in sub-Saharan Africa. Proc Natl Acad Sci U S A 107:19661–19666
Paul C, Knoke T (2015) Between land sharing and land sparing-what role remains for forest management and conservation? Int For Rev 17:210–230
Petersen B, Snapp S (2015) What is sustainable intensification? Views from experts. Land Use Policy 46:1–10
Phalan B, Onial M, Balmford A, Green RE (2011) Reconciling food production and biodiversity conservation: land sharing and land sparing compared. Science 333(6047):1289–1291
Pinho RC, Miller RP, Alfaia SS (2012) Agroforestry and the improvement of soil fertility: a view from Amazonia. Appl Environ Soil Sci 2012:1–11. https://doi.org/10.1155/2012/616383

Pretty JN (1997) The sustainable intensification of agriculture. Nat Res Forum 21(4):247–256
Pretty JN (2008) Agricultural sustainability: concepts, principles and evidence. Philos Trans R Soc 363(1491):447–465
Pretty JN, Bharucha ZP (2014) Sustainable intensification in agricultural systems. Ann Bot 114:1571
Pretty JN, Toulmin C, Williams S (2011) Sustainable intensification in African agriculture. Int J Agric Sustain 9(1):5–24
Raj A, Jhariya MK, Yadav DK, Banerjee A, Meena RS (2019a) Agroforestry: a holistic approach for agricultural sustainability. In: Jhariya MK, Banerjee A, Meena RS, Yadav DK (eds) Sustainable agriculture, forest and environmental management. Springer Nature Singapore Pte Ltd., 152 Beach Road, #21-01/04 Gateway East, Singapore 189721, Singapore, pp 101-131, 606. eISBN: 978-981-13-6830-1, Hardcover ISBN: 978-981-13-6829-5. https://doi.org/10.1007/978-981-13-6830-1
Raj A, Jhariya MK, Banerjee A, Yadav DK, Meena RS (2019b) Soil for sustainable environment and ecosystems management. In: Jhariya MK, Banerjee A, Meena RS, Yadav DK (eds) Sustainable agriculture, forest and environmental management. Springer Nature Singapore Pte Ltd., 152 Beach Road, #21-01/04 Gateway East, Singapore 189721, Singapore. pp 189–221, 606, eISBN: 978-981-13-6830-1, Hardcover ISBN: 978-981-13-6829-5. https://doi.org/10.1007/978-981-13-6830-1
Raj A, Jhariya MK, Yadav DK, Banerjee A (2020) Climate change and agroforestry systems: adaptation and mitigation strategies. Apple Academic Press Inc., CRC Press- A Taylor and Francis Group, US & Canada. ISBN: 9781771888226. p 383. https://doi.org/10.1201/9780429286759
Royal Society London (2009) Reaping the benefits: science and the sustainable intensification of global agriculture. The Royal Society, London, 72 p
Ruerd R, Lee D (2000) Combining internal and external inputs for sustainable intensification. International Food Policy Research Institute (IFPRI), Washington, DC, pp 1–2
Singh BR, McLaughlin MJ, Brevik EC (eds) (2017) The nexus of soils, plants, animals and human health. Catena-Schweizerbart, Stuttgart, p 156
Singh NR, Jhariya MK (2016) Agroforestry and agrihorticulture for higher income and resource conservation. In: Narain S, Rawat SK (eds) Innovative technology for sustainable agriculture development. Biotech Books, New Delhi, India pp 125–145. ISBN: 978-81-7622-375-1
Smith A, Snapp S, Chikowo R (2017) Measuring sustainable intensification in smallholder agro-ecosystems: a review. Glob Food Secur 12:127–138
Stachetti RG, Roberto MC (2018) Sustainability assessment of ecological intensification practices in coconut production. Agric Syst 165:71–84
Struik PC, Kuyper TW (2017) Sustainable intensification in agriculture: the richer shade of green- a review. Agron Sustain Dev 37:39
Thompson I, Mackey B, McNulty S, Mosseler A (2009) Forest resilience, biodiversity, and climate change: a synthesis of the biodiversity/resilience/stability relationship in forest ecosystems. Secretariat of the Convention on Biological Diversity, Montreal. Tech. Ser. No. 43
Tilman D, Balzer C, Hill J, Befort BL (2011) Global food demand and the sustainable intensification of agriculture. PNAS 108(50):20260–20264
Tittonell P (2014) Ecological intensification of agriculture-sustainable by nature. Curr Opin Environ Sustain 8:53–61. https://doi.org/10.1016/j.cosust.2014.08.006
Tittonell P, Giller KE (2013) When yield gaps are poverty traps: the paradigm of ecological intensification in African smallholder agriculture. Field Crop Res 143:76–90. https://doi.org/10.1016/j.fcr.2012.10.007
Tscharntke T, Klein AM, Kruess A, Steffan-Dewenter I, Thies C (2005) Landscape perspectives on agricultural intensification and biodiversity – ecosystem service management. EcolLett 8:857–874
Tsiafouli MA, The'bault E, Sgardelis SP, de Ruiter PC, van der Putten WH, Birkhofer K, Hemerik L, de Vries FT, Bardgett RD, Brady MV, Bjornlund L, Jørgensen HB,

Christensen S, Hertefeldt TD, Hotes S, Gera Hol WH, Frouz J, Liiri M, Mortimer SR, Setala H, Tzanopoulos J, Uteseny K, Pižl V, Stary J, Wolters V, Hedlund K (2015) Intensive agriculture reduces soil biodiversity across Europe. Glob Chang Biol 21:973–985

USAID (2013) Feed the Future. Program for sustainable intensification. http://feedthefuture.gov/sites/default/files/resource/files/ftf_factsheet_fsicsustainableint_nov2013.pdf. Accessed June 2015

van der Plas F, Manning P, Soliveres S, Allan E, Scherer-Lorenzen M, Verheyen K, Fischer M (2016) Biotic homogenization can decrease landscape-scale forest multifunctionality. Proc Natl Acad Sci U S A 113:3557–3562

Waste and Resources Action Programme (2011) Environmental audit committee: Written evidence submitted by WRAP. Retrieved from http://www.publications.parliament.uk/pa/cm201012/cmselect/cmenvaud/879/879vw20.htm

Wezel A, Soboksa G, McClelland S, Delespesse F, Boissau A (2015) The blurred boundaries of ecological, sustainable, and agroecological intensification: a review. Agron Sustain Dev 35:1283–1295

Xie H, Huang Y, Chen Q, Zhang Y, Wu Q (2019) Prospects for agricultural sustainable intensification: a review of research. Land 8(157):1–27

Ecological Intensification for Sustainable Agriculture in South Asia

6

Akbar Hossain, Sukamal Sarkar, Md. Atikur Rahman, Rajan Bhatt, Sourav Garai, Saikat Saha, Mst. Tanjina Islam, and Ram Swaroop Meena

Abstract

In South Asian countries huge population pressure, foster urbanization and industrialization lead to dwindle the agro-ecological resources like land, water, agroforestry, human and climatic stability. Agricultural intensification has been accompanied by a set of innovations, collectively referred to as the Green Revolution, which has increased food production significantly. However, the intensification poses a major threat to the physical environment such as the loss of natural resources, genetic diversity, land degradation and non-judicious application of water and nutrient. Recent evidence recommends that ecological intensification (EI) of distinctive agriculture particularly in rice-based cropping

A. Hossain (✉)
Bangladesh Wheat and Maize Research Institute, Dinajpur, Bangladesh

S. Sarkar · S. Garai
Department of Agronomy, Bidhan Chandra Krishi Viswavidyalaya, Nadia, West Bengal, India

M. A. Rahman
Species Research Center, Bangladesh Agricultural Research Institute, Bogra, Bangladesh

R. Bhatt
Regional Research Station, Kapurthala, Punjab Agricultural University, Ludhiana, Punjab, India

S. Saha
Nadia Krishi Vigyan Kendra, Bidhan Chandra Krishi Viswavidyalaya, Nadia, West Bengal, India

M. T. Islam
Department of Agronomy, Hajee Mohammad Danesh Science and Technology University, Dinajpur, Bangladesh

R. S. Meena
Department of Agronomy, Institute of Agricultural Sciences, Banaras Hindu University, Varanasi, Uttar Pradesh, India

M. K. Jhariya et al. (eds.), *Ecological Intensification of Natural Resources for Sustainable Agriculture*, https://doi.org/10.1007/978-981-33-4203-3_6

systems of South Asia can preserve the food production in addition with environmental welfares. Since EI of sustainable agriculture could organize the precise constituents of biodiversity and can be used to either balance synthetic-input as well as expand the productivity without adverse effect on agricultural production. Besides, it is also reported that the performance of organic and agro-ecological farming system is much better in case of ecosystem services relevant to climate change, carbon sequestration from a soil depth of 30 cm and other parameters like soil water holding capacity, etc. Therefore, scientists and policymakers consequently and progressively have emphasized the welfares of ecological-intensifying agriculture to a sociable way towards food, nutritional, environmental and livelihood security by assisting biodiversity and enhance the ecosystem services. This chapter highlights the available agro-ecological resources for improving crop productivity to obtain the goal of sustainable agricultural intensification without negotiating the agricultural outputs.

Keywords

Agriculture · Ecology · Environment · Food security · South Asia

Abbreviations

AEI	Agro-environmental indicators
AI	Agricultural intensification
C	Carbon
CA	Conservation agriculture
CH_4	Methane
CO_2	Carbon dioxide
EI	Ecological intensification
FAO	Food and Agricultural Organization
GDP	Gross domestic product
GHGs	Greenhouse gases
ICM	Integrated crop management
IGP	Indo-Gangetic plains
INM	Integrated nutrients management
IPCC	Intergovernmental Panel on Climate Change
IPNS	Integrated plant nutrition system
LCC	Leaf colour chart
N	Nitrogen
N_2O	Nitrous oxide
PA	Precision agriculture
RCTs	Resource conservation technologies
RUE	Resources use efficiency
RWCS	Rice wheat cropping systems

RYE	Rice-yield-equivalent
SA	South Asia
SID	Index of diversity
SOM	Soil organic matter
SSNM	Site-specific nutrient management
UAA	Utilized agricultural area

6.1 Introduction

World population is anticipated to grow by more than 9.1 billion between 2009 and 2050 and need to increase the overall nourishment about 70% between 2005–2007 and 2050 (FAO 2009). For example, for both food and animal feed, cereals are expected to extend approximately 3 billion tonnes in 2050 as compared to the demand of today's almost 2.1 billion tonnes (FAO 2009). At the same time, IPCC (2007) already projected that agriculture will face numerous constraints in the twenty-first century due to the changing climate (FAO 2009). Agro-ecology refers to an ecological concept and methodology that enhances agricultural productivity without altering the sustainability in the long term. It is considered as the beneficial for plant and soil biodiversity, natural resource conservation, recycling and climate change based on the adoption of conservation practices, nutrient restorative crop incorporation, and diversification of existing cropping system and optimization of resource use efficiency (RUE) (Kumar et al. 2020; Meena et al. 2018; Meena and Lal 2018). The term "agroecological practices" evolved in the 1980s to support the sufficient crop production to feed the world population, nutritional security and financial viability with the consideration of biodiversity conservation, quality life, social and climate-resilient. Hence, the aim and scope of agro-ecology are in a tune with sustainable agriculture. While to meet the food demand of increasing population, it is impossible to increase the food production through conventional intensification of agriculture without causing significant damage to the environment (Lundgren and Fergen 2011; Tschumi et al. 2015). Emergent proof recommends that ecological intensification (EI) of typical agriculture can defend the food production in additional with environmental welfares (Kleijn et al. 2019; Meena et al. 2020a, b). EI has been proposed as an environment-constructed substitute that accompaniments or changes *exterior involvement* such as agrochemicals, with production-supportive ecological procedures, to withstand agricultural productivity while mitigating the hostile effects on the environment (Bommarco et al. 2013). Since EI is established on the assumption that conveyance of *ecosystem services* is sub-optimal in high inputs farming (Jonsson et al. 2012; Poeplau and Don 2015; Venter et al. 2016), which organize the precise constituents of *biodiversity* and can be used to either balance synthetic-input as well as improve the agricultural output without adversely influence on production (Kleijn et al. 2019; Fig. 6.1).

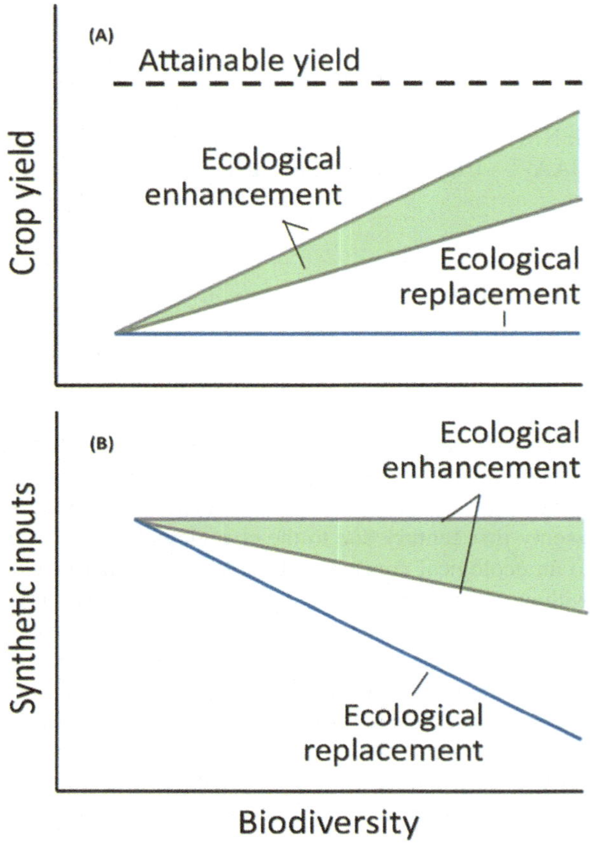

Fig. 6.1 Ecological intensification in more-input agricultural systems. Ecological replacement assumes that growing biodiversity can increasingly substitute artificial contributions through association with ecosystem facilities (**a**) while preserving continuous crop productivity (**b**). For easiness, linear associations are shown, while in the model, drenching curves are to be projected. (Adapted: Kleijn et al. 2019)

To be more convincing to farmers, scientific studies on EI need to address the costs and benefits that are most relevant to farmers. Potential costs of EI should be an integral component of research (Van der Horst 2011; Liebman et al. 2016). The benefits of EI generally improve with increased targeting of the specific species groups providing the bulk of the services to a particular crop (Tschumi et al. 2015). Although Garibaldi et al. (2017) revealed that proven benefits alone do not guarantee uptake of management practices. For example, due to a 15–16% cost saving, climate-smart agriculture (conservation tillage) in wheat has met with large-scale adoption in south Asia, but an adaption of the technology is limited in Mexico and Southern Africa not withstanding a proof of greater and extra stable yields both for maize and wheat (Erenstein et al. 2012).

In South Asian countries huge population pressure, foster urbanization and industrialization lead to dwindle the agro-ecological resources like land, water, agroforestry, human and climatic stability. Agricultural intensification has been accompanied by a set of innovations, collectively referred to as the "Green Revolution", which has increased food production significantly. However, the crop intensification postures a major threat to the physical environment such as the loss of natural resources, genetic diversity, land degradation and injudicious application of water and nutrient. Recent evidence recommends that EI of distinctive agriculture particularly in rice-based cropping systems of South Asia can preserve the food production in addition to environmental welfares. Since EI of sustainable agriculture could organize the precise constituents of *biodiversity* and can be used to either balance synthetic-input as well as expand the productivity without adversely affecting agricultural production (Kleijn et al. 2019; Meena et al. 2020). Therefore, scientists and policy-makers are consequently and progressively emphasized the welfares of ecological-intensifying agriculture to embrace EI as an ecologically sociable way towards food security by assisting the execution of biodiversity and increasing practices of ecosystem service (Blanco-Canqui et al. 2015; Potts et al. 2016). This chapter highlights the available agro-ecological resources for improving crop productivity to obtain the goal of agricultural sustainability without negotiating the agricultural outputs.

6.2 Intensive Use of Ecosystem Services and Its Adverse Effect on Agriculture

Ecosystem services simply referred to various types of benefits to humans gifted by the natural environment more precisely from properly working ecosystems (MEA 2005). Such ecosystems include agro-ecosystem, forest ecosystem and aquatic ecosystem, etc. These all afterwards contribute to the food ecosystems. Ecosystem services comprise four distinct units, viz., (1) Provisioning services, (2) Regulating services, (3) Cultural services, (4) Supporting services. First three services directly affect the people but the necessity of supporting services is to maintain other services (La Notte et al. 2017). On the other hand, rigorous use of these services leads to offer various benefits to mankind and in short, this is often called as intensive use of ecosystem services (MEA 2005).

The adverse effect refers to loss and damage occurring due to exhaustive utilization of ecosystem services. Due to excessive utilization of these ecosystem services, short term or long term effects have been reported in different research articles (Zommers et al. 2016; Parker et al. 2017). These ecosystem services play an important part in regulating climatic factors, human and natural systems also (Fig. 6.2).

6.2.1 Intensive Use of Ecosystem Services in Rice–Wheat Ecosystem

Traditional Rice–Wheat Cropping System (RWCS) has provided ecosystem services from ancient times but also have a major drawback in lodging disservices or losses and damages (Power 2010). In various ways, i.e. biodiversity loss, contamination of agrochemicals, pesticide poisoning in beneficial organisms and emissions of greenhouse gases are some of the examples in RWCS for accommodating adverse effects

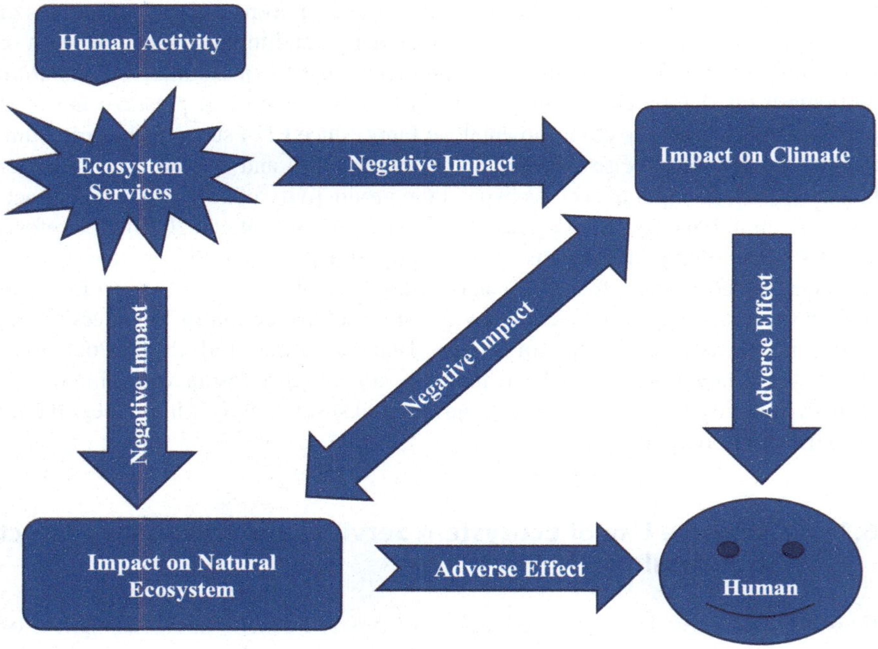

Fig. 6.2 Intra-relationship between ecosystem services and their possible impact on the natural ecosystem, climate and human

(Dale and Polasky 2007; Zhang et al. 2007). Most of the time, these adverse effects inevitably harm humans through the natural system (pesticide contamination in drinking water) or by manipulating climatic condition (global warming) (Power 2010).

6.2.2 Shortage in Freshwater

It is estimated that the use of irrigation water or water use for urban or industry, collected from rivers or lakes showing as doubled between periods of 1960 and 2000. By the by, 70% of freshwater used for food production purpose of different crops across the globe, RWCS shared the major portion (Du Preez et al. 2020). This excessive depletion tends to the shortage of available freshwater sources across the globe. Sometimes, this excess amount is emitted from a groundwater pool, but the most alarming scenario is the rate of groundwater recharge is slower than recharge (Du Preez et al. 2020). It is also estimated that water availability in India is showing a rapidly decreasing trend (Fig. 6.3). In 2001, per capita, water availability is only 1820 m^3 which is about 5177 m^3 in 1951 which can create an alarming situation in percent deficit in water. This situation can only manage by judicious and scientific

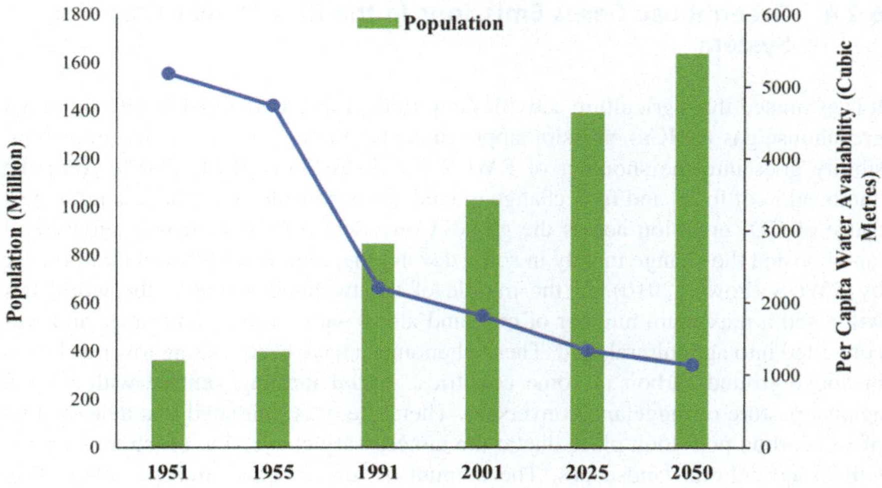

Fig. 6.3 Growth of population and declining trend in per capita water availability in India (Data source: National Commission for Women 2020)

use of our valuable water resource especially the use of water in the agricultural sector.

6.2.3 Nutrient Cycling

Agriculture has potential effects on biogeochemical cycles and the availability of nutrients in ecosystems from a local level to a global extent (Vitousek et al. 1997; Galloway et al. 2004). Nitrogen and phosphorus, namely two nutrients mostly inhibit the biological production in agricultural as well as natural ecosystems, which are heavily applied in RWCS (Vitousek et al. 1997). These nutrients are highly mobile, contaminated both groundwater and surface water, which resulted in numerous off-putting consequences for human health and on the environment also. It is estimated that 20% of N fertilizer applied in agricultural production run away into aquatic or marine ecosystems (Galloway et al. 2004). Nutrient loss in RW ecosystems mainly contaminates groundwater and increases pollution level in drinking water through augmented nitrate level (Bouwman et al. 2009). It is also reported that trade-offs in ecosystem services magnify manifold due to the impact of climate change in various parts across the globe (Power 2010).

6.2.4 Greenhouse Gases Emissions in the Rice–Wheat Cropping System

It is estimated that agriculture activities contributed an amount of 12–14% of global greenhouse gas (GHGs) emission approximately, amongst them maximum responsibility goes into the shoulder of RWCS (US-EPA 2006; IPCC 2007). After the combustion of fossil and fuel, change in land use is considered as the second largest cause of CO_2 emission across the globe. Conversion of this land into agricultural land boosted the change mostly in some developing countries followed traditionally by RWCS (Power 2010). In the middle of the twentieth century, the world has witnessed a maximum number of cropland along with pastures and rangeland was converted into agricultural land. These phenomena have been leading towards losses in above-ground carbon in some countries, started farming venture with RWCS against pasture or rangeland conversion. Therefore, it is estimated that a heavy loss of soil carbon pool took place due to the unnecessary conversion of natural ecosystem to agricultural landscapes. The amount of loss is approximately 30–50% in temperate regions over 50–100 years and in the tropics, 50–75% over a period from 20 to 50 years (IPCC 2007).

The contribution of agricultural activities in greenhouse emission is compiling in various ways. It is accredited that 49% of global methane (CH_4) emission and annually 66% of nitrous oxide (N_2O) emitted from agricultural production system across the globe and a large amount is shared by RWCS itself (Tubiello et al. 2013). Although there is great confusion in estimating agricultural contribution and total N_2O emission occur during the nitrogen cycle in the soil profile. But it is a recognized fact that the emission rate is increased significantly after nitrogen application to crops, most of the time when more doses will be applied than the amount a plant can be taken up in RWCS.

Nitrogen in various forms added to soils and only an amount of 50% N used as fertilizer is taken by the crop. Rest 45% is lost in GHGs emission and aquatic systems, only 2–5% is stored as nitrogen in the soil (Galloway et al. 2004). Besides, flooded cultivation in RWCS contributes a major portion of GHGs emissions through various CH_4- emitting soil microbes (UNEP 2016). Burning of crop residue practice after in rice–wheat ecosystems has been producing both CH_4 and N_2O up to a great extent.

In the global context of agricultural production, rice occupies the second largest area after wheat. More than one billion people consume rice as their staple food mostly in South and Southeast Asian countries. Rice crop occupies around 10% cultivated area across the globe and 20% of the global natural wetland area (Timsina and Connor 2001). With an increasing demand for boosting rice grain production due to rising population pressure in South Asian rice-growing countries, over-exploitation of natural resources is a major concern. It is also calculated that one of the dominant sources of methane emission to the atmosphere is rice fields (estimated as 15% of global methane emission (IPCC 1994)).

Rice cultivation includes the traditional practice of flood irrigation, unlike other crops. The available depth of water must be up to a depth of 4–6 cm. It is also

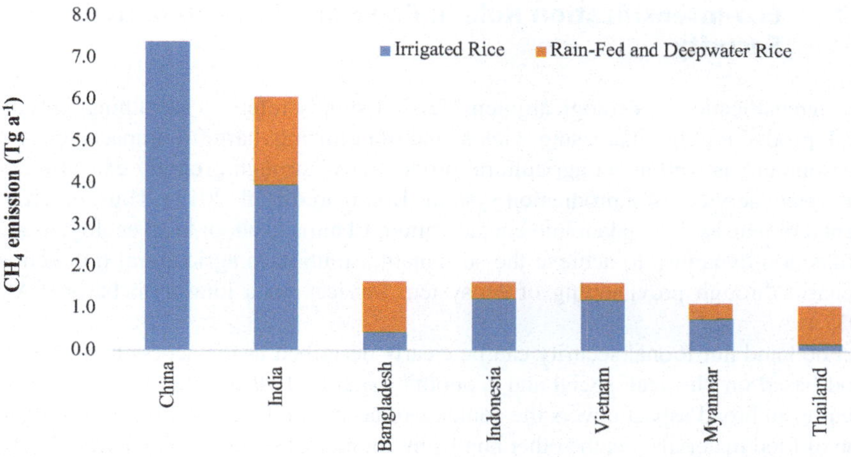

Fig. 6.4 Estimated methane emission from rice-growing ecology of different leading rice-producing countries (Data source: Yan et al. 2009)

reported that rice is a semi-aquatic crop and like another semi-aquatic crop it also produces methane and releases to the environment (Wassmann et al. 1993). Methane emission from rice fields has been identified and recognized as a major contributor to GHGs. Over the last three decades, considerable advancement has been made to estimate global CH_4 emission from rice fields. It is reported from different research articles that Southeast Asian countries contribute a major share in CH_4 emission from rice fields due to a large area in these countries are under RWCS (Yan et al. 2009). China and India, the most populous country across the globe, comprise major acreage under rice farming with an area of 20.0% and 28.5% of the total rice-growing area worldwide (FAOSTAT 2014). It is also reported that around 90% of the rice fields in China are adequately irrigated (Zhang et al. 2016), with different techniques and numerous strategies related to irrigation in recent decades (Chen et al. 2013). Annually these two countries contribute around 7.4 and 6.1 Tg a^{-1} of CH_4 from the different rice-growing ecologies (Fig. 6.4).

The hypothesis of ecosystem services lies in its potential to promote environmental conservation. Advantages derived from ecosystem services encourage understanding the humans' dependence on nature. But all human activities related to the ecosystem services may not be environmentally viable and ecologically sustainable. Thus, any intensive use of ecosystem services may seriously affect the components of ecosystem services, which ultimately impeded the human endeavour.

6.3 Eco-Intensification Role in Food and Environmental Security

Eco-intensification or ecological intensification simply refers to sustaining agricultural production by increasing yields and minimizing harmful impacts on the environment as well as on agricultural productivity, through properly execution of ecosystem services of a production system (Bommarco et al. 2013). Thus, environmental dimensions are taken into consideration, EI aims to adopt local ecologies and production dynamics to achieve the sustainable solution in agricultural production scenario through provisioning of ecosystem services on a long-term basis (FAO 2017).

Food and nutritional security can be clearly described as the access to nutritious food based on physical, social and economic criteria of all people over time. Food security in broad aspect covers the contexts of accessibility, availability and utilization of food materials. On the other hand, environmental security ascribes the threats occur through environmental dealings and individual trends, communities or nations. In other words, environment security is the viability of the environment to support life system across the globe with its different components (Mathews 1989). Environment security may cover an extensive assortment of issues or events related to the global environment: recent trends with climate change. Impact of individual divergence coupled with intercontinental relations among different nations on environmental issues also may be a topic for the centre of attraction (Brown 1977; Westing 1986).

Worldwide the scenario of food production has changed drastically. An increase in the amount of overall food production has increased manifold over past decades but food insecurity over a mass population across the globe has been rising day by day since the past two decades (FAO 2009). On the other hand, an abrupt undermining of critical natural resources to support the increasing demand for production of food grain leading towards restraining natural resources used for food grain production (Daily et al. 1998). This prime cause of uncontrolled depletion of different resources in one side hinders the future production scenario of food grains on the other side signalling towards a great danger on the environment. However, meeting the demand of agricultural foodstuffs for an ever-increasing population across the globe through the traditional way of intensification in crop rising is considered to be impossible without mounting significant damage on the environment (Foley et al. 2005; Tilman et al. 2011; Lundgren and Fergen 2011). Therefore, EI has been reported as an alternative way of crop production along with minimizing the risk associated with negative environmental impact also. Replacing external inputs like chemical fertilizers and chemical pesticides by using of nature-based substitutes for sustainable growth in agriculture production to make a proper blueprint for controlling environmental damages are the prime objectives behind EI (Cassman 1999; Bommarco et al. 2013). Besides, EI emphasizes on releasing of ecosystem services to optimize high utilization of external inputs and properly manage of specific biodiversity regulating components (Kremen et al. 2002; Jonsson et al. 2012; Ulén et al. 2010), which usually either complements through artificial inputs and help in boosting agricultural productivity or reduced negative costs over the environment through replacing of artificial inputs (Bommarco et al. 2013).

EI very often relates to practices associated with inputs and integration of multiple technologies through sustainable intensification strategies to sustain the agricultural production without creating a negative impact on our mother environment (Kassie et al. 2015). Practising of conservation agriculture, i.e. minimum soil tilth, proper rotation of field crops and intercropping with legume crops, the establishment of new crop varieties. Soil moisture conservation through mulching, integrated techniques for pest management along with the application of bio-fertilizers and organic fertilizers are some of the important evidence (Falconnier et al. 2018). Therefore, EI practices intend to increase both the per-unit production as well as the flexibility of agricultural production system by adding a range of methods in the process of agricultural production without or minimum depleting of the natural resource base. Many research articles also pointed out that the local specific practices and strategies related to EI depend upon the behaviour and choices of the farmers (Kassie et al. 2015; Ndiritu et al. 2014; Ortega et al. 2016). Integration of numerous components related to farming without harnessing the environment is a smouldering subject across the globe (Vanlauwe et al. 2014; Tittonell and Giller 2013; Falconnier et al. 2018). Practices, their situation specificity, farmers' adaptability and many other factors are also related to the worldwide EI process. However, systemic review and proper research investigation in this topic remain blank in different international research platforms (Franke et al. 2014; DuriauxChavarría et al. 2018; Cortner et al. 2019) and there must be a provision in onwards chapters for a thorough knowledge on EI for food and environmental security.

South Asian countries consist almost half of the world's inhabitants but producing a considerable amount of food grains within a limited land area (Xie et al. 2019). But, urbanization coupled with industrialization occupied the cultivable land leads to a decline in land area for agricultural practices. Besides, labour forces shortage due to migration threatens food security across different nations in this continent (Xie et al. 2019). All at once, sequentially to make a most of agricultural production application of synthetic artificial inputs like chemical fertilizers, pesticides are inevitable (Long et al. 2018). Excessive use of these agrochemicals however boosting the production but put a heavy load on the environment. As a result, payment for ecological cost has been increasing day by day. Therefore, intensive use of eco-friendly components and meet the food demand of an enormous population along with minimum payment of ecological costs are the major challenges now a day (Zuo et al. 2018; Jhariya et al. 2019a, b). Recent trends in researches also suggested that numerous phenomena include the provisioning of ecological services, the sustainability of agricultural production system and minimize the environmental pollution loads, systematic and comprehensive research on EI in South Asian countries (Yu et al. 2017). Therefore, categorization and summarizing the research outputs related to EI and its role in agricultural production system have great importance for further study in this arena.

6.4 Agricultural Intensification for Sustainability

6.4.1 Modern Agriculture

After a steady increase of agricultural productivity during twentieth century now become plateaus in many countries (Cassman et al. 2010) but the population is increasing consequently generating huge demand of fibre, food as well as bioenergy which create a pressure to increase agro-production from the same agricultural land (Godfray et al. 2010). Worldwide, arable land area has increased by 9% only since 1961 (Pretty 2008) but there is ample scope of increasing food production by minimum depletion of vital natural ecosystems. On the other hand, existing agricultural productivity is predicted to be limited in the near future due to climate change (Lobell and Field 2007).

Modern agriculture knows nothing but increasing production by manipulating or destroying biological functions of diverse communities of organisms and a raising application of energy and agrochemicals aiming to remove all sorts of limitations of plant productivity. Modern day's Agricultural Intensification (AI) models are ineffective and play detrimental role on the environment which leads to a loss in biodiversity and is not sustainable or ecological or eco-efficient to feed the future (Tilman et al. 2002). Besides, EI often promotes crop varieties with improving genetic material that helps in raising productivity, soil tillage mechanism to allow better root growth along with indiscriminate pesticide application for pest control (Tilman et al. 2001). Therefore, such healthy ecosystems which provide well-being to mankind globally get eroded due to change in climatic conditions, degradation and pollution in biotic organisms, heavy loss in biodiversity and so on (Hooper et al. 2005).

6.4.2 Challenges in Agriculture

Agriculture needs to become more productive to meet the future challenges based on climatic, economic and social criteria by achieving stability and resilience (Foley et al. 2005; Raj et al. 2020; Banerjee et al. 2020). It is possible for reducing overuse of nutrients and water to reduce the negative impact on the environment from crop production (Mueller et al. 2012). Decomposition of soil organic matter (SOM) and soil organisms, soil genesis, structure, moisture and soil mineral nutrients cycling including soil C are indispensable for crop yield which are directly interconnected with soil formation and nutrient cycling ecosystem. Soil as an irreplaceable component of different ecosystems provides several services including water storage and purification, C storage and gas regulation such as agricultural soils stored almost one-fifth of global C pool (Wood et al. 2000) and non-agricultural soils stored 80% global terrestrial C. On the other hand, it also promotes plant growth, pest and disease regulation, nutrient flow, root penetration service, gas exchange, soil water retention and erosion control. All of these processes are mediated by a large, diversified and more or less unexploited soil organisms like bacteria, fungi, protozoa, nematodes, earthworms, micro-symbionts (e.g. symbiotic nitrogen fixation), decomposers (e.g. nitrification and denitrification), elemental transformers, arthropods, pests and their natural enemies (Barrios 2007; Brussaard et al. 2004; Verbruggen et al. 2010). Landslide increase of food production by green revolution could not meet nutritional demand which is the key duty of agriculture but destroyed inherent species diversity of cultivated crops (DeFries et al. 2015). Production of nutrients in terms of food and assurance to access to nutritional diversity in a stable way with minimum disturbance of environment is the key role to play by modern day's agriculture (Wood 2018). Large-scale inception of pesticides since 1940 and their overuses and landscape changes led to the incidence of pest resurgence, development of secondary pests and failure in crop production due to devastating pest attack (Settle et al. 1996; Bommarco et al. 2011; Ekström and Ekbom 2011). There are several natural enemies for each pest species which control the pest population in the agricultural crop production system (Settle et al. 1996; Thies et al. 2011).

6.4.3 Climate Change

Average global temperature increment due to higher emission of GHGs in recent time warms the global climatic system that rises sea water temperature, glacier/ice sheet melting and thereby increases the sea level. Hydrological cycle alters as a consequence of global warming which in turns increases the evaporation, changes rainfall pattern, depletes the groundwater reserve, and increases the probabilities of extreme events such as drought, flood, heat-wave, cold wave and severe cyclonic events. Altering rainfall pattern and temperature fluctuation changes the activities of agricultural landscapes in overwhelming and often destructive ways. Climate change pertains to an increase in the atmospheric CO_2 concentration that entrapped the long-wave radiation within the troposphere and leads global warming. At present-day atmospheric CO_2 level covers around 397 ppm, significantly much higher than the pre-industrial level of 280 ppm and according to IPCC (2007) at the end of this era, the concentration level reaches double as compared with existing level. The agriculture sector has been contributed 24% of the total anthropogenic emission (IPCC 2007) which is consisted of CO_2, N_2O and CH_4, the three major GHGs. Energy use in farming activity and land management are the broad anthropogenic sources of GHGs emission from the agriculture sector. The cultivation of rice mainly in wetland conditions is the major source of emission of GHGs like CH_4 and N_2O. According to FAOSTAT (2013), global emissions increased from 0.37 to 0.52 $GtCO_2$-eq/year. Approximately, 94% emission from rice field comes from developing countries. Additionally, agricultural soil is also responsible for GHGs emission either directly or indirectly. N fertilization, residues burning and organic matter mineralization are considered as sources of direct emission. Indirect sources comprise nutrient leaching and deposition in the atmosphere. The agricultural land contributes 50% of the total N_2O emission. The huge application of synthetic fertilizers towards getting more products has been increasing emission more than nine-fold from 0.07 to 0.68 Gt

CO_2eq/year from 1951 to 2010 (Tubiello et al. 2013). In developing countries recently, crop residue burning as a part of clean cultivation contributes 0.5% of the total agricultural emissions. It has been projected that 0.6–2.5 °C temperature will increase by 2050 and subsequently 1.4–5.8 °C by 2100 (IPCC 2007). It was more awful for particularly south Asian countries, maybe suffered by an increase of 0.5–1.2 °C temperature in 2020, 0.88–3.16 °C by 2050, and 1.56–5.44 °C by 2080, will be dependent on the progress of future policies (IPCC 2007). The global warming effect is likely to be much higher in the winter season than summer, whereas the rainfall would be higher throughout the season except December to February (FCCC 2012). The lower attitude of Southern Asia would face the challenge of extreme temperature increase or macro-scale warming. In India, it was projected that climate change increases overall temperature by 2–4 °C with the least changes in the trend of rainfall at the next decade (Kavikumar 2010). The west coast, north-eastern states and some parts of Gujarat and Kerala may receive the higher rainfall by 6–8% of normal. A trend of higher temperature has been observed across the west coast, central, peninsular and north-eastern part of India. In contrary, the low-temperature trend has been monitored throughout the north-western and some parts of southern India.

6.4.4 Intensification Options

To meet these challenges EI may play a vital role in boosting crop production by minimizing the negative impact on the environment by integrating the components of natural ecosystems (Dore et al. 2011). By maintaining and enhancing ecosystem functions of feeding the world in future in a sustainable way EI is getting momentum in the scientific and development arena (Tittonell 2014) and they agreed that sustainable agriculture is needed to feed the present and next generation after keeping the earth green. To satisfy this aim of agriculture, EI is pivotal rather than AI as "win-win" or "win-neutral" balances between crop yield and ecosystem service outputs resulted in most EI systems (Garbach et al. 2017). EI generally provides enhancement of crop productivity through different supporting services for proper management of agricultural practices (Bommarco et al. 2012). Three pillars of EI mainly considered as increasing the yield potential of field crops, ensure in increasing input use efficiency and minimize the loss along with soil quality improve (Cassman 1999). Knowledge of EI and its goals should be implemented in the field level to make agriculture green. For sustaining agriculture, the pesticide should be replaced by biological, cultural and mechanical or integrated pest (insects, diseases and weeds) management methods to make a control over pests and substitute of inorganic fertilizers with manures, composts and nitrogen fixing legumes. A generous integration of ecosystem components mainly natural inputs must be incorporated in a holistic manner for making agriculture profitable along with highly input intensive enterprise through minimizing the use of inorganic inputs. It is proved that technologies and practices which are regenerative and resource-conserving mostly beneficial for the environment, farmers, communities and nation. Balanced use of inputs stated by the integrated plant nutrition system (IPNS) and targeted use of inputs indicated by integrated crop management (ICM) are conductive in use of external inputs and highlighting regenerative processes (Pretty 1997). Farmers having high input-oriented irrigated lands started to adopting regenerative technologies can improve or maintain yield successfully by reducing inputs (Kamp et al. 1993) and a similar result was found in the USA (Hewitt and Smith 1995) and Europe (Roling and Wagemakers 1997). Many farmers in the USA reduced 60–70% use of fertilizers, pesticides and total energy undergo in agricultural venture through adopting sustainable agricultural technologies but still, their yields are roughly comparable with the conventional agriculture (NAF 1994). Besides, it is also reported that the performance of organic and agro-ecological farming system is much better in case of ecosystem services relevant to climate change, C sequestration from a soil depth of 30 cm and other parameters like soil water holding capacity, etc. (Rossing et al. 2013). EI relies on controlling the ecosystem services provider components and measuring their direct and indirect contribution to agricultural production. These organisms provide supporting and regulating ecosystem services that can be incorporated into cropping systems to maximize production without/minimum impacting on climate that not means that anthropogenic inputs (e.g. inorganic fertilizers, pesticides, energy and irrigation) excluded (Cassman 1999). Farmers generally valued only ecosystem services (food, fibre and energy) but the importance of supporting (e.g. soil fertility) and regulating (e.g. pest control and crop pollination) services remains grossly undermined. The EI may help to increase the nutritional quality of food and stabilize crop yields over time by reducing soil degradation, increasing soil C storage, and water retention and coupling soil.

For sustainable EI of agriculture by ensuring minimal or no disturbance of natural ecological services and producing or maintaining yield goal, there are some technologies need to become forward like crop rotation, green manuring, compost as fertilizers, crop diversification, cover crop, no burning fields of forests, marigolds for nematode control, introducing legumes, nutrient recycling by creating crop-animal interactions, utilization of residual moisture and introduction of border crops/grasses (Shah 1994). Thus, the role of EI is to replace external inputs in agricultural production to optimize the role of ecosystem services (García-Palacios 2019). Mismanagement of ecological variables like overgrazing and intensive tillage practices may lead to significant loss of soil organic matter (Plaza-Bonilla et al. 2015). Therefore, besides soil water retention, other major ecological services especially soil C accumulation and N retention can be optimized by EI to boost crop production. On the other hand, N loss would be prevented by maintaining a tight soil N cycle after synchronization of soil N availability and plant N demand resulting in increasing N retention (Tilman 1998). Conservation agriculture and intercropping can reduce a list of activities like land degradation from repeated tillage operations (Kassam et al. 2012) enhancing soil C accumulation along with water retention capacity (Lal 2004; Plaza-Bonilla et al. 2015). N fixing woody species and cereals in alley cropping demanding few or no external inputs can be considered as another way to increase plant diversity, nutritional, wood and food production (Wood 2018) and soil organic matter content via intercropping reduces soil erosion which can be a sustainable tool for managing area-specific EI (Lal 1989). Therefore, EI increasing the demand for natural resource services and decline the application of synthetic fertilizers which creates beneficial impacts on plant competition (Brooker et al. 2015). This is more rigid at the time of atmospheric N fixing through legume crops in the intercropping system (Hauggaard-Nielsen and Jensen 2005).

Three pillar of conventional agriculture (CA) can do that in a sustainable way by crop rotation or diversification of crops, minimizing repeated soil tillage along with covering the soil through crop residues or by cover crops (FAO 2016). The reduction of soil compaction and increased in soil C in surface soil are ensured through the minimum soil disturbance of soil, adding more soil organic matter in soil (Hobbs et al. 2008). An integration of these components of CA can deliver important ecosystem services by reducing soil erosion, GHGs emissions and increasing soil C sequestration (Kassam et al. 2012). Recently, thrust has been given especially for small holding farmers to adopt CA to increase sustainable crop production, improve soil quality and reduction in dependencies on external input use. Globally declining yields of monoculture of many crops induces a suggestion to introduce crop diversification which may reduce indiscriminate use of inputs in traditional system, increases species diversity and enhances crop productivity due to complementarity in mechanisms in resource use and facilitative interactions between species or genotypes (Hooper et al. 2005).

Only superficial or catastrophically degradation of soil processes due to unsustainable agricultural practices like soil salinization, erosion and acidification was under discussion but below ground biological vital community structure remains undermined. External input dependent current agricultural lands are providing yield by diminishing microbial biomasses due to declines in SOM that fuel microbial communities (Paul et al. 1997) have long term negative effect on food security. Building up of SOM in the soil increases soil microbial activity, workability, water-holding capacity and pest control, improves soil fertility, decreases soil compaction, erosion, surface crusting and supporting in uninterrupted nutrient supply because most of the plant nutrients are bound to or released from SOM. More intensive agricultural lands will be less efficient in functioning due to decrease in SOM subsequently providing the reduced flow of soil services for both yield and sustainability (Pan et al. 2009). Thus, management of ecological services to increase SOM and ensuring diversification of crops by rotation will enhance sustainable agriculture. In addition to crop rotation by the inclusion of perennial grasses and legumes, adding of manure, residue management, reduced tillage (Paul et al. 1997) will decrease monoculture crop sequences and improve soil health. At least 10% yield of wheat will be reduced when wheat is planted just after wheat (Bennett et al. 2012) but the why is not known. It is thought that mono-crop will help to develop soil-borne pathogens and pests (Bennett et al. 2012), degrade soil fertility (Karlen et al. 1994) and decrease the efficiency of water-use (Roder et al. 1989) where break crop can minimize these incidences.

Biological control, landscape and crop rotation ecosystem services are conducive to reduce pest and use of pesticides by ensuring diversified crop rotation, creating and conserving of natural enemies and their habitats. Over 75% of food crops especially fruits and vegetables are directly or indirectly dependent on insect pollinators contributing significantly to vitamins and micronutrients genesis (Gallai et al. 2009) as well as food for human where bees are major (Klein et al. 2007). Though, the number of honeybees are increasing worldwide, but the number of flowering plants increasing faster than the previous (Aizen and Harder 2009). Which number of bees is needed for what species for which flowering plants and is there any synchronization on spatial and temporal distribution of bees and flowering plants and what will happen if modern agriculture is dependent on only one species in respect of climate change (Yachi and Loreau 1999) and is there any need to rear wild species of pollinators as an alternative option and is there any relationship of pollinator with habitat loss and fragmentation, agrochemicals, climate change, pests and pathogens (Potts et al. 2010; Schweiger et al. 2010). EI of pollinating services could be an only sustainable option such as long term management strategies like low tillage (Williams et al. 2010), provisioning of nesting resources (Steffan-Dewenter and Schiele 2008).

So, quantification of different ecosystem services, links between different ecosystem services and more specifically dynamics of different communities provides services need to be known to understand how they affect strength and flexibility of crop productivity that will help to estimate economic benefits and costs associated with EI. Multifunctional agriculture is a new system of agriculture that estimates these benefits and advice strategies to encourage land managers and farmers to support them (Renting et al. 2009). EI strategies beneficial for other services and vice versa but most of them remain inadequately understood.

With the level of intactness, difficulty and species richness of ecosystems generally service delivery increases (Díaz et al. 2006). Species communities generally formed by multiple drivers which operate through human-dominated landscapes (Cardinale et al. 2012) such as pollinator diversity increases crop yields (Hoehn et al. 2008) or with diversified crop rotations (Bennett et al. 2012). It is an exceptional challenge to ensure supply, steadiness as well as production with minimum effect on biodiversity and environment which is illustrated in resource management framework of "safe space" presented by Beddington et al. (2012) that how EI can be managed while avoiding potential negative trade-offs. This figure depicts how different strategies allow one to navigate into or widen a safe area where global food demands are met with minimum impact on the environment. Integration of expertise of scientists, ecologists and other experts should be combined to draw a needful conclusion for developing scales in managing of multiple ecosystem services globally. EI has to work together with other measures that reduce demands such as reducing food loss across the supply chain and by stepping down the food chain in global consumption to meet increasing food demands. A strategic role of EI based on local management interventions can move global crop production into the safe space globally. This framework to address similar food security issues in the livestock, forestry and fishery sectors is potentially portable.

Ecological intensification has the ability to provide globally environment-friendly agriculture than the AI does not. Understanding and providing ecological processes to the farmers those can able to secure biodiversity and food for human without damaging the environment in the long run, demands minute research. Farmers want to understand the economic opportunities of using different EI management tools replacing costly external inputs of modern agriculture.

6.5 Agro-ecological Resources for Sustainable Agriculture in South Asia

6.5.1 Land Resources

The land is a crucial resource for agricultural intensification. In developing countries of South Asia, there is no scope for the horizontal increase of arable land due to rapid urbanization. Maximum per capita land availability was estimated in India (0.16 ha), followed by Nepal (0.13 ha), Pakistan (0.13 ha), Bangladesh (0.06 ha) and Sri Lanka (0.05 ha) (FAOSTAT 2004). Rather than per capita land, biological productive land is more important to accomplish the aim of sustainability. In Bangladesh, the estimated productive land is 0.50 ha per person while in India it is 0.80 ha. Bangladesh uses the highest proportion of land (70%) in farming purposes followed by India (60%) (Fig. 6.5).

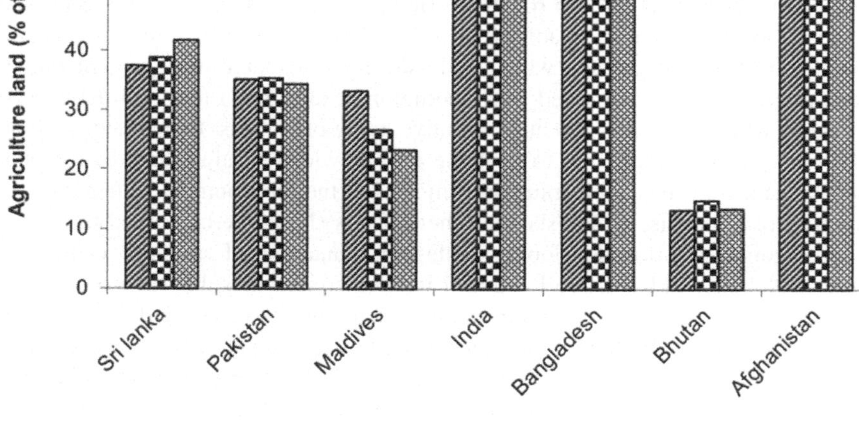

Fig. 6.5 Agricultural land share in South Asian countries

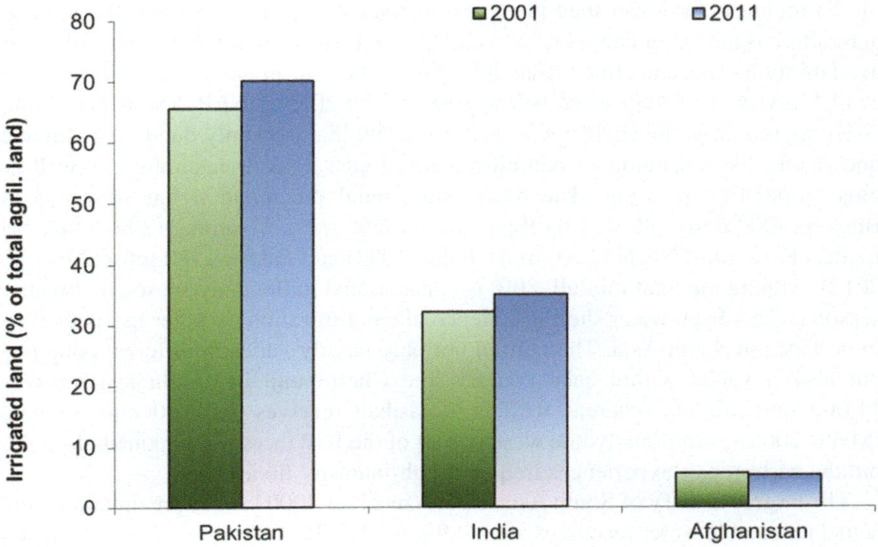

Fig. 6.6 Irrigated land out of the total in Pakistan, India and Afghanistan

While other South Asian countries like Nepal, Pakistan and Sri Lanka utilize only about 30% of the total land in agriculture purposes, and the proportion is likely to be increased. The maximum ratio of arable land to agricultural land is accounted for by India and Bangladesh. Incidence of irrigation (irrigated area as a percentage of arable land) increases continuously in all countries. The highest incidence of irrigation (83%) is accounted for Pakistan (Fig. 6.6) where the rapid improvement has depicted from Bangladesh, from the base at the 1960s to more than 50% in 2002. The spreading of the irrigated area over similar time duration has been relatively slow in India, from 11% to 33%. Rather than incidence of irrigation, Bangladesh has secured the first position in respect to irrigation intensiveness (165%), while India and Pakistan have 133% and 110%, respectively (Weligamage et al. 2002). This irrigation incidence and intensity indicate the country's land topography, soil characteristics, cropping pattern and cropping intensity such as northern and north-western states of India like Punjab, Himachal Pradesh and Haryana have greater irrigation intensity than the national average. Agricultural progress in South Asian countries has been shifted from external land-augmentation to internal land-augmentation.

6.5.2 Water Resources

The one-fourth of the world population belongs from the South Asian countries, while only about 4.5% (1945 BCM) of total world's annual renewable water resources are available in this region (Ali et al. 2012). Per capita, water availability in this region is far lower than the world average except in Nepal and Bhutan and agriculture is the major consumer of available freshwater. In India, the value of water availability has been declined from 4000 m^3 to 1869 m^3 in the last 20 years and the availability would likely to be below 1000 m^3 by 2025 (UNEP 2008). The South Asian region faces the challenge of water scarcity that is mainly due to the climatic uncertainty like changing in rainfall pattern, higher evapotranspiration as well as huge population pressure. The maximum annual precipitation has observed in Bhutan (4000 mm) followed by Bangladesh (2666 mm), Myanmar (2091 mm), Sri Lanka (1712 mm), Nepal (1500 mm), India (1083) and Pakistan (80 mm) (Ali et al. 2012). Among the total rainfall, 80% is concentrated in the rainy season and winter season suffers from water shortage. Hence, the optimization of water use is the key importance in South Asia. The rainfall not only largely varies in different countries but also it varies within each country like Cherrapunji in Meghalaya receives 11,000 mm rainfall, whereas western Rajasthan receives just 100 mm in India (MWR 2008). Additionally, the western part of the IGP faces with frequent drought, but the eastern part experiences frequent high-intensity floods.

The largest country of South Asia is India receives 4000 km^3 of precipitation, and actual renewable water resources are 1869 km^3 (MWR 2009). It has been estimated that in India, only about 1123 km^3 water, including 690 km^3 from surface water and 433 km^3 from groundwater is used for beneficial purposes (MWR 2009). Among the South Asian countries, India comes in the first position with respect to renewable water resources (1911 km^3). Bangladesh (1200 km^3) and Pakistan (223 km^3) come into the second position, followed by Bhutan and Sri Lanka, accounting 78.0 km^3 and 52.8 km^3, respectively (FAO 2011). Pakistan comes under the water stress category and in contrast, Bhutan is considered as a water-surplus category. Out of total internal RWR, almost 58% is withdrawn by the surface and groundwater together (UNEP 2008).

Overexploitation of groundwater though the millions of tube wells and salinity intrusion in India and Pakistan lead to drop the groundwater level in many areas. Shakir et al. (2011) reported that the water table decreased by an average rate of 40.0 $\pm$ 10 mm $year^{-1}$. The most serious challenge in farming activities is faced by the farmers in the rainfed region and almost 60% arable lands in SA are rainfed. Though the rainwater harvesting should be taken into consideration earnestly, but less than 30% rainfall is presently harvested in this region. With the consumption of 95% of total water, agriculture is known as the highest water consuming sector in SA. Rapid developmental pressure, resource degradation, ecological insecurity, failure in policy-making facilitate the threats and issues of water scarcity. However, SA has the potentiality to build-up water resources with consideration of climatic abnormalities and developmental economy.

The development phases of the irrigation system in South Asian countries are divided into three phases, namely, the Colonial era (1850–1940); the Cold War era (1950–1990) and the New Era of Globalization (1990 onwards). The last phase of irrigation development is a consequence of the shifting of agricultural practices from subsistence type to semi-commercial type and the primary aim of the nation has shifted towards food security with environmental sustainability.

6.5.3 Agroforestry

Agroforestry refers to alternate land use system where field crops are cultivated with perennial trees. The South Asian (SA) region is often recognized as the cradle of agroforestry in recognition of its long history of the practice of an array of systems under diverse agro-ecological conditions. The agroforestry system fulfils the basic needs of millions of smallholder farmers by supplying the food, fuelwood, fodder, plant-derived medicines and cash income (Jhariya et al. 2015; Singh and Jhariya 2016). In Indian sub-continent home gardening, silkworm rearing and the cultivation of lac insect were practiced from the Epic era. Agroforestry may be a great source of income for high-density population in SA, if properly utilize as it mitigates the demand of food and supplies the additional income from woods, honey and other forest products (Painkra et al. 2016). Additionally, sometimes animals can be reared in the combination with this system to properly utilize the spatial and temporal resources. More than 21% area of SA countries comes under this system, in which more than 10% area is under tree cover. The major agroforestry systems present in SA are agri-silviculture involving poplar (*Populus deltoides*) and *Eucalyptus* spp.; plantation crops include tea (*Camellia sinensis*), coffee (*Coffea* spp.), cacao (*Theobroma cacao* L.), and spices (black pepper, cardamom) with perennials trees; betel vine (Piper betel) + areca palm (*Areca catechu*); intercropping systems with coconut, para rubber, and other trees commercial crop production under the shade of trees in natural forests (e.g. cardamom) and homestead farming systems. The planting of trees on field bund to protect the crops as a natural wind barrier is also predominant in the subcontinent.

6.5.4 Crops and Diversification

The overall performance of agriculture is continuously improved in SA. Amongst different SA countries, Bangladesh and Pakistan are acknowledged as most rice-intensive and wheat-intensive countries, respectively. Additionally, HYVs of wheat and rice in SA countries lead to higher productivity and give the opportunity to high-value crop diversification. Diversity of agro-ecology in terms of land, climate, water resources permits farmers to cultivate a variety of crops, and associative commodities like livestock, fisheries, agroforestry, etc.

According to Joshi et al. (2014), the Simpson Index of Diversity (SID) has been increased from SA is gradually diversifying its crop sector to high-value fruits, vegetables, etc. Among the different countries Bangladesh, Bhutan and Nepal show less diversity than other countries. In Bangladesh, more than 75% area is under rice cultivation and the rest area is highly diversified. Introduction of dwarfing rice and wheat varieties in SA countries are considered as a major breakthrough enhancing the agricultural productivity and they are highly responsive to nutrient application especially N. The rice–wheat system is till now major livelihood support for millions of farm families. Rather than a rice–wheat system, in IGP, cotton–wheat, rice–maize, sugarcane–wheat are also predominant. However, the yield potential is categorized into two broad spectrums, i.e. the irrigated rice–wheat system is mainly confined into Pakistan, north-western India and some portion of eastern IGP, whilst a part Bangladesh, West Bengal, northern Bihar and eastern Uttar Pradesh of India, and the *Terai* region of central Nepal are habituated in rainfed wheat cultivation. Understanding the existing cropping pattern and the pace of diversification may lead to achieving the goal of sustainable agriculture.

6.5.5 Conservation Agriculture

Conservation agriculture (CA) refers to a suit technologies comprising least soil disturbance, soil cover, crop rotation and management of input and agrochemicals in a sustainable way. It is also well documented that CA is not only beneficial for ecological sustainability but also resulted in 33% yield enhancement in cereals crops over conventional agriculture. This concept was first introduced in SA during the 1980s by using zero tillage technology, first time used in Punjab, Pakistan. The conservation system spread rapidly throughout the SA and currently, it covers more than 5 million ha in the rice–wheat system (Friedrich et al. 2012). The success of CA is attributed to a multi-stakeholder approach and frequent farmer's participatory programme in SA countries. Integration of local manufactures with national partners has made available cheap, affordable and effective machineries used in CA. Zero-till wheat cultivation has been popularized in rice–wheat cropping system at a larger portion of north-western India, facilitating less labour and time, energy and cost involvement, and overall yield enhancement. Additionally, precision farming enhances the system productivity (7.4%), water-saving in wheat and rice accounting a value of 10–13% and 12–14%. The implementation of resource conservation yielded 0.5 million tons more wheat and saved 80 million US$ by lowering fuel consumption, tillage practices and input use.

6.5.6 Population and Economic Resources

South Asia region is covering 5,137,190 sq. km. area, 1653 million population and 450 Million US $ of gross domestic product (GDP) (Singh et al. 2014a, b). India as the largest country covers 64% of geographical area, 74.8% of the population and 81.3% of the share of GDP of the region. In contrast, Maldives and Bhutan share 1% of geographical area, population and GDP of the region (Singh et al. 2014a, b). About 23.24% of the world population belongs to the SA region and occupied only 2.62% of the world GDP. The majority of share (20%) in GDP comes from the agriculture sector in this region and a large proportion of the population depends on agriculture as their primary source of income. Share of the rural population in the total population varies between 66% in Pakistan and 86% in Nepal. Per capita income from agriculture is low mainly due to lower productivity and a higher proportion of farm families are belonging in poor economic condition. Agriculture farming in SA is dominated by smallholdings; as per example, Bangladesh holds less than 0.5 ha, whereas less than 1.0 ha for Sri Lanka and Nepal. In India, the average farm size is 1.41 ha (SAARC 2009).

Therefore, the comprehensible knowledge about the available agro-ecological resources, their present status and the climatic situation in SA countries would be helpful to resource conservation, precision application of resources and to prepare proper resilient planning against uncertainties.

6.6 Ecological Parameters for Intensification of Rice-Based Cropping Systems

Rice is considered as the staple food for more than half of the world population (Biswas et al. 2019). Rice-based cropping system refers to the farming practices, where rice is cultivated as a major crop with other suitable crops may be in intercropping, relay cropping mixed cropping or as a sole crop in the subsequent season. More intensified cropping system or in other words more production in per unit land or time is earnestly needed due to the debate on global food security (Teillard et al. 2012; Buckwell 2014). Recently, the term sustainable intensification (SI) has gained the popularity that conceptualizes the importance of system intensification and diversification without compromising with ecological balance (Garnett et al. 2013). Rice-based system intensification aims to achieve the maximum system productivity, utilization of available resources with minimum environmental impact and synergistic interaction with each other.

Ecological parameters for SI in rice-based cropping system vary with the geographical location, climatic situation, socio-economic background, and food and cultural habitat in the particular arena (Vliet et al. 2015). This chapter is focused on the EI of rice-based cropping system in SA, more precisely on two rice-dominating agriculture-based countries, namely, India and Bangladesh. Hot and sub-humid summer (March–May), humid rainy (June–September), cold and dry winter (October–February) with 1000–1200 mm annual average rainfall are the major characteristics of this region (Balasubramanian et al. 2013). Rainfall in rainy season is mainly happened by South-west monsoon, while winter rainfall is influenced by western disturbance. Recently, climatic variability delays the planting time, soil moisture, crop management practices and consequently affects the whole food systems (Aggarwal et al. 2004).

6.6.1 Major Rice-Based Production System

In SA especially in Indo-Gangetic plain rice, maize and wheat are major cereals contributing to livelihood security. These crops are cultivated either as a monoculture or in rotation or intercropping with suitable crops. Among the different cropping systems in irrigated lowland rice—rice is most predominant in tropical and sub-tropical regions, i.e. in Bangladesh, Eastern India, Eastern Nepal and some parts of South India (Yadav et al. 2019). Rather than rice–rice, rice–wheat, rice–

Table 6.1 Rice-based cropping system in South Asian countries

Irrigated condition	Upland condition	Rice under the integrated farming system
Totally rice	Rice–pulse (chickpea/lentil/pea)	Rice–fish–poultry
Rice–Rice–cereal		Rice–fish–duckery
Rice–Rice–pulses	Rice–oilseed (mustard/linseed)	
Rice–wheat–pulse	Rice–coarse cereals	
Rice–Toria–wheat	Rice–wheat	
Rice–wheat		
Rice–mustard		

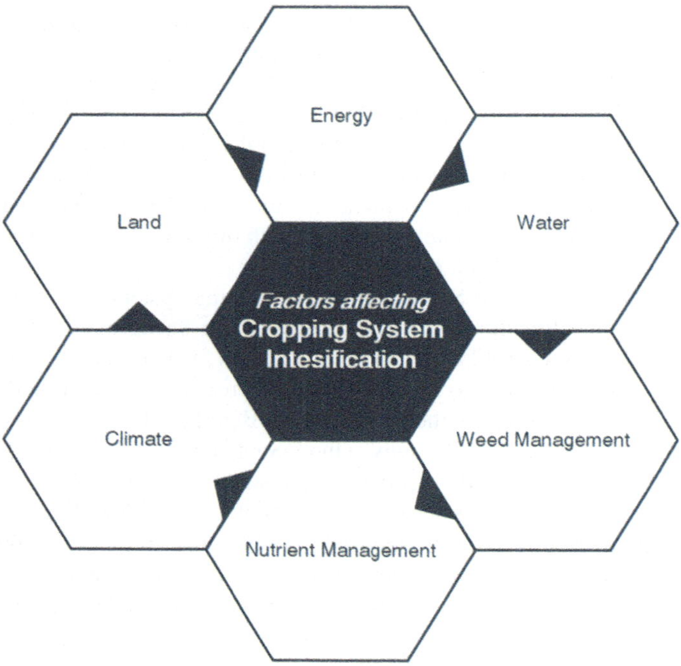

Fig. 6.7 Ecological indicator for intensification in rice-based cropping systems

maize, rice–rice/wheat–pulse are also cultivated in lowland condition. In upland condition rice–chickpea, rice–mustard/linseed, rice–barley are most common. Rice–fish–poultry/duckery is popular as an integrated cropping system. Major cropping systems are summarized in Table 6.1. Major ecological components (Fig. 6.7) for SI of rice-based cropping system are as follows:

6.6.2 Land Use Pattern

Land use comprises a number of crops grown, tillage practices and irrigation management, all of are considered as a part of Agro-Environmental indicators (AEI) (Oenema et al. 2011). Number of crops defines how many crops can be adjusted in total cropping system that attributes to the structural biodiversity of a cropping system. It also indicates better sustainability when the high-value crops are incorporated. Tillage practices indicate the percentage of the Utilized Agricultural Area (UAA) cultivated with conventional practices. Conservation agriculture has also been popularized for rice–wheat cropping system in IGP that includes minimum soil disturbance, land cover and crop rotation could be beneficial for soil health, reduce soil erosion and conserve for organic matter and soil C (Johansen et al. 2012). Two-wheel tractors (2WT), Versatile Multi-crop Planter (VMP) are promising machineries for CA in IGP in rainfed cropping system (Haque et al. 2017). Non-puddle transplanting with minimum tillage enhanced the SOC by 24% at 0–5 cm soil depth, as well as increased rice, yield up to 12%. Conservation tillage also reduced fossil fuel consumption by 85% and labour engagement by 50% in rice-based cropping system (Bell et al. 2019). In the rice–wheat cropping system, 25–30% land area is converted into zero tillage system. Irrigation management indirectly measures water consumption as the percentage of the UAA that is effectively irrigated.

6.6.3 Nutrient Management

Nutrient management is a key factor in terms of productivity and profitability of a cropping system. For adoption of successful cropping system, good knowledge about the nutrient requirement, efficiency, availability to next crops and their uptake pattern is necessary to optimize nutrient management schedule. Long term fertilizer trial on intensive rice based cropping system in IGP has resulted negative trend of crop productivity with existing recommended doses of fertilizers. Therefore, an alternative nutrient management strategy such as INM, organic farming is required to achieve the desired levels of crop productivity. The combination of organic with inorganic nutrient sources is more suitable as it can boost the SOC and soil health as well as enhance the system productivity (Hossain et al. 2016). The manure obtained from cow dung when applied in rice–wheat cropping system with NP fertilizers enhanced the SOC by 0.30 Mg ha^{-1} year^{-1}. INM with the incorporation of green manure or legume residue can save up to 80–120 kg N ha^{-1}.

Furthermore, 37% higher REY (rice equivalent yield) was obtained in rice–blackgram intercropping system when 15 kg N from FYM was supplemented with 20 kg N through urea and other nutrients were applied as per recommended dose (Mishra et al. 2012). Similarly, in a rice–rice cropping system, application of 50% RDN through green manuring of *Azolla* couple with chemical fertilizer in *kharif* season and subsequently supply of N solely from chemical sources to summer rice attributed higher yield and better soil health than farmers' practices (Mishra et al.

2017). Site-specific nutrient management (SSNM) uses of leaf colour chart or chlorophyll meter are the most promising approaches for need-based fertilizer recommendation in a particular location to optimize the fertilizer dose. Singh et al. (2015) observed that SSNM could increase REY up to 30% over the blanket recommendation. Moreover, incorporation of rice straw in the rice–wheat system instead of residue burning not only prevents the emission of CO_2 but also enhances the system yield. Use of mulching as a soil cover in CA leads to better soil organic matter formation, maintains soil temperature and reduces evapotranspiration loss. Therefore, from aforesaid it can be suggested that there is a huge scope to supply the plant nutrient especially N through integrated manner rather than chemical fertilizer solely. INM encourages the recycling of the resources, precision farming, lower soil-water pollution and land quality degradation.

6.6.4 Water Management

Being a water-loving crop, in rice-based cropping system water is a crucial indicator for determining the components crops, particularly in the dry season. Crop water management and duration of rice cultivars influence the water regimes for succeeding crops such as late harvesting of long duration rice reduces the soil residual moisture for the sowing of next crop. Pudding in rice is more water, capital and energy-intensive as well as creates subsurface hardpan that is not favourable for subsequent crops. Therefore, more water productive technology such as alternate wetting drying raised bed, aerobic rice is found promising as well. However, coarse texture soil with a favourable combination of puddling and water regimes attributed higher water and system productivity (Alam et al. 2017).

Additionally, life-saving irrigation at critical crop water requirement stage for dry season crops has been reported beneficial such as two supplemental irrigations at tasseling and dough stage in maize, pegging and pod development in groundnut, 50% flowering and grain filling in sunflower, crown root initiation and late jointing in wheat, stolon formation and tuberization in potato (Singh et al. 2014a, b). Moreover, rice–pulse–pulse cropping system results in higher crop water productivity than rice–cereals–pulse. Residue incorporation and surface mulching can retain higher moisture into soil (Mondal et al. 2018) than bare land or clean cultivation.

6.6.5 Weed Dynamics and Management

The efficiency of a cropping system and input use are well dependent on weed infestation. Relative weed density and distribution of weed flora are varied with cultural management practices, cropping pattern, choice of crops (Soni et al. 2012). The density of *Echinochloa colona* L. and *Cyperus iria* L. is predominant in the rice–wheat cropping system mainly due to minimum tillage/zero tillage, whereas *Avena ludoviciana* L. and *Chenopodium album* L. population are reduced by ZT. Another promising method of weed control is the preparation of stale seedbed in which water is applied prior to sowing that influences to emerge out all weeds, and then any post-emergence herbicide is sprayed to kill them. Senthilkumar et al. (2019) reported that adoption of stale seedbed using glyphosate application @ 1 kg ha^{-1} in direct-seeded rice at dry season reduced weed density; therefore, rice yield and net return were maximized than traditional practices. Hence, it is necessary to understand the nature of weed seed bank, existing weed flora and traditional cropping system in a particular location and the check the weeds beyond the economic threshold level.

6.6.6 Environmental Management

The cropping system includes double rice cultivation in a calendar year is being seen as a major culprit by higher methane emission under anaerobic condition. Additionally, 90% of the total crop residue comes from the rice-based cropping system (MNRE 2009). The on-farm burning of the residue to quick disposal releases a huge amount of GHGs, aerosols and another hydrocarbon (Jain et al. 2014) that might have a direct or indirect effect on the radiation balance, global warming potential and ozone layer depletion. Apart from these, a significant loss of major nutrients such as N, S, P and K (80%, 80%, 25%, and 21%, respectively) also happen during the time of burning (Yadvinder-Singh et al. 2005). However, Bhattacharyya et al. (2014) reported that rice ecosystem acts as a net agricultural sink rather than a polluter.

So, an alternate way of management practices such as diversification of rice–rice cropping system to low water requiring crops, alternate wetting drying instead of flooding, conservation tillage, and crop residue incorporation may improve the anthropogenic emissions. However, these resource conservation techniques are applicable for a specific location and particular situation and rarely investigated for a whole cropping system (Bhatt 2015). Therefore, there is a need to study to long term effect of resource conservation technologies (RCTs) on succeeding crops with consideration of farmer's livelihood and ecological balance.

6.6.7 Energy Management

The energy requirement for intensive rice-based cropping system mainly with cereals is increasing day by day. Shortage of groundwater leads to the use of centrifugal pumps instead of submersible for uplifting the water from a deeper layer but it involves more energy than a submersible. Additionally, free electricity for agriculture in some states influences the irrational expenses of energy (Bhatt et al. 2016) by farmers. Traditional practices like puddling, transplanting, conventional tillage, blanket fertilizer application, manual weeding and harvesting are attributed to higher energy involvement. Efficient cropping system including suitable crop choice and their management would be a sustainable approach for energy use. Rao et al. (2014) resulted in higher system energy efficiency in rice–maize cropping system followed by rice–mustard and rice–sunflower. The value of energy use efficiency was significantly higher in rice–*Lathyrus* cropping system than other rice-based non-legume systems as legumes require lower energy input (Rautaray et al. 2017). A whole cropping system is more dependent on the total input energy rather than use efficiency. Rice–potato–rice/sesame is most suitable for medium and large farmers as higher system productivity while rice–rapeseed mustard/pulse is suitable for small and marginal farmers involving moderate cost and higher energy use efficiency (Biswas 2017). Green manuring with *Sesbania* enhances the energy output in the rice–wheat cropping system by 10–14% over the rice–wheat summer ploughing system. However, it should be always kept into the mind that energy budgeting should compatible with monetary budgeting in a cropping system.

6.6.8 Productivity and Profitability

RYE is the most common indicator of system productivity. It converts the yield of different crops in a system into the equivalent of rice yield (Biswas et al. 2006). Whereas rice–fallow–-rice systems have an average RYE of about 10.6 t/ha, the rice–potato–rice system has represented 19.64 t ha^{-1} RYE in IGP. However, rice–potato–maize cropping systems depicted higher RYE (23.25 t/ha) than rice–potato–rice systems (Gatto 2020).

Profitability is a major consideration for farmers. It measures how much additional income is generated by adopting a cropping system than the sole crop. Net profit is measured in terms of gross income minus costs of production. The benefit-cost ratio is another method of profitability measurement across the whole cropping system (Roy et al. 2007).

6.7 Integrated Approaches for Eco-Intensification of Rice-Following Cropping Sequences

Rice-following cropping sequences because of conventional crop establishment methods, viz., flooded puddled transplanted rice suffering from a decline in water tables, decreasing soil quality, arising micronutrient deficiencies, stagnant yields because of conventional puddle transplanted rice establishment followed by the flooded conditions (Bhatt et al. 2016, 2019, 2020a, b). Rice shared a major part of the cereals water footprints, viz., 992 billion cubic metres per year (Gm^3 year^{-1}) than from the other cereals which further responsible for the lower water productivity of rice-based cropping systems (Mekonnen and Hoekstra 2011). The agricultural sector is the major consumer of irrigation water (with differential volumes of green and blue water) among different sectors, where water footprints needs to be cut down by 10–15% (Singh et al. 2010; Rost et al. 2008; Döll et al. 2012). Agricultural cropping sequences with rice constitute an important sequence which covers an area of about 24 million hectares (M ha) in China and India alone. In India, rice-based sequences particularly practised in the IGP because of the good quality of soil and better availability of the irrigation water (Singh et al. 2005). Among major Asian countries, viz., India, Nepal, Pakistan, Bangladesh, 12.3 M ha, 0.5 M ha, 2.2 M ha and 0.8 M ha is under rice-based cropping sequences which in totality constitutes around 85% of IGP (Ladha et al. 2003; Timsina and Connor 2001) which is responsible for the >45% of land productivity and satisfied 42% of the population with staple grains of the region (Jat et al. 2005). Some emerging issues because of faulty conventional practices for establishing rice crop particularly found to be responsible for many sustainability issues, because of lesser water and nutrient use efficiencies, the formation of hard plough pans, declining underground water tables and arising power crisis (Bhatt et al. 2016, 2019, 2020a, b). If this condition prevails in the region for some more decades, then only one-third of the population has an approach for the good quality water and that too in danger because of arising water demands by other competitive sectors, viz., industrial, urbanization, etc. (Wallace and Gregory 2002).

Ecological intensification (EI) involves sustainable use of the farmlands for multiplying the ecological processes intensity that favours land productivity, pest control, nutrient cycling and ecological pollution (Bommarco et al. 2013; Tittonell 2014; Cassman 1999; Pretty 1997). Further, as per Cassman (2017), it could also be pronounced as to improve the grains production potentials by carefully considering all the factors responsible for this production along with the soil quality. Hence, EI involves smart use of nature's role and performance for producing more and more grains from less and lesser resources, viz., land and water by improving the declining water as well as nutrient use efficiencies and which restricts the use of agricultural good quality land to the other uses, viz., urbanization, industrialization. The food deficient conditions could be easily eliminated by EI of rice-based cropping systems including R–W or R–M systems, an attempt made during the "Green revolution" and credit no doubt goes to shorter highly productive crop cultivars, assured irrigations facilities, availability of inputs, viz., organic and inorganic chemical fertilizers, better insecticides and pesticides (Mekonnen and Hoekstra 2011). Most important three elements of EI are mentioned below (Cassman 2017):

6.7.1 Shrinking the Exploitable Yield Gap

Deviation in the exploitable grain yield (i.e. 15–25% lesser than potential yields) reflects the degree of risk associated (inputs needed to move yields up), the response curve beyond 75% of the potential yield, and the ratio of commodity price to input costs (Lobell et al. 2009). Rice-based cropping sequence with wheat having a higher centre of gravity needs sufficient quantities of inputs, viz., fertilizers to achieve 85% of yield potential, which further resulted in lodging and thereby reducing yields and quality of grains so produced. Therefore, need-based fertilizer application particular of N is of utmost importance and in this regards, Punjab Agricultural University. Punjab, India already recommended leaf colour chart for the farmers (PAU 2020). In contrast to wheat, maize ears are located in the middle of the stalk having lesser susceptibility to lodging and the exploitable yield ceiling is likely closer to 85% (Grassini et al. 2011).

6.7.2 Improving Soil Quality

The second most important pillar of EI is the improved soil quality and special emphasis is provided to the most important properties like bulk density affecting other soil properties, viz., soil temperature, soil moisture dynamics, and soil aeration, infiltration, hydraulic properties, etc. which further influences use efficiencies of resources, viz., land, water or nutrients and thus affected could be potential yields. Further, the combined use of organic as well as inorganics inputs played key responsibility in this regard followed by green manuring, composting particularly rice compost, crop diversification with legumes, etc. As these practices enhance the soil organic C which further affected almost all of the soil physico-chemical properties by one or other way. An underpinning assumption is that a change in soil quality affects the relationship between yield and input requirements (Fig. 6.8). A reduction in soil quality means that increased external inputs are needed to overcome this degradation. Conversely, an increase in soil quality reduces input requirements and thus increases input use efficiency (Cassman 2017).

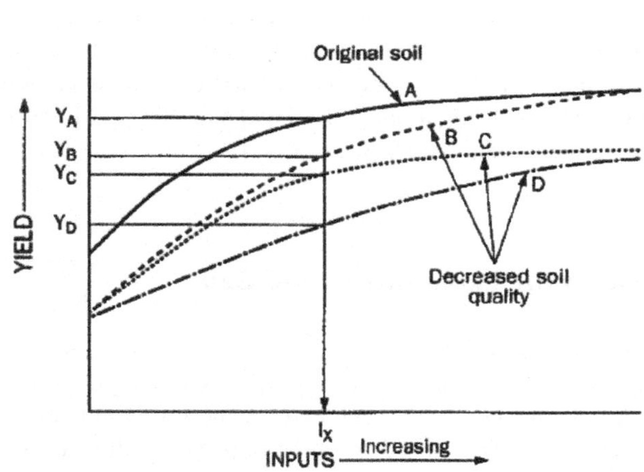

Fig. 6.8 Conceptual framework delineating the association between crop yields and input needs as prejudiced by soil quality. A reduce in soil quality from an initial state (curve A) can result in the need for superior inputs of energy, nutrients, water, seed and pest control measures to achieve the same yield. The slope and asymptote of the shifted response (shown by curves B, C, and D) depend on the type of soil deprivation and result in abridged input use efficiency, yield impending, or both. (Cassman 1999)

6.7.3 Precision Agriculture

The third pillar for the EI for the rice-based cropping systems is the precision agriculture (PA). Further, it depends on the several factors, viz., seed, fertilizer, lime, irrigation and pesticides (for the larger fields of the developed countries) and field-specific management for higher water as well as nutrient use efficiencies for having higher yields (in small fields of developing nations with smaller land holdings) (Cassman 2017). PA ensures optimum availability of nutrients and soil moisture sustainable crop production, and attack of insect-pest and diseases should be minimum or their effects should be minimized during the growing season. PA includes the correct dose, correct time and correct method of fertilizer applications for having higher nutrient use efficiency in a climate-smart way. Precision management of resources includes a site-specific approach and among different resource conservation technologies, all depend upon the soil textural class. For example, direct-seeded rice is success under medium to heavy textured soils, while failure techniques under the light-textured soils (Bhatt and Kukal 2015).

From 1993 and 2020, cereals demands increased to many folds for the ever-increasing population (1.2% per annum for wheat and rice, and 1.5% for maize) Table 6.2 (Rosegrant et al. 1998) which could be jumped to 882, 827 and 916 million metric tons of wheat, rice and maize, respectively, from 1997 to 2027 Hence, cereal production increases during the next 30 years of 44% for wheat, 43% for rice and 56% for maize, respectively.

Further, advanced cultivars which could stand in the stressed conditions, having higher harvest index and nutrient use efficiencies, reduced susceptibility to lodging, farmer's friendly also helps in the EI of the rice-based cropping systems. Judicious use of fertilizers based on the inherent soil supply and crop need must be there as on side it reduces the production of the GHGs while on the other, cut down the extra burden of the farmers practising the rice-based cropping systems. Government policies for providing subsidies on the resource conservation machines, viz., laser levellers, mechanical transplanters, happy seeders, seed cum fertilizers drills, zero-till seed drills must be there to attract the farmers to use these is a must. Further, the cooperative basis will also be helpful for the use of these machines by the poor farmers. The government must invest in the different irrigation project which further assured the irrigation for bumper yields by ensuring adequate moisture in the rhizosphere. Short duration cultivars will also be helped in enhancing the EI as

Table 6.2 Global harvested land of rice-based crops in 1967 and 1997, average growth rate since 1997 and the projected cereal demands from 1993 to 2020 (Source: http://apps.fao.org/)

	Harvested land, ha $\times 10^6$		Annual growth rate, % 1967–1997		
Crop	1967	1997	Yield	Production	Grain demand 1993–2020
Rice	128	150	1.9	2.5	1.19
Maize	112	142	1.8	2.6	1.49
Wheat	219	229	2.3	2.5	1.22

solo crop possible with the earlier wild cultivars while the use of new short-duration cultivars, fertilizers and assured irrigation facilities help to harvest two to three (wheat–legume–rice) crops annually. Harvesting two crops per year with rice, wheat or maize is now the most favourite among the farmers due to many factors; where conditions allow intensified cropping. But, now onwards, there is limited scope for the higher EI because of shrinking resources, reduced to satisfy the global burgeoning population. Further, under this attempt, the N fertilizers consumption in 2017 hiked to 9.13 times since 1960 (Saini and Bhatt 2020). Decreased N-fertilizers performance in terms of grain production up to 50% from 1960 to 2000 in cereals (Russenes et al. 2019) might be responsible for that. Further, injudicious fertilizers use claimed to one of the chief reason for the ecological pollution, which needs to addressed soon (Fowler et al. 2013; Neubauer and Megonigal 2015; IPCC 2014). Excessive use of N fertilizer blamed to be one of the main causes for higher production of N_2O from agricultural land (Bouwman et al. 2002), which further affected by its structure (Dobbie and Smith 2003) like poly-coated urea has lesser loss than conventional NFs), quantities (Ma et al. 2010) and application methods, viz., broadcasting or fertigation, etc. (Wu-xing 2010; Liu et al. 2011).

6.8 Conclusion

This chapter highlights the available agro-ecological resources for improving crop productivity to obtain the goal of agricultural sustainability without negotiating the agricultural outputs. In SA countries huge population pressure, foster urbanization and industrialization lead to dwindling the agro-ecological resources like land, water, agroforestry, human and climatic stability. Agricultural intensification has been accompanied by a set of innovations, collectively referred to as the Green Revolution, which has increased food production significantly. However, the intensification poses a major threat to the physical environment such as the loss of natural resources, genetic diversity, land degradation and injudicious application of water and nutrient. Further, climate change, changes in the dietary habits and food wastage trends hiked the goals of the productions to many folds for which EI of rice-based cropping systems is an answer. Recent evidence recommends that EI of distinctive agriculture particularly in rice-based cropping systems of SA can preserve the food production in additional with environmental welfares. Since EI of sustainable agricultute could organize the precise constituents of biodiversity and can be used to either balance synthetic-input as well as expand the productivity without adversely affect on agricultural production. Hence, three pillars of EI, viz., shrinking the exploitable yield gap, improving soil quality and PA must be addressed in a sustainable way which highlighted the number of questions, viz., the projection for increasing the yield potential of rice-based cropping systems, soil quality changes under intensively cultivated rice-based cropping sequences and its further consequences, adoption of resource conservation technologies, viz., short-duration cultivars, timely transplanting, laser levelling, tensiometer based irrigation, need-based N-fertilizers through LCC, raised beds, zero rather double zero tillage, etc., viz-*a*-viz., sustainable rice-based cropping systems.

6.9 Future Prospective

Further there is a need to address the issues of the new cultivars, seeding rate, sowing date, tillage method, nutrient quantities formulation-amounts-placement-timing, weed, insect pest, and disease control measures, use of organic nutrients, lime, other soil amendments and crop for meeting food requirements of the global population will require understanding the physiological basis of crop yield potential, the processes affecting or effecting the soil quality land productivity and environmental factors. Hence, there is a need to carry out more research on the new advances in the seed production, soil health maintenance, and resource conservation technologies, biotic and abiotic stress management to practice EI of the rice-based cropping systems for feeding the world but in a sustainable way and reduction of negative environmental impacts. Although scientists and policy-makers are consequently and progressively emphasized the welfares of ecological-intensifying agriculture to embrace EI as an ecologically sociable way towards food security by assisting the execution of biodiversity and increasing practices of ecosystem service.

Conflict of Interest Authors declare no conflict of interest.

References

Aggarwal PK, Joshi PK, Ingram JS, Gupta RK (2004) Adapting food systems of the indo-Gangetic plains to global environmental change: key information needs to improve policy formulation. Environ Sci Policy 7(6):487–498

Aizen MA, Harder LD (2009) The global stock of domesticated honey bees is growing slower than agricultural demand for pollination. Curr Biol 19:915–918

Alam MJ, Humphreys E, Sarkar MAR, Yadav S (2017) Intensification and diversification increase land and water productivity and profitability of rice-based cropping systems on the high Ganges River floodplain of Bangladesh. Field Crop Res 209:10–26

Ali H, Ahmad S, Mehmood MJ, Majeed S, Zinabou G (2012) Irrigation and water use efficiency in South Asia. Global Development Network, New Delhi. http://www.gdn.int/sites/default/files/WP/WP68-SA-Irrigation-Water_64f.pdf

Balasubramanian V, Adhya TP, Ladha JK (2013) Enhancing eco-efficiency in the intensive cereal-based systems of the indo-Gangetic Plains. Eco-efficiency: from vision to reality. CIAT, Cali, pp 99–115

Banerjee A, Jhariya MK, Yadav DK, Raj A (2020) Environmental and sustainable development through forestry and other resources. Apple Academic Press, New York. ISBN: 9781771888110. 400. https://doi.org/10.1201/9780429276026

Barrios E (2007) Soil biota, ecosystem services and land productivity. Ecol Econ 64:269–285

Beddington JR, Asaduzzaman M, Clark ME, Bremauntz AF, Guillou MD, Howlett DJ, Jahn MM, Lin E, Mamo T, Negra C, Nobre CA (2012) What next for agriculture after Durban? Forensic Sci 335(6066):289–290

Bell RW, Haque M, Jahiruddin M, Rahman M, Begum M, Miah MA, Islam M, Hossen M, Salahin N, Zahan T, Hossain MM (2019) Conservation agriculture for rice-based intensive cropping by smallholders in the Eastern Gangetic Plain. Agri 9(5):2–17. https://doi.org/10.3390/agriculture9010005

Bennett AJ, Bending GD, Chandler D, Hilton S, Mills P (2012) Meeting the demand for crop production: the challenge of yield decline in crops grown in short rotations. Biol Rev 87:52–71

Bhatt R (2015) Soil water dynamics and water productivity of rice-wheat system under different establishment methods. PhD thesis, Punjab Agricultural University, Ludhiana. https://krishikosh.egranth.ac.in/handle/1/5810015998

Bhatt R, Kukal SS (2015) Direct seeded rice in South Asia. In: Lichtfouse E (ed) Sustainable agriculture reviews, Sustainable agriculture reviews, vol 18. Springer, Cham, pp 217–252. https://doi.org/10.1007/978-3-319-21629-4_7

Bhatt R, Kukal SS, Busari MA, Arora S, Yadav M (2016) Sustainability issues on rice-wheat cropping system. Int Soil Water Conserv Res 4:68–83. https://doi.org/10.1016/j.iswcr.2015.12.001

Bhatt R, Kaur R, Gosh A (2019) Strategies to practice climate smart agriculture to improve the livelihoods under rice-wheat systems in South Asia. In: Sustainable soil and environmental management. Springer, Singapore, pp 29–72. https://doi.org/10.1007/978-981-13-8832-3_2

Bhatt R, Hossain A, Singh P (2020a) Scientific interventions to improve land and water productivity for climate smart agriculture in South-Asia. In: Agronomic crops: management practices, vol 2. Springer, Singapore, pp 499–558. https://doi.org/10.1007/978-981-32-9783-8

Bhatt R, Hossain A, Hasanuzzaman M (2020b) Adaptation strategies to mitigate the evapotranspiration for sustainable crop production: a perspective of Rice-wheat cropping system. In: Agronomic crops. volume 2: management practices. Springer, Singapore, pp 559–582. https://doi.org/10.1007/978-981-32-9783-8

Bhattacharyya P, Neogi S, Roy KS, Dash PK, Nayak AK, Mohapatra T (2014) Tropical low land rice ecosystem is a net carbon sink. Agric Ecosyst Environ 189:127–135

Biswas B (2017) Cropping system analysis for agricultural sustainability-productivity, economy, ecology and energy use efficiency. J Exp Biol Agric Sci 5(3):12–19

Biswas B, Ghosh DC, Dasgupta MK, Trivedi N, Timsina J, Dobermann A (2006) Integrated assessment of cropping systems in the eastern indo-Gangetic plain. Field Crop Res 99(1):35–47

Biswas B, Timsina J, Patra SR, De D, Mishra B, Chakraborti R, Patra A, Mahato B, Banerje P, Ghosh PK, Mukherjee J, Das S, Sarkar S, Adhikary S, Monda M, Kanthal S, Baidya S, Banerjee S, Mondal B, Ray BR (2019) Climate change-resilient rice production technology: a high yielding water efficient and remunerative option for south Asian farmers. Glob J Agric Allied Sci 1:20–29. https://doi.org/10.35251/gjaas.003

Blanco-Canqui H, Shaver TM, Lindquist JL, Shapiro CA, Elmore RW, Francis CA, Hergert GW (2015) Cover crops and ecosystem services: insights from studies in temperate soils. Agron J 107(6):2449–2474

Bommarco R, Miranda F, Bylund H, Björkman C (2011) Insecticides suppress natural enemies and increase pest damage in cabbage. J Econ Entomol 104:782–791

Bommarco R, Lundin O, Smith HG, Rundlöf M (2012) Drastic historic shifts in bumble-bee community composition in Sweden. Proc Royal Soc B: Biol Sci 279(1727):309–315

Bommarco R, Kleijn D, Potts S (2013) Ecological intensification: harnessing ecosystem services for food security. Trends Ecol Evol 28:230–238

Bouwman AF, Boumans LJM, Batjes NH (2002) Emissions of N_2O and NO from fertilized fields: summary of available measurement data. Global Biogeochem Cycles 116:6-1–6-13

Bouwman AF, Beusen AHW, Billen G (2009) Human alteration of the global nitrogen and phosphorus soil balances for the period 1970–2050. Global Biogeochem Cycles 23(4): GB0A04. https://doi.org/10.1029/2009GB003576

Brooker RW, Bennett AE, Cong WF, Daniell TJ, Hallett PD, Hawes C, Iannetta PPM, Jones HG, Karley AJ, Li L, McKenzie BM, Pakerman RJ, Paterson E, Schöb C, Shen J, Squire G, Watson

CA, Zhang F, Zhang J, White PJ (2015) Improving intercropping: a synthesis of research in agronomy, plant physiology and ecology. New Phytol 206:107–117

Brown L (1977) Redefining security, Worldwatch paper no. 14. Worldwatch Institute, Washington, DC

Brussaard L, Kuyper TW, Didden WA, De Goede RG, Bloem J (2004) Biological soil quality from biomass to biodiversity-importance and resilience to management stress and disturbance. In: Schjønning P, Elmholt S, Christensen BT (eds) Managing soil quality: challenges in modern agriculture. CABI Publishing, London, pp 139–161

Buckwell A (2014) The sustainable intensification of European agriculture. Public utility foundation for rural investment support for Europe (RISE). http://www.cap2020.ieep.eu/2014/7/28/sustainable-intensification-of-european-agriculture

Cardinale BJ, Duffy JE, Gonzalez A, Hooper DU, Perrings C, Venail P, Narwani A, Mace GM, Tilman D, Wardle DA, Kinzig AP (2012) Biodiversity loss and its impact on humanity. Nature 486:59–67

Cassman KG (1999) Ecological intensification of cereal production systems: yield potential, soil quality, and precision agriculture. Proc Natl Acad Sci USA 96(11):5952–5959. https://doi.org/10.1073/pnas.96.11.5952

Cassman KG (2017) Ecological intensification of maize-based cropping systems. Better Crops 101 (2):1–6

Cassman KG, Grassini P, van Wart J (2010) Crop yield potential, yield trends, and global food security in a changing climate. In: Handbook of climate change and agroecosystems, vol 3. Imperial College Press, London, pp 37–51

Chen H, Zhu QA, Peng C, Wu N, Wang Y, Fang X, Jiang H, Xiang W, Chang J, Deng X, Yu G (2013) Methane emissions from rice paddies natural wetlands, lakes in China: synthesis new estimate. Glob Chang Biol 19(1):19–32

Cortner O, Garrett RD, Valentim JF, Ferreira J, Niles MT, Reis J, Gil J (2019) Perceptions of integrated crop-livestock systems for sustainable intensification in the Brazilian Amazon. Land Use Policy 82:841–853

Daily G, Dasgupta P, Bolin B, Crosson P, Du Guerny J, Ehrlich P, Folke C, Jansson AM, Jansson BO, Kautsky N, Kinzig A (1998) Food production, population growth, and the environment. Science 281:1291–1292

Dale VH, Polasky S (2007) Measures of the effects of agricultural practices on ecosystem services. Ecol Econ 64:286–296. https://doi.org/10.1016/j.ecolecon.2007.05.009

DeFries R, Fanzo J, Remans R, Palm C, Wood S, Anderman TL (2015) Metrics for land-scarce agriculture. Science 349:238–240

Díaz S, Fargione J, Chapin FS III, Tilman D (2006) Biodiversity loss threatens human well-being. PLoS Biol 4:e277

Dobbie KE, Smith KA (2003) Nitrous oxide emission factors for agricultural soils in Great Britain: the impact of soil water-filled pore space and other controlling variables. Glob Chang Biol 9:204–218

Döll P, Hoffmann DH, Portmann FT, Siebert S, Eicker A, Rodell M, Strassberg G, Scanlon B (2012) Impact of water withdrawals from groundwater and surface water on continental water storage variations. J Geodyn 59–60:143–156

Doré T, Makowski D, Malézieux E, Munier-Jolain N, Tchamitchian M, Tittonell P (2011) Facing up to the paradigm of ecological intensification in agronomy: revisiting methods, concepts and knowledge. Eur J Agron 34:197–210

Du Preez CC, van Huyssteen CW, Kotzé E, van Tol JJ (2020) Ecosystem services in sustainable food systems: operational definition, concepts, and applications. In: The role of ecosystem services in sustainable food systems. Academic Press, Amsterdam, pp 17–42

DuriauxChavarría JY, Baudron F, Sunderland T (2018) Retaining forests within agricultural landscapes as a pathway to sustainable intensification: evidence from southern Ethiopia. Agric Ecosyst Environ 263:41–52

Ekström G, Ekbom B (2011) Pest control in agro-ecosystems: an ecological approach. Crit Rev Plant Sci 30:74–94
Erenstein O, Sayre K, Wall P, Hellin J, Dixon J (2012) Conservation agriculture in maize-and wheat-based systems in the (sub) tropics: lessons from adaptation initiatives in South Asia, Mexico, and southern Africa. J Sust Agric 36(2):180–206
Falconnier GN, Descheemaeker K, Traore B, Bayoko A, Giller KE (2018) Agricultural intensification and policy interventions: exploring plausible futures for smallholder farmers in southern Mali. Land Use Policy 70:623–634
FAO (2011) General summary Asia—AQUASTAT. Food and Agriculture Organization, Rome
FAO (2016) Conservation agriculture principles. http://www.fao.org/conservationagriculture/en/. Accessed 7 Jan 2019
FAO (2017) Ecological intensification in EU agriculture policy implications of research findings from project Liberation introduction. FAO, Rome
FAO (Food and Agriculture Organization) (2009) How to Feed the World: Global agriculture towards 2050. High level expert forum, 12–13 October 2009. FAO, Rome. Available http://www.fao.org/fileadmin/templates/wsfs/docs/Issues_papers/HLEF2050_Global_Agriculture.pdf
FAOSTAT (2004) FAOSTAT data 2004. Accessed Dec 2005
FAOSTAT (2013) FAOSTAT database. Food and Agriculture Organization of the United Nations. Available http://faostat.fao.org/
FAOSTAT (2014) Food and Agriculture Organization of the United Nations Statistics Division
FCCC (2012) Slow onset events-technical paper, FCCC/TP/2012/7. Framework Convention on Climate Change, United Nations
Foley JA, DeFries R, Asner GP, Barford C, Bonan G, Carpenter SR, Chapin FS, Coe MT, Daily GC, Gibbs HK, Helkowski JH (2005) Global consequences of land use. Science 309:570–574
Fowler D, Coyle M, Skiba U, Sutton MA, Cape JN, Reis S, Sheppard LJ, Jenkins A, Grizzetti B, Galloway JN, Vitousek P (2013) The global nitrogen cycle in the twenty-first century. Philos Trans Roy Soc B: Biol Sci 368(1621):20130164. https://doi.org/10.1098/rstb.2013.0164
Franke AC, Van Den Brand GJ, Giller KE (2014) Which farmers benefit most from sustainable intensification? An ex-ante impact assessment of expanding grain legume production in Malawi. Eur JAgron 58:28–38
Friedrich T, Derpsch R, Kassam A (2012) Overview of the global spread of conservation agriculture. Field Actions Sci Rep 6:2012
Gallai N, Salles JM, Settele J, Vaissière BE (2009) Economic valuation of the vulnerability of world agriculture confronted with pollinator decline. Ecol Econ 68:810–821
Galloway JN, Dentener FJ, Capone DG, Boyer EW, Howarth RW, Seitzinger SP, Asner GP, Cleveland CC, Green PA, Holland EA, Karl DM (2004) Nitrogen cycling: past, present, and future. Biogeochem 70:153–226
Garbach K, Milder JC, DeClerck FAJ, de Wit MM, Driscoll L, Gemmill-Herren B (2017) Examining multi-functionality for crop yield and ecosystem services in five systems of agroecological intensification. Int J Agric Sustain 15:11–28
García-Palacios P (2019) Ecological intensification of agriculture in drylands. J Arid Environ 167:101–105
Garibaldi LA, Gemmill-Herren B, D'Annolfo R, Graeub BE, Cunningham SA, Breeze TD (2017) Farming approaches for greater biodiversity, livelihoods, and food security. Trends Ecol Evol 32(1):68–80
Garnett T, Appleby MC, Balmford A, Bateman IJ, Benton TG, Bloomer P, Burlingame B, Dawkins M, Dolan L, Fraser D, Herrero M (2013) Sustainable intensification in agriculture: premises and policies. Science 341(6141):33–34
Gatto M (2020) Sustainable intensification of Rice-based systems with potato in eastern Indo-Gangetic Plains. Am J Potato Res 97:162–174. https://doi.org/10.1007/s12230-020-09764-6

Godfray HC, Beddington JR, Crute IR, Haddad L, Lawrence D, Muir JF, Pretty J, Robinson S, Thomas SM, Toulmin C (2010) Food security: the challenge of feeding 9 billion people. Science 327:812–818

Grassini P, Thornburn J, Burr C, Cassman KG (2011) High-yield irrigated maize in the Western US Corn Belt: I. On-farm yield, yield potential, and impact of agronomic practices. Field Crop Res 120:144–152

Hauggaard-Nielsen H, Jensen ES (2005) Facilitative root interactions in intercrops. Plant and Soil 274:237–250

Haque ME, Bell RW, Vance W (2017) Proceedings of the 2nd conference on Conservation Agriculture for Smallholders (CASH-II), 14–16 February, Mymensingh, Bangladesh, p 170

Hewitt TI, Smith KR (1995) Intensive agriculture and environmental quality: examining the newest agricultural myth. Henry Wallace Institute for Alternative Agriculture, Greenbelt

Hobbs PR, Sayre K, Gupta R (2008) The role of conservation agriculture in sustainable agriculture. Philosop Trans Royal Soc B: Biol Sci 363(1491):543–555

Hoehn P, Tscharntke T, Tylianakis JM, Steffan-Dewenter I (2008) Functional group diversity of bee pollinators increases crop yield. Proc Royal Soc B: Biol Sci 275(1648):2283–2291

Hooper DU, Chapin FS, Ewel JJ, Hector A, Inchausti P, Lavorel S, Lawton JH, Lodge DM, Loreau M, Naeem S, Schmid B (2005) Effects of biodiversity on ecosystem functioning: a consensus of current knowledge. Ecol Monogr 75:3–35

Hossain MS, Hossain A, Sarkar MA, Jahiruddin M, Teixeira da Silva JA, Hossain MI (2016) Productivity and soil fertility of the rice–wheat system in the High Ganges River Floodplain of Bangladesh is influenced by the inclusion of legumes and manure. Agric Ecosyst Environ 218:40–52. https://doi.org/10.1016/j.agee.2015.11.017

IPCC (1994) Climate change 1994. Cambridge University Press, Cambridge

IPCC (2007) Contribution of working group III to the fourth assessment report of the Intergovernmental Panel on Climate Change. Cambridge University Press, Cambridge

IPCC (2014) Climate change 2014: synthesis report. Contribution of working groups I, II and III to the fifth assessment report of the Intergovernmental Panel on Climate hange. IPCC, Geneva

Jain N, Bhatia A, Pathak H (2014) Emission of air pollutants from crop residue burning in India. Aerosol Air Qual Res 14(1):422–430

Jat ML, Singh S, Rai HK, Chhokar RS, Sharma SK, Gupta RK (2005) Furrow irrigated raised bed (FIRB) planting technique for diversification of rice-wheat system in indo-Gangetic Plains. Proc Japan Assoc Intern Collab Agric For 28:25–42

Jhariya MK, Bargali SS, Raj A (2015) Possibilities and perspectives of agroforestry in Chhattisgarh. In: Zlatic M (ed) Precious forests-precious earth. InTech, Croatia, pp 237–257. https://doi.org/10.5772/60841ISBN: 978-953-51-2175-6

Jhariya MK, Banerjee A, Meena RS, Yadav DK (2019a) Sustainable agriculture, forest and environmental management. Springer, Singapore, p 606. ISBN: 978-981-13-6830-1, ISBN: 978-981-13-6829-5. https://doi.org/10.1007/978-981-13-6830-1

Jhariya MK, Yadav DK, Banerjee A (2019b) Agroforestry and climate change: issues and challenges. Apple Academic Press, Oakville. ISBN: 978-1-77188-790-8 (hardcover), 978-0-42957-274-8 (E-book), p 335. https://doi.org/10.1201/9780429057274

Johansen C, Haque ME, Bell RW, Thierfelder C, Esdaile RJ (2012) Conservation agriculture for small holder rainfed farming: opportunities and constraints of new mechanized seeding systems. Field Crop Res 132:18–32

Jonsson M, Buckley HL, Case BS, Wratten SD, Hale RJ, Didham RK (2012) Agricultural intensification drives landscape-context effects on host–parasitoid interactions in agroecosystems. J Appl Ecol 49(3):706–714

Joshi PK, Gulati A, Birthal PS, Tewari L (2014) Diversification in South Asia patterns, determinants and policy implications. Econ Pol Wkly 39:2457–2467

Kamp K, Gregory R, Chowhan G (1993) Fish cutting pesticide use. ILEIA Newslett 2(93):22–23

Karlen DL, Varvel GE, Bullock DG, Cruse RM (1994) Crop rotations for the 21st century. Adv Agron 53:1–45

Kassam A, Friedrich T, Derpsch R, Lahmar R, Mrabet R, Basch G, González-Sánchez EJ, Serraj R (2012) Conservation agriculture in the dry Mediterranean climate. Field Crop Res 132:7–17
Kassie M, Teklewold H, Jaleta M, Marenya P, Erenstein O (2015) Understanding the adoption of a portfolio of sustainable intensification practices in eastern and southern Africa. Land Use Policy 42:400–411
Kavikumar KS (2010) Climate sensitivity of Indian agriculture: role of technological development and information diffusion. In: Lead papers of national symposium on climate change and rainfed agriculture, February 18–20, 2010, p 192. Indian Society of Dryland Agriculture, Central Research Institute for Dryland Agriculture, Hyderabad, India
Kleijn D, Bommarco R, Fijen TP, Garibaldi LA, Potts SG, van der Putten WH (2019) Ecological intensification: bridging the gap between science and practice. Trends Ecol Evol 34(2):154–166. https://doi.org/10.1016/j.tree.2018.11.002
Klein AM, Vaissiere BE, Cane JH, Steffan-Dewenter I, Cunningham SA, Kremen C, Tscharntke T (2007) Importance of pollinators in changing landscapes for world crops. Proc Royal Soc B: Biol Sci 274:303–313
Kremen C, Williams NM, Thorp RW (2002) Crop pollination from native bees at risk from agricultural intensification. Proc Natl Acad Sci 99(26):16812–16816
Kumar S, Meena RS, Jhariya MK (2020) Resources use efficiency in agriculture. Springer, Singapore. https://doi.org/10.1007/978-981-15-6953-1
La Notte A, D'Amato D, Mäkinen H, Paracchini ML, Liquete C, Egoh B, Geneletti D, Crossman ND (2017) Ecosystem services classification: a systems ecology perspective of the cascade framework. Ecol Indic 74:392–402
Ladha JK, Dawe D, Pathak H, Padre AT, Yadav RL, Bijay S, Yadvinder-Singh SY, Singh P, Kundu AL, Sakal R, Ram N, Regmi AP, Gami SK, Bhandari AL, Amin R, Yadav CR, Bhattarai EM, Das S, Aggarwal HP, Gupta RK, Hobbs PR (2003) How extensive are yield declines in long-term rice–wheat experiments in Asia? Field Crop Res 81:159–180
Lal R (1989) Agroforestry systems and soil surface management of a tropical alfisol. I: soil moisture and crop yields. Agr Syst 8:7–29
Lal R (2004) Soil carbon sequestration impacts on global climate change and food security. Science 305:1623–1627
Liebman M, Baraibar B, Buckley Y, Childs D, Christensen S, Cousens R, Eizenberg H, Heijting S, Loddo D, MerottoJr A, Renton M (2016) Ecologically sustainable weed management: how do we get from proof-of-concept to adoption? Ecol Appl 26(5):1352–1369
Liu A, Fong A, Beckett E, Yuan J, Tamae C, Medrano L, Maiz M, Wahba C, Lee C, Lee K, Tran KP, Yang H, Hoffman RM, Salih A, Miller JH (2011) Selective advantage of resistant strains at race levels of antibiotics: a simple and ultrasensitive color detection of antibiotics and genotoxic agents. Antimicrob Agents Chemother 55:1204–1210
Lobell DB, Field CB (2007) Global scale climate-crop yield relationships and the impacts of recent warming. Environ Res Lett 2:014002
Lobell DB, Cassman KG, Field CB (2009) Crop yield gaps: their importance, magnitudes, and causes. Annu Rev Env Resour 34:179–204
Long YQ, Wu WB, Yu QY (2018) Review on the research progress of intensive use of cultivated land. Nat Resour 33:337–350
Lundgren JG, Fergen JK (2011) Enhancing predation of a subterranean insect pest: a conservation benefit of winter vegetation in agroecosystems. Appl Soil Ecol 51:9–16
Ma BL, Wu TY, Tremblay N, Deen W, Morrison MJ, McLaughlin NB, Gregorich EG, Stewart G (2010) Nitrous oxide fluxes from corn fields. Glob Chang Biol 16:156–170
Mathews JT (1989) Redefining security. Foreign Aff 68(2):162–177
MEA (2005) Ecosystems and human well-being: synthesis. Island Press, Washington, DC
Meena RS, Lal R (2018) Legumes for soil health and sustainable management. Springer, Singapore, p 541. ISBN 978-981-13-0253-4 (eBook), ISBN: 978-981-13-0252-7(hardcover). https://doi.org/10.1007/978-981-13-0253-4_10

Meena RS, Kumar V, Yadav GS, Mitran T (2018) Response and interaction of *Bradyrhizobium japonicum* and Arbuscular mycorrhizal fungi in the soybean rhizosphere: a review. Plant Growth Regul 84:207–223

Meena RS, Kumar S, Datta R, Lal R, Vijaykumar V, Brtnicky M, Sharma MP, Yadav GS, Jhariya MK, Jangir CK, Pathan SI, Dokulilova T, Pecina V, Marfo TD (2020) Impact of agrochemicals on soil microbiota and management: a review. Land (MDPI) 9(2):34. https://doi.org/10.3390/land9020034

Meena RS, Lal R, Yadav GS (2020a) Long term impacts of topsoil depth and amendments on soil physical and hydrological properties of an Alfisol in Central Ohio, USA. Geoderma 363:1141164

Meena RS, Lal R, Yadav GS (2020b) Long-term impact of topsoil depth and amendments on carbon and nitrogen budgets in the surface layer of an Alfisol in Central Ohio. Catena 194:104752

Mekonnen MM, Hoekstra AY (2011) The green, blue and grey water footprint of crops and derived crop products. Hydrol Earth Syst Sci 15:1577–1600

Mishra A, Behera B, Pal AK, Mohanty SK, Rath BS, Subudhi CR, Nayak SC, Sahoo N (2012) Performance of rice and blackgram with different nutrient management practices in rainfed upland. ORYZA-An Intl J Rice 49(4):273–279

Mishra KN, Patra AK, Garnayak LM, Mohanty AK, Swain SK (2017) Long-term effects of integrated nutrient management on productivity and soil properties of rice (*Oryza sativa*)—rice cropping system in coastal Odisha. Ind J Agron 62(3):239–246

MNRE (2009). www.nicra.iari.res.in/Data/FinalCRM.doc

Mondal M, Gunri SK, Sengupta A, Kundu R (2018) Productivity enhancement of Rabi groundnut (*Arachishypogaea* L.) under polythene mulching and rhizobium inoculation under new alluvial zone of West Bengal. Int J Curr Microbiol App Sci 7:2308–2313. https://doi.org/10.20546/ijcmas.2018.709.286

Mueller ND, Gerber JS, Johnston M, Ray DK, Ramankutty N, Foley JA (2012) Closing yield gaps through nutrient and water management. Nature 490:254–257. https://doi.org/10.1038/nature11420

MWR (2008) Efficient use of water for various purposes. National Water Mission under National Action Plan on Climate hange. Ministry of Water Resources, New Delhi

MWR (2009) Ministry of Water Resources, Government of India. http://wrmin.nic.in/resource/default3.htm

NAF (1994) A better row to hoe: the economic, environmental and social impact of sustainable agriculture. Northwest Area Foundation, St Paul

National Commission for Women (2020) Water availability report. https://www.indiastat.com/table/per-capita-availability-data/24/water-supply/18198/365176/data.aspx. Accessed 17 Mar 2020

Ndiritu SW, Kassie M, Shiferaw B (2014) Are there systematic gender differences in the adoption of sustainable agricultural intensification practices? Evidence from Kenya. Food Policy 49:117–127

Neubauer SC, Megonigal JP (2015) Moving beyond global warming potentials to quantify the climatic role of ecosystems. Ecosystems 18:1000–1013. https://doi.org/10.1007/s10021-015-9879-4

Oenema O, Amon B, van Beek C, Hutchings N, Perez-Soba M, Procter C, Pietrzak S, Velthof GL, Vinther F, Wilson L (2011) Farm data needed for agri-environmental reporting. In: Selenius J, Baudouin L, Kremer AM (eds) Eurostat methodologies and working papers. Publications Office of the European Union, Luxembourg

Ortega DL, Waldman KB, Richardson RB, Clay DC, Snapp S (2016) Sustainable intensification and farmer preferences for crop system attributes: evidence from Malawi's central and southern regions. World Dev 87:139–151

Painkra GP, Bhagat PK, Jhariya MK, Yadav DK (2016) Beekeeping for poverty alleviation and livelihood security in Chhattisgarh, India. In: Narain S, Rawat SK (eds) Innovative technology

for sustainable agriculture development. Biotech Books, New Delhi, pp 429–453ISBN: 978–81–7622-375-1
Pan G, Smith P, Pan W (2009) The role of soil organic matter in maintaining the productivity and yield stability of cereals in China. Agric Ecosyst Environ 129:344–348
Parker HR, Boyd E, Cornforth RJ, James R, Otto FE, Allen MR (2017) Stakeholder perceptions of event attribution in the loss and damage debate. Clim Pol 17(4):533–550. https://doi.org/10.1080/14693062.2015.1124750
PAU (2020) The package of practices for the crops of Punjab Kharif 2020. Half yearly package. Punjab Agricultural University, Ludhiana
Paul EA, Paustian KH, Elliott ET, Cole CV (1997) Soil organic matter in temperate agroecosystems: Long-term experiments in North America. Boca Raton, CRC Press
Plaza-Bonilla D, Arrúe JL, Cantero-Martínez C, Fanlo R, Iglesias A, Álvaro-Fuentes J (2015) Carbon management in dry land agricultural systems. A review. Agron Sustain Dev 35:1319–1334
Poeplau C, Don A (2015) Carbon sequestration in agricultural soils via cultivation of cover crops—a meta-analysis. Agric Ecosyst Environ 200:33–41
Potts SG, Biesmeijer JC, Kremen C, Neumann P, Schweiger O, Kunin WE (2010) Global pollinator declines: trends, impacts and drivers. Trends Ecol Evol 25:345–353
Potts SG, Ngo HT, Biesmeijer JC, Breeze TD, Dicks LV, Garibaldi LA, Hill R, Settele J, Vanbergen A (2016) The assessment report of the intergovernmental science-policy platform on biodiversity and ecosystem services on pollinators, pollination and food production. Secretariat of the Intergovernmental Science-Policy Platform on Biodiversity and Ecosystem Services, Bonn, p 556. http://nora.nerc.ac.uk/id/eprint/519227/
Power AG (2010) Ecosystem services and agriculture: tradeoffs and synergies. Philos Transact Roy Soc B: Biol Sci 365(1554):2959–2971
Pretty JN (1997) The sustainable intensification of agriculture. Nat Res Forum 21:247–256. https://doi.org/10.1111/j.1477-8947.1997.tb00699.x
Pretty J (2008) Agricultural sustainability: concepts, principles and evidence. Philos Trans R Soc B Biol Sci 363(1491):447–465. doi.org/10.1098/rstb.2007.2163
Raj A, Jhariya MK, Yadav DK, Banerjee A (2020) Climate change and agroforestry systems: adaptation and mitigation strategies. Apple Academic Press, Palm Bay, p 383. https://doi.org/10.1201/9780429286759
Rao UA, Sridhar TV, Adi Lakshmi D, Raju SK (2014) Identification of viable rice based cropping systems for double cropped delta areas of Andhra Pradesh. Int J Sci Nat 5(3):512–514
Rautaray SK, Mishra A, Verma OP (2017) Energy efficiency, productivity and profitability of rice (*Oryza sativa* L.) based cropping systems for selected conservation practices. Archiv Agron Soil Sci 63(14):1993–2006
Renting H, Rossing WA, Groot JC, Van der Ploeg JD, Laurent C, Perraud D, Stobbelaar DJ, Van Ittersum MK (2009) Exploring multifunctional agriculture. A review of conceptual approaches and prospects for an integrative transitional framework. J Environ Manage 90:S112–S123
Roder W, Mason SC, Clegg MD, Kniep KR (1989) Yield-soil water relationships in sorghum soybean cropping systems with different fertilizer regimes. Agron J 81:470–475
Roling NR, Wagemakers MA (1997) Social learning for sustainable agriculture. Cambridge University Press, Cambridge
Rosegrant MW, Leach N, Gerpacio RV (1998) Alternative futures for world cereal and meat consumption. International Food Policy Institute, Washington, DC
Rossing WAH, Modernel P, Tittonell P (2013) Diversity in organic and agro-ecological farming systems for mitigation of climate change impact, with examples from Latin America. In: Fuhrer J, Gregory PJ (eds) Climate change impact and adaptation in agricultural systems. CAB International, Wallingford, pp 69–87
Rost S, Gerten D, Bondeau A, Lucht W, Rohwer J, Schaphoff S (2008) Agricultural green and blue water consumption and its influence on the global water system. Water Resour Res 44:W09405. https://doi.org/10.1029/2007WR006331

Roy SB, Saha NK, Kadian MS, Quiroz R, Ilangantileke S (2007) Improving the livelihood of farmers by intensifying the rice-potato-rice system through double-transplanting of rice in West Bengal, India. International potato center, natural resources management division working paper no. 2007-1. p 28. http://cipotato.org/wp-content/uploads/publication%20files/working%20papers/003830.pdf

Russenes AL, Korsaeth A, Bakken LR, Dörsch P (2019) Effects of nitrogen split application on seasonal N_2O emissions in southeast Norway. Nutr Cycl Agroecosyst 115(1):41–56. https://doi.org/10.1007/s10705-019-10009-0

SAARC (2009) Fifth meeting of SAARC committee on agricultural & rural development. South Asian Association for Regional Cooperation, Kathmandu

Saini J, Bhatt R (2020) Global warming-causes, impacts and mitigation strategies in agriculture. Curr J Appl Sci Technol 29:93–107. https://doi.org/10.9734/cjast/2020/v39i730580

Schweiger O, Biesmeijer JC, Bommarco R, Hickler T, Hulme PE, Klotz S, Kühn I, Moora M, Nielsen A, Ohlemüller R, Petanidou T (2010) Multiple stressors on biotic interactions: how climate change and alien species interact to affect pollination. Biol Rev 85:777–795

Senthilkumar D, Murali AP, Chinnusamy C, Bharathi C, Lavanya Y (2019) Stale seed bed techniques as successful weed management practice. J Pharmacogn Phytochem SP2:120–123

Settle WH, Ariawan H, Astuti ET, Cahyana W, Hakim AL, Hindayana D, Lestari AS (1996) Managing tropical rice pests through conservation of generalist natural enemies and alternative prey. Ecology 77(7):1975–1988

Shah P (1994) Village-managed extension systems in India: implications for policy and practice. In: Scoones I, Thompson J (eds) Beyond farmer first: rural People's knowledge, agricultural research and extension practice. Intermediate Technology Publications, London

Shakir AS, ur Rehman H, Khan NM, Qazi AU (2011) Impact of canal water shortages on groundwater in the lower Bari doab canal system in Pakistan. Pak J Eng Appl Sci 9:87–97

Singh NR, Jhariya MK (2016) Agroforestry and Agrihorticulture for higher income and resource conservation. In: Narain S, Rawat SK (eds) Innovative technology for sustainable agriculture development. Biotech Books, New Delhi, pp 125–145ISBN: 978-81-7622-375-1

Singh KK, Jat AS, Sharma SK (2005) Improving productivity and profitability of rice (*Oryza sativa*)-wheat (*Triticul aestivum*) cropping system through tillage and planting management. Indian J Agric Sci 75:396–399

Singh R, Kundu DK, Bandyopadhyay KK (2010) Enhancing agricultural productivity through enhanced water use efficiency. J Agric Phy 10:1–15

Singh D, Bhaskar BP, Baruah U, Sarkar D (2014a) Evaluation of rice (*Oryza sativa* L.) based cropping systems in major soil series of upper Brahmaputra Valley Asom. Global J Sci Frontier Res: D Agric Vet 14(6):64–69

Singh J, Singh A, Kaur H (2014b) Agricultural growth and inequality in South Asia. Int J Res Comm Econom Manag 4(5):52–58

Singh VK, Shukla AK, Dwivedi BS, Singh MP, Majumdar K, Kumar V, Mishra RP, Rani M, Singh SK (2015) Site-specific nutrient management under rice-based cropping systems in indo-Gangetic Plains: yield, profit and apparent nutrient balance. Agric Res 4(4):365–377

Soni M, Upadhyay VB, Singh P (2012) Weed dynamics and production efficiency of rice-based cropping system. Ind J Weed Sci 44(1):21–25

Steffan-Dewenter I, Schiele S (2008) Do resources or natural enemies drive bee population dynamics in fragmented habitats? Ecology 89:1375–1387

Teillard F, Allaire G, Cahuzac E, Léger F, Maigné E, Tichit M (2012) A novel method for mapping agricultural intensity reveals its spatial aggregation: implications for conservation policies. Agric Ecosyst Environ 149:135–143

Thies C, Haenke S, Scherber C, Bengtsson J, Bommarco R, Clement LW, Ceryngier P, Dennis C, Emmerson M, Gagic V, Hawro V (2011) The relationship between agricultural intensification and biological control: experimental tests across Europe. Ecol Appl 21:2187–2196

Tilman D (1998) The greening of the green revolution. Nature 396:211–212

Tilman D, Fargione J, Wolff B, D'Antonio C, Dobson A, Howarth R, Schindler D, Schlesinger WH, Simberloff D, Swackhamer D (2001) Forecasting agriculturally driven global environmental change. Science 292:281–284

Tilman D, Cassman KG, Matson PA, Naylor R, Polasky S (2002) Agricultural sustainability and intensive production practices. Nature 418:671–677

Tilman D, Balzer C, Hill J, Befort BL (2011) Global food demand and the sustainable intensification of agriculture. Proc Natl Acad Sci 108(50):20260–20264

Timsina J, Connor DJ (2001) Productivity and management of rice-wheat cropping systems, issues and challenges. Field Crop Res 69:93–132

Tittonell P (2014) Ecological intensification of agriculture — sustainable by nature. Curr Opin Environ Sustain 8:53–61

Tittonell P, Giller KE (2013) When yield gaps are poverty traps: the paradigm of ecological intensification in African smallholder agriculture. Field Crop Res 143:76–90

Tschumi M, Albrecht M, Entling MH, Jacot K (2015) High effectiveness of tailored flower strips in reducing pests and crop plant damage. Proc Roy Soc B: Biol Sci 282(1814):20151369. https://doi.org/10.1098/rspb.2015.1369

Tubiello FN, Salvatore M, Rossi S, Ferrara A, Fitton N, Smith P (2013) The FAOSTAT database of GHG emissions from agriculture. Environ Res Lett 8:1–11. https://doi.org/10.1088/1748-9326/8/1/015009

Ulén B, Aronsson H, Bechmann M, Krogstad T, Øygarden L, Stenberg M (2010) Soil tillage methods to control phosphorus loss and potential side-effects: a Scandinavian review. Soil Use Manage 26(2):94–107

UNEP (2008) Fresh water under threat, South Asia, vulnerability assessment of freshwater resources to environmental change. United Nations Environment Program, Nairobi

UNEP (2016) Food systems and natural resources. In: Westhoe H, Ingram J, van Berkum S, Ozay L, Hajer M (eds) A report of the working group on food systems of the inter-national resource panel. UNESCO, Paris

US-EPA (2006) Global anthropogenic non-CO_2 greenhousegas emissions: 1990–2020. United States Environmental Protection Agency, Washington, DC. EPA 430-R-06-003

Van der Horst D (2011) Adoption of payments for ecosystem services: an application of the Hägerstrand model. Appl Geogr 31(2):668–676

Vanlauwe B, Coyne D, Gockowski J, Hauser S, Huising J, Masso C, Nziguheba G, Schut M, Van Asten P (2014) Sustainable intensification and the African smallholder farmer. Curr Opin Environ Sustain 8:15–22

Venter ZS, Jacobs K, Hawkins HJ (2016) The impact of crop rotation on soil microbial diversity: a meta-analysis. Pedobiologia 59(4):215–223

Verbruggen E, Röling WF, Gamper HA, Kowalchuk GA, Verhoef HA, van der Heijden MG (2010) Positive effects of organic farming on below-ground mutualists: large-scale comparison of mycorrhizal fungal communities in agricultural soils. New Phytol 186:968–979

Vitousek PM, Aber JD, Howarth RW, Likens GE, Matson PA, Schindler DW, Schlesinger WH, Tilman DG (1997) Human alteration of the global nitrogen cycle: sources and consequences. Ecol Appl 7:737–750

Vliet VJ, Groot HLF, Rietveld P, Verburg PH (2015) Manifestations and underlying drivers of agricultural land use change in Europe. Landscape Urban Plan 133:24–36

Wallace JS, Gregory PJ (2002) Water resources and their use in food production. Aq Sci 64:363–375

Wassmann R, Papen H, Rennenberg H (1993) Methane emission from rice paddies and possible mitigation strategies. Chemosphere 26(1–4):201–217

Weligamage P, Barker R, Hussain I, Amarasinghe U, Samad M (2002) World irrigation and water statistics. International Water Management Institute, Colombo

Westing AH (1986) An expanded concept of international security. In: Westing AH (ed) Global resources and international conflict. Oxford University Press, Oxford

Williams NM, Crone EE, T'ai HR, Minckley RL, Packer L, Potts SG (2010) Ecological and life-history traits predict bee species responses to environmental disturbances. Biol Conserv 143:2280–2291

Wood SA (2018) Nutritional functional trait diversity of crops in South-Eastern Senegal. J Appl Ecol 55:81–91

Wood S, Sebastian K, Scherr SJ (2000) Agroecosystems. A joint study by International Food Policy Research Institute and World Resources Institute. International Food Policy Research Institute and World Resources Institute, Washington, DC. http://www.wri.org/wr2000, http://www.ifpri.org

Wu-xing L (2010) Effect of self-allelopathy on AOS of *Casuarina equisetifolia* forest seedling. Fujian J Agric Sci 25:108–113

Xie H, Huang Y, Chen Q, Zhang Y, Wu Q (2019) Prospects for agricultural sustainable intensification: a review of research. Landarzt 8(11):157. https://doi.org/10.3390/land8110157

Yachi S, Loreau M (1999) Biodiversity and ecosystem productivity in a fluctuating environment: the insurance hypothesis. Proc Natl Acad Sci U S A 96:1463–1468

Yadav GS, Das A, Lal R, Babu S, Datta M, Meena RS, Patil SB, Singh R (2019) Impact of no-till and mulching on soil carbon sequestration under rice (Oryza sativa L.)-rapeseed (Brassica campestris L. var. rapeseed) cropping system in hilly agro-ecosystem of the Eastern Himalayas, India. Agric Ecosyst Environ 275:81–92

Yadvinder-Singh, Bijay-Singh, Timsina J (2005) Crop residue management for nutrient cycling and improving soil productivity in rice-based cropping systems in the tropics. Adv Agron 85:269–407

Yan X, Akiyama H, Yagi K, Akimoto H (2009) Global estimations of the inventory and mitigation potential of methane emissions from rice cultivation conducted using the 2006 intergovernmental panel on climate change guidelines. Global Biogeochem Cycles 23:GB2002. https://doi.org/10.1029/2008GB003299

Yu ZN, Zhang XB, Wu CF (2017) Review on the progress of land science related research in the journals of nature and science from 2000 to 2016. China Land Sci 31:89–97

Zhang W, Ricketts TH, Kremen C, Carney K, Swinton SM (2007) Ecosystem services and dis-services to agriculture. Ecol Econ 64:253–260. https://doi.org/10.1016/j.ecolecon.2007.02.024

Zhang B, Tian H, Ren W, Tao B, Lu C, Yang J, Banger K, Pan S (2016) Methane emissions from global rice fields: magnitude, spatiotemporal patterns, and environmental controls. Global Biogeochem Cycles 30(9):1246–1263

Zommers Z, van der Geest K, De Sherbinin A, Kienberger S, Roberts E, Harootunian G, Sitati A, James R (2016) Loss and damage: the role of ecosystem services. United Nations Environment Programme, Nairobi, p 84

Zuo L, Zhang Z, Carlson KM, MacDonald GK, Brauman KA, Liu Y, Zhang W, Zhang H, Wu W, Zhao X, Wang X (2018) Progress towards sustainable intensification in China challenged by land–use change. Nat Sustain 1:304–313

Ecological Intensification for Sustainable Agriculture and Environment in India

7

Saikat Mondal and Debnath Palit

Abstract

Sustainable intensification (SI) of farming practices provides synergistic incentives for agricultural and natural resource outputs to co-produce. Performance and replacement are measures towards SI, but reconstruction of the program is necessary to achieve optimal outcomes when ecological and economic circumstances alter. This chapter addresses global development in SI in agriculture and food system through different aspects of SI. SI is commonly assumed to have the attributes of enhancing productivity and minimizing environmental degradation. The approaches and metrics for SI in agriculture are varied followed by calculation cases which primarily spread in Europe, Asia, and Africa. Key priorities include engaging in more ways of SI in farming processes, building ecosystems of agricultural expertise, and implementing policy steps to scale up SI. Several worldwide data demonstrating the capacity of SI's towards climate-smart agriculture and agricultural sustainability in southern part of Asia and Africa in particular. The literature has been extensively evaluated to determine the degree to which SI will raise productivity throughout South Asia, an area that is likely to pose some of the biggest hurdles to food security in the upcoming years. It observed that the yield gains from SI strategies were diverse, and the overall yield increase in all experiments was about 21%. It can be deduced that lasting intensification can address a turning point where it might be revolutionary. The present chapter deals with the significant social-ecological probabilities resulting from SI by pinpointing a spectrum of outcome mechanisms at the magnitude at which the intensification takes place, and to explore in detail the circumstances under which these different outcomes are likely to happen. We do

S. Mondal (✉)
Department of Zoology, Raghunathpur College, Purulia, West Bengal, India

D. Palit
Department of Botany, Durgapur Government College, Durgapur, West Bengal, India

M. K. Jhariya et al. (eds.), *Ecological Intensification of Natural Resources for Sustainable Agriculture*, https://doi.org/10.1007/978-981-33-4203-3_7

so by reviewing the research journals that evaluates both ecosystem services and well-being effects associated with intensifying agriculture.

Keywords

Agriculture · Food system · Food security · Natural resource · Sustainable intensification

Abbreviations

CA	Conservation agriculture
CGIAR	International Agricultural Science Consultative Group
CO_2	Carbon dioxide
FAO	Food and Agricultural Organization
GHG	Greenhouse gas
ha	Hectare
IAASTD	The International Assessment of Agricultural Information, Science and Production Technology
IFAD	International Fund for Agricultural Development
Ksh	Kenyan shilling
Mg	Milligram
Mha	Mega hectare
Mt.	Metric ton
N_2O	Nitrous oxide
OA	Organic agriculture
PPB	Participatory plant breeding
SCAR	Standing Committee on Agricultural Science
SDG	Sustainable development goals
SI	Sustainable intensification
SRI	System of rice intensification
UN	United Nations
UNCTAD	United Nations Conference on Trade and Growth
UNEP	United Nations Environment Program
US	United States

7.1 Introduction

A growing, more prosperous, and urban world's population is driving the demand for more sustainable food (Nature Editorial 2020). Global food production has seen significant rises over the last half century. World population has risen by 2.5 times since 1960, and yet food production per capita has expanded by 50% during the same span (FAOSTAT 2017). At about the same moment, data suggests that livestock is the single biggest source of habitat depletion, greenhouse gas (GHG) emissions, groundwater intake, nitrogen loading into the biosphere, and a big contributor of pesticide contamination (West et al. 2014). It is embodied in surface loss and destruction, river and sea runoff, aquifer depletion, and water forcing (Pywell et al. 2015). As a result, attempts have gone ahead to build manufacturing processes that at least lessen the footprint of harm per unit generated (Rockström et al. 2017). This drive for agricultural practices to provide adequate and healthy food without harming to the ecosystem and to make meaningful contributions to economic, social, and human resources was embodied in demands for a wide variety of productive agriculture (Benton 2015).

The twin dynamics of rising and rapidly affluent communities and deteriorating environmental effects have sparked concern in "sustainable intensification" (SI) (Nature Editorial 2020). Intensification (productivity improvements) without incorporating efficiency is the prevailing model for agricultural growth centers. In contemplating the climate, the traditional emphasis is not on finding synergies between intensification and development but on-harmful impacts. There is growing proof that biodiversity systems can increase efficiency by changing agricultural output factors, like changes from chemical fertilizers to natural manures, shifting from pesticides to biological enemies. In the 1980s, there was proof of harmony between stabilization and intensification: first used in connection with an African agriculture analysis (Pretty 1997a, b). Intensification had traditionally been linked with styles of farming which caused environmental damage (Collier and Soentaro 1973; Meena and Lal 2018). The use of these two words has been an effort to prove that positive outcomes, including more calories and improved environmental services, should not be mutually incompatible. Both may be accomplished by properly using ground, water, wildlife, energy, information and technology. In a range of main commissions, SI was also suggested to follow it as it rose from about ten papers a year before 2010 to more than 100 a year by 2015 (Gunton et al. 2016).

Fair intensification related to the farming is currently seemed to be the priority for driving global development strategies (DeClerck et al. 2016; Godfray et al. 2010; Rockström et al. 2017), which emerges as a core plan for stopping poverty which promoting sustainable usage of terrestrial habitats in the Sustainable Development Goals (SDGs) (UN 2015). SI is an essential but not adequate transformative factor in the broader food system. Changes in eating habits as well as decreases in food waste will contribute further to the overall resilience of food and farming processes (Benton 2015; Meena et al. 2018). Nevertheless the need for successful SI is imperative at the farm and landscape stage. Demand on developed farm land tends to expand. Reduces the wealth base by environmental destruction (Rockström et al. 2017; Hallman et al. 2007); expansion of industrial and road development consumes farm land (Francis et al. 2012) and the resulting severe weather produce additional pressures (IPCC 2014). Attempts to establish SI will contribute to both agricultural production and natural resource gains (Pretty and Bharucha 2014; Pretty et al. 2006a, b, 2011a, b). In less developed countries, the greatest rises in food production have prevailed, often beginning from a lower base of output. Systems also continued to see improvements in productivity in developing nations, mitigating the harm to environmental resources and also the crop and livestock yields (Reganold and Wachter 2016).

The global problem, though, is extremely important, planetary borders are under stress, world population should begin to rise from about 7600 million (2018) to 10,000 million by 2050 (UN Department of Economic and Social Affairs 2017), and consumption trends for some parts of the population converge on those characteristic of wealthy nations, which still leave some 800 million people starving globally. One problem focuses on size: can farming always produce adequate healthy food while at the same time developing natural resources while not losing other facets of well-being; and does this work on a size to help thousands of lives, reducing biotic diversity depletion and reduce GHG emissions. Another concern centers on how far wider improvements in the food system towards healthier diets could influence agricultural output requirements (Smith et al. 2015). Again, healthy diets may create improved market demand for smaller residues from pesticides. Because SI is a paragliding concept that incorporates a broad variety of various agricultural activities and technologies, the exact scope of current SI activities was largely unexplored. The cumulative social-ecological consequences resulting from agricultural intensification are identified in depth in this chapter by describing a variety of outcome scenarios at the scale at which the intensification happens, and discussing the circumstances in which such specific outcomes are likely to happen.

7.2 Origins of Sustainable Intensification

The British research organization the Royal Society has described SI as a method whereby the productivity of the crop increases with least impact on environment (The Royal Society 2009). It corresponds to parallel improvements in production per unit region, productivity of resource utilization, natural resources and environmental support flow, and decreases in adverse environmental effects, like GHG emissions (Pretty 1997a, b; Godfray et al. 2010). This concept has been commonly used for positive intensification. It is worth remembering, though, the usage of the word is not completely clear in the literature. Some studies appear to interchangeably use agricultural intensification and sustainable agriculture (IFAD 2011). Some use a somewhat different word, for example, SI of crop output, thereby restricting it to the crop sub-sector (FAO 2011). Similar terms like ecological intensification (EI) and agroecology are found in numerous literatures as well. The word was initially developed in the 1990s in order to address the farming system of smallholding farmers in African subcontinents associated with lesser production and depletion of resources (Pretty 1997a, b). In these early researches, there are many aspects worth mentioning. First, they both concentrated on smallholder development in the developed world (especially in Africa) and took a firmly pro-smallholder approach. This compares with later perceptions under which SI has been a paradigm for international agricultural growth. Second, the initial definition of SI put similar focus on sustainability and intensification; in addition, considering the deteriorated condition of several agricultural lands in Africa as well as other areas, sustainability has been seen as a criterion for intensification. Thirdly, focus was put on versatility and the need to adjust methods to the local environment rather than following a predetermined range of farming methods. Fourthly, strengthening the farmers' livelihoods has become a key priority of SI.

7.3 Present Concept

In earlier times which emerged 20 years back, the idea of prolonged intensification has only achieved popularity over the last 5 years or so. This period it has been backed by countries like the United Kingdom, the United States (US), and the African Union; foreign bodies like the Food and agriculture Organization (FAO) and the International Fund for Agricultural development (IFAD); Government bodies like the International Agricultural Science Consultative Group (CGIAR) and its 15 research centers along with various internationally reputed organizations (IFAD 2011; Foresight 2011; FAO 2011; CGIAR 2014; Jawahery 2015). SI is being extended today not only to Africa but to areas across the globe. Indeed, it has become a key topic that occurs regularly at international conventions and the preparation of agricultural production. The players having major role in terms of highlighting the concept from its origin point of view in Africa up to international topics related to agriculture are:

1. The Royal Society report: "Reaping the Benefits" (2009).
2. FAO's prior objective of SI in agricultural production up to 2010.
3. The UK report on the issue of food security and production in 2011.

The aforesaid factors highlighted the importance of SD and optimum use of resource in global food production with decent application of SI.

SI has now become a widely agreed paradigm for researchers all over the world, and the commonly accepted definition is to maximize output unit intake through reduced environmental impacts (Schiefer et al. 2015, 2018; Liao and Brown 2018). The primary aim is to enhance agricultural production and capital by raising yield per hectare (ha), thereby further achieving agricultural intensification (Buckwell et al. 2014). Therefore, the main aspects include attributes, concepts, and activities of SI in agriculture are outlined in Table 7.1.

7.4 Difference Between Ecological and Sustainable Intensification

To explore the newer aspect of agricultural intensification for satisfying the food demands of the globe and therefore, sustaining human civilization and ecosystem services to promote growth and development. In fact, two of these have gained traction in the literature of research and technology, namely SI (Pretty et al. 2011a, b, c) and EI (Dore et al. 2011). Are there essential variations in both of

Table 7.1 Different aspects of sustainable intensification

Characteristics	Concepts	Activities
Reinforce the usage of economic, social, and human properties and using the necessary equipment and inputs accessible to reduce harm to the environment	Decrease land usage and expand the use of renewable resources to boost demand, like energy and information (Godfray et al. 2010; Firbank et al. 2013)	Preservation of plant genetic capital and better varieties, water control, fertilizer application, insufficient drainage, excessive irrigation (FAO 2004)
Boost productivity without environmental damage and without developing more land when preserving natural resources (Pretty and Bharucha 2014; Baulcombe et al. 2009)	Boost resource quality, maximize the use of external resources, raising the adverse environmental effect of food processing, and close the export difference (Matson et al. 1997)	Integrated pest management (Pretty 1997a, b)
Regulation of livestock processing processes inputs and outputs to maximize efficiency or development while preserving network and ecological credibility (Gibon et al.1999)	Reduce food waste and increase productivity (Garnett et al. 2013)	Tillage maintenance, seed substitution, and usage of biofouling and residual plastic packaging to protect the land (Wezel et al. 2015)

these terms? Is EI always contributing sustainability? It is a big question that whether intensification would be ecological compatible or not? Various leading organizations such as the World Economic Council, FAO, the Montpellier Panel (2013), Global science and decision bodies, has recommended for the promotion of "Feed the Future" program through proper legal enforcement.

The word is now commonly used by major international donor organizations, even in the agribusiness community. It was observed that new dimension of intensification related to agriculture was found to be unsafe from ethical and environmental perspective. Further, it was not being able to fulfill the demand of the earth and having its negative consequences over the global people and ecosystem (Tilman et al. 2002; Kleijn et al. 2009; Godfray et al. 2010; Geiger et al. 2010; Barnosky et al. 2011). These are incontestable signals from reality. Such are indisputable real-life signs. We speak to a growing need for new ways of intensifying agriculture. The definition of EI is much older, and practical relations with ecosystem resources have been seen. It means generating more while growing differently and creating new products (Dore et al. 2011), while SI means extracting more production within the same land considering the environmental aspects along with natural resource and services of the ecosystem. Grassroots networks and awareness generation from environmental perspective happens to be the work done in form of word SI.

Therefore, the disparity between the two intensification criteria is not simply textual, which is closer to the old dichotomy of input technology and method technology (Vanloqueren and Baret 2009).To reality, however, EI does not quite present a centralized range of management strategies but instead alternate frameworks that adopt diverse forms across the world and combine culture and environment to varying degrees.

7.5 Sustainable Intensification: Then and Now

SI means growing yield rates from the same land area, thus raising the adverse impacts of farm development on the ecosystem and growing the delivery of environmental services. Although this concept sounds innocent enough, it has become a notorious word for prolonged intensification. It is an exciting new model for advocates to lead agriculture in an age of increasing food demand and lack of capital. It is an oxymoron among critics—a pretext to reinforce the new capitalist model of industrial farming with a green sugar wrapping (Cook et al. 2015).

SI has been granted growing prominence to deal with the issue of food security and crisis for the ever growing global population. SI originates 20 years back, but in the last 5 years it has come into fashion these days via the publishing of a number of high-level studies by prominent organizations. Today the definition is disputed, with the controversy arising primarily from the usage of the word by various parties, their respective fundamental objectives and principles, as well as conclusions regarding the nature of the problems it may tackle inside the broader food supply chain (Cook et al. 2015; Meena et al. 2020b, c).

None will question the requirement of improvement in the production sector through sustainable approach to maximize the productivity of resource usage. Further, the problem lies with the interpretation of the word in various ways. Risk is associated with intensification of surpassing the objectives of SD; secondly, there is a risk that sustainability would be so loosely defined without considering about the capital and incorporation of its socio-economic dimension. Hitherto, debates have concentrated mainly on technological approaches while ignoring social and political aspects that are important as application of the definition should be for the local stakeholders and not for the others. Thirdly, the scarcity of a viewpoint on the food system in most of the literature on SI has contributed to an exclusive emphasis on crop development, while food protection involves approaching the entire agricultural method, including animals, from a larger perspective on food systems (Banerjee et al. 2020; Raj et al. 2020).

7.6 Rationale for Sustainable Intensification

Rise in demand for produce of agricultural sector due to increasing population for diverse diets present an enormous threat to global farming. To resolve these problems, many writers asked for more intensification of the agricultural sector (Mueller et al. 2012; Lampkin et al. 2015). Whether to accomplish better grain, fiber, and fuel output in the immediate future is a matter of discussion, since "conventional intensification" requires concentrated usage of inputs to increase efficiency, while EI applies to alternate farming systems that value and preserve natural capital whereas producing appropriate farm profits (SCAR 2011; Malezieux 2012; Bommarco et al. 2013). Academics accept that intensification in cultivation has to be "sustainable," adding the phrase "SI" to the farming science and policy agenda. However, definitions of the idea of "SI" vary significantly, and the debate mostly focuses on growth, ignoring the endpoint of the supply chains consumption (Loos et al. 2014).

Global agricultural intensification took place mostly during Green Revolution, which multiplied the worldwide cereal production between 1950 and 2000, providing adequate amount of energy from food consumption across the globe of about 6 billion till twentieth century (Trewavas 2002). Nevertheless, Green Revolution developments are part of traditional intensification as they largely depends upon the agrochemical use as well as the resources (Nair 2014). Although extremely competitive in the short term, the traditional intensification model's large-scale monocultures have poor genetic diversity that causes the crops vulnerable to collapse caused by natural disasters and epidemics. Many scholars have well reported the detrimental impacts of modern agriculture on human safety and the ecosystem (Chappell and Lavalle 2011; Bommarco et al. 2013). So it can be argued that these activities are not consistent with sustainable agricultural intensification, especially in sight of the continuing climate change and dwindling natural resources.

Stoate et al. (2001) mentioned about the degenerative nature of monoculture practice after green revolution in terms of loss of biodiversity. This created the need of conservation of biodiversity, nature's intangible worth, and ecosystem functionality. First narrative advocates argue that the functionality of the ecosystem alone do not reasonably rationalize conserving biodiversity (Kleijn et al. 2015). In addition, it maintains the ecosystem services that are not accountable (De Groot et al. 2012; Meena et al. 2020a). Worldview explains the functional attributes of ecosystem from farmer perspective. On the other hand it perceived it as a contribution towards production by species (Balvanera et al. 2001; Isbell et al. 2011; Cardinale et al. 2012). The primary goal of this portion of biodiversity is EI (Bommarco et al. 2013). Monitarizing nature's inherent worth is difficult as it serves a collective utility (Losey and Vaughan 2006); calculating the expense of preserving biologically valuable organisms is comparatively simpler, but significant work holes, such as defining efficiencies and exchange-offs between various ecological services, still exist. Such knowledge is critical in the harmonization of environmental resources management with agriculture, for instance, by ecological reimbursement strategies (Lal et al. 2015).

7.7 Sustainability and Resilience

It is much more challenging to describe "sustainability" in reference to agriculture than "intensification" or "productivity" as it is related to the principles of the individual utilizing the word. Because SI relies on agricultural development, contributions to sustainability in the publications appear to relate predominantly to environmental and, in some instances, economic aspects. Nevertheless, sustainability is wider than that, as described in the Bruntland Report through various dimensions of sustainable development (United Nations World Commission on Environment and Development 1987).

The idea of sustainable agriculture and food processing encompasses four main concepts (The Royal Society 2009):

1. Persistence: The capability to keep creating desirable performance over lengthy stretches of time thus accorded consistency.
2. Resilience: The existence of the system with proper adjustment of the changes that has taken place in the surroundings.
3. Autarchy: Ability for producing desirable outcomes from inputs and services (production factors) obtained within main process limits.
4. Benevolence: Capacity to generate desirable resources while preserving the functionality of ecological assets and not inducing the loss of natural capital.

Pretty (2008) has provided a valuable foundation for the recognition of values with the capacity to improve productivity and resilience, which include:

1. Various processes in terms of biogeochemical cycling, species interactions, soil quality restoration needs to be integrated into the production system of the food.
2. Lower use of non-renewable resources which have negative impact in atmosphere and human health.
3. Recognizing the traditional knowledge of the farming community in terms of cultivation practices to develop self-resiliency of the agroecosystem as an alternative for human build agricultural system with allochthonous inputs.
4. Allow effective use of the mutual ability of citizens to involve for identifying possible problems such as pests, watersheds, drainage, forests, and loans.

7.8 Integrating Sustainable Intensification in the Agriculture and Food System

Efficient SI plays vital and tested contribution towards more efficient, competitive, and robust agriculture. However, it is vital to guarantee that its usage is maintained to strict standards and that it is not permitted to be a tool for supporting relatively high agricultural inputs or strategies that have negative impacts on local livelihoods. Although SI will render agriculture more competitive and efficient, it is only feasible to tackle the several other variables leading to food security by putting SI inside the broader context of sustainability.

It is necessary to note that several powerful organizations with control over the global food environment have supported SI. Agricultural production will be under progressively unfavorable situations in the coming decades due to land degradation, pollution, water shortage, weather patterns, and unstable commodities prices. Since agriculture is the biggest single reason for the decline of biodiversity and one of the key contributors (direct and indirect) to GHG emissions, the task of generating more food is probably to be a sustainability problem, instead of just intensiveness. The United Nations Conference on trade and development (UNCTAD), United Nations Environment Program (UNEP), and IFAD recently highlighted the need to concentrate more on biodiversity, demonstrating how existing food systems are at threat from the progressive loss of their own environmental quality and services (UNCTAD 2013; UNEP 2012).

An increasing body of research is exploring what in reality feels like continuous intensification. Their results suggest that efficiency can be improved using biologically, socially, and economically secure relatively low-input strategies. For example, one analysis analyzed 286 projects in 57 nations where farmers improved agricultural productivity by an average of about 80% while at the same time increasing the output of water usage, enhancing carbon sequestration, and reducing pesticide requirements (Pretty et al. 2006a, b). The forethought Project investigated 40 initiatives in 20 African nations where activities that could be called SI were implemented and identified significant economic gains for 10.39 million farmers and their communities, and also environmental changes with some 12.75 million ha of land (Pretty et al. 2011a, b, c). The Montpellier Panel report address a range of agroecological and genetic solutions to SI in Africa from the perspective of production of water resource at smaller level, various indigenous practices of agriculture, pests control practices, climate resilience agriculture practices, etc. which been shown to increase yields and farm income and at the same time offering numerous environmental benefits. This also addresses certain facets of the rural economy critical for improving living standards like diet, economies, and social capital (The Montpellier Panel 2013; Juma et al. 2013).

7.8.1 The System of Rice Intensification: A Significant Breakthrough for Sustainable Farming

The system of rice intensification (SRI) was initially created by a priest in Madagascar obtained from field activities aimed at addressing biophysical restrictions on Madagascar Plateau along with the perspective of the farming communities in rice cultivation from social and economic point of view. In this region farmers tend to suffer from seed scarcity as well as issues related to irrigation. They also suffer from monetary support due to their poor economic conditions (Zotoglo 2011; Kepha et al. 2014). The implementation of SRI has resulted in increased rice yields and other positive benefits. Nevertheless, SRI demands more and stronger skilled labor, attempting to make it less suitable for situations in which rice competes for labor with many other activities on or off the farm (Chapagain and Yamaji 2010; Thakur et al. 2013, 2014a, b). In order to allow single plants to grow a comprehensive root network, SRI employs very low intensity planting, allowing adequate nutrients for the crop plants to use. In most cases SRI was considered as technique but it can also be viewed as accumulation of methods and its sensible application depending upon the existing agroecological and other features of the concerned area (Stoop 2011).

Fundamentally, SRI activities establish more desirable soil-water-plant-atmosphere connections than those obtained with continuous supply of water along with low oxygen environment in the soil which is prevalent in the existing rice cultivation practices in wetland area. These activities would have positive impacts on advantageous soil biota and on the roots and canopies of rice plants (Thakur et al. 2014a, b). Grain yield increase through SRI may be due to improvement in the crop morpho-physiology both above and below the soil (Thakur et al. 2014a, b).

On over 4 million ha, nearly 10 million farmers of 50 nations implemented the SRI concept in their cultivation practice (Uphoff et al. 2015). SRI has stretched mostly from five Asian countries—China, India, Vietnam, Indonesia, and Cambodia. India has begun working on SRIs later than China and Indonesia, however, SRI took numerous pathways for information creation and distribution. An environmental problem, drought in 2002–03, has triggered SRI study in South India, with state agricultural institutions taking the lead in Tamil Nadu and Andhra Pradesh (Prasad 2006; Moore and Westley 2011). Recent data indicated that the SRI provides an alluring possibility to expand agricultural production per unit of water and increase productivity in Tamil Nadu (Nayar et al. 2020). The appraisal of the effect of SRI on rice output in Tamil Nadu indicates that SRI strategies can be used to yield considerably higher amounts of paddy. Within SRI, paddy production is greater due to the synergistic impact of youth seedlings, novel approaches of transplantation, robotic weeding and sporadic irrigation that collectively save significant water and electrical power on the farm scale. Such technique lowers the use of seeds and chemicals and the allocation of labor, thus reducing the cost of output on certain transactions and growing the incomes of farmers. The usage of SRI methods often limits the water use (Nayar et al. 2020).

7.8.2 Participatory Plant Breeding Builds Local Resilience and Knowledge

Participatory Plant Breeding (PPB), presently being used in many countries across the world, is a procedure of participatory crop improvement research. PPB enables producers and breeders to participate equally in making decisions at all stages, from identifying favorable traits and parent lines to assessing the consequent varieties. PPB allows the production of new varieties that are both higher yielding and more robust by utilizing native varieties that are also more tolerant and conventional high yielding varieties. It customizes crop breeding to different local conditions, significantly increases the level of acceptance of technology and, by its application, creates opportunities for conserving agro-biodiversity. A maize PPB program launched in 2000 has increased yields by 15–30% in the province of Guangxi in China. This, together with the availability of organically produced foods in the commercial centers of a town area have 30% rise of economic gain in comparison to non-PPB villages increasing hybrid maize and generated opportunities for the introduction of agroecological farming methods in PPB villages (such as the use of ducks to combat pests, intercropping, and use of bio-fertilizers) (Swiderska et al. 2011). In vulnerable condition crop diversification could be a suitable alternative for getting sustainable yield in comparison to Monocropping system. The former one has less chance of the failure of crop due to use of robust local variety. As a consequence they are well adapted with the local conditions and therefore more successful in raising and sustaining yields than standard improved varieties that are less sensitive to local conditions.

7.8.3 Localizing Knowledge, Training, and Incentives

Enhancing the efficiency and sustainability of a farming system at the same time requires both exposure to technology and resources and knowledge producing new inputs and techniques or new forms of utilizing them (The Montpellier Panel 2013). These can be given by scientists working in modern research programs and businessmen from the private sector, and can even be extracted from local and traditional expertise or from a mixture of the two by contact between farmers and researchers. Many agroecological methods are directly extracted from conventional knowledge (Silici 2014).

Nevertheless, in terms of the position of information and technology, supporters of economic intensification have different viewpoints and goals. For example, the "Reaping the Benefits" study by the Royal Society reflects on traditional research, and in specific biological science; it notes that no innovation can be omitted (The Royal Society 2009). In the other side, the FAO study Save and Expand focuses in small-scale farmers and illustrates knowledge-intensive production, local traditions, and organizations (FAO 2011).

The International Assessment of Agricultural Information, Science and Production Technology (IAASTD) is exceptional in the context of agricultural science evaluations, evaluating both structured science and technology and local and conventional information, covering not just production and efficiency, but also the versatility of farming. The World Bank and FAO launched this enormous regional consultative mechanism in 2002, and its several publications rely on the research of hundreds of specialists from all parts of the world. It sees the key task of agricultural expertise, science and technology as sustainably increasing the efficiency of agriculture in order to cater the requirements of small-scale farms in different ecosystems. It calls for expanded public and private involvement in agricultural education, research, and technology; establishment of policies and structures to promote them; revaluation of conventional and local information; and a multidisciplinary, comprehensive, and system-based approach to knowledge creation and sharing (Heinemann et al. 2009).

7.8.4 Enhancing Food Security Through Greater Equity, Access, and Control

Nevertheless, although with greater focus on biodiversity, environmental intensification cannot solve a number of other problems that threaten the food system, such as extremely unequal access and right to food, unregulated land ownership, enhanced demand for resource-intensive meat and dairy goods, food shortages and pollution, discriminatory trading laws and supply chains, and concentrations of control in the possession of a small number of multinational agribusiness companies. Therefore, while prolonged intensification has much significant to bring to existing discussions and strategies, it is not, nor is it supposed to be, an overall mechanism for solving food sustainability or the challenges of our modern food environment (Kumar et al. 2020; Jhariya et al. 2019a, b). For this rationale, the framework must be placed within a broader perspective of the food systems, as illustrated in Fig. 7.1. Four critical aspects of food protection are identified by the FAO (2006): availability, access, utilization, and stability.

7.8.5 Re-energizing Farming

More broadly, energy appears to be ignored in debates on environmental intensification and food security. In reality, because it is focused on large inputs of fossil fuel oil, it is unlikely how often farther the highly energy-intensive farming model predominant in developing countries will be sustained. Significant volumes of inorganic fertilizers extracted from fossil fuels are added to soils worldwide and are critical for sustaining productivity in large external input agricultural systems (Meena et al. 2020a). Pesticides are often generated from fossil fuels and their processing and application is extremely energy-intensive. Fossil fuels, though, are the main culprits of climate change, and are therefore susceptible to market fluctuations and disruptions of production. For all those purposes they ought to be buried in the field. The energy needs of smallholder farmers and corporations manufacturing, storing, and selling their goods are an overlooked region, but one that is essential to tackle if farmers are to increase their productivity. Many smallholder farmers across the globe are experiencing a massive gap in obtaining energy resources. This problem has to be tackled across the supply chain—from field to market—to improve food security (Best 2014).

7.9 Governance and Regulatory Frameworks

Although there is an immediate need to tackle the work deficiencies outlined above, policy change is much more important to turning food systems and value chains into higher sustainability. Advancement towards more sustainability parameters and measures requires joint research and policy objectives, some of which have been mentioned below.

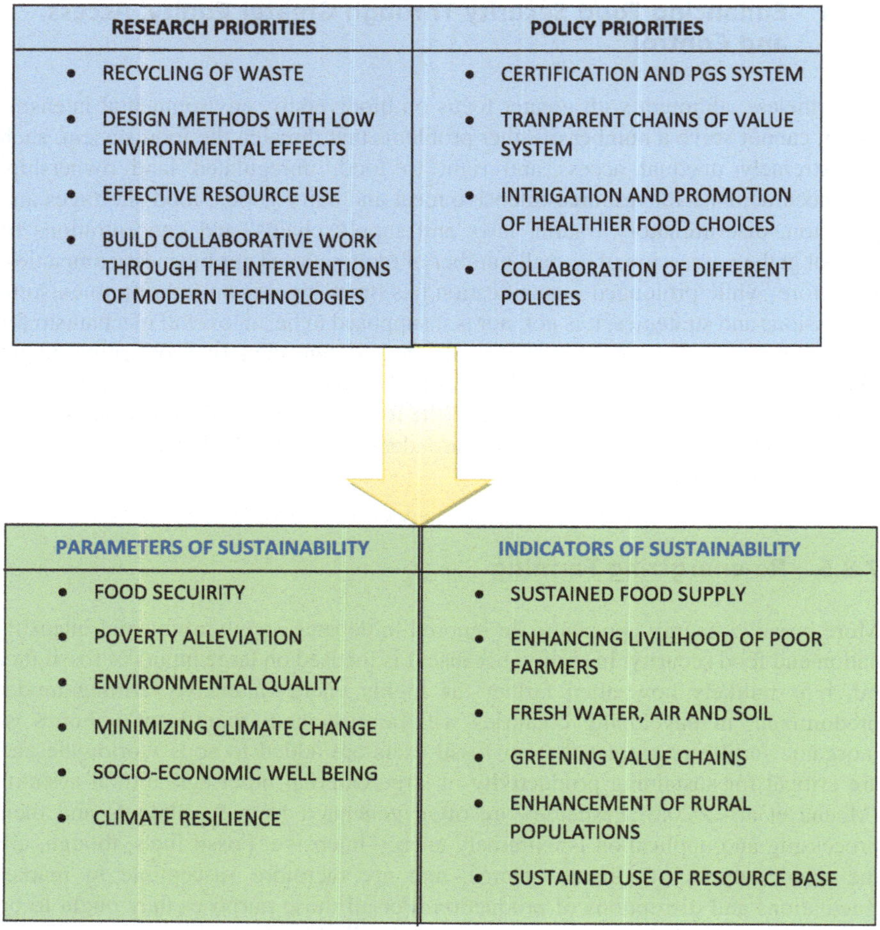

Fig. 7.1 Complex framework of sustainable intensification

7.9.1 Consumer Preferences and Cost Internalization

In the past half century urban diets have drastically changed, mainly in developing nations. The concept of "sufficiency" presented in the third foresight study of the European Commission's Standing Committee on Agricultural Science (SCAR) stresses that, in order to remain within the capability of the "World" framework, increased the stress of changing perception pattern and behavior to fulfill the demand for food as well as make structural modification in the food system. We must inform and empower customers to select safer and more nutritious products, enforce policy mechanisms that increase access to and availability of healthy products, and tackle the destructive impact of uncontrolled exchange that could promote behavioral change on the customer side (SCAR 2011).

Present policies and market conditions promote unhealthy agronomic activities by promoting large-scale development of single goods, which are priced at distortedly low rates at the detriment of the ecosystem and eventually humanity. In 1996, for example, Pretty et al. (2000) recorded overall external costs of UK farming to be £208 per ha of arable and/or permanent pasture. Another research showed that Switzerland's non-monetarized pesticide usage expenses contribute to at least 50–100 million Swiss francs a year (Zandonella et al. 2014). When these factors were internalized, traditional produce would become more profitable and more efficient and renewable. This will contribute to a move towards greater sustainable agricultural output (Fry and Finley 2005; Reisch and Gwozdz 2010).

7.9.2 Sustainability Standards and Participatory Guarantee Systems

Certification criteria (e.g. Rainforest Alliance, UTZ Accredited, Fair trade, etc.) combine sustainable development activities with restoration of habitats, and civil rights security. Compared with single-aspect requirements, multidimensional requirements (e.g. UTZ) achieve agroecosystem comprehensive sustainability by integrating cultural, social, and environmental parameters. Generally speaking, certification tends to operate well where supply/demand levels are small in product markets and quality premiums are substantial. Nevertheless, when supply/demand ratios grow, sustainability requirements are losing their appeal as approved commodity costs are still dropping. This places a question mark around the viability of quality requirements in the lengthy period. In some situations, it may be appropriate to nationalize the sustainability standards (Indonesia's oil palm sector).

Farmers typically gained from the principles of sustainability by having exposure to foreign markets and undergoing preparation, which increased food consistency and helped protect natural resources. Furthermore, the expected impact on rural livelihoods has been minimal, particularly for smallholder farmers (COSA 2013; Potts et al. 2014). Processors, distributors, dealers, and other participants in the supply chain enjoy comparatively greater advantages opposed to smallholders due to their better negotiating force (Bjorndal et al. 2014; Meybeck and Redfern 2014). In addition, the required conversion periods could deter traditional farmers from entering certification schemes. Thus, subsidies or incentive schemes for farmers in the "in-conversion phase" could encourage greater numbers of smallholders to adopt sustainable practices. In fact, the extra expenses of testing and registration tend to be a significant obstacle. Smallholders also will only profit by collective approval, through which cooperatives need to be established. Under this context the "Participatory Assurance Schemes (PGS)" alternate credential scheme was investigated. By focusing on smallholders and local customers, PGS has been popular in different areas of the tropics (Zanasi et al. 2009; Nelson et al. 2010). However, as well as viable policy and regulatory mechanisms are not yet in place, assessments on effective certification systems are required on a case-by-case basis, particularly since their performance depends on the background. Nations who were at the frontline of the implementation of environmental principles (e.g. Austria and Switzerland, where 19.5 and 12.2% of agricultural land were accredited organic farming, respectively) (Willer and Lernoud 2015) could provide positive examples and facilitate in the establishment of appropriate policy and regulatory systems in developing nations.

7.10 Different Technologies Contributing Sustainable Intensification

7.10.1 Conservation Agriculture

It was observed that there is significant contribution of organic farming towards fulfilling the SDGs. As per the records >70 million ha land area of 2.8 million producers is under organic farming (IFOAM 2020). Conservation Agriculture (CA), coupled with external inputs through integrated system of various biotic and abiotic assets acts effectively as a tool for management. Its vision is to enhance agricultural production by implementing economically, environmentally, and socially sustainable ways of agriculture (FAO 2015). It is popular practice for farmers, especially in temperate climates, to perform the tillage activity before sowing and then after sowing proper aeration in soil helps in weed eradication. Tillage activity has been reported to disintegrate the clay lump in the soil. This is, however, problematic in soils under the threat of drought and erosion as observed in sub-Saharan Africa may cause total destruction of the structure of soil leading to more evaporative water loss (Montpellier Panel 2013).

CA is a synthesis of three linked concepts represented by techniques: zero / minimal till, mulch cover, and mixed cropping. At the same time, CA promotes environmental conservation and boosts agricultural output by enhancing conditions for growth (FAO 2007). CA will boost efficiency of input usage, improve farm income, and maintain the natural resource base (FAO 2007), particularly when accompanied by the use of higher or enhanced seeds, water management, fertilizer micro-dosing, and integrated pest management (FAO 2015). As per the FAO (2018), CA is an agricultural method that encourages constant no or minimal soil disruption (i.e. no tillage for seeding and weeding), conservation of soil structure, and plant species diversification. It increases biodiversity and natural ecological processes above and below the land, thereby leading to enhanced quality and sustainability through the use of water and nutrients, to more efficient crop systems and to better and sustainable crop growth. CA is performed on about 125 M ha worldwide. The major countries that practice CA are the USA (26.5 M ha), Brazil (25.5 M ha), Argentina (25.5 M ha), Canada (13.5 M ha), and Australia (17.0 M ha). CA adoption in India is also in its early stages. The implementation of zero tillage and CA has grown in the last few years to include about 1.5 million ha (Jat et al. 2012). Different case studies related to the potentiality of CA is described in Table 7.2.

Table 7.2 Data showing benefits of conservation agriculture and its contribution towards sustainable intensification

Case report	References
In Zimbabwe, 8000 ranchers have adopted CA methodologies, going to result in a 67% increase in maize yields	Silici et al. (2011)
Concern Nationwide observed that 133 farmers performing CA produced an average maize yield of 2.8 tons per ha from 2005–2006	Harty et al. (2010)
Reduced labor time and less farm power requirements which decreased input costs, resulting in higher profits in Tanzania, Zambia, and Malawi	
In Zimbabwe's Matopo area, Christian Aid has finds that conservation farming (CA) strategies help farmers improve yields and preserve natural resources	Hobbs and Powell (2011)
In Zimbabwe, farmers have achieved yields between 2 and 6 times that of conventional farming practices, while at the same time benefiting from reduced labor and costs due to lower input levels	
Use CA crop rotation is a well-recognized solution to minimizing the production of pests and diseases that could propagate in the existence of crop residues	Nunez (2010)
Farmers in approximately 20 African countries encourage and endorse CA, including Algeria, Ghana, Kenya, Lesotho, Madagascar, Malawi, Morocco, Mozambique, Namibia, South Africa, Sudan, Swaziland, Tanzania, Tunisia, Uganda, Zambia, and Zimbabwe, with the new CA adoption estimates for annual crops in Africa (season 2015/16) totaling 1.5 M ha	Kassam et al. (2018)
CA has also been used in national agricultural policies and has gradually been accepted as a central element of climate-smart farming	FAO (2016a, b), Kassam et al. (2018)
Actually the total amount of African carbon sequestration due to 1.5 Mha CA implementation is much more than 5.6 metric ton (Mt) CO_2 per year. The future effect of CA's carbon sequestration activity is to improve this to 533 Mt. of CO_2each year, just over a hundredfold higher	FAO (2016a, b), Kassam et al. (2018)
CA strengthened with a compact cover crop and its corresponding thick residue cover can minimize losses from evaporation and capture nutrients, thus fostering productivity gains. CA may help to alleviate climate impacts and respond to impact of climate change like decreased rainfall in tropical and/or arid areas like South Africa	Deb O'Dell et al. (2020)
A systematic review leveraging 9686 combined site-year analyses reflecting various crop-system success metrics indicates substantial ($P < 0.05$) advantages while conservation agriculture aspect activities are either applied independently or in tandems. Zero tillage with residue preservation, for instance, had an average output benefit of about 6%, an improvement in water usage capacity of about 13%, an improvement in net economic gain of about 26% and a decrease of about 12–33% in global climate change potential, with much more beneficial response on loamy soils and maize-wheat regimes in South Asia	Jat et al. (2020)

7.10.2 Organic Agriculture

Organic agriculture (OA) was found to be a sustainable approach in terms of crop and livestock producing service (Montpellier Group 2013). OA usually has stringent restrictions on some aspects related to amount and nature of manure application with zero application of chemical fertilizer and pesticides. However, it emphasizes the use of naturally produce substances. The usage of genetically modified organisms (GMOs) is banned, so there is a prohibition on the daily use of medications, pesticides, so wormers in livestock production, so animal feed must be sustainable. The OA works by mimicking the natural processes of nutrient exchange between crop and soil ecosystem as well as make the system resilient and resistance for pests infestation and disease outbreaks. Through building natural resources in this manner, farms will be more durable and more competitive against shocks and stresses. Organic farming has been suggested as a climate change mitigation technique, owing to improved soil ability to offset nitrous oxide (N_2O) and carbon dioxide (CO_2) pollution by decreased soil degradation and therefore a higher ability to contain GHG emissions (Niggli et al. 2001). CO_2 levels in OA applications can even be smaller as pesticides and fertilizers produced from fossil fuels are not used (FAO 2015). Different case studies related to the potentiality of OA is described in Table 7.3.

7.10.3 Water Conservation and Harvesting

The storage and recycling of water lead to a balanced intensification by enabling productive usage of resources. This results in increased agricultural output across the year and increased resistance to drought, therefore improving livelihoods and food protection for peasants (Taddele Dile et al. 2013; Rockstrom et al. 2002). Taddele Dile et al. (2013) found that water harvesting fulfills the SI standards by (1) enhancing availability of water throughout dry spells; (2) enhancing agricultural yields for

Table 7.3 Data showing benefits of organic agriculture and its contribution towards sustainable intensification

Case report	References
The UNEP and the UNCTAD observed that crop yields and livestock productivity or overall food output improved in organic farms with an emphasis on food processing and that agricultural yields in organic systems appeared to be more robust when transferred from other low-input models	UNEP-UNTCAD Task Force (2008)
In tropical Africa organic conversion is consistent with yield rises rather than decreases	Gibbon and Bolwig (2007)
In Uganda, organic cotton production yields 1000–1250 kg per ha of cotton seed, providing 300–320 kg per ha of lint	Organic Research Centre (2015)
Organic cotton production grew mostly in India, China, and Kyrgyzstan, as well as in Turkey, Tanzania, and Tajikistan	FiBL and IFOAM (2020)

food safety; (3) trying to rehabilitate degraded land to reestablish biodiversity; (4) lessening the use of external inputs which have adverse environmental impacts; 5) enabling enhanced carbon sequestration in soils to deal with climate change; and (6) lowering downstream river pollution from upstream farmland nutrient releases. Various tactics for water management are performed throughout the planet, some of which are: contour farming, soil basins, planting holes, drip irrigation, and ridge binding. Case study related to the potentiality of water conservation and harvesting in tomato yields in South Africa reflected significant increase in production and yield through appropriate innovative approaches (Karlberg et al. 2007).

7.10.4 Precision Farming

Precision farming is an approach that utilizes inputs in accurate amounts to increase the average yields compared to the orthodox agricultural techniques. It is a holistic framework intended to increase production through the use of key knowledge, technology, and administration components in order to generate maximum efficiency, improve product quality, conserve energy and environment protection (Shibusawa 2002). Precision farming seeks to insure that inputs—including resources, chemicals, plants, or water—are utilized in a specific, preserving, efficient, and strategic way to maintain minimum carbon footprint (McBratney et al. 2005). Precision agriculture, which consists of implementing inputs (what is required) where and when appropriate, has been the third phase of the new agricultural transformation (the first one was mechanization and the second was indeed the green revolution with its genetic manipulation (Zhang 2019), and is now being strengthened with an improvement in farm information systems owing to the accessibility of greater data volumes. As early as October 2016, the United States Department of Agriculture (USDA) announced that precision agriculture innovations are growing net gains and gross income (Schimmelpfennig 2016). It recognizes the temporal and spatial variations of field-scale crop production (Wells and Dollarhide 1998), and accounts for these disparities by approaching input implementation to optimize yields (Adamchuk et al. 2004). To this end, it is necessary to minimize the amount of inputs required to achieve the specified output rates. Such approaches have been proposed in reaction to growing depletion of the ecosystem and increasing input costs that endanger food production across the world. Farms are usually larger in developed countries than in developing countries (10–1000 ha or more) and are better resourced to enable mechanized agricultural production networks. Enable to equipment and specialized technology ensures that agricultural precision rapidly requires emerging developments such as satellite imaging, radar, and geospatial methods (Tran and Nguyen 2006). Such technologies can be used by farmers to gather, evaluate, and plot data on production, environmental, and soil quality parameters in various sections of their fields and then set different fertilizer mixes in conjunction with soil requirements at particular places (Sonka et al. 1997).

Table 7.4 Data showing the benefits of precision farming

Case report	References
Yield Data from Southern Ontario related to the precision farming can be used to classify areas on the farm that are unprofitable over period and could therefore be freed up for conservation without any economic effect on the farmer	Capmourteres et al. (2018)
Under the leadership of the Agricultural Transformation Agency (ATA) in Ethiopia, farmers increasing hybrid maize were capable of achieving 6–8 tons/ha—exceeding the European average—when adding an appropriate NPK balance (Nitrogen, Phosphorus and Potassium). This was combined with Boron, which had been found to be defective in the area after soil tests	Montpellier Panel (2014)
In North Eastern parts of Karnataka state, India, the precision farming technology was implemented in farm and get higher net income	Shruthi et al. (2018)
In Tamil Nadu, India, the Tamil Nadu Agricultural University has initiated precision farming project in high-value crops such as tomatoes, brinjal, sugarcane, gherkins, etc. The project helps farmers to improve productivity and increase profitability as well as empower them socially, economically, and technically	Sangeetha et al. (2013), Ravikumar (2016)

In several developing nations, Western precision farming technology is used little to no. This is due to smaller field sizes, inadequate technological availability, finance capital and mentoring. Nonetheless, farmers are exploring the means and tools at their disposal to improve agricultural productivity and output, allow effective use of scarce capital and yield more (Tran and Nguyen 2006). Perhaps precision farming is even more important in which resource efficiency is propelled by a scarcity of resources without availability to alternatives. In the precision farming family there are many methodologies, from controlling and evaluating farm situations, like soil analysis, locating, and implementing a wide range of organic or inorganic inputs, for instance, by micro-dosing. Such measures can be more or less technically advanced, rendering precision agriculture widely available in numerous ways. Different case studies related to the potentiality of precision farming are described in Table 7.4.

7.10.5 Diversification

Diversity—calculated as species richness in its simplest type—is commonly regarded as a critical factor in preserving fairly healthy and robust environments (Mori et al. 2013). Agricultural processes are usually optimized by natural environments to increase crop or livestock growth, however, these generalization may render the system more susceptible to external stress and adversity. It may take several forms to diversify an agricultural sector, like intercropping, agroforestry, and advanced pest control (Jhariya et al. 2018a, b). Intercropping is the process of growing on a given piece of land two or more crops simultaneously (Montpellier Group 2013).

Diversification of farm output by intercropping will, if carefully managed, improve productivity, minimize occurrences of pests and diseases, increase the overall resilience of the environment, and promote the availability of environmental resources such as nutrient and soil water retention. Intercropping can deliver sustainable uses of assets.

Nevertheless, intercropping will dramatically increase yields in developed nations and for smallholder farmers without the economic and environmental risks related to traditional monocultures (Altieri 2002; Gliessman 2007a, b; Chappell and LaValle 2011; Noltze et al. 2013). In an appraisal of 286 initiatives applying development steps, especially diversification, to small-scale farms in developed nations, outputs for a range of agricultural system, and crop varieties was improved by an average of 79% (Pretty et al. 2006a, b). After all, where peasants cannot buy inputs like nitrogen fertilizer, incorporating plants that make nitrogen accessible to certain crops may be an efficient way for peasants to raise productivity. The *Tephrosia vogelii* shrub can grow really rapidly, grow up to 4 m long, fix nitrogen and will be used as green manure (World Agroforestry Centre 2009).

The *Tephrosia*-produced mulch has also been used on the coffee plots. *Tephrosia* intercropped with coffee generated 1.4–1.9 tons of biomass per ha during the first year, and added 42–57 kg of nitrogen per ha. Compared with conventional management approaches, this procedure raised coffee yields by 400–500 kg per ha. *Tephrosia* mulch was 87% as effective as inorganic fertilizer used under comparable conditions and provided a 30-day saving of working hours per ha relative to current farmers' management by minimizing weeding labor.

7.11 Why Sustainable Intensification Should Go Green?

Tittonell and Giller (2013) stressed that intensification of African smallholder agriculture is merely a requirement, whether traditional, environmentally sound or ecological. This is not really an easy statement: input prices are sometimes improperly functional and agricultural goods markets are not always working well; markets that crash when smallholders achieve yield increases; furthermore, smallholders also are devoid of costly investment in terms of seeds, required agrochemicals along with technological and scientific inputs. Further, improper availability of the labors may also tend to reduce the higher output. Knowledge and comprehension typically require knowledge unique to the web. For decades, farmers have established site-specific awareness and were able to operate with that awareness based on skill, practice and experiential learning. Intensification of the environment needs more qualitative information than traditional intensification.

EI may benefit from an ecological approach focused on characteristics, i.e. an approach centered on plant functional features in agroecology (Garnier and Navas 2012; Martin and Isaac 2015). Trait-based ecology tends to put capabilities at the forefront and generates a sustainable agroecosystem by integrating the characteristics necessary to achieve that stability in the easiest way possible. Ecological knowledge based upon the crop attributes can be applied in the production system at global level in various dimensions starting from field to landscape level to achieve sustainable intensification in a greener way. Additional choices may involve breeding systems (Lammerts van Bueren 2016) and crop architecture. Several of the EI concepts and hypotheses also lack facts. For instance, it is still hard to relate biodiversity directly to resource quality, yield increase or resilience. However, a recent study proposed that intercropping may play a prominent part in sustained intensification (Brooker et al. 2015). Another new studies through different nations found that crop diversification is an effective method for intensifying the climate (Gurr et al. 2016). A global committee of specialists agreed that in view of production, "diversified agroecological systems" would do as effectively as "modern food systems" (IPES-Food 2016). Sadras and Denison (2016) reported that there is a double obstacle to maximizing crops and crop systems (both natural and agronomic). It is difficult to refine specific crop traits or specific agronomic activities due to trade-offs and variability in the setting. They inferred that neither crop genetics nor crop management could be enhanced; thus it is challenging to eliminate agronomic obstacles for sustainable intensification. Additionally, several simply assume that a core factor in SI is sustainability by diversification (Garnett et al. 2013). There is definitely a need to give greater attention to the durability of agricultural production processes, like continuity, ability to adapt and versatility, by improving durability analysis to the instability of such processes (Ge et al. 2016; Urruty et al. 2016). Resilience thought, like ecological resilience and engineering resilience (Holling 1996), will involve the crop sector, the plant, the agroecosystem, and the food network, and requires to move beyond theoretical conceptualization into practical implementation (Walker and Cooper 2011; Baduhur et al. 2013; Anderson 2015). Disregarding trade-offs would undoubtedly render intensification less efficient, without take into consideration the incentives presented to render intensification more environmentally friendly.

7.12 The Social Dimensions of Sustainable Intensification

SI needs significant transitions in agricultural social and economic organization, based on fair allocation, individual empowerment, and equality. This would be difficult to make these improvements, since existing socio-technical structures are established in current organizations and retained preserved by powerful interest groups (Vanloqueren and Baret 2009; Struik et al. 2014; Fraser et al. 2016; IPES-Food 2016).

Science needs to pay attention far more than knowledge: values and beliefs are also at great risk, and they need to perform a significant role in the way forward decision-making. SI includes public discussions and open decision-making about what should and should be stepped up, provided that trade-offs are normal. Pushing for agricultural sustainability makes it compulsory that we first strengthen our trade-off research. Both now and in the ahead, the fundamental problem is: Will the right to food be achieved in a fair way? This right to food is directly related to the manner it is processed and also used: the privilege extends only because the food is sustainably grown and eaten. SI needs different sources of information and modern approaches in agronomy (Doré et al. 2011), but also education in social dimensions of sustainability, as SI is primarily about systemic collaboration, structural creativity, and proactive management, instead of becoming the sole domain of agronomic debate.

We would have to tell agricultural learners and the wider community that a much more standard (or value-laden) agronomy is imminent. We must rethink and reinterpret agronomy as a multidisciplinary subject, combining natural and social sciences, with modern curriculum creation that recognizes that farming puts tremendous strain on limited capital and that it requires remaining within planetary boundaries. Agronomy should provide alternatives and measure trade-offs such that society should make informed trade-off choices. Besides this difficulty of turning ourselves from the wrongdoers into professionals, we also have the issue that such a program is challenging to develop. Wals (2015) claimed that the uncertain complexity, difficulty, and perpetuation of sustainable development and education for sustainable growth may potentially become a justification for not participating in sustainability education. This tends to make sustainability teaching a dilemma but learning in sustainability is absolutely essential.

7.13 Policy Option

SI growth has started to occur through a large variety of agroecosystems level. Increasingly apparent are the advantages of both science and farmer research into technology and activities that integrate crops and animals with effective agroecological and agronomic leadership. The related construction of modern social networks helps in all knowledge transfer and trust builds between individuals and agencies. This will contribute to enhanced awareness and capability of farmers via the use of forums for collaboration with modern information technologies. Growing efficiency (Rockström et al. 2017), growing network complexity (Pywell et al. 2015), decreasing farmers' costs (Pretty et al. 2011a, b, c; Gurr et al. 2016), minimizing adverse externalities (Smith 2013; Gurr et al. 2016), and enhancing ecosystem resources (Sandhu et al. 2015; Gurr et al. 2016) is illustrated through the use of SI. There are also a number of possible incentives for peasants to follow SI strategies and for national government, private sector, and foreign organizations to provide policy support. SI needs expenditure to construct economic, social, and human resources, and is therefore not costless (FAO 2016a, b; Benton 2015).

However, state strategies for the SI remain badly defined or inefficient in certain circumstances. Farm subsidies in the EU have gradually moved to specific environmental results instead of export incentives, a trend that the UK Government aims to promote (Morris et al. 2017), but this still promises synergistic gains across entire ecosystems. Numerous nations have given specific public policy funding for the creation of social groups such as Landcare (Australia), Watershed Protection (India), Cooperative Forest Management (India, Nepal, Democratic Republic of Congo), Irrigation Consumer Groups (Mexico), and Farmers Field Schools (Indonesia,

Burkina Faso). In India's Andhra Pradesh state, the state government has made clear its commitment for zero-budget natural farming (ZBNF), targeting to touch 6 million peasants by 2027 (Kumar 2017); political promises have been established in Bhutan and the Indian states of Kerala and Sikkim to turn all property into organic farming; greening of the Sahel by means of agroforestry started when national tree ownership regulations were modified to favor local communities (Godfray et al. 2010). The 2016 No1 Core Paper in China stresses creativity, teamwork, greening, and exchanging as main components of a modern SI policy (Xinhua 2016). At the current period, customers are performing a growing position in direct interaction with peasants in wealthy countries, such as by community buying schemes, farmers' markets, and qualification schemes, which may in effect shift customer preference (Allen et al. 2017). With this increased recognition of the constructive roles that policymakers should serve in formulating opportunities and strategies, and promoting systems of agricultural information.

7.14 Research and Development

The factors influencing the implementation of SI can be defined as: socio-economic influences; own traits of the peasants and natural environmental conditions, as per established studies. The socio-economic factors influence opportunities from peasants and investors in many other financial entities to engage in sustainable macroscopic intensification Enhanced agricultural production levels hinder the implementation of SI by small-scale farmers (Schut et al. 2016). Woelcke's (2010) work in eastern Uganda, for example, found that encouraging small holders to shift towards sustainable crop development by addressing knowledge gaps, raising crop values, growing consumer access to information and that shipping costs was crucial. Work on the integrated crop-farming program in the Brazilian Amazon pastures by Cortner et al. (2019) found that consumer quality, food costs, pasture infrastructure, credit availability, and other conditions significantly hinder farmers' acceptance of the process. The Eastern African countries' agricultural policies, and Uganda's national development plan suggested that inadequate facilities like transport systems hindered the introduction of sustainable crop intensification by the agricultural attendees (Yami and Van 2017).

SI focuses on how peasants exist in, and rely on, natural, environmental, social, cultural, and political conditions to achieve sustainable lifestyle (Prager and Posthumus 2010). Currently several researchers are investigating the effect of various features of peasants on sustainable agricultural intensification from the peasants' viewpoint. Indicators representing peasants' features include mostly family size, age, class, standard of education, and extent of subsequent work, etc. (Ndiritu et al. 2014; Scherer et al. 2018).

Most research have also shown that climate change, height, slope, soil organic matter, precipitation, and soil quality often influence the implementation of sustainable agricultural intensification in terms of environmental factors (Scherer et al. 2015, 2016). Table 7.5 explains different influencing factors of agricultural SI in

Table 7.5 Influencing factors of agricultural sustainable intensification in different regions of the world

Regions	Influencing factors
Brazil	Market exposure, costs of goods, ranch equipment, access to finance, and a shortage of marketing choices (Cortner and coll 2019)
Malawi (Central and South)	Gender opinions, climate, population size, crop costs, size of the farmlands (Snapp et al. 2018; David et al. 2016).
Africa (East and West)	Market participation, agricultural extension programs, intelligence gaps, and the adoption of new technology (Kassie et al. 2015; Yami and Van 2017)
Uganda	Population strain, market knowledge, instability in markets, environmental degradation, poor yield, subsistence for farmers. (Rahn et al. 2018; Yami and Van 2017)
Eastern, Southern and Western Kenya	Population growth, quality of the soil, access to knowledge, behavior of the peasants, environment. Sexual identity, technological gaps, market access, quality of land, payment prospects, institutional support (Rolando et al. 2017; Ndiritu et al. 2014)
Germany	Population strain, appetite for food, soil organic matter, gradient, and depth of soil (Schiefer et al. 2015)
Sub-Saharan Africa	Soil productivity, demographic strain, output distance, approach to prosperity, consumer access (Vanlauwe et al. 2014)

various regions throughout the globe. The original meaning of sustainable agricultural intensification is complex, and creating an index framework is a widely utilized tool for calculating whether such a topic of research achieves sustainable agricultural intensification. Review of the assessment findings allows explaining the basic steps for achieving sustainable agricultural intensification and improves the percentage of sustainable reinforcement of research objects (Smith et al. 2017). Smith et al. (2017) subdivided SI appraisal indicators into five fields from the viewpoint of SI indicators: growth, economic sustainability, environmental sustainability, social sustainability, and individual well-being.

Stachetti and Roberto (2018) assessed the level of EI of coconut production in Brazilian farmlands from five factors: landscape ecology, excellence of the ecosystem, socio—cultural environment, economic values, and strategic planning. Mahon et al. (2018) established benchmarks for SI from a UK stakeholder viewpoint. They analyzed such subsystems from seven subsystems focused on widely agreed sustainability principles: resource structures, resource groups, policy, resource consumers, relationships, results, and the ecosystem. Liao and Brown (2018) explain the nexus between SI in agricultural sector, livelihood of farming community and maintaining ecosystem services. Slätmo et al. (2017) looked into the sustainability of farms. Snap et al. (2018) implemented the System for SI metrics to determine the importance of Maavian corn diversification and advanced strategies for soil fertility control in Africa. Ndiritu et al. (2014) used a multivariate likelihood (MVP) model to test if there are systemic gender disparities in Kenya's SI approaches.

David et al. (2016) used the studies of option (CE) method to determine the behavioral priorities of small holders by implementing adaptive intensification methods. Associated with different forms of field, Firbank et al. (2013) assessed the extent of SI via a comparison analysis of improvements in agricultural output and environmental parameters, i.e. increased food production without decreases in ecological services. Sánchez-Escobar et al. (2018) used expenditure return on education (EROI) and activity-based costing (ABC) approaches to calculate the extent of agricultural systems' SI at the farm level. Yami et al. (2017) used the Policy Agreement (PAA) methodology to determine the extent of government funding for sustainable crop intensification in eastern Africa.

The realistic ways of sustainable agricultural intensification in different regions and agricultural systems are heterogeneous because of the various environmental ecosystems and socio-economic factors in various regions (Weltin et al. 2018). For instance, in certain arid areas, instead of inputs like irrigation, fertilizer or better crops, inputs linked to animal welfare and social organizations may be the most suitable intensification type, creating an entirely different circumstance from other areas (Scherer et al. 2018). Even so, crop yields can be greatly improved in certain arid regions of North Africa and West Asia by the introduction of supplementary irrigation and good management techniques, thus leading to sustainable crop output intensification (FAO 2004). Since agroforestry complex management is commonly advocated as a model for achieving sustainable agricultural intensification in mountainous areas, it is seen as a remedy to the growing concerns of shortened harvest times and land loss (Nath et al. 2016; Raj et al. 2019a, b). Evaluating strategies to preserve and rebuild ecological resources as a core component of agroecosystems in the plateau areas is the path to achieving sustainable agricultural intensification (Rahn et al. 2018; Willy et al. 2019). Work on rice cultivation in the North China Plain of China indicates that precision nitrogen fertilizer management techniques raise rice yield by 2–3 times, while at the same time the negative effect of development on the climate, thereby offering a guideline for sustainable agricultural intensification (FAO 2004).

7.14.1 Recent Global Sustainable Intensification Research Scenario

While there is general consensus that satisfying food requirements without more destruction of natural habitats needs continuous intensification, there is very little debate about the work agenda required to accomplish this (Cassman and Grassini 2020). Assessing the capacity of current agricultural structures to be subjected to SI at regional and global scales offers insight into soil, water, and energy needs to ensure sufficient food supplies while resolving climate change and biodiversity issues (Cassman and Grassini 2020). A global outlook tends to highlight large patterns and factors of potential foreign food supplies and product markets, and in effect gives important insight to national research and development strategies since most nations still depend on imports to satisfy food requirements. Efficient national target setting recognizes the crops, crop systems, territories, and innovations most likely to progress SI provided landscape, land, and water management inheritances (Cassman and Grassini 2020). The most noticeable reported SI research is based on four key arguments:

- The first argues for an extended SI reach to involve activities throughout the food production system's entire supply chain to meet the nutritional and nutritious food needs through processes that enhance social-ecological resilience and improve natural resources inside the Earth system's safe operating room (Rockström et al. 2017).
- A second push aims to improve crops and cropping regimes in sub-Saharan Africa and Asia's low-income developing regions where existing yield deficits are important. And although SI is important for these areas, the technical obstacle is negated by the fact that present production mechanisms have very low yields, as they obtain little supplies of fertilizers, enhanced seed and pest control steps, and have therefore experienced so little intensification (Tilman et al. 2011).
- The third and increasing body of SI research represents a shift to increased crop diversity and crop frameworks as a part of enhancing human nutrition by producing more healthful crop species than principal cereals, and since increased crop diversity sometimes can lessen risk, improve soil health, and reduce fertilizer and pesticide requirements (Pretty et al. 2018; Mabhaudhi et al. 2019).
- A fourth focus involves research seeking to develop benchmarks for evaluating field or farm level development towards SI (Mabhaudhi et al. 2019; Gadanakis et al. 2015; Thomson et al. 2017). Most research of this sort relies on indicators correlated with fertilizer, oil, and water input-use efficiencies leading to a shortage of very well-defined environmental output thresholds. A latest analysis of maize projects in Malawi, in contrast to such production and productivity indicators, contained likelihood of crop loss, possibility of food sufficiency, and scores of women's crop and soil control options, each of these are essential concerns for smallholder farming in most sub-Saharan African nations (Snapp et al. 2018).

The most plausible outcome is that the global food environment will become increasingly globalized and exchange-dependent in 2050 owing to the exponential momentum of urban development, which is projected to grow from 55% of the world's population currently to approximately 70% by 2050 (UN Economics and Social Affairs Division 2018). Even though pursuing SI to guarantee world food protection by 2050 is a huge science obstacle, it is not really beyond grasp if there are excellently-prioritized national and international R&D strategies with a relentless emphasis on the dual goals of achieving major yield increases on established farmland coupled with substantial improvements in environmental efficiency that protects human and economic capital properly, which significantly reduces GHG pollution. A 50% rise in yields on current farmland and a 50% reduction in negative environmental external costs offer valuable initial goals for the development of national study portfolios for SI. The research needed will come across a broad variety of discipline like fundamental and analytical sciences that stretch well across conventional agricultural fields to include computer and analytical sciences, ecosystem ecology, and molecular biology, to mention a few. But securing the necessary degree of SI in national and international food production processes is only one element of the commitment to food security; it must be accompanied by social, political, and economic circumstances that improve the likelihood, sustainability, and sufficient food for everyone. While research and development on SI's biophysical properties is required, it is not enough as there is a vital need for effective policies, institutions, and trade agreements to guarantee that productive SI in terms of production and environmental objectives contributes to nature-saving land and an accessible and nutritious food supply to everyone (Byerlee et al. 2014; Phalan 2016).

Apart from scope and magnitude, one thing is for sure: without SI in the precise context of growing agricultural productivity on current farmland whereas dramatically simultaneously minimizing impacts on the ecosystem, it would be impossible to attain a food-safe planet without major biodiversity destruction and rapid environmental degradation (Khan et al. 2020a, b). Thus the value of sufficient expenditure and successful optimization of Research and development activities in order to achieve the necessary level of SI in food systems that contributes very much to the supply of human nutrition.

7.15 Critiques of Sustainable Intensification

Given the diverse meanings and uses of the word by various parties, it is not shocking that sustainability has been heavily criticized, especially its usage in recent years by foreign agencies, Western governments, academic institutions, and agribusiness firms, instead of the manner it was created initially. Nongovernmental groups focused on agriculture and food security are among its detractors. The core critiques are the latest SI definitions of:

1. Represent a productivist ideology that promotes development over certain food protection factors.
2. In fact, they are reinforcement of existing, strong external input solutions but with a glowing coating of sustainability.
3. Favor technical solutions though dedicating limited energy to lower agroecological manufacturing techniques.
4. Focusing exclusively on crop development rather than tackling the entire agricultural environment, including animals, which is a prerequisite for achieving food protection entails agroecology co-option, lacking its social and political aspects.
5. So narrowly define sustainability, disregarding its essential social and economic elements, such as livelihoods, education, social justice, and commercial feasibility.
6. SI does not entail a radically different strategy but instead an intense repackaging.
7. It offers an opportunity for policymakers and other actors to disregard litigious market concerns. Through creating a win-win situation under which yields on the same amount of land can be sustainably improved, there is no need to address

high amounts of demand for beef, milk, and other resource-intensive goods among the wealthiest components of the global population (Hird 2012).
8. Use of uncommon statements which lack social and political evaluation.
9. Another popular critique of the recent usage of SI is that it is available as the means to accomplish it to all forms of technology. This helps certain corporations and scientists to advocate new innovations such as biotechnology, which are viewed as inherently inconsistent with biodiversity by other organizations interested in food system issues (Collins and Chandrasekaran 2012; Parmentier 2014).

7.16 Future Prospect

Agricultural SI has grown to this day along with the discussion regarding the possibility of climate, atmosphere, and agriculture. The idea of agricultural intensification is developing and aims to harness the advantages of conventional agriculture and organic farming while mitigating its limitations. In fact, the disruption to the natural ecosystem and its utility mechanisms is minimized, thereby bridging the distance between the two and saving the world. SI is actually only being viewed as a generic term with little indicators, and is being used ambiguously at various rates (Petersen and Snapp 2015). Therefore, in order to effectively quantify SI and achieve agreement on the definition, it is necessary to create criteria and a coherent indexing framework for accurate calculation of sustainable intensification. The functional ways of sustainable agricultural intensification are heterogeneous in various regions and cropping practices, owing to the specific environmental environments and socio-economic factors in various regions (Weltin et al. 2018). For example, in certain arid regions, instead of inputs like irrigation, fertilizer or better crops, inputs relevant to animal welfare and social institutions may be the most effective ways of intensification, creating a very different scenario from other regions (Scherer et al. 2018).

However, crop yields can be increased significantly in the some arid regions of North Africa and West Asia by implementing complementary irrigation and proper management procedures, thereby making a contribution to a SI of crop production (FAO 2004). Since agroforestry complex management is commonly advocated as a model for achieving sustainable agricultural intensification in mountainous regions, it is seen as a remedy to the growing concerns of shortened fallow times and land loss (Nath et al. 2016). So we will be reflecting on the natural world and socio-economic developments of various areas in the future. Incorporating the idea of sustainable agricultural intensification, pick suitable sizes and traditional regions, then examine the methods, activities of patterns and the probability of model implementation in regions with specific socio-economic and natural environmental factors. Analysis of the driving factors of sustainable agricultural intensification is desperately required in the light of decreasing crop yields and crop production and tightening resource limitations.

Work on the affecting factor of SI needs to be improved, in particular the calculation of influencing factors like farm-scale land tenure protection, livelihood approaches, access to loans and industry, facilities, policy subsidies, and program architecture, so that it can expose the impacting process and make suitable recommendations. We will need to strengthen efforts to consistently determine the relevant geographic and detailed impacts of economic intensification in agriculture. The expense of the context of SI of agriculture and the inadequate short-term economic opportunities produced by SI will intensify the numerous agricultural uncertainties in operation. Building a risk reduction system and improving the blueprint of sustainable agricultural intensification accordingly is thus urgently needed. In the long term, the assessment of incentive structures should be strengthened to encourage heterogeneous growers to switch to sustainable agricultural intensification, studies on enhancing the social welfare of farmers, the procurement costs of foreign agricultural inputs, farmers' readiness to adapt farming practices, offering crop insurance schemes and exposure to information and the sector, defining the determinants of the insecurity of farmers at the household level and offering suitable opportunities are also priorities in further research.

7.17 Conclusion

SI is the way of boosting agricultural production to feed an increasing population while minimizing the farmer's environmental footprint. EI, as one of the three foundations of SI, is about protecting natural capital, being particular in the usage of resources towards sustainable ecology and economy. Although EI shows great potential, at least in the context of biodiversity and sustainability, these technologies are seldom brought to scale, owing in part to the significant energy, resources, and expertise they need. Thus, although agricultural ecology practice is essential to biodiversity change, the cycle of crop and livestock breeding and socio-economic intensification is just as significant. The aim is to achieve all of the targets and increasing the productivity. Innovation is required for all the activities mentioned in this summary to attain Sustainable Intensification, whether by creating new strategies, deeper comprehension of the influences, and local background of current practices or by promoting their acceptance and progress. Many barriers prohibit smallholders from implementing modern strategies and tailoring them towards their own climate. Some obstacles, like insecurity over land tenure, lack of economic capital, etc. More research is needed for implementing innovations, their usage and impact, the active involvement of smallholder families in the process to increase adoption rates, business growth for organic farming and, most of all, political leadership.

A lot of work has to be performed to make sure that agricultural systems should boost nutrient production globally while also ensuring positive effect on ecosystems and social capital. We infer that a change from production by replacement to reconstruction would be necessary, indicating that a phase of adaptation should be the definition and action of SI in agriculture, guided by a broad variety of actors collaborating in new economies of agricultural awareness. Farmers and community would also need to spend in SI, not only for the sake of survival but for living standards and productivity.

References

Adamchuk VI, Marx DB, Kerby AT, Samal AK, Soh LK, Ferguson RB, Wortmann CS (2004) Guided soil sampling for enhanced analysis of georeferenced sensor-based data, p 1–4. http://www.geocomputation.org/2007/2A-Apps_Environment_2/2A2.pdf

Allen JE, Rossi J, Woods TA, Davis AF (2017) Do community supported agriculture programmes encourage change to food lifestyle behaviours and health outcomes? New evidence from shareholders. Int J Agric Sustain 15:70–82

Altieri MA (2002) Agroecology: the science of natural resource management for poor farmers in marginal environments. Agric Ecosyst Environ 93:1–24

Anderson B (2015) What kind of thing is resilience? Politics 35:60–66

Baduhur AV, Ibrahim M, Tanner T (2013) Characterising resilience: unpacking the concept for tackling climate change and development. Clim Dev 5:55–65

Balvanera P, Daily GC, Ehrlich PR, Ricketts TH, Bailey SA, Kark S, Pereira H (2001) Conserving biodiversity and ecosystem services. Science 291(5511):2047

Banerjee A, Jhariya MK, Yadav DK, Raj A (2020) Environmental and sustainable development through forestry and other resources. Apple Academic Press, Boca Raton. ISBN: 9781771888110, p 400. https://doi.org/10.1201/9780429276026

Barnosky AD, Matzke N, Tomiya S, Wogan GOU, Swartz B, Quental TB, MarshallC MGJL, Lindsey EL, Maguire KC, Mersey B, Ferrer EA (2011) Has the Earth's sixth mass extinction already arrived? Nature 471:51–57

Baulcombe D, Crute I, Davies B, Green N (2009) Reaping the benefits: science and the SI of global agriculture. Royal Society, London

Benton TG (2015) In: Pritchard B, Ortiz R, Shekar M (eds) Handbook of food and nutrition security. Routledge, Abingdon

Best S (2014) Growing power: exploring energy needs in smallholder agriculture. IIED discussion paper. IIED, London

Bjorndal T, Child A, Lem A (2014) Value chain dynamics and the small-scale sector: policy recommendations for small-scale fisheries and aquaculture trade. FAO fisheries and aquaculture technical paper 581, Rome, Italy, p 112. http://www.fao.org/fileadmin/user_upload/fisheries/docs/Value_chain_ dynamics_and_the_small-scale_sector.pdf

Bommarco R, Kleijn D, Potts SG (2013) Ecological intensification: Harnessing ecosystem services for food security. Trends Ecol Evol 28:230–238

Brooker RW, Bennett AE, Cong WF, Daniell TJ, George TS, Hallett PD, Hawes C, Iannetta PPM, Jones HG, Karley AJ, Li L, McKenzie BM, Pakeman RJ, Paterson E, Schöb C, Shen J, Squire G, Watson CA, Zhang C, Zhang F, Zhang J, White PJ (2015) Improving intercropping: a synthesis of research in agronomy, plant physiology and ecology. New Phytol 206 (1):107–117

Buckwell A, Nordang-Uhre A, Williams A, Polakova J, Blum WE, Schiefer J, Haber W (2014) The SI of European Agriculture; a review sponsored by the RISE Foundation. The RISE Foundation, Brussels

Byerlee D, Stevenson J, Villoria N (2014) Does intensification slow crop land expansion or encourage deforestation? Glob Food Secur 3:92–98

Capmourteresa V, Adamsb J, Bergb A, Fraserb E, Swantonc C, Ananda M (2018) Precision conservation meets precision agriculture: A case study from southern Ontario. Agr Syst 167:176–185

Cardinale BJ, Duffy JE, Gonzalez A, Hooper DU, Perrings C, Venail P, Narwani A, Mace GM, Tilman D, Wardle DA, Kinzig AP, Daily GC, Loreau M, Grace JB, Larigauderie A, Srivastava DS, Naeem S (2012) Biodiversity loss and its impact on humanity. Nature 486:59–67

Cassman KG, Grassini P (2020) A global perspective on sustainable intensification research. Nat Sustain 3:262–268

CGIAR (2014) Addressing crosscutting issues. CGIAR. www.cgiar.org/ar2012/strong-progress/collaboration/addressing-crosscutting-issues

Chapagain T, Yamaji E (2010) The effects of irrigation method, age of seedling and spacing on crop performance, productivity and water-wise rice production in Japan. Paddy Water Environ 8:81–90

Chappell MJ, LaValle LA (2011) Food security and biodiversity: can we have both? An agroecological analysis. Agric Human Values 28:3–26

Collier WL, Soentaro WG (1973) Recent changes in rice harvesting methods. Some serious social implications. Bull Indon Econ Stud 9(2):36–45

Collins ED, Chandrasekaran K (2012) A Wolf in Sheep's Clothing? An analysis of the 'sustainable intensification' of agriculture. Friends of the Earth International, Amsterdam

Cook S, Silici L, Adolph B, Walker S (2015) SI revisited. IIED issue paper. IIED, London

Cortner O, Garrett RD, Valentim JF (2019) Perceptions of integrated crop-livestock systems for sustainable intensifcation in the Brazilian Amazon. Land Use Policy 82:841–853

COSA (2013) The COSA measuring sustainability report: coffee and cocaine 12 countries. The Committee on Sustainability Assessment, Philadelphia. http://thecosa.org/wp-content/uploads/2014/01/The- COSA-Measuring-Sustainability-Report.pdf

David LO, Kurt BW, Robert BR (2016) SI and farmer preferences for crop system attributes: evidence from Malawi's central and southern regions. World Dev 87:139–151

DeClerck FAJ, Jones SK, Attwood S, Bossio D, Girvetz E, Chaplin-Kramer B, Enfors E, Fremier AK, Gordon LJ, Kizito F, Lopez Noriega I, Matthews N, McCartney M, Meacham M, Noble A, Quintero M, Remans R, Soppe R, Willemen L, SLR W, Zhang W (2016) Agricultural ecosystems and their services: the vanguard of sustainability? Curr Opin Environ Sustain 23:92–99

DeGroot R, Brander L, vanderPloeg S, Costanza R, Bernard F, Braat L, Christief M, Crossmang N, Ghermandii A, Heina L, Hussain S, Kumar P, McVittie A, Portela R, Rodriguezg LC, Brinkm PT, Beukering PT (2012) Global estimates of the value of eco systems and their services in monetary units. Ecosyst Serv 1:50–61

Doré CK, Makowski D, Malézieux E, Munier-Jolain N, Tchamitchian M, Tittonell T (2011) Facing up the paradigm of ecological intensification in agronomy: revising methods, concepts and knowledge. Eur J Agron 34:197–210

FAO (2004) The ethics of sustainable agricultural intensification; ethics series. Food and Agriculture Organization of the United Nations, Rome, pp 3–5

FAO (2006) Policy brief on food security, vol 2. FAO, Rome

FAO (2007) Organic agriculture in Africa. FAO, Rome. www.unep.org/training/programmes/Instructor%20Version/Part_2/Activities/Huma n_Societies/Agriculture/Supplemental/Organic_Agriculture_in_Africa.pdf

FAO (2011) Save and grow: a policymakers guide to the SI of smallholder crop production. FAO, Rome

FAO (2015) Organic agriculture and climate change. FAO, Rome. www.fao.org/docrep/005/y4137e/y4137e02b.htm

FAO (2016a) Mid-term evaluation of the conservation agriculture scaling-up (CASU) project. FAO, Rome. http:// www.fao.org/3/a-bq888e.pdf

FAO (2016b) Save and grow: maize, rice and wheat—a guide to sustainable crop production. FAO, Rome

FAO (2018) SI of agriculture. http://www.fao.org/policy-support/policythemes/sustainable-intensification-agriculture/en/

FAOSTAT Database (2017) Food and Agriculture Organization, 2017. FAO, Rome

FiBL and IFOAM (2020) The world of organic agriculture statistics and emerging trends 2020. http://www.organic-world.net/yearbook/yearbook-2020.html

Firbank LG, Elliott J, Drake B (2013) Evidence of SI among British farms. Agric Ecosyst Environ 173:58–65

Foresight (2011) The future of food and farming. Final project report. The Government Office for Science, London

Francis CA, Hansen TE, Fox AA, Hesje PJ, Nelson HE, Lawseth AE, English A (2012) Farmland conversion to non-agricultural uses in the US and Canada: current impacts and concerns for the future. Int J Agric Sust 10L:8–24

Fraser E, Legwegoh A, Krishna KC, CoDyre M, Dias G, Hazen S, Johnson R, Martin R, Ohberg L, Sethuratnam S, Senyd L, Smithers J, Van Acke R, Vansteenkiste J, Wittman H, Yada R (2016) Biotechnology or organic? Extensive or intensive? Global or local? A critical review of potential pathways to resolve the global food crisis. Trends Food Sci Technol 48:78–87

Fry J, Finley W (2005) The prevalence and costs of obesity in the EU. Proc Nutr Soc 64:359–362

Gadanakis Y, Bennett R, Park J, Areal FJ (2015) Evaluating the SI of arable farms. J Environ Manage 150:288–298

Garnett T, Appleby MC, Balmford A, Bateman IJ, Benton TG, Bloomer P, Burlingame B, Dawkins M, Dolan L, Fraser D, Herero M, Hoffmann I, Smith P, Thornton PK, Toulmin C, Vermeulen SJ, Gadfray HCJ (2013) SI in agriculture: premises and policies. Science 341:33–34

Garnier E, Navas ML (2012) A trait-based approach to comparative functional plant ecology: concepts, methods and applications for agroecology. Agron Sustain Dev 32:365–399

Ge L, Anten NPR, VanDixhoorn IDE, Feindt PH, Kramer K, Leemans R, Meuwissen MPM, Spoolder H, Sukkel W (2016) Why we need resilience thinking to meet societal challenges in bio-based production systems. Curr Opin Environ Sustain 23:17–27

Geiger F, Bengtsson J, Berendse F, Weisser WW, Emmerson M, Morales MB, Ceryngier P, Liira J, Tscharntke T, Winqvist C, Eggers S, Bommarco R, Part T, Bretagnolle V, Plantegenest M, Clement LW, Dennis C, Palmer C, Onate JJ, Guerrero I (2010) Persistent negative effects of pesticides on biodiversity and biological control potential on European farmland. Basic Appl Ecol 11:97–105

Gibbon P, Bolwig S (2007) The economics of certified organic farming in tropical Africa: a preliminary analysis. SIDA DIIS working paper number 2007/3, Subseries on Standards and Agro-Food-Exports (SAFE) No. 7

Gibon A, Sibbald AR, Flamant JC (1999) Livestock farming systems research in Europe and its potential contribution for managing towards sustainability in livestock farming. Lives Prod Sci 61:121–137

Gliessman SR (2007a) Agroecology: ecological processes in sustainable agriculture, 2nd edn. Lewis Publisher, Boca Raton, New York

Gliessman SR (2007b) Agroecology: the ecology of sustainable food systems. CRC Press, New York

Godfray HCJ, Beddington JR, Crute IR, Haddad L, Lawrence D, Muir JF, Pretty J (2010) Food security: the challenge of feeding 9 billion people. Science 327:812–818

Gunton RM, Firbank LG, Inman A, Michael Winter DM (2016) How scalable is sustainable intensification? Nat Plants 2:16065

Gurr GM, Lu Z, Zheng X, Xu H, Zhu P, Chen G, Yao X, Cheng J, Zhu Z, Catindig JL, Villareal S, Chien HV, Cuong LQ, Channoo C, Chengwattana N, Lan LP, Hai LH, Chaiwong J, Nicol HI, Perovic DJ, Wratten SD, Heong KL (2016) Multi-country evidence that crop diversification promotes ecological intensification of agriculture. Nat Plants 2:16014

Hallmann CA, Sorg M, JongejansE SH, Hofland N, Schwan H, de Kroon H (2007) More than 75 percent decline over 27 years in total flying insect biomass in protected areas. PLoS One 12: e0185809

Harty M, Wagstaff P, Harper M, Chikate J (2010) Conservation agriculture: measuring the impact on livelihoods in Zimbabwe. Agriculture for Development, No. 11

Heinemann JA, Abate T, Hilbeck A, Murray D (2009) Agriculture at a crossroads: the synthesis report of the International Assessment of Agricultural Knowledge, Science and Technology for Development. World Bank and various UN Agencies

Hird V (2012) The politics of plenty. Food Ethics 7(2):19–21

Hobbs B, Powell S (2011) Healthy harvests: the benefits of sustainable agriculture in African and Asia. A Christian Aid Report, London

Holling CS (1996) Engineering resilience versus ecological resilience. In: Schulze PC (ed) Engineering within ecological constraints. The National Academic Press, Washington, pp 31–44

IFAD (2011) Rural poverty report: new realities, new challenges:new opportunities for tomorrow's generation. IFAD, Rome. www.cgiar.org/consortium-news/have-we-got-our-priorities-straight

IFOAM (2020) The world of organic agriculture statistics and emerging trends 2020. http://www.organic-world.net/yearbook/yearbook-2020.html

IPCC (2014) Summary for policymakers. In: Climate change 2014: Impacts, adaptation, and vulnerability. Part A: global and sectoral aspects. Contribution of working group ii to the fifth assessment report of the intergovernmental panel on climate change Field CB, Barros VR, Dokken DJ, Mach KJ, Mastrandrea MD, Bilir TE, Chatterjee M, Ebi KL, Estrada YO, Genova RC, Girma B, Kissel ES, Levy AN, Macracken S, Mastrandrea PR, White LL,(eds.). Cambridge University Press, Cambridge, 1–32

IPES-Food (2016) From uniformity to diversity: a paradigm shift from industrial agriculture to diversified agroecological systems. International Panel on Sustainable Food Systems. www.ipes-food.org

Isbell F, Calcagno V, Hector A, Connolly J, Harpole WS, Reich PB, Loreau M (2011) High plant diversity is needed to maintain ecosystem services. Nature 477:199–203

Jat, ML, Malik RK, Saharawat YS, Gupta R, Bhag M, Raj P (2012) Proceedings of regional dialogue on conservation agricultural in South Asia, New Delhi, APAARI, CIMMYT, ICAR, p 32

Jat ML, Chakraborty D, Ladha JK, Rana DS, Gathala MK, McDonald A, Gerard B (2020) Conservation agriculture for SI in South Asia. Nature Sustain 3(4):336–343

Jawahery AAH (2015) Undated on the post-2015 sustainable development goals, Blogpost of the International Fertilizer Industry Association. www.gpic.com/media/Pdf/IFA.pdf

Jhariya MK, Yadav DK, Banerjee A (2018a) Plant mediated transformation and habitat restoration: phytoremediation an eco-friendly approach. In: Gautam A, Pathak C (eds) Metallic contamination and its toxicity. Daya Publishing House, A Division of Astral International Pvt. Ltd, New Delhi, pp 231–247ISBN: 9789351248880

Jhariya MK, Banerjee A, Yadav DK, Raj A (2018b) Leguminous trees an innovative tool for soil sustainability. In: Meena RS, Das A, Yadav GS, Lal R (eds) Legumes for soil health and sustainable management. Springer., ISBN 978-981-13-0253-4 (eBook), ISBN: 978-981-13-0252-7 (Hardcover), Singapore, pp 315–345. https://doi.org/10.1007/978-981-13-0253-4_10

Jhariya MK, Banerjee A, Meena RS, Yadav DK (2019a) Sustainable agriculture, forest and environmental management. Springer, Singapore. eISBN: 978-981-13-6830-1, Hardcover ISBN: 978-981-13-6829-5, p 606. https://doi.org/10.1007/978-981-13-6830-1

Jhariya MK, Yadav DK, Banerjee A (2019b) Agroforestry and climate change: issues and challenges. Apple Academic Press, Boca Raton. ISBN: 978-1-77188-790-8 (Hardcover), 978-0-42957-274-8 (E-book), p 335. https://doi.org/10.1201/9780429057274

Juma C, Tabo R, Wilson K, Conway G (2013) Innovation for SIin Africa. The Montpellier Panel. Agriculture for Impact, London

Karlberg L, Rockström J, Annandale J, Martin Steyn J (2007) Low-cost drip irrigation—a suitable technology for southern Africa? An example with tomatoes using saline irrigation water. Agric Water Manag 89(1–2):59–70

Kassam A, Friedrich T, Derpsch R (2018) Global spread of conservation agriculture. Int J Environ Stud 76:29–51

Kassie M, Teklewold H, Jaleta M (2015) Understanding the adoption of a portfolio of SIpractices in eastern and southern Africa. Land Use Policy 42:400–411

Kepha GO, Bancy MM, Patrick GH (2014) Determination of the effect of the system of rice intensification (SRI) on rice Yields and water saving in Mivea irrigation scheme, Kenya. J Water Resour Protec 6:895–901

Khan N, Jhariya MK, Yadav DK, Banerjee A (2020a) Herbaceous dynamics and CO_2 mitigation in an urban setup—a case study from Chhattisgarh, India. Environ Sci Pollut Res 27 (3):2881–2897. https://doi.org/10.1007/s11356-019-07182-8

Khan N, Jhariya MK, Yadav DK, Banerjee A (2020b) Structure, diversity and ecological function of shrub species in an urban setup of Sarguja, Chhattisgarh, India. Environ Sci Pollut Res 27 (5):5418–5432. https://doi.org/10.1007/s11356-019-07172-w

Kleijn D, Kohler F, Baldi A, Batary P, Concepcion ED, Clough Y, Diaz M, Gabriel D, Holzschuh A, KnopE KA, Marshall EJP, Tscharntke T, Verhulst J (2009) On the relationship between farmland biodiversity and land-use intensity in Europe. Proc R Soc 276:903–909

Kleijn D, Winfree R, Bartomeus I, Carvalheiro LG, Henry M, Isaacs R, Klein AM, Kremen C, M'Gonigle LK, Rader R, Ricketts TH, Williams NM, Adamson NL, Ascher JS, Andras Baldi A, Batary P, Benjamin F, Biesmeijer JC, Blitzer EJ, Bommarco RB, Brand MR, Bretagnolle V, Button L, Cariveau DP, Chifflet R, Colville JF, Danforth BN, Elle E, MPD G, Herzog F, Holzschuh A, Howlett BG, Jauker F, Jha S, Knop E, Feon VL, Mandelik Y, May EA, Park MG, Pisanty G, Reemer M, Riedinger V, Rollin O, Rundlof M, Sardinas HS, Scheper J, Sciligo AR, Smith HG, Steffan-Dewenter I, Thorp R, Tscharntke T, Verhulst J, Viana BF, Vaissiere BE, Veldtman R, Ward KL, Westpha C, Potts SG (2015) Delivery of crop pollination services is an insufficient Argument for wild pollinator conservation. Nat Commun 6:1–8

Kumar VT (2017) Zero-budget nature farming. Department of Agriculture, Andhra Pradesh

Kumar S, Meena RS, Jhariya MK (2020) Resources use efficiency in agriculture. Springer, Singapore. eBook ISBN: 978-981-15-6953-1, Hardcover: 978-981-15-6952-4, p 760. https://doi.org/10.1007/978-981-15-6953-1

Lal R, Singh BR, Mwaseba DL, Kraybill D, Hansen DO, Eik LO (2015) SI to advance food security and enhance climate resilience in Africa. Springer, Cham

Lammerts van Bueren E (2016) Enhancing resilience through plant breeding requires an integrated and interdisciplinary approach. In: Davis K (ed) Cultivating resilience. Organic Seed Alliance, Port Townsend, pp 133–134

Lampkin NH, Pearce BD, Leake AR, Creissen H, Gerrard CL, Girling R et al (2015) The role of agroecology in sustainable intensification,In Report for the land use policy group. Organic Research Centre, Elm Farm and Game and Wildlife Conservation Trust, New Bury, p 163

Liao C, Brown DG (2018) Assessments of synergistic outcomes from SI of agriculture need to include smallholder livelihoods with food production and ecosystem services. Curr Opin Environ Sustain 32:53–59

Loos J, Abson DJ, Chappell MJ, Hanspach J, Mikulcak F, Metal T (2014) Putting meaning back into "sustainable intensification". Front Ecol Environ 12:356–361

Losey JE, Vaughan M (2006) The economic value of ecological services provided by insects. Bioscience 56:311–323

Mabhaudhi T, Chibarabada T, Chimonyo V, Murugani V, Pereira L, Sobratee N, Govender L, Slotow R, Modi AT (2019) Mainstreaming underutilized indigenous and traditional crops into food systems: a South African perspective. Sustainability 11:172

Mahon N, Crute I, Di BM (2018) Towards a broad-based and holistic framework of SI indicators. Land Use Policy 77:576–597

Malezieux E (2012) Designing cropping systems from nature. Agron Sustain Dev 32:15–29

Martin AR, Isaac ME (2015) Plant functional traits in agroecosystems: a blueprint for research. J Appl Ecol 52:1425–1435

Matson PA, Parton WJ, Power AG (1997) Agricultural intensification and ecosystem properties. Science 277:504–509

McBratney A, Whelan B, Ancev T (2005) Future directions of precision agriculture. Precision Agric 6:7–23

Meena RS, Lal R (2018) Legumes for soil health and sustainable management. Springer, Singapore, p 541. ISBN 978-981-13-0253-4 (eBook), ISBN: 978-981-13-0252-7(Hardcover). https://doi.org/10.1007/978-981-13-0253-4_10

Meena RS, Kumar V, Yadav GS, Mitran T (2018) Response and interaction of *Bradyrhizobium japonicum* and Arbuscular mycorrhizal fungi in the soybean rhizosphere: a review. Plant Growth Regul 84:207–223
Meena RS, Kumar S, Datta R, Lal R, Vijaykumar V, Brtnicky M, Sharma MP, Yadav GS, Jhariya MK, Jangir CK, Pathan SI, Dokulilova T, Pecina V, Marfo TD (2020a) Impact of agrochemicals on soil microbiota and management: a review. Land (MDPI) 9(2):34. https://doi.org/10.3390/land9020034
Meena RS, Lal R, Yadav GS (2020b) Long term impacts of topsoil depth and amendments on soil physical and hydrological properties of an Alfisol in Central Ohio, USA. Geoderma 363:1141164
Meena RS, Lal R, Yadav GS (2020c) Long-term impact of topsoil depth and amendments on carbon and nitrogen budgets in the surface layer of an Alfisol in Central Ohio. Catena 194:104752
Meybeck A, Redfern S (2014) Voluntary standards for sustainable food systems: challenges and opportunities. In: Paper presented at a workshop of the FAO/UNEP programme on sustainable food systems. FAO, Rome, Italy, June 2013. http://www.fao.org/3/a-i3421e.pdf
Moore ML, Westley F (2011) Surmountable chasms: networks and social innovation for resilient systems. Ecol Soc 16(1):5
Mori AS, Furukawa T, Sasaki T (2013) Response diversity determines the resilience of ecosystems to environmental change. Biol Rev 88:349–364
Morris C, Fish R, Winter M, Lobleyet M (2017) Sustainable intensification: the view from the farm. Asp Appl Biol 136:19–26
Mueller ND, Gerber JS, Johnston M, Ray DK, Ramankutty N, Foley JA (2012) Closing yield gaps through nutrient and water management. Nature 490:254–257
Nair RP (2014) Grand challenges in agroecology and land use systems. Front Environ Sci 2:1
Nath TK, Jashimuddin M, Kamrul HM (2016) The SI of agroforestry in shifting cultivation areas of Bangladesh. Agr Syst 90:405–416
Nature Editorial (2020) Green and grow. Nat Sustain 3(4):253
Nayar V, Ravichandran VK, Barah BC, Uphoff N (2020) Learnings from an irrigated agriculture management project in Tamil Nadu. Econ Pol Wkly 55(2):11
Ndiritu SW, Kassie M, Shiferaw B (2014) Are there systematic gender dierences in the adoption of sustainable agricultural intensification practices? Evidence from Kenya. Food Policy 49:117–127
Nelson E, Gomez Tovar L, Schwentesius R, Gomez Cruz MA (2010) Participatory organic certification in Mexico: an alternative approach to maintaining the integrity of the organic label. Agric Hum Values 27:227–237
Ngigi SN, Thome JN, Waweru DW, Blank HG (2001) Low-cost irrigation for poverty reduction. An evaluation of low-head drip irrigation technologies in Kenya. International Water Management Institute, Colombo
Noltze M, Schwarze S, Qaim M (2013) Impacts of natural resource management technologies on agricultural yield and household income: the system of rice intensification in Timor Leste. Ecological Economics, Ne Climate Economics 85:59–68
Nunez J (2010) Crop rotation as a method of disease control. Western Farm Press, Madera. westernfarmpress.com/management/crop-rotation-method-disease-control
O'Dell D, Eash NS, Hicks BB, Oetting JN, Sauer TJ, Lambert DM, Thierfelder C, Muoni T, Logan J, Zahn JA, Goddard JJ (2020) Conservation agriculture as a climate change mitigation strategy in Zimbabwe. Int J Agric Sustain 1:16
Organic Research Centre (2015) The role of SI and agroecology in achieving food security sustainably. www.organicresearchcentre.com/?go=Information%20and%20publications&page=2013Closing plenary
Parmentier S (2014) Scaling up agroecological approaches: what, why and how? Oxfam-Solidarity, Bruxelles

Petersen B, Snapp S (2015) What is sustainable intensification? Views from experts. Land Use Policy 46:1–10

Phalan B (2016) How can higher-yield farming help to spare nature? Science 351:450–451

Potts J, Lynch M, Wilkings A, Huppé G, Cunningham M, Voora V (2014) State of sustainability initiatives review 2014—standards and the green economy. International Institute for Sustainable Development/International Institute for Environment and Development, Winnipeg/London. www.iisd.org/sites/default/files/pdf/2014/ssi_2014.pdf

Prager K, Posthumus H (2010) Socio-Economic factors influencing farmers' adoption of soil conservation practices in Europe. Nova Science Publishers, Hauppauge

Prasad CS (2006) System of rice intensification in India: innovation history and institutional challenges. WWF-Dialogue Project at the International Crop Research Institute for Semi-Arid Tropics, Hyderabad

Pretty J (1997a) The SI of agriculture. Nat Resources Forum 2(4):247–256

Pretty JN (1997b) The SI of agriculture. In: Natural resources forum. Blackwell, Oxford

Pretty J (2008) Agricultural sustainability: concepts, principles and evidence. Philos Trans R Soc B 363:447–465

Pretty J, Bharucha ZP (2014) SI in agricultural systems. Ann Bot 114:1571–1596

Pretty JN, Brett C, Gee D, Hine RE, Mason CF, Morison JI, Raven H, Rayment M, Van Der Bijl G (2000) An assessment of the total external costs of UK agriculture. Agr Syst 65:113–136

Pretty JN, Noble AD, Bossio D, Dixon J, Hine RE, Penning de Vries FWT, Morison JIL (2006a) Resource-conserving agriculture increases yields in developing countries. Environ Sci Technol 4:1114–1119

Pretty J, Noble A, Bossio D, Dixon J, Hine R, Penning de Vries F, Morrison J (2006b) Resource conserving agriculture increases yields in developing countries. Environ Sci Technol 40:114–119

Pretty J, Toulmin C, Williams S (2011a) SI in African agriculture. Intern J Agric Sustain 9:5–24

Pretty J, Toulmin C, Williams S (2011b) SI in African agriculture. Int J Agric Sustain 9(1):5–24

Pretty JN, Toulmin C, Williams S (2011c) SI in African agriculture. Int J Agr Sustain 9:5–24

Pretty J, Benton TG, Bharucha ZP, Dicks LV, Flora CB, Godfray HCJ (2018) Global assessment of agricultural system redesign for sustainable intensification. Nat Sustain 1:441–446

Pywell RF, Heard MS, Woodcock BA, Hinsley S, Ridding L, Nowakowski M, Bullock JM (2015) Wildlife friendly farming increases crop yield: evidence for ecological intensification. Proc R Soc Lond B 282:20151740

Rahn E, Liebig T, Ghazoul J (2018) Opportunities for SI of ecoagro-ecosystems along an altitudinal gradient on Mt. Elgon, Uganda. Agric Ecosyst Environ 263:31–40

Raj A, Jhariya MK, Yadav DK, Banerjee A, Meena RS (2019a) Agroforestry: a holistic approach for agricultural sustainability. In: Jhariya MK, Banerjee A, Meena RS, Yadav DK (eds) Sustainable agriculture, forest and environmental management. Springer, Singapore, pp 101–131. https://doi.org/10.1007/978-981-13-6830-1

Raj A, Jhariya MK, Banerjee A, Yadav DK, Meena RS (2019b) Soil for sustainable environment and ecosystems management. In: Jhariya MK, Banerjee A, Meena RS, Yadav DK (eds) Sustainable agriculture, forest and environmental management. Springer, Singapore. eISBN: 978-981-13-6830-1, Hardcover ISBN: 978-981-13-6829-5, pp 189–221. https://doi.org/10.1007/978-981-13-6830-1

Raj A, Jhariya MK, Yadav DK, Banerjee A (2020) Climate change and agroforestry systems: adaptation and mitigation strategies. Apple Academic Press, New York. ISBN: 9781771888226, p 383. https://doi.org/10.1201/9780429286759

Ravikumar R (2016) Practicing precision agriculture in Dharmapuri district of Tamil Nadu: a case study. MPRA paper no. 73233. https://mpra.ub.uni-muenchen.de/73233/

Reganold JP, Wachter JM (2016) Organic agriculture in the twenty-first century. Nat Plants 2:15221

Reisch LA, Gwozdz W (2010) Einfluss des Konsumverhaltens auf dieEntwicklungvon Übergewicht bei Kindern. Bundesgesundheitsblatt Gesundheitsforschung Gesundheitsschutz 53:725–732
Rockstrom J, Baron J, Fox P (2002) Rainwater management for increased productivity among small-holder farmers in drought prone environments. Phys Chem Earth 27:949–959
Rockström J, Williams J, Daily G, Noble A, Matthews N, Gordon L, Wetterstrand H, DeClerck F, Shah M, Steduto P, de Fraiture C, Hatibu N, Unver O, Bird J, Sibanda L, Smith J (2017) SI of agriculture for human prosperity and global sustainability. Ambio 46:4–17
Rolando JL, Turin C, Ramírez DA (2017) Key ecosystem services and ecological intensification of agriculture in the tropical high-Andean Puna as acted by land-use and climate changes. Agric Ecosyst Environ 236:221–233
Sadras VO, Denison R (2016) Neither crop genetics nor crop management can be optimised. Field Crop Res 189:75–83
Sánchez-Escobar F, Coq-Huelva D, Sanz-Cañada J (2018) Measurement of SIby the integrated analysis of energy and economic flows: case study of the olive-oil agricultural system of Estepa, Spain. J Clean Prod 201:463–470
Sandhu H, Wratten S, Costanza R, Pretty J, Porter JR, Reganold J (2015) Significance and value of non-traded ecosystem services on farmland. Peer J 3:762
Sangeetha S, Ganesan R, Indumathy K, Jai Sridar P (2013) Exploring the extent of adoption of precision farming technologies in tomato cultivation. Indian J Agric Res 47(2):178–182
SCAR (2011) Sustainable food consumption and production in a resource—constrained world in European Commission–Standing Committee on Agricultural Research (SCAR), The 3rd SCAR foresight exercise, Brussels, pp 1–50
Scherer LA, Verburg PH, Schulp CJE (2018) Opportunities for SI in European agriculture. Glob Environ Chang 48:43–55
Schiefer J, Lair GJ, Blum WEH (2015) Indicators for the definition of land quality as a basis for the SI of agricultural production. Int Soil Water Conserv Res 3:42–49
Schiefer J, Lair GJ, Blum WEH (2016) Potential and limits of land and soil for SI of European agriculture. Agric Ecosyst Environ 230:283–293
Schimmelpfennig D (2016) Farm profits and adoption of precision agriculture. USDA 217:1–46
Schut M, Van A, Okafor P, C. (2016) SI of agricultural systems in the Central African Highlands: the need for institutional innovation. Agr Syst 145:165–176
Shibusawa (2002) Precision farming approaches to small farm agriculture. Agro-Chem Rep 2 (4):13–20
Shruthi K, Hiremath GM, Joshi AT (2018) An overview of use of precision farming technologies by the farmers—a case study of North Eastern Karnataka. Indian J Agric Res 52(1):93–96
Silici L (2014) Agroecology: what it is and what it has to offer. IIED, London
Silici L, Ndabe P, Friedrich T, Kassam A (2011) Harnessing sustainability, resilience and productivity through conservation agriculture: the case of Likoti in Lesotho. Int J Agric Sustain 9:137–144
Slätmo E, Maye D, Duncan J (2017) The framing of sustainability in sustainability assessment frameworks for agriculture. Sociol Rural 57:378–395
Smith P (2013) Delivering food security without increasing pressure on land. Glob Food Secur 2:18–23
Smith MR, Singh GM, Mozaffarian D, Myers SS (2015) Effects of decreases of animal pollinators on human nutrition and global health: a modelling analysis. Lancet 386:1964–1972
Smith A, Snapp S, Chikowo R (2017) Measuring SI in smallholder agro-ecosystems: a review. Glob Food Secur 12:127–138
Snapp SS, Grabowski P, Chikowo R (2018) Maize yield and profitability trades with social, human and environmental performance: is SI feasible? Agr Syst 162:77–88
Sonka ST, Bauer ME, Cherry EJ (1997) Precision agriculture in the 21st century: geospatial and information technologies in crop management. Executive summary. National Academies Press, Washington, p 168

Stachetti RG, Roberto MC (2018) Sustainability assessment of ecological intensification practices in coconut production. Agr Syst 165:71–84

Stoate C, Boatman ND, Borralho RJ, Carvalho CR, DeSnoo GR, Eden P (2001) Ecological impacts of arable intensification in Europe. J Environ Manage 63:337–365

Stoop W (2011) The scientific case for system of rice intensification and its relevance for sustainable crop intensification. Int J Agric Sustain 9(3):443–455

Struik PC, Kuyper TW, Brussaard L, Leeuwis C (2014) Deconstructing and unpacking scientific controversies in intensification and sustainability: why the tensions in concepts and values? Curr Opin Environ Sustain 8:80–88

Swiderska K, Song Y, Li J, Reid H, Mutta D (2011) Adapting agriculture with traditional knowledge. IIED briefing. IIED, London

Taddele Dile Y, Karlberg L, Temesgen M, Rockström J (2013) The role of water harvesting to achieve sustainable agricultural intensification and resilience against water related shocks in sub-Saharan Africa agriculture. Ecosyst Environ 181:69–79

Thakur AK, Rath S, Mandal KG (2013) Differential responses of system of rice intensification (SRI) and conventional flooded rice management methods to applications of nitrogen fertilizer. Plant and Soil 370:59–71

Thakur AK, Mohanty RK, Patil DU, Kumar A (2014a) Impact of water management on yield and water productivity with system of rice intensification (SRI) and conventional transplanting system in rice. Paddy Water Environ 12:413–424

Thakur AK, Singh R, Kumar A (2014b) The science behind the system of rice intensification (SRI). Research Bulletin No. 69, Directorate of Water Management (ICAR), Bhubaneswar, Odisha, p 58

The Montpellier Panel (2013) Sustainable intensification: a new paradigm for African agriculture. Agriculture for Impact, London

The Montpellier Panel (2014) No ordinary matter: conserving, restoring and enhancing Africa's soils. Agriculture for Impact, London

The Royal Society (2009) Reaping the benefits: science and the SI of global agriculture. RS policy document 11/09. The Royal Society, London

Thomson AM, Ramsey S, Barnes E, Basso B, Eve M, Gennet S, Grassini P, Kliethermes B, Matlock M, McClellen E, Spevak E, Snyder CS, Tomer MD, Kessel CV, West T, Wick G (2017) Science in the supply chain: collaboration opportunities for advancing sustainable agriculture. Agric Environ Lett 2:170015

Tilman D, Cassman KG, Matson PA, Naylor R, Polasky S (2002) Agricultural sustainability and intensive production practices. Nature 418:671–677

Tilman D, Balzer C, Hill J, Befort BL (2011) Global food demand and SI of agriculture. Proc Natl Acad Sci 108:20260–20264

Tittonell P, Giller KE (2013) When yield gaps are poverty tramps: the paradigm of ecological intensification in African smallholder agriculture. Field Crop Res 143:76–90

Tran DV, Nguyen NV (2006) The concept and implementation of precision farming and rice integrated crop management systems for sustainable production in the twenty-first century In: Trees, agro-forestry and multifunctional agriculture in Cameroon. Int J Agric Sustain 9:110–119

Trewavas A (2002) Malthus foiled again and again. Nature 418:668–670

UN Department of Economic and Social Affairs (2017) World population prospects revision. United Nations, New York

UNCTAD (2013) Trade and environment review (2013) wake up before it is too late: make agriculture truly sustainable now for food security in a changing climate. UNCTAD, Geneva

UNEP (2012) Avoiding future famines: strengthening the ecological foundation of food security through sustainable food systems. United Nations Environment Programme, Nairobi

UNEP-UNTCAD Task Force (2008) Organic agriculture and food security in Africa. Capacity Building Task Force on Trade, Environment and Development (CBTF)

United Nations (2015) Transforming our world: the 2030 agenda for sustainable development. United Nations, New York

United Nations Division of Economics and Social Affairs (2018) The 2018 revision of the world urbanization prospects. United Nations, New York

United Nations World Commission on Environment and Development (1987) Report of the world commission on environment and development: our common future. Oxford University Press, Oxford

Uphoff N, Fasoula V, Iswandi A, Kassam A, Thakur AK (2015) Improving the phenotypic expression of rice genotypes: rethinking "intensification" for production systems and selection practices for rice breeding. Crop J 3:174–189

Urruty N, Tailliez-Lefebvre D, Huyghe C (2016) Stability, robustness, vulnerability and resilience of agricultural systems- A review. Agron Sustain Dev 36:1–15

Vanlauwe B, Coyne D, Gockowski J (2014) SI and the African smallholder farmer. Curr Opin Environ Sustain 8:15–22

Vanloqueren G, Baret PV (2009) How agricultural research systems shape a technological regime that develops genetic engineering but locks out agroecological innovations. Res Policy 38:971–983

Walker J, Cooper M (2011) Genealogies of resilience: from systems ecology to the political economy of crisis adaptation. Secur Dialogue 42:143–160

Wals AEJ (2015) Beyond unreasonable doubt. Education and learning for socio-ecological sustainability in the anthropocene. Wageningen University, Wageningen, p 36

Wells KL, Dollarhide JE (1998) Precision agriculture: a field study of soil test variability and its effect on accuracy of fertilizer recommendations. Soil Sci News Views 19:5

Weltin M, Zasada I, Piorr A (2018) Conceptualising fields of action for sustainable intensification-A systematic literature review and application to regional case studies. Agric Ecosyst Environ 257:68–80

West PC, Gerber JS, Engstro MPM, Mueller ND, Brauman KA, Carlson KM, Cassidy ES, Johnston M, MacDonald GK, Ray DK, Siebert S (2014) Leverage points for improving global food security and the environment. Science 345:325–328

Wezel A, Soboksa G, Mcclelland S (2015) The blurred boundaries of ecological, sustainable, and agroecological intensification: a review. Agron Sustain Dev 35:1283–1295

Willer H, Lernoud J (2015) The world of organic agriculture- statistics and emerging trends 2015. FiBL-IFOAM report. Research Institute of Organic Agriculture (FiBL), Frick, and IFOAM Organic International, Bonn

Willy KD, Muyanga M, Jayne T (2019) Can economic and environmental benefts associated with agricultural intensification be sustained at high population densities? A farm level empirical analysis. Land Use Policy 81:100–110

Woelcke J (2010) Technological and policy options for sustainable agricultural intensification in eastern Uganda. Agric Econ 34:129–139

World Agroforestry Centre (2009) Agroforestry: a global land use. Annual Report 2008–2009. World Agroforestry Centre, Nairobi

Xinhua (2016) CPC and state council guide opinion on using new development concepts to accelerate agricultural modernisation and realise moderate prosperity society. http://news.xinhuanet.com/fortune/2016-01/27/c_1117916568.htm

Yami M, Van AP (2017) Policy support for sustainable crop intensification in Eastern Africa. J Rural Stud 55:216–226

Zanasi C, Venturi P, Setti M, Rota C (2009) Participative organic certification, trust and local rural communities development: the Case of Rede Ecovida. New Medit 2:56–64

Zandonella R Sutter D, Liechti R, VonStokar T (2014) Volkswirtschaftliche Kosten des Pestizideinsatzes In der Schweiz-Pilotberechnung. http://www.pronatura.ch/landwirtschaft?file=tl_files/dokumente_de/2_unsere_themen/landwirtschaft/pestizide/Studie%20Volkswirtschaftliche%20Kosten% 20Pestizideinsatzes%20Schweiz.pdf

Zhang Y (2019) The role of precision agriculture. Resource 19:9

Zotoglo K (2011) Training manual on system of rice intensification. Integrated Initiatives for Economic Growth (IICEM), Mali, pp 3–15

Mulching and Weed Management Towards Sustainability

8

Taher Mechergui, Marta Pardos, Manoj Kumar Jhariya, and Arnab Banerjee

Abstract

Mulching is a very effective technology for conservation of soil moisture as well it adds nutrient through decomposition which helps to increase the productivity of the crops. Initiation of planting process tends to be successful due to maintenance, management of weed as well as their monitoring. Weed control is known to favor survival and plant growth by suppressing competition for water, nutrients, light, and space. Many weed control strategies (manual, mechanical, chemical, etc.) have been adopted. However, weed control using mulching has occupied a particular role among weed control methods used until now not only due to the effectiveness of this technique but also to its ability to provide conducive microclimatic conditions (increase of temperature and soil moisture) around mulched plant while improving the structural stability and soil structure and nutrients availability. This may improve early survival and plant growth in forest plantations, which makes weed control using mulching merits consideration in afforestation programs. Mulching cost depends on type of mulch used. However, weed control using mulching is generally less costly compared to traditional

T. Mechergui (✉)
Faculté des Sciences de Bizerte, Laboratoire des Ressources Sylvo-Pastorales de Tabarka, Tabarka, Tunisie

M. Pardos
Forest Dynamics and Management Department, INIA, Madrid, Spain
e-mail: pardos@inia.es

M. K. Jhariya (✉)
Department of Farm Forestry, Sant Gahira Guru Vishwavidyalaya, Sarguja, Ambikapur, Chhattisgarh, India

A. Banerjee
Department of Environmental Science, Sant Gahira Guru Vishwavidyalaya, Sarguja, Ambikapur, Chhattisgarh, India

M. K. Jhariya et al. (eds.), *Ecological Intensification of Natural Resources for Sustainable Agriculture*, https://doi.org/10.1007/978-981-33-4203-3_8

techniques (manual, mechanical, chemical, etc.) due to the reduction of maintenance activities after plantation.

Keywords

Growth · Mulching · Nutrients availability · Soil structure · Survival · Weed control

Abbreviations

%	Percent
cm	Centimeter
FAO	Food and Agriculture Organization
g/m^2	Gram per meter square
kg/m^2	Kilogram per meter square
m	Meter
mg/kg	Milligram per kilogram
mm	Millimeter
ppm	Parts per million
UV	Ultraviolet
μm	Micrometer

8.1 Introduction

One of the main causes of woody plantation failure is the absence or irregularity of maintenance against natural herbaceous vegetation. It is always beneficial to eliminate or better avoid the appearance of spontaneous weeds in the immediate vicinity of plants because of competition that can occur at different levels: the consumption of water and nutrients, the occupation of airspace and underground (Jhariya and Singh 2020; Meena et al. 2018).

Since the root density of competing vegetation in the first centimeters of the soil can be up to 50 times that of young seedlings (Nambiar and Sands 1993), the suppression of weeds is essential to ensure the survival and development of woody plants. In order to increase the chances of successful fallowing forest plantation, experts agree to: (1) consider maximum weed control for at least the first 2–3 years after planting (Von Ahlten 1990; Albouchi and Abbassi 2000) and (2) to prefer weed control at the feet of plants (or on planting lines) during this phase of installation of young trees, rather than between lines of plantation. Weeding should occur during periods when vegetation competition is more active.

Studies on weed-tree competition have shown that: (1) the loss of tree growth is even stronger when weed is close to the plant (Frochot 1984; Davies 1987), (2) the adverse effects of weed on the tree are first played at soil level and are mainly attributed to competition for water, and (3) weed control on the root zone of plant promotes its growth.

In order not to hinder the resumption and development of young trees for at least 3 years, a minimum weed area of 1 m in diameter at the base of plants usually recommended (Davies 1987). For conifer species this diameter should be between 1.5 m (Fiddler and McDonald 1987) and 1.90 m (McDonald and Helgerson 1990).

The intervention strategies of the forest manager aim towards managing the unwanted plants with low cost technologies. Therefore, several processes and mechanisms are there to achieve the target of forest management through necessary weed control (Meena and Lal 2018). Harrowing only between plantation lines is not advised during the planting phase of seedlings as it concentrates its action in an area where most of the barely planted woody plants have not had time to develop their root system and where they do not yet suffer from competition phenomena for water (Van Lerberghe 2004a).

It is preferable to carry out manual or mechanical operations to maintain a minimum surface area of 1 m^2 around the base of young trees during periods when vegetation competition is most active, generally from March to September, during the first 3 years after planting. In this particular context of intervention, two traditional methods of weeding are conceivable, but they present certain drawbacks likely to limit their eventual implementation:

Manual hoeing poses no risk to the environment. It involves uprooting weeds using a manual tool (e.g., hoe). As the weeds grow rapidly, this treatment must take place at regular intervals during the active growth phase of unwanted plants. Costly along with lesser durability (Albouchi and Abbassi 2000), it requires careful man-power that ensures weeding close to plants without risk of injury. Long and tedious (Albouchi and Abbassi 2000), this work is conceivable only for very small areas that can be maintained by oneself.

Chemical weeding is a suitable alternative with efficient functioning and low cost technique to be used for control of unwanted plants (Van Lerberghe 2004b). However, it assumes a real competence for the choice, the dosage of products, the calculation of quantities of liquid or granules to be spread, the optimal date of treatment, and the adaptation to species to be protected and species to be destroyed (Van Lerberghe 2004a).

It is clear that the chemical fight is increasingly poorly considered by the public and that in the face of the risks of pollution and danger to health, the arsenal of active ingredients available is gradually reduced. In addition, the current context of forest certification encourages the adoption of plantation maintenance strategies respectful of the environment (Raj et al. 2020; Banerjee et al. 2020; Jhariya et al. 2019a, b).

Mulching is an effective method (McDonald and Helgerson 1990) and likely to respond to growing public concern for environmental protection. Mulching, used mainly in sericulture, horticulture, and market gardening, does not know the same extent in forestry. However, past studies conducted worldwide report, most often, the positive effect that exerts mulching on establishment and seedlings growth (Afocel 1978; Gallois et al. 1997; Leclerc 1997) and the realization at lower cost compared to traditional techniques for weed control (Afocel 1978; Leclerc 1997).

Mulching is the process of effective management of agroecosystem to reduce the negative consequences. In the absence of mulching the soil moisture gets depleted and therefore seed germination becomes problematic. Mulching often acts as a cover over the topsoil which reduces the impact of water (Rahma et al. 2019) and wind (Ranjan et al. 2017) erosion. The bare soil gets frequently eroded due to the absence of mulches. Further, fallow lands without mulching help to promote weed growth which reduces the nutrient pool of the soil. In these perspectives, mulches tend to conserve the soil nutrient by adding as biomass into the soil after their life cycle.

Thus, the objective of this chapter is to put emphasis on mulching use in forests and the role that may play in the conservation and sustainability of these natural resources.

8.2 Concepts of Mulching

Mulching is a technique used in planting and maintenance which consists of covering the soil surface with an organic or inorganic material (Fig. 8.1) continuously (film) or discontinuously (grains, fragments, etc.), in order to prevent the development of weeds (Fig. 8.2) that compete with young seedlings, limit soil

Fig. 8.1 Mulching practices of organic nature (**a**, **b**) and inorganic nature (**c**, **d**): (**a**) cork oak seedling mulched with Italian stone pine, (**b**) cork oak seedling mulched with lentisk; (**c**) cork oak seedling mulched with black plastic; and (**d**) cork oak seedling mulched with gravel (Mechergui 2008, 2016)

Fig. 8.2 Mulching effect on weed growth, 3 years after plantation: (**a**) mulched zeen oak seedling with Italian stone pine and (**b**) unmulched zeen oak seedling invaded by Montpellier cistus (Mechergui 2016)

water losses, and help conserve fresh soil, increase or regulate soil temperature, improve its structural stability and structure, and influence nutrients availability (Van Lerberghe and Gallois 1997).

8.3 Different Types of Mulches

The constituent materials of the mulches are organic, biodegradable, or inorganic, non-biodegradable, in particular synthetic ones to which one can add specific adjuvants stabilizing type dyes anti-UV, wax, or paraffin.

8.3.1 Plastic Mulch

Plastics belong to the chemical family of "polyolefins." The main plastics are polyethylene and polypropylene. Their intrinsic properties (lightness, unalterability, aspect, impermeability, thermal insulation, mechanical or chemical resistance, etc.) and their sales price will vary depending on the constituent polymer (polyethylene and polypropylene).

In order to improve or modify the intrinsic characteristics of these synthetic polymers, two types of products are mainly used in agricultural plastic products with prolonged use: the dyes and the anti-oxidant and anti-UV stabilizers.

A great number of plastics have a light transmission coefficient close to that of glass: they are naturally transparent. In order to modify their optical and radiative properties, they are colored with carbon black or titanium oxide (white).

Stabilizers enhance the resistance to heat and UV rays and therefore the durability of mulches. They avoid the weakening of the cohesion of the macromolecular structure of plastics by fragmentation of the carbonated chains and thus the modification of the mechanical properties (elongation, tensile strength). It must always be ensured that the plastic used is treated against UV, in order to avoid the risk of its photo-fragmentation in less than 6 months.

8.3.2 Organic Mulches

Organic mulches consist of vegetable fibers (leaves, branches, wood, bark, etc.). Used in the same way as plastic, they have, however, the main characteristic of being biodegradable that is to disappear in 2, 3 years or more (without leaving synthetic residues), but sometimes more quickly, after they have more or less fulfilled their protective mission. In some organic mulch stabilizing products such as wax or paraffin are frequently used in order to increase their durability. Unfortunately, these mulches are not always 100% natural, with the addition of non-biodegradable components such as tar (substance water repellent) or chemical resin. In theory, these materials are not colored.

Depending on their rigidity or thickness there are three main types of mulch, namely leaves, plates, and layers:

8.3.3 Mulch Sheets

Flexible and thin, these mulches fit well on the ground surface; they are inexpensive and easy to handle (Meena et al. 2020a, b). Depending on their composition, three categories of products are identified: plastic sheets, vegetable fiber sheets, and hybrid sheets.

Plastic Sheets

Often used in agriculture, these plastics are currently synthetic. If they all improve the availability of soil water, their effect on light, temperature, and weed development may vary with their color, opacity or thickness, water-tightness, durability (Fig. 8.3). These plastics are worked by appropriate techniques for two main types of products sold: films (polyethylene) and veils (polypropylene).

Polyethylene Films

They are usually sold in rolls of 0.80–3 m and 100–600 m long, more rarely in individual format. Their thickness is 20μm, 30μm, 50μm, and 80μm. The preferred quality is the standard quality (80μm) whose durability is greater than 2–3 years.

Fig. 8.3 Development of weeds around mulched cork oak seedling after the tear of sheet mulch (black polyethylene 50 μm thick) after only 1 year from plantation (Mechergui 2008)

Non-wettable and non-porous, they are impermeable to air and water. If they reduce water losses from soil by evaporation they also delay the replenishment of soil water reserves by intercepting a significant portion of rainfall. It must be so to avoid their use on already dry soils.

Some may be micro-perforated (holes 1 mm in diameter at 2500 holes/m^2), perforated (holes 1 cm in diameter at a rate of 250, 500, 750, or 1000 holes/m^2) to change their permeability. It is recommended to avoid the use of perforated films at the risk of seeing a development of weeds through the film. The micro-perforated films ensure a better distribution of rainwater while avoiding the appearance of weeds.

The biological effectiveness (survival and growth of plants) of plastic sheets and low cost make them the main type of mulch currently used.

Nonwoven Polypropylene Veils

A nonwoven veil polypropylene consists of polypropylene filaments distributed in the most isotropic possible way, hot pressed (thermo-welding) and whose diameter (<20–25 microns) is less than that of a hair. Packaged in reels 1–3 m wide and UV stabilized, their weight varies between 17 g/m^2 (standard quality) and 100 g/m^2. The preferred qualities are the heavy qualities (>30–50 g/m^2) for a durability greater than 2–3 years. The high porosity of the veils ensures their permeability to air and water. This permeability is a function of the weight of the veil; the higher the weight, the poorer the permeability. On the contrary, the higher the veil, the more the veil is thermal. Their porosity, finesse, and flexibility combined with mechanical performance often remarkable make it a promising product for arboreal use, despite a higher cost than a plastic film.

Vegetable Fiber Sheets

Woven or nonwoven, the models are varied. Sheets of newsprint, waxed kraft paper or waxed cardboard, cotton, etc., these mulches have the characteristic of being more or less rapidly biodegraded that is to say decomposed by living organisms of the soil. They do not leave unwanted residues in the environment, unlike plastics.

The most fragile sheets, especially, those with unpainted paper base, last less than a year. The incorporation or spraying of vegetable latexes or waxes can increase the resistance to degradability, the shelf life than being greater than 2 years. They are packaged in reels or in individual format whose weight can reach 2.5 kg/m^2.

The mulching effects on temperature and soil moisture rely upon materials that compose them. If they improve, like plastic sheets, the availability of soil water by reducing evaporation and limiting transpiration, they have the capacity to absorb more or less strongly rainwater by restoring to forest plant that (water) in excess.

This increase in moisture resistance slows down the warming of the soil because part of the incident solar radiation can be used to evaporate the absorbed water. Moreover, their generally pale color accentuates the reflection of the solar radiation towards the atmosphere with, as a result, a floor generally cooler than under the plastic sheets.

Hybrid Sheets

They consist of a combination of a plastic film, often propylene, aluminized or not with a cotton fabric, a sheet of paper, a layer of wax and/or paraffin. It should be noted that these products are not often found in agriculture.

8.3.4 The Plates

The plates are rigid and thick (>5 mm) (Fig. 8.4). They consist essentially of fibers or wood pulp, cork particles assembled by an organic binder or synthesis, associated or not with clay. Individually packaged, their shape is round, square, or octagonal, their diameter varying between 40 and 100 cm.

Their durability is greater than 2 years and their total opacity prevents any development of adventitious vegetation. Their high density contributes to greatly limit their penetration by the roots after germination of seeds located on the surface.

If we can blame them for not marrying the surface of the ground or for requiring a more thorough preparation of the soil, they gradually lose their rigidity after impregnation of rainwater.

Mulching and its various forms serve various purposes. For instance, plastic mulching preserves soil moisture, organic mulching leads to addition of organic matter, inorganic mulching performs various ecological services, and plate mulching inhibits the weed growth (Fig. 8.5). Further mulching activity leads to increase the yield under various agro-climatic conditions. Application of mulching should be done on scientific basis considering the types of crops, species, site conditions, soil types, agro-climate, and local ecological conditions.

Fig. 8.4 A young tree mulched with plate (https://www.filoche-et-ficelle.fr/paillage-naturel/368-442-paillage-jardin-disque.html)

8.4 Utility of Mulching

When using mulching in woody plantations, the forest manager/owner aims to:

- prevent the development of weeds that compete with young plants;
- limit soil water losses and help conserve fresh soil;
- increase or regulate the soil temperature, improve its structural configurations;
- influence soil nutrient availability and fertility. Thus, this soil cover or mulching is not limited only to the suppression of undesirable vegetation. It constitutes a real physical barrier to heat, water, and gas exchanges between the soil and the atmosphere of its immediate environment. It modifies production factors as important as temperature, humidity, soil structure, nutrient content, gas exchange,

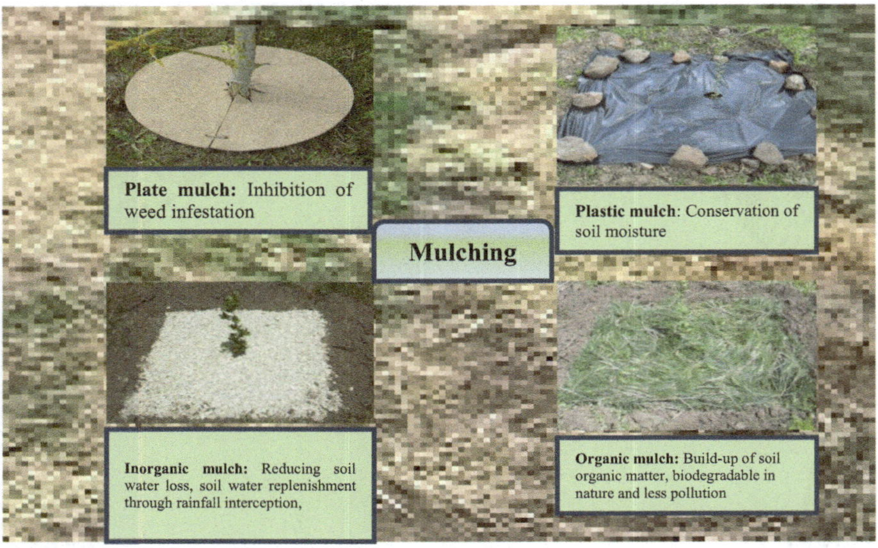

Fig. 8.5 Form of mulching and its benefit

and the root system, which exert a major influence on plant survival and growth (Table 8.1).

Thus, mulch is intended to mitigate the effects of certain adverse conditions on the development of plants: the competing vegetation, but also the cold, wind, driving rain, hail, excess solar radiation, and drought.

8.4.1 Inhibition of Competing Weeds

It is always beneficial to eliminate or better prevent the emergence of weeds in the immediate vicinity of plants because of the special neighborhood relationships that reflect the existence of competition.

This competition takes place at different levels:

- The consumption of water: water deficiency causes the woody plant to reduce the intensity of its metabolism and the constitution of its plant tissues (75–90% of their weight); it also hinders the circulation of the mineral substances of the raw and organic sap of the elaborated sap. In its acute form, the lack of water causes wilting of the tissues, which when irreversible leads to the death of the plant;
- The consumption of nutrients: nutrient deficiencies are frequently expressed by discolorations, defoliation, and loss of growth of the forest plant;

Table 8.1 Mulching technology and its application

S. No.	Technology used	Impact	Source
1	Mulching of 0.6 inch above ground	86% reduction in soil erosion	Borst and Woodburn (1942)
2	The use of various materials for covering the ground	It reduces the population of weed and retains moisture. It also helps in increasing yield	Kader et al. (2019)
3	Mulching	Reduces runoff water, increases soil infiltration capacity, and also regulates evapotranspiration	Rathore et al. (1998)
4	Weed mulching	Weeds used as mulching materials increase evapotranspiration, improve soil water infiltration rate, increase soil water level	Harris et al. (2004)
5	Straw mulch layer	Straw mulch comprising of 1.5 inch thickness has the potentiality to reduce evaporation up to 35%	Russell (1939)
6	Living mulches/turf	Improves the water retention capacity of the soil Turf mulches help to lower the surface temperature through evaporation	Kacinski (1951), George et al. (2003), Montague and Kjelgren (2004), Cregg and Dix (2001)
7	Black plastic mulch	Inhibits the exchange of water between soil and atmosphere	Banko and Stefani (1991)
8	Organic mulches/ geotextiles/plastics	Reduces evaporation and improves soil water retention	Lakatos et al. (2000)
9	Various types of organic mulching (i.e., leaf, jute, hay, straw, grasses, coir pith, barks, softwood, hardwood, etc.)	Organic mulches tend to conserve water. Under tropical condition organic mulches tend to reduce the soil temperature in comparison to bare soil	Lakatos et al. (2000), Oliveira and Merwin (2001), Rokich et al. (2002), Downer and Hodel (2001)
10	Mulching	By retaining water in the soil mulching reduces water demand for irrigation	Pfammatter and Dessimoz (1997)
11	Coarse organic mulches	It acts as sponge and helps to capture precipitation and irrigation water. It also reduces the amount of water runoff. It has been	Oliveira and Merwin (2001), Tilander and Bonzi (1997)

(continued)

Table 8.1 (continued)

S. No.	Technology used	Impact	Source
		reported that straw mulch of 0.6 inch thickness has the capability to reduce up to 43% of runoff Bark and jute help in aggregation of soil particles and improve the soil porosity Coarse mulches have the temperature regulation ability with organic system tends to reduce the temperature significantly	
12	Mulch protection	Mulching helps to protect the vegetal community from drought and cold stress	Smith (2000)
13	Living mulches	They provide absolute protection against erosion, hold the soil particle under steep slope Straw mulch reduces soil erosion more than 80% Logging debris helps to reduce runoff	Borst and Woodburn (1942), Rothwell (1978), Tanavud et al. (2001)
14	Chunky inorganic mulch	They are the most effective temperature regulators, while the synthetic ones are the poor performers	Montague and Kjelgren (2004), McGovern et al. (2002)
15	Increased soil nutrition through mulching	Variable results were obtained in terms of role of organic mulches towards nutrient content in soil. In this connection green materials and animal waste products perform better. Undecomposed bark or straw materials reduce the nitrogen level in soil water to avoid runoff pollution. Conversely, mulches with higher nitrogen improve the yield and productivity. Some of the low nitrogen level mulches such as bark, sawdust, and straw may also promote elevated	Downer and Hodel (2001), Pfammatter and Dessimoz (1997), Tilander and Bonzi (1997), Ansari et al. (2001), Pickering and Shepherd (2000), Szwedo and Maszczyk (2000)

(continued)

Table 8.1 (continued)

S. No.	Technology used	Impact	Source
		nutrient level in soil and vegetation	
16	Lowering salt and contamination of pesticide through mulching	Mulching helps to reduce the negative impact of salt toxicity on growth of plants. Further, they increase the desalinization of soil. Organic mulches act as metal binders and degraders of pollutants through associated microbial activity	Ansari et al. (2001), Gan et al. (2003)
17	Heavy metals binding through mulching	Organic mulches are found to be effective in removing heavy metals. For example, leaves of eucalyptus used as mulching purpose are found to be effective in removing lead. Further, compost and wood chips tend to reduce copper load in forest soils	Salim and El-Halawa (2002), Kiikkila et al. (2002)
18	Plant germination and development through mulching	Mulches have been reported to promote establishing woody and herb species. In this process they improve the germination of seeds, enhance seedling survival rate, help in root proliferations, and increase the success in transplanting materials. Mulching associated with manure or sawdust improves growth and development of oak species. Mulches also help in weed control after post-seeding conditions	Kacinski (1951), Rokich et al. (2002), Chalker-Scott (2007)
19	Growth and develop of root system through mulching	Mulches help proper growth and development of roots of plants and their subsequent stabilization. It was reported that the density and development of root was higher in organic mulching	Fausett and Rom (2001)

(continued)

Table 8.1 (continued)

S. No.	Technology used	Impact	Source
20	Growth performance of plant under mulching	Mulching has positive impact in terms of better growth of plants both under field and nursery conditions. Various growth attributes reflected positive results under mulching	Cahill et al. (2005), Downer and Hodel (2001), Pfammatter and Dessimoz (1997), Downer and Hodel (2001), Tilander and Bonzi (1997)
21	Disease reduction through mulching	Mulches arrest the splashing action of rain and irrigation water which may be the potential source of infection of diseases. Conversely, mulches help to reduce soil pathogens and promote beneficial colonization of the microbes in the soil. This therefore helps in disease reduction. Mulches have been reported to maintain optimum soil conditions to promote plant growth and development along with reduction in susceptibility towards diseases. Researchers have revealed that straw and wood chips mulch is very much effective towards inhibiting diseases	Downer and Hodel (2001)
22	Weed reduction through mulching	Research report reveals that mulching is very much effective towards reducing weed growth. Wilen et al. (1999) reported more than 90% weed reduction under laboratory conditions. This process takes place through inhibition of light by the mulches. Some mulches tend to have allelopathy impact over the weed population Crop residues and forest produce have been reported to have inhibitory effect on weed population	Wilen et al. (1999), Horowitz and Thomas (1994)

(continued)

Table 8.1 (continued)

S. No.	Technology used	Impact	Source
23	Reducing use of pesticide through biological control by mulching	Mulches help to reduce the stress of weed growth or other pest infestation either biologically or chemically. In this way they reduce the necessity of use of various agrochemicals in the form of pesticides and insecticides	Chalker-Scott (2007)
24	Scenic beautification through mulching	Mulches tend to have positive influence in terms of beautification of the surroundings by improving the local esthetics. For example, tumbled glass can be effective both in terms of esthetic pleasure and protection of soil	Chalker-Scott (2007)
25	Valuation of mulching	It is proven fact through research about the increasing yield and productivity of crop under mulching. Local based wood debris would be an effective mulching material in terms of cost reduction and giving more output. Research report reflects utilization of brush mulch in beautification and revegetation of roadside. Paddy straw under tropical conditions may tend to have higher benefit–cost ratio at field level application	Rothwell (1978)

- The absorption of solar energy: weeds can intercept visible light (white light) which provides the energy necessary for the formation of chlorophyll and photosynthetic activity;
- The occupation of airspace: the spatial development of a weed can induce a change in the circulation of water, air, light energy, thermal layers; too close proximity can result in physical damage (crushing or chafing) of forest plants;
- The occupation of underground space: weeds often have faster root development that gives them an advantage over plants in the competition for water, nutrients, and light. Soil cover removes competitive weeds, essentially by inhibiting light

source needed in physiological process as well as regulating growth to some extent. By definition, only mulch that is absolutely opaque to solar radiation is able to stifle the germination of annual and biennial weeds or the growth of perennials.

8.4.2 Limits Soil Water Losses and Helps to Conserve Soil

Water is essential for the constitution of the plants and the realization of all the physiological and biochemical processes necessary for their growth. In many climatic conditions, rainfall is insufficient or too irregular to cover their continuous water requirements during the growing season. The water improvement of the plant is done, via the roots, from the reserves stored in the porous spaces of the soil.

In general, mulch increases soil water availability by limiting plant transpiration and atmospheric evaporation:

- The water absorbed by the roots and conveyed in the plant can be evacuated by the leaves, in the atmosphere in the state of vapor, we speak of vegetal transpiration. It is by preventing competing vegetation that mulches reduce water losses due to weed transpiration;
- The atmosphere has an evaporating power that is explained by the fact that the air is only exceptionally saturated with water vapor: it is called atmospheric evaporation. By covering the soil, mulch prevents direct evaporation of water from its surface. Thus, the moisture of the soil decreases less rapidly in the ground than in bare soil (Henin and Monnier 1961).

This water saving strategy must also be related to:

- The shading effect of the cover: by intercepting the solar radiation, the mulch decreases, during the summer period, the energy available to vaporize the water;
- The wind protection effect: the wind can no longer move the layer of air above the ground containing the water vapor from the ground.

The influence of mulch is all the more sensitive in certain pedoclimatic conditions: low fertility soils with low water retention capacity (sandy, stony, superficial, very steep), hot and dry climates.

Laying mulches will be done on cool soil, after rainfall, for example, especially if the soil drains quickly. Avoid covering hydromorphic soil because mulch can cause temporary soil engorgement and poor aeration of plant roots.

8.4.3 Increases or Regulates Soil Temperature

The functioning and development of the root system of a plant are directly influenced by soil temperature. The temperature increase intensifies the absorption

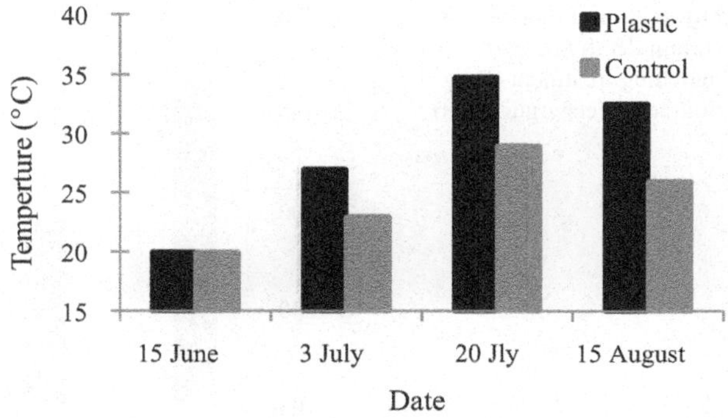

Fig. 8.6 Soil temperature at certain depth (10 cm) under mulching treatment (Data Sources: Mechergui 2008)

of water and nutrients by increasing the permeability of the root walls and decreasing the viscosity of the water. The warming of the soil will depend on its thermal properties (heat capacity and thermal conductivity), its humidity and external conditions.

By its mere presence on the soil, mulch constitutes a "screen" capable of modifying energy exchanges in both directions between the surrounding environment (atmosphere) and the plant, in particular with regard to the transmission of solar radiation and heat flux.

In general, mulches have an important effect on soil temperature compared to an unprotected surface, either by raising the maximum or minimum temperature, or by reducing daily fluctuations or again, by exerting both effects at the same time (Robitaille 1994). The sense and magnitude of the effects vary according to the nature (organic or inorganic), the constitution (color, thickness, perforation) of the mulch, and the time of year: the phenomena of absorption, reflection, and transmission of solar radiation are directly a function of the intrinsic properties of mulch. It is therefore necessary to know the thermal effects of marketed mulches in order to choose a product that is well adapted to the afforestation of former agricultural lands.

Agricultural plasticulture farming already provides several responses, particularly concerning the influence of color on the behavior of synthetic films. In general, black plastic, the most commonly used mulch in weed control increases both daily soil temperatures (Fig. 8.6) and moisture (Fig. 8.7). These positive temperature and moisture differences from uncovered soil remain sensitive throughout the year, creating good conditions of activity and root growth.

Organic mulches play mainly a role in temperature regulation. They reduce the changes in soil temperature induced by atmosphere by lowering maximum temperatures and/or raising minimum temperatures (Van Lerberghe and Gallois 1997).

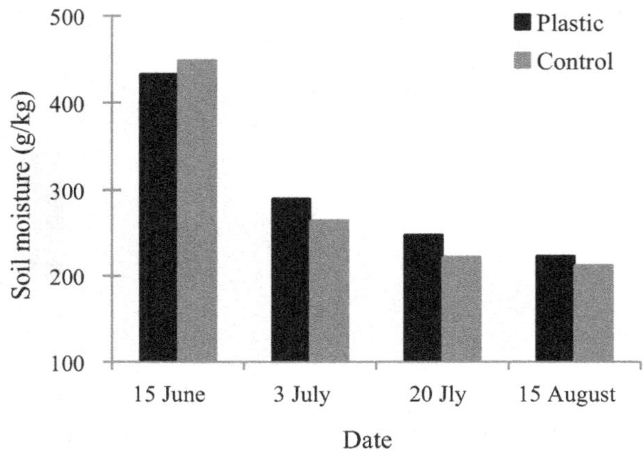

Fig. 8.7 Soil moisture at certain depth (10 cm) under mulching treatment (Data Sources: Mechergui 2008)

8.4.4 Improve the Structural Stability of the Soil

Mulches act as a protective screen against bad weather (beating rain and hail) while they preserve soil structure, even after soil preparation for planting. They serve as a shield on which the drops of water come to break: by crashing against the cover, they lose most of their energy. The structure in surface is thus much less degraded. In particular, the phenomena of flapping on fragile soils are avoided as silty soils with very fine sands are insufficiently provided with clays or humus. Under the effect of violent precipitations the fine elements are piled up between the aggregates. The soil compacts on the surface and becomes impervious, to the detriment of the supply of water and oxygen to roots.

8.4.5 Improve the Soil Structure

The improvement of the structure of a soil goes hand in hand with the increase of its rate of organic matter. Only the layers of decomposable organic material such as vegetable debris and crop residues (manure, straw, crushed green manure) are likely to improve soil properties by composting.

Surface composting consists of allowing organic matter to decompose on the surface of soil, which will gradually enrich the profile. Gradually disintegrated, they associate intimately with the earthy particles that transform into solid and stable aggregates, thus inducing an increase of the porosity in favor of a better circulation in the soil environment.

8.4.6 Impact on Soil Nutrient Availability and Its Fertility

Brought by some biodegradable mulches or simply present in the soil before mulching a plantation, the organic matter will be gradually decomposed and mineralized.

In general, mulch soil cover increases the rate of decomposition of organic matter and accelerates its mineralization by maintaining moisture and temperature conditions conducive to the activity of soil micro-organisms, which is desirable and beneficial for the growth of young woody stems.

Inducing different soil and temperature conditions in the soil, the different types of mulch have varying effects on the mineralization and nutrient availability, depending on the fermentable or non-fermentable nature of the constituent material and the duration of mulch. There is a lacuna regarding the influence of market mulch upon soil fertility, it has been shown that several behaviors are possible in mulches consisting of layers of organic material, in particular the availability of nitrogen and potassium in the soil.

Nitrates

Among the positive effects of weed control with mulching it is the enrichment of the soil with nitrates, either by increasing the biodegradation of humus by soil biota associated with increase of temperature or by the decomposition of the mulch (organic mulch). Indeed, organic mulches are often used to enrich the soil with mineral elements following their decomposition. By covering the soil with straw, Truax and Gagnon (1993) found that soil nitrate availability increased from 2.8 to 7.07 mg/kg.

Phosphorus and Potassium

Tuckey and Schoff (1963) tested decomposable and non-decomposable mulches. Decomposable materials included legume hay, peanut hulls, corn cobs, straw, and sawdust, where levels of phosphorus and potassium that can be used by plants were, respectively, of 27.5 and 63.3 ppm, 23 and 50.3 ppm, 26.7 and 44.3 ppm, 24.8 and 63 ppm, 23.2 and 51.7 ppm, while non-decomposable materials included foam rubber and gravel, under which phosphorus and potassium were, respectively, of 24.9 and 47.3 ppm, 21.5 and 47 ppm. Both levels of phosphorus and potassium are higher for all decomposable or non-decomposable mulch than for control treatment, where phosphorus and potassium were, respectively, of 19 and 43.7 ppm, although this increase being higher when the materials are decomposable.

8.5 Setting up of Mulch

For mulch to be well applied, it must be preceded by a preparation of the soil around the plant. In addition, to avoid being displaced by wind or runoff, mulch should be well fixed to the soil. However, it is necessary to grant a big importance for the surface of control of vegetation competition; according to Davies (1987), for the

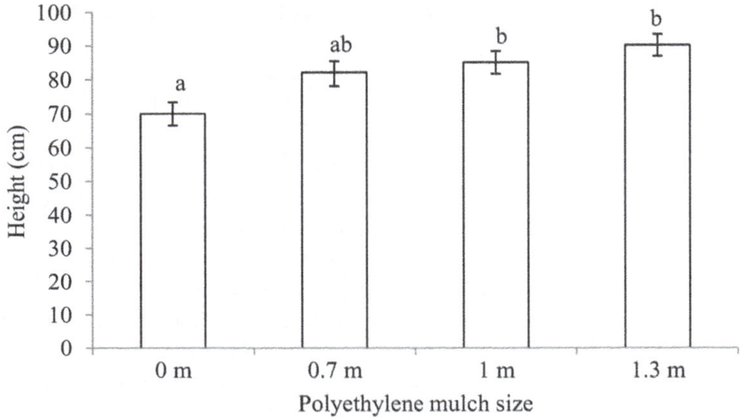

Fig. 8.8 Height growths according to polyethylene mulch size. Means having significant difference ($P \leq 0.05$) was represented by different letters (Data Sources: Mechergui 2008)

Fig. 8.9 Weed control, using black polyethylene mulch, and tree shelters (non-vented (left) and vented (right) tree shelters) used with combination (Adapted, Mechergui 2008)

control of competition vegetation to be effective, it must be applied on a minimum surface of $1m^2$. This has also been recently proven by Mechergui (2008), where the beneficial effect of vegetation control, using black polyethylene mulch, was efficient for a mulch size equal or greater than 1 m^2 (Fig. 8.8). This trend seems to be true when weed control is used alone. However, when used in combination with tree shelters (Fig. 8.9) the effectiveness of weed control applied on such a surface around plant may be dependent on used tree shelter type. For instance, for non-vented tree shelters a significant improvement of seedlings growth was achieved even under a weed control on a surface inferior to 1 m^2 (0.7 m^2), compared to these tree shelters type when used alone (Fig. 8.10). However, for vented tree shelters seedlings growth

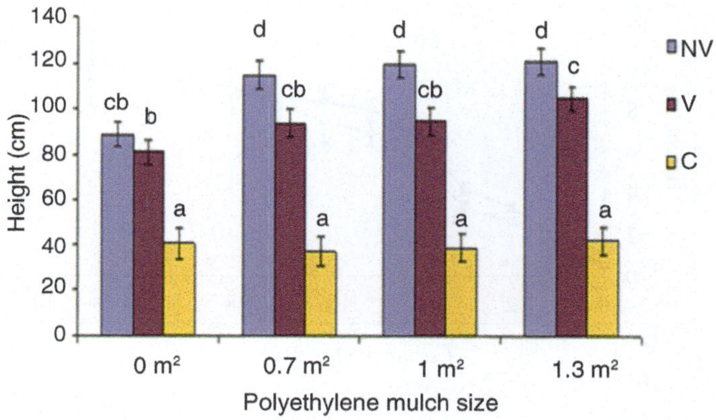

Fig. 8.10 Height growth of cork oak seedlings according to combination of tree shelter type (*NV* non-vented, *V* vented, *C* control) and polyethylene mulch size. Means marked with different letters varied significantly ($P \leq 0.05$) (Adapted, Mechergui 2008)

Table 8.2 Survival rate (%) of different tree species with or without black plastic mulch, 8 months after plantation (Albouchi and Abbassi 2000)

	Treatment		
	With mulch (1)	Without mulch (2)	Gain in survival (1)–(2)
Casuarina equiset	86	40	46
Casuarina glauca	98	90	8
Casuarina stricta	85	80	5
Casuarina torulosa	88	74	14
Cupressus sempervirens	99	86	13
Eucalyptus gomphocephala	84	59	25
Mean	90	71.5	18.5

was significantly improved when the weed control was performed on a surface of 1.3 m^2.

8.6 Influence of Mulching Over Growth and Survivability of Plant

Vegetative competition control using mulching has favorable effects on both survival (Table 8.2) and seedling growth (Figs. 8.11, 8.12). One year after implementation, the gain on the survival rate was 15% in *Quercus robur*, 12% in *Quercus petraea*, 51% in *Prunus avium*, 21% in *Robinia pseudoacacia*, and 33% in *Tilia platyphyllos* (Leclerc 1997). This author also reported a gain of 11% in *Alnus incana*, although this gain was not significant.

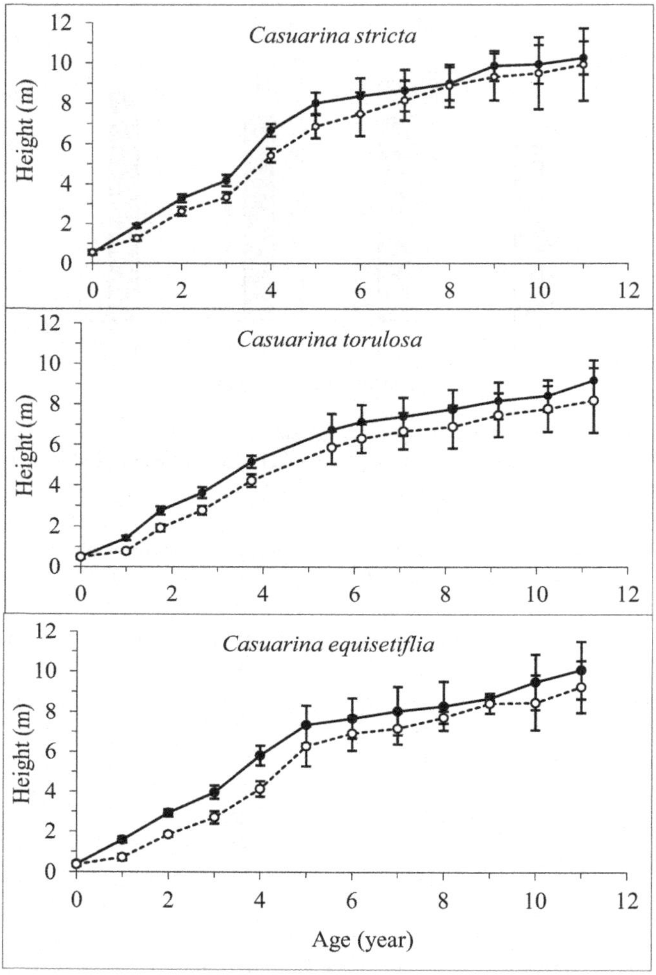

Fig. 8.11 Evolution, over time, of height growth *Casuarina stricta* Dry and, *Casuarina torulosa* Aiton and *Casuarina equisetifolia* Forst planted in windbreaks with (—) or without black plastic mulch (- - - - -) (Albouchi and Abbassi 2000)

After 1 year, Leclerc (1997) noted a significant gain in height of 16 cm in *Quercus robur* L., 12 cm in *Quercus petraea*, 51 cm in *Prunus avium*, 57 cm in *Robinia pseudoacacia*, and 31 cm in *Alnus incana* and *Acer campestre*. For the first three species, the total height of the mulched plants, measured after 3 years, was a little more than double the control plants. Gallois et al. (1997) reported an increase in height growth of 16 cm/year, on average, in mulched *Prunus avium* and *Acer campestre* comparatively to control, after 3 years of planting. In *Pinus pinaster*, gain in growth due to mulching was 4.4–7.2 cm after 1 and 2 years, respectively, and varied from 40 to 85 cm after 7 months in various species of *Populus* (Afocel 1978).

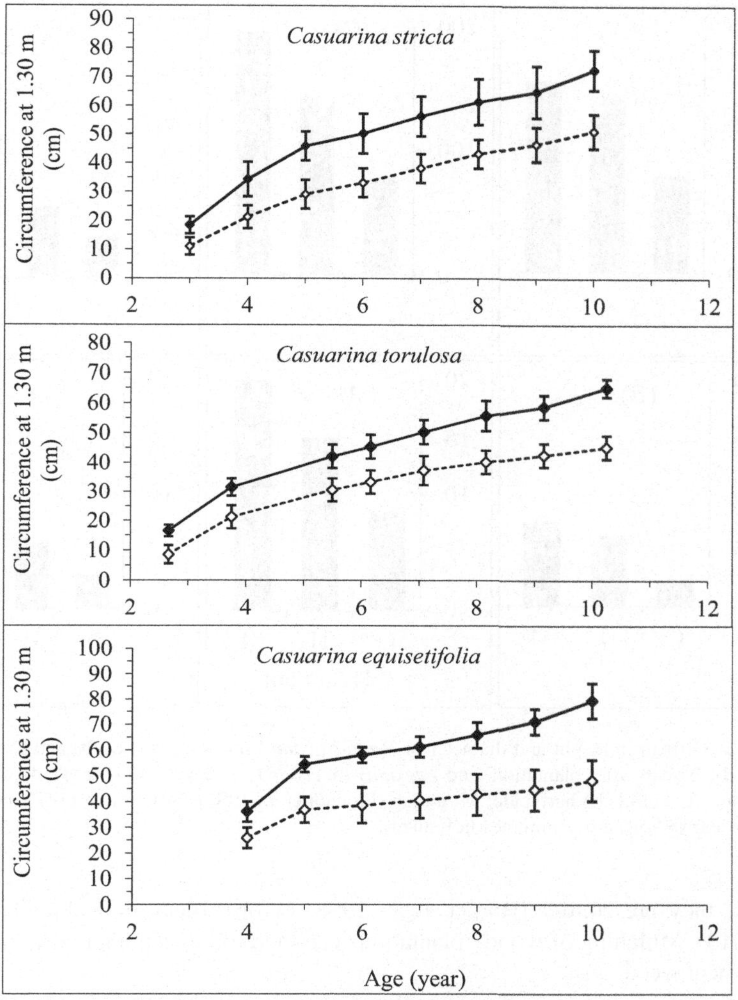

Fig. 8.12 Evolution, over time, of the circumference *Casuarina Stricta* Dry and *Casuarina torulosa* Aiton and *Casuarina equisetifolia* Forst planted in windbreaks with (—) or without black plastic mulch (- - - - -) (Albouchi and Abbassi 2000)

Beneficial effects of mulching are, in fact, clearer and more durable on growth in circumference measured at breast height (1.30 m) than on height growth; this is because circumference of forest seedlings seems to be more sensitive to competition than their height (Albouchi and Abbassi 2000).

The comparison between weeding using mulching and chemical weeding which is suitable techniques as it has received criticism due to its non-eco-friendliness as previously mentioned shows that height and diameter growths of mulched trees are often comparable and even greater than that measured on trees weed chemically,

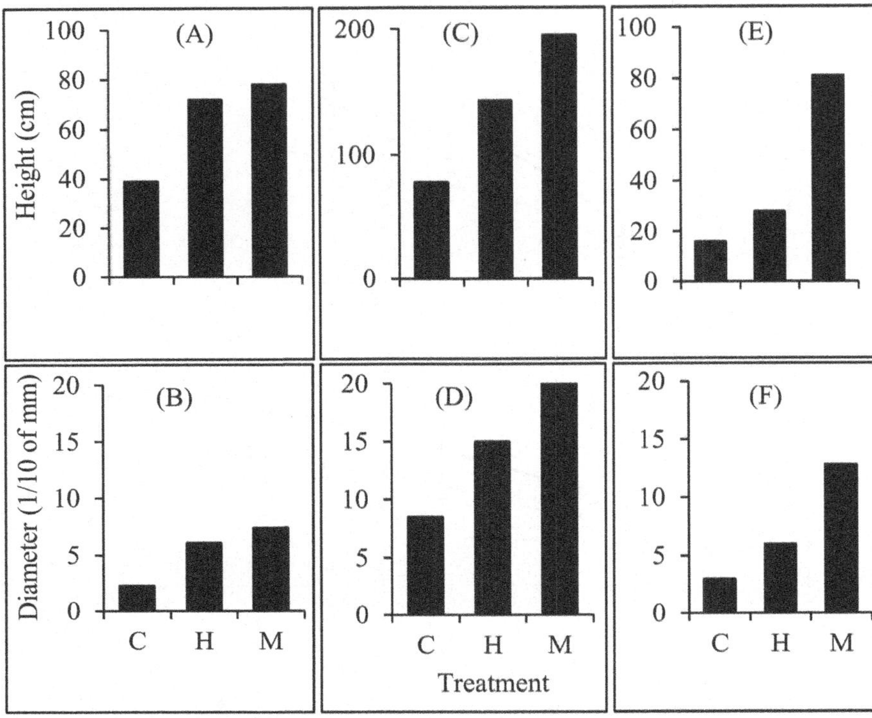

Fig. 8.13 Growth in height and diameter of *Prunus avium* L.in Foug (54) (**a**, **b**) and Commercy (55) (**c**, **d**), 2 years after plantation, and *Fraxinus excelsior* L. in Parroy (54) (**e**, **f**), 4 years after plantation. C: control; H: herbicide; M: mulch (Frochot et al. 1992). Note: Foug (54), Commercy (55), and Parroy (54) are communes in France

whether they are conifer (Harper et al. 1998) or broad-leaved (Robitaille 2003) (Fig. 8.13). Mulching of woody plantations can therefore be an alternative solution to chemical weeding.

Improvement of survival and seedlings growth may be attributed to optimum growth conditions offered by mulching (Robitaille 1994). Indeed, many advantages are offered by mulching: (1) it contributes significantly towards soil moisture loss reduction, allowing plant to better take advantage of rainwater and water use (Awe et al. 2015; Jemai et al. 2013; Li et al. 2015), (2) it enhances temperature of upper soil layer. This therefore regulates seed germination physiology as well as seedling growth (Zhang et al. 2009; Siczek et al. 2015), (3) it changes the soil organic matter and its associated soil fertility and quality (An et al. 2015; Huo et al. 2017; Muñoz et al. 2017), (4) it retains the soil organic carbon equilibrium and promotes nutrient recycling (Naab et al. 2015; Wang and Xing 2016), (5) it favors soil metabolism and enzyme activities (Masciandaro et al. 2004; Elfstrand et al. 2007), (6) it suppresses weed growth (Jabran et al. 2015; Splawski et al. 2016; Nawaz et al. 2017).

Recently, worldwide forest resources assessment which is coordinated by FAO (2015) found that forest area passed from 31.6% to 30.6% between 1990 and 2015, a loss of 1% in 25 years. The conservation and sustainability of this natural resource lie partly in the planting of degraded forest sites. However, the success of a forest plantation requires, among others, the control of competing vegetation for at least the first 2–3 years, corresponding to the delicate phase of installation of young trees (Von Ahlten 1990). Encouraging results obtained for establishment and plant growth of many forest species highlight the crucial role that mulching can play as a promising tool for natural resource management and sustainability. Indeed, a review conducted by Wagner and Robinson (2006) compiling 60 research findings related to weed control across the globe showed, in 75% of cases, a 30% to 500% increase in productivity in terms of volume of timber when vegetative competition was controlled. It is therefore very important to consider vegetation management, especially using mulching, in the forest plantations.

8.7 Challenges in Mulching Techniques

Mulching technology has several benefits along with various challenges and issue which may reduce the utility of mulching. Research work has reported various form of negative interaction in terms of acidification, competition, disease, and pest outbreaks besides economic input, etc. (Table 8.3).

Considering the negative impacts one needs to take care about proper method of application, types of materials used for mulching, time of mulching, rate of mulching, compatibility of the specific technology, etc., and other associated scientific facts related to mulching (Chalker-Scott 2007).

8.8 Mulching and Sustainability

Addressing sustainability in agroecosystem and other land-use requires development of eco-friendly technologies to improve the health of the system (Khan et al. 2020a, b). For instance, mulching can be effectively utilized for maintenance of soil conditions and health leading to sustainable production. It has been found that mulching performs diverse ecological functions in terms of soil conservation and management which can be sustainably addressed through mulching (Ngosong et al. 2019).

Development of suitable techniques and methodology is required in order to achieve sustainability functions on case to case basis depending upon the local ecological conditions. It can be used as a suitable sustainable approach for conservation of soil and water (Meena et al. 2020; Kumar et al. 2020). It also helps proper utilization of nutrient and other resources within agroecosystem which reflects proper management and conservation of natural resources (Kader et al. 2017).

At a stress mulching performs various functions in an integrated way which makes it a sustainable tool for soil management (Fig. 8.14). Mulching tends to

Table 8.3 Challenges of mulching

S. N.	Impact	Remarks	References
1	Acidification	Mulches of organic nature such as bark and wood chips tend to acidify the soil. Some other research study reveals that organic mulches may convert the soil pH towards alkaline in nature. The nature of acidification through organic mulch depends upon the phenolic acid content	Pickering and Shepherd (2000)
2	Allelopathy	Allelopathy is a significant problem that influences the regulation of weeds seed germination and growth through organic mulches. Specific chemical such as juglonic acid is secreted from black walnut (*Juglans nigra*) which may inhibit the understory vegetation. Ground vegetation without proper development of the root system may be affected by the allelopathy impact of the organic mulches	Harris et al. (2004)
3	Competition	Resource partitioning becomes a critical issue between the living mulch and local inhabitants. Scientific reports reveal the involvement of the grass hampering the growth of tree species, inhibitory role of turf due to their competitive nature which severely affects the shrub and tree population	Chalker-Scott (2007)
4	Chemical contamination	Woody mulches can be a potential source of environmental contaminants as revealed from a study conducted in Florida about arsenic accumulation in woody mulches. Similar type of reports was reported for wood mill waste which can negatively impact the upper storey of the vegetation	Townsend et al. (2003)
5	Disease	Mulches made up of materials containing disease pores or pathogens may spread the disease to other organisms. Therefore, before application of mulches heat treatment is mandatory	Brantley et al. (2001), Davis et al. (2005)
6	Flammability	Research reports reveal the flammable nature of mulches. Among them rubber mulches have high flammability. After rubber mulch various forms of organic mulches	Brantley et al. (2001)

(continued)

Table 8.3 (continued)

S. N.	Impact	Remarks	References
		containing water reflect flammable nature	
7	Nitrogen deficiency	It has been reported that woody mulches due to their high C:N ratio tend to cause nutrient deficiency for the plants. However, proper investigation in this matter has nullified the impact of woody materials over nutrient deficiency	Szwedo and Maszczyk (2000), Pickering and Shepherd (2000)
8	Pests	Organic mulches in association with wood products may attract pests. Mulches with high nutrient content may be highly fruitful for termite population to grow and infest the vegetation	Long et al. (2001)
9	Weed contamination	Mulches can be a potential source of weed contamination by carrying weeds seed and spores with them. It is seen that untreated mulch usually contents pathogens and helps in spreading of diseases. In this category crop residues and uncomposted material are the potential carriers of weed seeds	Zaragoza et al. (1995), Horowitz and Thomas (1994), Chalker-Scott (2007)

preserve soil moisture, reduces erosivity through wind and water, adds organic matter after their decomposition and many more fruitful functions for soil ecosystem.

8.9 Conclusion

The elimination of competition vegetation is a prerequisite for a good recovery and a uniform growth of seedlings. To be efficient, weed control must be applied to a minimal area of 1 m^2, and when the competition vegetation is more active. Many weed control methods are employed (manual, mechanical, chemical, etc.). Mulching has gained, however, considerable popularity due to its ability to create conducive conditions around mulched plant including especially microclimatic conditions (temperature, soil moisture). These benefits which cannot be offered by another technique of weed control favor the establishment and plants growth.

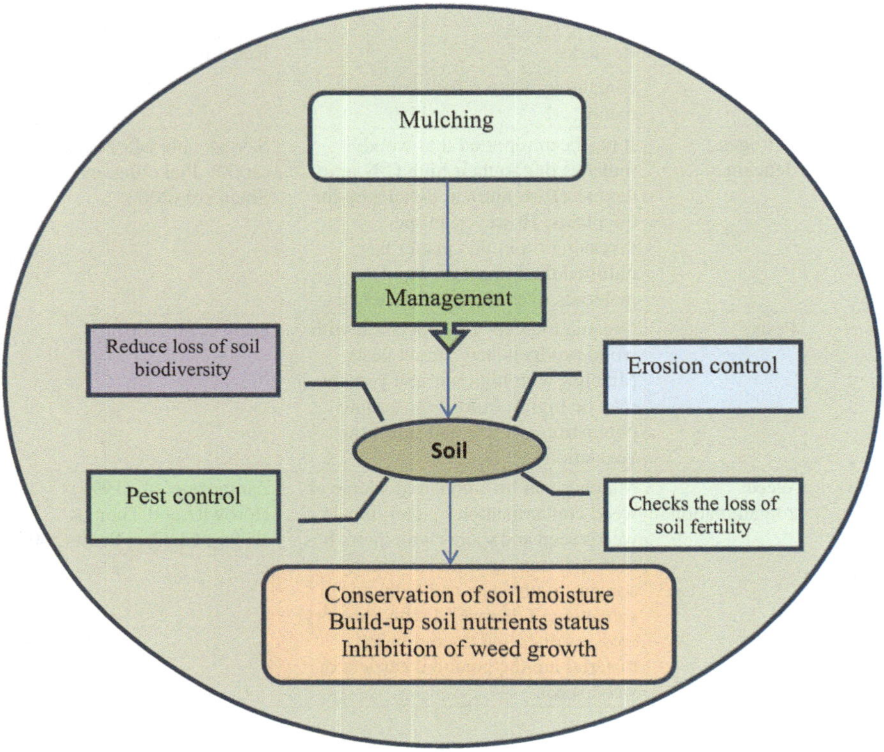

Fig. 8.14 Mulching towards sustainability

8.10 Future Perspective

Mulching is a promising tool in terms of soil moisture conservation as well as prevents soil erosion due to wind and water. Various technologies have been adopted across the globe in relation to mulching methods and techniques which help to maintain the soil resource as well as boost up the agricultural sustainability.

From future perspective screening of suitable techniques of mulching on the basis of habitat conditions needs to be formulated for sustainable management of soil and agroecosystem. Under arid climatic condition moisture stress is a significant factor hampering the agricultural productivity. In this context potentiality of mulching needs to be explored as it is an efficient conservator of soil moisture. Further, screening of suitable species for effective mulching needs to be explored and implemented depending upon the ecosystem types.

Moreover, plant species used for mulching purpose often utilize some portion of soil nutrient pool which may hamper the crop production and productivity. This is also a serious issue which needs to be explored properly from future perspective for sustainable production of agricultural crops. Further, the cost-effectiveness of each technique adopted under mulching needs to be analyzed properly through cost–benefit analysis. Therefore, based on the results, recommendation and extension of suitable technology of mulching can be formulated and implemented under diverse situations.

References

Afocel A (1978) Le paillage plastique des plantations forestières. Informations-Forêt Fasc 113 (3):155–162

Albouchi A, Abbassi M (2000) Effet du paillage plastique noir sur la survie et la croissance de six espèces forestières plantées en brise-vent en région semi-aride. Annales de l'institut national de recherches en génie rural, eaux et forêts 4:40–61

An T, Schaeffer S, Li S, Fu S, Pei J, Li H, Zhuang J, Radosevich M, Wang J (2015) Carbon fluxes from plants to soil and dynamics of microbial immobilization under plastic film mulching and fertilizer application using 13C pulse-labeling. Soil Biol Biochem 80:53–61

Ansari R, Marcar NE, Khanzada AN, Shirazi MU, Crawford DF (2001) Mulch application improves survival but not growth of *Acacia ampliceps* Maslin, *Acacia nilotica* (L.) Del. and Conocarpus lancifolius L. on a saline site in southern Pakistan. Intern For Rev 3:158–163

Awe GO, Reichert JM, Timm LC, Wendroth OO (2015) Temporal processes of soil water status in a sugarcane field under residue management. Plant Soil 387:395–411

Banerjee A, Jhariya MK, Yadav DK, Raj A (2020) Environmental and sustainable development through forestry and other resources. Apple Academic Press, Boca Raton. ISBN: 9781771888110, p 400. https://doi.org/10.1201/9780429276026

Banko TJ, Stefani MA (1991) Effects of container medium peat content and bed surface on plant growth during capillary irrigation. J Environ Hort 9:33–36

Borst HL, Woodburn R (1942) The effect of mulching and methods of cultivation on runoff and erosion from Muskingum silt loam. J Agric Eng 23:19–22

Brantley EA, Davis DD, Kuhns LJ (2001) Influence of mulch characteristics on sporulation by the artillery fungus *Sphaerobolus stellatus*. J Environ Hort 19:89–95

Cahill A, Chalker-Scott L, Ewing K (2005) Wood-chip mulch improves plant survival and establishment at no-maintenance restoration site (Washington). Ecol Restoration 23:212–213

Chalker-Scott L (2007) Impact of mulches on landscape plants and the environment—a review. J Environ Hort 25(4):239–249

Cregg BM, Dix ME (2001) Tree moisture stress and insect damage in urban areas in relation to heat island effects. J Arboriculture 27:8–17

Davies RJ (1987) Trees and weeds-weed control for successful tree establishment. In: Forestry commission handbook. HMSO, London, p 36

Davis DD, Kuhns LJ, Harpster TL (2005) Use of mushroom compost to suppress artillery fungi. J Environ Hort 23:212–215

Downer J, Hodel D (2001) The effects of mulching on establishment of Syagrus romanzoffiana. Cham. Becc., Washingtonia robusta H. Wendl. and Archontophoenix cunninghamiana (H. Wendl.) H. Wendl. & Drude in the landscape. Sci Hortic 87:85–92

Elfstrand S, Båth B, Mårtensson A (2007) Influence of various forms of green manure amendment on soil microbial community composition, enzyme activity and nutrient levels in leek. Appl Soil Ecol 36:70–82

FAO (2015) Évaluation des ressources forestières mondiales 2015: Comment les forêts de la planète changent-elles? Deuxième édition. FAO, Rome. www.fao.org/3/a-i4793f.pdf

Fausett JB, Rom CR (2001) The effects of transitioning a mature high-density orchard from standard herbicide ground-cover management system to organic ground-cover management systems. Arkansas Agric Expt Sta Res Series 483:33–36

Fiddler G, McDonald P (1987) Alternative treatments for releasing conifer seedlings: a study update. In: Proceedings of the 8th annual forest vegetation management conference, Sacramento, CA, pp 64–69
Frochot H (1984) Influence de Festucapratensis sur le développement de jeunes peupliers. 7ème colloque international sur l'écologie, la biologie et la systématique des mauvaises herbes. Columa-ewers pp 307–313
Frochot H, Levy G, Lefèvre Y, Wehrlen I (1992) Amélioration du démarrage des plantations de feuillus précieux: cas du frêne en station à bonne réserve en eau. Rev For Fr XLIV:61–65
Gallois F, Schmutz T, Basin P (1997) Nouveaux matériaux de paillage : premiers résultats d'essai en milieux ouverts. Forêt-entreprise 116:34–41
Gan J, Zhu Y, Wilen C, Pittenger D, Crowley D (2003) Effect of planting covers on herbicide persistence in landscape soils. Environ Sci Technol 37:2775–2779
George AP, Nissen RJ, Mowat A, Collins RJ, Collins R (2003) Innovative production systems for non-astringent persimmon. Acta Hortic 601:151–157
Harper GJ, Comeau PG, Biring BS, Reid WJ, Fielder P (1998) A comparison of mulch mat and herbicide treatments for reducing grass competition in the wet warm interior Douglas-fir subzone. Extension Note, 27, British Columbio, p 7
Harris RW, Clark JR, Matheny NP (2004) Arboriculture: integrated management of landscape trees, shrubs, and vines, 4th edn. Prentice Hall, Upper Saddle River, p 578
Henin S, Monnier G (1961) Mécanisme de l'action d'une couverture de paille sur le bilan de l'eau du sol. C R Acad Sci t 152:939–941
Horowitz M, Thomas JM (1994) Couverture du sol pour la gestion des mauvaises herbes (Soil cover for weed management), Maitrise des adventices par voie non chimique. Communications de la quatrieme conference internationale I.F.O.A.M. 2nd ed, pp 149–154
Huo L, Pang H, Zhao Y, Wang J, Lu C, Li Y (2017) Buried straw layer plus plastic mulching improves soil organic carbon fractions in an arid saline soil from Northwest China. Soil Tillage Res 165:286–293
Jabran K, Mahajan G, Sardana V, Chauhan BS (2015) Allelopathy for weed control in agricultural systems. Crop Prot 72:57–65
Jemai I, Ben Aissa N, Ben Guirat S, Ben-Hammouda M, Gallali T (2013) Impact of three and seven years of no-tillage on the soil water storage, in the plant root zone, under a dry subhumid Tunisian climate. Soil Tillage Res 126:26–33
Jhariya MK, Singh L (2020) Herbaceous diversity and biomass under different fire regimes in a seasonally dry forest ecosystem. Environ Dev Sustain 22:1–19. https://doi.org/10.1007/s10668-020-00892-x
Jhariya MK, Banerjee A, Meena RS, Yadav DK (2019a) Sustainable agriculture, forest and environmental management. Springer, Singapore. eISBN: 978-981-13-6830-1, Hardcover ISBN: 978-981-13-6829-5, p 606. https://doi.org/10.1007/978-981-13-6830-1
Jhariya MK, Yadav DK, Banerjee A (2019b) Agroforestry and climate change: issues and challenges. Apple Academic Press, Palm Bay. ISBN: 978-1-77188-790-8 (Hardcover), 978-0-42957-274-8 (E-book), p 335. https://doi.org/10.1201/9780429057274
Kader MA, Senge M, Mojid MA, Ito K (2017) Recent advances in mulching materials and methods for modifying soil environment. Soil Tillage Res 168:155–166
Kader MA, Singha A, Begum MA, Jewel A, Khan FH, Khan NI (2019) Mulching as water-saving technique in dry land agriculture. Bulletin Nat Res Centre 43:1–6
Kacinski NA (1951) Posev duba v mikroponizenija kak sredstvo borjby s zasuhoi na svetlokastanovyh pocvah (Sowing oak in micro-depressions as a means of combating drought on light chestnut soils). Pocvoved 10:585–603
Khan N, Jhariya MK, Yadav DK, Banerjee A (2020a) Herbaceous dynamics and CO_2 mitigation in an urban setup- a case study from Chhattisgarh, India. Environ Sci Pollut Res 27(3):2881–2897. https://doi.org/10.1007/s11356-019-07182-8

Khan N, Jhariya MK, Yadav DK, Banerjee A (2020b) Structure, diversity and ecological function of shrub species in an urban setup of Sarguja, Chhattisgarh, India. Environ Sci Pollut Res 27 (5):5418–5432. https://doi.org/10.1007/s11356-019-07172-w

Kiikkila O, Derome J, Brugger T, Uhlig C, Fritze H (2002) Copper mobility and toxicity of soil percolation water to bacteria in metal polluted forest soil. Plant and Soil 238:273–280

Kumar S, Meena RS, Jhariya MK (2020) Resources use efficiency in agriculture. Springer, Singapore. eBook ISBN: 978-981-15-6953-1, Hardcover: 978-981-15-6952-4, p 760. https://doi.org/10.1007/978-981-15-6953-1

Lakatos T, Buban T, Muller W, Polesny F, Verheyden C, Webster AD (2000) Effectiveness of different groundcover materials to preserve soil water content in a young apple orchard. Acta Hortic 525:425–426

Leclerc D (1997) Le paillage en forêt. ONF Bulletin Tech 32:39–46

Li X, Jin M, Huang J, Yuan J (2015) The soil–water flow system beneath a cotton field in arid north-West China, serviced by mulched drip irrigation using brackish water. Hydrgeol J 23:35–46

Long CE, Thorne BL, Breisch NL, Douglass LW (2001) Effect of organic and inorganic landscape mulches on subterranean termite (Isoptera: Rhinotermitidae) foraging activity. Environ Entomol 30:832–836

Masciandaro G, Ceccanti B, Benedicto S, Lee HC, Cook HF (2004) Enzyme activity and C and N pools in soil following application of mulches. Can J Soil Sci 84:19–30

McDonald P, Helgerson O (1990) Mulches aid in regenerating California and Oregon forests: past, present and future. USAD Forest Service Gentech Rep PSW-123:20

McGovern RJ, McSorley R, Bell ML (2002) Reduction of landscape pathogens in Florida by soil solarization. Plant Dis 86:1388–1395

Mechergui T (2008) Effets de l'utilisation des abris-serres et du paillage plastique sur l'installation et la croissance de plants de chêne-liège dans la région de Nefza (Nord-Ouest de la Tunisie) [mémoire de mastère]. Faculté des Sciences de Bizerte

Mechergui T (2016) Régénération artificielle du chêne-liège (Quercus suber L.) et du chêne-zeéen (Quercus canariensis Willd.): impacts du paillage et des abrisserres sur l'installation, la croissance et le développement architectural des plants. Tunisie: Thèse de doctorat. Faculté des Sciences de Bizerte, Tunisie, p 154

Meena RS, Lal R (2018) Legumes for soil health and sustainable management. Springer, Singapore, p 541. ISBN 978-981-13-0253-4 (eBook), ISBN: 978-981-13-0252-7(Hardcover). https://doi.org/10.1007/978-981-13-0253-4_10

Meena RS, Kumar V, Yadav GS, Mitran T (2018) Response and interaction of *Bradyrhizobium japonicum* and Arbuscular mycorrhizal fungi in the soybean rhizosphere: A review. Plant Growth Regul 84:207–223

Meena RS, Kumar S, Datta R, Lal R, Vijaykumar V, Brtnicky M, Sharma MP, Yadav GS, Jhariya MK, Jangir CK, Pathan SI, Dokulilova T, Pecina V, Marfo TD (2020) Impact of agrochemicals on soil microbiota and management: a review. Land (MDPI) 9(2):34. https://doi.org/10.3390/land9020034

Meena RS, Lal R, Yadav GS (2020a) Long term impacts of topsoil depth and amendments on soil physical and hydrological properties of an Alfisol in Central Ohio, USA. Geoderma 363:1141164

Meena RS, Lal R, Yadav GS (2020b) Long-term impact of topsoil depth and amendments on carbon and nitrogen budgets in the surface layer of an Alfisol in Central Ohio. Catena 194:104752

Montague T, Kjelgren R (2004) Energy balance of six common landscape surfaces and the influence of surface properties on gas exchange of four containerized tree species. Sci Hortic 100:229–249

Muñoz K, Buchmann C, Meyer M, Schmidt-Heydt M, Steinmetz Z, Diehl D, Thiele-Bruhn S, Schaumann GE (2017) Physicochemical and microbial soil quality indicators as affected by the agricultural management system in strawberry cultivation using straw or black polyethylene mulching. Appl Soil Ecol 113:36–44

Naab JB, Mahama GY, Koo J, Jones JW, Boote KJ (2015) Nitrogen and phosphorus fertilization with crop residue retention enhances crop productivity, soil organic carbon, and total soil nitrogen concentrations in sandy-loam soils in Ghana. Nutr Cycl Agroecosyst 102:33–43

Nambiar EKS, Sands R (1993) Competition for water and nutrients in forests. Can J For Res 23:1955–1968. https://doi.org/10.1139/x93-247

Nawaz A, Farooq M, Lal R, Rehman A, Hussain T, Nadeem A (2017) Influence of sesbania brown manuring and rice residue mulch on soil health, weeds and system productivity of conservation rice-wheat systems. Land Degrad Dev 28:1078–1090

Ngosong C, Justin O, Tening AS (2019) Mulching: a sustainable option to improve soil health. In: Soil fertility management for sustainable development, pp 231–249. https://doi.org/10.1007/978-981-13-5904-0_11

Oliveira MT, Merwin IA (2001) Soil physical conditions in a New York orchard after eight years under different groundcover management systems. Plant and Soil 234:233–237

Pfammatter W, Dessimoz A (1997) Influence de l'irrigation et de la couverture du sol sur le developpement et le rendement de jeunes pommiers (Influence of irrigation and ground cover on development and yields of young apple trees). Revue Suisse de Viticulture, d'Arboriculture et d. Horticulture 29:301–304

Pickering JS, Shepherd A (2000) Evaluation of organic landscape mulches: composition and nutrient release characteristics. Arboricultural J 23:175–187

Rahma AE, Warrington DN, Lei T (2019) Efficiency of wheat straw mulching in reducing soil and water losses from three typical soils of the loess plateau, China. Int Soil Water Conserv Res 7:335–345

Raj A, Jhariya MK, Yadav DK, Banerjee A (2020) Climate change and agroforestry systems: adaptation and mitigation strategies. Apple Academic Press, Palm Bay. ISBN: 9781771888226, p 383. https://doi.org/10.1201/9780429286759

Ranjan P, Patle GT, Prem M, Solanke KR (2017) Organic mulching—a water saving technique to increase the production of fruits and vegetables. Curr Agri Res 5(3):371–380

Rathore AL, Pal AR, Sahu KK (1998) Tillage and mulching effects on water use, root growth, and yield of rain-fed mustard and chickpea grown after lowland rice. J Sci Food Agric 78:149–161

Robitaille D (1994) Bilan annuel des expérimentations sur les paillis forestiers dans la station de la forêt de Drmmondville. Rapport interne MRN. Québec, p 142

Robitaille D (2003) Protection des plantations de feuilles contre la végétation concurrente dans les friches herbacées: plantation de Beaumont-résultats de sept ans. Mémoire de recherche forestière n°142

Rokich DP, Dixon KW, Sivasithamparam K, Meneyka KA (2002) Smoke, mulch, and seed broadcasting effects on woodland restoration in Western Australia. Restor Ecol 10:185–194

Rothwell RL (1978) Erosion control on forest roads. Agric Forestry Bull 1:29–32

Russell JC (1939) The effect of surface cover on soil moisture losses by evaporation. Proc Soil Sci Soc Am 4:65–70

Salim R, El-Halawa RA (2002) Efficiency of dry plant leaves (mulch) for removal of lead, cadmium and copper from aqueous solutions. Process Safety Environ Prot 80:270–276

Siczek A, Horn R, Lipiec J, Usowicz B, Łukowski M (2015) Effects of soil deformation and surface mulching on soil physical properties and soybean response related to weather conditions. Soil Tillage Res 153:175–184

Smith MW (2000) Cultivar and mulch affect cold injury of young pecan trees. J Amer Pomology Soc 54:29–33

Splawski CE, Regnier EE, Harrison SK, Bennett MA, Metzger DJ (2016) Weed suppression in pumpkin by mulches composed of organic municipal waste materials. Hort Sci 51:720–726

Szwedo J, Maszczyk M (2000) Effects of straw-mulching of tree rows on some soil characteristics, mineral nutrient uptake and cropping of sour cherry trees. J Fruit Ornamental Plant Res 8:147–153

Tanavud C, Kheowvongsri P, Yongchalermchai C, Leowarin W, Densrisereekul O, Bennui A, Murase J, Kimura M (2001) Effects of land use patterns on soil and water quality in Khlong U-Taphao Basin. Thai J Agric Sci 34:15–31

Tilander Y, Bonzi M (1997) Water and nutrient conservation through the use of agroforestry mulches, and sorghum yield response. Plant and Soil 197:219–232

Townsend TG, Solo GH, Tolaymat T, Stook KK (2003) Impact of chromated copper arsenate (CCA) in wood mulch. Sci Total Environ 309:173–185

Truax B, Gagnon D (1993) Effects of straw and black plastic mulching on the initial growth and nutrition of butternut, white ash and bur oak. For Ecol Manage 57:17–27. https://doi.org/10.1016/0378-1127(93)90159-K

Tuckey LM, Schoff EL (1963) Influence of different mulching materials upon the environment. Proc Amer Soc Hort Sci 82:69–82

Van Lerberghe P (2004a) Les paillis biodégradables ont-ils un avenir en plantations ligneuses ? Forêt-entreprise 157:20–21

Van Lerberghe P (2004b) Le paillage des plantations ligneuses, une alternative au désherbage chimique. Forêt-entreprise 157:22–26

Van Lerberghe P, Gallois F (1997) Les objectifs culturaux du paillage et ses conséquences. Forêt-entreprise 116:26–30

Von Ahlten FW (1990) Guide révisé relatif à la plantation des bois durs sur les terres agricoles abandonnées au sud de l'Ontario, Forêt Canada-Région de l'Ontario, Centre de foresterie des grands lacs, Sault-Sainte-Marie, p 90

Wagner RG, Robinson AP (2006) Critical period of interspecific competition for four northern conifers: 10-year growth response and associated vegetation dynamics. Can J For Res 36:2474–2485

Wang X, Xing Y (2016) Effects of mulching and nitrogen on soil nitrate-N distribution, leaching and nitrogen use efficiency of maize (*Zea mays* L.). PLoS One 11:e0161612

Wilen CA, Schuch UK, Elmore CL (1999) Mulches and subirrigation control weeds in container production. J Environ Hort 17:174–180

Zaragoza C, Moya S, Martinez G (1995) Efectos de las coberturas organicas a base de cortezas de pino y restos de poda en un Huerto de frutales (effects of mulches based on pine bark and pruning residues in a fruit orchard), pp 83–290. Proceedings of 1995 congress, Spanish Weed Science Society, Huesca

Zhang S, Lövdahl L, Grip H, Tong Y, Yang X, Wang Q (2009) Effects of mulching and catch cropping on soil temperature, soil moisture and wheat yield on the loess plateau of China. Soil Tillage Res 102:78–86

Vertical Greenhouses Agro-technology: Solution Toward Environmental Problems

9

Evgeniya P. Klyuchka and Marko Petkovic

Abstract

Population growth and urbanization, climate change, and the environmental disadvantages of traditional agriculture have reached a critical limit. Global processes can reduce the amount of industrial waste and find environmentally friendly ways of recycling, abandon hazardous food products, solve problems in the market for organic products, and reduce food waste. World forums discuss topics such as sustainable development theory, environmental rents, the prospects for the green revolution, and the 4.0 science and technology revolution industry. Greenhouse productions are aimed at solving some of the environmental and food problems. Greenhouse production could be divided into three groups: the technology of growing plants without soil, the practical application of LED lighting systems, and the capabilities of digital IT technologies. Greenhouse production is located on a "scale of comparison," from the simple technologies without soil in a house, office, on roofs, on the street, all the way to the system with various microclimate systems, a system of nutrient solutions, heating and air conditioning systems, humidification and dehumidification systems, lighting systems, gas generation systems, monitoring, and control systems, and other microclimate systems. The production efficiency would be increased by technical equipment and electrical installations, improving biotechnological methods and rising energy costs. The higher the manufacturability of greenhouse production, the higher the energy intensity of the process of growing plants. We are introducing digital IT technologies and are approaching the extreme point of the "scale of comparison" on which cyber-physical systems are located. Greenhouse

E. P. Klyuchka
Don State Technical University, Rostov-on-Don, Russian Federation

M. Petkovic (✉)
Faculty of Agronomy, University of Kragujevac, Čačak, Serbia
e-mail: marko.petkovic@kg.ac.rs

M. K. Jhariya et al. (eds.), *Ecological Intensification of Natural Resources for Sustainable Agriculture*, https://doi.org/10.1007/978-981-33-4203-3_9

technologies have varying degrees of success. Israel has no fertile soil, half of the territory in the form of a desert, and a lack of fresh water. However, Israel produces 17% of fruits and vegetables from all agricultural products. The agricultural sector represents approximately 5% of the population, which satisfies Israel's needs for agricultural products by 92%. The success of Israel crop production is based on greenhouses, hydroponic systems in the field; drip irrigation of plants through a network of flexible tubes; plant breeding; digital IT technologies of phyto-monitoring, etc. The global achievement of greenhouse production in Japan is the practical application of new concepts: the Internet of Things (IoT) and cyber-physical systems (CPS). The greenhouse business in Japan has confirmed environmental safety, the prospect of growing clean organic products, and making a profit.

Keywords

Agro-technology · Cyber-physical systems · Urbanized agricultural production · Vertical greenhouse

Abbreviation

ACPS	Agricultural Cyber-Physical Systems
ADM	The Australian grain producer
AgTech	Agriculture Technology
BSG	Boston Consulting Group
CPS	Cyber-Physical Systems
EEC	European Economic Community
EU	European Union
FAO	Food and Agriculture Organization
FinTech	Financial Technology
LED	Light Emitting Diodes
NOSB	National Organic Standards Board
UN	United Nations
USA	United States of America

9.1 Introduction

Population growth, urbanization, climate change, and a decrease in the quality of ecology have revealed problems in traditional agriculture. Traditional agriculture has reached the limit of increasing the amount of food through crop production and land exploitation. Land and water resources are declining in quantity and quality (FAO 2020).

World processes should lead to new trends for maintaining a delicate ecological balance, such as the desire to reduce the amount of industrial waste and the search for environmentally friendly ways of disposal, stabilize food security and self-sufficiency in food, reduce food waste and solve problems in the market for organic products (Raj et al. 2020; Banerjee et al. 2020; Jhariya et al. 2019a, b). In this regard, it is clear that the most discussed topics at world forums are: the theory of sustainable development, environmental rent, the prospects of the Green Revolution, and the scientific and technological revolution Industry 4.0. Thus, these large-scale trends require new innovative methods of food production, which is city farming. It is necessary to pay attention to the preparation of public opinion and the upbringing of a completely new young generation on current world trends and global events (CAP 2020).

The current global trend, which is the subject of debate, is the technology of urban agriculture or AgTech (agriculture technology) urbanized agricultural production. The development of AgTech technologies as a combination of innovative, highly effective agrarian practices for the production of crop products is directly related to the environmental and food safety of the country, and the health of the nation.

The need for AgTech development is recognized at the world level following paragraph 95 of the implementation plan of the New Urban Development Program, the Quito Declaration on Ecologically Sustainable Cities and Communities for All has been drawn up (Habitat III Conference 2016). The "New Urban Development Program" was adopted at the UN-Habitat III Conference on Housing and Sustainable Urban Development and approved by the UN General Assembly. A 197 UN member states have declared a commitment to "support agriculture and farming in urban settings."

The number and scope of vertical farm projects in different countries of the world over the past 10 years, indeed, have begun to acquire impressive proportions, which speaks not only of the viability of this technology, but it also affects many areas: environmental, economic, social, market. In particular, the organization of logistics, infrastructure, food prices, the diet of the population, and much more (Kozai et al. 2020; Meena et al. 2018).

Among such economically and socially significant projects are the company's farms: Pasona in city offices (Japan); AeroFarms in New Jersey; 9-story farm in Dronten (Netherlands); Plenty Farms (Seattle and San Francisco); Bowery (North Carolina and New York); Farm one (New York); Panasonic (Singapore); a vertical farm at the Paignton Zoo (UK); Farm-360 (Indianapolis); Metro Group installations (Europe); Urban Crops (Kortrijk, Belgium), La Caverne (Paris, France); AeroFarms, SunDrops and Farms Badia Farms (Dubai, Saudi Arabia); 142-m skyscraper farm in Linkoping (Sweden); Plantagon Agritechture and Sweco Architects (Sweden); Mirai Corp (Tokyo metropolitan area, Japan); vertical farms in Target chain stores and IKEA hydroponic installations.

Greenhouse technologies have varying degrees of success. For example, Israel has no fertile soil, half of the territory in the form of a desert, and a lack of fresh water. However, Israel produces 17% of fruits and vegetables from all agricultural products. The agricultural sector accounts for approximately 5% of the population, but the work of these people meets the needs of Israel for agrarian products by 92%. The unique experience of Israel agriculture has confirmed the success of the application of technologies for growing plants without soil in open rocky areas. The success of Israel crop production is based on: Israel greenhouses, hydroponic systems directly in the field; drip irrigation of plants through a network of flexible tubes; plant breeding; digital IT technologies of phyto-monitoring, and more. Another example illustrates the high-tech greenhouses that Japanese scientists created. The global achievement of greenhouse production in Japan is the practical application of new concepts: the Internet of Things (IoT) and cyber-physical systems (CPS). The greenhouse business in Japan has confirmed environmental safety, the prospect of growing clean organic products, and making a profit.

The idea of vertical farms is promising, giving a reason to rely on significant results that the construction of such facilities can lead to several certain and uncertain consequences. Therefore, the present chapter describes the role of vertical green agro-technology in solving environmental problems.

9.2 Global Problems of Modern Greenhouse Agricultural Complexes

According to forecasts, by 2050, the World's population will reach 9.7 billion people, with 70% of the people living in urban conditions (United Nations 2016). Based on the reports of the FA (FAO 2020), the following conclusions could be drawn. The growth and aging of the World's population lead to an increasing concentration of the people in cities (Meena et al. 2020a, b). According to the UN forecast, in the future, the share of urban residents in the World will increase steadily, according to various sources, from 55% in 2016 to 60% by 2030 and 70% by 2050. Twenty-five percentage of fertile land has already degraded, which directly affected 15% of the World's population; it would also be expected that by 2030 another 2.4% of highly productive areas would be "absorbed" by growing megacities. A report released by the UN on 27 November 2007, on environmental changes that have occurred since 1987, noted that, by general definition, humanity is in a state of ecological crisis, and there are no signs of its easing (Kantor 2013).

The indicative trend line was determined using data from the FAO (FAO 2018), with the observed difference correlates with the global recession of the early 1990s (Fig. 9.1, Benke and Tomkins 2017).

Pesticides that are used in traditional agriculture negatively affect water quality and reduce the number of water resources, pollute the air, contribute to the extinction of living organisms, pollution and deterioration of food quality, and the accumulation of industrial waste and household garbage (Meena et al. 2020). Traditional agriculture is the reason that the soil is degraded and depleted of nutrients, pollutes the environment with herbicides and nitrates, is the cause of death of animals and insects. It is understood that it is the economically developed countries that cause the most significant harm to the environment, consuming more than other countries, both natural raw materials and finished products, polluting the planet with

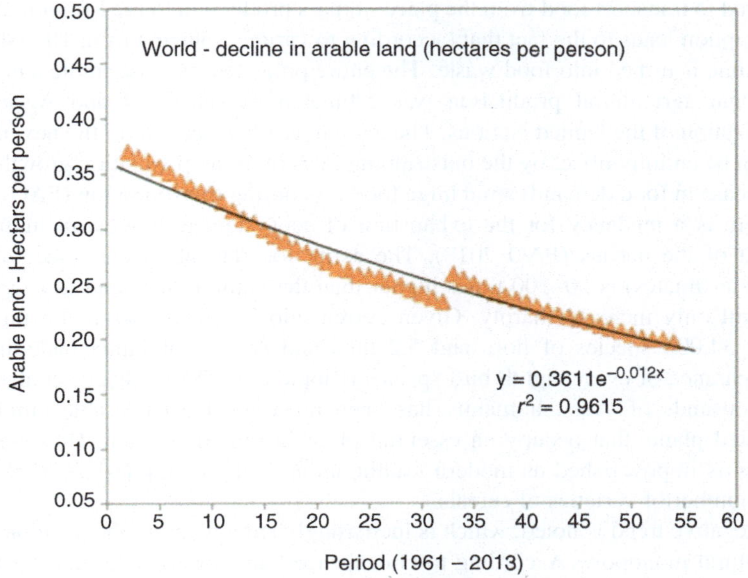

Fig. 9.1 Reduction of stocks of arable land in the world (Benke and Tomkins 2017)

production and consumption waste. There was an understanding that progress can be achieved by using predatory natural resources, cheap dirty technology, which only leads to a decrease in environmental safety (Meena and Lal 2018; Kumar et al. 2020).

Currently, economists around the world are discussing the theory of sustainable development, in which it is possible to ensure reproductive capital and environmental protection, due to the invested rent from natural resources, defined as the difference between the market price of supply and the marginal cost of its use (Dixon et al. 2003; Schulze et al. 2015). Products obtained in traditional agriculture are much more expensive, but not much expensive because no one currently takes into account all costs associated with depreciation and restoration of natural resources, which have the property of limitation and depletion.

A direct consequence of human activities is climate change on the earth (carbon dioxide emissions, global warming). The climate becomes the primary limiting (limiting) factor in the development of traditional agricultural production, so this industry becomes unstable, located in a high-risk area of manufacturing. Today we received a high harvest, and next year it may not be at all.

Territories with mass food production are far from consumer markets, this leads to the formation of large amounts of food waste, to the loss of food in the production logistics chain: storage, sorting, packaging, rejection, and failure during long-term transportation and the period of sale of products. At the same time, there is a catastrophic decline in quality due to the use of barbaric technologies, the transformation of immature crop products to a state of external commercial attractiveness.

The need to transport food from the place of mass production to the location of mass consumption leads to the fact that (according to various sources) from 10 to 40% of its volume is turned into food waste. The entire projected increase in the amount of traditional agricultural production was estimated by the Food and Agriculture Organization of the United Nations. The amount is 1.5% per year in the next decade and can be entirely offset by the outstripping rates of urban population growth (with an increase in food demand) amid huge food loss during transportation (FAO 2020).

There is a tendency for the exhaustion of ecosystem resources, including the support of the oceans (FAO 2018). The extinction rate of species (according to various estimates) is 50–100 times higher than the natural ones, and as suggested, they will only increase sharply. Given current global trends, extinction threatens almost 34,000 species of flora and 5.2 thousand species of fauna including the disappearance of every eighth bird species (Ripple et al. 2020; Briggs et al. 2015). For thousands of years, humanity has been breeding a considerable number of cultivated plants that occupy an essential place in our food chain. However, this treasury is impoverished as modern traditional agriculture emphasizes a relatively small number of varieties of cereals.

A negative trend is noted, which is increasingly strengthening the position of the agricultural monopoly. According to Greenpeace (Greenpeace International Public Environmental Organization 2020), four corporations—ADM, Bunge, Cargill, and Dreyfus—control about 80% of the global grain trade. Seed behemoth Monsanto's consolidation in the production and sale of seed with the chemical giant Bayer is currently being published. The turnover of both monopolists exceeds $ 62 billion and has long dominated the food industry in America with genetically modified seeds and gamma toxic pesticides. According to ETC Group estimates, which use only official data voiced, Monsanto controls more than 30% of seeds worldwide, in the USA this monopolist controls 93% of the soybeans market and more than 70% of cotton sales and judging by the trends in the world, these numbers will only grow. The corporations are under total control over how food is produced and what the world's population eats. The complete absorption of the global food market by single players threatens the implementation of environmental innovations and developments, biodiversity, and, ultimately, the food security of each country.

An increase in population, extensive urbanization, and an increase in industrial capacity pose a threat to environmental well-being and global biodiversity (Khan et al. 2020a, b; Raj et al. 2018). These are the reasons why tremendous pressure is created on the market for organic products. There is a real process of changing consumer preferences toward the priority of "healthy," "natural," "organic" food. Consumers are increasingly looking for and buying products: without stabilizers and flavor enhancers; not frozen ripe, not harvested for ripening during transportation and storage; grown without chemicals, pesticides, herbicides, antibiotics, steroids; with high transparency of production and supply chain. In this regard, development issues such as an alternative system of uninterrupted supply or self-sufficiency of cities with foodstuffs, and food security in the whole country are raised with particular urgency (Gorchakov and Durmanov 2002; Gorchakov 2004, 2009).

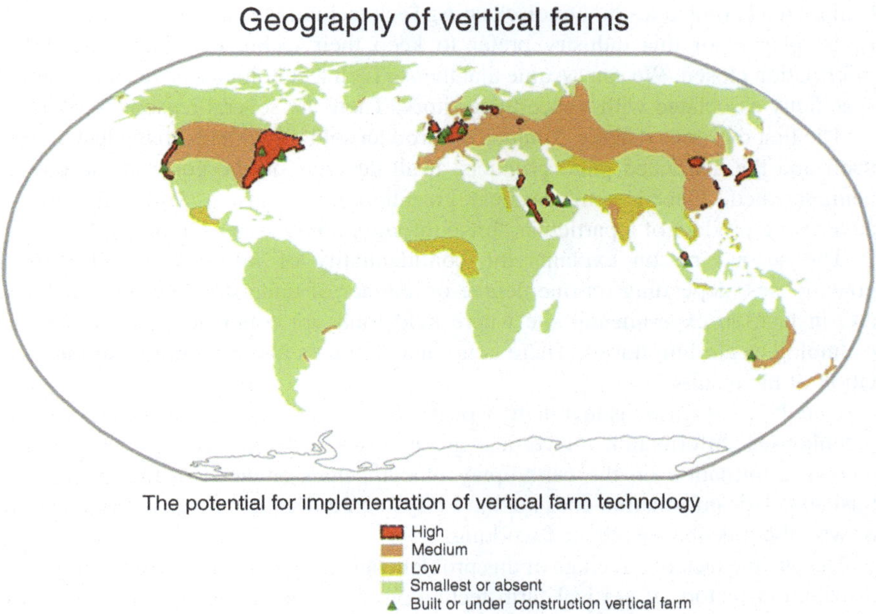

Fig. 9.2 The schematic map and the geography of vertical farms AgTech (Gres et al. 2019)

The EU and the USA were the first to start the process of consolidating the status of organic food at the level of legislation, certification, and standardization. In 1991, the European Council of Ministers adopted Agricultural Regulation (EEC) № 2092/91 on organic farming and the corresponding labeling of agricultural products and food products. The introduction of these rules as part of the reform of the Common Agricultural Policy of the EU (Common Agricultural Policy) represents the previous process's completion. Organic agriculture received official recognition (CAP 2020; US Department of Agriculture 2020).

Organic food research in the USA has been underway for over 20 years. National Organic Standards Board (NOSB), on 1 November 2017, decided to make eligible the container greenhouses that use hydroponics and aeroponics technology for certification of products as organic (National Council of Organic Standards NOSB 2020). Vertical farms function on the territories is shown in Fig. 9.2 (Tornaghi 2014; Gres et al. 2019). Currently, vertical farms have proven their viability.

According to a study by the Swiss financial institution (Union Financial institution of Switzerland 2020), innovations in food and agriculture, the vertical greenhouse production market will have an annual average annual growth rate of 39.6% over the next 5 years. The global market will reach $ 1.1 billion by 2024, compared with $ 1.480 million in 2019. Some of the key players in the market of vertical farmers and plants: According to the report, they have recently absorbed significant amounts of investment capital and will become a market of $ 700 billion by 2030. Private commercial companies are currently promoting the ideas of AgTech.

Commercial companies do not disclose AgTech technology, which is a trade secret. Large players of this industry prefer to keep their technology and profitability information closed. We can assume that the AgTech technologies have an individual specificity associated with a specific territory. There are several reasons for this.

The first reason is that the commercial product sold might be both the technology itself and the cultivated plant products. It all depends on the goals of the project being created. Indeed, some vertical greenhouses are just a demonstration, an advertising product of a particular "breakthrough, innovative technology."

The second reason explains the confidentiality of information; all vertical greenhouses, depending on one degree or another of understanding the problems, are, in fact, an experimental site where field trials are conducted, and innovative technologies are introduced. There is no unambiguous understanding and interpretation of the results.

And the third reason is that in the world practice, there are no statistics on vertical greenhouses. Information on vertical greenhouses is promoted, but we regularly receive information on the bankruptcy of companies engaged in the greenhouse business. This information does not have a system; it does not give a realistic picture of why the greenhouse stopped working.

Despite the lack of coverage of this problem, an attempt was made to highlight the distribution factors of vertical farm technology in various countries of the World (Kemp and Loorbach 2003; Lovell 2010; Miller 2011; Kapelyuk and Aletdinova 2017; Iconographer 2018). The fourth level of the territories is mostly poorly populated or not related to Oecumene (Antarctica, Amazon, and others). The third level of the regions is traditional agricultural areas (a frequently observed phenomenon of pseudo-urbanization), such as Russia, India, Egypt, or Tropical Africa. The second level of the territories is territories with a relatively high level of socioeconomic development and a developed agricultural sector (France, Italy, and others).

Finally, the first level of territories includes territories corresponding to a particular set of conditions, such as a high level of urbanization, the presence of large agglomerations or megalopolises and solvent population, a high standard of living and socioeconomic indicators, and a lack of highly productive agricultural land near agglomerations/megacities. There is a request for the implementation of projects from citizens and an understanding of the benefits of these innovations among the population. The presence of advanced institutes of science and social networks, with the help of which the accumulation of knowledge and the relay of knowhow is possible. Territories of the Eastern and Western coasts of the USA, coastal highly urbanized areas of the PRC, South Korea, Japan, Singapore, the United Arab Emirates, the coast of Saudi Arabia, Northern Europe, and others are more consistent with similar conditions.

The fourth industrial revolution, Industry 4.0, is democratizing the production of greenhouse plants. Knowledge becomes available with the development of information, communications, the Internet, and the means of production are becoming less and less expensive, more and more efficient and manageable. The basis of this global technological restructuring is preceded by investments in new agricultural technologies comparable, for example, with investments in the field of innovative financial technologies FinTech.

Over the past decade of emerging technology, the problems of transition to sustainable development have gained growing attention. Sustainable development is a process in which "the satisfaction of present needs does not compromise future generations 'capacity to meet their own needs'"(Pfeiffer 2017). Moreover, without radical changes in production business processes, such a transition is impossible. These changes are primarily due to the introduction of technology 4.0.

Today, scientists agree that the industry has created the necessary preconditions for the fourth industrial revolution when highly digitized processes are integrated with the internet and smart technologies (Rojko 2017). The BCG Consulting Company identifies nine critical technologies aimed at forming the fourth industrial revolution: autonomous robotics, simulation, horizontal and vertical system integration, augmented reality, the Internet of things, cloud computing, additive manufacturing, cyber security, and big data. This change also includes more global trends, the introduction of the digital economy, CPS, smart cities, smart buildings, intelligent greenhouses, and others.

Industry 4.0 will allow production costs to be reduced by 10–30%, logistics costs by 10–30%, and quality management costs by 10–20% (Rojko 2017; Lisovsky 2018). Industry 4.0 innovations are designed to minimize the time taken to market new products on the market, improve the quality of consumer service, and allow more effective use of resources (Lee et al. 2015; Trachuk and Linder 2018).

The most successful innovation growth strategy in South Korea was "Strategy 3.0," which describes smart plants as production systems, where all business processes are automated and incorporated into a single information system. The operation of the plants is ensured by the CPS, which allows the creation of virtual twins. CPS is designed to incorporate devices that are directly connected with the external world and existing processes by utilizing data collection and processing facilities over the Internet. Current CPS research focuses primarily on principles, emerging technology, information architecture design, existing problems, and new avenues of growth within the context of Industry 4.0. An integrated model for CPS into production processes is proposed via smart communication, data translation into information, virtual space transfer, knowledge management, and system configuration.

9.3 Concepts of Vertical Greenhouse Agro-Technology

Today, new technologies appear in the protected soil industry related to the creation of high-tech automated phytocomplexes for intense energy and resource-saving production. They are gaining wide popularity around the world and are called "urban farms" or "vertical greenhouses AgTech." Phytocomplexes include energy-efficient vegetative lighting systems of various modifications and original design solutions. These parameters are in conjunction with multiple microclimate systems for automated maintenance. The technology for growing plants on the shelves using electric lighting (or light culture) is currently the most advanced when growing plants require a strict balanced diet and additional artificial lighting. This environment is optimal when growing seedlings of vegetables, flowers, lettuce, green crops, medicinal plants (Benke and Tomkins 2017; Blok et al. 2017; Beacham et al. 2019).

The AgTech vertical greenhouse technology combines several independent areas, and if at least three of the six mentioned characteristics are present, then we can attribute this technology to city farming. Vertical greenhouses have a distinctive feature: technological surfaces are located one above the other and form a multi-level design. The main principle is to grow the most substantial amount of plant products in the smallest area, due to compaction and placement of levels vertically. Application of technologies for growing plants without soil: hydroponics, aeroponics, aggregatoponics, and many hybrid technologies.

The use of artificial light sources applies to either for illumination in conditions of decreasing natural light and reducing daylight hours, or light culture technology when plants are grown using only artificial light. LED lighting systems are considered the most promising, which reduce the energy intensity of the entire production (in comparison with low and high-pressure discharge lamps), and are also a tool for influencing the biochemical composition of plants, shortening the growing season and increasing the harvests.

Production should be waste-free and environmentally friendly. Therefore, it includes closed-loop recycling systems, as well as systems for cleaning and preparing the air and water environment for plants. Methods are needed that control what photocomplexes give into the environment and what they take from the environment (the urban environment is quite aggressive).

The use of renewable energy sources: solar panels, bioreactors, wind turbines, the use of rainwater, condensate, and more. Autonomy and independence from external sources of energy, water supply, sewage, and more are needed.

Automation/robotization of all production processes: technological, economic, logistics, and others. High expectations are laid on artificial intelligence, which should give an accurate answer to the choice and optimization of a particular technology in specific working conditions.

Vertical greenhouses AgTech of city farming have a rather diverse form of self-expression. The critical point for characterizing a vertical greenhouse is the degree of innovation of the technologies used. We will create a conditional classification according to the above-voiced characteristics (Fig. 9.3, Beacham et al. 2019). The lowest level (fourth) is made up of a home/office environment. It is the closed living microclimate and working rooms of vegetative lighting installations. At the same time, they can diversify the diet with organic plant products and carry an aesthetic burden. They can combine groundless technologies, artificial lighting systems, partial automation of individual systems.

The third level consists of industrial vertical greenhouses, which solve the difficult task of integration in an urban environment (Fig. 9.4, Vertical Farm in Romainville, France 2020). These are greenhouses in non-functioning industrial buildings and skyscrapers, on the roof, in basements, in utility rooms, as well as the task of growing greenhouse plants in a small area with a seal of planting material

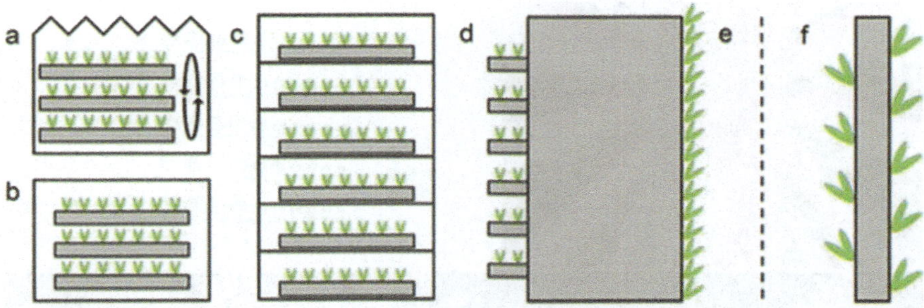

Fig. 9.3 Structural diagram of options (a–f) for placing plants in greenhouses of the third level, denoting different forms, i.e., options for placing plants in greenhouses of the third level (Beacham et al. 2019)

Fig. 9.4 Vertical farm in Romainville (France) (http://ifarmproject.ru/verticalfarmphotonews)

due to the location of technological surfaces one above the other. The main feature of these greenhouses is the likeness of the "classic" greenhouse technologies that are used in extensive agricultural holdings. In addition to the techniques of soilless cultivation, LED lighting systems, automation of production, there must be control over the incoming and outgoing environment. The use of this technology does not aggravate the ecological state of cities and does not call into question the issue of organic products.

The second level consists of modular or container installations (Fig. 9.5, Japanese company Mirai Mirai Corp 2020). It is necessary to solve many issues related to autonomy, which involves the use of renewable energy sources, closed water supply, and more. Only artificial light is used. It is advisable to use fully automatized control systems for artificial microclimate.

Finally, the first level (highest) consists of fully robotic cyber-physical systems (Figs. 9.6, 9.7, 9.8, 9.9, 9.10, 9.11, and 9.12), the technology product of the fourth industrial revolution Industry 4.0. The Farmbot system is a unique representative of the cyber-physical system. Farmbot-3 is an open-source digitally controlled

Fig. 9.5 Japanese company Mirai Mirai Corp (http://3476.jp/en/about/profile.html)

Fig. 9.6 French Startup Agricool (2020) (https://propozitsiya.com/francuzkiy-startap-stvoryuie-vertikalni-fermi-; http://green-city.su/gorodskie-fermy-potreblyayut-na-90-menshe-resursov/)

numerical control (CNC) computer. The Farmbot system is attractive in terms of availability and is entirely autonomous because solar panels power it.

The goal of robotics is to outsource all harvesting work, which accounts for 20% of all agricultural work, to robots to increase agricultural efficiency. Automation of

Fig. 9.7 Australian Container Farm Primary Module (2020) (Module http://www.modularfarms.com.au/primary-module-2/; Module http://www.modularfarms.com.au/primary-module-2/#)

Fig. 9.8 California (USA) Company (Fodder Works) (2020) (https://www.fodderworks.net/products/fodderworks-instruction-manual)

Fig. 9.9 Farmbot-3 (2020) (https://farm.bot/)

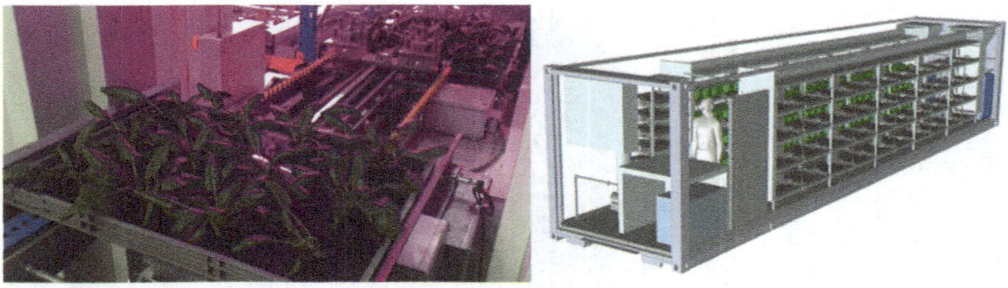

Fig. 9.10 Belgian company Urban Crop Solutions (2020) (https://urbancropsolutions.com/ru)

Fig. 9.11 American Robot Greenhouse Firm Iron Ox (2020) (http://www.iksmedia.ru/news/5536161-Robotizirovannaya-ferma-sposobna.html)

collection will save 20% of farm labor. Calculations showed that about 40% of labor costs are taken up by plants and 40% by calibration, sorting, and packing of tomatoes. For harvesting, using robots is relatively simple. The main advantage of robots is the ability to work at night. And the people who come to work in the morning have to pack the tomatoes collected at night. High technologies have been introduced in the greenhouse complex since its construction: computer control of temperature, humidity, lighting, irrigation, fertilizer, and carbon dioxide supply. The main problem was how to develop the "picking" robot to determine the degree of tomato ripeness. To do this, the designers focused on identifying the robot color and shade of the tomato. The work taught to record the image of the fetus and compare it with the reference image of a ripe tomato, which is introduced by farm workers.

Fig. 9.12 Japanese Robotic Greenhouse by technology company Spread Co (2020) (https://www.agritecture.com/blog/2018/12/2/japan-plants-to-open-worlds-largest-automated-leaf-vegetable-factory)

In agriculture, robotization has not begun yet: the financing problem holds back—it is not always possible to convince the investor of the effectiveness of the project. However, according to experts from the Massachusetts Institute of Technology, this is one of the ten most promising areas of robotics.

Robotization of the greenhouse development should be integrated into cyber-physical systems (CPS). CPS is computer systems focused on and reliant on the combination of computing and physical components. New CPSs will be coordinated, distributed, and linked, and must be reliable and responsive. CPS technology can change the way people communicate with the systems they have developed, just as the Internet has changed the way people interact with knowledge (National Science Foundation 2020). More precisely, the CPS is a system in which computing/information processing and physical processes are tightly combined and inseparable from a behavioral point of view. The system functionality and characteristic are manifested as a result of the interaction between physical and computing objects. Computers, networks, devices, and their environments have been interacting with physical properties, consume resources, and contribute to the overall behavior of the system.

In the twenty-first century, the next generation of agricultural enterprises must incorporate the vision, demands, and intellect of the "consumer" in the supply chain. It will be a performance-oriented organization that responds rapidly to consumer needs and minimizes resource use, maximizes environmental sustainability, and economic competitiveness. To get such a high level of results, it is necessary: preparation (training, education) of a new generation for new conditions, development (research, modeling) of innovative technologies and business development on these technologies. We have global work to prepare specialists with a high environmental culture who will see the prospects for the development of new technologies and be able to apply these technologies for the benefit.

Currently, vegetative plants on the market have been implemented in which plants and crops are controlled by computer algorithms. Agricultural Cyber-physical Systems (ACPS) provide an opportunity not only to copy experiments easily, but also to collect, analyze, and study the data obtained to reveal new features and

Fig. 9.13 The Grocon Pixel building is an example of a green office building and greenhouse (Melbourne, Australia) (https://stephenvaradyarchitraveller.com/2018/02/11/melbourne-pixel-building-australia/)

patterns. ACPS can help with plant optimization methods that provide more autonomous, efficient, and intelligent plant growth models by integrating robotic control loops in innovative vegetation devices with an artificial microclimate. At present, safe environment systems are based on open-source principles. The choice in the review, we focused on openly accessible technologies that can be implemented independently.

The Grocon Pixel building is an example of a green office building in Carleton, near the central business district of Melbourne (Fig. 9.13, Grocon Pixel green office building and greenhouse 2020). The multi-level building in the background features roof-mounted wind turbines and customizable side panels to monitor the effects of solar radiation. Once converted to a vertical truss, solar panels can be used on roofs and sidewalls, and high-performance LED sources are used for indoor lighting. A million-liter reservoir is being installed in the nearby Lincoln Square for local storm water reuse. The construction of the dwelling of a residential tower with a vertical truss is demonstrated.

9.3.1 Advantages of Vertical Greenhouses Agro-Technology

The main advantages of the vertical greenhouses AgTech are:

- Protecting crops and plants from harsh weather is one of the essential benefits of vertical farming. This advantage makes it possible to obtain year-round crops of certified organic products and growing plants in any extreme conditions of deserts, the North, military bases, drilling rigs, and more.
- In the "walking" accessibility from the consumer, these provide tremendous benefits for improving the quality of plant products and thereby contribute to a healthy lifestyle of the nation. This fact gives a significant advantage in the fight against food waste.
- Saving water resources and non-productive costs. A comparative analysis of vertical and traditional agriculture shows that the first method significantly

reduces water consumption. Almost 70% of the World's drinking water is used in conventional agriculture, while vertical agriculture needs only 20–40% of water resources, and in the future, this indicator will only decrease (FAO 2018). Plants grown in closed conditions trigger the evaporation process. This process allows the reuse of water for irrigation purposes. Water consumption is approaching a minimum level. Thus, vertical farms contribute to the rational use of natural resources.

- Constant control of the internal environment eliminates the need for financial costs for the purchase of pesticides and insecticides. Since all crops are cultivated in a controlled environment, the chances of pests, insects, and diseases are minimal.
- Lack of food waste. The risk of product spoilage is extremely small or equal to zero. A feature of vertical farming is that the crop is consumed immediately after harvest. The need for transportation costs is excluded since all crops are intended for consumption within the city.
- Vertical city farming greenhouses are needed to solve environmental problems. With a stable and constant supply of plant products to the urban population, the tension on traditional agriculture is relieved. It enables you to maximize the available land, growing more crops in a smaller area with a highly efficient system, with an opportunity to preserve some of the soil for conservation, to raise reserves, and much more.

9.3.2 Disadvantages of Vertical Greenhouses Agro-Technology

Opponents question the potential profitability of vertical farming. The primary and main negative argument not in favor of vertical greenhouses is that technology opponents put forward relatively high capital and operating costs. But in this matter, not everything is so simple. For the period only after 1960, food production in the world increased 2.5 times, water consumption two times, deforestation three times. As the population grows, the availability of agricultural land decreases: in 1980, 0.3 ha of arable land fell per capita in 2011, 0.24 ha (Final Report Vertical Farm 2.0 2015). The reduction of land resources as a global trend is due to the rejection of productive land for enterprises, cities, and other settlements, the development of the transport network. Vast areas of cultivated land are being lost because of erosion, salinization, waterlogging, desertification, physical, and chemical degradation.

Attention should be focused on the problems with world land, water, and forest resources. Over the past 10–15 years, awareness of environmental protection and to the development of new principles of state regulation to achieve more sustainable economic growth has noticeably increased.

One of the debated issues that affect the theory of sustainable development is the concept of natural rent, that is, payment for the use of natural resources. Scientists argue: the current rejection of the scientific definition of the nature and extent of natural rent leads to significant losses in the national economy. The primary purpose of calculating natural rents is the selection of unique macroeconomic indicators that would allow us to assess the dynamics of changes in the physical state of the environment. Natural rent should serve to coordinate the pace of macroeconomic development, improve the environmental situation, as well as fluctuations in the cost parameters of human and natural resources that are part of the full composition of national wealth. The general meaning of these discussions is the need to take into account the environmental costs of amortization and restoration of natural resources when calculating production efficiency. Further, this means only one thing that the cost of production in traditional agriculture is much higher and incomparably more significant than that presented in official reports.

The theory of sustainable development is gaining strength and significance, against the background of environmental degradation, depletion of natural resources, excessive pollution, which indicates failures in the market mechanism. The market allows you to evaluate only one function of the environment: the provision of natural resources. Another feature is not taken into account: the assimilation of waste and pollution, the performance of environmental functions (aesthetic, recreational, etc.). The feature that is related to the environment does not find its adequate reflection in market valuation.

Thus, if we consider the costs of building and operating vertical greenhouses against the background of a decrease in environmental well-being, quality of life, food security, and public health, then these expenses will seem not so big. But this is what concerns the comparison of greenhouse technology and traditional "classical" agriculture.

Currently, the elaboration of the successful theory of AgTech vertical greenhouse production is practically based on vegetative lighting and irradiation equipment in conditions of artificial controlled parametric microclimate.

Despite considerable experience in growing plants under artificial lighting, there is currently no single view of the optimal intensity (power) of illumination, the spectral composition of radiation, the dose of plant irradiation, photoperiod, etc. In almost every case, a peculiar lighting system is created that best meets the physiological needs of the cultivated plant (Fredani 2010; 6th Global Botanic Gardens Congress Geneva, Switzerland 2017). This direction is at the stage of formation and development.

9.4 Global Scenario of Vertical Greenhouse Agro-Technology

The production of vegetables in the open field remains a difficult problem for the production of environmentally quality plant products, which requires high costs (both physical and financial), time, and environment. The use of enclosed structures of protected soil (greenhouses, hotbeds) for the off-season growing of plants significantly facilitates by reducing adverse environmental factors. All the contradictions that arise during the creation and current operation of biotechnological systems have not been resolved. The more technologically advanced greenhouse production is the higher the energy intensity, the energy consumption of the process of growing greenhouse products. So far, no one has managed to create an innovative developer of greenhouse technology, which is fully recouped due to the significant energy demand. The main contradictions are as follows:

- Firstly, at the present stage of technical development, not a single greenhouse has solved the problem of enormous heat losses. At low ambient temperatures, there is a need for energy costs to maintain the optimum temperature inside the greenhouse. When the external temperature rises, a "greenhouse effect" appears, and the need arises to divert excess energy. Thus, either energy is expended for a favorable temperature environment or there is a need to get rid of excess energy with transoms, curtains, ventilation, air conditioning, which incurs additional costs.
- Secondly, with all the variety of lighting developments with various light sources, the question of lighting systems for greenhouses that fully satisfy the needs of the plant has not been resolved. The use of artificial irradiation for their illumination only increases the cost of greenhouse products. The principal contradiction is that the plant's light environment requires specific characteristics in terms of light intensity, spectral composition, and duration of exposure. Only 10% of the optical energy is used by plants in the process of photosynthesis in a controlled closed biotechnological system. The remaining 90% is a technological loss. Plants do not adequately absorb light energy, do not use all the energy in the biomass that they consumed, and have maximum active photosynthesis of 3–6% of total solar radiation. Closed systems of artificial microclimate allow plants to increase the efficiency of photosynthesis by up to 10%. The plant has internal processes: reflection and transmission of light energy, as well as a competing process in the form of photo-respiration and photosynthesis. Photosynthesis may be ineffective in certain conditions. Excessive light energy is dissipated to avoid damage to the photosynthetic apparatus. Excessive light energy is dissipated in the form of heat (non-photochemical quenching) and emits in the form of chlorophyll fluorescence (Ke 2001; Medvedev 2012).
- Thirdly, the environmental component means the use of large areas of land, light pollution from luminous agricultural complexes, the absence of closed systems and recycling systems. Automation of individual systems and inconsistency of work between systems is a massive waste of natural resources. This production requires a high level of specialists since decision making in the technological process depends on the expert assessment of a specialist. The low level of automation of labor-intensive processes requires manual labor.

Based on the previous, it is recognized that substantial greenhouse complexes have developed their resource for further technological and technical development. Such an organization of the greenhouse business incurs significant energy losses, considerable financial costs, and poses a severe environmental problem. Future technologies in greenhouse production should be based on the foundation and applying essential principles, such as a high environmental friendliness (closed-loop technology), the development of autonomy (renewable energy sources), energy-saving at all levels of production (technology of LED lighting systems), reduction in energy intensity (technology of soilless plant growing), automation, and robotization of closed systems of greenhouse production.

9.5 Technologies in Vertical Greenhouses

Vertical farming or vertical greenhouses AgTech there is a practice of growing crops at vertically located levels. Vertical greenhouses, as a rule, combine the most advanced technologies: soilless plant growing, LED lighting systems, automation/robotics, and many processes (constructive, engineering, technological, economic, and many others). The use of these technologies has now led to a more than tenfold increase in crop yields compared to traditional farming methods. The modern concept of vertical farming was proposed by Dickson Despommier, professor of public and environmental health at Columbia University (Despommier 2009, 2011). There are several developing concepts for vertical farming: Peyton (Devon, England) (2020); Israel (Green Zionist Alliance 2020), Singapore (Association for Vertical Farming in Singapore 2020), Baltimore (Company Gotham Greens–Baltimore, Maryland, USA 2020), Germany (Das Münchner AgTech-Unternehmen Agrando 2020), London (London-based startup Wefarm 2020), Japan (AgTech Innovation in Japan 2020), USA (American startups Freight Farms 2020); Brisbane (Eat Street's Modular Farm Brisbane in Australia 2020) and others.

9.5.1 Technologies for Growing Plants Without Soil

Hydroponics uses the ability of plants to consume nutrients dissolved in water. Natural soil acts as a source of minerals, but the presence of soil is not necessary for plant life. We know that plants feed on mineral salts in the form of electrically charged ions. There is currently no accurate and complete classification of hydroponics. We offer conditional classification by review and analysis of scientific literature (Fig. 9.14). Specialists distinguish three main methods of hydroponics: substrate culture (hydroponics and substrates); water culture (hydroponics); air culture (aeroponics).

We classify hydroponics according to the following criteria: environment around the roots, with or without substrate. Hydroponics has a feature of the environment around the roots: aquatic environment; the environment that forms the fog; an environment with an organic substrate; medium with an artificial substrate; combined medium.

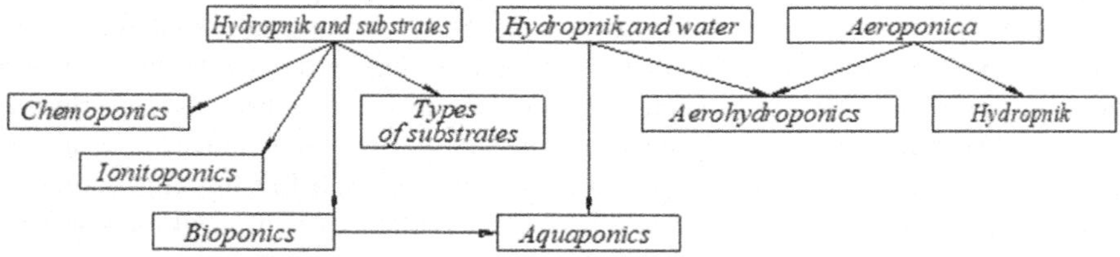

Fig. 9.14 Classification of technologies for growing plants without soil

Hydroponics is divided into active and passive. Active hydroponics uses a pump when water enters the reservoir with roots. Passive hydroponics uses the property of a substrate to raise moisture the voids and tubules with an artificial or organic substrate due to the surface tension force.

Aeroponics has a distinctive feature when there is no substrate. Aeroponics can be reversible and irreversible. Aeroponics has a reversible nutrient solution supply system, where the solution circulates continuously and flows back to the original reservoir. Aeroponics has a non-reversible nutrient solution supply system, where the solution is used once, so there is no need to control the parameters of the nutrient solution.

Aero-hydroponics is classified by a technical solution. Technical solutions may use a water pump, air pump, whirlpool ultrasonic generator.

Chemoculture, or culture of dry salts, is called a viable method in which the rooting of plants occurs in an organic substrate, which is saturated with nutrient solutions. Periodically, such a system is moistened with a nutrient solution. The main advantage of such a system is the possibility of the horizontal and vertical laying of the substrate on the plane. Chemoculture allows you to grow plants in adverse conditions (if there is no constant watering). Today, quite different methods have been created for introducing nutrients into horizontal and vertical chemoculture systems.

Ionitoponics is a method of growing plants in which ion-exchange materials are used as a substrate. Ionitoponics has the main difference from other methods; hydroponics uses substrates that can retain nutrient components for a long time. Therefore, you can water the plants directly with plain water without the additional introduction of macro and trace elements. However, watering should be required because the exchange of ions occurs in the aquatic environment. The intensity of the transfer of nutrients from the substrate to the decay products of the plant depends on many external factors: temperature, plant lighting, air humidity, and others.

Bioponics is organic hydroponics, in particular, the so-called aquaponics. Bioponics uses large microbial populations, which become the main obstacle to pathogens. Roots protect beneficial organisms with proper oxygenation, save water and fertilizer, and increase good quality food production.

Thus, the classification of hydroponics is the objective of the study of many scientific teams. This topic does not have a final methodological development. The rating and the effectiveness of certain types of hydroponics wait for further scientific study.

9.5.2 Hydroponics and Aeroponics

In recent decades, technology for growing plants without soil has taken many forms: hydroponics (growing plants in a nutrient aqueous solution and their roots are fixed in an inorganic permeable substrate), aquaponics (a type of hydroponics in which nutrients are extracted from fish waste), and aeroponics (growing plants from freely hanging roots in the air, which are periodically sprayed with a nutrient solution) (Kozai 2013; Schnitzler 2013; Lakhiar et al. 2018; Imran et al. 2018).

Table 9.1 The attractiveness of hydroponics technology

Techniques	Harvest (crop) (%)	Ripening speed (%)	Water consumption (%)
Aeroponics	130	150	10
Hydroponics	110	120	30
Traditional farming	100	100	100

Advantages of soilless plant cultivation: an order of magnitude higher yield per hectare; 10 times less water per unit of output is required; 5 times less fertilizer per unit of production; increased protection against diseases; no crop damage; and much more. The ratio of capital and operational costs an essential decision in the choice of technology. The more complex the technology, the more expensive it is. At present, hydroponics occupy a strong position, the main advantage of which is the simplicity of technical implementation, which makes it easier to control, operate, repair, cheaper to repair and replace components, and much more. Hydroponics does not cause crop loss if a failure occurs in the water supply system (Kozai et al. 2015).

In comparison with traditional hydroponics, aeroponics does not require any liquid or solid medium for growing plants (Kozai et al. 2020). Alternatively, a liquid solution of nutrients produces dust in the air chambers where the root portion of the plant is put through the nozzles. Of course, aeroponics is the most sustainable cultivation technology without the use of soil, since it uses up to 90% less water than the most effective conventional hydroponic systems and, most importantly, does not require replacing the nutrient medium. Aeroponics does not require additional aeration (the effect is absent when the roots are suffocating), there is no need for cleaning after using the nutrient solution (if correctly calculated, it is entirely absorbed by the roots). Aeroponics optimizes greater air access for more successful plant growth, unlike substrate methods. The plant in the aeroponic apparatus has 100% access to CO_2, which contributes to the accelerated growth of the plant. In addition, the absence of a nutrient medium requires aeroponic systems to follow a vertical structure that further saves energy as gravity naturally absorbs excess fluid.

In contrast, conventional horizontal hydroponic systems often require water pumps to control the excess solution. Water that circulates in hydroponics gives additional weight, which is reflected as a contribution to the material consumption of the structure (Table 9.1). Therefore, hydroponics is usually used in single-shelf technology (Benke and Tomkins 2017; Goddek et al. 2019a, b).

In general, hydroponics occupies a leading position in the city-farming market, which is estimated at $ 2 billion. Aeroponic systems have not received widespread use, since it requires more complicated calculations, more subtle and time-consuming settings, and therefore, more qualified maintenance personnel. Still, they are beginning to attract considerable attention. In the next 5 years, experts predict a faster growth of aeroponics, aerohydroponics, and aquaponics technologies because they are more economical and efficient than hydroponics systems (Battaglia 2017; Sidorenkov et al. 2017; Mytton-Mills 2018).

Automated vertical farms intend to transform the world of agriculture, and this is not surprising because advanced hydroponics and aeroponics technologies bypass

Fig. 9.15 The essence of "Hyponics" is aeroponics and computer technology (https://auto-grow.ru/assets/images/tickets/1776/cbafba0f04887c9b575327efc8bfc894d82bcd0c.pdf)

conventional crops are pioneers in change. There are many companies creating farms the size of a standard shipping container. Such farms can be used by novice entrepreneurs for the restaurant business because, in this way, it is possible to save on the transportation of plants and serve only fresh products on the table.

The "Hyponics" is an example of high technology. It is the most progressive subspecies of aerohydroponics, which appeared relatively recently due to the improvement of classical hydroponic methods. Today there are already melons, from one vine of which more than 90 fruits weighing 1.2 kg are collected, and a cucumber tree, giving 3000 cucumbers, and 6-m sugarcane. In 3 months, up to 4000 fruits ripen on this beautiful tomato. From one melon vine, the Japanese harvest 90 fruits with an average weight of 1.2 kg. And their "cucumber-tree" gives more than 3000 fruits. Natural sugarcane, instead of 2–3 m, almost two times faster than under normal conditions, grows to 6 m. They become even that papaya growth was not worse than in the tropics, although it does not grow in the Japanese climate.

The inventor of "Hyponics," Shigeo Nozawa, president of Kyowa, spent 20 years experimenting. The essence of this method is the use of aeroponics and computer technology (Fig. 9.15). The mixture is continuously circulating, feeding plants only with those substances that it needs, precisely in this period of development, and their concentration is several times higher than the plant could get from the soil (Hydropon East Magazine 2012).

Special equipment creates optimal temperature, humidity, and other characteristics, increasing the speed of development, and the ripening of fruits increases by 3–4 times. It took 1000 experiments and many years of painstaking work to develop nutritional compounds for each culture. The company keeps in strict confidence its solution formulas. It is known that no stimulants, hormonal, and other drugs that artificially accelerate growth are not used in them. The fruits grown on Hyponics are not inferior in quality to ordinary fruits and no danger or side effects for humans. According to the employees of the Japanese company Kyowa, using "Hyponics," you can grow any plant. An example of this is cucumber bush, which gives more than 3000 fruits, or a melon vine, with which up to 90 fruits are harvested with an average weight of 1.2 kg. A sugarcane reached 6 m in height, while in natural conditions, it does not grow more than 2–3 m; they began to grow tropical papaya in the Haydon way, as well as pumpkins in various colors and tobacco (Division of Hyponica 2020).

9.5.3 LED Lighting Systems

Currently, LEDs have no alternative. However, the LEDs have not reached the limit of their capabilities and are in the process of constant updating. LEDs have several unique advantages and many disadvantages.

LED is a semiconductor light source containing one or several crystals of the Crystals to be in one body with a lens. The lens forms the light flux. The LED has the main feature: the light that LED emits is in a narrow range of the spectrum. The range of radiation of the LED depends on the chemical composition of semiconductors.

The principle of operation of the LED based on electrical luminescence: cold emission occurs when the LED electric current flows. The electric current in the LED passes through the p-n junction in the forward direction, charge carriers of the semiconductor (electrons and holes) recombine with the emission of photons. Semiconductor materials have different efficiency, which emits light during the recombination. Semiconductors with different wavelengths create the LEDs from UV to mid-IR. LEDs have directional radiation through a plastic lens covering the LED chip.

All semiconductors are light-emitting diodes, currents of fear because the resistivity of semiconductors depends on temperature. The resistance of semiconductor decreases if the temperature rises, and there is an avalanche process which increases the electric current. The LED will burn if the electric current will increase. The lifetime of LEDs depends on temperature, while the quality of light depends on the lifetime of the LED. Illumination decreases with time, and labor is the aging (degradation) of the lamp.

The beautiful life of a typical LED lamp is 5000–100,000 h. Other manufacturers indicate more modest numbers, only 35,000 h. The lifetime of LED is the time that the device operates until the moment of failure, and this is not the final breakdown, a drop in performance below a certain level. Some manufacturers believe this threshold the reduction of the luminous flux by 30% of the nominal value, the other 50%. These data generally are not reported in promotional materials and in the documentation for luminaires that do not allow the buyer to make the right choice (Nikiforov 2015; Sun et al. 2017).

Developments in the LED market are developing very quickly. The LEDs that did not meet expectations, removed from production, and start a new, better one. Testing of the LED is carried out in extreme conditions (current strength and the temperature of the crystal are at the limit of acceptable values) within a relatively short period. The results of the tests extrapolated independence on a longer time interval for normal operating conditions (Liu et al. 2016).

Individual single LED is not used directly. The LED is part of a lighting unit (lamp, reflector), which produces the assembly of a LED matrix. The LED light has lighting equipment: driver block power supply, stabilizer. Manufacturers of lighting devices used to indicate the lifetime of the LEDs under normal conditions. The real power supply system cannot provide normal conditions. In the power system, there is always some voltage drop, voltage fluctuations, harmonics, etc. LED lights still work less of a lifespan that the manufacturer promises. The lifetime of the LED lamp depends on the driver (pulsed power supply) and secondary optics. The heat-dissipating system in LED lighting uses a ribbed aluminum radiator (Nikiforov 2005). Lenses on the secondary optics in LED lamps are usually made of plastic, which over time, becomes cloudy. The reflectors are mostly made of plastic, coated with a thin layer of metal. Here it may have the effect of tarnishing the metal surface. These problems are solved by the use of modern materials and the sealing of the luminary (Nikiforov 2009; Nikiforov 2012).

LEDs have many advantages relative to other light sources. The power of LEDs is the efficiency of conversion of electricity to light. Modern commercial LEDs have the indicator light output 170 LM/W, industrial 110 LM/W, and every year they improve.

LED lighting system effectively adjusts light modes in the spectral composition and intensity. Other light sources are not flexible regulation of the light conditions. The LED lighting system provides energy savings on lighting (not only material things but also environmental safety). Every kilowatt of energy that saves the LED reduces the combustion of fossil fuels, harmful emissions in the air, and no need to build new power stations, electric networks, and infrastructure. LED lamps have the advantage of simple disposal in comparison with other lights.

Technologies that grow plants and use LEDs in an artificial microclimate have a promising future. One of them is the opportunity to obtain a certified organic product and raw materials, year-round to grow crops of high productivity, to abandon herbicides and pesticides, and to reduce environmental pressure on natural resources and more.

The practical implementation of the lighting plants with LEDs in a synthetic controlled microclimate is faced with many difficulties (Horibe et al. 2018; Kozai et al. 2019; Kozai 2019). We were discussing the unresolved question, such as the effective range of optical energy for plant growth. The range for adequate growth and development of plants varies across crops, types, varieties, hybrids, according to the vegetation periods. The range should change depending on the vegetative period of plant development, changes of microclimate parameters, and other settings. Currently, there is no understanding based on what position the action of light fields; there is no clear theory, which can be used to reasonably select the optimal parameters of a bright environment for artificial microclimate (Donskoy et al. 2019; Klyuchka et al. 2019a, b; Klyuchka and Lukyanov 2019).

Spectrum is only one of the characteristics of the light environment and light environment is part of an artificial microclimate. An adjustable parametric climate, as habitat for plants, is a complex multifactorial object for the simulation (Klyuchka et al. 2018a, b). The plant as a research object can be divided into vegetative part and root part. The vegetative part of the plant responds to crucial factors: light, temperature, humidity, and gas composition of the air. The root part of the plant responds to the water temperature, water acidity (pH), the concentration of salts, oxygen content, and conductivity. Both plant parts are interdependent and react to internal changes with each other. Every single unit of these factors is a tool to encourage, influence, impact on plants. There are dynamics at the time of the processes. An essential biological element is the progressive growth and development of plants, the changes in the environment under the action of the life processes of plants (Li et al. 2018; Klyuchka et al. 2019a, b; Klyuchka and Lukyanov 2019).

Currently, there is no understanding based on what position the action of light fields; there is no clear theory, which can be used to reasonably select the optimal parameters of a bright environment for artificial microclimate.

Problems that associate artificial light microclimate and photobiological processes of plants impact on modern greenhouse technology. Modernization of greenhouse production takes place only at the level of individual systems. It makes a slight win for reducing energy consumption. The breakthrough could provide the studies that establish a definite relationship between microclimate parameters of internal processes in plants. The research will give an accurate prediction of the operation of the cultivation of greenhouse plants, increasing the productivity of plants, to obtain the specific biochemical composition of plant materials in conditions of artificial microclimate. Research is aimed at building the biotech systems to a new level, the so-called CPS. CPS is physical and engineered systems whose operations are monitored, coordinated, and integrate computing and communication core. Cyber-physical processes require dynamic-programming in real-time (Kozai et al. 2020; Niu et al. 2020; Takagaki et al. 2020).

9.5.4 Researches on LED Lighting Systems in Greenhouse

Electrical energy passes through several stages in the power supply system, such as generation, transmission, distribution, and consumption. The lighting system is a consumer of electrical energy, that is, part of the power supply system. The lighting system also creates a light environment for plants in greenhouses. The lighting system is an intermediary between the power supply system and the lighting environment for plants. The quality of electrical energy determines the quality of the artificial microclimate in greenhouses. The task of the quality of electric power is not unique. Consumers of electric energy must use electrical equipment that does not violate the operating mode of the entire network or consumers of electric energy should use additional material to eliminate negative phenomena. The negative aspects of the power supply system are as follows: the state of the network from a variable load schedule, reactive power, voltage drop or surge, asymmetry of the network, non-sinusoidal shape of the current and voltage curves, and more.

The energy component in the cost structure in the production of greenhouses (according to various sources) approaches 45% and can increase up to 65% depending on the climate. An event that reduces the consumption of electrical energy

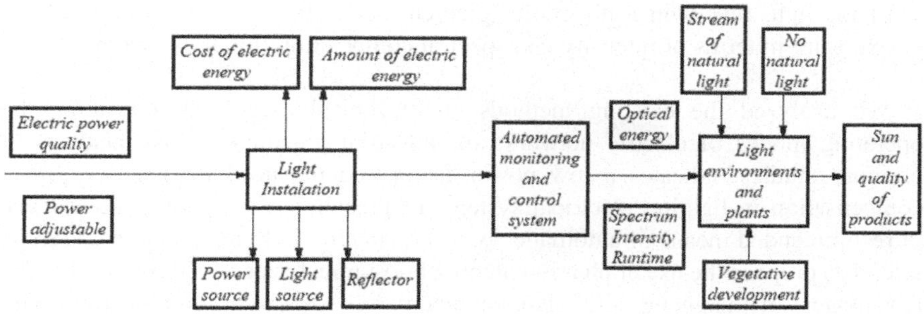

Fig. 9.16 A conceptual model of the organization of the light environment in a biotechnological system

is called energy-saving. For example, reducing the use of electric power for lighting is the replacement of gas-discharge light sources with LED light sources. Figure 9.16 shows the primary factors that affect the light environment of greenhouses.

An analysis of the literature found that the parameters and modes of electric lighting networks of greenhouses have several specific properties. The main feature is a weak correlation between load curves and reactive power. The reason for this property is associated with a significant degree of uncertainty about the operation of lighting electrical networks. Weather and climatic conditions (excess or lack of sunlight) create work uncertainty. The reason for this property limits the possibility of using traditional methods of voltage regulation and reactive power compensation in networks using batteries of statistical capacitors.

The conceptual model provides an analysis of factors that affect the light environment of the greenhouse (Fig. 9.16). The conceptual model illustrates a complex objective. The conceptual model is formed by functional modeling. Functional modeling consists of breaking the object under study into functional blocks. Function blocks detect points in the functional diagram. The aspects of the functional diagram are characterized by the optimization criterion and the main process parameters. The choice of lighting installation depends on several groups of factors:

- Economic factors: the amount of equipment used, electricity consumption, and electricity tariff;
- Power supply factors: power quality and power loss; equipment used for power control and voltage regulation;
- Factors that influence the choice of a light source according to light, technical and electrical, operational parameters;
- Factors that relate to the choice of schemes for switching on the lighting system, lighting equipment;
- Factors that influence the choice of an automation system, control, measurement, registration, control, dimming (adaptive dynamic lighting);
- Factors that influence the choice of lighting method (constant, variable, pulsed, multi-spectral, and others). The choice significantly affects the spatial distribution

of the light flux with a directivity gradient determined by the duration of the exposure in terms of intensity and spectral composition of optical energy.

We analyzed the existing methods and technical means of controlling the operating modes of electric networks of industrial greenhouses. An analysis of previous studies indicates a low power factor and the need for reactive power compensation in lighting electrical systems of greenhouses. It shows the absence of recommended means of automatic control regulation in lighting networks. Some scientists propose the use of phase-switching boost transformers. This device has the advantages of magnetic and semiconductor technology in combination with connecting secondary windings to various phases of the supply network, which allows a relatively simple and effective way to solve the problems of voltage regulation and compensation of reactive power. Booster transformer can create a large number of combinations of modules and phases of additional EMF. This property makes it possible to control the voltage and reactive power modes, and as a result, control the parameters of the electric lighting networks in the conditions of weak correlation of the responsive power load graphs.

Another problem with lighting electrical networks is a large proportion of non-linear loads exceeding the linear component. The nonlinear load is the source of higher harmonics in the electrical grid. Sources of nonlinear load are electric drives with electronic speed converters, air conditioning, and ventilation systems, single-phase consumers, high-pressure discharge lamps with electronic ballasts. Sources of nonlinear load are office equipment part of an automated control system, tracking, control: computers, servers, uninterruptible power supplies, LED light sources. In the future, the problem of nonlinear loads will only grow and increase.

Non-linear load distorts the shape of the current and voltage curves. Deformation of the shape of the current and voltage curves affects the reduction of the power factor of electrical equipment; for prolonged heating and to increase losses in the network; to rapid aging of insulation; to reduce the useful life; to the false triggering of the microprocessor relay protection system. Some scientists note the malfunctioning of measuring devices and monitoring devices. Problems for green-house enterprises increase capital costs for the replacement of electrical equipment and additional operating expenses.

Currently, scientists are developing organizational and technical measures to improve the reliability of the power supply system and the quality of electricity. LED lamps and single-phase power supplies are energy-saving technologies, but the main types of distorting loads in the electric network depend on LED equipment. Therefore, the problem is not trivial; the method of optimal design for lighting systems, which allows obtaining new structures of lighting networks. New arrangements of lighting networks should solve the problem of a rational configuration of electrical equipment for lighting networks of greenhouses to create an adjustable lighting environment for plants.

Thus, the potential for optimizing and improving the artificial lighting system for plants can be realized in the following areas:

- The power supply for lighting systems is important for a lighting system. The quality of power supply cannot be ignored; operational indicators of light sources depend on it; therefore, the characteristics of the light environment in which the plants are forced to be. The quality of the power supply determines the pulsation and duration of the entire lighting system.
- The electrical connection diagram is the main point in the design of lighting systems. A factor that takes into account the location of the light source relative to the plants in the volume of the greenhouse affects the intensity of the light environment, the number of necessary lighting devices, and energy consumption.
- To focus on the spatial redistribution of optical energy through various lighting methods (variable, pulsed, multi-spectral, and other lighting methods), which reduce the number of equipment in use, thereby reducing the load on the entire power supply system.
- Automated control systems use the principle of the consistency interdependence of the microclimate subsystems (innovative information and software technologies).
- Information and software technologies for the development of the concept of biotechnological feedback using methods of functional diagnostics of the physiological activity of plants.

9.6 Modern Vertical Greenhouses of the Third Level

Vertical greenhouses of the third level of constructive and technical implementation have a reasonably broad interpretation and, in a sense, they use "classical" greenhouse technologies: soilless plant growing, artificial light sources, automation of microclimate systems, and much more. The difference is that vertical greenhouses need to solve the issue of integration in the urban environment, to use a minimum area due to the multi-level design of technological surfaces. And also use closed process cycles. And so the estate, the success of a particular greenhouse depends on a combination of many individual factors. The analysis of vertical greenhouses among themselves, as well as vertical greenhouses in comparison with the "classical" greenhouse technologies, can be found in large numbers in the English-language scientific literature.

A study on the concept of a vertical farm optimization model, with excellent increasing conditions in a managed environment, improved energy, and waste flow, automation systems as well as modular system applications, was published in January 2017 (Final Report Vertical Farm 2.0 2015). The study Vertical Farm 2.0 (VF 2.0, Fig. 9.17) is the product of a parallel engineering (CE) workshop at the German Aerospace Center (DLR) in Bremen, together with the Association for Vertical Agriculture (AFV).

The purpose of the report was to develop a basic functional scenario in all areas from design, technical equipment, technological process, logistics process, and more. The VF 2.0 module has four development levels (2 for leafy greens and 2 for grape growing). On the ground floor, there is a center for processing, sorting,

Fig. 9.17 Basic functional model of the vertical greenhouse VF 2.0 (http://www.fao.org/3/a-i4904e.pdf%20Accessed%2013%20Feb%202020)

and storage of finished plant products. The project was provided by the possibility of adding a level of a roof greenhouse for expanding the production. Each cultivation module is designed as autonomously as possible as an independent unit, independently of other modules.

The following technologies were taken as the base model: hydroponics on a multi-rack installation, LED lighting system, varieties of lettuce, and tomatoes were selected. Conventional agricultural technologies were made in the calculation. Automated technology Nutrient Film Technique (NFT) (which is used by the American company AeroFarms) was chosen for the production of plants. The plant's lighting system and air management system cover almost 98% of the energy requirement (Table 9.2). Reducing energy costs is the main objective of reducing manufacturing costs. Vertical farms become a safe, economically feasible, and environmentally efficient way of generating a substantial amount of food for people in large cities.

The process design involves greenhouse development, the climate control system components, the power supply system element (NDS), and the structure itself. Unit inputs include seeds, electricity (light and heat/cooling), carbon dioxide, and nutrient irrigation water (Klyuchka et al. 2020). Reversible standards are water coming from the drain or wastewater vapor and excess heat in the air management system (AMS). The final results obtained through the crop are waste (uneatable material) and product (eatable material). Separate chapters include a detailed analysis of lighting requirements, climatic conditions. Each subsystem is associated with a part of the overall process.

Table 9.2 Total energy consumption per subsystem on the baseline scenario

Subsystem/relevant area	Yearly energy demand [kWh]	Distribution of energy use (%)	Chapter reference
Illumination system	12,152,463	69.9	7
Air management system	4,853,038	27.9	8
Nutrient delivery system	69,739	0.4	6
Plant health monitoring system	106,872	0.6	9
Horticulture operations	36,500	0.2	5
Growth floors core area	24,094	0.1	11
Ground floor core	15,067	0.1	11
Ground floor working area	125,008	0.7	11
Total	17,382,422		

The concept of the vegetative and generative stage, where most structures are linked either through the associated components or through the object itself, is a good example. The primary aim of this analysis is to evaluate the total value of VF. A well-designed roadmap of the information is necessary to achieve this goal regarding essential aspects in each subsystem to avoid bottlenecks that can potentially slow down the process of both design and operation.

The aim of the basic modules VF 2.0 is to achieve the most efficient yield, food quality standards, and protection, rather than the maximum return at the optimum cost. According to a study (Zeidler et al. 2013), waste disposal within the program would require high costs, which will minimize the effectiveness of the project. The device will use the currently available heat and water recovery technologies. Recycling of purified water for plants will be carried out. The heat from the LED lamps is removed from the production facilities and transported to other parts of the building using a heat pump. A tank can be created to store this energy for future applications.

Despite considerable experience in growing plants under artificial lighting, there is currently no single view of the optimal intensity (power) of illumination, the spectral composition of radiation, a dose of plant irradiation, photoperiod, etc. In almost every case, a peculiar lighting system is created that best meets the physiological needs of the cultivated plant. This direction is at the stage of formation and development.

The concept of the base module VF 2.0 uses artificial lighting. Designers admitted, based on the opinion of optimistic scientists, that the optimal light spectrum and photoperiod (Sabzalian et al. 2014; Kozai et al. 2015) maximize the yield of plant products. The model chosen is the Heliospectra Light Bar V101G-L which is configured for water cooling and designed explicitly for VF applications. It generates a spectrum of light precisely designed to enhance photosynthesis.

The result of the initial project of the VF 2.0 module was the construction of a real building measuring 75 × 35 meters, which is a particular starting point for further research and implementation of the most promising innovative technologies that city farming needs so much. The experiment is currently ongoing, and the results for a functioning VF 2.0 base model have not been published.

9.7 Vertical Farm 2.0

The final report, VF 2.0 on the design of a functionally and economically feasible vertical farm, as an innovative technology for sustainable agriculture, gave an ambiguous assessment and some conflicting conclusions. The purpose of the report was the demonstration of how realistic it is to implement this project in the current economic situation. Of course, in the project, the calculation was carried out both for capital and operational costs for equipment, so the warranty period is given for quite a long 30 years. A long-term warranty does not require the cheapest machine, which offers a guarantee of high reliability and fast high-quality repairs. At the same time, an extended warranty period provides for the modernization of gradual re-equipment of production with the advent of new technologies.

Moreover, in practice, multi-level vertical greenhouses have demonstrated economic success in Japan and Singapore, for example, where products are merchandised in a different retail segment at a slightly higher price than the EU or US markets.

The killer's criterion in this equation is not the technology itself, but the restrictions on reducing the cost of electricity and labor, which make up the majority of operating expenses, probably increasing these costs several times per 1 m^2. VF 2.0 needs additional expenditure without any difference in the production of a net economic benefit relative to regular plant food. This attempt exposes a fractured plan to address the crucial problems of food production today.

Another conclusion is that urban vertical greenhouses should demonstrate the full range of the most advanced agricultural technology in a controlled environment (CEA), which implements the principles of a closed cycle. The VF 2.0 report conducted a comparative analysis of a multi-layer growing system with large-scale thermal air masses in a greenhouse in terms of using recycling systems. For example, in an air movement control system combined with precise touch control of air quality, the VF 2.0 manufacturer gives a clear advantage in optimizing environmental parameters. There are several HVAC solutions in terms of climate control applications by creating overlapping sensor technologies to create multiple data points.

Automation using technology would be the most logical development direction to reduce labor costs. The report emphasized that any technology has an adaptation period when testing and debugging take place after installation. The period of transition to full production capacity may be delayed due to weak bottlenecks in technology not taken into account. It would be appropriate automation with the help of artificial intelligence (AI), which can be of great importance in the role of an advanced, informed expert, without spending years on the training of personnel and highly skilled specialists. The potential for machine learning and personalized feedback from sensors is essential and can further lead to the production of automated VF 2.0.

The concept, when it comes to the design project of the vertical greenhouse VF 2.0, is very inventive and unique, with such a mix of experts from various fields and young entrepreneurs. As a result, the AVF has no precedent in the agriculture of foreign organizations representing this form of operation and membership.

Existing typical agricultural organizations tend to focus primarily on the rural, regional representation of farmers or on the form of organization of production, which focuses mostly on national issues or, at best, on particular supply chains. Searching for versions to different types of organizations or networks that specialize in technologies of this size, AVF is the only foreign association with 300 members from both sides of the CEA and beyond. Therefore, AVF needs to be sure that this is the impetus for the development of innovative solutions in the field of food production. This project has the distinctive character of AVF, which supports new food production on a large scale and with high energy in this expanding market. Based on these efforts, the framework for the next step was successfully laid to begin work on the design of the new VF 3.0 base model.

Concepts related to the incorporation of vertical greenhouses in the city reflect valid information that is debated in small academic communities. Such principles (such as CPUL) and the underlying theories have not yet entered the general public consciousness or the political discourse; they have not yet reached the mainstream or do not reflect the opinion of the majority. The difference between existing standards and the concept of vertical greenhouses is the most critical obstacle to more extensive transformation and system integration-now and in the future. Global work is needed at all levels of state, scientific, educational, public, and so on.

Generally speaking, the understanding of new goods and innovations is critical for further implementation. Innovation, such as vertical greenhouses, depends on its social acceptability, especially in the early stages (Specht et al. 2016). "Acceptance" is defined as "the process or fact of receiving something as adequate, valid or appropriate" (Oxford Dictionary Acceptance 2014). Therefore, one of the specific goals of such research is to analyze people's attitudes to certain new technologies, especially those associated with risks. The widespread occurrence of risks and low public acceptance of innovations in biotechnology, energy production, GMOs, carbon capture, and storage, as well as precision farming, organic farming, and conservation agriculture is perceived. As for vertical greenhouses, concerns are raised about integration, the use of public resources, accessibility, technical complexity, and aesthetics of vertical greenhouse projects. Perceived risks are associated with possible low-quality products and potential health risks associated with urban pollution. Finally, the ecological and economic balance is being questioned (Kozai et al. 2019; Kozai 2019; Kozai et al. 2020; Niu et al. 2020).

Another type of potential threat applies to the implemented or suggested production. The crucial factor in this context is the stakeholder belief that the technologies used in vertical greenhouses (i.e., soil-free cultivation technologies) are unnecessarily complicated. This tool primarily relates to more technologically sophisticated systems, such as autonomous, automated, closed vertical greenhouse systems. This tool refers mainly to more technically advanced systems, such as independent, automatic, closed vertical greenhouse systems. The use of soilless cultivation or methods that use synergies between urban agriculture and buildings (for example, by combining heat, water cycles, waste, etc.) is also covered by this tool. Since sophisticated technologies are associated with high costs, it can be assumed that the technology of vertical greenhouses can contribute to higher property prices and, thus, change the surroundings.

The perceived high complexity and high operating costs of vertical greenhouses also pose the perceived risk of pursuing large enterprises as a profitable but unsustainable business. For example, vertical greenhouses are managed for profit without integrating social or other functions. This category of risks is directly related to the search for specialists who could design, implement a technical project, and maintain the technical part separately and separately greenhouse plants, and much more.

The development of vertical greenhouse technologies is associated with a diverse set of risks, according to many stakeholders. The main threats were related to urban integration of vertical greenhouses, the production system, the food itself, the environmental balance, and economic indicators. However, it has been shown that there are many risks associated with a lack of expertise, non-integrative policy-making, inadequate transfer of research findings to the general public, and a lack of ongoing demonstration projects (Kozai 2013, 2018; Kozai et al. 2019, 2020). Also, comparing the results in the available literature that current practice and market data negate some of the perceived risks. Further research should focus on the creation, dissemination, and dissemination of new data to raise awareness and knowledge through pilot and demonstration projects (He et al. 2019; Fang and Chung 2019; Paponov et al. 2020).

The German Center for Research and Innovation in Tokyo (DWIH Tokyo), together with the French Embassy in Japan and the Japanese Agency for Science and Technology (JST) invited 16 speakers from Germany, France, and Japan to Tokyo to discuss how artificial intelligence can help solve environmental problems. More than 150 participants took part in the one-day event. The panelists came up with four different points of view. At the first meeting, they presented the experience of developing policies about environmental issues and the digital transition process and their relationship to each country's national policies. Speakers emphasized the enormous potential of AI to promote more resilient societies, especially in the energy and mobility sectors. However, they also indicated a need to realize this potential. It is necessary to change the attitude of the population (Udaltsova et al. 2015; Niu et al. 2020). The second session discussed the use of artificial intelligence in agriculture and land use. Agriculture is one of the main factors contributing to global warming, and demand will grow as the world's population grows. Artificial intelligence applications such as robotic tractors or drones to test crop growth can significantly increase productivity and labor efficiency, and natural resource management will reduce greenhouse gas emissions. It was emphasized that environmental benefits must be economically attractive, which is still not always the case.

The third session brought artificial intelligence from rural areas to cities under the theme of applying artificial intelligence to Smart Cities. Thanks to AI, the mobility of connected, autonomous, typical, and electrical (CASE) can soon become a reality, and people will be able to use applications to compare the costs of various routes, for example, with a car-sharing service with independent driving. AI can also be used in "smart buildings" that are looking for ways to reduce energy consumption independently. However, the innovative systems that lay the foundation for smart cities vary widely between countries. It is also reflected in the question of who owns the enormous amounts of data processed in a smart city that affects personal data, legal business data, and government data. It was noted that Japan and Europe share similar positions regarding data privacy.

In the last session, examples of already used AI applications in the field of environmental protection were presented. For example, a demonstration project of a virtual power plant (VPP), implemented jointly by a German and Japanese company, the use of AI for climate service products, AI for studying the impact of climate change on ocean resources, and machine learning in the context of environmental hazard monitoring. Throughout the conference, the importance of the international and intersectoral exchange of AI and climate change issues has become apparent. Both transitions (digital and environmental) will occur at the global level and will affect almost all walks of life.

It is a vivid illustration of promising areas in all spheres of human life, including AgTech vertical greenhouses.

9.8 Research and Development Toward Greenhouse Agro-Technology

AgTech is an industry that grows plants in an artificial microclimate and has many prospects for the future. AgTech receives a certified organic product, allows high productivity all year long, eliminates the need for herbicides, pesticides, and reduces environmental stress on natural resources. Generally, it is attractive for investment. The practical implementation of the technology faces many difficulties in growing plants in an artificially controlled microclimate.

The biologist studies the processes of plant physiology at the level of molecules, cells, tissues, at the level of an individual organ, the whole organism (phylogenesis). Scientists with biological education direct their eyes to the internal processes of plant systems, and they ignore the technical component of the experiment. Biologists cannot make the right choice of a light source for many factors: light, technical, electrical, photometric, and other characteristics.

Engineers simulate the artificial environment in which greenhouse plants are forced to be. Scientists with technical education are aimed at the implementation of various electrical technologies, engineering systems, devices, process automation, instrumentation, and the use of information technology. Engineers do not genuinely understand the features of a biological object. Engineers research the development of specific analysis methods:

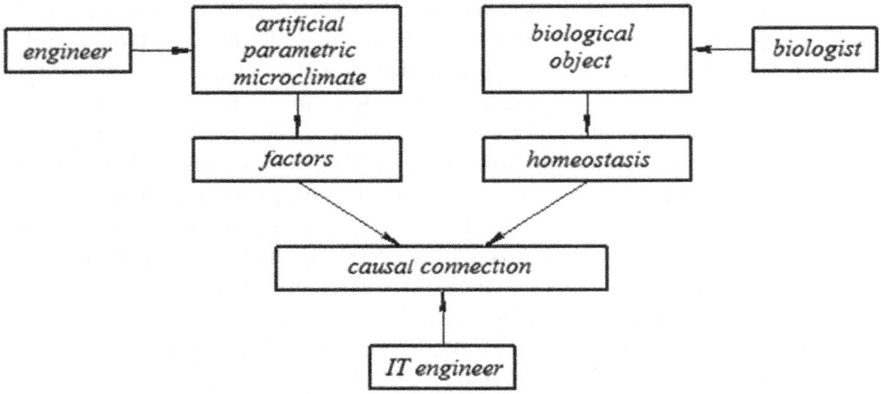

Fig. 9.18 Biotechnological system model

- Methods of analysis of electrical technologies which are methods of stimulating plants and seeds with fields of various nature: optical, laser, electric, magnetic, electromagnetic, acoustic and other areas,
- Methods of analysis of engineering systems that create an artificial microclimate and are a way of influencing plants,
- Analysis methods of instrumentation, which is an indirect tool for assessing the effectiveness of impact on plants.

An engineer strives to create an artificial microclimate with stable parameters that vary in a particular range. At first glance, the task is quite simple and is to create a favorable living environment. To assess the effectiveness of microclimate systems, engineers choose to increase the productivity of photosynthesis. However, the daily intensity of photosynthesis depends on many factors: on the environment, on the state of the plant at a given time, on the internal biological rhythm during the day. Engineers conclude the single results of increasing the productivity of photosynthesis. Plant physiology experts emphasize that photosynthesis productivity does not always correlate with overall plant productivity (Chikov 1997; Chikov et al. 2012).

Information technology experts are trying to unite biologists and engineers. Information technology specialists are trying to establish a connection between various systems: biological, technical, information. Information technology specialists solve specific analytical problems of a practical application: collecting information and processing results. The next part of the paper is the interpretation of the results and the determination of the causal relationship.

The model of the biotechnological system of artificial microclimate (Fig. 9.18) contains several studied objects:

- Parametric artificial microclimate as a technical complex system with many heterogeneous components;
- Plants like a complex living system;

- Information control and measurement system that establishes the relationship between the microclimate and the process of growing plants.

We conducted an analytical study on scientific works (231 scientific articles) that study the effect of LED lighting systems on plants in an artificial microclimate. The experiments are organized by scientific laboratories and production sites at various levels. The question that currently has no solution and causes discussion: the effective spectrum of optical energy for plant growth. An analysis of scientific research methodologies does not provide a universal and accurate answer. The lighting spectrum varies by plant cultures, varieties, hybrids, and by vegetative periods of development. The spectrum is part of an adjustable microclimate. The spectrum efficiency varies depending on temperature changes, on the gas composition of the air, the formation, and quality of the nutrient solution for the roots and other factors.

Currently, there is no sense: what is the basis of the position of the action of light fields of different spectra. There is no theory with the help of which it is possible to substantiate the choice of optimal parameters of the light-medium for an artificial microclimate. There are no criteria for assessing the effectiveness (the same for all) of the processes of plant growth and development. Scientific research methods need serious revision.

At present, the norm is accepted as optimal for plants separately for each element of the light-medium: norm in the spectrum, the norm in intensity, the norm in time of exposure. The patterns of the individual parts of the light-medium are chosen empirically (random nature), according to the reference and scientific literature on the results of past experiments on various crops and plant varieties. There is no information about the optimal ratios (proportions) between the elements of the light-medium inside the whole norm. The optimal norms are not the same for different rates (different dimensions) of the aspects of the light-medium. The ratio of the light environment elements is an independent factor that forms the light habitat of plants. Currently, this approach is poorly researched and not practiced. The problem in which it is necessary to find the optimal ratio of the elements of the light-medium is a more complex and time-consuming task.

Our hypothesis is as follows. The ratio of the elements of the light-medium (spectrum, intensity, exposure time) within the whole norm is a separate acting factor that affects all structural and service systems of plants.

A significant advantage of LED lighting modules over other lighting systems is the simple technical implementation of the control of optical energy by spectrum, by intensity, by duration. The exposure dose during the operation of the light installation is determined as the product of the intensity "E" on the exposure time "τ": $H = E \times \tau$. Thus, the same exposure dose can be obtained at different depths and exposure times. The LED lighting system provides a unique opportunity to use the power of the LED, which is an individual value and can vary. The power of the LED is regulated through changes in current and is determined according to Ohm's law: $P = I \times U$, where "I" is the current strength, and "U" is the voltage of the LED. A monochromatic LED of a particular spectrum determines the level of the experiment, where the intensity and time of exposure vary.

Efficiency criteria are the length of the roots and the length of the seedlings. A crucial qualitative indicator of the development of the seedling is the proportional ratio of the root length to the length of the seedling, thus harmonizing the development. The mass of seedlings gives indirect information about the leaf area of the seedling and the diameter of the stem. The graph is formed according to the results of an experiment from a family of transfer characteristics, which are the dependencies of the relative increase in parameters on exposure doses for various ratios of the elements of the light-medium. The analysis is carried out according to the graph of the curve family. The choice is made relative to the level of increase to determine the exposure dose. The dose level of exposure is determined based on two conditions. In the first condition, the level of exposure should be significant, or the relative increase in the parameter under study should be higher than the relative error of its determination. In the second condition, the line of the significant level must cross as many transfer characteristics as possible so that the corresponding spectral sensitivity curve consists of a more substantial number of points. Each monochromatic LED's definition with a different wavelength and exposure doses can increase germination and primary parameters of germination. On the graph of the transfer characteristics, draw a horizontal line for the selected level of growth and, at the initial intersection with the transfer characteristic for a given wavelength, determine the exposure dose. Analysis of all levels of research will allow you to create a "light recipe." The results are generated and accumulated in the database in the control and measuring climate control system.

We want to highlight several aspects that are not taken into account in many scientific studies.

9.8.1 Parametric Artificial Microclimate

We will show an ideal model for a biotechnological system for growing plants. As an object of study, the plant can be divided into vegetative and root parts. The vegetative part of the plant responds to the primary factors: light, temperature, humidity, gas composition of the air. The root part of the plant responds to water temperature, water acidity (or pH), salt concentration, oxygen content, and electrical conductivity. Both parts of the plants are interdependent and react to each other's internal changes.

The main factors of the microclimate interact with background factors: pressure, fluctuations in parameters, and so on. Each individual of these factors is an instrument of influence on plants. Thus, we create for each element a separate microclimate system: a control system for the light environment, a temperature control system, a humidity control system, a carbon dioxide concentration control system, and a power supply system.

At first glance, the task is simple: each microclimate system forms an optimal controlled parameter. But such a concept contradicts the common goal of the entire closed biotechnological system of plant growing: increasing the productivity of a biological object for the lowest energy costs. A biotechnological system that has many various components is determined by the "direction of the goal" of this system. In other words, we can organize a complex structure of individual elements, but the functionality of a complex system determines the relationship between these elements.

All factors interact with each other; therefore, it is necessary to take into account the dynamics of ongoing processes. The dynamics of the processes are stable, periodic, or oscillatory. For example, the work of light sources increases the temperature and humidity of the environment. As a result, the transpiration process in plants increases. Another example, the balance between carbon dioxide and oxygen is not stable due to the vital activity of plants. Another factor: uniform distribution of parameters in the volume of the artificial system. Uniform distribution of parameters provides circulation (speed of air, heat, humidity, gas, nutrient solution). An essential biological factor, such as the progressive growth and development of plants, is a change in the habitat under the influence of vital plant processes. Daily fluctuations in the process of photosynthesis and biological rhythms of all internal procedures of plants depend on optimal environmental conditions and will always be present.

The use of optical energy of a specific spectrum for the intensification of plant growth demonstrates positive experience. However, the practice of their application reveals that the biological effect is lower than expected and is unstable. This fact can be explained by the fact that technological, biological, and engineering issues are created for lighting systems that are isolated and not deep enough.

Thus, the scientific idea is as follows. The optimal ratio of the characteristics of the controlled light environment (spectrum, intensity, exposure time), which stimulates or inhibits the development of plant seeds during germination, is characterized by the functional spectral sensitivity of the seeds. Patterns provide the basis for creating a synthetic spectrum of LED arrays and make the optimal selection of the characteristics of the light-medium for a given culture. Models provide an idea of how to control the process of seed germination, which will increase germination, reduce germination time, and improve the quality of commodity seedlings of medicinal and green crops.

9.8.2 Plants as a Complex Living System

The plant plays two roles in scientific research. The first role threatens the plant as an object of study. That is, we are interested in variability, plasticity, adaptation of the living system, the law of irritability. In the second role, the plant as an assessment of the quality of the effectiveness of the technical system, to obtain a crop of high productivity with specific attributes of biochemical composition. We note: an artificial microclimate for plants is created through a long chain of experiments using the "trial and error" method.

The plant as a biological object is plastic, and the plant adapts to any proposed conditions to realize its genetic potential. The biological curve of plants clearly shows that any factor can pass from a positively stimulating to a limiting factor of both low and high intensity. It has been established that the range of positive effects of microclimate factors is quite narrow, unstable, and depends on other environmental factors. And here lies the tremendous potential for improving the efficiency of greenhouse production.

We must consider concepts such as ontogenesis and morphogenesis. Ontogenesis is the development of an individual organism, while the general laws of morphogenesis relate to the formation and differentiation of organs and tissues. Morphogenesis events (stable patterns) are embedded in ontogenesis events (individual development). The problem is to separate the personal characteristics of plants, seeds and highlight the models of morphogenesis of a particular crop and variety. Therefore, a qualitative research result should be based on a large sample of events. The flow of information from the many ongoing researches is not clear. In the future, we must learn to compare, evaluate the microclimate parameters with the parameters of the functional state of plants for a qualitative assessment of the cause–effect relationship. Scientists must obtain information, store, process, interpret, and apply in practice, as well as repeatedly verify the reproducibility of experiments and establish a pattern.

An artificial microclimate, as a habitat for plants, is a complex multifactorial object for modeling. The specific features of the artificial microclimate are reflected in the subjective individual reaction of plants. The potential of plants is difficult to assess with the advent of ever new varieties and hybrids that are obtained using biotechnology methods. We concluded that the properties of plants deprive the versatility of the parameters of the artificial microclimate. Each plant needs a particular personal environment for artificial microclimate.

9.8.3 Information Control and Measuring System

Information technology should bring together biologists and engineers to solve specific analytical problems of a practical application. Artificial microclimate creates a habitat for a biological object using engineering equipment, which is structured into separate subsystems of lighting, heating, air conditioning, ventilation, watering, power, gas composition, and more. The microclimate subsystems are served by automated control systems through instrumentation to coordinate work.

The empirical models describing biomass growth due to photosynthesis productivity depending on many factors have disadvantages. In essence, they require a large number of scientific multifactor experiments. The organization of multivariate analyses is a process that involves a lot of time, money, and high accuracy. The determination of the best parameters of a light source is carried out using various methods, including both mathematical (calculation methods for determining optical energy by quantity and quality) and physiological (by analyzing the reaction of plants to the lighting spectrum) methods (Nikiforov 2018; Good et al. 2019). But none of the methods can be adopted unambiguously since the reproducibility of scientific research and the quality of the predictability of the result remain very low.

Biotechnological systems for growing plants in an artificial microclimate are among the complex systems. Such complex systems include biological and technical objects with the presence of many components that are combined by a single technological process. Biotechnological systems require organized management. Control methods for closed technical systems are practically transferred to complex biotechnological systems of artificial microclimate.

A task that solves the problem of increasing the efficiency of the management process requires management based on a context-sensitive language. Currently, several insurmountable contradictions interfere with organizing the management of complex systems in a context-sensitive style. This circumstance forces us to use problem-oriented words that are close to nature, with further formalization following the requirements of the technical base of its presentation.

The control problem is related to the concept of biotechnological feedback. Biotechnological feedback is a combination of signal and structural information exchange between the components of a complex system. The idea of biotechnological feedback requires some extraordinary approaches to understanding the management of a complex system. In this case, it is necessary to take into account the features of components with fundamentally different properties. Management in technical systems is usually organized according to judgments based on mathematical modeling. The construction of a technical system with biological components imposes restrictions on the entire system, both of a technical nature and many other limitations. Evaluation of the biological part in a complex system raises the question of the appropriateness of applying a control action at any given time. Now the problems of the biotechnological system are being solved by increasing the complexity of the whole system by creating new subsystems and not established new connections.

An analysis of the work showed that the control problem boils down to isolating the physical signal with subsequent actions at the signal feedback level. The specificity of the biological component is found in a functional state, which is a reaction to past actions of the system, or as consequences of delayed response. Thus, an operational state does not make sense without a comparative analysis of possible functional states in the context of environmental monitoring in an artificial microclimate. An indicator of the assessment of the habitat is plants, on which all changes are reflected. Biotechnological feedback is a source for the formation of databases in a biotechnological system. Management boils down to the need for decision making through the use of systems with databases (knowledge). Biotechnological feedback is an unknown unexplored topic for the practical application of methods that provide an integrated assessment of the quality of the environment regarding the functional state of plants.

Currently, microscopy and analytical methods remain quite popular, which destroys the object as a result of taking a biological sample for analysis. These methods are aimed at assessing the performance of specific plant organs, which are not an integral assessment of the functional state of a biological object. In the last 10 years, studies of functional diagnostic methods have spread in practice. Functional diagnostics is an objective assessment that detects deviations from the norm based on measurements of physical, chemical, or other objective indicators using instrumental or laboratory research methods. An accurate estimate of the environment in an artificial microclimate and the plant's functional state obtains reproducible results. It establishes laws and patterns that can be applied to a wide range of plant objects. Signals of various functional origins can significantly improve the accuracy of early diagnosis, and therefore quickly record the change in the operational state of a biological purpose. Thus, functional diagnostics as the basis of biotechnological feedback is an indicator of the adaptive capabilities of plants through signs of mutual relations between various components of the system.

Currently, there are many methods related to the concepts of the functional state of a biological object (express diagnostics) for assessing the holistic image of plants: the radiography, the method of slow induction of chlorophyll fluorescence, hyperspectral analysis, phenotyping (pattern recognition), the method of acoustic response, technique determination of electric biopotential, electrography, and other methods. But microscopy and analytical methods remain quite popular, which destroys the object as a result of taking a biological sample for analysis. These methods are aimed at assessing the performance of specific plant organs, which are not an integral assessment of the functional state of a biological object. In the last 10 years, methods of functional diagnostics have spread in the practice of research. Functional diagnostics is an objective assessment that detects deviations from the norm based on measurements of physical, chemical, or other objective indicators using instrumental or laboratory research methods. An accurate evaluation of the environment in an artificial microclimate and the functional state of the plant obtains reproducible results and establishes the legality to be applied to a wide range of plant objects. Signals of various functional origins can significantly improve the accuracy of early diagnosis, and therefore quickly record the change in the functional state of a biological purpose. Thus, functional diagnostics as the basis of biotechnological feedback is an indicator of the adaptive capabilities of plants through signs of mutual relations between various components of the system.

The lack of understanding of the internal processes associated with the creation of an artificial microclimate and its effect on plant photobiological processes is reflected in current high-level greenhouse technologies. Modernization of greenhouse production occurs only at the level of individual systems, which makes a slight gain in terms of reducing the energy intensity of the process of growing greenhouse plants, increasing plant productivity, obtaining a particular biochemical composition of plant materials. A breakthrough can provide research to establish an unambiguous relationship between microclimate parameters and internal processes in plants. Based on these studies, biotechnological systems of a new level, the so-called CPS, can be built. CPS is physical and engineering systems whose operations are controlled, coordinated, and integrated by the computing and communication core. Cyber-physical processes will require dynamic self-programming in real-time.

The uniqueness of the situation at the moment is as follows:

- Lack of specialists who combine biological knowledge, technical knowledge, and knowledge of information technology.

- Lack of a simple understanding of the effect of controlled light on the internal processes of plants, including the impact of stimulating the process of germination of plant seeds.
- LED technology is developing rapidly. LED light sources make it possible to form a light-medium by spectral properties in the range from near IR to near UV with sufficient radiation intensity. At the same time, the efficiency of LED sources is continuously increasing. The research spectrum should cover LED not only light sources, but also LED power sources and the quality of the power supply system. LED lighting systems await further development.
- Microprocessor control systems allow you to effectively manage and monitor all parameters of the artificial microclimate and light environment. Precise control systems for plant growth dynamics await further development.
- Methods for determining the functional state of plants in the process of plant development make it possible to assess the influence of stimulating factors through real-time assessment of the functional activity of seeds. Analysis of the process of seed germination can be carried out without waiting for the appearance of external phenotypic characters. Methods for determining the functional state of plants are aimed at finding an integral parameter that is a consolidator of all other influences and factors. The essential indicator tracks the impact of all elements of artificial microclimate. The control and measuring system regularly sends data that is generated, analyzed, and creates a greenhouse operation system. Methods for determining the functional state of plants are waiting for further development and practical application.

9.9 Policies and Legal Framework Toward Greenhouse Agro-Technology

The goal of the AgTech greenhouse agricultural technology policy is to create such institutional conditions that will allow us to form a competitive product and enhance the competitiveness of agricultural enterprises. The main instrument of such a policy should be the digital economy. The digital economy is an adaptive cyber-physical system of systems. The digital economy allows at every moment the most rational use of the resources at its disposal for the fullest possible satisfaction of the needs of its participants. The core elements of the digital economy are constantly transforming integrated product and service systems (PSS).

The European Union (EU) Commission is a prime example of addressing climate and environmental issues (Brussels, December 11, 2019). The EU is trying to create an industrial strategy within the framework of the European Green Deal project to solve the double problem of green and digital transformation. The principal objective of the new political program will be to stimulate the development of leading markets for a climate-neutral in the EU and beyond. The new project should unite citizens in all their diversity, with national, regional, local authorities, civil society, and industry, working closely with EU institutions and advisory bodies. Digital technology is an essential tool for achieving the sustainable development goals of

The European Green Deal. The Commission aims to examine measures to ensure that digital technologies, such as artificial intelligence, 5G, cloud, and peripheral computing, and the Internet of things, can accelerate and maximize the impact of policies to combat climate change and protect the environment. The practice of the EU Commission has no precedent in the scale of the work done and in the number of countries that are united by one goal.

The European Green Deal project was the result of past work. A group of scientists from three international institutes held more than 20 months of various kinds of events: 18 official negotiations, bilateral meetings, and meetings of working groups of interested parties, after which an agreement was drawn up on June 28, 2017. The new organic regulation is EU regulation 2018/848. In 2017, organic regulation was finally agreed upon after one of the longest stages of debate and negotiation that the EU has seen. The European Council said that over the next 10 years, the commission would assess the compatibility of this practice with the principles of organic production. The commission will be able to use the results of the analysis, after which the commission will be able to submit an appropriate legislative proposal.

Organic food research in the USA has been underway for over 20 years. The USA National Organic Standards Board (NOSB), on November 1, 2017, decided that hydroponic and container greenhouses would be eligible for organic certification. The Straits Times (June 25, 2019) reported that Sky Greens, an island city, the vertical farm, was awarded the world's first national standard for organic vegetables grown in urban settings. The equivalence rules between organic in the USA and the EU have been controversial. All US and EU trade agreements now have another area in which manufacturing differences exist.

Thus, scientific research in the field of AgTech should develop in parallel with the development of concepts such as organic food products, functional food products, and the status of vertical farms at the level of international standards. AgTech standards will provide a fundamental foundation for building policies to address climate and environmental issues.

9.10 Conclusion

The evolution of greenhouse production is at the beginning of the journey, we must hurry. Over the next 50 years, scientists predict rapid climate change. Scientists estimate that when the air temperature rises for every 1 °C, 10% of agricultural land will be lost. Governments will face enormous challenges in the future related to safe food and water supply. Vertical farming can be potentially useful to increase food production and promote sustainable urban management.

Thus, we believe that there are no terrible greenhouse technologies. An essential factor in the success of technology is external conditions: the technological and intellectual level of the country; a unique way of thinking to combine technical and biological systems; non-standard methods of research and design of closed systems.

9.11 Future Roadmap Toward Greenhouse Agro-Technology and Sustainability

An optimistic forecast is provided by a parallel engineering workshop (CE) at the German Aerospace Center (DLR) in Bremen, together with the Association for Vertical Agriculture (AVF) (Final Report Vertical Farm 2.0 2015). The workshop brought together experts from around the world to share their experience and knowledge in collaboration. The seminar aimed to develop a basic functional scenario in all areas for design, technical equipment, technological process, logistics process, and more. Searching for equivalents of new types of organizations or platforms specializing in innovations of this scale, AVF is currently the only international organization with 300 members from all sides of CEA and beyond.

An analysis of the conclusions that scientists gave at the events could be the basis of the roadmap of the future on the path to agricultural greenhouse technologies and sustainable development:

- A vertical farming project can be seen as a model for simulating future projects. The main objective of the project is to formulate a criterion for comparing technologies and a methodology for assessing the effectiveness of agricultural technologies.
- The project plays the role of a platform on which the most advanced technologies were introduced or, for comparison, several different techniques.
- The project must have an open information platform with the same projects in different countries. An open-source concept would contribute to the accumulation of data for analysis and the search for patterns.
- The project should be developed continuously in the framework of the digital economy, logistics, business projects, and much more.
- The project should solve problems that are related to the issues of big data, artificial intelligence, robotics, and much more.

References

6th Global Botanic Gardens Congress Geneva, Switzerland (2017). https://www.bgci.org/our-work/services-for-botanic-gardens/bgci-congresses/bgci-global-botanic-garden-congresses/. Accessed 10 Mar 2020

AgTech Innovation in Japan (2020). https://blog.agthentic.com/agtech-innovation-in-japan-25d733f9d815. Accessed 10 Mar 2020

American Robot Greenhouse Firm Iron Ox (2020). http://www.iksmedia.ru/news/5536161-Robotizirovannaya-ferma-sposobna.html. Accessed 10 Mar 2020

American startups Freight Farms (2020). https://www.freightfarms.com. Accessed 10 Mar 2020

Association for Vertical Farming in Singapore (2020). https://www.hortidaily.com/article/9195770/japanese-vertical-farm-spread-brings-large-scale-expertise-to-the-avf/. Accessed 10 Mar 2020

Australian Container Farm Primary Module (2020). http://www.modularfarms.com.au/primary-module-2/. Accessed 10 Mar 2020

Banerjee A, Jhariya MK, Yadav DK, Raj A (2020) Environmental and sustainable development through forestry and other resources. Apple Academic Press, Palm Bay, p 400. https://doi.org/10.1201/9780429276026

Battaglia D (2017) Aeroponic gardens and their magic: plants/persons/ethics in suspension. Hist Anthropol 28(3):263–292. https://doi.org/10.1080/02757206.2017.1289935

Beacham AM, Vickers LH, Monaghan JM (2019) Vertical farming: a summary of approaches to growing skywards. J Hortic Sci Biotechnol 94(3):277–283. https://doi.org/10.1080/14620316.2019.1574214

Belgian company Urban Crop Solutions (2020). https://urbancropsolutions.com/ru. Accessed 10 Mar 2020

Benke K, Tomkins B (2017) Future food-production systems: vertical farming and controlled-environment agriculture. SSPP 13(1):13–26. https://doi.org/10.1080/15487733.2017.13940548

Blok C, Jackson BE, Guo X, de Visser PHB, Marcelis LFM (2017) Maximum plant uptakes for water, nutrients, and oxygen are not always met by irrigation rate and distribution in water-based cultivation systems. Front Plant Sci 8(562):1–15. https://doi.org/10.3389/fpls.2017.00562

Briggs S, Kennel C, Victor D (2015) Planetary vital signs. Nat Clim Change 5:969–970. https://doi.org/10.1038/nclimate2828

California (USA) company (Fodder Works) (2020). http://robotrends.ru/pub/1705/robot-dlya-vertikalnyh-ferm-nakormit-lyudyay-i-zhivotnyhhttps://www.fodderworks.net/products/fodderworks-instruction-manual. Accessed 10 Mar 2020

CAP – European Commission Common Agricultural Policy Union Common Agricultural Policy (2020). https://www.ec.europa.eu/agriculture/organic/home_en Accessed 10 March 2020. Accessed 10 Mar 2020

Chikov VI (1997) Ассоциация фотосинтеза с продуктивностью растений. Соровский учебный журнал 2:23–27 (in Russ.). http://www.pereplet.ru/nauka/Soros/pdf/9712_023.pdf. Accessed 23 Mar 2020

Chikov VI, Salyakhova GA, Safiullina GF, Zamalieva FF (2012) Чиков Фотосинтез, транспорт ассимиляции и продуктивность растений картофеля, выращенных в различных условиях освещения. Сельскохозяйственная биология 1:72–77. UDK 635.21/.24:57.043:581.132 (in Russ.). https://cyberleninka.ru/article/n/fotosintez-transport-assimilyatov-i-produktivnost-u-rasteniy-kartofelya-vyraschennyh-pri-raznoy-osveschennosti/viewer. Accessed 23 Mar 2020

Company Gotham Greens – Baltimore Maryland USA (2020). https://www.prnewswire.com/news-releases/from-steel-to-sustainability-gotham-greens-opens-greenhouse-in-baltimore-301011101.html. Accessed 10 Mar 2020

Das Münchner AgTech-Unternehmen Agrando (2020). https://www.presseportal.de/pm/141290/4519043. Accessed 10 Mar 2020

Despommier D (2009) The Rise of Vertical Farms. Sci Am 301(5):80–87. https://doi.org/10.1038/scientificamerican1109-80

Despommier D (2011) The vertical farm: controlled environment agriculture carried out in tall buildings would create greater food safety and security for large urban populations. J Verbr Lebensm 6:233–236. https://doi.org/10.1007/s00003-010-0654-3

Division of Hyponica (2020). http://www.kyowajpn.co.jp/hyponica/index.html. Accessed 10 Mar 2020

Dixon J, Beckes J, Hamilton C, Kant A, Lutz E, Peggiola S, Hee J (2003) Новыйвзгляднабогатствонародов–Индикаторыэкологическиустойчивогоразвития. pp 128 (in Russ.). http://documents.worldbank.org/curated/ru/155071468739138996/pdf/170461RU1ver1Exp10the0Measure0of0Wealth.pdf. Accessed 12 Mar 2020

Donskoy YD, Lukyanov AD, E Klyuchka, Petkovic M (2019) Research of the effect of discrete light sources on seeds of vegetable and green cultures and the possibility of their approximation to modified sunlight. In: Conference: proceedings of September 2019 IEEE east west design & test symposium (EWDTS), Batumi, Georgia. https://doi.org/10.1109/EWDTS.2019.8884391

Eat Street's Modular Farm Brisbane in Australia (2020). https://theweekendedition.com.au/food-drink/eat-street-northshore-modular-farm/. Accessed 10 Mar 2020

Fang W, Chung H (2019) Bioponics for lettuce production in a plant factory with artificial lighting. Acta Hortic 1227:593–598. https://doi.org/10.17660/ActaHortic.2018.1227.75

FAO (2018) The state of world fisheries and aquaculture 2018 - achieving the sustainable development goals. Rome. http://www.fao.org/3/i9540ru/I9540RU.pdf. Accessed 15 Feb 2020

FAO (2020) The Food and Agriculture Organization of the United Nations. http://www.fao.org/home/en/. Accessed 08 Mar 2020

Farmbot-3 (2020). https://farm.bot/. Accessed 23 Feb 2020

Final Report Vertical Farm 2.0 (2015) Climate-smart agriculture: a call for action, Food and Agriculture Organization of the United Nations. http://www.fao.org/3/a-i4904e.pdf Accessed 13 Feb 2020. Accessed 10 Mar 2020

Fredani K (2010) Vertical plant production as a public exhibit at Paignton Zoo. In: Proceedings of the 4th global botanic gardens congress. pp 1–8. https://www.bgci.org/files/Dublin2010/papers/Frediani-Kevin.pdf. Accessed 23 Feb 2020

French startup Agricool (2020). https://propozitsiya.com/francuzkiy-startap-stvoryuie-vertikalni-fermi-dlya-viroshchuvannya-polunic, http://green-city.su/gorodskie-fermy-potreblyayut-na-90-menshe-resursov/. Accessed 23 Feb 2020

Goddek S, Joyce A, Kotzen B, Dos-Santos M (2019a) Aquaponics and global food challenges. In: Goddek S, Joyce A, Kotzen B, Burnell G (eds) Aquaponics food production systems. Springer, Cham, pp 3–17. https://doi.org/10.1007/978-3-030-15943-6_1

Goddek S, Joyce A, Kotzen B, Burnell GM (2019b) Aquaponics food production systems: combined aquaculture and hydroponic production technologies for the future. Springer, Cham. https://doi.org/10.1007/978-3-030-15943-6

Good LA, Kalashnikova EA, Tarakanov IG (2019) Влияниесветаразногоспектральногодиапазонанаморфогенезежевики и малины*in vitro*. Лесохозяйственная информация 2:97–102. UDC: 634.7/57.085.23 (in Russ.). https://doi.org/10.24419/LHI.2304-3083.2019.2.09

Gorchakov YV (2004) Development trends and market aspects of global organic farming. Barnaul, Az Buka, p 256

Gorchakov YV (2009) Agritourism in Europe and the USA: the experience of farmers. Bull Veg Grow 3:38–43

Gorchakov YV, Durmanov DN (2002) World organic farming of the XXI century. PAIMS, p 402

Green Zionist Alliance GZA (2020) Bold resolutions for 36th world Zionist congress green prophet. Impact news for the Middle East. https://www.greenprophet.com/2010/06/green-zionist-alliance-gza-resolutions/. Accessed 10 Mar 2020

Greenpeace International Public Environmental Organization (2020). https://greenpeace.ru. Accessed 10 Mar 2020

Gres RA, Kazachkova YS, Kholmatov SR (2019) Spatial patterns of the distribution of technologies of vertical farms and their impact on the geography of agriculture. Rostov Sci J 1:437–444. https://www.elibrary.ru/item.asp?id=36833776

Grocon Pixel Green Office Building and Greenhouse (2020) (Melbourne, Australia). https://stephenvaradyarchitraveller.com/2018/02/11/melbourne-pixel-building-australia. Accessed 10 Mar 2020

Habitat III Conference (2016). https://www.un.org/sustainabledevelopment/en/habitat3/. Accessed 13 Feb 2020

He D, Kozai T, Niu G, Zhang X (2019) Light-emitting diodes for horticulture. In: Li J, Zhang GQ (eds) Light-emitting diodes. Solid state lighting technology and application series, Springer, Cham, pp. 513–547. doi:https://doi.org/10.1007/978-3-319-99211-2_14

Horibe T, Imai S, Matsuoka T (2018) Effects of light wavelength on daughter cladode growth and quality in edible cactus Nopalea cochenillifera cultured in a plant factory with artificial light. J Hortic Res 26(2):71–80. https://doi.org/10.2478/johr-2018-0018

Hydropon East Magazine (2012) Hyponika: new technologies in hydroponics. https://www.hydroponeast.com/. Accessed 10 Mar 2020

Iconographer OG (2018) Экоархитектуравертикальныхфермкакноваятипологияагропромышленныхзданийгородскогохозяйствабудущего. Bull Samara Sci Cent Russ Acad Sci 3 (60):34–41 (in Russ.). https://cyberleninka.ru/article/n/ekoarhitektura-vertikalnyh-ferm-kak-novaya-tipologiya-agropromyshlennyh-zdaniy-gorodskogo-hozyaystva-buduschego. Accessed 12 Feb 2020

Imran AL, Gao J, Tabinda NS, Farman AC, Noman AB, Waqar AQ (2018) Monitoring and control systems in agriculture using intelligent sensor techniques: a review of the aeroponic system. J Sens 2018:8672769. https://doi.org/10.1155/2018/8672769

Japanese company Mirai Mirai Corp (2020). http://3476.jp/en/about/profile.html. Accessed 10 Mar 2020

Japanese Robotic Greenhouse by technology company Spread Co (2020). https://www.agritecture.com/blog/2018/12/2/japan-plants-to-open-worlds-largest-automated-leaf-vegetable-factory

Jhariya MK, Banerjee A, Meena RS, Yadav DK (2019a) Sustainable agriculture, forest and environmental management. Springer, Singapore, p 606. https://doi.org/10.1007/978-981-13-6830-1

Jhariya MK, Yadav DK, Banerjee A (2019b) Agroforestry and climate change: issues and challenges. Apple Academic Press, Palm Bay, p 335. https://doi.org/10.1201/9780429057274

Kantor EL (2013) Rent and economic valuation of natural resources. Theor Econ 1:7–13. https://cyberleninka.ru/article/n/renta-i-ekonomicheskaya-otsenka-prirodnyh-resursov

Kapelyuk ZA, Aletdinova AA (2017) Вертикальноесельскоехозяйствокакноваяконцепцияразвитияаграрногосектора. Интернет-журнал "Науковедение" 9(6):1–7 (in Russ.). https://cyberleninka.ru/article/n/vertikalnoe-selskoe-hozyaystvo-kak-novaya-kontseptsiya-razvitiya-agrarnogo-sektora. Accessed 11 Mar 2020

Ke B (2001) Photosynthesis: photobiochemistry and photobiophysics. Springer, Dordrecht. https://doi.org/10.1007/0-306-48136-7

Kemp R, Loorbach D (2003) Governance for sustainability through transition management. In: EAEPE 2003 Conference, November 7–10. Maastricht, the Netherlands. pp 1–26. https://eaepe.org/content/documents/ConferenceArchive/EAEPE_Confernece2003_Programme.pdf. Accessed 12 Mar 2020

Khan N, Jhariya MK, Yadav DK, Banerjee A (2020a) Herbaceous dynamics and CO_2 mitigation in an urban setup- a case study from Chhattisgarh, India. Environ Sci Pollut Res 27(3):2881–2897. https://doi.org/10.1007/s11356-019-07182-8

Khan N, Jhariya MK, Yadav DK, Banerjee A (2020b) Structure, diversity and ecological function of shrub species in an urban setup of Sarguja, Chhattisgarh, India. Environ Sci Pollut Res 27 (5):5418–5432. https://doi.org/10.1007/s11356-019-07172-w

Klyuchka E, Lukyanov A (2019) The biotechnological method of medicinal plants cultivation with Aeroponics and LED lamps. XXIV Conf Biotechnol Int Particip 606:721–729. https://www.afc.kg.ac.rs/files/data/sb/zbornik/Zbornik_radova_2_-_SB2019.pdf

Klyuchka E, Radin V, Groshev M, Kambulov S (2018a) Problems of modeling of complex technological systems of greenhouse production. MATEC Web Conf 226:1–6. https://doi.org/10.1051/matecconf/201822602020

Klyuchka E, Radin V, Groshev L, Maksimov V (2018b) Modelling a complex technical system of greenhouse production: the foundations of an interdisciplinary approach. MATEC Web Conf 226:1–6. https://doi.org/10.1051/matecconf/201822602019

Klyuchka E, Kuznetsov D, Dudnik V, Lukyanov A, Gaponov V (2019a) New methods of seeds functional state and activity control for the development of the biotechnical feedback concept. AIP Conf Proc 2188(1):030015. https://doi.org/10.1063/1.5138408

Klyuchka E, Donskoy D, Lukyanov A, Tregubenko A, Slashchev I (2019b) Simulation of light environment in a climatic chamber (phytotron). AIP Conf Proc 2188(1):030013. https://doi.org/10.1063/1.5138406

Klyuchka E, Lukyanov A, Donskoy D, Petkovic M (2020) Digital roentgenography is an innovative method of assessing seed quality. In: XXV conference on biotechnology faculty of agronomy Čačak, Serbia, Proceedings. pp 563–569. http://www.afc.kg.ac.rs/files/data/sb/zbornik/Zbornik%20radova%20-%20SB2020%20-%202a.pdf

Kozai T (2013) Plant factory in Japan - current situation and perspectives. Chron Horticult 53 (2):8–11. https://www.researchgate.net/publication/260871244_Plant_Factory_in_Japan__Current_situation_and_perspectives. Accessed 15 Feb 2020

Kozai T (2018) Benefits, problems and challenges of plant factories with artificial lighting (PFALs): a short review. Acta Hortic 1227:25–30. https://doi.org/10.17660/ActaHortic.2018.1227.3

Kozai T (2019) Towards sustainable plant factories with artificial lighting (PFALs) for achieving SDGs. Int J Agr Biol Eng 12(5):28–37. https://doi.org/10.25165/j.ijabe.20191205.5177

Kozai T, Niu G, Takagaki M (2015) Plant factory: an indoor vertical farming system for efficient quality food production. Elsevier, Amsterdam. https://www.elsevier.com/books/plant-factory/kozai/978-0-12-801775-3

Kozai T, Yumiko A, Hayashi E (2019) Towards sustainable plant factories with artificial lighting (PFALs): from greenhouses to vertical farms. Burleigh Dodds Ser Agric Sci. https://doi.org/10.19103/AS.2019.0052.06

Kozai T, Niu G, Takagaki M (2020) Plant factory: an indoor vertical farming system for efficient quality food production, 2nd edn. Elsevier, Amsterdam. https://doi.org/10.1016/C2018-0-00969-X

Kumar S, Meena RS, Jhariya MK (2020) Resources use efficiency in agriculture. Springer, Singapore, p 760. https://doi.org/10.1007/978-981-15-6953-1

Lakhiar IA, Jianmin G, Syed TN, Chandio FA, Buttar NA, Qureshi WA (2018) Monitoring and control systems in agriculture using intelligent sensor techniques: a review of the aeroponic system. J Sens 2018:1):1–1)18. https://doi.org/10.1155/2018/8672769

Lee J, Bagheri B, Kao HA (2015) A cyber-physical systems architecture for industry 4.0-based manufacturing systems. MFGLET 3:18–23. https://doi.org/10.1016/j.mfglet.2014.12.001

Li Q, Li X, Tang B, Gu M (2018) Growth responses and root characteristics of lettuce grown in aeroponics, hydroponics, and substrate culture. Horticulturae 4(4):35. https://doi.org/10.3390/horticulturae4040035

Lisovsky AL (2018) Оптимизациябизнес- процессовдляперехода к устойчивомуразвитию в условияхчетвертойпромышленнойреволюции. SDRM 4:10–19 (in Russ.). https://doi.org/10.17747/2078-8886-2018-4-10-19

Liu S, Liu W, Ji W, Yu J, Zhang W, Zhang L, Xie W (2016) Top-emitting quantum dots light-emitting devices employing microcontact printing with electric field-independent emission. Sci Rep 6(22530):1–9. https://doi.org/10.1038/srep22530

London-Based Startup Wefarm (2020). https://www.eu-startups.com/2019/10/london-based-agtech-startup-wefarm-secures-e11-7-million-to-scale-its-smallholder-agricultural-ecosystem/. Accessed 05 Mar 2020

Lovell ST (2010) Multifunctional urban agriculture for sustainable land use planning in the United States. Sustainability 2(8):2499–2522. https://doi.org/10.3390/su2082499

Medvedev SS (2012) Физиологиярастений: учебник. BHV-Petersburg, St. Petersburg, p 512 (in Russ.). https://bhv.ru/product/fiziologiya-rastenij-uchebnik/. Accessed 15 May 2020

Meena RS, Lal R (2018) Legumes for soil health and sustainable management. Springer, Singapore, p 541. https://doi.org/10.1007/978-981-13-0253-4_10

Meena RS, Kumar V, Yadav GS, Mitran T (2018) Response and interaction of *Bradyrhizobium japonicum* and Arbuscular mycorrhizal fungi in the soybean rhizosphere. Rev Plant Growth Regul 84:207–223

Meena RS, Kumar S, Datta R, Lal R, Vijaykumar V, Brtnicky M, Sharma MP, Yadav GS, Jhariya MK, Jangir CK, Pathan SI, Dokulilova T, Pecina V, Marfo TD (2020) Impact of agrochemicals on soil microbiota and management: a review. Land (MDPI) 9(2):34. https://doi.org/10.3390/land9020034

Meena RS, Lal R, Yadav GS (2020a) Long term impacts of topsoil depth and amendments on soil physical and hydrological properties of an Alfisol in Central Ohio, USA. Geoderma 363:1141164

Meena RS, Lal R, Yadav GS (2020b) Long-term impact of topsoil depth and amendments on carbon and nitrogen budgets in the surface layer of an Alfisol in Central Ohio. Catena 194:104752

Miller A (2011) Scaling up or selling out. A critical appraisal of current developments in vertical farming. MSc thesis, Carleton University Ottawa, Canada. https://curve.carleton.ca/system/files/etd/748f35ee-2799-4ff4-80b6-de3032dd65f7/etd_pdf/a275c00b3631e12bc0a228ccf79bb6cd/miller-scalinguporsellingoutacriticalappraisalof.pdf

Mytton-Mills H (2018) Reimagining resources to build smart futures: an agritech case study of aeroponics. In: Dastbaz M, Naudé W, Manoochehri J (eds) Smart futures, challenges of urbanisation, and social sustainability, pp 169–191. https://doi.org/10.1007/978-3-319-74549-7_10

National Council of Organic Standards NOSB (2020). https://www.ams.usda.gov/rules-regulations/organic/nosb. Accessed Mar 10 2020

National Science Foundation (2020) USA: National Science Foundation. 2415 Eisenhower Avenue, Alexandria, Virginia 22314. http://www.nsf.gov. Accessed 10 Mar 2020

Nikiforov SG (2005) Почемумногиесветодиодыневсегдаработаюттак, какхотятихпроизводители. Компоненты и технологии 7:16–24 (in Russ.). https://cyberleninka.ru/article/n/pochemu-mnogie-svetodiody-ne-vsegda-rabotayut-tak-kak-hotyat-ih-proizvoditeli/viewer. Accessed 17 Feb 2020

Nikiforov SG (2009) Исследованиеновогосемействамощныхсветодиодов CREE XLamp XP-E. Светодиоды, светодиодныекластеры и сборки 2:20–22 (in Russ.). http://www.led-e.ru/assets/files/pdf/2009_2_20.pdf. Accessed 17 Feb 2020

Nikiforov SG (2012) Актуальностьисследования и необходимостьсовершенствования методовизучениядеградациипараметровсветодиодовнаосноветвердыхрастворовAlGaInP араметровсветодиодовнаосноветвердыхрастворовAlGaInP и AlGaInN. Полупроводниковаясветотехника 1:36–37 (in Russ.). https://www.led-e.ru/order.php?year=2012&number=1. Accessed 17 Feb 2020

Nikiforov GG (2015) Прогнозсрокаслужбы и измененияпараметровпромышленныхсветодиодовпринаработке с помощьюфотометрическогометода. Инновации и инвестиции 1:152–156 (in Russ.). http://arhilight.ru/images/prez/arhilight_buklet_2018.pdf. Accessed 17 Feb 2020

Nikiforov SG (2018) Современное состояние оценки и современные методы измерения параметров приборов для тепличного освещения, Полупроводниковая светотехника 2:68–75. (in Russ.). https://www.elibrary.ru/item.asp?id=32839543. Accessed 17 Feb 2020

Niu G, Kozai T, Sabeh N (2020) Physical environmental factors and their properties. In: Kozai T, Niu G, Takagaki M (eds) Plant factory: an indoor vertical farming system for efficient quality food production, 2nd edn. Academic Press, Cambridge, MA, pp 185–195. https://doi.org/10.1016/B978-0-12-816691-8.00011-X

Oxford Dictionary Acceptance (2014). www.oxforddictionaries.com/de/definition/englisch/acceptance. Accessed 25 Feb 2020

Paponov M, Kechasov D, Lacek J, Verheul MJ, Paponov IA (2020) Supplemental light-emitting diode inter-lighting increases tomato fruit growth through enhanced photosynthetic light use efficiency and modulated root activity. Front Plant Sci 10(1656):1–14. https://doi.org/10.3389/fpls.2019.01656

Peyton (Devon, England) (2020). https://foodtank.com/news/2016/09/englands-paignton-zoo-feeding-animals-from-the-ground-up/. Accessed 07 Mar 2020

Pfeiffer S (2017) The vision of "Industrie 4.0" in the making—a case of future told, tamed, and traded. NanoEthics 11:107–121. https://doi.org/10.1007/s11569-016-0280-3

Raj A, Jhariya MK, Harne SS (2018) Threats to biodiversity and conservation strategies. In: Sood KK, Mahajan V (eds) Forests, climate change and biodiversity. Kalyani Publisher, New Delhi, pp 304–320

Raj A, Jhariya MK, Yadav DK, Banerjee A (2020) Climate change and agroforestry systems: adaptation and mitigation strategies. Apple Academic Press, Palm Bay, p 383. https://doi.org/10.1201/9780429286759

Ripple WJ, Wolf C, Newsome TM, Barnard P, Moomaw WR (2020) World scientists' warning of a climate emergency. BioSci 70(1):8–12. https://doi.org/10.1093/biosci/biz088

Rojko A (2017) Industry 4.0 concept: background and overview. IJIM 11(5):77–90. https://doi.org/10.3991/ijim.v11i5.7072

Sabzalian MR, Heydarizadeh P, Zahedi M, Boroomand A, Agharokh M, Sahba MR, Schoefs B (2014) High performance of vegetables, flowers and medicinal plants in a red-blue LED incubator for indoor plant production. Agron Sustain Dev 34:879–886. https://doi.org/10.1007/s13593-014-0209-6

Schnitzler WH (2013) Urban hydroponics for green and clean cities and for food security. Acta Hortic 1004:13–26. https://doi.org/10.17660/actahortic.2013.1004.1

Schulze E, Pakhomova NV, Nesterenko NY, Krylova Y, Richter KK (2015) Традиционное и органическоесельскоехозяйство: анализсравнительнойэффективности с позицииконцепцииустойчивогоразвития. ВестникСанкт-ПетербургскогоУниверситета 5(4):4–39 (in Russ.). https://cyberleninka.ru/article/n/traditsionnoe-i-organicheskoe-selskoe-hozyaystvo-analiz-sravnitelnoy-effektivnosti-s-pozitsii-kontseptsii-ustoychivogo-razvitiya. Accessed 07 Mar 2020

Sidorenkov VA, Losev OD, Anchutin VA (2017) Мехатронноеустройстводлявыращиваниярастенийнакосмическихаппаратах и станциях. Символнауки, 12:32–37 (in Russ.). https://cyberleninka.ru/article/n/mehatronnoe-ustroystvo-dlya-vyraschivaniya-rasteniy-na-kosmicheskih-apparatah-i-stantsiyah. Accessed on 07 Mar 2020

Specht K, Siebert R, Thomaier S (2016) Perception and acceptance of agricultural production in and on urban buildings (Z Farming): a qualitative study from Berlin, Germany. Agric Hum Values 33:753–769. https://doi.org/10.1007/s10460-015-9658-z

Sun B, Fan X, Ye H, Fan J, Qian C, Van Driel W, Zhang G (2017) A novel lifetime prediction for integrated LED lamps by electronic-thermal simulation. Reliab Eng Syst Safe 16:14–21. doi: https://doi.org/10.1016/j.ress.2017.01.017

Takagaki M, Hara H, Kozai T (2020) Micro- and mini-PFALs for improving the quality of life in urban areas. In: Kozai T, Niu G, Takagaki M (eds) Plant factory: an indoor vertical farming system for efficient quality food production, 2nd edn, pp 117–128. https://doi.org/10.1016/B978-0-12-816691-8.00006-6

Tornaghi C (2014) Critical geography of urban agriculture. PIHG 38(4):551–567. https://doi.org/10.1177/0309132513512542

Trachuk A, Linder N (2018) Четвертаяпромышленнаяреволюция: каквлияетинтернетвещейнавзаимодействиепромышленныхкомпаний с партнерами? Стратегические решения и риск-менеджмент 3:16–29 (in Russ.). https://doi.org/10.17747/2078-8886-2018-3-16-29

Udaltsova NL, Kozhanov EN, Gorbulina DV (2015) Инновационныйуспехяпонии: мифилиреальность? Вопросыинновационнойэкономики 5(2):37–46 (in Russ.). https://doi.org/10.18334/inec.5.2.440

Union Financial institution of Switzerland (2020). https://www.swissfinancecouncil.org/en/. Accessed 08 Mar 2020

United Nations (2016) The world's cities in 2016. Department of Economic and Social Affairs, Population Division – World Urbanization Prospects: the 2014 Revision. http://www.un.org/en/development/desa/population/publications/pdf/urbanization/the_worlds_cities_in_2016_data_booklet.pdf. Accessed 09 Mar 2020

US Department of Agriculture (2020). https://www.usda.gov. Accessed 08 Mar 2020

Vertical Farm in Romainville, France (2020). http://ifarmproject.ru/verticalfarmphotonews. Accessed 23 Feb 2020

Zeidler C, Schubert D, Vrakking V (2013) Feasibility study: vertical farm EDEN. Institute of Space Systems, German Aerospace Center. Bremen. fhttps://www.researchgate.net/publication/259899768_Feasibility_Study_Vertical_Farm_EDEN. Accessed 23 Feb 2020

Bioremediation of Lead Contaminated Soils for Sustainable Agriculture

10

Dyhia Boukirat and Mohamed Maatoug

Abstract

Metal pollutants such as lead, zinc, and copper are among the most widespread pollutants in the environment and especially in our soils, their origin is diverse and their persistence is worrying. In Algeria, the main sources of pollution are petroleum derivatives and industrial effluent, and studies have found metal levels up to 2712 ppm for Pb (Lead), 910 ppm for Zn (Zinc), and 10.18 ppm for Cd (Cadmium). It has become essential to control this pollution and find sustainable solutions for soil remediation and conservation. Several studies are directed towards new solutions for soil decontamination, bioremediation, and other promising techniques which are economical and above all more respectful of the environment. One of these innovative techniques is phytoremediation, the use of the capacity of the vegetation to bio-remediate varying concentrations of heavy metals for the rehabilitation of contaminated soils. However, these so-called hyper-accumulating plants have relatively low biomass, slow growth, and different rates of accumulation depending on the species and the metallic element. Various reports reflected the positive influence of earthworms on plant biomass and their indispensable role as a soil engineer. The aim of this chapter is to study the possibility of using this macro-invertebrate (*Lumbricus sp.*) to increase the efficiency of phytoremediation processes of *Hordeum vulgare*.

D. Boukirat (✉)
Department of Natural and Life Sciences, Tissemsilt University, Tissemsilt, Algeria

Laboratory of Agro Biotechnology and Nutrition in Semi-arid Zones, Faculty of Natural and Life Sciences, Tiaret University, Tiaret, Algeria

M. Maatoug
Laboratory of Agro Biotechnology and Nutrition in Semi-arid Zones, Faculty of Natural and Life Sciences, Tiaret University, Tiaret, Algeria

M. K. Jhariya et al. (eds.), *Ecological Intensification of Natural Resources for Sustainable Agriculture*, https://doi.org/10.1007/978-981-33-4203-3_10

Keywords

Barley · Bioaccumulation · Earthworms · Lead · Phytoremediation · Soil pollution

Abbreviations

BCF	Bioconcentration factors
CCME	Canadian Council of Ministers of the Environment
Cd	Cadmium
CEC	Cation-exchange capacity
Co	Cobalt
Cr	Chromium
Cu	Copper
DOC	Dissolved organic carbon
DW	Dry weight
EDTA	Ethylene diamine tetraacetic acid
EW	Earthworm
Fe	Iron
Hg	Mercury
$mg{\cdot}g^{-1}$	Milligram per gram
Mo	Molybdenum
Ni	Nickel
OECD	Organization for Economic and Cooperation and Development
OM	Organic matter
Pb	Lead
pH	Potential hydrogen
ppm	Part per million = $\mu g.g^{-1}$ = $mg.kg^{-1}$
SD	Standard deviation
Se	Selenium
SOM	Soil organic matter
t/y	ton per year
UE	European Union
Zn	Zinc
$\mu g.g^{-1}$	micro gram per gram

10.1 Introduction

Anthropogenic activities are responsible for the contamination of soils with various components. Industrial activities, agriculture, and automobile traffic are the main sources of contamination by various pollutants (Almehdi et al. 2019; Azizollahi et al. 2019; Kumari and Dey 2019; Colandini 1997; Raj et al. 2020; Banerjee et al. 2020), among them heavy metals whose levels in the soil receive great attention due to their negative consequences on biota as Pb (lead) and Cd (cadmium) (Li et al. 2009; Cheng and Wong 2002; Li NY et al. 2009).

Lead tends to have maximum toxic effect; its concentrations in lithosphere are aggravated by anthropogenic activity (Baize 1997). Lead is harmful to living entity at trace amounts and can also be toxic (Uzu et al. 2009). Pb in soil environment destroys the ecosystem homeostasis, inhibits plant growth leading to environmental degradation and human prosperity (Mishra et al. 2006; Zeng et al. 2006).

In Algeria, petroleum derivatives and industries are the population sources. There are relatively few data on soil pollution by heavy metals; no inventory on metal polluted soils has been done so far. However, there are some studies on metallic pollutants in the soil of various sources (road traffic and industrial effluent, etc.), for example, in the region of Tiaret—Algeria most of the soils which are near roads are contaminated by heavy metals and specifically by lead with concentration ranging from 20 $mg.g^{-1}$ to 4780 $mg.g^{-1}$.

Soil is the principal natural resource of humanity (Singh et al. 2020); its remediation has become necessary to conserve soils and water resources, the use of biological treatment (plants, bacteria, and invertebrates) to decontaminate soils is very promising (Jhariya et al. 2019a, b). However, bioremediation processes are still limited. The use of earthworms in bioremediation processes can increase its effectiveness (Eslami et al. 2019; Meena et al. 2018).

The introduction of the macro-invertebrates such as earthworms into soil contaminated with metals has been suggested to improve the phytoremediation capacity of vegetation (Lemtiri et al. 2016; Jusselme et al. 2012). Several experiments have been carried out using the association (plant–earthworms) for soil decontamination; earthworm activities enhance the metal availability as well as regulate soil structure and quality.

Earthworms are identified as ecosystem engineers for their functional role on long time span on properties of soil (physical, chemical, and biological) (Edwards and Bohlen 1996; Blouin et al. 2013; Bityutskii et al. 2016; Singh et al. 2016). They can be extremely affected by soil pollution and accumulated contaminants in their bodies. These characteristics, among others, have allowed their use as indicators of the contamination of the soil (Lanno et al. 2004; Xiao et al. 2006; Meena and Lal 2018). Edwards and Bohlen (1992) underline the interest of using earthworms in the biomonitoring of soil quality (De Vaufleury et al. 2013). It can be an advantage in soil remediation (Singh et al. 2020).

This chapter deals with the role of earthworms on lead phytoremediation and soil sustainability.

10.2 Soil Contamination by Lead: A Worldwide Problem

Natural presence of metals is very common. Certain elements (zinc and copper) are required for the biota in low doses. Human activities are the source of soil contamination by various organic and inorganic components. Heavy metals produced by industrial activities, agriculture, and road traffic have reached high levels in the soil,

Table 10.1 Standards for maximum levels in soils of some toxic metals (mg.kg^{-1} DW) enacted by the EU and various OECD countries (Ramade 2011)

Country	Metal	Maximum levels in soils
EU	Cd	3
	Cr	200
	Cu	140
	Pb	300
	Zn	300
Canada	Cd	3
	Cr	70
	Cu	150
	Pb	375
	Zn	600
France	Cd	2
	Cr	150
	Cu	100
	Pb	100
	Zn	300
GB	Cd	3
	Cr	400
	Cu	135
	Pb	300
	Zn	300
Holland	Cd	0.8
	Cr	100
	Cu	36
	Pb	85
	Zn	140
Sweden	Cd	0.4
	Cr	30
	Cu	40
	Pb	40
	Zn	75
Swiss	Cd	0.8
	Cr	75
	Cu	50
	Pb	50
	Zn	200

even toxic for some, in many regions (Ha et al. 2011; Jiang et al. 2012; Adriano 2001; Alkorta et al. 2004) and it has become more severe than other soil pollution (Demarco et al. 2019; Gómez-Garrido et al. 2018; Guo et al. 2019; Jeelani et al. 2018; Liu et al. 2018).

The contamination occurs when the concentration of an element in the environment increases beyond the values usually encountered (Table 10.1) (Alloway 1997).

Table 10.2 Global lead emission to land from all sources (Nriagu and Pacyna 1988 cited in Thornton 1995)

Sources	Emissions to land (t/y)
Metals and mining industry	329,100–791,000
Use and elimination of Pb	215,800–461,700
Other sources	57,190–303,700
Atmospheric fallout	202,000–263,000
Total	804,090–1,819,400

The toxicity of metals depends on their content in the medium and their chemical form, adding to this the physicochemical and mineralogical characteristics of the environment which influences the mobility and bioavailability of these elements and their accumulation by living beings (bacteria, plants, animals, and humans) (Baize 1997; Ebadi and Hisoriev 2017; Galal et al. 2019; Nan et al. 2019; Mai et al. 2019). Among the metallic elements mentioned, lead is toxic from the trace stage; it is probably the most known pollutant in public opinion. Its accumulation and transfer constitute a risk for public health, animals, and plants, but also for the natural environment as a whole. The estimation of heavy metal contents in soils is a crucial step in order to assess the risks and better remedy them (Blum et al. 1997; Meena et al. 2020a, b).

Lead is naturally present in the earth's crust, its average abundance is estimated at around 15 ppm (Kabata-Pendias and Pendias 2001), generally in poorly soluble form. Inorganic derivatives are present in waters, sediments, soils, atmosphere and possibly in micro-traces in living organisms. Uncontaminated soils would contain 10 to 30 $mg.kg^{-1}$ (Nriagu 1978; Baize 2002). Various forms of Pb are found in soil particles (Raskin and Ensley 2000).

Accumulation of Pb into the soil occurs mainly in the surface horizons (Sterckeman et al. 2000; Huynh 2009) and more precisely in horizons rich in OM (organic matter). The Pb contents then decrease more deeply (De Abreu et al. 1998; Huynh 2009). This is explained by the fact that the Pb is not very mobile. Being mainly associated with clays, oxides, iron hydroxides, and OM, it is only mobile when it forms soluble organic complexes and/or the soil has exceeded its absorption capacity for Pb (Morlot 1996; Raskin and Ensley 2000).

According to Ramade (1993), anthropogenic contributions of Pb are ten times greater than natural contributions (volcanism, erosion, wildfires, etc.), 99.7% of atmospheric Pb emissions are of anthropogenic origin, and a large part related to automobile traffic (Colandini 1997). The level of Pb in the soil surface has reached values of around 2% of DW (dry weight) in the soil at several sites in the world (Kabata-Pendias and Pendias 2001). Pb levels in soil are from various sources (Table 10.2). In Algerian soils several studies have shown the extent of this pollution and road traffic seems to be the main source of metallic pollution of the soils (Table 10.3).

Table 10.3 Heavy metals concentration in Algerian soils (ppm)

Origin of the pollution	Pollutant	Mean	Range	Author
Road traffic	Pb	1714,4	845,6–2712	Maatoug et al. (2013)
	Zn	666,5	179–910	Zerrouki (2014)
	Cu	3,18	0,8–13	
	Pb	643,61	–	Bouras et al. (2010)
	Zn	836,93	–	
Atmospheric fallout in the ground (industrial zone)	Pb	3,42	0–13.7	Benahmed et al. (2016)
	Zn	7,45	0–21.96	
Industrial effluent	Pb	–	7.99–39.14	Kebir and Bouhadjera (2011)
	Zn	–	21.15–731.08	
	Cu	–	4.32–15.32	
	Cd	–	1.89–10.18	

The evaluation of lead concentrations in soil, which may be harmful or even toxic to plants, is difficult; several authors have estimated these concentrations between 100 and 500 ppm (Kabata-Pendias and Pendias 2001). Pb disrupts the physiological mechanisms of plants and affects their growth and morphogenesis. It inhibits germination, causes the formation of small plants, and affects the morphology of the roots. Inhibition of the division of cells and their elongation are the most often reported phenomenon to explain its effects on the roots (Kopittke et al. 2007; Patra et al. 2004; Seregin and Ivanov 2001; Makowski et al. 2002).

Lead concentrations toxic to plants vary according to the studies; it strongly depends on the species of plant studied, but also on the culture method and environmental factors. Rooney et al. (1999) showed that for concentrations of metals extractable with EDTA in a soil reaching 800 mg.kg^{-1}, the growth of Raygrass was not affected. Päivöke (2002), on the other hand, showed that the toxicity of Pb to peas is dependent upon plant age and Pb concentration in the soil. However, he reports harmful effects on plant growth for lead concentrations below 500 mg.kg^{-1}. Liu et al. (2003) demonstrated that the sensitivity or tolerance of plants to lead was cultivar-dependent; the toxicity strongly depends on the behavior of plants towards this metal.

10.3 Earthworms for Improving Phytoremediation Process

Although phytoremediation has several advantages, natural process often carried out on-site and less costly than other technologies (Jhariya et al. 2018a, b). The majority of hyper-accumulative plants used for phytoremediation have a low biomass production and a slow growth rate, and in some cases trace elements have a harmful effect on them (phytotoxicity, inhibition of growth and metabolism, etc.); these negative effects depend on the element and its level in the soil (Salt et al. 1998; Shah and Nongkynrih 2007; Singh et al. 2003).

Table 10.4 Abundance and biomass of soil fauna in temperate regions (Gobat et al. 2010)

Group	Individuals/m^2	Biomass (g/m^2)
Protozoa	10^7–10^{11}	6 à ▯30
Nematodes	1 à 30 millions	1–30
Acari	20,000–400,000	0.2–0.4
Collembola	20,000–400,000	0.2–0.4
Myriapods		
Diplopods	20–700	0.5–12.5
Chilopods	100–400	1–10
Isopods	Up to 1800	Up to 4
Insect larvae	Up to 500	4.5
Earthworms	50–400	20–400

However, a plant applied for metal phytoremediation should be able to accumulate high concentrations but also reflect higher growth and development (Lemtiri et al. 2016; Banerjee et al. 2018). To overcome these drawbacks earthworms have been found to increase the phytoextraction of pollutants (Lemtiri et al. 2016; Jusselme et al. 2012, 2015; Huynh 2009).

Earthworms can increase plant biomass, in some cases tend to adjust in soils with high level of mental concentration and increase the availability of pollutants; their activities must be considered in phytoremediation strategies (Jana 2009; Azhar-u-ddin et al. 2020). In fact, earthworms positively affect plant biomass in 75% of cases (an increase in aboveground and root biomass is observed) (Brown et al. 1999; Jana 2009), and the growth of hyper-accumulating plants (Jusselme et al. 2015; Huynh 2009) and have a favorable impact on the soil and its organisms.

Earthworms are soil engineering organisms; they alter soil resource availability for organisms (Jones et al. 1994; Lavelle 1996). They tend to be important organisms in soil (Huynh 2009) and are known for their important biological functions (Table 10.4). Incorporation of OM in soils promotes proper biogeochemical cycling and maintenance of favorable physicochemical conditions for biota (Brown et al. 1999, 2000; Lavelle 1988; Lavelle et al. 1998, 2006; Blouin et al. 2013; Meena et al. 2020; Kumar et al. 2020). Ecological category of earthworms determines their functionality in soil environment (Bouche 1977).

However, it seems that all earthworms increase heavy metals availability in soil and influence their mobility (Sizmur and Hodson 2009). Significant differences in bioaccumulation have been reported between eco-physiologically distinct earthworm species (Morgan et al. 1986, 1993, 1999; Beyer et al. 1987; Morgan and Morgan 1992, 1999; Van Vliet et al. 2005; Kamitani and Kaneko 2007). The endogens and the anecics are the most influential among the various earthworm species (Bouche 1977).

Earthworms have an impact on the dynamics of metals (Wen et al. 2004; Devliegher and Verstraete 1997 in Huynh 2009). They raise the availability of metals due to the decomposition of OM (Wen et al. 2004). Further, it also affects their distribution in soils (Zorn et al. 2005). Thus, the presence of earthworm *Lumbricus terrestris* increases the availability of copper by 6% (Devliegher and Verstraete 1995 in Huynh 2009). However, Abdul Rida (1996) considers that the impact of earthworms reflects lesser bioaccumulation in plants under low pollution level.

The effect of heavy metals on earthworms varies according to species, their ability to adapt to pollution, stage of development, diet, and place of life (Depta et al. 1999). Earthworms tend to be most sensitive than invertebrates towards metals (Bengtsson et al. 1992).

Metal accumulation in earthworms varies as per its chemical nature and soil properties and is generally higher than in other animal species (Depta et al. 1999; Beyer et al. 1982).

Earthworms can act on the growth and metals uptake by plants, either directly by modifying the physicochemical parameters of the soil (pH and the DOC level) or indirectly by modifying the communities of microorganisms present in soils (Gaur and Adholeya 2004; Jusselme et al. 2015; Eisenhauer et al. 2009; Ortiz-Ceballos 2007; Sizmur and Hodson 2009; Ma et al. 2006a).

Several works report the earthworm impact on metal accumulation of plants (Table 10.5) as a part of phytoextraction process (Jana 2009). For example, the *Pheretima* (*Metaphire*) genus increases the bioavailability of zinc in artificially contaminated soils and increases the aboveground and root biomass as well as the zinc concentrations of two plants *Lolium multiflorum* and *Brassica sp.* (Wang et al. 2006). The same positive effect on biomass and accumulation is observed for *Metaphire gillelmi* earthworms and the plant ryegrass (*Lolium multiflorum*) in soils artificially contaminated with copper and enriched in OM by the addition of straw (Jana 2009).

The tropical species *Pontoscolex corethrurus* also improves the phytoextraction of Pb by the plant *Lantana camara* in soils artificially doped with 500 and 1000 $\mu g.g^{-1}$of Pb (Duarte et al. 2012; Huynh 2009), and that the soluble and exchangeable fraction of Pb significantly decreases under earthworm presence in natural soil contamination (Duarte et al. 2012).

Another study, in microcosms on the soil of an old mining site contaminated by several metals (Pb, Zn, Cd, and Cu) involving the earthworm *Eisenia fetida* and plants (*Zea mays*) and (*Hordeum vulgare*), reports an increased rate of accumulation of these elements in both plant species and in all of their organs (Jana 2009).

10.4 Phytoextraction of Lead by *Hordeum vulgare* Under Controlled Conditions

Barley, considered as a hyper-accumulator plant of heavy metals (Morel 1997). It accumulates trace elements to different degrees depending on the metal and its level in the soil. In an agricultural soil contaminated with lead, barley accumulates up to 2% of the Pb in soil (Maatoug et al. 2013).

Table 10.5 Various publications which have studied (directly or indirectly) the effects of earthworms on the phytoremediation of soils polluted by heavy metals

Earthworms	Plants	Heavy metals	Authors
Eisenia fetida	*Vicia faba*, *Zea mays*	Cu, Zn, Pb, Cd	Lemtiri et al. (2016)
	Brassica juncea L.	Cd	Kaur et al. (2018)
	Neyraudia reynaudiana	Pb, Zn	Li et al. (2018)
	Vetiver grass	Cd	Wu et al. (2020)
	Stachys inflata	Pb, Zn	Mahohi and Raiesi (2019)
	Zea mays, *Hordeum vulgare.*	Pb, Zn, Cd, Cu	Ruiz et al. (2009)
	Phaseolus vulgaris L.	Se	Azhar-u-ddin et al. (2020)
	Paulownia tomentosa, *Cytisus scoparius*	Pb, Cr, Cd, Zn, Cu, Ni	Macci et al. (2012)
Eisenia andrei	*Canavalia ensiformis*	Cu	Santana et al. (2019)
Pheretima guillelmi	*Leucaena leucocephala*	Pb, Zn	Ma et al. (2003)
Pheretima sp.	*Lolium multiflorum*, *Brassica juncea*	Zn	Wang et al. (2006)
	Lolium multiflorum, *Brassica juncea*	Cd	Yu et al. (2005)
Lumbricus terrestris	*Pennisetum purpureum*	Pb	Das and Osborne (2018)
	Zea mays *Hordeum vulgare*	Pb, Zn, Cu	Ruiz et al. (2011)
Lumbricus rubellus	*Zea mays L.*	Cd	Aghababaei et al. (2014)
Lumbricus sp.	*Hordeum vulgare*	Pb	Boukirat et al. (2017)
Pontoscolex corethrurus	*Lantana camara*	Pb	Jusselme et al. (2012)
	Acacia mangium	Pb, Ni, Cr	Bongoua-Devisme et al. (2019)
Metaphire guillelmi	*Lolium multiflorum*	Cu	Wang et al. (2007)

Pb unlike other elements (Zn, Cu, etc.) is not an essential element because although present in plants, it does not participate in any known physiological or biochemical function (Marschner 1995). It is absorbed by the plant according to a different uptake pathway than that of the essential elements (Lemtiri et al. 2016).

The accumulation of Pb by barley is influenced by several parameters related to the soil but also to the form in which the pollutant is found. Zerrouki (2014) reported that parameters such as pH, CEC, and clay content have a significant effect on the bioaccumulation of metals by barley. However, the interaction between the different metals and the forms in which they are present in the soil can modify their bioavailability for the plant.

Table 10.6 Descriptive statistics of lead concentrations (μg.g^{-1} DW) in soil–plant system

Attributes	*N*	Mean ± SD	Min	Max
Pb-soil	16	331.88 ± 66.95	200.00	430.00
Pb-barley	16	59.44 ± 13.16	38.00	75.00
Control-soil	4	50.00 ± 21.60	20.00	70.00
Control-barley	4	11.75 ± 3.10	9.00	16.00

We studied the bioaccumulation of barley under controlled conditions in artificially contaminated soils by lead 331.88 ± 66.95 μg.g^{-1}, the concentrations accumulated by barley were 59.44 ± 13.16 μg.g^{-1}, these concentrations greatly exceed those found in barley cultivated under the same conditions on an unpolluted soil 50.00 ± 21.60 μg.g^{-1} with a concentration in the order of 11.75 ± 3.10 μg.g^{-1}, (Table 10.6) (Boukirat et al. 2017).

The accumulation of lead by barley was the subject of a study in soils with a high polymetallic load (Pb, Cu, Zn). Maatoug et al. (2013) reported that for lead concentrations of 1714.39 ± 512.62 μg.g^{-1}, barley accumulates up to 36.28 ± 14.90 μg.g^{-1}. The removal of trace element by plants is related to the supply of soil (Nguyen 2007). The correlation between soil lead and lead accumulated by barley is positive and highly significant ($r = 0.688^{**}$, $p < 0.001$) (Boukirat et al. 2017). An increasing trend between metal levels in the soil and the concentration accumulated in plants was recorded after an artificial contamination of a healthy soil by saline solutions of metals (Mahler et al. 1978; Mitchell 1978). In general, the Pb concentrations of a plant are closely correlated to the Pb concentrations of the soil.

Barley accumulates 18.11% of lead from the soil under controlled conditions (Fig. 10.2), a high rate compared to that found by Maatoug et al. (2013) which is 2%. The form in which lead is added to the soil may explain the large amount accumulated. Diehl et al. (1983) found that Pb particles in soils are rapidly converted into water-soluble compounds, readily disposable to plants.

Soil physicochemical parameters influence metals mobility in the soil and their bioavailability but also their bioaccumulation by plants (Boukirat et al. 2017; Maatoug et al. 2013). The constituents of the soil, in particular clays and OM can interact with metals through different chemical interactions, all these interactions limit the bioavailability of these metals in the soil (Tanner and Headley 2011).

Soil pH regulates the absorption of Pb at variable concentrations in various species (Seiler and Paganelli 1987) and clays which are fine fractions which intervene mainly during the phenomena of retention and fixation of heavy metals (Sanders 1983).

Lead solubility and mobility in soil depend on the type of Pb compound that has been added to the soil. For example, Pb nitrates are very soluble and will be easily leached from the soil (CCME 1999) and easily bioaccumulated by the plant (Boukirat et al. 2017).

10.5 Bioaccumulation of Lead by Earthworms (*Lumbricus sp.*)

Earthworms are known for their bioaccumulation capacity (Hopkin 1989) and for their important role in soil remediation (Singh et al. 2020). They are constantly in contact with the soil, whether in their dermis or during ingestion of soil. Earthworms tend to accumulate pollutants of the soil environment. Differences in metal concentrations between eco-physiologically different species of earthworms are reported by Morgan et al. (1993). Also depending on their ecological class (endogeic, epigeic or anecic), earthworms are more or less sensitive to metallic trace elements (Tomlin 1992).

Earthworms that live in soils contaminated by metals, mainly from anthropogenic sources, accumulate high concentrations of heavy metals (Ireland 1983; Morgan and Morgan 1988; Dai et al. 2004). Metal bioaccumulation in their tissues relies on the species and the characteristics of their living environment, in particular the composition of the soil and its pH (Van Gestel and Ma 1988; Morgan and Morgan 1991, 1999). Morgan et al. (1993) emphasize the importance of earthworms determining biotic factors, which regulates metal accumulation. However, the study of the bioavailability of pollution in terrestrial environments by measuring the bioaccumulation of metals by earthworm species is quite difficult since the bioavailability data are specific to each species (De Vaufleury et al. 2013).

Nahmani et al. (2007) mentioned difficulty in establishing simple statistical relationships between the metals concentrations in organisms and those of abiotic components of the environment. Thereby, better understanding of the potential of different species of earthworms to bioaccumulate pollutants according to the nature of the metals and their concentration is primordial (De Vaufleury et al. 2013).

Earthworms that live in soils polluted by metals due to the proximity of motorways (Gish and Christensen 1973) or mining (Ireland 1975; Dai et al. 2004) or the spreading of waste (Helmke et al. 1979) have much higher levels of heavy metals than those which develop in unpolluted areas.

On soils artificially contaminated with lead 342.50 ± 54.59 $\mu g.g^{-1}$, the earthworms *Lumbricus sp.* accumulate approximately 24.01 ± 10.97 $\mu g.g^{-1}$ (Table 10.7). Higher accumulation compared to their concentrations 8.20 ± 0.48 $\mu g.g^{-1}$ in unpolluted soils whose concentration is <100 $\mu g.g^{-1}$. The concentrations of lead in earthworm tissue are positively correlated with Pb-soil concentrations with a correlation coefficient ($r = 0.92^{**}$, $p < 0.000$) (Boukirat et al. 2017).

An increase in the concentrations of Pb in the earthworm *Eisenia fetida* exposed to the highest contamination (2000 $mg.kg^{-1}$) is observed, but this remains much lower than that of the environment (Scaps et al. 1997). These results agree with those of Grelle and Descamps (1998) and Boukirat et al. (2017).

The ability of earthworms to accumulate Pb makes them useful indicators of lead pollution of the soil. Usually the lead concentrations in the soil exceed that of earthworms. On the other hand, in certain situations, high concentrations of Pb combined with low calcium content in an acidic soil can induce a greater accumulation of lead coming from the soil in earthworms (Ireland 1979).

Table 10.7 Descriptive statistics of lead concentrations (μg.g^{-1} DW) in soil–earthworms system

Attributes	*N*	Mean ± SD	Min	Max
Pb-soil	16	342.50 ± 54.59	250.00	440.00
Pb-EW	16	24.01 ± 10.97	12.55	54.54
Control-soil	4	70.00 ± 16.33	50.00	90.00
Control-EW	4	8.20 ± 0.48	7.50	8.59

In *Aporrectodea tubulata* the Pb concentration varies from 5 mg.kg^{-1} to 12 mg.kg^{-1} in a pH range between 7.1 and 4.9 (Beyer et al. 1987).

The accumulation of Cd and Zn in the body of earthworms is variable; it depends on the species, the soil nature, duration of the experiments (De Vaufleury et al. 2013). The low bioavailability of metals (Pb, Zn, and Cu) from contaminated soil at alkaline pH has been confirmed in *Lumbricus terrestris* in soils that are not rich in nutrients with the absence of toxic effects (Kennette et al. 2002). The variation in soil pH also influences the accumulation of lead by earthworms *Lumbricus sp.* (Boukirat et al. 2017).

The nature of the soil is of primordial importance in the bioaccumulation and toxicity of metals in earthworms (Van Gestel 1992). Clays, due to their properties, play a very important role in the availability of heavy metals. Li and Li (2000) showed that clay minerals can adsorb metals and immobilize them. The metals can also form an organometallic complex by complexing with soil organic matter (SOM) (Lamy 2000). Heavy metals toxicity depends also on the nature of the clay and SOM. Lock and Janssen (2001) showed that for *Enchytraeus albidus*, the toxicity of Zn and Cd depends on the nature of the clay used (kaolinite, illite, and montmorillonite) and on OM (fallen leaves of various dead trees or stems of nettles or reeds).

The accumulation of metals occurs when the concentration factor in the soil is >1 (Van Hook 1974). The accumulation rates in earthworms vary depending on the heavy metal studied 16.0 for Cd, 4.1 for Zn, 1.1 for Cu, 0.5 for Ni, and 0.4 for Pb (Abdul Rida 1996; Van Hook 1974; Van Rhee 1977; Czarnowska and Jopkiewicz 1978; Kennette et al. 2002).

The level of lead accumulated by *Lumbricus sp.* is 7.24% of soil lead, 24.01 ± 10.97 μg.g^{1} DW, BCF: 0.07. Compared to the results found in other studies on the accumulation of heavy metals by earthworms low values are recorded. However, a monometallic pollution combined with specific behaviors for each species and between the different ecological categories can explain the low bioconcentration factor (BCF) (Boukirat 2018).

Morgan et al. (1999) report that earthworms (*Lumbricus rubellus*) survive in the soil of the metalliferous site of Rudry, South Wales (GB) are exposed to concentrations of Pb (2337 μg.g^{-1} DW), Zn (5902 μg.g^{-1} DW), and Cd (604 μg.g^{-1} DW). Their tissues contain substantial amounts of the three metals (Cd: 1212 μg.g^{-1} DW, BCF: 2.01; Zn: 2470 μg.g^{-1} DW, BCF: 0.42; and Pb: 892 μg.g^{-1}DW, BCF: 0.38) (De Vaufleury et al. 2013).

Several studies have shown properties of soil influence Pb absorption by *Eisenia andrei* (Smith et al. 2012; Peijnenburg et al. 1999; Spurgeon and Hopkin 1995; Bradham et al. 2006; Luo et al. 2014).

Higher Pb concentrations in earthworm tissue are associated with high clay and silt levels, and high lead concentrations in soil, combined with acidic pH (Boukirat 2018). Luo et al. (2014) reported the bioavailability of Pb and its toxicity on earthworms. He further reported that Pb concentrations in earthworm tissues reflect positive correlation with total soil Pb concentration and the available fraction, the content of sand and Fe. Negative correlation was found between earthworm accumulation with the content of clay, silt, and CEC. They also point out that the concentrations of Pb in earthworms are well predicted by the total concentrations of Pb and silt rate in the soils.

Numerous studies have made it possible to identify the parameters on which the bioaccumulation process depends: trace element and its concentration in the soil, the species and the ecological category of the earthworm, the physical and chemical properties of the soil (Ma 1982; Marino et al. 1992; Abdul Rida 1996; Ireland 1979; Ash and Lee 1980; Ireland and Richards 1981; Smith 1996; Jusselme et al. 2013; Boukirat et al. 2017). According to De Vaufleury et al. (2013) it is primordial to study the factors that control the mobilization and absorption of the compounds in other terrestrial animal species with more varied lifestyles and anatomical and physiological characteristics.

10.6 Impact of the Plant/Earthworm Association on Soil Bioremediation

The positive effect of earthworms on vegetation was confirmed by the study of Wollny (1890) on cereals and legumes. He observed an increase of 35–50% in grain yield and 40% for straw (Jana 2009), combined with their impact on soil fertility and the bioavailability of nutrients, their use as reinforcement for soil phytoremediation seems promising.

The interaction between plants and earthworms varies the concentrations of lead accumulated by each of them (Boukirat et al. 2017). For example, Lemtiri et al. (2016) found that adding earthworms to contaminated soil with Pb, Zn, and Cu does not affect the concentrations of metals in plants. On the other hand, Wen et al. (2004) and Azhar-u-ddin et al. (2020) reported an increase in the concentration of metals in plants in the presence of earthworms.

The influence of *Hordeum vulgare* and *Lumbricus sp.* association on lead bioaccumulation by each of them was studied in a greenhouse experiment with different arrangements: (1) with or without inoculation of earthworms; (2) with or without the presence of the plant (Boukirat et al. 2017) their interaction is represented in (Fig. 10.1). The concentrations of Pb accumulated by the association of barley and earthworms are 38.00 ± 5.68 μg.g^{-1} and 26.01 ± 6.66 μg.g^{-1}, respectively, and for polluted soils it ranged from 500 μg.g^{-1} to 2000 μg.g^{-1}, a

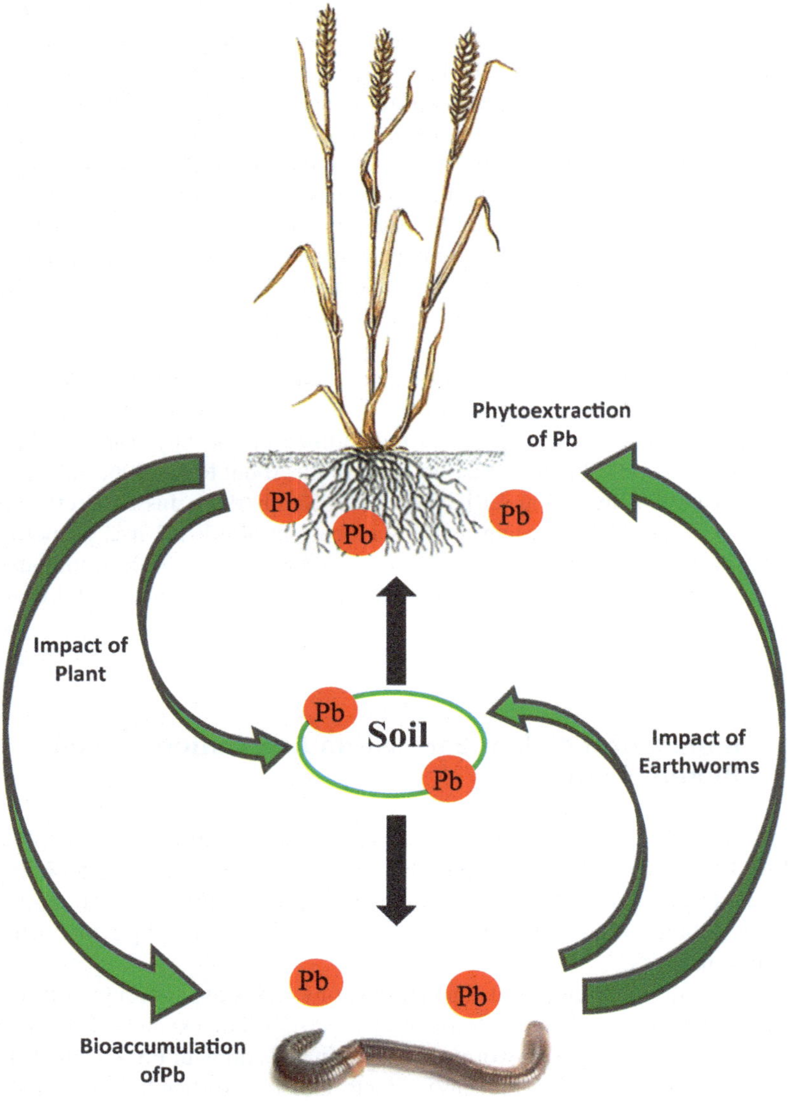

Fig. 10.1 Diagram representing soil, plant, and earthworms interactions under lead pollution

strong accumulation relative to the control soil (Pb <100 $\mu g.g^{-1}$) 20.25 ± 2.25 $\mu g.g^{-1}$ and 12.84 ± 1.09 $\mu g.g^{-}1$, respectively (Table 10.8).

The association of *Hordeum vulgare* and the earthworm (*Lumbricus sp.*) showed an accumulation by *Hordeum vulgare* of 12.91% of soil lead and by *Lumbricus sp.* of 8.78% of soil lead (Fig. 10.2). The presence of earthworms reduced the phytoextraction of Pb by barley by 5.2% compared to the results found in the

Table 10.8 Descriptive statistics of lead concentrations (μg.g^{-1} DW) in the soil–plant–earthworms system

Attributes	*N*	Mean ± SD	Min	Max
Polluted soil				
Pb-soil	16	316.25 ± 87.24	150,00	430,00
Pb-barley	16	38.00 ± 5.68	31.00	50.00
Pb-EW	16	26.01 ± 6.66	16.53	39.00
Control soil				
Pb-soil	4	67.5 ± 9.57	60,00	80,00
Pb-barley	4	20.25 ± 2.25	17.00	23.00
Pb-EW	4	12.84 ± 1.09	11.36	13.93

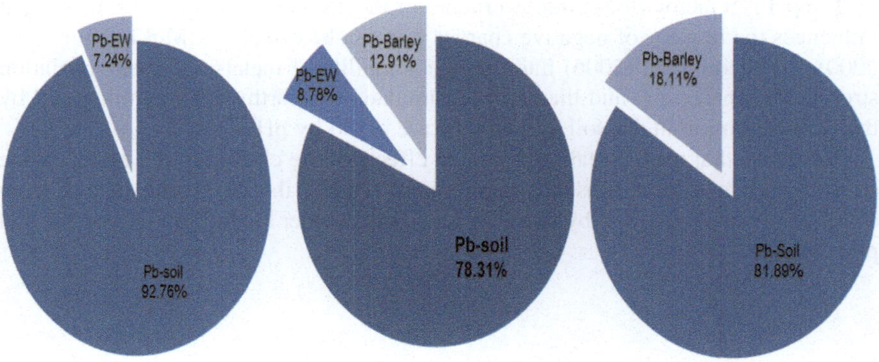

Fig. 10.2 Percentage of lead accumulated by barley and earthworms in soil–earthworms system, soil–plant system, and soil–plant–earthworms system

absence of earthworms and the concentration of lead in earthworms increased by 1.54%.

The bioavailability of metals in the soil influences their bioaccumulation (Brown 1995), and biotic and abiotic factors can influence this bioavailability in the polluted environment. Romheld (1986) and Hinsinger (2001) pointed out that the exchange activity of the root scan changes the pH. Furthermore, it is well known that earthworms strongly influence soil physical and chemical properties. Hence, the importance of considering their role in the bioavailability of metals is thus justified (Huynh 2009).

Soil characteristics can change the bioavailability of metals by modifying the speciation of metals and/or by modifying the adsorption of soil particles (Spurgeon and Hopkin 1996; Alloway 1995a, b; Van Gestel 1992). In theory, the decrease in the pH of a metal-rich soil should increase their bioavailability.

The presence of earthworms in the soil significantly lowers the pH (Cheng and Wong 2002; Huynh 2009; Yu and Cheng 2003; Boukirat 2018). The decrease in soil pH improves the phytoextraction process (Sanders et al. 1986).

However, other studies point out that earthworms increase the pH of the soil (Cheng and Wong 2002; Hu et al. 1998; Salmon 2001; Wen et al. 2006; Udovic and Lestan 2007). Except for the study of Lemtiri et al. (2016) on the earthworm *E. fetida,* who reported that the presence of the earthworm did not affect the pH values. These differences observed between the different studies may be due to the earthworms used (species, ecological category, etc.), the type of soil, and the different plants.

The activity of earthworms on the soil causes changes in soil physical and chemical properties. The association of both earthworms and plant in the same soil induces at a much lower pH than that observed when they are separated (Boukirat 2018). Soil pH is a crucial factor that affects adsorption and desorption behavior and therefore metals bioavailability in the soil, and it is important to assess the pH changes induced by the activity of earthworms (Lemtiri et al. 2016; Wen et al. 2004).

The pH is a factor whose role is crucial for the mobility of metal ions, because it influences the number of negative charges that can be dissolved (McLaughlin et al. 2000). Spurgeon et al. (2006) find that the solubility of metals and their speciation strongly depend on pH and that their accumulation in earthworms is influenced by their concentration in the soil and, in the case of Cd by pH.

The cause that may explain the limited effects of the change in pH on the tissue concentration of earthworms was explained by Oste et al. (2001) who suggested an effect of pH on absorption by the skin, and, on the other hand, the influence of soil particles ingested.

10.6.1 Lead Concentration in the Plant Between Presence/Absence of Earthworms

The presence of earthworms significantly affects the lead concentrations in the plant ($P < 0.001$) (Fig. 10.3) as the plant *Hordeum vulgare* accumulates more lead alone than in the presence of the earthworm *Lumbricus sp.* 59.44 ± 13.16 $\mu g.g^{-1}$ and 38.00 ± 5.68 $\mu g.g^{-1}$, respectively. The presence of earthworms reduced 5.2% phytoextraction of lead by barley compared to the results found in the absence of earthworms. The earthworm seems to retain a part of the lead by accumulating it and makes it less available for barley. A decrease in metal concentration of the plant *Leucaena leucocephala* which is cultivated in soil that contains earthworms was also observed by Sizmur and Hodson (2009). Earthworms can in certain situations decrease the availability of metals in polluted soils, which has been confirmed by some studies (Ma et al. 2000, 2006b; Liu et al. 2005; Lukkari et al. 2006; Duarte et al. 2012).

The decrease in lead concentrations accumulated by barley can be advantageous if taken from a health standpoint. The use of barley as a fodder plant and for human consumption represents a risk of bioconcentration of this metal in the food chain through direct or indirect way on human health.

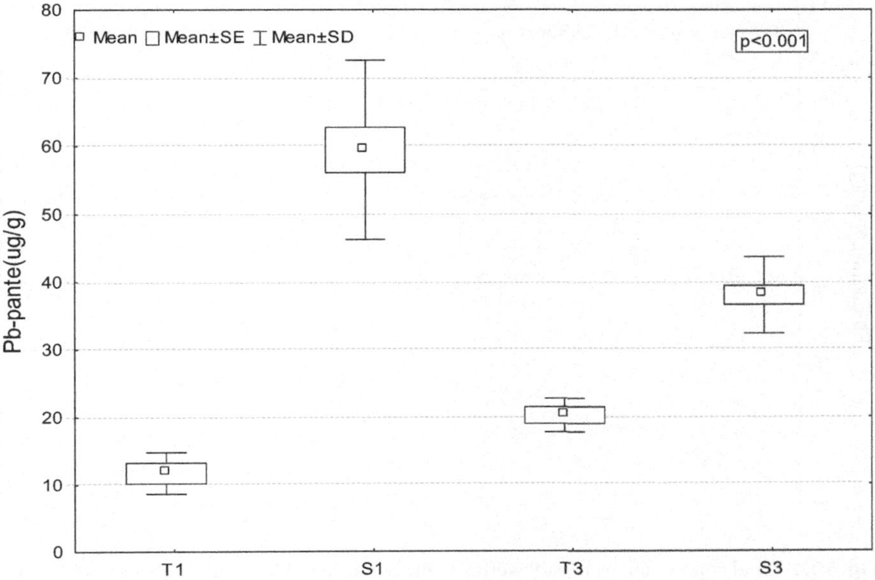

Fig. 10.3 Lead concentrations in the plant in the presence (S3-T3) and absence (S1-T1) of earthworms; **S1**: Plant + Pb polluted soil; **S3**: Plant + Earthworms + Pb polluted; **T1**: Plant + Unpolluted soil; **T3**: Plant + Earthworms + Unpolluted soil

10.6.2 Lead Concentration in Earthworms Between Presence/Absence of the Plant

The presence of the plant significantly affects lead concentrations in earthworms ($P < 0.001$) (Fig. 10.4). The concentrations recorded in earthworm tissue suggest that the presence of the plant *Hordeum vulgare* increases these concentrations by 1.54%, 26.01 ± 6.66 μg.g^{-1} in the presence of the plant against 24.01 ± 10.97 μg.g^{-1} in its absence. In contrast Lemtiri et al. (2016) found that the addition of *Vicia faba* or and *Zea mays* reduced Pb and Cd accumulation in earthworm tissues. Also Elyamine et al. (2018) reported a significant decrease in Cd concentrations of earthworms in the presence of plants.

Earthworms immobilize heavy metals by accumulating them in their bodies, making them less available to other soil organisms. They also reduce the toxicity of the soil (Eslami et al. 2019).

The impact of earthworm activities increases the availability of metals in contaminated soils (Wen et al. 2004; Coeurdassier et al. 2007; Udovic and Lestan 2007; Jusselme et al. 2013). Wen et al. (2004) observed that the earthworms increase the concentration of metal in plants. The earthworm *Pontoscolex corethrurus* increases also the bioavailability of Pb compared to soils without earthworms and enhances its absorption by the plant (Jusselme et al. 2012). In contrast Ma et al.

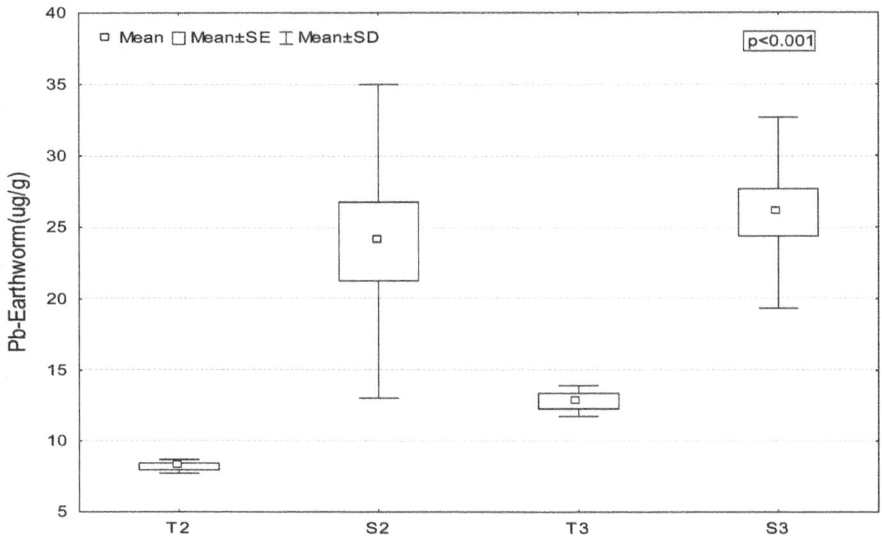

Fig. 10.4 Lead concentrations in earthworms in the presence (S3-T3) and absence (S2-T2) of the plant; **S2**: Earthworms + Pb polluted soil; **S3**: Plant + Earthworms + Pb polluted; **T2**: Earthworms + Unpolluted soil; **T3**: Plant + Earthworms + Unpolluted soil

(2006b) found that the effects of the earthworm *P. guillelmi* decreased the concentration of extractable Pb and Zn.

These different observations can be explained by a different impact depending on the earthworm, the plant, the metal and their interaction.

10.7 Role of Earthworms in Soil Sustainability

Soil is a fundamental and important element for the survival and the development of human beings (Qayyum et al. 2020). Soil sustainability is an important and complex challenge which is influenced by soil degradation and agricultural pollution (Zentner et al. 2004; Raj et al. 2019a, b). Soil management is crucial for improving productivity of crops and to maintain environmental sustainability (White et al. 2012). Studying the impact of earthworms on soil and plants is essential for understanding the biological functioning of soils and to support practices that are based on this functioning (Hedde 2018). Earthworms are very sensitive to their environment, that is why they are good bio-indicators of soil management (Paoletti et al. 1991).

Earthworms (annelids, oligochaetes) represent a major component of soil invertebrates which have a major impact on soil characteristics (physical, chemical, and biological), ensuring the maintenance of the soil's capacity to deliver ecosystem services (Lavelle 2002; Lavelle et al. 2006). They are also sensitive to the physical and chemical modifications of soil parameters (Paoletti et al. 1991).

The beneficial role of earthworms in soil sustainability and plant growth, and the amount of research done on them make them more promising for assessing the sustainability of the environment (Paoletti 1999).

Their importance for soils was first cited by Darwin: "*...the land was regularly ploughed and still continued to be thus ploughed by earthworms*" (Darwin 1881 in Pelosi 2008). He emphasized their role in soil formation, dynamics, and fertility (Brown et al. 2003a; Jana 2009).

Nowadays, earthworms are considered engineers of the terrestrial ecosystem (Jones et al. 1994); they change the availability of resources for other organisms (Pelosi 2008). Their populations vary from a few individuals in a square meter of soil to more than 1000 individuals (Edwards 2004 in Da Silva 2013).

Earthworms play an important role in their environment through various physicochemical and biological mechanisms, making it possible to improve fertility and preserve soil structure (Stork and Eggleton 1992; Lavelle et al. 1997) which is not without importance for the growth of the plant (Huynh 2009). Also, by affecting both physical and chemical properties of the soil (Edwards 2004) they modify the biotope of microbial communities (Lavelle and Gilot 1994; Lavelle et al. 1997). They also play a major role in the incorporation of OM in soils, the functioning of biogeochemical cycles, and the maintenance of favorable physicochemical conditions for plants and other soil organisms especially microorganisms (Brown et al. 1999, 2000; Lavelle 1988; Lavelle et al. 1998, 2006; Edwards 2004; Blouin et al. 2013; Da Silva 2013). The decrease in their activities in the soil can disrupt the recycling of OM (Cluzeau et al. 1987).

Earthworms play a key role in soils because they participate in the physical, chemical, and biological dynamics of the soil (Pelosi 2008; Singh et al. 2016). The absence of earthworms blocks OM on the surface; in fact their activity allows the fragmentation and mineralization of OM in the soil and therefore the recycling of nutrients and makes them more available to plants (Jana 2009; Brown et al. 2003b, 2004; Chauhan 2014; Datta et al. 2016).

Their main functions (Fig. 10.5) consist in: (1) Decomposition of OM (Datta et al. 2016) (2) The creation of galleries which improve the porosity and the aeration of the soil (Lavelle 1997; Fonte et al. 2009), the infiltration of water (Pelosi 2008; Datta et al. 2016) and facilitate root penetration (Jegou et al. 2002) (3) The excretion of dejection (deposited on the walls of the galleries) or turricules (discharges present on the ground surface) (Huynh 2009; Pelosi 2008). The successive passages of earthworms in the soil cover the galleries of mucus and excrement rich in nutrients (Sims and Gerard 1985; Binet and Curmi 1992; Huynh 2009), especially in nitrate (NO^{3-}), ammonium (NH^{4+}), and organic carbon (Bhatnagar 1975), thus promoting the development of a large bacterial microflora throughout the gallery walls, resulting in an increase in respiratory activities and enzymatic mineralization (Binet 1993; Huynh 2009; Fonte et al. 2009). The turricules also present higher concentrations of potassium, calcium, magnesium, phosphorus (Lavelle and Martin 1992).

However, earthworm impact on soil varies according to their ecological category. The endogens and the anecics are the most influential (Bouche 1977).

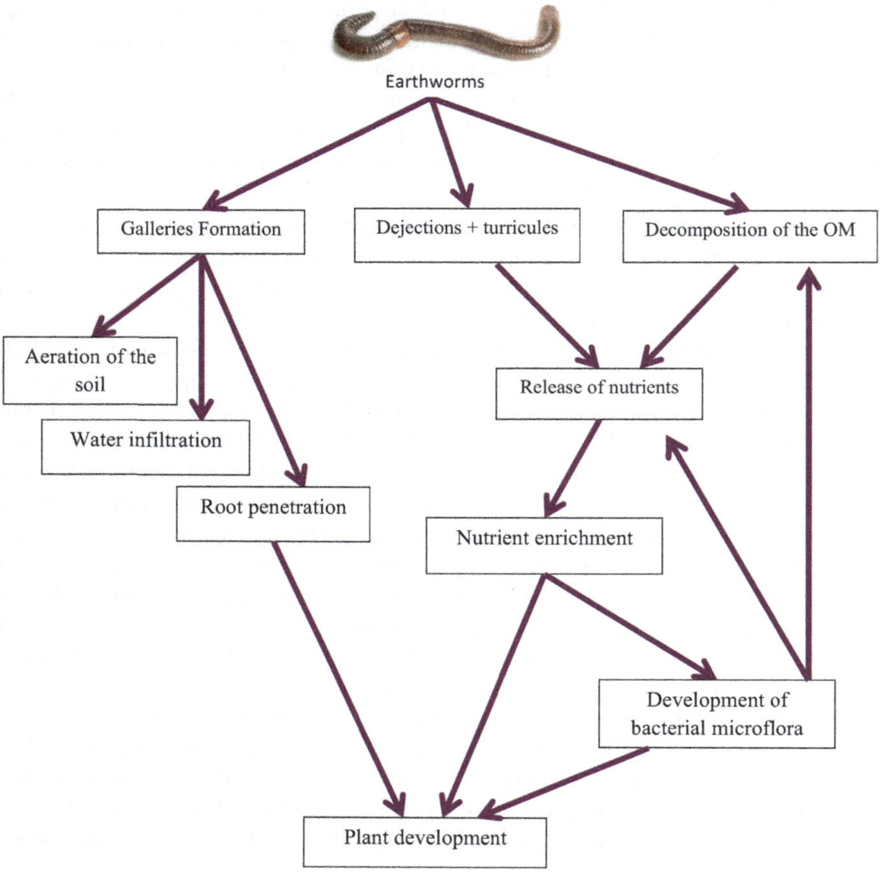

Fig. 10.5 Functions of earthworms in the soil

The class of epigeic described by Bouche (1977) concerns the earthworms living in the soil surface in organic heaps (manure, compost, leaf litter, bark). These earthworms due to their diet based on dead OM (fragments of plants, fungi) will be responsible for the decomposition and fragmentation of dead plant matter (Puga Freitas 2012; Da Silva 2013).

Anecics are large individuals, living in permanent vertical galleries which open on the surface (Bouche 1977). They are characterized by a strong activity in the soil, they feed on OM in an intermediate state of decomposition, and ensure its mixing with the mineral fraction of the different horizons of the soil (Huynh 2009; Da Silva 2013; Pelosi 2008; Puga Freitas 2012). Due to the construction of their galleries, they allow the development of hot spots rich in water and oxygen where the bacterial and fungal populations will be able to develop (Jana 2009). The anecics are sensitive to mechanical work on the soil and to inputs (Da Silva 2013).

The endogeic species (Bouche 1977) represent the most important group; they constitute 20–50% of the biomass of earthworms in the soil. Endogeics live permanently in the ground where they dig horizontal temporary galleries highly branched. They feed on the OM contained in the soil, and influence the soil microorganisms (Da Silva 2013; Huynh 2009; Jana 2009; Pelosi 2008). Endogeics spend the most time in the soil and are therefore likely to be most affected by soil contamination (Da Silva 2013).

Several studies have identified the effects of earthworms on plant growth and development (Scheu 2003; Brown et al. 1999, 2003b, 2004; Jana 2009). These studies have shown a positive impact on the aboveground biomass in 75% of the cases for tropical earthworms and 79% of the cases for temperate earthworms (Puga Freitas 2012). An augmentation in the biomass of aerial and root parts and in seed yield is also observed by Brown et al. (1999).

Earthworms participate in the release of substances such as vitamins and proteins that stimulate plant growth (Edwards and Bohlen 1996 in Pelosi 2008) and interactions with beneficial microorganisms (Jana 2009). PGPB (Plant Growth Promoting Bacteria) that synthesize substances similar to plant hormones, whose populations and activity increase in turricules (Pederson and Hendrikson 1993). The origin of these substances, produced indirectly by microorganisms associated with the digestive tract of earthworms or their structures (turricules, galleries) seems specific to the plant and earthworm species studied (Cluzeau et al. 2005).

Earthworms impact the activity and abundance of bacteria populations (Puga Freitas 2012; Jana 2009; Fonte et al. 2009) such as nitrifying and denitrifying bacteria (Wu et al. 2012; Parkin and Berry 1999; Businelli et al. 1984). The bacterial populations are involved in the solubilization and mineralization of inorganic phosphate (Rodriguez and Fraga 1999; Wu et al. 2012) as well as on the solubilization of potassium (Wu et al. 2012).

Several studies have also shown the effect of earthworms on reducing the severity of many diseases (Stephens et al. 1993, 1994; Stephens and Davoren 1995, 1997; Clapperton et al. 2001; Wolfarth et al. 2011). These effects are often associated with better plant nutrition or an earthworm predation effect (Puga Freitas 2012).

Earthworms play a central role in improving and maintaining the productivity of agro-systems. However, these organisms directly or indirectly suffer the consequences of soil management methods (Pelosi 2008). The level of their population is sensitive to cultivation techniques, the products that are applied, and the quality and quantity of the carbon resource, which are important characteristics of the sustainability of agro-systems (Bockstaller et al. 2008). Earthworms are considered to be practical indicators of soil fertility and the sustainability of agricultural systems (Richards et al. 2007; Moonen and Bàrberi 2008; Paoletti 1999).

10.8 Bioremediation and Agricultural Sustainability

Sustainable agriculture is a system of agricultural practices which aims to ensure the objectives of sustainable development, the management of agricultural lands must be able to offer the best possible production, durably, without harming the environment, the quality of the soil, and its long-term production capacity to meet the needs of future generations. Lal (1998) defines sustainable agriculture as the ability of agricultural systems to remain productive, efficiently and indefinitely.

The agriculture of tomorrow must be respectful of the environment without harming economic profitability (White et al. 2014), and must have as main concern better management of natural resources and preservation of soil and public health (Chowdhury et al. 2008; Khan et al. 2020a, b).

The soil is a complex and fragile ecosystem which represents a life support for most living organisms. Intensive agriculture, modern farming techniques, the use of pesticides, fertilizers, etc., deteriorate the soil and unbalance its proper functioning. The quality of the soil is thus threatened, and is in constant decline due to the overexploitation of the soil and the pollution generated by the excessive use of chemicals. The pollutants accumulate in the soil; contaminate surface and groundwater and the food chain. They can negatively affect soil biodiversity, plant growth and their production capacity.

The preservation of soil quality and biodiversity is fundamental to ensure a sustainable agricultural system (Datta et al. 2016; White et al. 2014). To preserve the soil it is necessary to have recourse to remediation methods which disturb the soil less and are able to maintain soil quality and biodiversity. Bioremediation is a decontamination technique which uses the capacity of certain living organisms as plants and bacteria, to accumulate or degrade pollutants. Bioremediation do not have environmentally harmful processes or negative impacts compared to other techniques (chemical or physical treatments) (Juwarkar et al. 2010). It is a safe and environmentally friendly method and cost effective. Its goal is to decontaminate the soil and improve its health (Purohit et al. 2018) and it has advantageous effects on soil fertility and structure (Juwarkar et al. 2010).

There are many methods of bioremediation and they can be divided into two categories: In situ bioremediation which involves the treatment of soil directly at the level of contaminated sites (Talley 2005; Azubuike et al. 2016). It is a less expensive method and less disruptive to the environment (Vidali 2001). The ex situ bioremediation which involves excavating contaminated soils from sites and moving them to another site for treatment. This method presents a risk of contamination of other sites (Purohit et al. 2018; Srivastava 2015; Azubuike et al. 2016). It can lead to health risks and safety risks due to the excavation and displacement of the contaminated soil (Juwarkar et al. 2010; Talley and Sleeper 2006).

Bioremediation is a good alternative for soil decontamination; however, its effectiveness depends on multiple factors, one of them is the pollutant. Not all the pollutants can be removed from soil; some of them are biodegradable or bio-transformable but pollutants such as heavy metals are persistent. Phytoremediation is a suitable technique for this type of pollutant which is an in situ *bioremediation* technique that uses plants capacity to bioaccumulate pollutants and is considered to be a major tool in bioremediation of heavy metals and sustainable agriculture (Belliturk et al. 2015).

Several researches made it possible to develop new methods of bioremediation to improve the capacity of remediation, among them the association of two organisms such as plant and bacteria. The use of combined technologies such as fertilized-assisted remediation (Sun et al. 2018) and transgenic organisms (Gerhardt et al. 2017) is important to develop in order to respond to the current and future need for remediation of polluted sites, and to ensure a healthy environment.

To ensure sustainable agricultural production, the soil must be preserved (its structure, quality, biodiversity, etc.) and not just decontaminated. Phytoremediation is a solution that meets the need to decontaminate the soil, preserve it, and maintain a healthy environment. However, the problem which arises for the rehabilitation of agricultural soils is that they are constantly used while these techniques take time and mobilize the land.

Sustainability cannot be achieved if agricultural practices overexploit agricultural ecosystems and abuse the use of chemicals (Saber 2001). Whatever the performance of the soil remediation techniques used and their effectiveness, the sustainability of agricultural systems cannot be reached without changing the current mode of exploitation of agricultural lands, the excessive use of chemicals, and if we do not opt for a more responsible and environmentally friendly soil management (sustainable management of soil).

10.9 Research and Development Activities: Case of Algeria

Bioremediation is an old concept that was used from the earliest civilizations to recycle water resources by exploitation of the marshes. However, the first installations which testify to the voluntary use of a wetland for water purification date from the end of the nineteenth century. Bioremediation has become a current trend due to its simplicity, efficiency, and safety (Singh et al. 2020).

In 1869, the work of Mille and Durand on the degradation of pollutants from wastewater by agricultural crops made it possible to set up in Paris one hectare of "model garden" open to the public in order to convince farmers of the benefits of this type of installation. In 1875, the cities of Moscow and Berlin set up similar treatment systems on an area of 25,000 ha for the Berlin region. In 1901, the city of San Antonio set up an artificial lake (Lake Mitchell) with an area of 275 ha with the aim of treating its wastewater (Boutin et al. 2000). At the same time, the city of Munich developed a similar project by installing a 233 ha water body to create a biological treatment plant. This technique makes it possible to reproduce the mechanisms of degradation, transformation, and recycling of OM thanks to the purifying properties of microscopic algae (Boutin et al. 2000).

Vascular plants floating or rooted (macrophytes) will not be involved in treatment until much later. Their depolluting effect has been known empirically for a very long time, but it was not until the 1950s that German researchers began to study this phenomenon scientifically. Dr. Kathe Seidel of the Max Planck Institute undertook the first scientific work followed by experiments using an artificial wetland pilot for the treatment of wastewater (Campbell and Ogden 1999). These works, presented for the first time in 1953, showed that certain plant species such as marsh rush (*Scirpus lacustris*) could have a real purifying activity. However, it was not until 1972 that the first functional station was built in Othfresen, Germany (Boutin et al. 2000). This process was introduced in France in the 1980s and one of the first installations, still in service today, was carried out in the commune of Pannissières in the Jura.

Since the 1970s, phytoremediation has been booming. The use of plants to eliminate, in particular, metallic trace elements and pesticides, is attracting growing interest and has been the subject of numerous studies. The nature of the pollutant and its accessibility for plants are decisive in the choice of the phytoremediation process to be used for optimal treatment.

In 2010 the Algerian Ministry of Higher Education and Scientific Research (General direction of scientific research and technological development) launched a national research program (NRP) and more than 2000 research projects have been selected by the intersectoral commissions including more than 80 thematic projects "Environment and Promotion of Sustainable Development." These projects must contribute to the universal knowledge of science and contribute to a better classification of Algeria in the field of experimental and fundamental sciences. This program aims to strengthen research on the vulnerability issues of "biological systems" through a social and economic approach. The main objective is to reduce the cost of transfer of pollution control technologies, as well as taking charge of the management of the effects on populations and ecosystems. All this is in order to develop a national environmental policy within companies, and to develop their capacities for observation, forecasting, and scientific investigation.

From the numerous scientific works carried out by Algerian researchers, the use of plants, microorganisms, and macro-invertebrates as biological means of depollution and rehabilitation of soils contaminated by hydrocarbons and heavy metals becomes more and more encouraging.

The capacity of some species to tolerate and accumulate certain pollutants opened up new areas of research into soil treatment involving bioremediation. Among the current concerns in Algeria and in particular that of SONATRACH (Algerian petroleum company), soil pollution by petroleum derivatives which are one of the principal sources of pollution in the region, the development of a biological treatment for contaminated soils is a well-founded and promising solution, which will allow real management of soil pollution at affordable costs. For this purpose, several studies have been carried out, among them: Ferradji et al. (2014) who studied the degradation of petroleum and naphthalene by *Streptomyces* spp. in the surface soils of the Mitidja plain (North of Algeria); Meliani and Bensoltane (2016) on the application of *Pseudomonas* biofilm in the absorption of heavy metals; and Hamoudi-Belarbi et al. (2018) on the potentials of biostimulation of carob croquettes and carrot peel waste, for the bioremediation of soils polluted by crude oil.

Benchouk and Chibani (2017) also showed that the bioremediation of artificially contaminated soils by bacterial strains and the consortium showed a very great effect on the degradation of hydrocarbons. They also carried out mutagenesis on the strains: *Candida sp., Bacillus sp.,* and *Pseudomonas putida*. To improve the degradation of hydrocarbons on soil samples from the Arzew oil refinery (West Algeria), which showed that the mutagenesis of strains developed a greater or reduced capacity for degradation of petroleum and diesel (*Pseudomonas aeruginosa* representing the highest capability of degradation with 80.86% of hydrocarbons biodegradation).

Heavy metals from road traffic also represent a major source of soil contamination. The study of Maatoug et al. (2013) showed that barley has the capacity to decontaminate the soil located near the highways of the Tiaret region (West Algeria).

Soil remediation aims to preserve soil health and sustainability and to protect the organisms (flora and fauna) that live there as well as humans, by protecting one of the most important resources (the soil) that provides various services. The aim of soil remediation is also to reduce the transfer of pollutants to the groundwater and their bioconcentration in trophic networks (biomagnification). Physicochemical depollution methods have several disadvantages: they are expensive and difficult to implement (Gadd 2000; Salt et al. 1995; Raskin et al. 1994). Further, they strongly disturb the biological activity of soils and alter their physical structure (Huynh 2009). For this purpose, biotechnologies (bioremediation, phytoremediation) seem to be an interesting alternative, these techniques are more respectful of the environment, preserve the characteristics of the soil, and inexpensive.

It should be noted that understanding the biogeochemical processes related to the bioremediation of polluted soils requires a multidisciplinary approach based on both analytical (identification and monitoring of contaminants, modification of soil quality) and biological aspects. This last point implies in-depth knowledge both at the soil level (identification and monitoring of microorganisms of interest which are involved in bioremediation) and at the plant level (characterization of molecules exuded at the rhizospheric level, adaptation and monitoring of plant growth).

10.10 Policy Strategy and Legal Framework for Soil Protection in Algeria

Since the end of the nineteenth century, the industrial sector has grown without concern for the release of toxic elements into the soil. In fact, soil has long been considered a renewable resource, even inexhaustible on the scale of human generations, capable of receiving without consequence rejections from our activities. We now know that the soils are characterized by a fragile and vulnerable balance. Today, this heritage is threatened both by the heavy heritage of the past and by the extension of the areas devoted to industrial development (Lecomte 1998).

In Algeria, agriculture is subject to other physical constraints which burden its natural potential and weigh on the ecological balance of the different natural regions. Indeed, despite the efforts made by the State in the fields of water and soil conservation, soil erosion and pollution continues to increase and water resources are seriously affected by overexploitation or salinization. Desertification is a threat to the 32 million hectares of rangelands and the forest cover of northern Algeria is permanently exposed to natural risk (fire) or to anthropogenic pressure (deforestation-clearing).

In this context, Algeria has implemented, since the 2000s, a policy aimed at improving national food security, the development of certain priority agricultural sectors and land development, including the National Agricultural Development Program (Bessaoud et al. 2019).

The commitment of the Algerian government, for a rational management of natural resources is evident, having regard to the strengthening of the legislative and institutional framework and to the numerous programs launched in the field of environmental education, promotion of renewable energies, fight against poverty, protection of soil and biodiversity, and this, integrated in a three-dimensional approach combining economic, social, and environmental considerations.

The 2010–2014 programs are part of this sustainable development approach and strengthen the intersectoral and participatory approach to planning and implementing integrated natural resource management. It allows the management of issues related to the preservation and sustainable use of biodiversity, soil degradation, water management, and/or the stabilization of greenhouse gases in the atmosphere (National Report of Algeria 2011). Indeed, law No. 83-03 on environmental protection (1983) aims to implement a national environmental protection policy aimed at: protecting, restructuring, and enhancing natural resources, soil conservation/soil improvement (Algerian official journal, www.jordp.dz).

In terms of thinking for the future, the improvement, in practice, of the integrated management of natural resources, in particular the soils affected by pollution, would require significant changes in approaches and interventions, which should be based on more coherent and effective legal and institutional arrangements. Successful water and soil conservation measures also require removing barriers to long-term investment. Without land and financial security, can the farmers concerned engage in long-term activities to maintain soil fertility? This should also call for greater integration of environmental policies (forests, waters, soils, biodiversity, etc.) and agricultural and rural development (Plan Bleu 2003).

10.11 Conclusion

Earthworms play an important role in the soil through various mechanisms, physicochemical and biological, which improve fertility and preserve soil structure and therefore its sustainability. Their proposal as a means to improve the phytoremediation efficiency can be beneficial for maintaining soil fertility and its rehabilitation. Several studies report the effectiveness of the earthworm/plant association for soil remediation, on the one hand, by increasing the bioavailability and phytoextraction of metals, and, on the other hand, by increasing plant biomass. The interaction between these two organisms differs from one species to another (plant and earthworms).

The activity of the earthworm *Lumbricus sp.* reduces the bioaccumulation of lead by *Hordeum vulgare* by 5.2%, and the concentrations recorded in earthworm tissue suggest that the presence of the plant increases these concentrations by 1.54%. The use of another species of earthworm (for example: an eco-physiologically different species) can have a different impact on the phytoextractor behavior of the barley, hence the interest in extending these studies to other species. This study highlights the influence of earthworm activities on the bioaccumulation of lead by barley but also that of the plant on earthworms. The introduction of earthworms into polluted soils represents a great advantage for maintaining soil quality and better remediation; large-scale inoculations of earthworms can be imagined in the soil to improve soil fertility and sustainability. The potential of earthworms must be fully studied for better sustainable soil management.

10.12 Future Perspective

This research opens up many perspectives in the field of phytoremediation and in other fields. Complementary works to this study are envisaged; study the evolution of the earthworm populations in different soils for a better understanding of their behavior towards pollution and cultural practices. Further, study on the physiology and biochemistry of barley to better understand the effect of earthworm activity on this plant and the mechanisms involved in its response. Also a study of the impact of earthworms on the evolution of bacterial communities in metals polluted soils which play a role in the availability of metals for plants needs to be explored.

A study on the possibility of increasing earthworm population in the soil or of introducing them on a large-scale is to be considered to improve soil fertility and durability.

On the other hand, these bioremediation tools have indeed become essential to guide the policy of soil remediation for sustainable agriculture. However, bioremediation studies are still too ad hoc and limited to meet all of the needs.

The industrial and socio-development of Algeria has not always taken into account its impact on the quality of the environment and on the conservation of the environment and natural resources. However, from the year 2010, this gap was filled by the definition and the implementation of a policy of environmental protection, and in particular of fight against the pollution of agricultural soils, aiming in particular at:

- The creation of a multidisciplinary research network in a context of sustainable development,
- Determining the impact of certain micro-pollutants whose concentrations in the soil are difficult to quantify directly,
- Raising public awareness, in particular farmers about soil pollution,
- The development of a preservation strategy with regard to soil pollution in agricultural land,
- Exchange of experiences in the field of bioremediation of soil pollution.

It is crucial to assess the threats linked to soil pollution and to identify contaminated sites in order to remediate them and to ensure soil sustainability.

References

Abdul Rida AMM (1996) Concentrations et croissance de Lombriciens et de plantes dans des sols contaminés ou non par Cd, Cu, Fe, Pb et Zn: Interactions plant-sol-lombricien. Soil Biol Biochem 28:1037–1044

Adriano DC (2001) Trace elements in terrestrial environments: biochemistry, bioavailability and risks of metals. Springer, New York

Aghababaei F, Raiesi F, Hosseinpur A (2014) The significant contribution of mycorrhizal fungi and earthworms to maize protection and phytoremediation in Cd-polluted soils. Pedobiologia 57 (4–6):223–233. https://doi.org/10.1016/j.pedobi.2014.09.004

Alkorta I, Hernandez-Allica J, Becerril JM, Amezaga I, Albizu I, Garbisu C (2004) Recent findings on the phytoremediation of soils contaminated with environmentally toxic heavy metals and metalloids such as zinc, cadmium, lead and arsenic. Environ Sci Biotechnol 3:71–90

Alloway BJ (1995a) The mobilisation of trace elements in soils. In: Proceeding contaminated soils, 3rd international conference on the biogeochemistry of trace elements, Paris, France, 15–19 May. p 133

Alloway BJ (1995b) Heavy metals in soils. Blackie Academic & Professional, London

Alloway BJ (1997) The mobilisation of trace elements in soils. Contaminated soils. INRA, Paris, pp 133–145

Almehdi A, El-Keblawy A, Shehadi I, El-Naggar M, Saadoun I, Mosa KA, Abhilash PC (2019) Old leaves accumulate more heavy metals than other parts of the desert shrub Calotropis procera at a traffic-polluted site as assessed by two analytical techniques. Int J Phytoremediation 21:1254–1262. https://doi.org/10.1080/15226514.2019.1619164

Ash CPJ, Lee D (1980) Lead, cadmium, copper and iron in earthworms from roadside sites. Environ Pollut 22A:59–67

Azizollahi Z, Ghaderian SM, Ghotbi-Ravandi AA (2019) Cadmium accumulation and its effects on physiological and biochemical characters of summer savory (Satureja hortensis L.). Int J Phytoremediation 21:1241–1253. https://doi.org/10.1080/15226514.2019.1619163

Azhar-u-ddin HJ-C, Gan X, He S, Zhou W (2020) Interactive effects of earthworm *Eisenia fetida* and bean plant Phaseolus vulgaris L on the fate of soil selenium. Environ Pollut 2020. https://doi.org/10.1016/j.envpol.2020.114048

Azubuike CC, Chikere CB, Okpokwasili GC (2016) Bioremediation techniques–classification based on site of application: principles, advantages, limitations and prospects. World J Microbiol Biotechnol 32(11):180. https://doi.org/10.1007/s11274-016-2137-x

Baize D (1997) Teneurs totales en ETM dans les sols (France). INRA, Paris, p 408

Baize D (2002) Les éléments traces métalliques dans les sols. INRA éditions, Paris

Banerjee A, Jhariya MK, Yadav DK, Raj A (2018) Micro-remediation of metals: a new frontier in bioremediation. In: Hussain C (ed) Handbook of environmental materials management. Springer, Cham. https://doi.org/10.1007/978-3-319-58538-3_10-1

Banerjee A, Jhariya MK, Yadav DK, Raj A (2020) Environmental and sustainable development through forestry and other resources. Apple Academic Press, Palm Bay, p 400. https://doi.org/10.1201/9780429276026

Bellitürk K, Shrestha P, Görres JH (2015) The importance of phytoremediation of heavy metal contaminated soil using vermicompost for sustainable agriculture. J Rice Res 03:02. https://doi.org/10.4172/2375-4338.1000e114

Benahmed M, Dellal A, Hellal B (2016) Mobilite Du Plomb Et Du Zinc Issus De Retombees Atmospheriques Dans Le Sol: Cas De La Zone Industrielle De Tiaret, Algerie. Eur Sci J 12 (18):131–140. https://doi.org/10.19044/esj.2016.v12n18p131

Benchouk A, Chibani A (2017) Petroleum-hydrocarbons biodegradation by pseudomonas strains isolated from hydrocarbon-contaminated soil. J Fundam Appl Sci 9(2):713–726. https://doi.org/10.4314/jfas.v9i2.7

Bengtsson G, Ek H, Rundgren S (1992) Evolutionary response of earthworms to long term metal exposure. Oikos 63:289–297

Bessaoud O, Pellissier J-P, Rolland J-P, Khechimi W (2019) Rapport de synthèse sur l'agriculture en Algérie. [Rapport de recherche]. CIHEAM-IAMM, Montpellier, p 82

Beyer WN, Chaney RL, Mulhern BM (1982) Heavy metal concentrations in earthworms from soil amended with sewage sludge. J Environ Qual 11:381–385

Beyer WN, Hensler G, Morre J (1987) Relation of pH and other soil variables to concentrations of Pb, Cu, Zn, Cd and Se in earthworms. Pedobiologia 30:167–172

Bhatnagar T (1975) Lombriciens et humification: un aspect nouveau de l'incorporation microbienne d'azote induite par les vers de terre. In: Kilbertus G, Reisinger O, Mourey A, da Fonseca JAC (eds) Humification et Biodégradation. Pierron, Sarreguemines, pp 169–182

Binet F (1993) Dynamique des peuplements et fonctions des lombriciens en sols cultivés tempérés. Thèse de Doctorat, Université de Rennes, p 1

Binet F, Curmi P (1992) Structural effects of Lumbricus terrestris (Oligochaeta: Lumbricidae) on the soil-organic matter system: micromorphological observations and autoradiographs. Soil Biol Biochem 24:1519–1523

Bityutskii N, Kaidun P, Yakkonen K (2016) Earthworms can increase mobility and bioavailability of silicon in soil. Soil Biol Biochem 99:47–53

Blouin M, Hodson ME, Delgado EA, Baker G, Brussaard L, Butt KR, Dai J, Dendooven L, Peres G, Tondoh JE, Brun JJ (2013) A review of earthworm impact on soil function and ecosystem services. Eur J Soil Sci 64:161–182

Blum WEH, Brandstetter A, Wenzel WW (1997) Trace element distribution in soils as affected by land use. In: Adriano DC, Chen Z-S, Yang S-S, Iskadar IK (eds) Biogeochemistry of trace elements. Science Reviews, Northwood, p 61

Bockstaller C, Galan M-B, Capitaine M, Colomb B, Mousset J, Viaux P (2008) Comment évaluer la durabilité des systèmes en production végétale? In: Systèmes de culture innovants et durables. Educagri, Dijon, pp 29–51

Bongoua-Devisme AJ, Akotto OF, Guety T, Kouakou SAAE, Ndoye F, Diouf D (2019) Enhancement of phytoremediation efficiency of Acacia mangium using earthworms in metal-contaminated soil in Bonoua, Ivory Coast. Afr J Biotechnol 18(27):622–631. https://doi.org/10.5897/ajb2019.16852

Bouche MB (1977) Stratégies lombriciennes. Soil organisms as components of ecosystems. Swed Nat Sci Res Council Ecol Bull Stockh 25:122–132

Boukirat D (2018) Bioremédiation d'un sol agricole pollué par le plomb à l'aide de l'interaction des macro-invertébrés terrestres (vers de terre : *Lumbricus sp*) et des céréales (orge commune: *Hordeum vulgare*), Thèse de doctorat de l'université de Tiaret, Algérie

Boukirat D, Maatoug M, Zerrouki D, Lahouel H, Heilmeier H, Kharytonov M (2017) Bioremediation of agricultural soil contaminated with lead using interaction: common barley *Hordeum Vulgare* and earthworm *Lumbricus Sp.* INMATEH Agric Eng 51(1):133–142

Bouras S, Maatoug M, Hellal B, Ayad N (2010) Quantification of ground pollution by lead and zinc in relation to the road traffic (cases of the town of Sidi Bel Abbes, Algeria west). Les Technol Lab 5(20):11–17

Boutin C, Esser D, Molle P, Lienard A (2000) March. Les filtres et lits plantés de roseaux en traitement d'eaux usées domestiques. Perspectives pour le traitement d'eaux pluviales. Manuscrit auteur, publié dans "Les rendez-vous du GRAIE. Le traitement des eaux par macrophytes. Perspectives pour les eaux pluviales, Le Bourget-du-Lac: France

Bradham KD, Dayton EA, Basta NT, Schroder J, Payton M, Lanno RP (2006) Effect of soil properties on lead bioavailability and toxicity to earthworms. Environ Toxicol Chem 25:769e775

Brown GG (1995) How do earthworms affect microflora and faunal community diversity? Plant Soil 170:209–231

Brown G, Pashanasi B, Villenave C, Patron J, Senapati BK, Giri S, Barois I, Lavelle P, Blanchart E, Blakemore RJ, Spain AV, Boyer J (1999) Effects of earthworms on plant production in the tropics. In: Lavelle P, Brussaard L, Hendrix PF (eds) Earthworm management in tropical agroecosystems. CABI Publishing, Wallingford, pp 87–147

Brown GG, Barois I, Lavelle P (2000) Regulation of soil organic matter dynamics and microbial activity in the drilosphere and the role of interactions with other edaphic functional domains. Eur J Soil Biol 36:177–198

Brown GG, Feller C, Blanchart E, Deleporte P, Chernyanski SS (2003a) With Darwin, earthworms turn intelligent and become human friends. Pedobiologia 47:924–933

Brown GG, Benito NP, Pasini A, Sautter KD, Guimaraes M, Torres E (2003b) No-tillage greatly increases earthworm populations in Parana state, Brazil. Pedobiology 47:764–771

Brown GG, Edwards CA, Brussaard L (2004) How earthworm affect plant growth: burrowing into the mechanisms. In: Edwards CA (ed) Earthworm ecology. CRC Press, Boca Raton, pp 13–49

Businelli M, Perucci P, Patumi M, Giusquiani P (1984) Chemical composition and enzymic activity of some worm casts. Plant Soil 80:417–422

Campbell C, Ogden MR (1999) Constructed wetlands in the sustainable landscape. Wiley, Hoboken, 270 pp

CCME (Canadian Council of Ministers of the Environment) (1999) Canadian soil quality guidelines for the protection of environmental and human health: Lead (1999). In: Canadian environmental quality guidelines, 1999. Canadian Council of Ministers of the Environment, Winnipeg

Chauhan RP (2014) Role of earthworms in soil fertility and factors affecting their population dynamics: a review. Int J Res 1(6):642–649

Cheng J, Wong MH (2002) Effects of earthworms on Zn fractionation in soils. Biol Fertil Soils 36:72–78

Chowdhury A, Pradhan S, Saha M, Sanyal N (2008) Impact of pesticides on soil microbiological parameters and possible bioremediation strategies. Indian J Microbiol 48(1):114–127

Clapperton MJ, Lee NO, Binet F, Conner RL (2001) Earthworms indirectly reduce the effects of take-all (*Gaeumannomyces graminis* var. tritici) on soft white spring wheat (*Triticum aestivum* cv. Fielder). Soil Biol Biochem 33:1531–1538

Cluzeau D, Lebouvier M, Trehen P, Bouché MB, Badour C, Perraud A (1987) Relations between earthworms and agricultural practices in the vineyards of Champagne. Preliminary results. In: On earthworms. Mucchi, Modena, pp 465–484

Cluzeau D, Blanchart E, Peres G, Ablain F, Cuendet G, Fayolle L, Lavelle P (2005) Faune du sol et Lombriciens dans les sols tempérés agricoles. In: Girard M-C, Walter C, Rémy J-C, Berthelin J, Morel J-L (eds) Sols et environnement 2e cycle. Cours, exercices et études de cas. Dunod, Paris, pp 386–407

Coeurdassier M, Scheifler R, de Vaufleury A, Crini N, Saccomani C, Du Mont LS, Badot PM (2007) Earthworms influence metal transfer from soil to snails. Appl Soil Ecol 35:302–310

Colandini V (1997) Effets des structures réservoirs à revêtement poreux sur les eaux pluviales: qualité des eaux et devenir des métaux lourds. Thèse doctorat., Université de Pau et des pays de l'Adour, p 162

Czarnowska K, Jopkiewicz K (1978) Heavy metals in earthworms as an index of soil contamination. Pol J Soil Sci II:57–62

Da Silva E (2013) Interactions "sol - vers de terre" et dynamique du mercure en Guyane française. Earth Sciences. Universite Paris-Est. French. <NNT: 2013PEST1162>. <tel-00944329>

Dai J, Becquer T, Rouiller JH, Reversat G, Bernhard-Reversat F, Nahmani J, Lavelle P (2004) Heavy metal accumulation by two earthworm species and its relationship to total and DTPA-extractable metals in soils. Soil Biol Biochem 36:91–98

Darwin C (1881) The formation of vegetable mould through the action of worms with observations on their habits. Murray, London, 298 pp

Das A, Osborne JW (2018) Enhanced lead uptake by an association of plant and earthworm bioaugmented with bacteria. Pedosphere 28(2):311–322. https://doi.org/10.1016/s1002-0160(18)60021-9

Datta S, Singh J, Singh S, Singh J (2016) Earthworms, pesticides and sustainable agriculture: a review. Environ Sci Pollut Res 23(9):8227–8243. https://doi.org/10.1007/s11356-016-6375-0

De Abreu CA, de Abreu MF, Andrade JC (1998) Distribution of lead in the soil profile by DTPA and Mehlich-3 solution. Bragantia 57:185–882

De Vaufleury A, Gimbert F, Gomot L (2013) Bioaccumulation, Bioamplification des polluants dans la faune terrestre, Un outil pour la biosurveillance des écosystèmes. ADEME, Paris, pp 46–90

Demarco CF, Afonso TF, Pieniz S, Quadro MS, Camargo FADO, Andreazza R (2019) Phytoremediation of heavy metals and nutrients by the Sagittaria montevidensis into an anthropogenic contaminated site at Southern of Brazil. Int J Phytoremediation 21:1145–1152. https://doi.org/10.1080/15226514.2019.1612843

Depta B, Koscielniak A, Rozen A (1999) Food selection as a mechanism of heavy metal resistance in earthworms. Pedobiologia 43:608–614

Devliegher W, Verstraete W (1995) *Lumbricus terrestris* in a soil core experiment: effects of nutrient-enrichment processes (NEP) and gut-associated processes (GAP) on the availability of plant nutrients and heavy metals. Soil Biol Biochem 28:489–496

Devliegher W, Verstraete W (1997) Microorganisms and soil physico-chemical conditions in the drilosphere of *Lumbricus terrestris*. Soil Biol Biochem 29:1721–1729

Diehl KH, Rosopulo A, Kreuzer W, Judel GK (1983) Das Verhalten von Bleitetraalkylen im Boden und deren Aufnahme durch die Pflanzen. Z Pflanzenernähr Bodenkd 146(5):551–559

Duarte AP, Melo VF, Brown GG, Pauletti V (2012) Changes in the forms of lead and manganese in soils by passage through the gut of the tropical endogeic earthworm (*Pontoscolex corethrurus*). Eur J Soil Biol 53:32–39

Ebadi AG, Hisoriev H (2017) Metal pollution status of Tajan River – northern Iran. Toxicol Environ Chem 99:1358–1367. https://doi.org/10.1080/02772248.2017.1345191

Edwards CA (2004) Earthworm ecology, 2nd edn. St. Lucie Press, Boca Raton

Edwards CA, Bohlen PJ (1992) The effects of toxic chemicals on earthworms. Rev Environ Contam Toxicol 125:23–99

Edwards CA, Bohlen PJ (1996) Biology and ecology of earthworms, 3rd edn. Chapman & Hall, London

Eisenhauer N, Konig S, Sabais A (2009) Impacts of earthworms and arbuscular mycorrhizal fungi (*Glomus intraradices*) on plant performance are not interrelated. Soil Biol Biochem 41:561–567

Elyamine AM, Moussa MG, Ismael MA, Wei J, Zhao Y, Wu Y, Hu C (2018) Earthworms, rice straw, and plant interactions change the organic connections in soil and promote the decontamination of cadmium in soil. Int J Environ Res Public Health 15:2398. https://doi.org/10.3390/ijerph15112398

Eslami A, Eftekhari A, Siavoshi M (2019) Investigating the interaction of silica and earthworm on biological refining of heavy metal nickel in dwarf Lilyturf (*Ophiopogon japonicus*). Albanian I Agric Sci 18(4):99–104

Ferradji FZ, Mnif S, Badis A (2014) Naphthalene and crude oil degradation by biosurfactant producing Streptomyces spp. isolated from Mitidja plain soil (north of Algeria). Int Biodeter Biodeg 86:300–308. https://doi.org/10.1016/j.ibiod.2013.10.003

Fonte SJ, Winsome T, Six J (2009) Earthworm populations in relation to soil organic matter dynamics and management in California tomato cropping systems. Appl Soil Ecol 41:206–214

Gadd JM (2000) In: Raskin I, Ensley BD (eds) Phytoremediation of toxic metals: using plants to clean up the environment. Wiley, New York, p 304

Galal TM, Shedeed ZA, Hassan LM (2019) Hazards assessment of the intake of trace metals by common mallow (Malva parviflora K.) growing in polluted soils. Int J Phytoremediation 21:1397–1406. https://doi.org/10.1080/15226514.2018.1524842

Gaur A, Adholeya A (2004) Prospects of arbuscular mycorrhizal fungi in phytoremediation of heavy metal contaminated soils. Curr Sci 86(4):528–534. http://www.currentscience.ac.in/Downloads/article_id_086_04_0528_0534_0.pdf

Gerhardt KE, Gerwing PD, Greenberg BM (2017) Opinion: taking phytoremediation from proven technology to accepted practice. Plant Sci 256:170–185. https://doi.org/10.1016/j.plantsci.2016.11.016

Gish CD, Christensen RE (1973) Cadmium, nickel, lead and zinc in earthworms from roadside soil. Environ Sci Technol 7:1060–1062

Gobat JM, Aragno M, Matthey W (2010) Le sol vivant: bases de pédologie, biologie des sols. PPUR Presses Polytechniques Et Universitaires Romandes, Lausanne, p 817

Gómez-Garrido M, Mora Navarro J, Murcia Navarro FJ, Faz Cano Á (2018) The chelating effect of citric acid, oxalic acid, amino acids and Pseudomonas fluorescens bacteria on phytoremediation of Cu, Zn, and Cr from soil using Suaeda vera. Int J Phytoremediation 20:1033–1042. https://doi.org/10.1080/15226514.2018.1452189

Grelle C, Descamps M (1998) Heavy metal accumulation by *Eisenia fetida* and its effects on glutathione-S-transferase activity. Pedobiologia 42:289–297

Guo Z, Gao Y, Cao X, Jiang W, Liu X, Liu Q, Chen Z, Zhou W, Cui J, Wang Q (2019) Phytoremediation of Cd and Pb interactive polluted soils by switchgrass (*Panicum virgatum* L.). Int J Phytoremediation 21:1486–1496. https://doi.org/10.1080/15226514.2019.1644285

Ha NTH, Sakakibara M, Sano S, Nhuan MT (2011) Uptake of metals and metalloids by plants growing in a lead-zinc mine area, Northern Vietnam. J Hazard Mater 186:1384–1391

Hamoudi-Belarbi L, Hamoudi S, Belkacemi K (2018) Bioremediation of polluted soil sites with crude oil hydrocarbons using carrot peel waste. Environments 5(11):124. https://doi.org/10.3390/environments5110124

Hedde M (2018) Indicateurs basés sur la faune des sols: des outils pour l'agriculture innovante ? Innov Agron 69:15–26. hal-02002765

Helmke PA, Robarge WP, Korotev RL, Schomberg PJ (1979) Effects of soil-applied sewage sludge on concentration of elements in earthworms. J Environ Qual 8:322–327

Hinsinger P (2001) Bioavailability of trace elements as related to root-induced chemical changes in the rhizosphere. In: Gobran GR, Wenzel WW, Lombi E (eds) Trace elements in the rhizosphere. CRC Press, Boca Raton, pp 25–41

Hopkin SP (1989) Ecophysiology of metals in terrestrial invertebrates. Elsevier Applied Science, London

Hu F, Wu XQ, Li HX, Wu SM (1998) Effects of earthworm and ants on the properties of red soils. In: Research on the red soil ecosystem. China Agricultural Science and Technology Publishing House, Beijing, pp 276–258 (in Chinese)

Huynh TMD (2009) Impact des métaux lourds sur les interactions plante/ver de terre/microflore tellurique. Océan, Atmosphère. Université Paris-Est, 2009. French. <tel-00486649>

Ireland MP (1975) The effect of earthworm Dendrobaena rubida on the solubility of lead, zinc and cadmium in heavy metal contaminated soils in Wales. J Soil Sci 26:313–318

Ireland MP (1979) Distribution of essential and toxic metals in the terrestrial gastropod Arion ater. Environ Pollut 20:271–278

Ireland MP (1983) Heavy metal uptake and tissue distribution in earthworms. In: Satchell JE (ed) Earthworm ecology from Darwin to Vermiculture. Chapman and Hall, London, pp 247–265

Ireland MP, Richards KS (1981) Metal content after exposure to cadmium of two species of earthworms of known differing calcium metabolic activity. Environ Pollut 26A:69–78

Jana U (2009) Etude des interactions entre la plante *Arabidopsis thaliana* (L.) Heynh et les vers de terre *Aporrectodea caliginosa* (Savigny) : Application à la phytoremédiation de l'arsenic et de l'antimoine, these de doctorat de l'université Paris-Est

Jeelani N, Yang W, Qiao Y, Li J, An S, Leng X (2018) Individual and combined effects of cadmium and polycyclic aromatic hydrocarbons on the phytoremediation potential of *Xanthium sibiricum* in co-contaminated soil. Int J Phytoremediation 20:773–779. https://doi.org/10.1080/15226514.2018.1425666

Jegou D, Brunotte J, Rogasik H, Capowiez Y, Diestel H, Schrader S, Cluzeau D (2002) Impact of soil compaction on earthworm burrow systems using X-ray computed tomography: preliminary study. Eur J Soil Biol 38:329–336

Jhariya MK, Yadav DK, Banerjee A (2018a) Plant mediated transformation and habitat restoration: phytoremediation an eco-friendly approach. In: Gautam A, Pathak C (eds) Metallic contamination and its toxicity. Daya Publishing House, A Division of Astral International, New Delhi, pp 231–247

Jhariya MK, Banerjee A, Yadav DK, Raj A (2018b) Leguminous trees an innovative tool for soil sustainability. In: Meena RS, Das A, Yadav GS, Lal R (eds) Legumes for soil health and sustainable management. Springer, Singapore, pp 315–345. https://doi.org/10.1007/978-981-13-0253-4_10

Jhariya MK, Banerjee A, Meena RS, Yadav DK (2019a) Sustainable agriculture, forest and environmental management. Springer, Singapore. https://doi.org/10.1007/978-981-13-6830-1

Jhariya MK, Yadav DK, Banerjee A (2019b) Agroforestry and climate change: issues and challenges. Apple Academic Press, Palm Bay, p 335. https://doi.org/10.1201/9780429057274

Jiang ZF, Huang SZ, Han YL, Zhao JZ, Fu JJ (2012) Physiological response of Cu and Cu mine tailing remediation of Paulownia fortunei (Seem) Hemsl. Ecotoxicology 21:579–767

Jones CG, Lawton JH, Shachak M (1994) Organisms as ecosystem engineers. Oikos 69:373–386

Jusselme MD, Poly F, Miambi E, Mora F, Blouin M, Pando A, Rouland-Lefèvre C (2012) Effect of earthworms on plant Lantana camara Pb-uptake and on bacterial communities in root-adhering soil. Sci Total Environ 416:200–207

Jusselme MD, Miambi E, Mora P, Diouf M, Rouland-Lefevre C (2013) Increased lead availability and enzyme activities in root-adhering soil of *Lantana camara* during phytoextraction in the presence of earthworms. Sci Total Environ 445–446:101–109

Jusselme MD, Miambi E, Lebeau T, Rouland-Lefevre C (2015) Role of earthworms on phytoremediation of heavy metal-polluted soils. In: Heavy metal contamination of soils. Soil biology, vol 44. Springer, Cham. https://doi.org/10.1007/978-3-319-14526-6_15

Juwarkar AA, Singh SK, Mudhoo A (2010) A comprehensive overview of elements in bioremediation. Rev Environ Sci Bio/Technol 9(3):215–288. https://doi.org/10.1007/s11157-010-9215-6

Kabata-Pendias A, Pendias H (2001) Trace elements in soils and plants, 3rd edn. CRC Press, Boca Raton, pp 222–234

Kamitani T, Kaneko N (2007) Species-specific heavy metal accumulation patterns of earthworms on a floodplain in Japan. Ecotoxicol Environ Saf 66:82–91

Kaur P, Bali S, Sharma A, Vig AP, Bhardwaj R (2018) Role of earthworms in phytoremediation of cadmium (Cd) by modulating the antioxidative potential of *Brassica juncea* L. Appl Soil Ecol 124:306–316. https://doi.org/10.1016/j.apsoil.2017.11.017

Kebir T, Bouhadjera K (2011) Effects of heavy metals pollution in soil and plant in the industrial area, West Algeria. J Korean Chem Soc 55(6):1018–1023. https://doi.org/10.5012/jkcs.2011.55.6.1018

Kennette D, Hendershot W, Tomlin A, Sauvé S (2002) Uptake of trace metals by earthworm *Lumbricus terrestris* L. in urban contaminated soils. Appl Soil Ecol 19:191–198

Khan N, Jhariya MK, Yadav DK, Banerjee A (2020a) Herbaceous dynamics and CO_2 mitigation in an urban setup- a case study from Chhattisgarh, India. Environ Sci Pollut Res 27(3):2881–2897. https://doi.org/10.1007/s11356-019-07182-8

Khan N, Jhariya MK, Yadav DK, Banerjee A (2020b) Structure, diversity and ecological function of shrub species in an urban setup of Sarguja, Chhattisgarh, India. Environ Sci Pollut Res 27 (5):5418–5432. https://doi.org/10.1007/s11356-019-07172-w

Kopittke PM, Colin JA, Kopittke RA, Menzies NW (2007) Toxic effects of Pb2+ on growth of cowpea (*Vigna unguiculata*). Environ Pollut 150:280–287. https://doi.org/10.1016/j.envpol.2007.01.011

Kumari R, Dey S (2019) A breakthrough column study for removal of malachite green using coco-peat. Int J Phytoremediation 21:1263–1271. https://doi.org/10.1080/15226514.2019.1633252

Kumar S, Meena RS, Jhariya MK (2020) Resources use efficiency in agriculture. Springer, Singapore, p 760. https://doi.org/10.1007/978-981-15-6953-1

Lal R (1998) Soil quality and agricultural sustainability. In: Lal R (ed) Soil quality and agricultural sustainability. Ann Arbor Press, Chelsea, pp 3–12

Lamy I (2000) Réactivité des matières organiques des sols vis-à-vis des métaux. Journées nationales de l'étude des sols. p 22

Lanno R, Wells J, Conder J, Bradham K, Basta N (2004) The bioavailability of chemicals in soil for earthworm. Ecotoxicol Environ Saf 57:39–47

Lavelle P (1988) Earthworms and the soil system. Biol Fertil Soils 6:237–251

Lavelle P (1996) Diversity of soil fauna and soil function. Biol Int 33:3–16

Lavelle P (1997) Faunal activities and soil processes: adaptive strategies that determine ecosystem function. Adv Ecol Res 27:93–132

Lavelle P (2002) Functional domains in soils. Ecol Res 17:441–450

Lavelle P, Gilot C (1994) Priming effects of macroorganisms on microflora: a key process of soil function? In: Ritz K, Dighton J, Giller K (eds) Beyond the biomass. Wiley-Sayce, Chichester, pp 176–181

Lavelle P, Martin A (1992) Small-scale and large-scale effect of endogeic earthworms on soil organic matter dynamics in soil of humid tropics. Soil Biol Biochem 24:1491–1498

Lavelle P, Bignell D, Lepage M, Wolters V, Roger P, Ineson P, Heal OW, Ghillion S (1997) Soil function in a changing world: the role of invertebrate ecosystem engineers. Eur J Soil Biol 33:159–193

Lavelle P, Pashanasi B, Charpentier F, Gilot C, Rossi JP, Derouard L, Andre J, Ponge JF, Bernier N (1998) Large-scale effects of earthworms on soil organic matter and nutrient dynamics. In: Edwards CA (ed) Earthworm ecology. St. Lucie Press, Boca Raton, pp 103–122

Lavelle P, Decaëns T, Aubert M, Barot S, Blouin M, Bureau F, Margerie P, Mora P, Rossi J-P (2006) Soil invertebrates and ecosystem services. Eur J Soil Biol 42:S3–S15

Lecomte P (1998) Les sites pollués: Traitement des sols et des eaux souterraines. TEC & DOC, Lavoisier, Paris

Lemtiri A, Lienard A, Alabi T, Brostaux Y, Cluzeau D, Francis F, Colinet G (2016) Earthworms *Eisenia fetida* affect the uptake of heavy metals by plants *Vicia faba* and *Zea mays* in metal-contaminated soils. Appl Soil Ecol 104:67–78. https://doi.org/10.1016/j.apsoil.2015.11.021

Li LY, Li RS (2000) The role of clay minerals and effect of H+ ions on removal of heavy metal (Pb2 +) from contaminated soil. Can J Geotech/Rev 37:296–307

Li M, Liu ZT, Xu Y, Cui YB, Li DS, Kong ZM (2009) Comparative effects of Cd and Pb on biochemical response and DNA damage in the earthworm *Eisenia fetida* (Annelida, Oligochaeta). Chemosphere 74:621–625

Li NY, Li ZA, Zhuang P, Zou B, McBride M (2009) Cadmium uptake from soil by maize with intercrops. Water, Air, and Soil Pollution 199:(1-4):45–56

Li Y, Luo J, Yu J, Xia L, Zhou C, Cai L, Ma X (2018) Improvement of the phytoremediation efficiency of Neyraudia reynaudiana for lead-zinc mine-contaminated soil under the interactive effect of earthworms and EDTA. Sci Rep 8(1):1–10. https://doi.org/10.1038/s41598-018-24715-2

Liu J, Li K, Xu J, Zhang Z, Ma T, Lu X, Yang J, Zhu Q (2003) Lead toxicity, uptake, and translocation in different rice cultivars. Plant Sci 165:793–802

Liu X, Hu C, Zhang S (2005) Effects of earthworm activity on fertility and heavy metal bioavailability in sewage sludge. Environ Int 31(6):874–879

Liu X, Cao L, Zhang X, Chen J, Huo Z, Mao Y (2018) Influence of alkyl polyglucoside, citric acid, and nitrilotriacetic acid on phytoremediation in pyrene-Pb co-contaminated soils. Int J Phytoremediation 20:1055–1061. https://doi.org/10.1080/15226514.2018.1460305

Lock K, Janssen CR (2001) Effect of clay and organic matter type on the ecotoxicity of zinc and cadmium to the potworm *Enchytraeus albidus*. Chemosphere 44:1669–1772

Lukkari T, Teno S, Väisänen A, Haimi J (2006) Effects of earthworms on decomposition and metal availability in contaminated soil: microcosm studies of populations with different exposure histories. Soil Biol Biochem 38(2):359–370

Luo W, Verweij RA, van Gestel CAM (2014) Determining the bioavailability and toxicity of lead contamination to earthworms requires using a combination of physicochemical and biological methods. Environ Pollut 185:1–9. https://doi.org/10.1016/j.envpol.2013.10.017

Ma WC (1982) The influence of soil properties and worm-related factors on the concentration of heavy metals in earthworms. Pedobiologia 24:109–119

Ma Y, Dickinson NM, Wong MH (2000) The effect of earthworm inoculation on metal bioavailability: potential use for phytoremediation of Pb/Zn mine spoils. In: Proceedings of Remade Lands 2000, international conference on the remediation and management of degraded lands, Fremantle, Western Australia. pp 33–34

Ma Y, Dickinson N, Wong M (2003) Interactions between earthworms, trees, soil nutrition and metal mobility in amended Pb/Zn mine tailings from Guangdong, China. Soil Biol Biochem 35:1369–1379

Ma JF, Tamai K, Yamaji N, Mitani N, Konishi S, Katsuhara M, Ishiguro M, Murata Y, Yano M (2006a) A silicon transporter in rice. Nature 440:688–691

Ma Y, Dickinson NM, Wong MH (2006b) Beneficial effects of earthworms and arbuscular mycorrhizal fungi on establishment of leguminous trees on Pb/Zn mine tailings. Soil Biol Biochem 38:1403–1412

Maatoug M, Amirat M, Zerrouki D, Ait Hammou M (2013) Decontamination of agricultural soil polluted with lead using the common barley (Hordium vulgare). Arab Gulf J Sci Res 31 (1):23–35

Macci C, Doni S, Peruzzi E, Ceccanti B, Masciandaro G (2012) Bioremediation of polluted soil through the combined application of plants, earthworms and organic matter. J Environ Monit 14 (10):2710. https://doi.org/10.1039/c2em30440f

Mahler RJ, Bingham FT, Page AL (1978) Cadmium-enriched sewage sludge application to acid and calcareous soils: effect on yield and cadmium uptake by lettuce and chard. J Environ Qual 7:274–281

Mahohi A, Raiesi F (2019) Functionally dissimilar soil organisms improve growth and Pb/Zn uptake by Stachys inflata grown in a calcareous soil highly polluted with mining activities. J Environ Manag 247:780–789. https://doi.org/10.1016/j.jenvman.2019.06.130

Mai X, Luo D, Wei L, Liu Y, Huang X, Wu Q, Yao G, Liu G, Liu L (2019) Evaluation method for the measuring comprehensive suitability of chelating agents: a study of the temporal dynamics of heavy metal activation. Int J Phytoremediation 21:1415–1422. https://doi.org/10.1080/15226514.2019.1633262

Makowski D, Wallach D, Tremblay M (2002) Using a Bayesian approach to parameter estimation; comparison of the GLUE and MCMC methods. Agronomie 22:191–203

Marino F, Ligero A, Diaz Cosin DJ (1992) Heavy metals and earthworms on the border of the road next to Santiago (Galicia, Northwest of Spain). Initial results. Soil Biol Biochem 24:1705–1709

Marschner H (1995) Mineral nutrition of higher plants, 2nd edn. Academic press, London

McLaughlin MJ, Zarcinas BA, Stevens DP, Cook N (2000) Soil testing for heavy metals. Commun Soil Sci Plant Anal 31:11–14. https://doi.org/10.1080/00103620009370531

Meena RS, Lal R (2018) Legumes for soil health and sustainable management. Springer, Singapore, p 541. https://doi.org/10.1007/978-981-13-0253-4_10

Meena RS, Kumar V, Yadav GS, Mitran T (2018) Response and interaction of *Bradyrhizobium japonicum* and Arbuscular mycorrhizal fungi in the soybean rhizosphere: a review. Plant Growth Regul 84:207–223

Meena RS, Kumar S, Datta R, Lal R, Vijaykumar V, Brtnicky M, Sharma MP, Yadav GS, Jhariya MK, Jangir CK, Pathan SI, Dokulilova T, Pecina V, Marfo TD (2020) Impact of agrochemicals on soil microbiota and management: a review. Land (MDPI) 9(2):34. https://doi.org/10.3390/land9020034

Meena RS, Lal R, Yadav GS (2020a) Long term impacts of topsoil depth and amendments on soil physical and hydrological properties of an Alfisol in Central Ohio, USA. Geoderma 363:1141164

Meena RS, Lal R, Yadav GS (2020b) Long-term impact of topsoil depth and amendments on carbon and nitrogen budgets in the surface layer of an Alfisol in Central Ohio. Catena 194:104752

Meliani A, Bensoltane A (2016) Biofilm-mediated heavy metals bioremediation in PGPR pseudomonas. J Bioremed Biodegr 7:370. https://doi.org/10.4172/2155-6199.1000370

Mille A, Durand C (1869) Utilisation et épuration des eaux d'égout de Paris. Dunod, Paris

Mishra S, Srivastava S, Tripathi RD, Kumar R, Seth CS, Gupta DK (2006) Lead detoxification by coontail (*Ceratophyllum demersum* L) involves induction of phytochelatins and antioxidant system in response to its accumulation. Chemosphere 65:1027–1039

Mitchell RL (1978) Cobalt in soil and its uptake by plants. Paper presented at 9th Simposio Int. di Agrochimica, Punta Ala, Argentina, 2. p 521

Moonen A-C, Bàrberi P (2008) Functional biodiversity: an agroecosystem approach. Agric Ecosyst Environ 127:7–21

Morel JL (1997) Bioavailability of trace elements to terrestrial plants, Chapter 6. In: Tarradellas J, Bitton G, Rossel D (eds) Soil ecotoxicology. Lewis Publishers, CRC Press, Boca Raton, pp 141–176

Morgan JE, Morgan AJ (1988) Earthworms as biological monitors of cadmium, copper, lead and zinc in metalliferous soils. Environ Pollut 54:123–138

Morgan JE, Morgan AJ (1991) Differences in the accumulated metal concentrations in two epigeic earthworm species (*Lumbricus rubellus* and *Dendrodrilus rubidus*) living in contaminated soils. Bull Environ Contam Toxicol 47:296–301

Morgan JE, Morgan AJ (1992) Heavy metal concentrations in the tissues, ingesta and faeces of ecophysiologically different earthworm species. Soil Biol Biochem 24:1691–1697

Morgan JE, Morgan AJ (1999) The accumulation of metals (Cd, Cu, Pb, Zn and Ca) by two ecologically contrasting earthworm species (*Lumbricus rubellus* and *Aporrectodea caliginosa*): implications for ecotoxicological testing. Appl Soil Ecol 13:9–20

Morgan AJ, Morris B, James N, Morgan JE, Leyshon K (1986) Heavy metals in terrestrial macroinvertebrates: species differences within and between trophic levels. Chem Ecol 2:319–334

Morgan AJ, Morgan JE, Turner M, Winters C, Yarwood A (1993) Metal relationships of earthworms, ecotoxicology of metals. In: Dallinger R, Rainbow PS (eds) Invertebrates. SETAC, Lewis Publ, Boca Raton, pp 333–358

Morgan AJ, Stürzenbaum SR, Winters C, Kille P (1999) Cellular and molecular aspects of metal sequestration and toxicity in earthworms. Invert Reprod Dev 36:17–24

Morlot M (1996) Aspects analytiques du plomb dans l'environnement. Ed Lavoisier, Paris

Nahmani J, Hodson ME, Black S (2007) A review of studies performed to assess metal uptake by earthworms. Environ Pollut 145:402–424

Nan G, Guo L, Gao Y, Meng X, Zhang L, Song N, Yang G (2019) Speciation analysis and dynamic absorption characteristics of heavy metals and deleterious element during growing period of Chinese peony. Int J Phytoremediation 21:1407–1414. https://doi.org/10.1080/15226514.2019.1633261

National Report of Algeria (2011) 19ème session de la Commission du Développement Durable des Nations Unies (CDD-19) Mai 2011. 40 p

Nguyen C (2007) La libération de composés organiques par les racines (rhizodéposition): modélisation et impact sur la biodisponibilité des éléments minéraux pour les plantes, Thèse de doctorat, Nancy, pp 100–111

Nriagu JO (1978) In: Nriagu JO (ed) The biogeochemistry of lead in the environment. Elsevier Biomedical Press, Amsterdam

Nriagu JO, Pacyna JM (1988) Quantitative assessment of worldwide contamination of air, water and soils by trace metals. Nature 333(12):135–138

Ortiz-Ceballos A (2007) Mycorrhizal colonization and nitrogen uptake by maize: combined effect of tropical earthworms and velvetbean mulch. Biol Fertil Soils 44:181–186

Oste LA, Dolfing J, Ma WC, Lexmond TM (2001) The effect of beringite on Cd an Zn uptake by plants and earthworms: more than a liming effect? Environ Toxicol Chem 20:1339–1345

Päivöke AEA (2002) Soil lead alters phytase activity and mineral nutrient balance of Pisum sativum. Environ Exp Bot 48:61–73

Paoletti MG (1999) The role of earthworms for assessment of sustainability and as bioindicators. Agric Ecosyst Environ 74:137–155

Paoletti MG, Favretto MR, Stinner BR, Purrington FF, Bater JE (1991) Invertebrates as bioindicators of soil use. Agric Ecosyst Environ 34:341–362
Parkin TB, Berry EC (1999) Microbial nitrogen transformations in earthworm burrows. Soil Biol Biochem 31:1765–1771
Patra M, Bhowmick N, Bandopadhyay B, Sharma A (2004) Comparison of mercury, lead and arsenic with respect to genotoxic effects on plant systems and the development of genetic tolerance. Environ Exp Bot 52:199–223
Pederson JC, Hendrikson NB (1993) Effect of passage through the intestinal tract of detritivore earthworms (Lumbricus spp.) on the number of selected gram-negative and total bacteria. Biol Ferti soil 18:227–232
Peijnenburg W, Baerselman R, de Groot AC, Jager T, Posthuma L, Van Veen RPM (1999) Relating environmental availability to bioavailability: soil type- dependent metal accumulation in the oligochaete *Eisenia andrei*. Ecotoxicol Environ Saf 44:294–310
Pelosi C (2008) Modélisation de la dynamique d'une population de vers de terre Lumbricus terrestris au champ. Contribution à l'étude de l'impact des systèmes de culture sur les communautés lombriciennes. Earth Sciences. AgroParisTech, 2008. French. <NNT: 2008AGPT0057>. <tel-00336523>
Plan Bleu (2003) Centre d'activités régionales. Les Cahiers du Plan Bleu 2 - Sophia Antipolis, Les menaces sur les sols dans les pays méditerranéens. Plan Bleu, Marseille, 80 p
Puga Freitas R (2012) Effet du ver de terre Aporrectodea caliginosa sur la croissance des plantes, leur développement et leur résistance aux pathogènes. Université Paris-Est, Paris
Purohit J, Chattopadhyay A, Biswas MK, Singh NK (2018) Mycoremediation of agricultural soil: bioprospection for sustainable development. In: Mycoremediation and environmental sustainability. Springer, Cham, pp 91–120. https://doi.org/10.1007/978-3-319-77386-5_4
Qayyum S, Khan I, Meng K, Zhao Y, Peng C (2020) A review on remediation technologies for heavy metals contaminated soil. Cent. Asian J Environ Sci Technol Innov 1:21–29
Raj A, Jhariya MK, Yadav DK, Banerjee A, Meena RS (2019a) Agroforestry: a holistic approach for agricultural sustainability. In: Jhariya MK, Banerjee A, Meena RS, Yadav DK (eds) Sustainable agriculture, forest and environmental management. Springer, Singapore, pp 101–131. https://doi.org/10.1007/978-981-13-6830-1
Raj A, Jhariya MK, Banerjee A, Yadav DK, Meena RS (2019b) Soil for sustainable environment and ecosystems management. In: Jhariya MK, Banerjee A, Meena RS, Yadav DK (eds) Sustainable agriculture, forest and environmental management. Springer, Singapore, pp 189–221. https://doi.org/10.1007/978-981-13-6830-1
Raj A, Jhariya MK, Yadav DK, Banerjee A (2020) Climate change and agroforestry systems: adaptation and mitigation strategies. Apple Academic Press, Palm Bay, p 383. https://doi.org/10.1201/9780429286759
Ramade F (1993) Dictionnaire Encyclopédique de l'Ecologie et des Sciences de l'Environnement. Edition Edisciences International, Paris
Ramade F (2011) Introduction à l'écochimie: Les substances chimiques de l'écosphère à l'homme. Lavoisier, Paris, p 376
Raskin I, Ensley BD (2000) Phytoremediation of toxic metals; using plants to clean up the environment. Wiley, New York
Raskin I, Kumar NPBA, Dushenkov S, Salt DE (1994) Bioconcentration of heavy metal by plant. Curr Opin Biotechnol 5:285–290
Richards S, Hewson K, Moller H, Wharton D, Campbell H, Benge J, Manhire J (2007) Soil biota as indicators of soil quality in organic and integrated management kiwifruit orchards in New Zealand. Acta Hortic 753:627–632
Rodriguez H, Fraga R (1999) Phosphate solubilizing bacteria and their role in plant growth promotion. Biotechnol Adv 17:319–339
Romheld V (1986) pH-Veränderungen in der Rhizosphäre verschiedener Kulturpflanzenarten in Abhängigkeit vom Nährstoffangebot. Potash Rev 55:1–8

Rooney CP, McLaren RG, Cresswell RJ (1999) Distribution and phytoavailability of lead in a soil contaminated with lead shoot. Water Air Soil Pollut 116:535–548

Ruiz E, Rodrıguez L, Alonso-Azcarate J (2009) Effects of earthworms on metal uptake of heavy metals from polluted mine soils by different crop plants. Chemosphere 75(8):1035–1041

Ruiz E, Alonso-Azcarate J, Rodrıguez L (2011) *Lumbricus terrestris* L. activity increases the availability of metals and their accumulation in maize and barley. Environ Pollut 159:722–728

Saber MSM (2001) Clean biotechnology for sustainable farming. Eng Life Sci 1(6):217. https://doi.org/10.1002/1618-2863(200112)1:6<217::AID-ELSC217>3.0.CO;2-Y

Salmon S (2001) Earthworm excreta (mucus and urine) affect the distribution of springtails in forest soils. Biol Fertil Soils 34:304–310

Salt DE, Blaylock M, Kumar NPBA, Dushenkov S, Ensley BD, Chet I, Raskin I (1995) Phytoremediation: a novel strategy for the removal of toxic metals from the environment using plants. Biotechnology 5:285–290

Salt DE, Smith RD, Raskin I (1998) Phytoremediation. Plant Physiol Plant Mol Biol 49:643–668

Sanders JR (1983) The effect of pH on the total and free ionic concentrations of manganese, zinc and cobalt in soil solutions. J Soil Sci 34:315–323

Sanders JR, McGrath SP, Adams TM (1986) Zinc, copper and nickel concentrations in ryegrass grown on sewage sludge-contaminated soils of different pH. J Sci Food Agric 37:961–968

Santana NA, Ferreira PAA, Tarouco CP, Schardong IS, Antoniolli ZI, Nicoloso FT, Jacques RJS (2019) Earthworms and mycorrhization increase copper phytoextraction by Canavalia ensiformis in sandy soil. Ecotoxicol Environ Saf 182:109383. https://doi.org/10.1016/j.ecoenv.2019.109383

Scaps P, Grelle C, Descamps M (1997) Cadmium and lead accumulation in the earthworm *Eisenia fetida* (Savigny) and its impact on cholinesterase and metabolic pathway enzyme activity. Comp Biochem Physiol 116C:233–238

Scheu S (2003) Effects of earthworms on plant growth: patterns and perspectives. Pedobiologia 47:846–856

Seiler JR, Paganelli D (1987) Photosynthesis and growth response of red spruce and loblolly pine to soil-applied lead and simulated acid rain. For Sci 33(3):668–675

Seregin IV, Ivanov VB (2001) Physiological aspects of cadmium and lead toxic effects on higher plants. Russ J Plant Physiol 48(4):523–544

Shah K, Nongkynrih JM (2007) Metal hyperaccumulation and bioremediation. Biol Plant 51:618–634

Sims RW, Gerard BM (1985) Earthworm. Keys and notes for identification and study on the species. Brill and Backhuys, London

Singh OV, Labana S, Pandey G, Budhiraja R, Jain RK (2003) Phytoremediation: an overview of metallic ion decontamination from soil. Appl Microbiol Biotechnol 61:405–412

Singh S, Singh J, Vig AP (2016) Earthworm as ecological engineers to change the physico-chemical properties of soil: soil vs vermicast. Ecol Eng 90:1–5. https://doi.org/10.1016/j.ecoleng.2016.01.072

Singh SI, Singh S, Bhawana, Vig AP (2020) Earthworm-assisted bioremediation of agrochemicals. In: Agrochemicals detection, treatment and remediation. Butterworth-Heinemann, Oxford, pp 307–327. https://doi.org/10.1016/b978-0-08-103017-2.00013-1

Sizmur A, Hodson ME (2009) Do earthworms impact mobility and availability in soil? A review. Environ Pollut 159:1981–1989. https://doi.org/10.1016/j.envpol.2009.02.029

Smith SR (1996) Agricultural recycling of sewage sludge and the environment. CAB, International, Wallingford

Smith BA, Greenberg B, Stephenson GL (2012) Bioavailability of copper and zinc in mining soils. Arch Environ Contam Toxicol 62:1–12

Spurgeon DJ, Hopkin SP (1995) Extrapolation of the laboratory-based OECD earthworm toxicity test to metal-contaminated field sites. Ecotoxicology 4:190–205

Spurgeon DJ, Hopkin SP (1996) The effects of metal contamination on earthworm populations around a smelting works: quantifying species effects. Appl Soil Ecol 4:147–160

Spurgeon DJ, Lofts S, Hankard PK, Toal M, McLellan D, Fishwick S, Svendsen C (2006) Effect of pH on metal speciation and resulting metal uptake and toxicity for earthworms. Environ Toxicol Chem 25:788–796

Srivastava S (2015) Bioremediation technology: a greener and sustainable approach for restoration of environmental pollution. In: Applied environmental biotechnology: present scenario and future trends. Springer, New Delhi, pp 1–18. https://doi.org/10.1007/978-81-322-2123-4_1

Stephens PM, Davoren CW (1995) Effect of the lumbricid earthworm Aporrectodea trapezoides on wheat grain yield in the field, in the presence or absence of Rhizoctonia solani and Gaeumannomyces graminis var. tritici. Soil Biol Biochem 28:561–567

Stephens PM, Davoren CW (1997) Influence of the earthworms Aporrectodea trapezoides and A. rosea on the disease severity of Rhizoctonia solani on subterranean clover and ryegrass. Soil Biol Biochem 29:511–516

Stephens PM, Davoren CW, Doube BM, Ryder MH, Benger AM, Neate SM (1993) Reduced severity of rhizoctonia solani disease on wheat seedlings associated with the presence of the earthworm aporrectodea trapezoides (lumbricidae). Soil Biol Biochem 25:1477–1484

Stephens PM, Davoren CW, Doube BM, Ryder MH (1994) Ability of the lumbricid earthworms Aporrectodea rosea and Aporrectodea trapezoides to reduce the severity of take-all under greenhouse and field conditions. Soil Biol Biochem 26:1291–1297

Sterckeman T, Douay F, Proix N, Fourrier H (2000) Vertical distribution of Cd, Pb and Zn in soils near smelters in the north of France. Environ Pollut 107:377–389

Stork NE, Eggleton P (1992) Invertebrates as determinants and indicators of soil quality. Am J Altern Agric 7:38–47

Sun S, Sidhu V, Rong Y, Zheng Y (2018) Pesticide pollution in agricultural soils and sustainable remediation methods: a review. Curr Pollut Rep 4(3):240–250. https://doi.org/10.1007/s40726-018-0092-x

Talley J (2005) Introduction of recalcitrant compounds. In: Jaferey W, Talley L (eds) Bioremediation of recalcitrant compounds. CRC, Boca Raton, pp 1–9

Talley WF, Sleeper PM (2006) Roadblocks to the implementation of biotreatment strategies. Ann N Y Acad Sci 829:16–29

Tanner CC, Headley TR (2011) Components of floating emergent macrophyte treatment wetlands influencing removal of stormwater pollutants. Ecol Eng 37(3):474–486

Thornton I (1995) Metals in the global environment: facts and misconceptions. ICME, Ottawa

Tomlin AD (1992) Behaviour as a source of earthworm susceptibility to ecotoxicants. In: Greig-Smith PW et al (eds) Ecotoxicology of earthworms. Intercept, Hants, pp 116–128

Udovic M, Lestan D (2007) The effect of earthworms on the fractionation and bioavailability of heavy metals before and after soil remediation. Environ Pollut 148:663–668

Uzu G, Sobanska S, Aliouane Y, Pradere P, Dumat C (2009) Study of lead phytoavailability for atmospheric industrial micronic and sub-micronic particles in relation with lead speciation. Environ Pollut 157:1178–1185

Van Gestel CAM (1992) The influence of soil characteristics on the toxicity of chemicals for earthworms: a review. In: Greig-Smith PW, Becker H, Heimbach EPJ, F. (eds) Ecotoxicology of earthworms. Intercept Press, Hants, pp 44–54

Van Gestel CAM, Ma WC (1988) Toxicity and bioaccumulation of chlorophenols in earthworms, in relation to bioavailability in soil. Ecotoxicol Environ Saf 15:289–297

Van Hook RI (1974) Cadmium, lead and zinc distributions between earthworms and soils: potential for biological accumulation. Bull Environ Contam 12:509–512

Van Rhee JA (1977) Effects of soil pollution on earthworms. Pedobiologia 17:201–208

Van Vliet PCJ, Van der Zee SEATM, Ma WC (2005) Heavy metal concentrations in soil and earthworms in a floodplain grassland. Environ Pollut 138:505–516

Vidali M (2001) Bioremediation: an overview. Pure Appl Chem 73(7):1163–1172

Wang D, Li H, Wei Z, Wang X, Hu F (2006) Effect of earthworms on the phytoremediation of zinc-polluted soil by ryegrass and Indian mustard. Biol Fertil Soils 43:120–123

Wang D, Li H, Hu F, Wang X (2007) Role of earthworm-straw interactions on phytoremediation of Cu contaminated soil by ryegrass. Acta Ecol Sin 27(4):1292–1299

Wen B, Hu X, Liu Y, Wang W, Feng M, Shan X (2004) The role of earthworms (*Eisenia fetida*) in influencing bioavailability of heavy metals in soils. Biol Fertil Soils 40:181–187

Wen B, Liu Y, Hu X, Shan X (2006) Effect of earthworms (*Eisenia fetida*) on the fractionation and bioavailability of rare earth elements in nine Chinese soils. Chemosphere 63:1179–1186

White PJ, Crawford JW, Alvarez MCD, Moreno RG (2012) Soil management of sustainable agriculture. Appl Environ Soil Sci 2012:850739. https://doi.org/10.1155/2012/850739

White PJ, Crawford JW, Díaz Álvarez MC, García Moreno R (2014) Soil management for sustainable agriculture 2013. Appl Environ Soil Sci 2014:536825. https://doi.org/10.1155/2014/536825

Wolfarth F, Schrader S, Oldenburg E, Weinert J, Brunotte J (2011) Earthworms promote the reduction of Fusarium biomass and deoxynivalenol content in wheat straw under field conditions. Soil Biol Biochem 43:1858–1865

Wollny E (1890) Untersuchungen über die beeinflussung der Fruchtbarkeit der Ackerkrume durch die Tätigkeit der Regenwürmer. Forsh Agrik Physik 13:381–395

Wu F, Wan Judy Hon C, Wu S, Wong M (2012) Effects of earthworms and plant growth–promoting rhizobacteria (PGPR) on availability of nitrogen, phosphorus, and potassium in soil. J Plant Nutr Soil Sci 175:423–433

Wu Y, Chen C, Wang G, Xiong B, Zhou W, Xue F, Qi W, Qiu CS, Liu Z (2020) Mechanism underlying earthworm on the remediation of cadmium-contaminated soil. Sci Total Environ 728:138904. https://doi.org/10.1016/j.scitotenv.2020.138904

Xiao N-W, Song Y, Ge F, Liu X-H, Ou-Yang Z-Y (2006) Biomarkers responses of the earthworm *Eisenia fetida* to acetochlor exposure in OECD soil. Chemosphere 65:907–912

Yu XZ, Cheng JM (2003) Effect of earthworm on bioavailability of Cu and Cd in soils. Acta Ecol Sinica 23:922–928 (in Chinese)

Yu X, Cheng J, Wong M (2005) Earthworm–mycorrhiza interaction on Cd uptake and growth of ryegrass. Soil Biol Biochem 37:195–201

Zeng LS, Liao M, Chen CL, Huang CY (2006) Effects of lead contamination on soil microbial activity and rice physiological indices in soil-Pb-rice (*Oryza sativa* L.) system. Chemosphere 65:567–574

Zentner RP, Campbell CA, Biederebeck VO, Selles F, Lemke R, Jefferson PG, Gan Y (2004) Long-term assessment of management of an annual legume green manure crop for fallow replacement in the brown soil zone. Can J Plant Sci 83:475–482

Zerrouki D (2014) Phytoremédiation d'un sol agricole contaminé par les métaux lourds (Pb, Zn et Cu) en milieu routier à l'aide de l'orge *Hordeum vulgare.L* et de tournesol *Helianthus annus. L.* thèse de doctorat de l'université de sidi bel Abbes

Zorn MI, Van Gestel CAM, Eijsackers H (2005) The effect of two endogeic earthworm species on zinc distribution and availability in artificial soil columns. Soil Biol Biochem 37:917–925

Pollination and Ecological Intensification: A Way Towards Green Revolution 11

I. Merlin Kamala and I. Isaac Devanand

Abstract

Coping with the negative influence of indigenous agricultural technologies, sustainable intensification has emerged to replace the term 'green revolution' and is widely used in agricultural sector. Sustainable ecological intensification, minimizing the agricultural inputs by maximizing ecosystem services is desperately required to feed the increasingly demanding human population. Enhanced yields of food with increased nutrition from the same land surface by supporting biodiversity and ecosystem service refers to ecological intensification. Inventions in agricultural sector are heading towards green revolution, whereby agricultural production is drastically increased with minimal inputs to feed the ever-increasing world population. To attain sustainability it is essential to enhance the pollinator services. Pollination is needed for plants to reproduce and set seeds, which is traditionally aided by honey bees or other insects as pollinators for centuries. Pollinators are vital in determining the fertility of plant and are keystone process in any ecosystems. Pollinators are responsible for the food crop production meeting human diet, but their population is constantly deteriorating. They not only contribute to our food supply, but are the crucial fragment of biodiversity that all forms of life rest on. Technology has aided in agricultural crop production by inventing different techniques and equipment's to support food production. Traditional ecological structure is modified leading to several ill effects that drastically affect the pollination and bee population. Since bee pollination is an important aid for any crop production, the tragic decline in bee population has forced to artificially pollinate the crops which are labour intensive and

I. M. Kamala (✉)
Department of Agricultural Entomology, Tamil Nadu Agricultural University, Coimbatore, Tamil Nadu, India

I. I. Devanand
Department of Agricultural Extension, Annamalai University, Chidambaram, Tamil Nadu, India

M. K. Jhariya et al. (eds.), *Ecological Intensification of Natural Resources for Sustainable Agriculture*, https://doi.org/10.1007/978-981-33-4203-3_11

economically not feasible. There is a wide expansion in cultivation of insect-pollinated crops lately, which leads to huge increase in demand of 300% for pollination services. As per various scientific reports, the economics of pollination services is about more than 200 billion dollar or approximately 10% of world's food production. Pollinator-dependent crops have widened their range, increasing demand for pollination services up to 300% worldwide. The quality and quantity of crop production deter due to extinction of bees. Research should focus on encouraging bee keeping for sustained global growth of food production. Reducing the usage of chemical pesticides and leaning of plant-based botanicals for crop protection in an organic manner of crop production and planting more bright coloured flowers in the vicinity will add for bees to thrive.

Keywords

Ecological intensification · Ecosystem · Food · Honeybees · Pollination · Pollinators

Abbreviations

FAO	Food and Agriculture Organization
NAS	National Academy of Sciences
NRC	National Research Council
IPCC	Intergovernmental Panel on Climate Change
IPBES	Intergovernmental Science-Policy Platform on Biodiversity and Ecosystem Services

11.1 Introduction

Recently, for achievement of higher agricultural outcome with reduced negative influence on ecosystem, sustainable agricultural crop intensification has attracted worldwide attention. Globally, there is an increasing concern on enhancing the ecological services in agricultural crop intensification, commonly termed as ecological intensification (Xie et al. 2019). Ecological intensification aims to coordinate ecological services to sustain agricultural crop productivity to minimize the adverse effect on environment (Kremen et al. 2000; Meena et al. 2018).

Ecosystem amenities are the set of assorted ecological functions profiting human beings, by providing ecosystem conservation (Heal 2000; Raj et al. 2020; Banerjee et al. 2020). Ecosystem services are positive services that nature provide free of cost, which include eco-provisions like air, water and food, adaptable provisions like air, water and climate regulation, cultural provisions like recreation and tourism, supporting provisions like soil formation, soil decomposition, plant photosynthesis, plant respiration and plant pollination (Daily 1997). It is a nature-based alternative to high-input agriculture, which impacts severely on the harmony of the environment. Meeting the food demand to feed the ever-increasing global population by conventional crop production imposes severe damage to the environment and impacts on the biodiversity (Allen-Wardell et al. 1998; Jhariya et al. 2019a, b; Raj et al. 2018). Ecological intensification is based on the delivery of eco-provisions in conventional agriculture, with ecological replacement of artificial inputs, thereby reducing environmental costs without negatively influencing crop production. Sustainable intensification practices focus to maximize agricultural productivity and resilience using a variety of specific measures by conserving the natural resource base (Kassie et al. 2015; Meena and Lal 2018).

Over decades, research works on ecological intensification is well strengthened, with studies supporting and enhancing the ecosystem services for improved agricultural production and environmental safety. The benefits of ecologically intensifying agriculture are highlighted in increasing number of researches relying greatly of safeguarding biodiversity by ecological services (Benadi et al. 2013). Ecological intensification is thus embraced as an environmentally friendly means of food security.

A significant ecosystem service in agricultural crop production is pollination, which is ultimate for the fertilization of agricultural crops that flower, contributing to one-third of the food production (Meena et al. 2020a, b). The process of pollination is a vibrant ecological amenity contributed by a wide group of animals. Pollination by living organisms, *viz.*, animals and insects is a significant package profiting the environment (Deuri et al. 2018). The largest contributors of this essential ecosystem services are the insects including honeybees, butterflies, beetles, wasps, flies and moths (Klien et al. 2007). A global estimate portrays that around 88% of all flowering plants and around 33% of the vegetarian food supplies count on pollinators for their successful stand. Nearly three-quarters of all living flowering plants receive pollination services from animals. Bees are the chief group of pollinators of agricultural crops (Chittka and Thomson 2001). A diversified variety of insects and other animals and birds provide pollination service collectively contributing to human diet. Less tangible, but significant is the cultural significance of pollinators worldwide (Potts et al. 2016). Majority of world's plants are partially reliant on cross-pollination by biotic and abiotic components (Ollerton et al. 2011) for food production, many important medicines, foods, building materials and fibres derived from wild plants (IPBES 2016). Many living things on earth contribute to pollination, of which insects play a key role, which include honey bee, bumble bee, wasp, ant, flies, hoverflies, butterflies, moths, beetles, etc. Vertebrate pollinators include bats and birds, non-bat mammals such as monkey, lemur, etc., and some lizards pollinate certain plants. Human beings also aid in artificial pollination.

Food is essential for life on earth, and so are the pollinators; maintaining both are of prime concern in agriculture. This book chapter highlights the value of pollination services, reasons for their decline and ways to mitigate the decline to enhance pollinator services.

11.2 Pollination

The backbone of composite ecosystem is pollination (Heithaus 1974), which is the base for agricultural production (Bommarco et al. 2013). The transmission of pollens from the male part, anther, to the female part, stigma is referred to as pollination (Fig. 11.1). Fertilization follows pollination, the chief process in plants, sexual reproduction cannot not take place in other case (Muchhala et al. 2008).

Pollination is a process that leads to the formation of plants that grow into new plants. As pollination is decisive, both for reproduction and evolution of flowering plants this involves the pollen movement to stigma of a flower. Pollination is the base for formation of seeds and fruits, either by itself within a flower, or with the aid of water and wind and through the act of vectors (Losey and Vaughan 2006).

11.2.1 Importance of Pollination

Pollination enables fertilization and sexual reproduction, fruit and seed production with the aid of two factors, pollen and stigma. Reproduction is the life process which helps an organism to breed its own offspring, which involves a lot of events and pollination is one among them with respect to plants (Ingram et al. 1996).

Angiosperms have several different plant parts that are important in pollination. Flowers have male organs, stamens that produce a sticky powder called pollen and female the pistil. The top of the pistil, stigma, is sticky in nature (Allen-Wardell et al. 1998). The base of the pistil produces seeds in the ovule. The pollen of a flower must

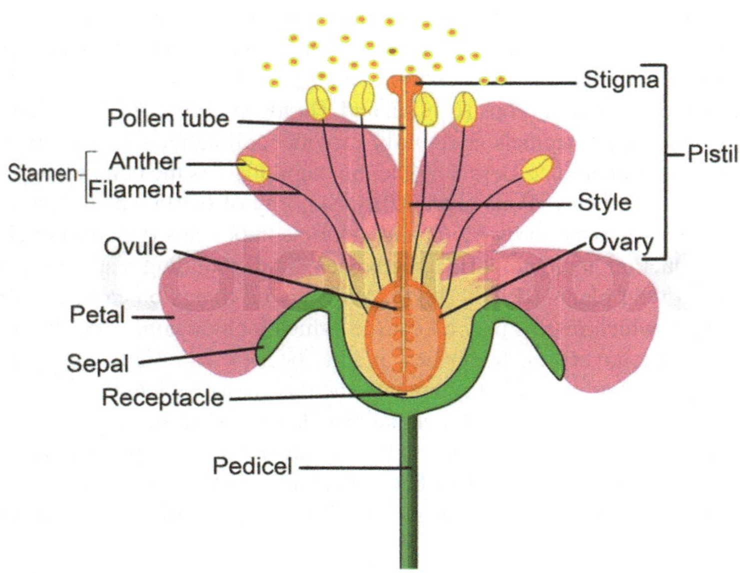

Fig. 11.1 Process of pollination

be moved from a stamen to the stigma to get pollinated. The pollination happening between pollen and stigma of same flowers within a plant is called self-pollination and that between different flowers within a plant is called cross-pollination. Cross-pollination causes efficient fertilization and produces stronger plant (Kevan and Randolf 2012).

11.2.2 Types of Pollination

The sexual reproduction of plants involves the pollen transfer from a flower to stigma of other plants through pollination. Pollination is of two types, *viz.*, abiotic and biotic pollination (Fig. 11.2). Abiotic pollination happens without the action of living organisms, *viz.*, wind and water. Biotic pollination involves pollen transmission with the aid of living organisms, a common form of pollination (Armbruster 1993). Around 90% of estimated pollination of all flowering plants is through living organisms (Abramovitz 1998). Pollination is achieved manually in some exceptional cases, by hands, known as hand pollination or mechanical pollination.

Pollination can be of two types: self-pollination and cross-pollination (Fig. 11.3).

Self-Pollination: Pollen transmission from the stigma of identical or diverse flower within the similar plant species is termed as self-pollination, which is commonly observed in dioecious plants, which has both the sexual parts on the same flower (Moldenke 1976; Barrett 2010). Self-pollinating plants have less dependence on the external factors for pollination. The anthers and stigma will be of similar lengths to facilitate the pollen transmission in self-pollinating flowers. Self-pollination occurs by two means:

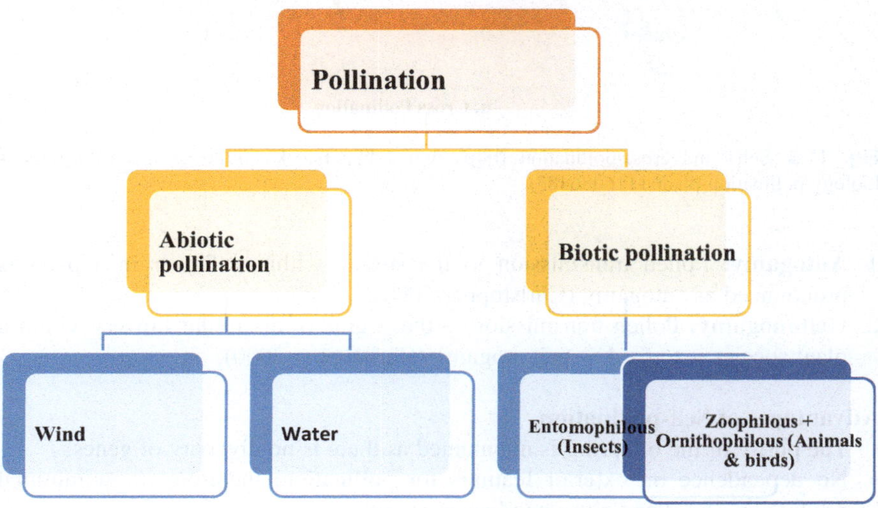

Fig. 11.2 Types of pollination

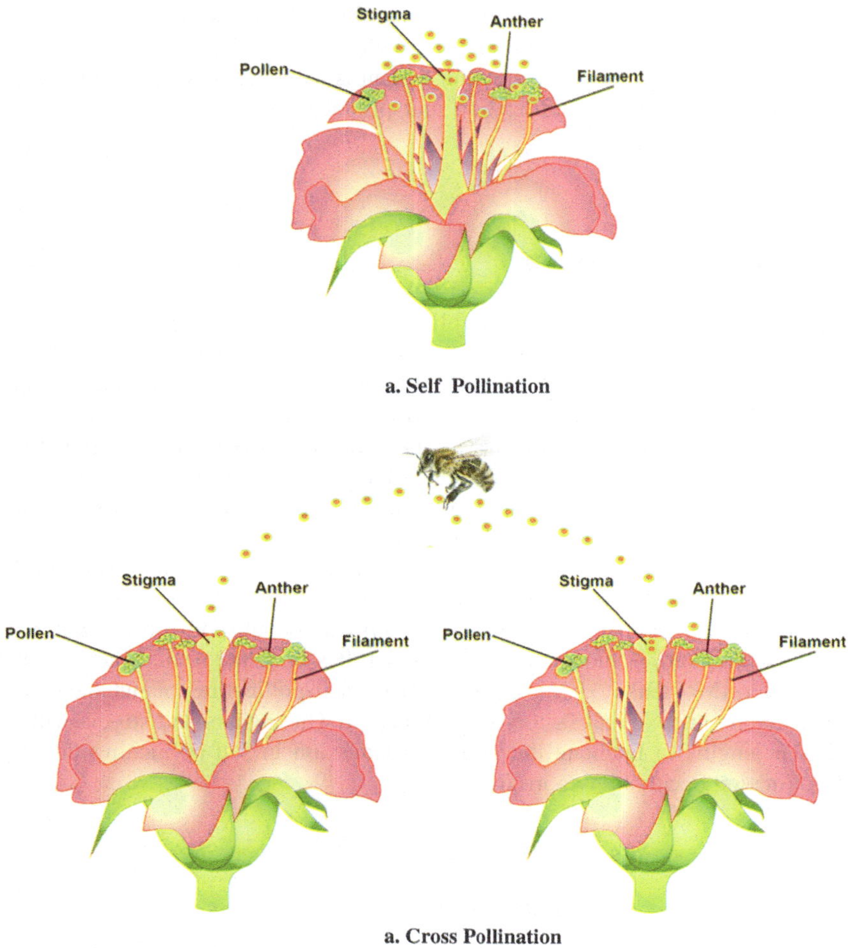

Fig. 11.3 Self- and cross-pollination (https://www.shutterstock.com/image-vector/illustration-biology-pollination-plant-1116086483)

1. **Autogamy:** Pollen transmission to the stigma within a flower in a plant is pronounced as autogamy (Christopher 2000).
2. **Geitonogamy:** Pollen transmission to the stigma of dissimilar flowers within a plant species is termed as geitonogamy (Christopher 2000).

Advantages of Self-pollination

- The purity of the offspring is maintained as there is no diversity of genes.
- No dependence on exterior features for pollination; therefore, even minimal production of pollens gives good success rate.
- It ensures that recessive characters are eliminated (Shore and Spencer 2011).

Disadvantages of Self-pollination

- No new features are introduced into the lineage as there is no mixing up of genes.
- It reduces the vigour and vitality of the race as no novel character is introduced.
- The immunity to diseases will be reduced as no new characters are introduced in the resultant offsprings' (Skogsmyr and Lankinen 2002).

Cross-Pollination: Pollen transmission from anther of a flower to stigma of another flower is termed as cross-pollination, where both the flowers involved differ genetically. Plants depend on an external factor for pollen transmission which includes animals, insects, birds, bats, marsupials, snails, water, wind, etc. Cross-pollination can be of different types based on the agent of pollination (Free 1993).

Hydrophilous Flowers: Pollination happens by means of water in hydrophilous flowers. These flowers are usually tiny and inobtrusive to other agents. They lack scent or bright featured petals and the pollen has adaptations to float in water. Water pollinated plants have tiny, inobtrusive male flowers that produce huge quantity of pollens that drift in water where they are stuck by the bigger fluffy stigmas of female flowers (Du and Wang 2014).

Anemophilous flowers: Pollination happens with the aid of wind in anemophilous flowers. These flowers are small and inconspicuous, have dull or no petals and very light to be easily carried by the wind. The stamens produce enormous quantity of pollen, when exposed to the wind the pollen is easily dispersed, while the female flowers are feathery and have large surface areas for being pollinated. The pollen grains of these flowers are very light in weight, non-sticky and occasionally winged (Du and Wang 2014).

Zoophilous flowers: Pollination occurs by means of pollinating agents like bats, birds, animals and human beings in zoophilous flowers. The zoophilous flowers have pollen designed to stick on animal body surface, to facilitate easy transportation from flower to flower (Nicole 2015).

Entomophilous flowers: Pollination is mediated by insects like honey bees, beetles, butterflies and moths in entomophilic flowers. Attractive flowers with bright coloured petals with luring fragrance attract the insect visitors to them. Insects have to perch on these flowers through broad stigmas or anthers. The nectar from the flower tempts bees, butterflies and other insects, which eventually help the flower to pollinate. Those flowers bear spiny pollens or pollens with extensions that help to stick on to insect body (Nicole 2015) (Fig. 11.4).

Ornithophilous flowers: Pollination happens with the aid of birds in ornithophilous flowers. Bird pollination is mediated very rarely. The birds commonly employed in pollination are humming birds, spider hunters, sunbirds, honeycreepers and honeyeaters (Nicole 2015).

Merits of cross-pollination: Cross-pollination is highly beneficial in offspring production of plants as they introduce new genetic factor in their heredity. It offers resistance to diseases and changes in the environment. The seeds produced are good in vigour and vitality. The presence of any recessive character in the lineage will be removed as a result of genetic recombination. It is the only option for reproduction in unisexual plants (Vaudo et al. 2015).

Fig. 11.4 Entomophilous pollinators

Demerits of cross-pollination: There is a heavy wastage of pollen grains for fertilization to be assured. There will be greater chances for potential characters to be eliminated and unnecessary characters to get in due to recombination of the genes (Vaudo et al. 2015).

11.2.3 Role of Pollinators

Pollination is the key process where fertilized plants are able to produce offsprings. An effectively fertilized flower will produce seeds, ensuring production of new generation of plants. Plants employ several tactics to ensure pollen is carried from flower to flower (Will 1983). A wide array of plants relies on routes for pollination such as water, wind and other living organisms. Some plants trust on wind and water, while most flowering plants fertilize through animal pollination. Around 75% fruit and vegetables in daily human diet require animal pollination (William and Anderson 1974). Pollinators powerfully impact environmental associations and stability, ecological maintenance, heritable dissimilarity among plants, floral diversity, specialization and evolution. Research estimates 200,000–400,000 diverse animal species aid in pollination, from birds to bats, marsupials to monarch (Naug 2009). Insects are chief animal pollinators, the most important are bees. Insect's role in reproduction of plants was initially found by Joseph Kolreuter at the University of Karlsruhe, Germany in the eighteenth century (NRC 2007). Charles Darwin widely attempted this initiation and usage of managed pollinators to improve fruit and seed set, which flourished gradually since then.

Plants and pollinators are reciprocally benefitted in pollination process. Pollination aids plants in seed set that is obligatory for plant reproduction, while pollinators obtain nectar and pollen rewards from the visiting flowers. Sugary nectar gives carbohydrates, while pollen offers other nutrients like proteins, vitamins, minerals, fats and other micronutrients to the visitors. Somewhere between 75 and 95% of the plants that flower on earth need help through pollinators. Plant visitors provide their amenities to more than 180,000 diverse species of plants and around 1200 food crops signifying that one out of every third mouth of food consumed by humans comes from pollinators (Buchmann and Nabhan 1996).

Honey bees are the first organism recognized among pollinators. However, many other organisms including insects, *viz.*, solitary bees, butterflies, flies, beetles, some birds and bats also contribute to plant pollination. Globally, more than 300,000 species of plants that bloom require living organism for pollination. The tremendous floral blends in every ecosystem care the corresponding diversity of pollinators by offering them nectar and pollen, a massive stream of them is insects. An estimation reports, around 1000 vertebrates and at least 16,000 different species of bees globally to serve pollination services (Rogers and Charlotte 2012).

11.2.4 Types of Pollinators

Honey bees are widely known pollinators of flowering plants, but a vast array of other organisms equally contributes to pollinate plants across the globe (Wilcock and Neiland 2002). The list of crops pollinated by diverse pollinators are listed in Table 11.1.

The Bees

Honey bees (*Apis cerana indica*): Honey bees serve as the most momentous insect visitor, with several specialist adaptations for lifting pollen with the aid of plumose hairs on their bodies (Roubik 1995). They work tirelessly pollinating a variety of crops. Honey bees are being important for crop production as they are able to be domesticated and managed to crops in need of pollination. Probably the most well-known pollinator, honey bees are credited with pollinating much of the human diet (Vasanthakumar et al. 2018). Berries, pears, apples, citrus, melons, peas and beans are just a few of the foods that would not exist without honeybees. In fact, about one-third of the food consumed by human beings is pollinated by honey bees (Hung et al. 2018).

Western honey bee (*Apis mellifera*): *Apis mellifera* is another efficient pollinator in agricultural systems worldwide boosting seed and fruit production (Moritz et al. 2005; Schmidt and Johnson 1984). The western honey bee provides valuable pollination services for a wide variety of food crops (Calderone 2012), and positions as the recurrent species of insect visitor (Garibaldi et al. 2019).

Bumble bee, *Bombus impatiens*: Bumblebees are efficient pollinators, increasing crop production (Libbrecht and Keller 2015). Many agricultural and horticultural crops are dependent on bumblebees for natural pollination. Crops like peppers, blueberries, tomatoes, strawberries, cane berries, cucumbers, melons and squash are benefitted by bumble bees through pollination (FAO 2008).

Stingless bee: Stingless bees (Apidae: Meliponini), tiny honey bees are common flower visitors and effective pollinators of several plants. Studies state that they are regular visitors of flowers of around 90 flowering plants and effective pollinators of

Table 11.1 List of diverse plants pollinated by diverse pollinators

S. No.	Type of pollination		Plants pollinated
1	Abiotic pollination	Wind	Rice, corn, rye, barley, oats, pines, spruces, firs, cattails
2		Water	Pond weed, eel-grass, Hydrilla, Coontail, *Vallisneria*
	Biotic pollination	Honey bee, *Apis cerana indica*	Apple, mango, kiwi, pear, plum, peach, guava, pomegranate, rambutan, nectarine, apricot, avacado, passion fruit, custard apple, cherry, lychee, strawberry, raspberry, starfruit, grapes, persimmon, cashew, durian, limes, alfalfa, okra, eggplant, onion, allspice, lima bean, kidney bean, green bean, coffee, walnut, cotton, cucumber, beet, mustard seed, rape seed, broccoli, cauliflower, cabbage, Brussel sprouts, turnip, Congo beans, sword beans, chilli peppers, papaya, safflower, sesame, clover, tamarind, cocoa, vanilla, tomato
3		*Apis mellifera*	Onion, strawberry, cashew, beet, strawberry, celery, papaya, safflower, watermelon
4		Bumble bee	Kiwi, strawberry, flax, lupine, cherry, plum, rose, pear, eggplant, clover, blue berry
5		Stingless bee	Eucalyptus, tamarind, citrus, fennel, coriander, eggplant, fennel, cherry, avocado, cardamom, rubber, sweetpepper, castor, pigeon pea, strawberry, peach, plum, guava, roseapple, sunflower, litchi, rambutan, breadfruit, jackfruit
6		Wasp	Orchid, fig
7		Butterfly	Alyssum, Aster, Beebalm, Butterfly bush, Calendula, Cosmos, Daylily, Fennel, Globe thistle, Goldenrod, Hollyhock, Lavender, Marigold, Oregano, Verbena, Zinnia
8		Moth	Morning glory, tobacco, Yucca and Gardenia
9		Beetle	Magnolias, pond lilies, goldenrods, Spiraea
10		Flies	Fennel, coriander, caraway, kitchen onion, parsley, carrots
11		Hoverfly	*Epipactis veratrifolia*, oilseed rape, *Brassica napus*
12		Mosquito	*Platanthera obtusata*
13		Thrips	*Echium plantagineum, Arctostaphylos pungens*
14		Ant	*Microtis parviflora, Euphorbia seguieriana, Alyssum purpureum, Frankenia thymifolia*
15		Bat	Guava, banana, cocoa, mango, fig, dates, cashew, peach, evening primerose, goldenrod, tobacco, Yucca, honeysuckle, marigold
16		Bird	*Erythrina, Bignonia, Strelitzia, Tecoma*
17		Lizard	*Metrosideros excelsa*, cactus
18		Squirrel	*Mucuna macrocarpa*

Source: Roubik (2018)

around 9 plant species (Wille et al. 1983). They are generalist visitors of flowers for nectar than pollen. Stingless bees are adjustable to any ecosystem, quick learner to navigate the floral localities (Heard 1999). Stingless bees prefer small to medium sized flowers, thick inflorescences, flowers with corolla tubes shorter and wider for bees to enter. They also favour trees with white, yellow flowers and mild shaded flowers (Cortopassi-Laurino et al. 1991).

Solitary bees: Solitary bees are non-social bees not living in colonies, including the carpenter bee, leaf-cutter bee and the orchard mason bee. They do not make honey, but visit flowers for nectar and serve as great pollinators (Wood et al. 2017). This group, as the name implies, lives alone, scavenging flowers for pollen and nectar and in the process pollinating several flowers of crops. Solitary bees are wild pollinators, pollinating a wide variety of wild plants (Everaars 2012).

Butterflies, Moths and Skippers: Butterflies, moths and skippers belong to order Lepidoptera, adding visual plea with diverse scientific issues (Subba and Meera 1984). They use their long proboscis to suck nectar from flowers, in the process, the pollen grains adhere to butterfly body parts such as proboscis, head, thorax, legs and wings, transmitted to stigma as they move to another flower, thereby pollinating them (Dotterl et al. 2006). However, reliable cases of butterflies pollinating plants are rare, except *Caesalpinia pulcherrima*, butterflies are primary pollinators and the pollen is carried primarily on their wings (Cruden and Wand Hermann-Parker 1979).

Pollinating beetles and bugs: Beetles were marked to be the earliest pollinators to evolve ages ago along with flowering plants. They are primeval pollinators and do not possess specific adaptations for carrying pollen (Wang et al. 2013). Their pollinating efficiency is more pronounced in flowers having sticky pollens which can easily cling to their hard exoskeletons. Plants with large flowers which emit decay like bad odours attract them, which eventually will be pollinated. Around 400,000 species of beetles, accounting for 25% of all known insects are beetles (Grimaldi and Engel 2005). Beetles encounter pollination of plants in tropics and arid desert regions of the world (Krell 2006).

Thrips: Thrips (Thysanoptera) are tiny primitive insects and known for their existence as pollinator of tiny flowers (Reitz 2009). Thrips are tiny insects, feeding on plant saps, tissues and pollen. They are recorded as oldest pollinators on the planet, identified by their tiny fossils. There are an estimated 6000 species of thrips (Peñalver 2012).

Hoverflies and other flies: Hoverflies act as pollinators of many flowering plants all around the world. Syrphid flies are visitors to a wide array of agricultural crops for pollen, nectar and honey dew (Rotheray and Gilbert 2011) visited by hoverflies are usually open and cup-shaped (Branquart and Hemptinne 2000; Gilbert 1975, 1981). Generally, hoverflies begin life as aquatic or semi-aquatic larva in stagnant water sources in cavities of decaying trees (Chapelin-Viscardi et al. 2015). The Belted Hoverfly, *Volucella zonaria* is a significant pollinator (Wratten et al. 1995; Hogg et al. 2011).

Wasps: Many wasps nectar on plants and thereby support in pollination (Jander and Herre 2010). Some plants are pollinated solely by wasps such as the Fig trees of the tropics, which are dependent on tiny fig wasps for pollination (Kislev 2006).

Mosquitoes: Mosquitoes was first determined as pollinators in the early 1970s. The mosquito, *Aedes communis* was an important pollinator of blunt-leaf orchid, *Platanthera obtusata*. In fact, many mosquitoes around the world pollinate small flowers that live in wetter environments (Rader 2015). When mosquitoes are not feeding blood, they feed on plant nectars to fuel their flight. *Aedes communis* feeds nectar from the floral spur of the *Platanthera obtusata*, where its eye naturally comes into contact with the pollinium a cluster of pollen (Smith and Snow 1976). As the mosquito moves on to another flower to feed on nectar, it touches the stigma of that flower, and the flower is pollinated (Fang 2010).

Ant: Workers of many ant species visit plants that flower to collect nectar and aid in pollination. Ants are generally reported to be nectar thieves as they are tiny in size, and are able to exploit sugary nectar without touching the pollens or stigma. Their smooth cuticle prevents pollen to adhere to the body (Barth 1991). Some other ant species are large and hairy and aid in pollination. Ant pollination of flowers is rarely reported (Proctor et al. 1996), 30 reports of insect pollination include the role of ants in pollination (Peakall et al. 1991). Discretely marked ants of *Tapinoma erraticum* and *Formica emarginatus* visit nectaries of *Euphorbia seguieriana*, and thereby aid in pollination is reported (Wolf and Wehner 2000).

Mammals: Mammalian pollinators are more abundant in tropics, where they aid to pollinate large trees (Johnson et al. 2001). Mammals such as marsupials, rodents and rats, sugar gliders, squirrels and lemurs, which do not have flying mechanism, have been identified to visit flowers and aid in pollination (Carthew and Goldingay 1997).

Bats: Bats are also important pollinators with some special strengths. Fruit bats are significant among mammalian pollinators as they are able to move to large distances and visit a greater number of flowers as they fly across the forest ecosystems (Burger 2005). A rare long-snouted bat (*Platalina genovensium*) present in Peru uses its elongated nose and tongue to feed on pollen and nectar inside deep or narrow flowers. The long-snouted bat has potential to hold large amounts of pollen on its unique nose and distribute it over greater distances. The long-snouted bat helps in pollination and distribution of seeds of native Peruvian plant, *Weberbauerocereus weberbaueri*.

Reptiles: Many reptiles help pollinate flowers in diverse ecosystems. In desert ecosystem, some lizards visit cacti flowers to drink the nectar, to satisfy thirst due to lack of water, and, in the process, collect pollen and transport from plant to plant ensuring its pollination (Jones and Stanley 2001). The endangered *Roussea* flower is pollinated solely by reptile species, the endemic blue-tailed day-gecko, *Phelsuma cepediana* in tropical forests of Mauritius (Johnson et al. 2001).

Birds: Birds are very important pollinators of wildflowers globally, which are brightly coloured with funnel shaped corolla. Humming birds, honey eaters and sunbirds are key pollinator birds, pollinating on deep-throated flowers. Humming birds contribute in pollinating a greater number of plants which they feed upon whilst hovering (McDade and Kinsman 1980). Honeycreepers and honeyeaters are important bird pollinators in Hawaii and Australia. In addition, brush-tongued parrots and sunbirds also contribute to pollination (Wester and Johnson 2017). Ornithophilous plants generally produce glowingly coloured flowers, but often lack odour, since birds are underprivileged to sense smell. The flowers are often long and tubular, with lots of nectar, and are stout enough for the birds to perch on and feed nectar (Bertin 1982).

Obligate Mutualism: A wide array of plants is visited by diversity of pollinators, but some plants depend on sole single species to visit to pollinate them. The Yucca plant (*Yucca elata*), hinge totally on a solitary species, *i.e.* Yucca moth (*Tegeticula maculata*) to pollinate. These cases of certain flowers depending on specific pollinators are called obligate mutualisms (Gallen 1999).

11.2.5 Value of Pollination Services

Pollination by living organism is an imperative ecosystem service; around 35% of worldwide crop-based food production profit from them (Ghazoul 2005). Major array of pollinators are wild, including approximately 20,000 species of bees, flies, butterflies, moths, wasps, beetles, thrips, birds, bats and other vertebrates. Many pollinators are managed; they include a few species of bees including the eastern or Indian honey bee (*Apis cerana indica*), the western honey bee (*Apis mellifera*), few bumble bees, some stingless bees and a few solitary bees (Gibbs et al. 2015). Honeybee farming provides an imperative cause of income for rural farming community across the globe (Painkra et al. 2016). *Apis mellifera* is the most extensively managed pollinator in the world, with an appraised 1.6 million tonnes of honey being produced from 81 million hives annually around the world (Klein et al. 2007).

Estimation of the cost of pollination services in agriculture or to envisage their role in food security is not an easy task (Melathopoulos et al. 2015); *Apis mellifera* as a single bee species is directly or indirectly accountable for production of over 90 crops in USA, collectively contributing to one-third of the American diet (Goulson et al. 2015). Honey bees are recognized as a premier in managed pollination throughout the world due to certain unique biological attributes: large perennial colonies with thousands of workers effectively foraging nectar and pollen, monocultures for honey bees to utilize a single plant species, communication behaviour to navigate particular locations easily, increased possibility of cross-fertilization and a preference for nesting in wooden substitutes for easy transportation (Ghazoul 2005). The managed pollination service provided by honey bees across the agricultural spectrum has been valued in to be more than $15 billion in US (NAS 2007). The value of pollination varies both quantitatively and qualitatively by crop; honey bees are responsible for pollinating 100% of almonds, more than 90% of commercial vegetation production of cucumber, carrot, broccoli, onion, pumpkin, etc. (Morse and Calderone 2000; Dams 1978). Honey bees indirectly profit the cattle industry by contributing the bulk of pollination services in alfalfa and clover production, the main sources for cattle fodder, seed production of soybean, olives and grapes, used in cattle feed (Soares-Filho et al. 2014). Pollinators are multi-beneficiary to human beings, beyond food contribution, assisting directly to medicinal drugs, fibres (cotton, jute and sunhemp), biofuels (Canola, *Jatropha* and palm oil) and in constructions (timber wood from agroforestry).

11.3 Pollinators and Pollination

11.3.1 Pollination Syndrome of Flowers

Pollination syndromes are phenotypic qualities and their blends theorized to adapt flowers for pollination by specific living organisms. They are hypothesized to increase the degree of attention of specified pollinators. They were initially characterized in the 1870s and refined in the twentieth century (Banskota et al. 2001). Pollinators are attracted to specific features of a flower, which is called pollindrome syndrome. The set of flower features like flower colour, shape, size, type of nectar and fragrance that fascinate pollinators is called the flower's pollination syndrome (Steinhauer and Rennich 2014). The shape of many flowers and scent from flowers have been adapted to match the shape and size of the pollinator's body part used to get nectar and their attraction to a particular flower. Conventionally floral evolution and multiplicity were constructed on the outlook of focused eco-communications between flowers and their specific visitors (Fenster et al. 2004). Pollination syndromes state the co-evolution of plants and pollinators in nature. Pollinators select for specific flowers based on their pollination syndrome and both the flower and pollinator adapt to each other's changes over time (Murren 2012). In pollination syndromes, plants and pollinators deed mutually to get benefitted. Pollinators exploit the association for their fertilization. They feed on the flowers for carbohydrates from nectar and protein and vitamins from pollen grains as their rewards. The plant assures its pollination and reproduction by providing nectar and the pollen (Anderson and Johnson 2009).

While there are a range of features that make up a flower's pollination syndrome, the colour preferences of pollinators with their preferred flowers take major lead (Caruso et al. 2003). Different plants produce flowers of different colours, which reflect light of different wave lengths (Gomez et al. 2003). Living organisms can distinguish colour of specific wavelength; therefore, different kinds of animals will be attracted to flowers of different colours (Briscoe and Chittka 2004). All organisms have different visible colour spectrums. Therefore, the colour of a flower fits the visible spectrum of their specific pollinator. Bees are likely to visit yellow, violet or blue flowers (Rausher 2008). Flies prefer brown or purple flowers, while butterflies desire pink flowers and moths on white flowers. Birds are attracted to flowers that are red (Whittall and Hodges 2007). Flower fragrance triggers a key role, where the pollinators are betrayed. Flower odours attract specific pollinators. Flies likes the smell of decaying food and some plants smell like rotting meat to attract flies (Hans and Thomas 2004).

11.3.2 Perception of Flowers by Pollinators

The remarkable variations of floral traits as colour, fragrance and structure are established as signals for dynamic mutualism between plants and visitors, especially insects (Galen and Kevan 1980). Yet, the intricacy of both the individual flowers and their ecological networks has made it tough to designate floral qualities and perceive associations between flowers and the visitors that visit them frequently (Klarhe et al. 2011). Pollinators use floral characters such as colour, scent, shape and size to navigate their food sources and to distinguish their specific plant species between diverse flowers (Chittka and Raine 2006; Knudsen and Tollsten 1993; Hoballah et al. 1992). Even though ultimate flower choice of pollinator's undoubtedly relies on a blend of stimuli, several researches suggest that few visitors count strongly on brightness to decide on their forage (Dafni et al. 1990; Heiling et al. 2003). A combination of floral structure shape and size, floral rewards pollen and nectar, colour and scent of the flower, colour perception by good vision and nutrient preference by pollinators will result in a specific pollination syndrome (Table 11.2) (Dötterl et al. 2014).

Colour

Visual information is a requisite for pollinators to trace, navigate and intermingle with flowers. Diurnal pollinators have sophisticated vision of colour that covers a diverse range of spectrum than human vision (Fenster et al. 2004). The diversified pollinators use different visual models for different pollen and nectar source (Troje 1993). The best studied species on vision of pollinators are the western honey bee, *Apis mellifera* and the bumblebees *Bombus terrestris*, which stake an analogous eye vision (Briscoe and Chittka 2004). The visual of spectrum of the honey bees, serving as a model, series between 300 and 700 nm with the photoreceptors peaking in blue, green and UV region of the spectrum corresponding to 344, 438 and 560 nm, respectively (Rausher 2008). The spectrum of the pollinators studied is trichromatic system, but certain pollinators share dichromatic system including beetles and flies and tetrachromatic system including butterflies (Briscoe and Chittka 2004). Colour preferences foreseeing pollination syndromes do not match for diverse pollinator groups (Lunau and Maier 1995), which in most cases are exhibited at a genus or species level rather than the group (Raine et al. 2006). Some research suggested deviations in brightness of flowers are not linked with dissimilar pollinator groupings (Cooley et al. 2008).

Scent

Floral scent is predicted to have a major role in pollinator–plant natural interaction. Flower fragrance is highly variable and plays a crucial role in pollinator's selection of flowers (Miyake and Yafuso 2003; Salzmann et al. 2007). Plants publicize their floral rewards by visual syndromes, *viz.*, shape, size and colour, and olfactory syndromes, *viz.*, scent-signals; nevertheless, fragrance as lure for pollinators is seen in some pollination systems (Dotterl et al. 2006). Generally, the olfactory stimulation of flower is thought to have specific supplement than visual facets

Table 11.2 Flower perception by pollinators

Insects	Colour	Nectar source	Odour	Nectar	Pollen	Flower shape
Honey bees	Bright white, yellow, blue or UV	Present	Fresh, mild, pleasant	Usually present	Limited; often sticky and scented	Shallow; have landing platform; tubular
Butterflies	Bright, including red and purple	Present	Faint but fresh	Ample; deeply hidden	Limited	Narrow tube with spur; wide landing pad
Moth	Pale and dull red, purple, pink or white	Absent	Strong sweet; emitted at night	Ample; deeply hidden	Limited	Regular; tubular without a lip
Flies	Pale and dull to dark brown or purple; flecked with translucent patches	Absent	Putrid	Usually absent	Modest in amount	Shallow; funnel like or complex and trap-like
Beetles	Dull white or green	Absent	None to strongly fruity or fetid	Sometimes present; not hidden	Ample	Large bowl-like, Magnolia
Bat	Dull white, green or purple	Absent	Strong musty; emitted at night	Abundant; somewhat hidden	Ample	Regular; bowl shaped — closed during day
Birds	Scarlet, orange, red or white	Absent	None	Ample; deeply hidden	Modest	Large funnel like; cups, strong perch support

Source: Krishna and Keasar (2018)

(Dobson 1994). Attraction of specific pollinators through fragrance depends on the concentration, configuration and period of secretion of odour. Statistical observation of pollinator fauna with scent features advises that both diurnal (Pombal and Morellato 2000) and nocturnal floral guests have diverse preferences for definite odours (Raguso and Willis 2005; Riffell et al. 2008), specially for aromatic components (Dobson 2006). Similarly, studies on behavioural choice assays using olfactometer display that bees, butterflies and moths were able to categorize both qualitative (Cunningham et al. 2004, 2006) and quantitative features of floral scent composition (Andersson 2003; Andersson and Dobson 2003; Wright et al. 2005).

Floral visitors desire to visit scented flowers over odourless ones (Knudsen et al. 1999; Kunze and Gumbert 2001; Ashman et al. 2005). The presence of volatile organic compounds (VOC's) in floral scent is the source of attraction of fly pollinators in *Ceropegia dolichophylla*. The flowers of the plant is found to emit volatile compounds, *viz.*, tridecane, pentadecene, and pentadecadiene components, which are the source of attraction of fly pollinator (Ollerton et al. 2011).

Structure of Flower

The diverse forms and structure of flowers have been recognized to be a mode of pollinator attraction and is configured into the perception of floral syndromes (Faegri and van der Pijl 1979; Proctor et al. 1996) where, flower presentation (Weberling 1989) and petal structure and texture (Kevan and Lane 1985) play a key role.

The flower and inflorescence size influence the rates of visitation by pollinators (Bell 1985), with great variation (Dafni 1990). Flower visitors have specific favouritism for floral patterns and structure (Lehrer and Bischof 1995). Beetles choose deeply grooved flowers, while honeybees desire flowers with broken outlines (Dafni and Kevan 1996; Herrera 1993). Honeybees favour larger flowers with more nectary source over smaller ones. Bees recognize several assorted flower cues, *viz.*, scent, colour, shape, size or symmetry for visiting decisions (Giurfa et al. 1996; Menzel and Shinida 1993). Pollinators displayed preferences for flower shapes, symmetry, sizes on different spatial frequencies, lowering for viewing images from a long distance and higher for viewing at close range, representing the resolution of bee vision (Vorobyev et al. 1997).

11.3.3 Mechanism of Pollination by Honey Bees

Majority of flowering plants and honey bee foragers have developed a complex mutualism for years. Research data depicts that over 80% of flowering plants are entomophilous, relying on insects for pollination among which bees play a major role (Tsukada and Tsukada 1986). The efficiency of honeybees in pollination is due to their mass foraging, their individual physique and their behaviour of foraging on a single plant species at a time. The bees look for their food source in flowers, nectar or pollen. Nectar is a sugary liquid produced by plants to lure visitors and pollen to ensure the next generation of plants (Wilson and Price 1983).

Bee pollinated flowers are designed in a way that visiting bee is supposed to brush against the flower's anthers bearing pollen or bear some special mechanism to release the anthers to cover the bee with pollen. Honey bees are gifted with hairy structures on legs effective enough to hold pollen while visiting a flower and while moving to the next flower, bees arrange them in the pollen baskets present in their hind legs (Rhodes 2002). Dry pollen grains cannot be held by bees and to avoid falling off on flight, the bee will regurgitate some nectar to stick pollen, which adds sweeter taste while feeding pollen balls collected by bees (Bloch 2010). The bees that lack pollen baskets utilize the abdominal hairs for transportation of pollen (*Osmia* bees and leaf-cutter bees) (Gaus and Larsen 2009). A honeybee usually visits 100–1000 flowers in a trip of half to four hours, with 7–14 trips per day. The honeybees are great pollinators pollinating wide number of plants, with some bee species are expertized in pollinating specific plant species. Honey is probably the most natural food on the globe, which human beings receive by the ecological services of bee to plants, the ecosystem and ultimately the human beings (Luig and Peterson 2005).

11.3.4 Pollinators and Biodiversity

Pollinators and flowering plants coexist, as both depend on each other for their survival. Biodiversity rate, the number of varied plant and animal species present in unit area, being higher is associated to the positive aspects of the environment and a stable ecosystem. A stable ecosystem creates the possibility of development of natural ecological niches. A stable ecosystem is created by the active services of pollinators, for fertilization, seed production and offspring development (Conradt and Roper 2005). In an ecosystem, some flower blooms every month, bimonthly, quarterly, half-yearly, annually, others every second or third year, whereas plants belonging to same species bloom at similar time and hours. Flowers at bloom are the basic requirement for any pollinator, for access to food all year round for their effective pollination service (Gallai 2009). Plants reproduce by simplest vegetative reproduction, where the same trait can be reproduced without any differential characteristics (DiPasquale et al. 2010). To overcome to environmental changes there need to be genetically different plants, which are better adapted due to special genetic constitutions (Maini et al. 2010). Cross-pollination of plants by bees involves mixing of genes, by which plants benefit by producing genetically different offsprings. In this case, there is a greater chance for at least some of the offspring to survive in the competition of life (Kremen and Chaplin-Kramer 2007). Pollinators find their role in this regard and therefore the ecosystem service given by pollinator is of chief concern.

11.4 Pollination and Food Production

Plants are perilous components for sustainable functioning of ecosystems provisioning food, shelter and other resources for a wide array of living organisms (Foley et al. 2012). Pollination is vital to life on our planet, without them food production is under question. Pollination plays key role in amendable provision to the environment and is crucial in relation to earth's ecology. Globally 80% of the plants that flower in the world depend upon visitors for fertilization (Gill 1991) and 90% them depend on the living organisms for pollen transfer (Donkersley et al. 2012).

Pollinators have prospered for long years on earth, ensuring food security and nutrition, maintaining biodiversity and lively ecosystems for living organisms

(Díaz 2015). The world's food crops for human consumption depend on pollinators for sustained production. The diverse range of food commodities that human beings consume is largely owed to animal pollinators (Moss and Schneider 2000). Pollinators are indispensable for vitamin, mineral and micronutrient rich cereals, fruits, vegetables, seeds and nuts in daily diet (Novais et al. 2016).

Sexual reproduction of plants is reliant on a living partner for assistance through pollination, the drive of pollen grains to receptive stigmatic surfaces of flower. More than 150,000 species were estimated to contribute to pollination; the massive population are insects (Levy 2011). Pollinators donated about $170 billion for global crop production, 10% amounting for human food consumption (Gallai et al. 2009). Although wind-pollinated grains such as rice, wheat and corn contribute to the bulk of human calories, animal pollination, specifically insects, is unreasonably imperative in global production of fruits, vegetables, edible oil crops and nuts (Potts et al. 2010). Moreover, worldwide production of agricultural crops pollinated by insects has dramatically amplified over centuries augmenting the demand of agricultural pollination services and their critical shortages in the foreseeable future (Aizen and Harder 2009).

Animal pollination contributes 5-8% of current global crop production, with annual market worth of $235 billion–$577 billion. The rank and value of pollination differs significantly among crops economies and growing regions (Morse and Calderone 2000).

11.4.1 Pollination and Agriculture

The majority of crops that provide our nutrition (fruits, vegetables and nuts) are cross pollinated by living organisms (Ivey et al. 2003). Human diets would be severely limited without the aid of pollinators, and they could not get the required vitamins and minerals for healthy living. However, some species of food crop does not depend on living external agents for pollination services; wheat a nutritious food crop is wind-pollinated (Gilliam et al. 1989). Roughly 75% of the globally raised crops depend to some degree on plant visitors to bear their output (Free 1993; William and Anderson 1974; Roubik 1995; Delaplane and Mayer 2000). Pollinators promote yield of crops quantitatively as well as qualitatively (Morandin and Winston 2006; Klatt 2014).

More than half of world food stream is facilitated by animal-pollinated crops including cotton, sunflower, flax, potato, onion, chilli, coriander, okra, pumpkin, cocoa, coffee and vanilla (Klein et al. 2007). Plants with medicinal properties also need pollination for seed set. The medicinal plant golden seal, known for respiratory and gastrointestinal diseases, sage, known for its antiseptic properties and dandelion a remedy for gastric troubles and heartburn are animal-pollinated plants (Levy 2011). Accordingly, pollination not only reinforces many ayurvedic medicines but is also vital in modern medications as up to 25% are derivatives of plants.

Apart from agricultural crop ecosystems, roughly 80% of the wild plants found in wild forest ecosystems require animal-mediated pollination (Breeze 2011). Plants are the foundation of food chains for terrestrial living things. The plant vegetation, fruits and seeds are fed by herbivores, which were hunted by carnivores. Besides, plants provide shade, shelter and nesting provisions for diverse living organism (Glavan and Bozic 2013). Accordingly, to sustain the assortment of global ecosystems, vigorous pollinator inhabitants are required to ensure production of offsprings.

The major cash crops that yield profitably in quantity as well as quality, which are leading in export including coffee, cocoa and almonds, offer employment and income for millions of people (Cooley et al. 2008). Pollinator-dependent food crops donate to healthy human diet in uplifting nourishment by supplying significant extent of nutrients, vitamins and minerals (Cooley et al. 2008).

11.4.2 Pollination in Higher Plants/Forestry

Agroforestry is the integration of trees, shrubs and herbaceous crops along with livestock in agricultural production system (Schoeneberger et al. 2017). Agroforestry provides pollinator food, habitat and breeding sites by adding structural and functional diversity to agricultural crops (Jose 2009; Singh and Jhariya 2016). Agroforestry provides protection for soil, reduces chemical and energy inputs, increases water use efficiency and improves air and water quality, sequesters carbon and enhances biodiversity (Jhariya et al. 2015, 2018a, b; Raj et al. 2019a, b). Agroforestry is seldom executed with respect to crop pollination services, but there are prospects in incorporating these services when scheming multifunctional practices (Udawatta et al. 2019). Agroforestry practices generally provide functions for pollinators such as providing habitats including foraging resources and nesting or egg-laying sites and enhancing site and landscape connectivity.

The foraging resources of pollinators in Agroforestry are as follows:

Pollen and nectar: Pollinators require a diversity of flowering plants over the foraging season to provide nectar and pollen resources to meet their nutritional need (IPBES 2016). Hedgerows, windbreaks and other linear agroforestry practices are used by bees (Morandin and Kremen 2013), butterflies (Varah et al. 2013), moths (Dover et al. 1997) and flies (Cole et al. 2017). According to the available resources agroforestry plays an important role in providing suitable floral resources, but if floral resources are limited, pollinator activity will also be limited. Some of the tree species that provide profuse nectar with high sugar content are maple (*Acer* sp.), cherry (*Prunus* sp.), willow (*Salix* sp.), horse chestnut (*Aesculus* sp.) (Stubbs et al. 1992; Loose et al. 2005). Pollen is a protein rich food reserve utilized by pollinators as an energy source. Key agroforestry species that offer pollen with high concentration of amino acids, sterols and other nutrients for bees and other pollinators include chestnut, willow, maple, cherry, plum, etc. (Ostaff et al. 2015; Tasei and Aupinel 2008; Russo and Danforth 2017; Filipiak 2019). Few pollinators are pollen specific, depend wholly on shrubs and trees such as willow, blueberry, roses, etc. (Fowler 2016; Dötterl and Vereecken 2010). Trees like willow, maple and *Prunus* sp. offer more than 90% of pollen collected by pollinators in the USA (Wood et al. 2018). The density of flowers and nectar availability in consequence can be higher in trees than herbs. During the peak flowering season, the grey willow produces 334,178 flowers/ m^2 and the nectar productivity is 3612 kg/ha/year (Baude et al. 2016). Management practices such as pruning and trimming of trees and hedge rows cause negative impact on pollinators.

Resins and Oils: Pollinators collect resins and floral oils from some kind of trees to assist in nest construction and provisioning immatures (Policarová et al. 2019; Portman et al. 2019; Cane et al. 2007). Some tunnel-nesting pollinators seal their nest with the aid of resins (Wcislo and Cane 1996), some others use tree resin, grains and other debris to create nest cells (O'Brien 2018). The best source of resin is poplar tree (*Populus* sp.), Polar tree is often used in agroforestry practices as they have fast growth rate, low shading effect, marketable products and low disease and pest complaints. Pine (*Pinus* sp.), beech (*Fagus* sp.), elm (*Ulmus* sp.) and horse chestnut are few other tree crops offering resins (Greenaway et al. 1990; Drescher et al. 2019). Some pollinators like honey bees use resins to make propolis to seal holes in their hives (Simone-Finstrom and Spivak 2010). Propolis has antibacterial properties that help to reduce disease transmission and parasite invasion (Simone-Finstrom et al. 2017).

Microclimate Modification: Pollinator behaviour and their services are strongly influenced by weather conditions such as temperature, wind speed, precipitation. Agroforestry increases air movement, modifies temperature and increases precipitation. Elevated temperature rather than low temperature favours pollinators activity in fruit growing region (Baldwin 1988; Norton 1988).

The horizontal and vertical structures and shaded sites within tree canopy offer diversity of niches for pollinators to get thermo regulated (IPBES 2016). Landscapes with higher proportion of hedgerows and semi-natural habitats enhance microclimatic situations (Hill and Webster 2016). Trees and shrubs can be used to shade beehives; woody plantings give maximum shade in summer. Agroforestry practices reduce wind speeds and thus increase pollinator efficiency and allow pollinators to forage during wind events that normally reduce or prohibit foraging (Pinzauti 1986).

Nesting and Egg-Laying Sites: Suitable nesting and egg-laying sites are important components of pollinator habitat. Pollinators benefit mostly from flower-rich foraging areas if nesting and breeding facilities are around (Sardiñas et al. 2016; Potts et al. 2005). Agroforestry enhances the presence of trees and shrubs that provide protected ground-nesting areas that have limited soil disturbance. Hedge rows provide suitable ground-nesting habitat for pollinators. Bumble bees construct nests in small cavities or old rodent burrows, underground or around fallen litter (Svensson et al. 2000; Kells and Goulson 2003). Hedge rows and other agroforestry practices can provide egg-laying sites, larval host plants and overwintering sites for lepidopteran butterflies (Macdonald et al. 2018).

11.5 Pollination, Pollinators and their Challenges

11.5.1 Pollinator Decline

Pollinators are key components of global biodiversity, extending vibrant ecological amenities to agricultural crops (Potts et al. 2010). Wild and managed pollinators face numerous stresses, which hinder their active involvement. Managed pollinator species such as honey bees, bumble bees, stingless bees and wild bees grieve due to pesticide exposure which severely affects their population (Meena et al. 2020). Parasites cause several diseases and pose a huge decline in their masses. Decline in floral resources due to increased human activity poses another threat (Goulson 2008). Widespread habitat demolition restricts nesting sites for wild pollinators. Those stressors interact synergistically to yield more negative effects on pollinator health. Recent declines in all categories of pollinator population have rightly raised concerns on global food security (NAS 2007).

The current global population of honeybees, the chief pollinator that pollinate the majority of agricultural crops, is under misery due to tragic losses (Delaplane 2011). The disappearance of many *Apis mellifera* colonies initiated in the USA with around 30% of beekeepers losing up to 3/4th of their colonies (Stokstad 2007). In extreme cases, the bees just left and never returned.

Many mammal and bird pollinators are also experiencing decline rate. An international rating declares that 45 bat species, 36 non-flying mammals, 26 humming birds and 7 sunbirds, which are known to pollinate plants are vanishing (Dafni et al. 1990). The lesser long nosed bat which lives in Mexico deserts, the chief pollinator for night-blooming plant such as agave cacti, is a significant endangered mammal pollinator (Delaplane and Mayer 2000). Globally 35% of plants depend on pollinators for crop production accounting as much as $577 billion a year (Allsopp et al. 2008); therefore, their decline severely affects agricultural economy. Many pollinator species are threatened with extinction, among which around 2000 avian species including humming birds contribute to 16% of vertebrates under threat of extinction. Extinction risk for insect pollinators is not as well demarcated, nevertheless some bees and butterflies are warned of high level of threat with at least 9% of bee and butterfly species at risk (Ashman 2004). The grounds for pressure of extinction of these insects include tough agricultural practices that eliminate wildflowers that provide food for pollinators (Buchmann et al. 1997). Agricultural farming also reveals pollinators to pesticides, and bees are under attack from parasites and pathogens, as well. The decline does not correspond to a sole cause, but comprises urbanization and change in land-use pattern for agriculture resulting in loss and deprivation of natural surroundings. In addition, intensive agriculture leads to consistent landscapes and lessening of diversity in flora, and minimized food and habitat resources (Johansen and Mayer 1990). Pesticides and other contaminants distress pollinators. Pesticides and other contaminants affect pollinators directly as contact poison (insecticides and fungicides) and indirectly as systemic poison (herbicides) (Desneux et al. 2007). Recently, the decline in both wild and domesticated pollinators and parallelly threat of plants that rely upon them is clearly evidenced. Without an international effort, the pollinators that promote the growth of food crops worth million dollars annually will face extinction (Barnett et al. 2007).

11.5.2 Causes for Decrease in Bee Population

Pollinators are fronting a multitude of pressures; thereby their population is declining around the world. Pollinator decline can interrupt the mutualistic relations between plants and pollinators, and potentially affect the plant populations (Rebekka et al. 2016). Human beings have had a dramatic impact on the ecosystems, which directly affects pollinator species.

Loss of Habitat

The main threats faced by pollinators are habitat loss, land degradation and fragmentation. Urbanization, fragmentation and a shift of large-scale agricultural lands to buildings reduce food and nesting sources and loss of natural territories for pollinators (Kluser and Peduzzi 2017). The native vegetation is replaced by roadways, lawn, constructions, etc., hence pollinators lose the food and nesting sites that are necessary for their survival (Bradbear 2003). Land use for substitute means may interfere with the foraging, overwintering behaviour and specific habitat needs of pollinators. Due to urbanization, hard metal surfaces replace vegetations, thereby limiting habitats for ground dwelling pollinators. Fragmented habitat breaks pollinator survival needs (Goulson 2008). The segregation of pollinators, which has the effect of reducing population sizes and corroding their genetic pools, is a consequence by loss of habitat (Zayed and Packer 2005). Diversity of pollinators and rate of crop visitation diminish with increasing isolation from their natural habitats (Ricketts et al. 2008). Pollinators require native vegetation and blooming plants for effective foraging to assist in pollination.

Fragmentation

Habitat fragmentation is one of the most illusory forms of ecological deprivation and is measured to be one of the utmost threats to diverse community of living organisms on the ground (Beverly and Erik 1993; Khan et al. 2020a, b). Landscape disturbance results in habitat loss and disruptions of habitat configuration. Landscape disturbance principally influences three components of plant–pollination interactions: pollinator mass, pollinator's movement and plant population (Johnson and Steven 2004). The interdependency between plants and animals essential for pollination systems also has the possibility to drive cumulative effects of fragmentation, originating adverse response loops between organism and plants they pollinate (Hadley and Betts 2011). The richness of pollinators as well as their contribution in the seed set in many species of flowering plants is greatly reduced due to fragmentation (Donaldson et al. 2002).

Agricultural Chemicals

Pesticides, as such and in grouping with other contaminants, have had a distressing consequence on flower visitors. Pesticidal usage lessens flower visitors directly through contact poisoning, and also contaminated their food sources acting as systemic poison, consequentially early mortality, reduced mobility and behavioural disorders were noticed (Boulter et al. 2006).

The concentration of toxin and exposure level of pesticides cause risk to pollinators (Yoder et al. 2013). Pesticides, predominantly insecticides, have a broad range of fatal effects on pollinators (Bradbear 2003). Certain pesticides may persist in the atmosphere for a prolonged period and infest the flower visitors for long generations (Schmehl et al. 2015). Insecticides applied to plants contaminate pollen grains stick on pollen surface and are lethal to pollinators (Ingram et al. 1996). Insecticides also hinder the capability of flower visitors to navigate to floral sources by distracting the odour (Mullin et al. 2010). Herbicides affect pollinators indirectly by destroying weed plants and other non-crop flora, which benefit pollinators (Edwards and Abivardi 1998).

Invasive Species

Deliberate or accidental introduction of foreign plant species poses another severe threat to pollinators (Enserink 1999). Pollinators require specific forage to provide nutrition and shelter for visitors. Plants native to a region is required by pollinators, which may be displaced by the introduction of new species (Eilers 2011). Foreign plant species may attract pollinators away from native species by delivering better forage options (Ebert 2011). The introduction of new species can encourage competition for foraging and nesting behaviour of pollinators with native species. The introduction of new species can encourage competition for foraging and nesting behaviour of pollinators with native species. The native pollinators have been suppressed after introduction of European honey bees from Eurasia to other continents (Buchmann 1996). The Western honey bee, *Apis mellifera* was introduced as primary pollinator of introduced invasive weeds in New Zealand, thereby, strengthen the colonization of invasive weeds (Goulson 2008). Furthermore, imported pollinators sometimes act as carriers of parasites and deadly diseases, which then spread to native mass of pollinators those deficit resistances to introduced pathogens. The bumblebee colonies showed massive decline, when imported to USA in 1998 for pollination in green houses, due to the carriage of parasites along with them, which infested native bees like *Bombus occidentalis*.

Climate Risks: Disruption of Pollination Timing

Climate change also alters the climatic shield of pollinator species. Most pollinator species are movable and adaptable to climatic variations, and, therefore, arrangement of plant and pollinator assemblages is likely to alter in according to weather change. Plant visitors in tropic regions live at their thermal optimum, and some species migrate to colder region or perish on further warming in tropics (Deutsch 2008). The existence of blooming plants may alter to northern region or higher elevations due to global warming, out of synchronization with their visitors (IPCC 2007). The distribution and range of pollinators may be varied; pollinators adapted to warmer temperature may enlarge their northward swift, replacing other pollinators. The timing of pollination is resoluted by climatic factors like temperature, humidity and availability of moisture (Cleland et al. 2007). The phenological response of plant and pollinators needs to remain synchronized for plant–pollinator relationship to be viable; but the lifecycles of pollinators are altered by climatic variations distressing the synchronization between plant and pollinator (Ellis et al. 2015). The adults of pollinators would not have emerged in abundance during exact pollination period due to altered lifecycle of pollinators in response to climatic variations. Kudo et al. (2004) reported that plants flower much earlier in alpine ecosystem, but the pollinator emergence is distressed due to warming temperature; thereby disrupting pollination.

11.5.3 Economic and Ecological Consequences of Pollinator Declines

Economics of pollination facilities offers statistics on the economic consequences of shortages in pollination due to decline of pollinators and contributes to decision-making in formulating alternative mitigation strategies (Bauer and Wang 2010). Pollination is an ecological amenity and acts as production package and an income generator. The wild pollinators visit a wide range of flowering plants and contribute as ecosystem facility, while the bee colonies are rented or acquired to supplement pollination as a production practice. Bee farming by using the tamed bee colonies indicates the market demand suggesting that natural pollination of crops is not sufficient to meet agriculture's pollination needs (Gill et al. 2016).

Honey bees are significant pollinators for vast number of plants, in taking nectar, assembling pollens with ease and foraging numerous floral sources in a single trip. Honey bees, and to a lesser extent bumble and stingless bees are favoured among growers as they are manageable and having large populations in colonies (Delaplane and Mayer 2000; Delaplane et al. 2010). Roughly 4000 wild bee species in the USA contribute to pollination service in the USA (Cane and Payne 1988,1990, 1993; Cane 1994), but their offer for commercial agriculture is not yet explored (Kremen et al. 2002; Spivak et al. 2011).

The population of wild bees is constantly declining as they undergo health issues as wild bees (Cameron et al. 2011). Wild bees undergo health issues as managed bees, and their populations have similarly declined. Wild bees contribute as efficient pollinators to numerous agricultural crops including pumpkin, tomato, cranberry, etc. With sufficient natural wild floral resources and nesting grounds they substitute in the face of managed pollinator decline (Spivak et al. 2011; Adamson et al. 2012).

Pollination by living organism is essential for fertilization of around 70% of plants that bloom in the world and more than 30% of world crops depend on honeybees for fertilization. The likelihood of lack of diet in the nonappearance of pollinators is less for cereals as typical diet sources are pollinated by wind. Possibility of less crop production is expected for healthy diet, *viz.*, fruits, vegetables as well as meat and dairy products, as they are chiefly anchored honey bees (Spivak et al. 2011). Approximately 35% of the global crop production is reliant on biotic pollination. The domain devoted for bee pollinated crop is increasing worldwide, meanwhile the usage of managed hives has lessened to half the quantity (Spivak et al. 2011).

The ecological amenity of fertilizing an agricultural crop to aid in food production by wild and managed plant visitors has shown stable deterioration lately (Allen-Wardell et al. 1998). The causes of this fall comprise agricultural intensification, crop monocultures, pest, diseases and parasites, usage of pesticides and other chemicals, land-use and urbanization and reduced floral sources for pollinators (Biesmeijer et al. 2006). Our understanding of pollination as a perilous component in the world's food supply, although the loss of pollination services has been confined to local cases (Michener 2000). Thus, attention has to be raised to maintain pollination services in agricultural management (Balmford et al. 2002).

Economic valuation of pollination is an important approach to uplift the attention of farming community and policy makers in sustainable services of pollinators to the environment. As the services of wild pollinators to plants cannot be graded to market value, their cost of pollination services is not considered in decision-making. Nevertheless, approximations for commercial worth of pollination provision differ extensively (Richards 1993; Costanza et al. 1997). Accordingly, the method of valuation should be reviewed and the understanding of current pollination value should be analysed. The statistics of pollination amenity is scale dependent (Kevan and Phillips 2001). The service supports farmer income in farming level, whereas the facility is essential for warranting supply of food at national level. The surpluses of the pollination amenity for both producers as well as consumers vary at these scales and diverse evaluation approaches have to be modulated (Hanley and Spash 1998).

It is of need for scientists and policy makers to be appraised on probable economic consequences as agricultural crop production is confronted with the crisis of pollinator decline. An extensive series of profits was offered to human beings by plant visitors as they provide a consistent and assorted source of fertilization producing seeds and fruits, sustaining ecoservices and biodiversity, monetary products from honey bees like honey, royal jelly, bee wax, etc. The aids from pollinators can be economically quantified to meet the consequences of their abundance and fall to population (Crane 1999). Existing scenario of economic predictions and indicators lacks to impose the full range of benefits from pollinator abundance and diversity, and the total value of managed pollinators (Ellis et al. 2015). These failures can result in minimal pollinator mediated gains that decisions on land-use management are based on market value and economic indicators from a social perspective (Roubik 1996). Definitely, deteriorations in pollinator richness and assortment have altered the paybacks they offer to human beings.

Knowledge to undertake remedial activities on failure in market and economic indicator is offered by economic valuation of pollinator derived benefits (Forup 2003). Consequently, nature's eco amenities are appreciated by human beings either by direct or indirect means (Ghazoul 2002). Usage of well-informed methodologies and justified criteria for clearing the benefits of pollination is defined as economic evaluation (Dafni 1990). The explicit data on pollinations worth generates statistics for farm managers, small-scale industrial managers, consumers and policy makers to set policies and decisions in favour of pollinator sustainability (Corbet 2000).

The consequences of pollinator decline and upliftment has multiple effects influencing the productivity of crops and other pollinator products such as honey. They can be treasured in both economic and non-economic means (Gikungu 2006). The economic means evaluated the market value of agricultural crop productivity and worth of by-products (honey, royal jelly, wax) from managed pollinators. The non-monetary means include improved nutrition quality of food sources, better productivity and better-quality offspring of plants for next generation (Nabhan and Buchmann 1997).

Consequently, it would be a vigorous tactic to encourage pollinator-friendly agricultural practices in crop production (Ellis 2014). An extinct knowledge on the relative extend of crop yield due to pollinator exists, but there are uncertainties that the relation between crop yield, agricultural inputs and pollination service vary for different crops.

A substantial risk is experienced economically with unstable pollinator masses than stable diverse groups (Steffan-Dewenter 2003). There is no extinct evaluation methodology for economic quantification of pollination amenity. The existing biological and economic statistics are insufficient for accurate estimation of economic value of minimal and maximal population of pollinators (Sessions 2000). Decision-making by policy makers through cost–benefit analyses and risk analyses aids in economic valuation of pollination facility (Thompson 2006).

11.6 Protecting Pollinators

The best means to maintain a healthy pollinator inhabitant is to guarantee an assorted and diverse floral plant community. Flowering plants that offer abundance of pollen and nectar should be raised in enormous quantity. Restoration practices are to be done to confirm a constant food supply for pollinators (Danforth et al. 2006). An enriching pollinator habitat should be created with varied floral resources. Mutually the pollinators in abundance will yield a more healthy and fertile land cover.

11.6.1 Protect Existing Forages for Pollinators

Existing pollen and nectar sources that offer nutritional reward to attract pollinators should be identified and protected. The blooming of efficient forage plants should be monitored at dawn and mid-day to the intensity of active pollinators. Among the whole pollinators, honey bees and bumble bees are considered effective indicators of flowering plants (Praz et al. 2008).

11.6.2 Ensure Flowering Throughout the Year

Pollinators have seasonal foraging activity. Bees are most active throughout the year, but lengthier in mild climates. Therefore, a palette of flowering plants offering a food source for diverse community of pollinators throughout the season is essential (FAO 2016). Bumble bees, one most efficient crop pollinators, freshly emerge, overwinter and bumble bee queens successfully establish their colonies and foraging sources are available early in the season (Vanderplanck et al. 2014). In addition, some solitary bees produce multiple generations each year, leading to greater populations in the summer; Plants with bloom throughout the season is mandatory for their forage. Late season blooms of goldenrod and asters benefit some wild and managed honey bees (Roulston and Cane 2002). If constant foraging of pollinators is minimized, planting annuals in borders, boundaries and ground covers will supplement floral sources.

11.6.3 Plant More Flowering Plants to Enhance Forage

Varied types of plants, *viz.*, herb, shrub and trees, that bloom abundance of flowers throughout the growing season should be planted (Kariyat et al. 2014). Native wild forage with diverse colour, shape, size and fragrant flowers will encourage the activity of numerous group of pollinators (Michener 2000). Trees with heavy blooms invite mammalian pollinators.

11.6.4 Protect Nesting Sites

Apart from food sources for foraging, pollinators require shelter to rest. The best approach for protecting pollinators is by providing potential nesting areas (Roulston et al. 2008). The nest of a pollinator is the home, where from they wander for nectar and pollen. The closer their habitats are located to their food sources, the lesser the energy pollinators need to spend travelling back and forth; therefore, energy can be utilized for their offspring production, thereby more foraging pollinator population can be produced (Li et al. 2014). Moreover, if the habitats of pollinators are far, they have to travel long distance for foraging, where they may encounter pesticides and other hazards (Hanley et al. 2008).

11.6.5 Protection from Pesticides

Pesticides are chemicals used in agriculture to manage insect pests, pathogens and weeds. Practice of pesticides should be well-adjusted with the status of nourishing healthy populations of crop pollinators that can be minimized by pesticide application. Pesticides use must be balanced against the importance of maintaining healthy populations of crop pollinators that can be damaged by pesticide applications (Chagnon 2015). Best agricultural practices will reduce the side effects of pesticide and offer refuge for native pollinators. Herbicides will destroy plants that provide flowers when crops are not in bloom, thereby pollinators are forced to forage widely for food (Henry et al. 2012). Foraging to far distances will expose them to more threats and so they produce reduced offspring. Insecticides are used to control insects; but were toxic to pollinators based on the active ingredient and formulation applied (Mao et al. 2013). Foraging pollinators are poisoned by pesticides as they absorb toxins through cuticle, suck pesticide sprayed nectar and gather contaminated pollen (Pettis et al. 2013). These toxins will pass to larval stage of pollinators in the nest by means of feeding (Chittka 1992). Pollinators entangled with pesticides will find it difficult to navigate back to their shelter. When there is need for pesticide spray, botanicals like neem-based pesticides, microbial biopesticides such as *Bacillus* can be utilized, which are safe for pollinators. Impact of herbicides can be reduced by spot treating only deadly weeds. Insecticides with less toxicity by careful choosing of active ingredients can be helpful for pollinators (Maxim et al. 2013). Pesticide application techniques that minimize drift, scheduling application when bees are less active during dusk will avoid pollinator loss.

11.7 Policy Formulation towards Eco-intensification of Pollination

Promotion of policies for ecological intensification on pollination services should be formulated which is lacking, though framed in few regions across the world.

11.7.1 Enact Pollinator-Friendly Pesticide Policies

Majority of pesticides may cause pollinator decline by contact or systemic action. Herbicides eliminate weeds that aid pollinators to forage and give shelter. Fungicides cause sub-lethal effects on pollinators (Morse and Calderone 2000). Neonicotinoids cause severe pollinator decline especially honey bees. Regulatory policies for pesticide mitigation include pesticide restriction, ban, especially neonicotinoid group. Few toxic pesticides like endosulfan is permanently banned and few are temporarily regulated (Heiling et al. 2003). Policies on restricted usages of pesticide are another approach to limit pesticide application.

11.7.2 Conserve and Enhance Pollinator Habitat Policies

Regulations to conserve the prevailing habitats and acclimatizing new habitats are chief factors in preserving the pollinators and provide crop pollination services (Lunau and Maier 1995). Policies on conservation strategies, subsidies for habitat enrichment, invoking endangered species laws should be regulated to conserve nesting sites for pollinators. Plants for shelter can be planted as border crops, row crops in farmers premises and in roadsides, public places, etc.

11.7.3 Regulating Incentives for Ecosystem Services by Pollinators

Policies for incentives for raising foraging crops for pollinators should be framed. Recently, economic valuation of profits for pollination services is offered by Government Agencies and Organizations (Pattinson 2012).

11.7.4 Ensure Participation and Empowerment of Diverse Stakeholders

Policies on involvement of rural and indigenous public, ensuring participation of diverse stalk holders are some important elements in encouraging pollinator activity. Indigenous rural people around the world have knowledge on pollinator biodiversity and their conservation measures. Raising awareness and tapping local knowledge are important factor to reverse pollinator crisis.

11.8 Ecological Intensification of Pollination Towards Green Revolution

11.8.1 Pollinator and Agroecological Intensification

Ecological intensification is a knowledge-intensive process that requires ideal organizing of nature's ecological functions and biodiversity to advance the performance efficiency of our agricultural system (Winfree et al. 2011).

In agricultural sector, pollinators contribute to productivity of agricultural, horticultural and forage crops by improving seed vigour. The managed pollinators such as honey bee yield nutritious honey which favours health and other co-products such as bee wax, royal jelly and propolis (Table 11.3).

In addition, improved biodiversity and landscape modification with flowering plants such as parks and other recreation centres improve recreational activities thereby add income of a nation (Calderone 2012). Improved agricultural practices in restoration of floral resources and conservation activities to protect the flowering plants and crops enhance the health of a nation (Table 11.4).

11.8.2 Sustainable Use of Pollinators in Farming

Crop pollination failures of several highly localized crops have been recorded worldwide, bringing the attention of pollination issues. Due to urbanization and allied activities the wild foragers activity is minimized due to lack of food and shelter. The managed bee population is also rapidly declining and more crops grown under intensive greenhouse systems will prevent the loss of pollination services. Unfortunately, the level of capacity to manage pollination services and the awareness of their importance is lacking (Geslin 2013). The penalties of pollinator declines

Table 11.3 Honey bee products and their utilities

S. No	Product	Origin	Main ingredients	Primary use
1	Honey	Flower nectar, honey dew from aphids	Sugar, water, pollen, protein, enzymes and proteins	Consumption as a food source
2	Wax	Wax producing glands of worker bees	Myricin (a waxy substance which forms the less soluble part of beewax)	Cosmetics, pharmaceuticals and candles
3	Propolis	Resin from trees	A biocidal compound that suppresses micro-organisms	Dermal and internal application in herbal treatments
4	Pollen	Flower anther	Protein, B vitamins and amino acids	A food additive
5	Royal jelly	Glands in the throats of worker bees	Carbohydrates, proteins, B vitamin, sugar and water	Applications in herbal medicine
6	Venom	Abdominal glands of female bees	A variety of toxic proteins (melittin, apamin and others) which act as neurotoxins.	Treatment for rheumatism and sciatica in 'apitherapy'

Source: Schmider and Escaille (2013)

Table 11.4 Ecosystem services on pollination

S. No	Provisions	Ecosystem service
1.	Agriculture	Agricultural crop pollination Forage pollination Fruit and vegetable production Seed set
2.	By-products	Honey, bee wax, royal jelly
3.	Cultural	Recreation, tourist location
4.	Environment	Restoration/conservation of habitats

are likely to influence the production of vitamin-rich crops like fruits and vegetables, leading to progressively unstable diet and health problems. Sustainable yield of agricultural and horticultural crops under agricultural development is critically important to health, nutrition, food security and improved income for farmers (Brittain 2010). In the past, pollination was aided by nature, without any cost to human communities, but increased acreage of farm fields amplified the usage of agricultural chemicals mounting pressure on pollinators. The process of securing pollinators for their service to agriculture has received renewed interest to help nature to provide pollination services.

For sustainable use of pollinators in farming, on a global scale, the convention of biological diversity has identified the importance of pollinators with the establishment of international initiative for the conservation and sustainable use of pollinators, facilitated and coordinated with FAO.

Biodiversity in agricultural landscapes encourages pollination services and serves as a critical form of insurance for pollinators. Pollinator-friendly pest management practices such as minimizing the use of farm chemicals through organic farming, integrated pest management or biopesticides are recognized as having benefits for keeping pollinators in the crop fields. In agro-ecosystems, pollinator-friendly management practices should be identified that serve to enhance yields, quality, diversity and resilience of cropping systems:

- Managing flower-rich cropping systems, ensure hedge rows, field margins and forage
- Cultivating shade trees
- Managing for pollinator nest sites
- Preserving wild habitat
- Reducing application of pesticides and associated risks
- Establishing landscape configurations that favour pollination services (Billeter 2008).

11.9 Pollination as an Ecological Intensification Towards Green Revolution

The Green revolution, which flourished in mid-twentieth century, where agricultural technologies with farm machinery, artificial fertilizers and breeding high yielding varieties increased crop productivity (Tester and Langridge 2010; Kumar et al. 2020). Green revolution has increased double and triple fold increase in cereals, especially wheat in India (Tilman et al. 2002). Agricultural green revolution is still going on across world, as influenced by agricultural development of technologies and agricultural ecological intensification.

Improved technologies in agriculture give positive loop with ever-increasing population of the world. With each invention of technologies, agriculture can feed more populations and with ever-increasing population, agriculture has to feed even more mouths (Rosegrant et al. 2001). The global population is expected to reach 10 billion by the twenty-first century (Williamsm and Kremen 2007), and, therefore, the food demands to feed the growing population will continue to increase for some decades (Godfray et al. 2010).

A major goal of ecological intensification towards pollinator-dependent crops is to encourage crop pollinators for improved crop productivity towards another green revolution. This can be achieved by conserving or restoring natural habitats and enriching biodiversity in the wider landscapes (Kleijn et al. 2019). Theories on ecology predict that species richness of pollinators on crops is determined by the size of local pollinator species pool, which chiefly depends on the quantity of resources available in near their habitats (Mandelik et al. 2012). Several factors contribute in decline of pollinators; therefore, it is our responsibility to improve pollinator habitats and protecting them.

The diversity of pollinators is quite extensive, but sensitive to pesticides. Thus, it is very important to limit the exposure of pollinators to pesticides and harmful chemicals. An effective general rule is to avoid spraying when flowers are in bloom (Mao et al. 2013). In addition, planting a variety of flowers throughout the agricultural crop vicinity, in between rows, in borders, as mixtures improves pollinator activity, leading to improved crop productivity, leading to another green revolution (Willmer 2011).

Pollinators require food, shelter and breeding sites; they are winged creatures so the required resources can be located within the field/landscape vicinity or spread over a local area. However, their flight mechanism is limited; therefore, local diversified resources is an important element in pollinator health. The landscape surrounding agricultural/horticultural crops has tremendous capacity for pollinator promotion (Klein et al. 2006). Multifunctional landscape elements such as managed field margins benefit the pollinator by increased habitat and food provisions and also serve to improve the connectivity of green infrastructure, the value of biodiversity.

11.10 Conclusion

Mainstream of world's flowering plants needs pollinators for seed and fruit production. A vast array of diverse pollinators is present across the world, but bees are primary pollinators of vegetable crops and fruit trees. Human diet will be limited if sufficient pollinators are not available globally. Therefore, there is a need to protect the pollinator population from factors that limit their efficiency. Exposure of pollinators to pesticides is the main cause of pollinator decline, and, therefore, farmers should be cautious to spray when honey bees are around or blooming period. Programmes on reduced pesticide usage by organic farming, diversified cropping such as intercrops, row crops, border crops and practicing integrated pest management tactics.

Protection of foraging and nesting habitats of native pollinators will increase the abundance and diversity of pollinator. Sustainable pollination service can be assured by enriched forages with blooming plants and trees in pollinator's vicinity. This ensures plant pollination; seed setting and sustainable perpetuation of native wild flora and restoration of native species will be achieved. Enhancing pollinator diversity by enriched forage and habitat spaces is a great opportunity to educate the community about the value of pollinator and their ecosystem services.

11.11 Future Roadmap

Future research should be on expanding and refining pollinator conservation tactics and to mitigate pollinator decline strategies. To raise the pollinator population and to tackle their decline, policies for improving their effectiveness should be framed. Improving the scientific information related to pollinators by increased investigation on the position and tendencies of pollinators, by constant monitor of the pollinators and the pollination service they offer, pollinator pressures, their importance and inter-linkages, ecological and economic importance of pollinators and their risks. To mitigate the causes for decline of pollinators, the conservation and restoration of various pollinator inhabitants in rural and urban areas should be promoted, lessening of pollinator risks including pesticides should be minimized through organic farming and integrated pest management techniques and exploring ways to further enhance risk assessment of pesticide on bees. The adverse impressions of invasive alien plant species on pollinator should be addressed in a proper manner. To promote alertness and advance team work besides informations' sharing, employment of rural and urban communities in the conservation of pollinators, knowledge distribution and virtuous practices among shareholders should be eased. Local, regional and national pollinator strategies should be developed on pollinators activity in collaboration with farming community involved in managed hives pollination as an ecological service.

References

Abramovitz JN (1998) Putting a value on nature's 'free' services. World Watch (January–February):10–19

Adamson NL, Roulston TH, Fell RD, Mullins DE (2012) Wild bees pollinating crops through the growing season in Virginia, USA. Environ Entomol 41:813–821

Aizen MA, Harder LD (2009) The global stock of domesticated honey bees is growing slower than agricultural demand for pollination. Curr Biol 19(11):915–918

Allen-Wardell G, Bernhardt P, Bitner R, Burquez A, Buchmann S, Cane J, Cox PA, Dalton V, Medellin P, Medellin-Morales S, Nabhan GP, Pavlik B, Tepedino V, Torchio P, Walker S (1998) The potential consequences of pollinator declines on conservation of biodiversity and stability of food crop yields. Conserv Biol 12:8–17

Allsopp MH, de Lange WJ, Veldtman R (2008) Valuing insect pollination services with cost of replacement. PLoS One 3:e3128

Anderson B, Johnson SD (2009) Geographical covariation and local convergence of flower depth in a guild of fly-pollinated plants. New Phytol 182:533–540

Andersson S (2003) Foraging responses in the butterflies Inachis io, Aglais urticae (Nymphalidae), and *Gonepteryx rhamni* (Pieridae) to floral scents. Chemoecology 13:1–11

Andersson S, Dobson HEM (2003) Behavioral foraging responses by the butterfly *Heliconius melpomene* to *Lantana camara*. J Chem Ecol 29:2303–2318

Armbruster WS (1993) Evolution of plant pollination systems – hypotheses and tests with the neotropical vine, *Dalechampia*. Evolution 47:1480–1505

Ashman TL (2004) Pollen limitation of plant reproduction: ecological and evolutionary causes and consequences. Ecology 85:2408–2421

Ashman TL, Bradburn M, Cole DH, Blaney BH, Raguso RA (2005) The scent of a male: the role of floral volatiles in pollination of a gender dimorphic plant. Ecology 86:2099–2105

Baldwin CS (1988) The influence of field windbreaks on vegetable and specialty crops. Agric Ecosyst Environ 22:191–203

Balmford A, Bruner A, Cooper P, Costanza R, Farber S, Green RE, Jenkins M, Jefferiss P, Jessamy V, Madden J, Munro K, Myers N, Naeem S, Paavola J, Rayment M, Rosendo S, Roughgarden J, Trumper K, Turner RK (2002) Why conserving wild nature makes economic sense. Science 297:950–953

Banerjee A, Jhariya MK, Yadav DK, Raj A (2020) Environmental and sustainable development through forestry and other resources. Apple Academic Press Inc., CRC Press – A Taylor and Francis Group, US & Canada, p 400. https://doi.org/10.1201/9780429276026

Banskota AH, Yasuhiro T, Shigetoshi K (2001) Recent progress in pharmacological research of propolis. Phytother Res 15(7):561–571. https://doi.org/10.1002/ptr.1029
Barnett EA, Charlton AJ, Fletcher MR (2007) Incidents of bee poisoning with pesticides in the United Kingdom. Pest Manag Sci 63:1051–1057
Barrett SCH (2010) Darwin's legacy: the forms, function and sexual diversity of flowers. Philos Trans R Soc Lond B Biol Sci 365(1539):351–368. https://doi.org/10.1098/rstb.2009.0212
Barth FG (1991) Insects and flowers. Princeton University Press, Princeton, NJ
Baude M, Kunin WE, Boatman ND, Conyers S, Davies N, Gillespie MAK, Morton RD, Smart SM, Memmott J (2016) Historical nectar assessment reveals the fall and rise of floral resources in Britain. Nature 530:85–88
Bauer D, Wang I (2010) Economic consequences of pollinator declines: a synthesis. Agric Resour Econ Rev 39(3):368–383. https://doi.org/10.1017/S1068280500007371
Bell G (1985) On the function of flowers. Proc R Soc London B 224:223–265
Benadi G, Thomas H, Poethke H (2013) When can plant-pollinator interactions promote plant diversity? Am Nat 182(2):131–146. https://doi.org/10.1086/670942
Bertin RI (1982) Floral biology, hummingbird pollination and fruit production of trumpet creeper (*Campsis radicans*, Bignoniaceae). Am J Bot 69:122–134
Beverly JR, Erik SJ (1993) Pollination biology in tropics. Curr Sci 65(3):273–277
Biesmeijer JC, Roberts SP, Reemer M, Ohlemueller R, Edwards M, Peters T, Schaffers A, Potts SG, Kleukers R, Thomas CD, Settele J, Kunin WE (2006) Parallel declines in pollinators and insect-pollinated plants in northwest Europe. Britain and the Netherlands. Science 313:5785
Billeter R (2008) Indicators for biodiversity in agricultural landscapes: a pan-European study. J Appl Ecol 45(1):141–150
Bloch G (2010) Industrial apiculture in the Jordan valley during Biblical times with Anatolian honeybees. Proc Natl Acad Sci U S A 107(25):11240–11244. https://doi.org/10.1073/pnas.1003265107
Bommarco R, Kleijn D, Potts SG (2013) Ecological intensification: Harnessing ecosystem services for food security. Trends Ecol Evol 28:230–238
Boulter SL, Kitching RL, Zalucki JM, Goodall KL (2006) Threats to pollination systems. In: Reproductive biology and pollination in rainforest trees: techniques for a community-level approach. Cooperative Research Centre for Tropical Rainforest Ecology and Management Rainforest CRC, 5356, Cairn, Australia
Bradbear N (2003) Bees are diligent pollinators of fruit and seed crops. In: Beekeeping and sustainable livelihoods. Food and Agriculture Organisation of the United Nations
Branquart E, Hemptinne JL (2000) Selectivity in the exploitation of floral resources by hoverflies (Diptera: Syrphinae). Ecography 23(6):732–742
Breeze TD (2011) Pollination services in the UK: how important are honeybees? Agric Ecosyst Environ 142(3–4):137–143
Briscoe AD, Chittka L (2004) The evolution of color vision in insects. Annu Rev Entomol 46:471–510
Brittain CA (2010) Impacts of a pesticide on pollinator species richness at different spatial scales. Basic Appl Ecol 11(2):106–115
Buchmann SL (1996) Competition between honey bees and native bees in the Sonoran desert and global bee conservation issues. In: Matheson A, O'Toole C, Buchmann S, Westrick P, Williams I (eds) The Conservation of Bees. Academic Press, New York
Buchmann SE, Nabhan GP (1996) The forgotten pollinators. Island Press, Washington, DC
Buchmann SL, Gary P, Nabhan S (1997) The forgotten pollinators. Island Press, Washington, DC
Burger BV (2005) Mammalian semiochemicals. In: Chemistry of pheromones and other semiochemicals. II, vol 240. Springer, Heidelberg, Germany, pp 231–278
Calderone NW (2012) Insect pollinated crops, insect pollinators and US agriculture: trend analysis of aggregate data for the period 1992–2009. PLoS One 7(5):e37235

Cameron SA, Lozier JD, Strange JP, Koch JB, Cordes N, Solter LF, Griswold TL, Robinson GE (2011) Patterns of widespread decline in North American bumble bees. Proc Natl Acad Sci U S A 108:662–667
Cane JH (1994) Nesting biology and mating behavior of the Southeastern blueberry bee, *Habropoda laboriosa* (Hymenoptera: Apidae). J Kans Entomol Soc 67:236–241
Cane JH, Payne S (1988) Foraging ecology of the bee *Habropoda laboriosa* (Hymenoptera: Anthophoridae), an oligolege of blueberries (Ericaceae: Vaccinium) in the South Eastern United States. Ann Entomol Soc Am 81:419–427
Cane JH, Payne S (1990) Native bee pollinates rabbit eye blueberry. Highl Agric Res 37:4
Cane JH, Payne S (1993) Regional, annual, and seasonal variation in pollinator guilds: Intrinsic traits of bees (Hymenoptera: Apoidea) Underlie their patterns of abundance at *Vaccinium ashei* (Ericaceae). Ann Entomol Soc Am 86:577–588
Cane JH, Griswold T, Parker FD (2007) Substrates and materials used for nesting by North American Osmia bees (Hymenoptera: Apiformes: Megachilidae). Ann Entomol Soc Am 100:350–358
Carthew SM, Goldingay RL (1997) Non-flying mammals as pollinators. Trends Ecol Evol 12 (3):104–108
Caruso CM, Peterson SB, Ridley CE (2003) Natural selection on floral traits of Lobelia (Lobeliaceae): spatial and temporal variation. Am J Bot 90:1333–1340
Chagnon M (2015) Risks of large-scale use of systemic insecticides to ecosystem functioning and services. Environ Sci Pollut Res 22:119–134. https://doi.org/10.1007/s11356-014-3277-x
Chapelin-viscardi JD, Tosser V, Maillet-Mezeray J, Sarthou V (2015) Contribution à la connaissance de la consommation de pollen par six espèces de Syrphes auxiliaires en milieux agricoles (Diptera Syrphidae). L'Entomologiste 71(3):169–178
Chittka L (1992) The colour hexagon: a chromaticity diagram based on photoreceptor excitations as a generalized representation of colour opponency. J Comparative Physiol A 170:533–543
Chittka L, Raine N (2006) Recognition of flowers by pollinators. Curr Opin Plant Biol 9:428–435. https://doi.org/10.1016/j.pbi.2006.05.002
Chittka L, Thomson JD (2001) Cognitive ecology of pollination. Cambridge University Press, Cambridge, UK
Christopher E (2000) Contributions of autogamy and geitonogamy to self-fertilization in a mass-flowering, clonal plant. Ecology 81(2):532–542. https://doi.org/10.2307/177446
Cleland EE, Chuine I, Menzel A, Mooney HA, Schwartz MD (2007) Shifting plant phenology in response to global change. Trends Ecol Evol 22(7):357–365
Cole LJ, Brocklehurst S, Robertson D, Harrison W, McCracken DI (2017) Exploring the interactions between resource availability and the utilisation of semi-natural habitats by insect pollinators in an intensive agricultural landscape. Agric Ecosyst Environ 246:157–167
Conradt L, Roper TJ (2005) Consensus decision making in animals. Trends Ecol Evol 20 (8):449–456. https://doi.org/10.1016/j.tree.2005.05.008
Cooley AM, Carvallo G, Willis JH (2008) Is floral diversification associated with pollinator divergence? Flower shape, flower colour and pollinator preference in Chilean *Mimulus*. Ann Bot 101:641–650
Corbet S (2000) Conserving compartments in pollination webs. Conserv Biol 14(5):1229–1231
Cortopassi-Laurino M, Knoll FRN, Ribeiro MF, van Heemert C, de Ruijter A (1991) Food plant preferences of Friesella schrottkyi. Acta Hort 288:382–385
Costanza R, d'Arge R, de Groot R, Farber S, Grasso M, Hannon K, Limburg S, Naeem RV, O'Neill J, Paruelo RG, Raskin P, Sutton van den Belt M (1997) The value of the world's ecosystem services and natural capital. Nature 387(6230):253–260
Crane E (1999) The world history of beekeeping and honey hunting. Routledge, London, p 682
Cruden R, Wand Hermann-Parker SM (1979) Butterfly pollination of *Caesalpinia pulcherrima*, with observations on a psychophilous syndrome. Ecology 67:155–168
Cunningham JP, Moore CJ, Zalucki MP, West SA (2004) Learning, odour preference and flower foraging in moths. J Exp Biol 207:87–94

Cunningham JP, Moore CJ, Zalucki MP, Cribb BW (2006) Insect odour perception: recognition of odour components by flower foraging moths. Proc Royal Soc London B 273:2035–2040

Dafni A (1990) Advertisement, flower longevity, reward and nectar protection in Labiatae. Act Hort 288:340–346

Dafni A, Kevan PG (1996) Floral symmetry and nectar guides: Ontogenetic constraints from floral development, colour pattern rules and functional significance. Bot J Linn Soc 120:371–377

Dafni A, Bernhardt P, Shmida A (1990) Red bowl-shaped flowers: convergence for beetle pollination in the Mediterranean region. Israel J Bot 39:81–92

Daily GC (1997) Nature's services: societal dependence on natural ecosystems. Island, Washington, DC

Dams LR (1978) Bees and honey-hunting scenes in the mesolithic rock art of Eastern Spain. Bee World 59(2):45–53. https://doi.org/10.1080/0005772X.1978.11097692

Danforth BN, Sipes S, Fang J, Brady SG (2006) The history of early bee diversification based on five genes plus morphology. Proc Natl Acad Sci U S A 103(41):15118–15123

Delaplane KS (2011) Understanding the impact of honey bee disorders on crop pollination. In: Sammataro D, Yoder JA (eds) Honey bee colony health: challenges and sustainable solutions. CRC Press, Boca Raton, FL, pp 223–228

Delaplane KS, Mayer DF (2000) Crop pollination by bees. CABI Publishing, NY

Delaplane KS, Thomas PA, McLaurin WJ (2010) Bee pollination of Georgia crop plants. In: University of Georgia cooperative extension cooperative extension bulletin 1106

Desneux N, Decourtye A, Delpuech JM (2007) The sublethal effects of pesticides on beneficial arthropods. Annu Rev Entomol 52:81–106

Deuri A, Rahman A, Gogoi J, Borah P, Bathari M (2018) Pollinator diversity and effect of *Apis cerana* F. pollination on yield of mango (*Mangifera indica* L.). J Entomol Zoology Stud 6 (5):957–961

Deutsch CA (2008) Impacts of climate warming on terrestrial ectotherms across latitude. Proc Natl Acad Sci 105(18):6668–6672

Díaz (2015) The IPBES Conceptual Framework – connecting nature and people. Curr Opin Environ Sustain 14:1–16

DiPasquale GM, Salignon S, LeConte LP, Belzunces A, Cecourtye A, Kretzschmar S, Suchail JLB, Alaux C (2010) Influence of pollen nutrition on honey bee health: do pollen quality and diversity matter? PLoS One 8(8):e72016

Dobson HEM (1994) Floral volatiles in insect biology. In: Bernays EA (ed) Insect-plant Interactions. CRC Press, London, Tokyo, pp 47–81

Dobson HEM (2006) Relationship between floral fragrance composition and type of pollinator. In: Dudareva N, Pichersky E (eds) Biology of floral scent. CRC Press, Taylor & Francis Group, Boca Raton, pp 147–198

Donaldson J, Nanni I, Zachariades C, Kemper J (2002) Effects of habitat fragmentation on pollinator diversity and plant reproductive success in renosterveld shrublands of South Africa. Conserv Biol 16(5):1267–1276

Donkersley P, Rhodes G, Pickup RW, Jones KC, Wilson K (2012) Honeybee nutrition is linked to landscape composition. Ecol Evol 4(21):4195–4206

Dötterl S, Vereecken NJ (2010) The chemical ecology and evolution of bee–flower interactions: a review and perspectives. Can J Zool 88:668–697

Dotterl S, Jurgens A, Seifert K, Laube T, Weissbecker B, Schutz S (2006) Nursery pollination by a moth in *Silene latifolia*: the role of odours in eliciting antennal and behavioural responses. New Phytol 169:707–718

Dötterl S, Glück U, Jürgens A, Woodring J, Aas G (2014) Floral reward, advertisement and attractiveness to honey bees in dioecious *Salix caprea*. PLoS One 9:e93421

Dover JW, Sparks TH, Greatorex-Davies JN (1997) The importance of shelter for butterflies in open landscapes. J Insect Conserv 1:89–97

Drescher N, Klein AM, Schmitt T, Leonhardt SDA (2019) A clue on bee glue: New insight into the sources and factors driving resin intake in honeybees (*Apis mellifera*). PLoS One 14:e0210594

Du ZY, Wang QF (2014) Correlations of life form, pollination mode and sexual system in aquatic angiosperms. PLoS One 9(12):e115653

Ebert A (2011) Nectar for the taking: the popularization of scientific bee culture in England, 1609–1809. Agric Hist 85:322–343

Edwards PJ, Abivardi C (1998) The value of biodiversity: where ecology and economy blend. Biol Conserv 83(3):239–246

Eilers EJ (2011) Contribution of pollinator-mediated crops to nutrients in the human food supply. PLoS One 6(6):e21363. https://doi.org/10.1371/journal.pone.0021363

Ellis H (2014) Spoonfuls of honey. A complete guide to honey's flavours & culinary uses with over 80 recipes. Pavilion Books, London

Ellis AM, Samuel S, Myers R, Taylor H (2015) Do pollinators contribute to nutritional health? PLoS One 10(1):e114805. https://doi.org/10.1371/journal.pone.0114805

Enserink M (1999) Predicting invasions: biological invaders sweep. Science 285(5435):1834–1836

Everaars J (2012) The response of solitary bees to landscape configuration with focus on body size and nest-site preference. Dissertation (Helmholtz-Zentrum für Umweltforschung – UFZ) 7/2012:1–146

Faegri K, van der Pijl L (1979) The principles of pollination ecology. Pergamon Press, Oxford

Fang J (2010) Ecology: a world without mosquitoes. Nature 466:432–434

FAO (2008) The value of bees for crop pollination. http://www.fao.org/3/i0842e/i0842e09.pdf. Accessed 15 Apr 2019

FAO (2016) FAO's global action on pollination services for sustainable agriculture. http://www.fao.org/pollination/projects/en/. Accessed 17 Mar 2016

Fenster CB, Armbruster WS, Wilson P, Dudash MR, Thomson JD (2004) Pollination syndromes and floral specialization. Ann Rev Ecol Systematics 35:375–403

Filipiak M (2019) Key pollen host plants provide balanced diets for wild bee larvae: a lesson for planting flower strips and hedgerows. J Appl Ecol 56:1410–1418

Foley K, Fazio G, Jensen AB, Hughes WHO (2012) Nutritional limitation and resistance to opportunistic *Aspergillus* parasites in honey bee larvae. J Invertebr Pathol 111:68–73

Forup ML (2003) The restoration of plant-pollinator interactions. PhD thesis, University of Bristol

Fowler J (2016) Specialist bees of the Northeast: host plants and habitat conservation. Northeastern Naturalist 23:305–320

Free JB (1993) Insect pollination of crops. Academic Press, London-New York

Galen C, Kevan PG (1980) Scent and color, floral polymorphisms and pollination biology in *Polemonium viscosum*. Nutt Am Midland Naturalist 104:281–289

Gallai N (2009) Economic valuation of the vulnerability of world agriculture confronted with pollinator decline. Ecol Econ 68(3):810–821. https://doi.org/10.1016/j.ecolecon.2008.06.014

Gallai N, Salles J, Settele J, Vaissière BE (2009) Economic valuation of the vulnerability of world agriculture confronted with pollinator decline. Ecol Econ 68:810–821

Gallen C (1999) Why do flowers vary? The functional ecology of variation in flower size and form within natural plant populations. BioSci 49:631–640. https://doi.org/10.2307/1313439

Garibaldi L, Pérez-Méndez N, Garratt M, Gemmill-Herren B, Miguez F, Dicks L (2019) Policies for ecological intensification of crop production. Trends Ecol Evol 34:10

Gaus H, Larsen H (2009) Pollination of fruit trees – Fact Sheet No.7.002. Colorado State University

Geslin N (2013) Plant pollinator networks along a gradient of urbanisation. PLoS One 8(5):e63421

Ghazoul J (2002) Flowers at the front line of invasion? Ecol Entomol 27:638–640

Ghazoul J (2005) Business as usual? Questioning the global pollination crisis. Trends Ecol Evol 20 (7):367–373

Gibbs HK, Rausch L, Munger J, Schelly I, Morton DC, Noojipady P (2015) Brazil's Soy Moratorium. Science 347(6220):377–378

Gikungu MW (2006) Bee diversity and some aspects of their ecological Interactions with plants in a successional tropical community. Ph.D. Thesis, University of Bonn, Germany

Gilbert LE (1975) Ecological consequences of a coevolved mutualism between butterflies and plants. In: Gilbert LE, Raven RH (eds) Coevolution of animals and plants. University of Texas Press, Austin and London, pp 210–240

Gilbert LE (1981) Foraging ecology of hoverflies: morphology of the mouthparts in relation to feeding on nectar and pollen in some common urban species. Ecol Entomol 6(3):245–262. https://doi.org/10.1111/j.1365-2311.1981.tb00612.x

Gill RA (1991) The value of honeybee pollination to society. Acta Horticulturae 288:62–68

Gill R, Baldock K, Brown M, Cresswell J, Dicks L, Fountain M, Garratt M, Gough L, Heard M, Holland J, Ollerton J, Stone G, Tang C, Vanbergen A, Vogler A, Woodward G, Arce A, Boatman N, Brand-Hardy R, Potts S (2016) Protecting an ecosystem service: approaches to understanding and mitigating threats to wild insect pollinators. Adv Ecol Res 54:135

Gilliam M, Prest DB, Lorenz BJ (1989) Microbiology of pollen and bee bread: taxonomy and enzymology of molds. Apidology 20:53–68

Giurfa M, Eichmann B, Menzel R (1996) Symmetry perception in an insect. Nature 382:458–461

Glavan G, Bozic J (2013) The synergy of xenobiotics in honey bee *Apis mellifera*: mechanisms and effects. Acta Biologica Slovenica 56:11–25

Godfray HC, Beddington JR, Crute IR, Haddad L, Lawrence D, Muir JF, Pretty J, Robinson S, Thomas SM, Toulmin C (2010) Food security: the challenge of feeding 9 billion people. Science 327:812–818

Gomez FG, David ER, Ramiro QL, María PBG, Patricia TG, Reyes TG, Cristina RP, Goulson D (2003) Effects of introduced bees on native ecosystems. Annu Rev Ecol Evol Syst 34:1–26

Goulson D (2008) Decline and conservation of bumble bees. Ann Rev Entomol 53:191–208

Goulson D, Nicholls E, Botías C, Rotheray EL (2015) Bee declines driven by combined stress from parasites, pesticides, and lack of flowers. Science 347(6229):1255957

Greenaway W, Scaysbrook T, Whatley FR (1990) The composition and plant origins of propolis: a report of work at Oxford. Bee World 71:107–118

Grimaldi D, Engel MS (2005) Evolution of the insects. Cambridge University Press, Cambridge, UK

Hadley A, Betts M (2011) The effects of landscape fragmentation on pollination dynamics: absence of evidence not evidence of absence. Biol Rev Camb Philos Soc 87:526–544. https://doi.org/10.1111/j.1469-185X.2011.00205.x

Hanley N, Spash CL (1998) Cost-benefit analysis and the environment. Edward Elgar Publishing, Cheltenham, UK

Hanley ME, Franco M, Pichon S, Darvill B, Goulson D (2008) Breeding system, pollinator choice and variation in pollen quality in British herbaceous plants. Funct Ecol 22:592–598. https://doi.org/10.1111/j.1365-2435.2008.01415

Hans P, Thomas S (2004) Flowers, faeces and cadavers: natural feeding and laying habits of flesh flies in Thailand (Diptera: Sarcophagidae, *Sarcophaga* spp.). J Nat Hist 38:1677–1694. https://doi.org/10.1080/0022293031000156303

Heal G (2000) Nature and the marketplace: capturing the value of ecosystem services. Island, Covelo, CA

Heard TA (1999) The role of stingless bees in crop pollination. Annu Rev Entomol 44:183–206

Heiling AM, Herberstein ME, Chittka L (2003) Pollinator attraction: crab-spiders manipulate flower signals. Nature 421:334

Heithaus ER (1974) The role of plant-pollinator interactions in determining community structure. Ann Missouri Bot Garden 61:675–691

Henry M, Beguin M, Requier F, Rolling O, Odoux JF, Aupinel P, Aptel J, Tchamitchian S, Decourtye A (2012) A common pesticide decreases foraging success and survival in honey bees. Science 336:348–350

Herrera CM (1993) Selection on floral morphology and environmental determinants of fecundity in hawkmoth-pollinated violets. Ecol Monogr 63:251–275

Hill DB, Webster TC (2016) Apiculture and forestry (bees and trees). Agr Syst 1995(29):313–320

Hoballah ME, Stuurman J, Heath RR, Landolt PJ, Dueben B, Lenczewski B (1992) Identification of floral compounds of night-blooming jessamine attractive to cabbage looper moths. Env Entomol 21:854–859

Hogg BN, Bugg RL, Daane KM (2011) Attractiveness of common insectary and harvestable floral resources to beneficial insects. Biol Control 56(1):76–84

Hung KJ, Kingstin JM, Albrecht M, Holway DA, Kohn JR (2018) The worldwide importance of honey bees as pollinators in natural habitats. Proc R Soc B 285(1870). https://doi.org/10.1098/rspb.2017.2140

Ingram M, Nabhan G, Buchmann S (1996) Our forgotten pollinators: protecting the birds and bees. Global Pesticide Campaigner 6(4) www.pmac.net/birdbee.html

IPBES (2016) The assessment report of the Intergovernmental Science-Policy Platform on Biodiversity and Ecosystem Services on pollinators, pollination and food production. In: Potts SG, Imperatriz-Fonseca V, Ngo HT (eds) Secretariat of the Intergovernmental Science-Policy Platform on Biodiversity and Ecosystem Services. IPBES, Bonn, Germany, p 552

IPCC (2007) Climate change 2007: synthesis report. In: Core Writing Team et al (eds) Contribution of working groups I, II and III to the fourth assessment report of the Intergovernmental Panel on Climate Change. IPCC, Geneva, Switzerland, p 104

Ivey CT, Martinez P, Wyatt R (2003) Variation in pollinator effectiveness in swamp milkweed, *Asclepias incarnata* (Apocynaceae). Am J Bot 90:214–225

Jander K, Herre E (2010) Host sanctions and pollinator cheating in the fig tree-fig wasp mutualism. Proc Royal Soc B: Biol Sci 277(1687):1481–1488. https://doi.org/10.1098/rspb.2009.2157

Jhariya MK, Bargali SS, Raj A (2015) Possibilities and perspectives of agroforestry in Chhattisgarh. In: Zlatic M (ed) Precious forests-precious earth. InTech, Croatia, Europe, pp 237–257. https://doi.org/10.5772/60841

Jhariya MK, Yadav DK, Banerjee A (2018a) Plant mediated transformation and habitat restoration: phytoremediation an eco-friendly approach. In: Gautam A, Pathak C (eds) Metallic contamination and its toxicity. Daya Publishing House, A Division of Astral International Pvt. Ltd, New Delhi, pp 231–247

Jhariya MK, Banerjee A, Yadav DK, Raj A (2018b) Leguminous trees an innovative tool for soil sustainability. In: Meena RS, Das A, Yadav GS, Lal R (eds) Legumes for soil health and sustainable management. Springer, pp 315–345. https://doi.org/10.1007/978-981-13-0253-4_10

Jhariya MK, Banerjee A, Meena RS, Yadav DK (2019a) Sustainable agriculture, forest and environmental management. Springer Nature Singapore Pte Ltd, Singapore, p 606. https://doi.org/10.1007/978-981-13-6830-1

Jhariya MK, Yadav DK, Banerjee A (2019b) Agroforestry and climate change: issues and challenges. Apple Academic Press Inc., CRC Press – a Taylor and Francis Group, US & Canada, p 335. https://doi.org/10.1201/9780429057274

Johansen CA, Mayer DF (1990) Pollinator protection: a bee and pesticide handbook. Wicwas Press, Cheshire, Connecticut, USA

Johnson S, Steven J (2004) The consequences of habitat fragmentation for plant–pollinator mutualisms. Int J Trop Insect Sci 24:29–43. https://doi.org/10.1079/IJT20049

Johnson SD, Pauw A, Midgley J (2001) Rodent pollination in the African Lily, *Massonia depressa* (Hyacinthaceae). Am J Bot 88:1768–1773

Jones GD, Stanley DG (2001) The uses of pollen and its implication for entomology. Neotrop Entomol 30(3):11–14

Jose (2009) Agroforestry for ecosystem services and environmental benefits: an overview. Agr Syst 76:1–10

Kariyat RR, Scanlon SR, Moraski RP, Stephenson AG, Mescher MC, De Moraes CM (2014) Plant inbreeding and prior herbivory influence the attraction of caterpillars (*Manduca sexta*) to odors of the host plant *Solanum carolinense* (Solanaceae). Am J Bot 101:376–380

Kassie M, Teklewold H, Jaleta M (2015) Understanding the adoption of a portfolio of sustainable intensification practices in eastern and southern Africa. Land Use Policy 42:400–411

Kells AR, Goulson D (2003) Preferred nesting sites of bumblebee queens (Hymenoptera: Apidae) in agroecosystems in the UK. Biol Conserv 109:165–174
Kevan PG, Lane MA (1985) Flower petal microtexture is a tactile cue for bees. Proc Natl Acad Sci U S A 82:4750–4752
Kevan PG, Phillips TP (2001) The economic impacts of pollinator declines: an approach to assessing the consequences. Conserv Ecol 5(1):8
Kevan PG, Randolf M (2012) The plight of pollination and the interface of neurobiology, ecology and food security. Environmentalist 32:300–310. https://doi.org/10.1007/s10669-012-9394-5
Khan N, Jhariya MK, Yadav DK, Banerjee A (2020a) Herbaceous dynamics and CO_2 mitigation in an urban setup – a case study from Chhattisgarh, India. Environ Sci Pollut Res 27 (3):2881–2897. https://doi.org/10.1007/s11356-019-07182-8
Khan N, Jhariya MK, Yadav DK, Banerjee A (2020b) Structure, diversity and ecological function of shrub species in an urban setup of Sarguja, Chhattisgarh, India. Environ Sci Pollut Res 27 (5):5418–5432. https://doi.org/10.1007/s11356-019-07172-w
Kislev M (2006) Early domesticated fig. Jordan Valley Sci 312(5778):1372–1374. https://doi.org/10.1126/science.1125910
Klarhe U, Gurba A, Herman K, Saxenhofer M, Bossolini E, Guerin PM, Kuhlemeier C (2011) Pollinator choice in *Petunia* depends on two major generic loci for floral scent production. Curr Biol 21:730–739
Klatt BK (2014) Bee pollination improves crop quality, shelf life and commercial value. Proc Biol Sci/Royal Soc 281(1775). https://doi.org/10.1098/rspb.2013.2440
Kleijn D, Bommarco R, Fijen TPM, Garibaldi LA, Potts SG, van der Putten WH (2019) Ecological intensification: bridging the gap between science and practice. Trends Ecol Evol 34:154–166
Klein AM, Vaissiere BE, Cane JH, Steffan-Dewenter I, Cunningham SA, Kremen C, Tscharntke T (2006) Importance of pollinators in changing landscapes for world crops. Proc R Soc B 3721:1–11. https://doi.org/10.1098/rspb.2006.3721
Klein BE, Vaissiere JH, Cane I, Steffan-Dewenter SA, Cunningham C, Kremen T, Tscharntke (2007) Importance of pollinators in changing landscapes for world crops. Proceedings of the Royal Society B, Biological Sciences 274 (1608):303–313
Kluser S, Peduzzi P (2017) Global pollinator decline: a literature review. UNEP/GRID-Europe
Knudsen JT, Tollsten L (1993) Trends in floral scent chemistry in pollination syndromes: floral scent composition in moth-pollinated taxa. Bot J Linnean Soc 113:263–284
Knudsen JT, Andersson S, Bergman P (1999) Floral scent attraction in *Geonoma macrostachys*, an understorey palm of the Amazonian rain forest. Oikos 85:409–418
Krell FT (2006) Fossil record and evolution of Scarabaeoidea (Coleoptera: Polyphaga). Coleopt Bull 60:120–143. https://doi.org/10.1649/0010-065X(2006)602.0.CO;2
Kremen C, Chaplin-Kramer R (2007) Insects as providers of ecosystem services: crop pollination on pest control. Royal Entomol Soc:349–382
Kremen C, Niles JO, Dalton MG, Daily GC, Ehrlich PR, Fay JP, Grewal D, Guillery RP (2000) Economic incentives for rain forest conservation across scales. Science 288:1828–1832
Kremen C, Williams NM, Thorp RW (2002) Crop pollination from native bees at risk from agricultural intensification. Proc Natl Acad Sci U S A 99:16812–11681
Krishna S, Keasar T (2018) Morphological complexity as a floral signal: from perception by insect pollinators to co-evolutionary implications. Int J Mol Sci 19(6):1681. https://doi.org/10.3390/ijms19061681
Kudo G, Nishikawa Y, Kasagi T, Kosuge S (2004) Does seed production of spring ephemerals decrease when spring comes early? Ecol Res 19:255–259
Kumar S, Meena RS, Jhariya MK (2020) Resources Use Efficiency in Agriculture. Springer Nature Singapore Pte Ltd, Singapore, p 760. https://doi.org/10.1007/978-981-15-6953-1
Kunze J, Gumbert A (2001) The combined effect of color and odor on flower choice behavior of bumble bees in flower mimicry systems. Behav Ecol 12:447–456
Lehrer M, Bischof S (1995) Detection of model flowers by honeybees: the role of chromatic and achromatic contrast. Naturwissenschaften 82:145–147

Levy S (2011) What's best for bees. Pollinating insects are in crisis. Understanding bees' relationships with introduced species could help. Nature 479:164–165

Li C, Xu B, Wang Y, Yang Z, Yang W (2014) Protein content in larval diet affects adult longevity and antioxidant gene expression in honey bee workers. Entomol Exp Appl 151:19–26. https://doi.org/10.1111/eea.12167

Libbrecht R, Keller L (2015) The making of eusociality: insights from two bumble bee genomes. Genome Biol 16:75

Loose JL, Drummond FA, Stubbs C, Woods S (2005) Conservation and management of native bees in cranberry. Maine Agricultural & Forest Experiment Station, Orono, ME

Losey JE, Vaughan M (2006) The economic value of ecological services provided by Insects. BioScience 56:311–323

Luig J, Peterson K (2005) Human impacts on pollinators and pollination services – ALARM Socioeconomic Working Paper, p 24. http://seit.ee/files/SEIT-DP1-0311-2005.pdf

Lunau K, Maier EJ (1995) Innate colour preferences of flower visitors. J Comp Physiol A 177:1–19

Macdonald KJ, Kelly D, Tylianakis JM (2018) Do local landscape features affect wild pollinator abundance, diversity and community composition on Canterbury farms? NZ J Ecol 42:262–268

Maini S, Medrzycki P, Porrini C (2010) The puzzle of honey bee losses: a brief review. Bull Insectol 63(1):153–160

Mandelik Y, Winfree R, Neeson T, Kremen C (2012) Complementary habitat use by wild bees in agro-natural landscapes. Ecol Appl 22:1535–1546

Mao W, Schuler MA, Berenbaum MR (2013) Honey constituents upregulate detoxification and immunity genes in the western honey bee *Apis mellifera*. Proc Natl Acad Sci 110:8842–8846

Maxim L, Jeroen P, van der S (2013) Seed-dressing systemic insecticides and honeybees. In: European Environment Agency (ed) Late lessons from early warnings: science, precaution, innovation. European Environment Agency (EEA) report 1/2013, Copenhagen, pp 401–438

McDade LA, Kinsman S (1980) The impact of floral parasitism in two neotropical hummingbird pollinated plant species. Evolution 34:944–958

Meena RS, Lal R (2018) Legumes for Soil Health and Sustainable Management. Springer, Singapore, p 541. https://doi.org/10.1007/978-981-13-0253-4_10

Meena RS, Kumar V, Yadav GS, Mitran T (2018) Response and interaction of *Bradyrhizobium japonicum* and Arbuscular mycorrhizal fungi in the soybean rhizosphere: a review. Plant Growth Regul 84:207–223

Meena RS, Kumar S, Datta R, Lal R, Vijaykumar V, Brtnicky M, Sharma MP, Yadav GS, Jhariya MK, Jangir CK, Pathan SI, Dokulilova T, Pecina V, Marfo TD (2020) Impact of agrochemicals on soil microbiota and management: a review. Land (MDPI) 9(2):34. https://doi.org/10.3390/land9020034

Meena RS, Lal R, Yadav GS (2020a) Long term impacts of topsoil depth and amendments on soil physical and hydrological properties of an Alfisol in Central Ohio, USA. Geoderma 363:1141164

Meena RS, Lal R, Yadav GS (2020b) Long-term impact of topsoil depth and amendments on carbon and nitrogen budgets in the surface layer of an Alfisol in Central Ohio. Catena 194:104752

Melathopoulos AP, Cutler GC, Tyedmers P (2015) Where is the value in valuing pollination ecosystem services to agriculture? Ecol Econ 109:59–70

Menzel R, Shinida A (1993) The ecology of flower colours and the natural colour vision of insect pollinators: The Israeli flora as a case study. Biol Rev 68:81–120

Michener CD (2000) The bees of the world. Johns Hopkins University Press, Baltimore and London

Miyake T, Yafuso M (2003) Floral scents affect reproductive success in fly-pollinated Alocasia odora (Araceae). Am J Bot 90:370–376

Moldenke AR (1976) Pollination ecology as an assay for ecosystemic organization-e-convergent evolution in Chile and California. Phytologia 42:415–454

Morandin LA, Kremen C (2013) Bee preference for native versus exotic plants in restored agricultural hedgerows. Restor Ecol 21:26–32
Morandin LA, Winston ML (2006) Pollinators provide economic incentive to preserve natural land in agroecosystems. Agric Ecosyst Environ 116:289–292
Moritz RF, Härtel S, Neumann P (2005) Global invasions of the western honeybee (*Apis mellifera*) and the consequences for biodiversity. Ecosci 12:289–301
Morse RA, Calderone NW (2000) The value of honey bee pollination in the United States. Bee Culture 128:1–15
Moss RH, Schneider SH (2000) Uncertainties in the IPCC TAR: recommendations to lead authors for more consistent assessment and reporting. In: Pachauri R, Taniguchi T, Tanaka K (eds) Guidance papers on the cross cutting issues of the third assessment report of the IPCC. World Meteorological Organization, Geneva, pp 33–51
Muchhala N, Caiza A, Vizuete JC, Thomson JD (2008) A generalized pollination system in the tropics: bats, birds and Aphelandra acanthus. Ann Bot 103:1481–1487. https://doi.org/10.1093/aob/mcn260
Mullin CA, Frazier M, Frazier JL, Ashcraft S, Simonds R, van Engelsdorp D, Pettis JS (2010) High levels of miticides and agrochemicals in North American apiaries: implications for honey bee health. PLoS One 5(3):e9754. https://doi.org/10.1371/journal.pone.0009754
Murren CJ (2012) The integrated phenotype. Integr Comp Biol 52:64–76
Nabhan GP, Buchmann SL (1997) Services provided by pollinators. In: Daily G (ed) Nature's services. Island Press, Washington, DC, pp 133–150
NAS (2007) Status of pollinators in North America. National Academies Press, Washington, DC
Naug D (2009) Nutritional stress due to habitat loss may explain recent honey bee colony collapses. Biol Conserv 142:2369–2372
Nicole W (2015) Pollinator power: nutrition security benefits of an ecosystem service. Environ Health Perspect 123:A210–A215
Norton RL (1988) Windbreaks: benefits to orchard and vineyard crops. Agric Ecosyst Environ 22:205–213
Novais SMA, Nunes CA, Santos NB, D'Amico AR, Fernandes GW, Quesada M (2016) Effects of a possible pollinator crisis on food crop production in Brazil. PLoS One 11(11):e0167292. https://doi.org/10.1371/journal.pone.0167292
NRC (2007) Status of pollinators in North America. National Academies Press, Washington, DC. www.nap.edu/catalog.php?record_id=11761
O'Brien M (2018) Notes on Dianthidium Simile (Cresson) (Hymenoptera: Megachilidae) in Michigan. Gt Lakes Entomol 40:1
Ollerton J, Winfree R, Tarrant S (2011) How many flowering plants are pollinated by animals? Oikos 120:321–326
Ostaff DP, Mosseler A, Johns RC, Javorek S, Klymko J, Ascher JS (2015) Willows (*Salix* spp.) as pollen and nectar sources for sustaining fruit and berry pollinating insects. Can J Plant Sci 95:505–516
Painkra GP, Bhagat PK, Jhariya MK, Yadav DK (2016) Beekeeping for poverty alleviation and livelihood security in Chhattisgarh, India. In: Narain S, Rawat SK (eds) Innovative technology for sustainable agriculture development. Biotech Books, New Delhi, India, pp 429–453
Pattinson D (2012) Pre-modern beekeeping in China: a short history. Agric Hist 86:235–255
Peakall R, Handel SN, Beattie AJ (1991) The evidence for, and importance of, ant pollination. In: Huxley CR, Cutler DF (eds) Ant–plant interactions. Oxford University Press, Oxford, pp 421–429
Peñalver (2012) Thrips pollination of Mesozic gymnosperms. PNAS Early Edition
Pettis JS, Lichtenberg EM, Andree M, Stizinger J, Rose R, van Engelsdorp D (2013) Crop pollination exposes honey bees to pesticides which alters their susceptibility to the gut pathogen *Nosema ceranae*. PLoS One 8(7):e70182
Pinzauti M (1986) The influence of the wind on nectar secretion from the melon and on the flight of bees: the use of an artificial wind-break. Apidologie 17:63–72

Policarová J, Cardinal S, Martins AC, Straka J (2019) The role of floral oils in the evolution of apid bees (Hymenoptera: Apidae). Biol J Linn Soc 128:486–497
Pombal ECP, Morellato LP (2000) Differentiation of floral color and odor in two fly pollinated species of Metrodorea (Rutaceae) from Brazil. Plant Systematics Evol 221:141–156
Portman ZM, Orr MC, Griswold TA (2019) Review and updated classification of pollen gathering behavior in bees (Hymenoptera, Apoidea). J Hymenopt Res 71:171–208
Potts SG, Vulliamy B, Roberts S, O'Toole C, Dafni A (2005) Role of nesting resources in organising diverse bee communities in a Mediterranean landscape. Ecol Entomol 30:78–85
Potts SG, Biesmeijer JC, Kremen C, Neumann P, Schweiger O, Kunin WE (2010) Global pollinator declines: trends, impacts and drivers. Trends Ecol Evol 25:345–353
Potts SG, Imperatriz-Fonseca V, Ngo HT, Aizen MA, Biesmeijer JC, Breeze TD (2016) Safeguarding pollinators and their values to human well-being. Nature 540:220–229
Praz CJ, Müller A, Dorn S (2008) Specialized bees fail to develop on non-host pollen: do plants chemically protect their pollen? Ecology 89:795–804. https://doi.org/10.1890/07-0751.1
Proctor M, Yeo P, Lack A (1996) The natural history of pollination. Harper Collins Publishers, New York
Rader R (2015) Non-bee insects are important contributors to global crop pollination. Proc Natl Acad Sci 113(1):146–151. https://doi.org/10.1073/pnas.1517092112
Raguso RA, Willis MA (2005) Synergy between visual and olfactory cues in nectar feeding by wild hawkmoths, *Manduca sexta*. Anim Behav 69:407–418
Raine NE, Ings TC, Dornhaus A, Saleh N, Chittka L (2006) Adaptation, genetic drift, pleiotropy, and history in the evolution of bee foraging behavior. Adv Study Behav 36:305–354
Raj A, Jhariya MK, Harne SS (2018) Threats to biodiversity and conservation strategies. In: Sood KK, Mahajan V (eds) Forests, climate change and biodiversity. Kalyani Publisher, India, pp 304–320
Raj A, Jhariya MK, Yadav DK, Banerjee A, Meena RS (2019a) Agroforestry: a holistic approach for agricultural sustainability. In: Jhariya MK, Banerjee A, Meena RS, Yadav DK (eds) Sustainable agriculture, forest and environmental management. Springer Nature Singapore Pte Ltd, Singapore, pp 101–131. https://doi.org/10.1007/978-981-13-6830-1
Raj A, Jhariya MK, Banerjee A, Yadav DK, Meena RS (2019b) Soil for sustainable environment and ecosystems management. In: Jhariya MK, Banerjee A, Meena RS, Yadav DK (eds) Sustainable agriculture, forest and environmental management. Springer Nature Singapore Pte Ltd, Singapore, pp 189–221. https://doi.org/10.1007/978-981-13-6830-1
Raj A, Jhariya MK, Yadav DK, Banerjee A (2020) Climate change and agroforestry systems: adaptation and mitigation strategies. Apple Academic Press Inc., CRC Press – a Taylor and Francis Group, US & Canada, p 383. https://doi.org/10.1201/9780429286759
Rausher M (2008) Evolutionary transitions in floral color. Int J Plant Sci 169(10):1086/523358
Rebekka S, Totland R, Láza Ro A (2016) Experimental simulation of pollinator decline causes community – wide reductions in seedling diversity and abundance. Ecology 97(6):1420–1430
Reitz SR (2009) Biology and ecology of the Western flower thrips (Thysanoptera: Thripidae): the making of a pest. Florida Entomologist 92:7–13
Rhodes J (2002) Cotton pollination by honeybees. Aust J Exp Agric 42:513–518. https://doi.org/10.1071/EA01063
Richards KW (1993) Non-Apis bees as crop pollinators. Rev Suisse Zool 100:807–822
Ricketts TH, Regetz J, Steffan Dewenter I, Cunningham SA, Kremen C, Bogdanski A, Gemmil-Herren B, Greenleaf SS, Klein AM, Mayfield MM, Morandin LA, Ochieng A, Viana BF (2008) Landscape effects on crop pollination services: are there general patterns? Ecol Lett 11:499–515
Riffell JA, Alarcón R, Abrell L, Davidowitz G, Bronstein JL, Hildebrand JG (2008) Behavioral consequences of innate preferences and olfactory learning in hawkmoth-flower interactions. PNAS 105:3404–3409
Rogers J, Charlotte S (2012) Here is my honey-machine: Sylvia Plath and the mereology of the Beehive. Rev of English Stud 63:293–310

Rosegrant MW, Paisner MS, Meijer S, Witcover J (2001) 2020 global food outlook: trends, alternatives, and choices. A 2020 vision for Food, Agriculture and Environment Initiative. International Food Policy Research Institute, Washington, DC

Rotheray GE, Gilbert F (2011) The natural history of hoverflies. Forrest Text, Tresaith

Roubik DW (1995) Pollination of cultivated plants in the tropics. Food Agric. Org. U.N, Rome

Roubik DW (1996) African honeybees as exotic pollinators in French Guiana. In: Matheson A, Buchmann SL, O'Toole C, Westrich P, Williams IH (eds) The conservation of bees. Academic Press, London, pp 173–182

Roubik D (2018) The Pollination of Cultivated Plants. A Compendium for Practitioners. Volume 2.

Roulston TH, Cane JH (2002) The effect of pollen protein concentration on body size in the sweat bee, *Lasioglossum zephyrum* (Hymenoptera: Apiformes). Evol Ecol 16:49–65. https://doi.org/10.1023/A:1016048526475

Roulston TH, Cane JH, Buchmann SL (2008) What governs protein content of pollen: pollinator preferences, pollen-pistil interactions, or phylogeny? Ecol Monographs 70:617–643

Russo L, Danforth B (2017) Pollen preferences among the bee species visiting apple (*Malus pumila*) in New York. Apidologie 48:806–820

Salzmann CC, Nardella AM, Cozzolino S, Schiestl FP (2007) Variability in floral scent in rewarding and deceptive orchids: the signature of pollinator-imposed selection? Ann Bot 100:757–765

Sardiñas HS, Ponisio LC, Kremen C (2016) Hedgerow presence does not enhance indicators of nest-site habitat quality or nesting rates of ground-nesting bees. Restor Ecol 24:499–505

Schmehl DR, Peter EA, Teal JL, Frazier CM, Grozinger S (2015) Genomic analysis of the interaction between pesticide exposure and nutrition in honey bees (*Apis mellifera*). J Insect Physiol 71:177–190

Schmider F, Escaille T (2013) Pollinators and agriculture: agricultural productivity and pollinator protection. In: Brochure – European Land Owners Organisation (ELO) and European Crop Protection Association (ECP). https://www.ecpa.eu/sites/default/files/Pollinators%20brochure_BàT2.pdf

Schmidt J, Johnson BE (1984) Pollen feeding preference of *Apis mellifera*, a polylectic bee. Southwest Entomol 9:41–47

Schoeneberger MM, Bentrup G, Patel-Weynand T (2017) Agroforestry: enhancing resiliency in U.S. agricultural landscapes under changing conditions. U.S. Department of Agriculture, U.S. Forest Service, Washington, DC

Sessions LA (2000) A floral twist of fate. Nat Hist 109(7):38

Shore J, Spencer B (2011) The effect of pollination intensity and incompatible pollen on seed set in *Turnera ulmifolia* (Turneraceae). Can J Bot 62:1298–1303. https://doi.org/10.1139/b84-175

Simone-Finstrom M, Spivak M (2010) Propolis and bee health: the natural history and significance of resin use by honey bees. Apidologie 41:295–311

Simone-Finstrom M, Borba RS, Wilson M, Spivak M (2017) Propolis counteracts some threats to honey bee health. Insects 8:46

Singh NR, Jhariya MK (2016) Agroforestry and agrihorticulture for higher income and resource conservation. In: Narain S, Rawat SK (eds) Innovative technology for sustainable agriculture development. Biotech Books, New Delhi, India, pp 125–145

Skogsmyr I, Lankinen A (2002) Sexual selection: an evolutionary force in plants. Biol Rev 77 (4):537–562. https://doi.org/10.1017/S1464793102005973

Smith GR, Snow GE (1976) Pollination ecology of *Platanthera* (*Hebenaria*) *ciliaris* and *Platanthera blepharoglottis* (Orchidaceae). Bot Gaz 137:133–140

Soares-Filho B, Rajão R, Macedo M, Carneiro A, Costa W, Coe M, Rodrigues H, Alencar A (2014) Cracking Brazil's forest code. Science 344(6182):363–364

Spivak M, Mader E, Vaughan M, Euliss NH (2011) The plight of the bees. Environ Sci Tech 45:34–38

Steffan-Dewenter I (2003) Importance of habitat area and landscape context for species richness of bees and wasps in fragmented orchard meadows. Conserv Biol 17:1036–1044

Steinhauer NA, Rennich K (2014) A national survey of managed honey bee 2012–2013 annual colony losses in the USA: results from the Bee Informed Partnership. http://hdl.handle.net/1957/49861

Stokstad E (2007) The case of the empty hives. Science 316:970–972. https://doi.org/10.1126/science.316.5827.970

Stubbs CS, Jacobson HA, Osgood EA, Drummond FA (1992) Alternative forage plants for native (wild) bees associated with lowbush blueberry, Vaccinium spp. Maine Agricultural & Forest Experiment Station, Orono, ME, p 57

Subba R, Meera B (1984) Butterflies and pollination biology. Proc Ind Acad Sci (Anim Sci) 93 (4):391–396

Svensson B, Lagerlöf J, Svensson BG (2000) Habitat preferences of nest-seeking bumble bees (Hymenoptera: Apidae) in an agricultural landscape. Agric Ecosyst Environ 77:247–255

Tasei JN, Aupinel P (2008) Nutritive value of 15 single pollens and pollen mixes tested on larvae produced by bumblebee workers (*Bombus terrestris*, Hymenoptera: Apidae). Apidologie 39:397–409

Tester M, Langridge P (2010) Breeding technologies to increase crop production in a changing world. Science 327:818–822

Thompson DM (2006) Detecting the effects of introduced species: a case study of competition between *Apis* and *Bombus*. Oikos 114:407–418

Tilman D, Cassman KG, Matson R, Naylor L, Polasky S (2002) Agricultural sustainability and intensive production practices. Nature 418:671–677

Troje N (1993) Spectral categories in the learning behaviour of blowflies. Zeitschrift für Naturforschung 48:96–104

Tsukada MSS, Tsukada Y (1986) Oldest primitive agriculture and vegetational environments in Japan. Nature 322:632–634

Udawatta RP, Rankoth LM, Jose S (2019) Agroforestry and biodiversity. Sustainability 11 (10):2879. https://doi.org/10.3390/su11102879

Vanderplanck M, Moerman R, Rasmont P, Lognay G, Wathelet B, Wattiez R, Michez D (2014) How does pollen chemistry impact development and feeding behaviour of polylectic bees? PLoS One 9:e86209. https://doi.org/10.1371/journal.pone.0086209

Varah A, Jones H, Smith J, Potts SG (2013) Enhanced biodiversity and pollination in UK agroforestry systems. J Sci Food Agric 93:2073–2075

Vasanthakumar S, Aruna R, Srinivasan MR (2018) Pollination efficiency of Indian Honey bees, *Apis cerana indica* (Fabricius) in mango orchard. In: Conference proceedings: international conference on biocontrol and sustainable insect pest management (ICBS 2018), Killikulam, Tamil Nadu, India, pp 579–582

Vaudo AD, Tooker JF, Grozinger CM, Patch HM (2015) Bee nutrition and floral resource restoration. Curr Opin Insect Sci 10:133–141

Vorobyev M, Gumbert A, Kunze J, Giurfa M, Menzel R (1997) Flowers through insect eyes. Isr J F Sci 45:93–101

Wang B, Ma JYM, Kenna D, Yan EV, Zhang HC, Jarzembowski EA (2013) The earliest known longhorn beetle (Cerambycidae: Prioninae) and implications for the early evolution of Chrysomeloidea. J Syst Palaeontol 12(5):565–574. https://doi.org/10.1080/14772019.2013.806602

Wcislo WT, Cane JH (1996) Floral resource utilization by solitary bees (Hymenoptera: Apoidea) and exploitation of their stored foods by natural enemies. Annu Rev Entomol 41:257–286

Weberling F (1989) Morphology of flowers and inflorescences. Cambridge University Press, Cambridge

Wester P, Johnson SD (2017) Importance of birds versus insects as pollinators of the African shrub *Syncolostemon densiflorus* (Lamiaceae). Bot J Linnean Soc 185(2):225–239. https://doi.org/10.1093/botlinnean/box054

Whittall JB, Hodges SA (2007) Pollinator shifts drive increasingly long nectar spurs in columbine flowers. Nature 47(7):706–709

Wilcock C, Neiland R (2002) Pollination failure in plants: why it happens and when it matters. Trends Plant Sci 7:270–277

Will MF (1983) Plant reproduction ecology. Wiley-Interscience, New York, p 218

Wille A, Orozco E, Raabe C (1983) Polinizaci´on del chayote Sechium edule (Jacq.) Swartz en Costa Rica. Rev Biol Trop 31:145–154

William PN, Anderson L (1974) Insect pollinators frequenting strawberry blossoms and the effect of honey bees on yield and fruit quality. J Am Soc Hort Sci 99(1):40–44

Williamsm NM, Kremen C (2007) Resource distributions among habitats determine solitary bee offspring production in a mosaic landscape. Ecol Appl 17:910–921

Willmer P (2011) Pollination and floral ecology. Princeton University Press, New Jersey

Wilson MF, Price PW (1983) The evolution of inflorescence size in *Asclepias* (Asclepiadaceae). Evolution:495–511

Winfree R, Gross BJ, Kremen C (2011) Valuing pollination services to agriculture. Ecol Econ 71:80–88

Wolf H, Wehner R (2000) Pinpointing food sources: olfactory and anemotactic orientation in desert ants, Cataglyphis fortis. J Exp Biol 203:857–868

Wood TJ, Holland JM, Goulson D (2017) Providing foraging resources for solitary bees on farmland: current schemes for pollinators benefit a limited suite of species. J Appl Ecol 54:323–333

Wood TJ, Kaplan I, Szendrei Z (2018) Wild bee pollen diets reveal patterns of seasonal foraging resources for honey bees. Front 6:210

Wratten SD, White AJ, Bowie MH, Berry NA, Weigmann U (1995) Phenology and ecology of hoverflies (Diptera: Syrphidae) in New Zealand. Environ Entomol 24:565–600

Wright GA, Lutmerding A, Dudareva N, Smith BH (2005) Intensity and the ratios of compounds in the scent of snapdragon flowers affect scent discrimination by honeybees. J Comparative Physiol A 191:105–114

Xie H, Huang Y, Chen Q, Zhang Y, Qing W (2019) Review prospects for agricultural sustainable intensification: a review of research. Landarzt 8:157. https://doi.org/10.3390/land8110157

Yoder JA, Jajack AJ, Rosselot AE, Smith TJ, Yerke MC, Sammataro D (2013) Fungicide contamination reduces beneficial fungi in bee bread based on an area-wide field study in honey bee, *Apis mellifera*, colonies. J Toxicol Environ Health Part A 76:587–600

Zayed A, Packer L (2005) Complementary sex determination substantially increases extinction proneness of haplodiploid populations. Proc Natl Acad Sci U S A 102:10742–10746

Ecosystem Services of Himalayan Alder 12

Zahoor ul Haq, Shujaul Mulk Khan, Sayed Afzal Shah, and Abdullah

Abstract

Alnus nitida (Himalayan Alder) is a monophyletic species of family Betulaceae. It is distributed in the mountainous ranges of Hindu Kush and western Himalayan of the Sino-Japanese belt. Family Betulaceae has a cosmopolitan distribution with 2 genera (*Betula* and *Alnus*) and 95 species. Previously, the family was divided into Betulaceae (*Alnus*, *Betula*) and Corylaceae (*Carpinus*, *Corylus*, *Ostrya,* and *Ostryopsis*). Various species of the family are used for different purposes such as timber, lumber, and household utensils production across the globe. Many species have been studied against for their therapeutic potential against various ailments, i.e. obesity, cancer, tuberculosis. Stem bark of *Betula utilis* is used for spiritual purpose as well as to treat various diseases. Himalayan Alder is among those species that offers numbers of ecosystem services. *A. nitida* being a riparian species is cultivated by farmers along the stream sides of their fields to control soil erosion. Wood is used by local people for the formation of different pots. Nodulation and biomass production are also the prominent characteristics of the family. Some studies revealed that *Alnus* acts as an indicator species of eastern North America for dry period. Ethno-ecological surveys have revealed that Alder is preferred by local communities in several ways such as fuel, fodder, and construction purposes. The species is ecofriendly to form a phyto-social association with about 146 species in its geographical range in Pakistan. The nitrogen fixation capability makes it more suitable for agroforestry and inter-cropping. Alder can be used as an alternate to discourage shift cropping and to enhance the

Z. ul Haq (✉)
Department of Plant Sciences, Quaid-i-Azam University, Islamabad, Pakistan

Department of Botany, Shaheed Benazir Bhutto University, Sheringal, Wari Campus, Sheringal, Pakistan

S. M. Khan · S. A. Shah · Abdullah
Department of Plant Sciences, Quaid-i-Azam University, Islamabad, Pakistan

M. K. Jhariya et al. (eds.), *Ecological Intensification of Natural Resources for Sustainable Agriculture*, https://doi.org/10.1007/978-981-33-4203-3_12

ecofriendly and high yield techniques of inter-cropping. Alder has prominent role in the absorption of heavy metals and boost up water quality. Since few years population of this important species is continuously decreasing and facing problem of extinction due to drought, developmental projects, deforestation, and other anthropogenic activities. This chapter provides a baseline for further comprehensive studies on its molecular genetics, phyto-chemistry, and conservation priorities for this and many other associated species.

Keywords

Himalayan Alder · Morphological diversity · Ecosystem services · Conservation · Threats

Abbreviations

C	Carbon
CCA	Canonical correlation analysis
Cm	Centimeter
CO_2	Carbon dioxide
DCA	Detrended correspondence analysis
Mm	Millimeter
N. and S. America	Northern and Southern America

12.1 Introduction

Alnus nitida commonly known as Himalayan Alder (Shrol: Hinku; Geeray: Pashto). It belongs to family Betulaceae and characterized by unique roots with nodules. Alder is a tree species of riparian habitat and grows up to 22 m or more in height. Shoots are pubescent during young stages and become glabrescent when the plant gets old. The floral characteristics show that male flowering catkin reaches size up to 20 cm in length, bracts is 1.1 mm long, tepals are oblong-obovate to spathulate, and filament is 1 mm long. Female cones are woody and range from 3–2.9 in length and 1–1.2 cm in width. Fruiting scale is 5-lobed and 5–6 mm in length. Nut is 2.3–4 mm long with less leathery wings. The flowering period is from September to October. Leaves are elliptical to ovate (5–15 × 2.9–14 cm), acuminate, and glabrous (Perveen and Qaiser 1999) (Fig. 12.1).

According to Furlow (1990), about 40 species of Alder are distributed in Bhutan, China, Bangladesh, Europe, India, Korea, Japan, Nepal, America, and Northern Hemisphere. Only 18 out of 35 species were documented from Asia (Kennedy et al. 2010); 8 N. America (Lin et al. 2015); 4 Europe, 1 South America, and 2 species from Mexico (Chen 2004). The subspecies of *Alnus* were differing from each other based on fruiting bodies and seed morphology. According to Navarro

Botanical name: *Alnus nitida*(Spach) Endl.

Local Name: Kashmiri and Hindku: Shrol; Pashto, Geeray

Synonym: *Clethropsis nitida* Spach

Family: Betulaceae

Flowering & Fruit season: September to October

Type:'in Emodo', *Jacquemont* (P)

Fig. 12.1 *Alnus nitida* in full bloom at riverside of Madyan (Swat), Pakistan

et al. (2003) the species are differentiated from each other based on seed size. In the subgenus *Alnobetula*, the seeds are borne on the twigs (Haq et al. 2020a, b). In China, there are five out of ten endemic species (Smith 1999). This family has higher rate of endemism in the eastern Asia. This region also hosts *Ostryopsis*, *Betula* (*Chinenses* and *Betulaster*): *Alnus* (*Cremastogyne, Acanthochlamys, Carpinus, Corylus,* and *Distegocarpus)* (Chen et al. 1998a, b). The Province Sichuan and its neighboring provinces are considered the hotspots for the extant species of this family. In 1873, genus *Ostryopsis* which is endemic to China was introduced. All six genera and 52 species are native to the above-mentioned region. Paleo-climatic and Paleo geographic characteristics revealed that, in the past, the Eurasian continent was divided into Late Cretaceous and Tertiary (Sun 2016).

Alnus nitida is the only single species reported from genus *Alnus* in Pakistan. In this chapter we have focused on the ethno-botany, pharmacology, photochemistry, ecology, taxonomy, and biomass and ecosystem services of family Betulaceae with special reference to Himalayan Alder.

12.2 Botanical Description and Phytosociology of Alder

The first section of this chapter highlights the detail description of family Betulaceae and *Alnus nitida* species. Along with that, it also throws light on the association of a single species of family Betulaceae with other species. The second portion contains ecosystem services of Alder plant. This chapter as a whole presents a complete package from occurrence of species, distribution, taxonomical variation, ecological linkages with associated species to ecosystem services.

Phyto-sociology is the study of vegetation in relation to its surrounding biotic and abiotic factors. This science is playing key role in quantification of ecological system services, vegetation mapping, and biodiversity conservation (Khan et al. 2013; Meena et al. 2018; Meena and Lal 2018). It is also used as a tool in the form of habitat for community and species level (Lin et al. 2015). Recently, Haq et al. (2019a, b) worked on *A. nitida* to explore vegetation structure, role of climatic and edaphic factors on the distribution, and formation of plant zonation in western Himalayas and Hindu Kush region of the Sino-Japanese belt. They recorded 146 plant species in association with *A. nitida* in the region. The reported species belong to 106 genera and 47 families. All associated plant species in different stations were classified through two-way cluster analysis into three major zones or communities, i.e., (1) *Celtis caucasia, Rubus fruticosus, Chenopodium murale,* (2) *Eucalyptus camaldulensis, Arundo donex, Mentha arvensis,* and (3) *Platanus orientalis, Saccharum munja, Oxalis corniculata* (Fig. 12.2). This association of Alder highlights is friendly nature towards other associated species in its vicinity.

12.3 Herbarium Specimen Recorded from Himalayas

Himalayan Alder is distributed in different areas of the Himalaya from Iran to Nepal. Numbers of herbaria in the region were visited to document herbarium records of Himalayan Alder. It was very interesting that specimens from the time of Indo-Pak were preserved in botanical centers of Pakistan. The herbarium visited were National herbarium of Pakistan (RAW), Quaid-i-Azam University herbarium also called Herbarium of Islamabad (ISL), University of Malakand Herbarium (BGH), Islamia college University Herbarium (ICP), Shaheed Benazir Bhutto University Herbarium. The 100-year-old specimen was recorded from Kashmir. The details are mentioned in the snapshot. Kulu (India) 1936, Rajanpur Kashmir 1917, Iran, India, China, and Nepal host more than one species but on the same belt Pakistan host only a single species of *A. nitida* distributed in area of 650 kilometer along the western Himalayas.

It is an enigma that the species has great variation in size of female and male catkin from zone to zone in Pakistan (Haq et al. 2020a, b). They investigated the reason to find whether it is due to variation in edaphic, environmental,

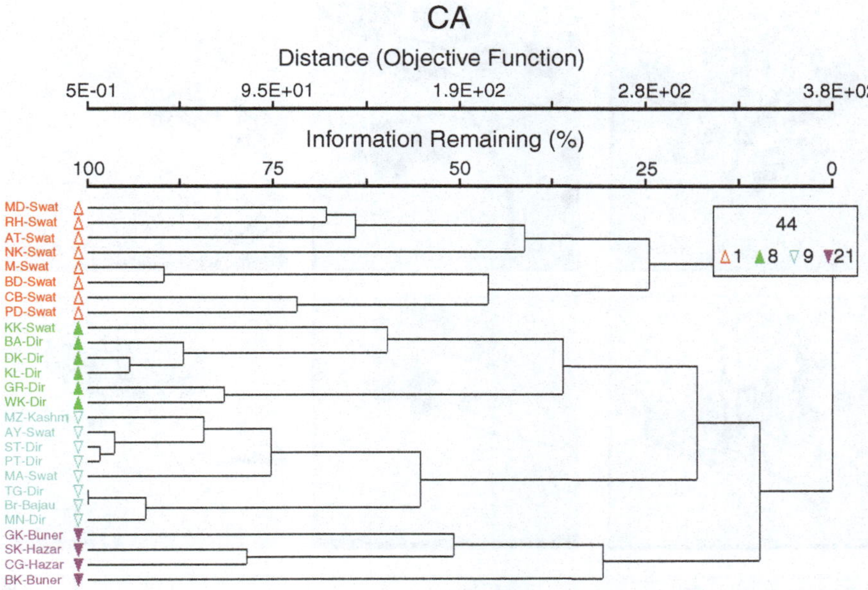

Fig. 12.2 Cluster dendrogram of 134 Quadrats based on Sorenson measures

micronutrients, and heavy metals. Their findings compile report of the only species of *A. nitida* along the western Himalayas in Pakistan (Fig. 12.3).

12.4 Taxonomy of Alder

12.4.1 Background of Family Betulaceae

The family Betulaceae consists of deciduous and few semi-deciduous shrubs and trees. Leaves are sessile having two or three scales. Leaves are dentate, serrate, rarely entire, or incised. Male inflorescence is cylindrical, elongate, and pediculate with overlapping bracts, having three to four bracteoles, sepals are 4 lobed, stamens one to four. The female inflorescence presents one or two in racemose or panicle, ellipsoid, or ovoid. Bracts are woody with numerous apexes 5-lobed. Nutlets on each bract are two in number. Flowering season is spring with some exceptions (Chen 2004). Family Betulaceae containing two genera and 95 species distributed in temperate and arctic zones of Asia, Europe, N. and S. America. This species also includes *Corylus* (hazel), *Betula* (birch), and *Carpinus* (hornbeam) (Li et al. 2004). Betulaceae is represented in Pakistan by both *Alnus* and *Betula* (Perveen and Qaiser 1999). Betulaceae is placed under the order Fagales (Perveen and Qaiser 1999). Geographically, Takhtajan (1980) classified it as a species of Holarctic Kingdom (the largest of all floristic kingdoms occupies more than half of the world terrestrial

Fig. 12.3 Photographs of herbarium specimen collected from Himalayas, i.e., Poonch (Kashmir) 1952; Muzzafarabad 1962; Kulu (India) 1936; Rajanpur Kashmir 1917; Hazara 1970; Buner 1998; Swat 1963; Dir 1998; Kohistan 1999; Buner 2001; Swat 2003; Swat 2007, respectively

kingdoms) and in separate monotypic order Betulales. Few species inhabit in subtropical regions, e.g., *Betula platyphylla. A. glutinosa* (L.) only occurs in Africa while *A. accuminata* in Argentina and America. The Betulaceae sensu lato was included in six extant genera. Authors divided the family Betulaceae into two tribes

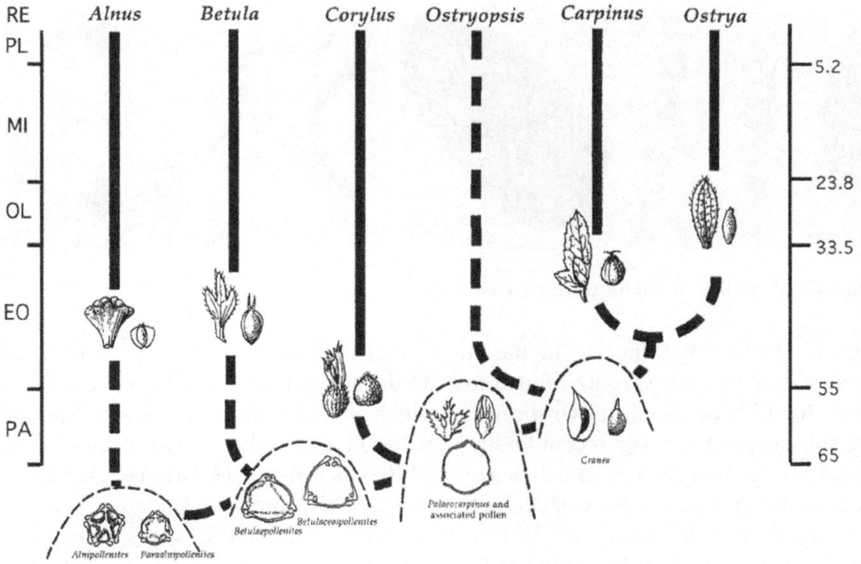

Fig. 12.4 Phylogenetic relationships within Betulaceae (Chen et al. 1998a, b)

or sub-families (Takhtajan 1980). In 1967, Hutchison ranked it in family—Betulaceae sensu stric to (*Alnus* & *Betula*) and Corylaceae (*Carpinus*, *Ostrya*, *Corylus,* and *Ostryopsis*). However, Abbe (1935) declares it as monophyletic family on the basis of morphology (Metcalfe and Chalk 1950), growth habit (Kikuzawa 1982), wood anatomy (Hall 1952), embryology, and S-type sieve tube (Behkne 1973). Fig. 12.4 shows the phylogeny and systematic evolution of different genus of family Betulaceae (Lachman et al. 2018).

12.4.2 Himalayan Alder Taxonomic Variations

Betula and *Alnus* are separated from each other by unique characters of nodules (Chen 2004). The reported species of *Alnus* in the world are up to 35, five in Europe, 18–23 in Asia, and nine in new world (Furlow 1990). The close geographic distribution and closely related subspecies of *Alnus* have led to confusion in taxonomical classification. Majority of *Alnus* species has been divided on the basis of palynology and morphology (Chen 2004). Few species have been reclassified based on DNA sequence (Navarro et al. 2003) and few other scientists recognized subgenus Gymnothyrsus and Alnobetula (Yu et al. 2007) declared the *Alnus* genus as monophyletic group. Chen (2004) introduced three subgenera, i.e., *Clethropsis*, *Alnobetula,* and *Alnus*. This classification trace back to species level. *Alnus viridis* complex and *Alnus incana* complex are further divided by Li et al. (2004) and they arranged three subgenera of *Alnus*: Alnobetula, Clethropsis, and *Alnus*. According to

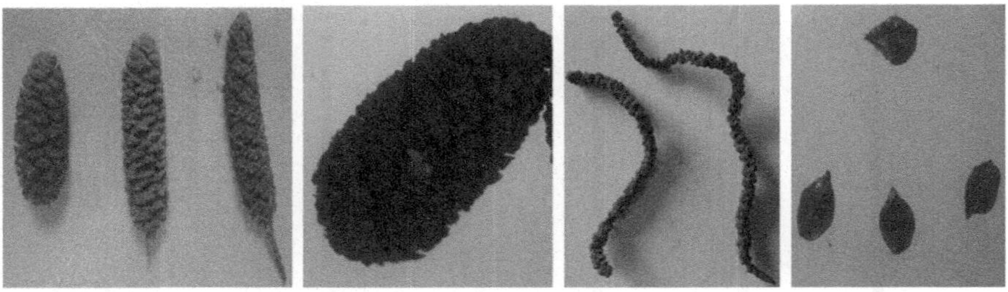

Fig. 12.5 Variation in catkins of Himalayan Alder

Furlow (1990), two species in the *Alnobetula* subgenus are shrubs. The three subspecies of *Clethropsis* are limited to Asia, Japan, and the USA. Previous reports show that flora of Iran host two species of which two are considered as varieties and one subspecies. However, recent findings increased the number to five species. After studying cone morphology and 28 leaves of 140 Alder plants, 11 different characters were recorded new, i.e., blade shape at the base, velocity intensity, leaf angle in apex, serration type, cone shape, and leaf hairs on both surfaces. The new species were *Alnus subcordata* varieties allocated to *villosa* and vice versa (Fig. 12.5). The morphological characters of the two newly introduced species consist of *Alnus dolichocarpa* and *Alnus djavanshirii* were most similar to that of *Alnus subcordata var. subcordata.* On the other hand, three subspecies of *Alnus glutinosa* were differentiated by similar character. To confirm the new taxa DNA bar-coding techniques were used to confirm the new taxa in Iran (Colagar et al. 2016). In Himalayan region the variation among catkin and leaf size of Alder still indicating chances of new species discovery.

12.4.3 Palynological Aspects

Palynology is considered the only genuine technique to identify fossil, as no other proper discovery happened yet (Blackmore et al. 2003). Various scientists had studied different aspects of Alder, but ample work is needed to be done on the genus *Alnus* (Lacourse 2007). *Alnus viridis* type and *Alnus rubra* type were classified based on pollen study (Arsenault et al. 2007). These morphotypes were based on European and American Alder pollens (Mayle et al. 1993). This pollen morphology provides base for differentiation of species or subspecies. The *A. incana* subsp. *tenuifolia* is classified into subspecies based on pollen morphological variations. The same thing happened for the pollen of two shrubs Alders (*A. incana* subsp. *tenuifolia* and *A. viridis* subsp. *Sinuate*) where they were also classified on the base of pollen study (Heusser 1969). Three species of *Alnus*; *A. djavanshirii* Zare, *A. orientalis* Decne, and *A. dolichocarpa.* Zare and Amini (2012) dissected into two subspecies *A. glutinosa* Gaerth subsp. *Antitaurica* Yaltrik and *A. glutinosa* Gaerth subsp. *Glutinosa* based on the supporting characters of palynology.

12.5 Ecosystem Services of Alder

Sustainability in forest ecosystem needs diversity of both flora and fauna. Biodiversity conservation and its ecosystem services studies have gain prominent importance and keen interest issues in last few decades (Raj et al. 2018; Jhariya et al. 2019a, b; Khan et al. 2020a, b). The diversity maintains and sustains greater nutrients supply, ecosystem stability, and plant communities (Raj et al. 2020; Banerjee et al. 2020). Diversity is a phenomenon because it can be linked to maturity, stability, evolutionary time, maturity, and predation pressure. It is also an important key to sustain thick forest, its niches, and habitats (Van Wieren 1996) (Fig. 12.6).

12.5.1 Ethno Ecological Importance

Different parts of *Alnus* are used by traditional communities for medicinal purposes for centuries (Sun 2016). Bark of *Betula utilis* is used for spiritual purpose as well as to treat different diseases (Fig. 12.7). *A. nitida* is used for the formation of different

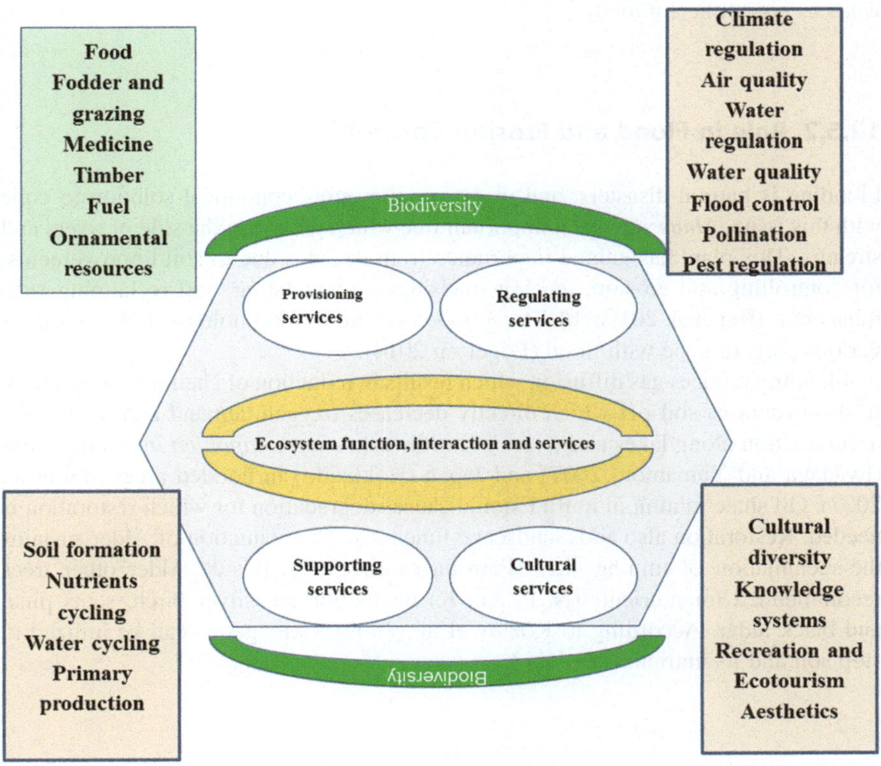

Fig. 12.6 Ecosystem services by plant species (Khan et al. 2013)

Fig. 12.7 Author interviewing medicinal plant experts (Mazharul Haq, Fazal-i-Subhan and Sirajul Haq) regarding ethno botanical importance of Himalayan Alder

handicrafts, pots, dolls, and furniture (Austin 2004). According to local inhabitants of the western Himalaya, i.e., Swat, Dir, Bajaur, and Buner in Pakistan, animal skins are salted and spread over the branches of *Alnus* for better dryness. Alder product particleboards fulfill the EN standard due to its elasticity and static binding (Nemli et al. 2003). The shoot bark is utilized for the treatment of diabetes mellitus by local *vaids* as economical remedy.

12.5.2 Role in Flood and Erosion Control

Flooding is natural disasters, and plants are the only economical solution to cope with this issue. *Alnus* spp. is an important tree which grows on the side of rivers and streams. This plant has gained importance from decades due to soil improvements, for controlling soil erosion, fencing roadsides, re-vegetating, and reclaiming strip mine areas (Raj et al. 2019a, b). The *A. japonica* shows morphological changes in its various parts to cope with flood (Edgerton 2014).

Flooding reduces gas diffusion which results in reduction of chemical production. It also decreases soil pH which directly decreases oxygen demand and phytotoxic accumulation along lakes and rivers of northwest Asia, *A. japonica* in swamp areas (Iwanaga and Yamamoto 2007) and Japan (Hokkaido) in flooded areas (Yu et al. 2007). Oil shale mining in north Estonia causes degradation for which restoration is needed. Restoration also adds landscape functions. The plantation of Alder sustains the reclamation of mining sites (Parrotta et al. 1997). Beside Alder other trees recommended for nutrients (N, P, and K) fluctuation are silver birch, scots pine, and black alder. According to Kozlov et al. (2009) *Alnus* genus can be utilized to stop soil and its nutrients erosion.

12.5.3 Alder and its Phytochemicals

Genus *Alnus* provides numerous services in the form of chemicals such as yashabushiketol and dihydroyashabushiketol, flavonoids, triterpenes and methanols, etc. Alder is used to treat tuberculosis. *A. incana* buds' decoction is used for lungs pain (Wopara et al. 2020) means that it has some special types of chemical compounds with therapeutic potential against T.B. A precise summary of published information regarding phytochemicals present in *A. nitida* is given in Table 12.1.

12.5.4 Role in Heavy Metals Accumulation

The term heavy metal is used (chemical point of view) for transition metals with gravity above 5 and atomic mass over 20. From biological point of view, the metalloids and metals that are or can be harmful both for animals and plants at low concentration. Mainly the metals are divided into essential and non-essential types. Non-essential metals, i.e., Pb, Hg, Se, As, and Cd, do not play significant physiological functions in most of the cases. Essential elements are required for metabolism, i.e., Cu, Ni, Zn, Mo, Co, and Fe (Rascio and Navari-Izzo 2011; Meena et al. 2020b, c). Environmental pollution due to disturbance of biogeochemical cycles and rapid industrialization creates an alarming situation in the form of heavy metal pollution in the environment (Khan et al. 2013). Both, anthropogenic and natural activities add heavy metals into the surrounding environment.

Natural activities are soil erosion, weathering, and volcanoes. The anthropogenic activities are mining, smelting, use of pesticides. Heavy metal accumulation potential of Himalayan Alder was assessed in different countries. As compare to *Salix* and *Acer. Alnus glutinosa* the plant species grow vigorously in highly alkaline anthropogenic sediment (Sarma 2017). The *Alnus nitida* heavy metal accumulation potential was studied first time from Pakistan by Haq et al. (2020a, b). They summarized that the plant species has great capability of bioaccumulation for heavy metal in hilly areas. *Alnus nitida* being a riparian species play important role in water filtration and quality enhancement and make it drinkable. Unfortunately, in recent decade the Himalayan Alder population is eroding and facing conservation problems. Conservation in its natural habitat through afforestation and controlling deforestation is essential for future generation, otherwise the recent speed of deforestation will be led it to extinction from western Himalayan regions of Pakistan (Haq et al. 2020a, b) (Fig. 12.8).

12.5.5 Role in Nitrogen Fixation

Alnus nitida being a riparian species is characterized by well-developed nodulation system. The nodules host nitrogen fixing bacteria. Nitrogen fixing bacteria fix atomic nitrogen into molecular nitrogen and boost up soil fertility. Soil fertility not only fulfills nutritional requirements of *Alnus nitida* but also of other associated species

Table 12.1 Alder species usage, chemical compounds, and methodology

S. no.	Plant name	Part use	Chemical name	Usage	Methodology	Reference
1	*Alnus incana* (L.) Moench	Cones/ bark	Triterpenoids	Scrofula caused by tuberculosis	Poultice usage on diabetes person effected area	Kaul (2011)
		Bark	Triterpenes	Tuberculosis (TB)	Betulin: anti-mycobacterial constituents in the bark of *A. incana* the functionality	Kaul (2011)
2	*Alnus nepalensis* (D. Don)	Leaves	Diarylheptanoids	Antifilarial activity	Lymphatic filariasis	Yadav et al. (2013)
		Leaf, roots, and bark	Taraxeryl acetate	Decoction used for wounds and cuts as a hemostatic	In dysentery, stomachache, and diarrhea	Pande et al. (2006)
3	*Alnus sieboldiana* (Matsum.) Murai	Flower	Dihydroyashabushiketol			Joo et al. (2009)
		Flowers and young shoots	Tannins and flavonoids	Used in infection, injury	Tumor necrosis	Barbara et al. (1994)
4	*Alnus glutinosa* (L.) Gaertn.	Bark	Methanol	Oregonin	Antimicrobial activity	Amin et al. (2016)
		Bark	Triterpenoids, diarylheptanoids, and flavonoids	Antioxidative activities and anti-inflammatory	Fever, diarrhea, and alcoholism	Kikuzaki et al. (1991)
5	*Alnus japonica* (Thunb.) Steud.	Bark	Diarylheptanoid	Atopic dermatitis	Fever, diarrhea, hemorrhage, and alcoholism	Joo et al. (2009)
		Leave	1,7-diarylheptanoids, yashabushiketol		Leaves have potent cytotoxic activities against murine human SNU-C1 and B16 melanoma cells gastric cancer cells	Choi et al. (2008)
		Bark	Diarylheptanoids	Anticancer	Mouse and human cancer cell lines	Choi et al. (2008)
		Bark	Diarylheptanoids	Obesity	The inhibitory activities of report of melanogenesis and cyclooxygenaxe-2 expression	Hussain and Kline (2004)

6	*Alnus hirsuta* Turcz. f.	Bark	Diarylheptanoid xyloside, oregonin		Antibiotic	Li (2015)
7	*Alnus rubra* Bong.	Leaves	Flavonoids and triterpenoids		To control HIV-1 virus replication and related enzymes	Yu et al. (2007)
8	*Alnus firma* Siebold & Zucc.	Young shoots			Two novel hydroxyketones, i.e., yashabushiketol and dihydroyashabushiket	Yu et al. (2007)
		Bark		Pharmacological use	Stop LPS-induced NF-*k*B activation and NO and TNF-a production	Jin et al. (2007)
9	*Alnus hirsute* Kunth	Male flower	Diarylheptanoids oregonin		Control 12-O-tetradecanoylphorbol-1, 3-acetate (TPa)-cyclo-oxygenase- 2 (COX-2) expression in immortalized human chest epithelial MCFIOA cells	Lee et al. (2000)
10	*Alnus sieboldiana* Kunth	Male flower	Diarylheptanoids oregonin		Showed significant inhibitory effect on 12-O-tetradecanoylphorbol-1, 3-acetate (TPa)-cyclo-oxygenase- 2 (COX-2) expression in immortalized human chest epithelial MCFIOA cells	Lee et al. (2000)

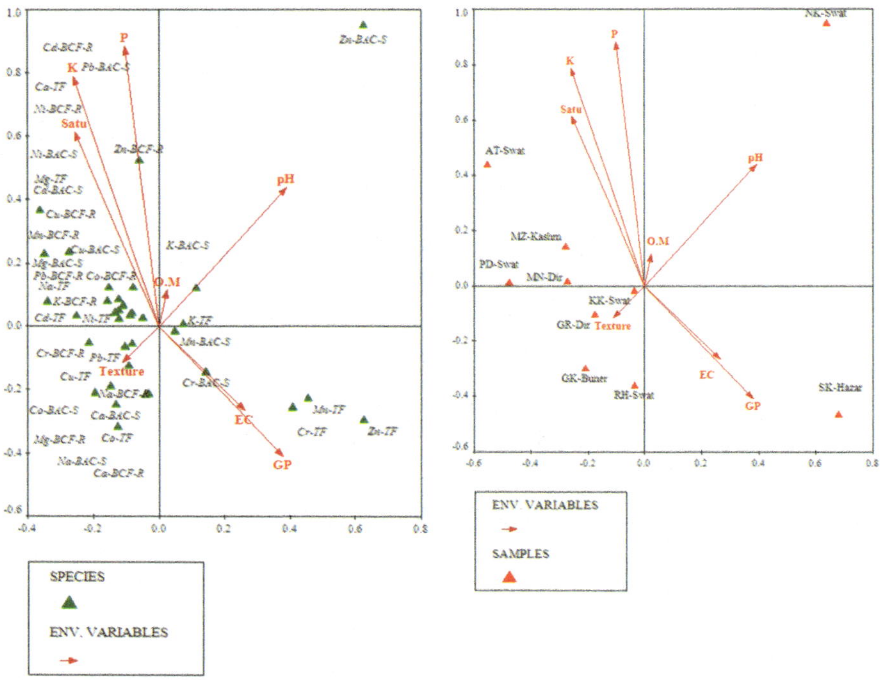

Fig. 12.8 Heavy metals accumulation in different parts of *Alnus nitida* in the Sino-Japanese belt

(Romero et al. 2004). There are two ways adopted by plants to fix nitrogen. Few plants are bestowed with special nodules such as legumes while some form mycorrhizal association (Table 12.2). *Alnus* is also among those species which fix atomic nitrogen into molecular form. Different species were considered and researched in this regard. *A. crispa* root nodules have the capability of nitrogen fixation. It utilizes insoluble metals through inoculated mycorrhiza (Becerra et al. 2005). *A. nepalensis* forms an association with *Frankia* spp. to form nitrogen fixing actinorhizal nodules (Chaia et al. 2010). On the other hand, PGPR are bacteria in soil that help in root colonization and stimulate growth. They adopt different mechanism like release of plant hormones (gibberellin, auxin, cytokinin, or ethylene) that directly stimulate growth (Kloepper et al. 2004). Indirect mechanisms involve production of metabolites that affect other factors in rhizosphere, resulting to enhanced growth of plant. The best known mechanisms in this group are inhibition of deleterious rhizobacteria and plant pathogens, and the release of either siderophores and/or antibiotics or lytic enzymes or HCN. *Alnus* species can fix nitrogen and provide utmost benefit to disturbed soil (Chen 1994). Black alder forms an association with bacteria and fungi for nitrogen fixation. *Alnus* with fungal associated root helps to make the uptake of nutrients more feasible. This association helps the plant to survive in fragile environmental conditions (Berliner and Torrey 1989). The mycorrhizal association is advantageous for plant growth in nutrient poor soil (Smith

Table 12.2 Alder species and its relationship with nodulated species

S. no.	Plant name	Nodulated species	Locality	Growing area	Reference
1	*Alnus incana* ssp. *Rugosa*	*Frankia*	New York Adirondack Mountains	Wetlands low accumulated area of NO3 and NH4	Kiernan et al. (2003)
		Frankia	Southern and eastern part of France	Poor calcareous soils	Normand et al. (2018)
2	*Alnus rubra* Bong.	*Frankia*	Washington	Maintain calcium and pH level.	Hilger et al. (2000)
		Frankia	Greenhouse	Cadmium effect, the nitrogen fixation	Lee et al. (2000)
		Frankia	USA	Forest floor	Neilson and Doudoroff (1973)
3	*Alnus glutinous* Kunth	*Frankia*	Netherlands	Waterlogged soils	Wolters et al. (1997)
4	*Alnus nepalensis* D. Don	*Frankia*	Eastern Himalayas	Different seasons study, i.e., rainy, summer, winter	Sharma et al. (1995)
		Streptomyces alni	China	ISP media 2, 3, 4, 5 and 7 and yeast-starch medium (DSMZ medium 1027). The type strain, D65T	Liu et al. (2009)
		Frankia	Kalimpong forest division of the Eastern Himalayas	Temperate forests	Sharma et al. (1995)
5	*Alnus tenuifolia* Nutt.	C2H2 assay	Lake Tahoe basin of California and Nevada	In vitro	Fleschner et al. (1976)
6	*Alnus acuminata* Kunth	*Frankia* and mycorrhizal fungi	Andes	Terrestrial processes	Becerra et al. (2009)

1999). Fleschner et al. (1976) worked over the springs where primary production in castle Lake of California was supported by *Alnus tenuifolia* growing on the shore.

12.5.6 Role as Biomass Producer

Biomass is alternate source of energy or other form of power. High and fast-growing plant species are the organic sources of energy. Alder tree is also among such species. *Alnus* is one of those trees having productive life cycle with maximum biomass and other uses. Alder tree is the best woody plant to rotate its life cycle. Alder produces 1.9–2.2 times more wood than Aspen and Birch at the age of 15–20 years (Daugavietis et al. 2009). Alder is mainly harvested for pulp, firewood, and timber. Alder species can be used for management in agroforestry, watershed protection, land reclamation, and erosion control (Becerra et al. 2005; Jhariya et al. 2018a, b). Due to vigorous growth Alder is usually considered as plant of top priority for cultivation and maximum biomass producer in different forests zones across the globe. Factors affecting site conditions for growth of black alder will result in progressive effects on wood production and forest growth. However, on contrary, increasing disturbance risks and drought will cause severities, which should be acknowledged in forest management practice (Socha and Ochał 2017). *Alnus* species are landscape plants, for example, *A. glutinosa* shows maximum stress of soil moisture. It was noted that in early Pleistocene era and Late Pliocene era the Nepalese alder more spread to eastern side as compare to the western side of the west. The *A. nitida* in Kashmir was migrated in latter stages. In arid mountains, the alder dominated the riparian habitat for landscape, it is valued due to its soil stabilizing capability, tree structure, shade in cool streams, host to bacteria and fungi and improve fish habitat. Except all these, it is a great source of fodder (Li 2015).

The above qualities favor the plant to survive better than any other angiosperm. That's why, Grey Alder at the age of 16-year yield 2.8 times more wood than Birch (Daugavietis et al. 2009). Proper management of Grey Alder can provide more biomass as compare to aspen, birch, and black alder. The same experiment was carried out by Saiz et al. (2006). The *A. viridis* observed with estimated production of 6.18 t/ha/year, of which 61.5% was leaf mass, 21.6% to stem, and 16.8% is branch-wood mass. The mass production of *A. viridis* is high as compare to scrub woodland due to its unique colonal growth quality (Wiedmer and Senn-Irlet 2006). The *A. glutinosa* at the age of 14-year may give 20–60 mg/ha as compare to other local plants (Bohanek and Groninger 2004). *A. incana* biomass at the age of 12 years was recorded up to 68.8 t ha^{-1} (Saiz et al. 2006) (Fig. 12.9).

12.5.7 Alder as a Source of Energy/Fuel Wood/Firewood

Woods are sustainable and vital sources of energy in the rural as well as urban areas of developing countries. The Agency for International Development (AID) has specially designed an international forum to aware the people regarding the

Fig. 12.9 *Alnus nitida* (1) Seedling of *Alnus* (2) Female catkin (3) Stump (4) cut down branches (5) Male catkins (6) Hide for drying

importance of plantation, fuel wood, and forestry. Fuel wood plantations are not only source of energy but also bring employment and environmental protection resources an area. Oil products, resin, dyes, paper, and green manure are the secondary benefits of plantation. Nearly 80% of Indian rural people depend on fuel wood for getting energy (Kataki and Konwer 2002).

Alder is used by local communities in Himalayan region for fuel purposes. They are unaware regarding the importance of it for medicinal uses and other services of the species, that is why they only prefer to use it for fuel purposes (Haq et al. 2020a, b). Important species growing in the Himalayan ranges and used by local for fuel purposes are *Celtis australis*, *M. nigra*, *Morus alba*, *Pinus roxburghii*, *Diospyros lotus,* and *Alnus nitida*. The market value of different species is different, few species ranges from 400–450 per mound, and other ranges from 450–600 per mound. Sohail et al. (2020) surveyed different markets and found that *A. nitida* local market rate is 420 per mound. According to wood sellers; the price of wood per mound depends on the ash amount and burning duration. If the wood of a species burns for long time, the highest will be the price and vice versa. Himalayan Alder is a source of livelihood and income for indigenous communities. Rapid and unplanned cutting may lead to extinction. Sustainable usage maybe adopted by the local inhabitants to maintain its availability for ever.

12.5.8 Alder as a Source of Fodder

Alnus is believed to be a delicious food and prefer by buffalos and cows. But some informants also reported that it is not a pleasant food for cattle like goat and sheep and they prefer other plants such as *Morus spp., Ailanthus spp., Melia spp.* etc. than Alder. Shepard usually visits the meadows with grass instead of trees dominant area, as many trees are not a hearty fodder for their cattle (Haq et al. 2020a, b). The green and fresh leaves are eaten by few mammals. By acquiring from local Shepard the reason of not eating shoots, they told me that shoots may be bitter when it is fresh. However, local dairy owners prefer to cut fresh leaves of Alder as fodder which effects its population. In meadow of Horret Wood *Alnus glutinosa* population reduce due to excessive grazing. Grazing pressure effect the new seedlings of *Alnus* to become a mature tree. New seedlings are very much sensitive to grazing. Alder has a very little germination capability which restricts its population. Therefore, suitable and alternate season for grazing is recommended in Alder hosting the areas (Vinthert 1983).

12.5.9 Role of Alder in Carbon Storage and Sequestration

Carbon sequestration is one of the ecosystem services provided by Alder species. The carbon sequestration is usually defined via population size, cultural perceptions, and consumption patterns. Carbon sequestration is the storage of carbon by a tree per year. Alder is evaluated by different scientist regarding carbon accumulation. Frouz et al. (2009) in Dakota (USA) designed an experiment with three stand design based on supply, demand, and anthropogenic activities to find the carbon accumulation ability of Alder. There results indicate that Alder was one among the trees to accommodate 1735.69 million kg carbon. According to Uri et al. (2014) the carbon accumulation capacity of Alder ranges from 0.60 ± 0.09 to 2.31 ± 0.23 t/ha/year. The biomass of a tree is usually linked to the amount of carbon absorbed from the environment (Zhao and Sander 2015). The young and old trees of Alder have the same potential to accumulate carbon. Plantation of Alder along polluted and industrial zones across the globe will help to reduce flow of carbon towards atmosphere which will ultimately lead to pollution control.

12.5.10 Productivity of Alder

Local folks in different areas of the world utilize their expertise to showcase Alder species in different ways for business purposes to enhance their economic conditions. Alder sapwood is used in industry to produce low-cost furniture, idols, timber, meter box, bulb holder, electric wiring support chips, socket box and junction box, etc. The age, buying cost, wood size, and selling cost overall affect the cost of final product of Alder. The log of 85 cm is first hollowed and then rubber or leather is attached to make it musical instruments: *Damru, Sarangi,* and *Madal* (Lohmus et al. 1996). The toys and goddesses are also the products of Alder. Chip piece and board for learner, electric wiring chips, mirror frame, bulb holder are the different products of Alder. The switch board made up of Dry Alder support electrical appliance as dryness make it insulator. Alder wood is excellent source of different designs of school benches and shelves of cupboard with long lasting products (Sharma et al. 2008). The productive nature of Alder made it a cash crop for the locals and farmers. They grow at least 4–5 trees in their fields for making different goods and earn penny (Fig. 12.10). Alnus log is fully utilized by carpenter to make different valuable items for sale. To enhance the livelihood of poor farmers in Himalayan region government should grow maximum Alder tree saplings that would be freely distributed to farmers to grow it in their respective areas.

12.6 Alder Based Agroforestry for Resources Conservation and Ecological Sustainability

Alnus nitida forms symbiotic association with *Frankia alni*, actinomycetes, and nitrogen fixing bacteria. Beside fixation of nitrogen these association helps in the provision of ecosystem services, i.e., biomass production, soil fertility, climax structure, successional trends, and productivity. Commercial production of plants with functional roots nodules is important in two reasons (Malézieux 2012). The grower of *A. nitida* uses relatively less N fertilizers in their fields which reduce production cost and environmental problems caused due to the runoff of water during irrigation. Second, the soil in which *Alnus* is placed may not be compatible for *Frankia*, but *Alnus* cultivation has the capability to overcome this possible barrier to N fixation in landscape (Kimmins 2011).

Its N fixation capability has made it more suitable for species for agroforestry and inter-cropping; one of the oldest practices is shift farming which is in practice from centuries. One of the oldest practices is shift farming which is in practice from centuries. In shift farming, an area with forest cover is cut down into a land with no tree. Then people wait for many years to fully demolish the nutrients so that it becomes part of the soil. This technique enhances the crop production as the soil

Fig. 12.10 Products of *Alnus* (Chhetri and Gauchan 2008)

receives ample amount of nutrients. In shift farming an area with forest cover is cut down into a land with no tree. Then they wait for few years to fully demolish the nutrients in the soil. This technique enhances the crop production as the soil receives ample amount of nutrients. The people from different tribes of Nagaland used to practice it to get maximum yield and fertile soil (Rathore et al. 2010). But this practice is now ban in Pakistan, India, and British due to few reasons. The forests

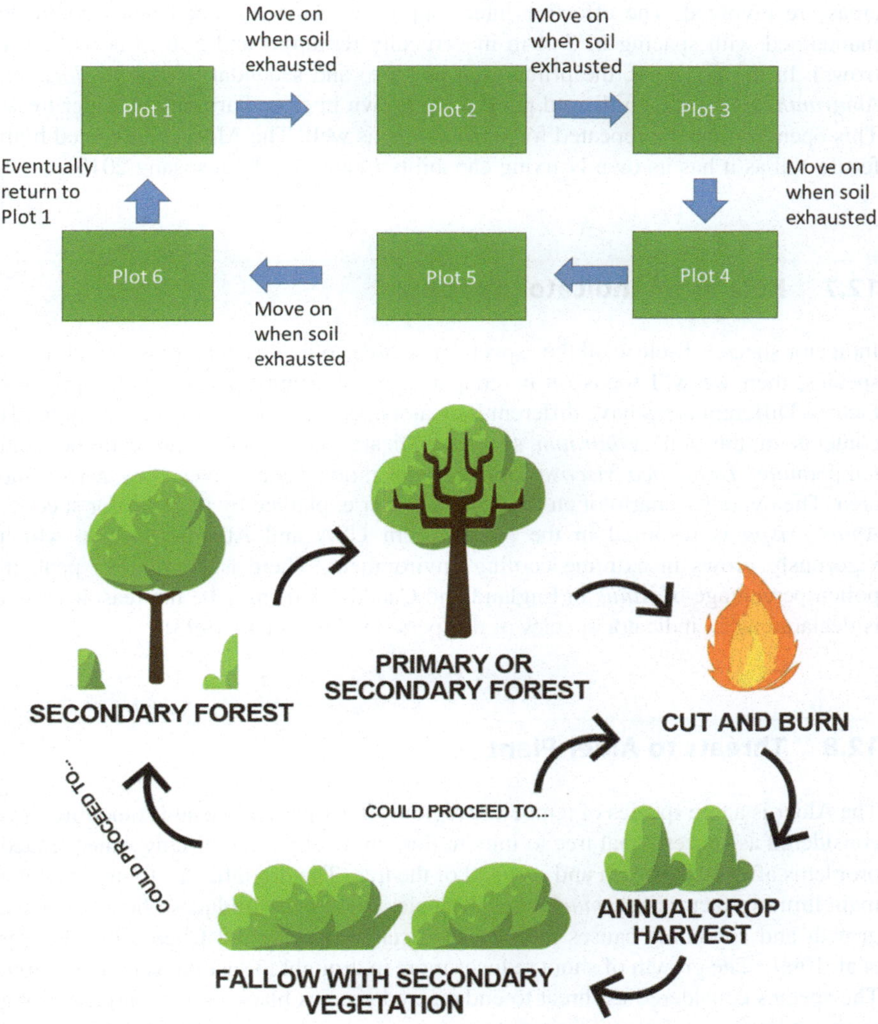

Fig. 12.11 Shift cropping technique used by the farmers to get fertile soil (Liang et al. 2009)

burn may affect the surrounding precious timbers, grasses, birds' nests, and wood land. It is too hard to calculate the tax by government from the area. The most important point is the deforestation which affects our environment and climate at local as well as global levels (Fig. 12.11).

There are many alternative solutions to address this issue. One of these solutions is intercropping. In the case of intercropping two or more crops are cultivated in an area with harmony to each other (Kehie et al. 2017). This technique is usually termed as agroforestry. Alder based agroforestry is an old age system of farming. It is in practice from centuries indigenous communities especially in India and some other areas are involved. The effective intercropping of Alder in crops fields could be maintained with spacing of 3–4 m in vertically (column) and 5–6 m horizontally (rows). In the first year, the primary crops (rice) and secondary crops (*Colocasia*, *Amaranthus*, tapioca, chilli, and potato) are grown in close harmony to Alder trees. This operation maybe repeated for second year as well. The Alder do not need high fertile soil as it has its own N-fixing capability (Yano and Lanusosang 2013).

12.7 Role as an Indicator Species

Indicator species disclose all the aspects of an area. When we talk about an indicator species, then we will focus on its edaphology, environmental and anthropogenic factors. Different areas have different indicators, *Urtica dioica* indicates the high pH contents in the soil, *Viburnum* species indicate snow factor and extreme cold temperature, *Dodonaea viscosa* indicates the moon soon waves flow across the area. The overall scenario of an area can be better explained by its indicator species. *Alnus crispa* is recorded in the Northeastern USA and Atlantic Canada which vigorously grows in extreme cooling environment. There is continuous peak in pollen percentage of *Alnus* in England and Canada. This may be the reason why it is declared as an indicator species of dry period (Mayle et al. 1993).

12.8 Threats to Alder Plant

The Alder is a tree species of temperate regions. It is riparian tree and sometimes it is considered as successional tree to initiate flora in nearby rivers. Many water related problems affect the growth and survival of the tree. The drought was found to be the main limiting factor as *A. maritima* subspecies *georgiensis* reduces photosynthesis, growth and sometime causes the leaf senescence (Zhang and Chen 1991; Packer et al. 1999). The growth of shoot reduces more in drought to sustain water hydration. The species considered as threat to endangered species black grouse based on long term landscape evolution (*Tetrao tetrix* L.) in subalpine belts (Wada et al. 1998) because it produces dense shrub cover (Kikuzaki et al. 1991). In subalpine regions anthropogenic activities are considered as the main reason for treeless land (Brunner and Fairbrothers 1979). *A. nitida* growth and photosynthesis reduce in drought but ample water supply increases its growth (Holtmeier 2009). *A. nitida* is facing water drying conditions in research area. Beside drought factor that *A. nitida* is facing restriction not only in research area but also due to cutting and deforestation, constructions, etc. The genus is ethno-botanically important as the *Betula* species stem bark is used to write spiritual versus for the cure of various diseases. They collect the bark and left the tree barks less; this ultimately leads to the drying of Alder tree. The unchecked cutting of *A. nitida* stem used to make various pots, dolls and furniture is also a threat to the species extinction.

12.9 Anthropogenic Activities and Alder Population Destruction

Human population continuously increases. It is believed that urbanization is an alarming threat for forest conservation. Indigenous communities depend on forest resources for fuel, timber, food, ploughing, and thatching on the forest ecosystem. Alder is also among such species occurring in great abundance in subalpine meadow. Overexploitation of Alder erodes its density and leads towards its extinction. Therefore, its conservation and sustainable utilization in its natural habitat are essential. Keep in mind sustainability here we present an example of French Alps, hosting a great diversity of population and condensed cover of Himalayan Alder. They use Himalayan Alder and its associated species in a sustainable way. However, few scientists claim that Himalayan Alder reduction occurs in subalpine meadows due to its topography and geology. Brunner and Fairbrothers (1979) recorded that anthropogenic pressure is responsible for the reduction of its population (David 2010). Construction of mega projects, roads, industries, and restaurants across the Sino-Japanese belt are the eroding factors for Alder population. Various areas of Himalayan, i.e., Kashmir, Gilgit Baltistan, and Chitral are important spots for tourism. Every year tourists go to visit the beautiful valleys, waterfall, and National Parks across of these areas. Therefore, local people construct hotels and restaurants on the bank of rivers to facilitate them, where they cut down the vigorous population of *A.nitida* and other plants species in its surrounding (Haq et al. 2020a, b).

12.10 Conservation Status of Himalayan Alder

Species extinction is an alarming threat for biodiversity conservation. The protection of threatened species and its conservation are major challenges for conservationists in present scenario (Deb 2017). It is estimated that many species are going to extinct at the rate of 100 to 1000 times faster than any geological time. The same rate of extinction will lead to disappear up to 100,000 species in near future. Anthropogenic activities, habitat destruction (Luetz 2017), and climate changes (Alam and Ali 2010) are the main causes for their erosion. The IUCN Red list categories are designed to identify the conservation status of any species across the world. It also provides explicit objectives and framework for species extinction and defines the status as well as assign priorities for its conservation and protection (Duckworth et al. 2012). The three main categories to check the status of plants are Threatened, Extinct, and Least concern. *A. nitida* has been declared by IUCN red list as least concern (LC) (Shaw et al. 2014). While Haq et al. (2020a, b) categorized it as endangered in few division of Pakistan according to Criteria A, B, C, and D. Due to recent climatic changes and population increase revisiting and documentation of the Alder are required to find its status in different areas of the globe.

12.10.1 Tissue Culture

The most promising technique for sustainability and conservation of Alder is tissue culture. *Alnus* tissue culturing is an important technique to increase its population. *Alnus* being an important plant species provides numerous services. *Alnus* forms an association with *Actinomyces* which forms nodules with its roots and fixing atmospheric N into molecular N, increases soil fertility and biomass. Forest breeding and silviculture require ample amount of *Alnus* in forest ecosystems (Kopelman 2000). To fulfill this demand tissue culture could be an important step (Richard and Tonnel 1987). Protoplast fusion will be homozygous breeding line. Tissue culturing is possible from protoplast of *A. glutinosa* (L.) Gaertn and cell wall of *A. incana* (L.) Moensch. This modern technique of tissue culture will help in cryopreservation of Alder for future studies, genetic record, and conservation.

12.10.2 Regeneration and Plantation

The species facing threats in different region of the world due to various factors, i.e., construction of tourist facilitation centers, hotels, housing schemes, fodder, furniture, water supply scheme (Nalla) are the major threats responsible for Alder population destruction. Mining degraded land in Estonia needs restoration (Kaar and Tomberg 2006). The restoration or afforestation of trees is usually designed for long term to sustain usage of resources (Chambers et al. 1994; Kumar et al. 2020; Meena et al. 2020a). On one side restoration of flora is a challenge due to multifunctional land area while on the other side it provides a base to landscape survival (Groot 2006). Scientist highly recommends reclamation on mining sites (Kuznetsova et al. 2010).

In Himalayan range it is recommended to develop tourism spots where least species may affect. Different economic routes affect the vigorous population of Alder and other species. Housing societies should be ban in those areas where natural environment is affected. The species is facing vulnerable status in western Himalayas (Haq et al. 2019a, b). If government, non-government organization, and local societies are not concern about the population or conservation status of the Alder in different areas of the world and especially in Himalayas, then plant scientists and conservationists should come forward to aware the local community regarding the present status of the plant in the area, future availability, and its effect on the local societies. To demolish a species means, we are putting ourselves away from the free of cost services provided by nature. The Himalayan Alder is one of the very special species which provide wood, fodder, furniture, and heavy metals accumulation facilities. We should take special care of it to keep continue its services for ever.

12.11 Conclusion

Himalayan Alder is ecologically and economically important Holarctic plant species distributed in different areas of Himalaya. It is a multipurpose plant providing ecosystem services to the community. The local people are using it for fuel, fodder, and construction purposes and the thick forest has reduced to sparsely distributed specimens. The species is facing conservation problems due to over exploitation its population decreasing continuously. Nowadays, the species is categorized against IUCN categories and criteria's critically endangered in different habitats due to habitat loss, anthropogenic activities, urbanization, developmental projects, and drought. The plant has great capacity in the accumulation of heavy metals and has been declared as the best bio-accumulator of heavy metals in hilly areas. The species absorb heavy metals from water and make it more suitable for drinking. Conservation strategies, awareness, afforestation, and reforestation will save Himalayan Alder for future generation otherwise the recent speed of deforestation will be led it to extinction from western Himalayan regions.

12.12 Future Perspective

Himalayan Alder is an economical, medicinal, agro-forestry, and fertility sustainer species found in Himalayan region. Many aspects of the species have been untouched and covered by different scientists across the globe. But still we have many aspects of the species to be discovered. Pharmaceutical industries should be involved in phyto-chemical analysis for the extraction of more chemicals because it contains number of chemicals extracted time by time (Sajid et al. 2016). Various chemicals isolated from Alder are used for multiple diseases. Many spiritual beliefs consider its shoots and bark very important. The species also has great variability in cone size and morphology. Therefore, its genetic studies are recommended to evaluate the macro-morphological difference, either it is due to environmental factors or it is a separate new or subspecies. Deforestation of *A. nitida* reducing its population with alarming speed, inhabitant of the area should be educated to conserve its population. Afforestation of *A. nitida* on riverside will be helpful to enhance water quality and purity through its heavy metals accumulation capacity (Haq et al. 2019a, b). Besides practicing ban technique of shift cropping, inter-cropping in agroforestry of Himalayan Alder will provide benefits to the inhabitants of Himalayan region (Uri et al. 2011). Due to multiple benefits of Alder tissue culturing technique may be applied for vigorous germination and propagation.

References

Abbe EC (1935) Studies in the phylogeny of the Betulaceae. 1. Floral and inflorescence anatomy. Bot Gaz 97:1–67

Alam J, Ali SI (2010) Contribution to the red list of the plants of Pakistan. Pak J Bot 42 (5):2967–2971

Amin A, Chollet S, Angelis A, Borie JN, Nuzillard A, Skaltsounis R, Reynaud SC, Gangloff J, Renault J, Hubert J (2016) Bioactivity-guided identification of antimicrobial metabolites in *Alnus* glutinosa bark and optimization of oregonin purification by centrifugal partition chromatography. J Chromatogr B 1029–1030:121–127

Arsenault A, Clague JJ, Mathewes RW (2007) Late Holocene vegetation and climate change at Moraine Bog, Tiedemann Glacier, southern Coast Mountains, British Columbia. Can J Earth Sci 44:707–719

Austin DF (2004) Florida ethnobotany. CRC Press, Boca Raton

Banerjee A, Jhariya MK, Yadav DK, Raj A (2020) Environmental and sustainable development through forestry and other resources. Apple Academic Press, Palm Bay, p 400. https://doi.org/10.1201/9780429276026

Barbara JAJ, Smith WB, Gamble JR, Van Ostade X, Vandenabeele P, Tavernier J, Fiers W, Vadas MA, Lopez AF (1994) Dissociation of TNF-_ cytotoxic and proinflammatory activities by p55 receptor- and p75 receptor-selective TNF-_ mutants. EMBO J 13:843–850

Becerra A, Zak MR, Horton TR, Micolini J (2005) Ectomycorrhizal and arbuscular mycorrhizal colonization of *Alnus* acuminate Calilegua National Park (Argentina). Mycorrhiza 15:525–531

Becerra AG, Menoyo E, Lett I, Li CY (2009) Alnus acuminata in dual symbiosis withFrankia and two different ectomycorrhizal fungi (Alpova austroalnicola and Alpova diplophloeus) growing in soilless growth medium. Symbiosis 47(2):85–92

Behkne HD (1973) Sieve-tube plastids of Hamamelididae. Taxon 22:205–210

Berliner R, Torrey JG (1989) On tripartite Frankia–mycorrhizal associations in the Myricaceae. Can J Bot 67(6):1708–1712

Blackmore S, Steinmann JAJ, Hoen PP, Punt W (2003) The northwest European pollen flora, 65, Betulaceae and Corylaceae. Rev Palaeobot Palynol 123:71–98

Bohanek JR, Groninger JW (2004) Productivity of European black alder (*Alnus*glutinosa) interplanted with black walnut (Juglans nigra) in Illinois, U.S.A. Agrofor Syst 64:99–106

Brunner F, Fairbrothers DE (1979) Serological investigations of the Corylaceae. Bull Torrey Bot Club 106:97–103

Chaia EE, Wall LG, Huss-Danell K (2010) Life in soil by the actinorhizal root nodule endophyte *Frankia*. A review. Symbiosis 51(3):201–226

Chambers JC, Brown RW, Williams BD (1994) An evaluation of reclamation success on Idaho's phosphate mines. Restor Ecol 2:4–16

Chen ZD (1994) Phylogeny and phytogeography of the Betulaceae. Acta Phytotaxon Sin 32:1–32, 101–153 (in Chinese with English summary)

Chen ZD (2004) Phylogeny and phytogeography of the Betulaceae. Acta Phytotaxon Sin 32:1–32, 101–153

Chen J, Karchesy JJ, Gonzalez-Laredo RF (1998a) Phenolic diarylheptenones from Alnus rubra bark. Planta Med 64:74

Chen ZD, Wang XQ, Sun HY, Han Y, Zhang ZX, Zou YP, Lu AM (1998b) Systematic position of the Rhoipteleaceae: evidence from nucleotide sequences of *rbcL*gene. Acta Phytotaxon Sin 36:1–7

Chhetri RB, Gauchan DP (2008) Traditional knowledge on wood processing of Utis in Panauti of Kavrepalanchowk district, Nepal. Ind J Tradit Knowl 7(1):112–115

Choi SE, Kim KH, Kwon JH, Kim SB, Kim HW, Lee WM (2008) Cytotoxic activities of diarylheptanoids from *Alnus* japonica. Arch Pharmacol Res 31:1287–1289

Colagar AH, Yousefzadeh H, Shayanmehr F, Jalali SG, Zare H, Tippery NP (2016) Molecular taxonomy of Hyrcanian *Alnus* using nuclear ribosomal ITS and chloroplast trnH-psbA DNA barcode markers. Syst Biodivers 14(1):88–101

Daugavietis M, Daugaviete M, Bisenieks J (2009) Management of grey alder (Alnus incana moench) stands in Latvia. In: Engineering for rural development, Jelgava. pp 229–234

David F (2010) Expansion of green alder (*Alnus* alnobetula [Ehrh] K. Koch) in the northern French Alps: a palaeoecological point of view. CR Biol 333:424–428

Deb D (2017) Folk rice varieties, traditional knowledge and nutritional security in South Asia. Agroecology ecosystems, and sustainability in the tropics. Studera Press, Delhi, pp 118–126

Duckworth JW, Batters G, Belant JL, Bennett EL, Brunner J, Burton J, Hedges S (2012) Why South-East Asia should be the world's priority for averting imminent species extinctions, and a call to join a developing cross-institutional programme to tackle this urgent issue. Surv Perspect Integr Environ Soc 5(2):1327

Edgerton EA (2014) Prevention and management of aquatic invasive plants in Texas. Doctoral dissertation

Fleschner MD, Delwiche CC, Goldman CR (1976) Measuring rates of symbiotic nitrogen fixation by *Alnus tenuifolia*. Am J Bot 63(7):945–950

Frouz J, Pižl V, Cienciala E, Kalčík J (2009) Carbon storage in post-mining forest soil, the role of tree biomass and soil bioturbation. Biogeochemistry 94(2):111–121

Furlow JJ (1990) The genera of Betulaceae in the southeastern United States. J Arnold Arbor 71:1–67

Groot R (2006) Function-analysis and valuation as a tool to assess land use conflicts in planning for sustainable, multi-functional landscapes. Landsc Urban Plan 75:175–186

Hall JW (1952) The comparative anatomy and phylogeny of the Betulaceae. Bot Gaz 113:235–270

Haq ZU, Khan SM, Abdullah, Asadullah, Razzaq A, Ahmad Z (2019a) An evaluation of plant zonation and conservation status of *Alnus nitida*; a monophyletic species of the Sino-Japanese belt. J Anim Plants Sci 30(4):2020

Haq Z, Khan SM, Abdullah, Razzaq A, Manan F, Rasheed S, Ahmad Z (2019b) Heavy metals uptake ability from water by the Himalayan alder growing in riparian habitat of Sino Japanese regions in Pakistan. Pure Appl Biol 9(1):704–713

Haq ZU, Khan SM, Abdullah, Asadullah, Razzaq A, Manan F, Kamran S, Rasheed S, Ahmad Z (2020a) Heavy metals uptake ability from water by the Himalayan alder growing in Riparian habitat of Sino Japanese regions in Pakistan. Pure Appl Biol 9(1):704–713

Haq ZU, Khan SM, Abdullah, Asadullah, Razzaq A, Ahmad Z (2020b) Conservation status' assessments of the Himalayan Alder through IUCN red list categories and criteria. Case Stud Sino-Jpn Belt Pak 14(1):11–21

Heusser CJ (1969) Modern pollen spectra from the Olympic Peninsula, Washington. Bull Torrey Bot Club 96:407–417

Hilger HA, Wollum AG, Barlaz MA (2000) Landfill methane oxidation response to vegetation, fertilization, and liming. J Environ Qual 29(1):324–334

Holtmeier FK (2009) Mountain timberlines: ecology, patchiness, and dynamics, vol 36. Springer, Dordrecht

Hussain I, Kline JN (2004) DNA, the immune system, and atopic disease. J Investig Dermatol Symp Proc 9(1):23–28

Iwanaga F, Yamamoto F (2007) Growth, morphology and photosynthetic activity in flooded *Alnus japonica* seedlings. J For Res 12:243–246

Jhariya MK, Yadav DK, Banerjee A (2018a) Plant mediated transformation and habitat restoration: phytoremediation an eco-friendly approach. In: Gautam A, Pathak C (eds) Metallic contamination and its toxicity. Daya Publishing House, A Division of Astral International Pvt. Ltd, New Delhi, pp 231–247

Jhariya MK, Banerjee A, Yadav DK, Raj A (2018b) Leguminous trees an innovative tool for soil sustainability. In: Meena RS, Das A, Yadav GS, Lal R (eds) Legumes for soil health and sustainable management. Springer, Singapore, pp 315–345. https://doi.org/10.1007/978-981-13-0253-4_10

Jhariya MK, Banerjee A, Meena RS, Yadav DK (2019a) Sustainable agriculture, forest and environmental management. Springer, Singapore, p 606. https://doi.org/10.1007/978-981-13-6830-1

Jhariya MK, Yadav DK, Banerjee A (2019b) Agroforestry and climate change: issues and challenges. Apple Academic Press, Palm Bay, p 335. https://doi.org/10.1201/9780429057274

Jin WY, Cai XF, Na MK, Lee JJ, Bae KH (2007) Diarylheptanoids from *Alnus hirsuta* inhibit the NF-kB activation and NO and TNF-_ production. Biol Pharm Bull 30:810–813. https://doi.org/10.1248/bpb.30.810

Joo SS, Kim SG, Choi SE, Kim YB, Park HY, Seo SJ, Choi YW, Lee MW, Lee DI (2009) Suppression of T cell activation by hirsutenone, isolated from the bark of *Alnus japonica*, and its therapeutic advantages for atopic dermatitis. Eur J Pharmacol 614:98–105

Kaar E, Tomberg E (2006) Recultivation of the quarry spoil. In: Valgma I (ed) 90 years of oil shale mining in Estonia. Proceedings of the conference of Estonian miners 2006 by Estonian mining society. Tallinn University of Technology, Tallinn, pp 78–83 (in Estonian)

Kataki R, Konwer D (2002) Fuelwood characteristics of indigenous tree species of north-East India. Biomass Bioenergy 22(6):433–437

Kaul P (2011) Nutritional potential, bioaccessibility of minerals and functionality of watermelon (*Citrullus vulgaris*) seeds. LWT-Food Sci Tech 44(8):1821–1826

Kehie M, Khamu S, Kehie P (2017) Indigenous alder based farming practices in Nagaland, India: a sustainable agricultural model. J Tradit Folk Pract 5(2):82–152

Kennedy PG, Schouboe JL, Rogers RH, Weber MG, Nadkarni NM (2010) *Frankia* and *Alnus rubra* canopy roots: an assessment of genetic diversity, propagule availability, and effects on soil nitrogen. Microbial Ecol 59(2):214–220

Khan SM, Page SE, Ahmad H, Harper D (2013) Sustainable utilization and conservation of plant biodiversity in montane ecosystems: the western Himalayas as a case study. Ann Bot 112 (3):479–501

Khan N, Jhariya MK, Yadav DK, Banerjee A (2020a) Herbaceous dynamics and CO_2 mitigation in an urban setup- a case study from Chhattisgarh, India. Environ Sci Pollut Res 27(3):2881–2897. https://doi.org/10.1007/s11356-019-07182-8

Khan N, Jhariya MK, Yadav DK, Banerjee A (2020b) Structure, diversity and ecological function of shrub species in an urban setup of Sarguja, Chhattisgarh, India. Environ Sci Pollut Res 27 (5):5418–5432. https://doi.org/10.1007/s11356-019-07172-w

Kiernan BD, Hurd TM, Raynal DJ (2003) Abundance of *Alnus incana* ssp. rugosa in Adirondack Mountain shrub wetlands and its influence on inorganic nitrogen. Environ Poll 123:347–354

Kikuzaki H, Usuguchi J, Nakatani N (1991) Constitutents of Zingiberaceae. I. Diarylheptanoids from the rhizomes of ginger (Zingiber officinale Roscoe). Chem Pharm Bull 39(1):120–122

Kikuzawa K (1982) Leaf survival and evolution in Betulaceae. Ann Bot 50:345–354

Kimmins JH (2011) Balancing act: environmental issues in forestry. UBC Press, Vancouver

Kloepper JW, Ryu CM, Zhang S (2004) Induced systemic resistance and promotion of plant growth by *Bacillus* spp. Phytopathology 94(11):1259–1266

Kopelman PG (2000) Obesity as a medical problem. Nature 404(6778):635–643

Kozlov M, Zvereva E, Zverev V (2009) Impacts of point polluters on terrestrial biota: comparative analysis of 18 contaminated areas, vol 15. Springer, Dordrecht

Kumar S, Meena RS, Jhariya MK (2020) Resources use efficiency in agriculture. Springer, Singapore, p 760. https://doi.org/10.1007/978-981-15-6953-1

Kuznetsova T, Rosenvald K, Ostonen I, Helmisaari HS, Mandre M, Lõhmus K (2010) Survival of black alder (*Alnus*glutinosa L.), silver birch (Betula pendula Roth.) and scots pine (*Pinus sylvestris* L.) seedlings in a reclaimed oil shale mining area. Ecol Eng 36(4):495–502

Lachman S, Boekholdt SM, Luben RN, Sharp SJ, Brage S, Khaw KT, Wareham NJ (2018) Impact of physical activity on the risk of cardiovascular disease in middle-aged and older adults: EPIC Norfolk prospective population study. Eur J Prev Cardiol 25(2):200–208

Lacourse T (2007) Environmental change controls postglacial forest dynamics through interspecific differences in life-history traits. Ecology 90:2149–2160

Lee MW, Kim JH, Jeong DW, Ahn KH, Toh SH, Surh YJ (2000) Cyclooxygenase-2 inhibitory effect of diarylheptanoids from the barks of *Alnus* hirsuta var. sibirica. Bio Pharm Bull 23:517–518

Li H (2015) Anti-mycobacterial natural products from Canadian medicinal plants: identification, confirmation of source and investigation of mode of action. Doctoral dissertation, University of New Brunswick

Li M, Chen X, Li X, Ma B, Vitányi PM (2004) The similarity metric. IEEE Trans Inf Theory 50(12):3250–3264

Liang L, Shen L, Yang W, Yang X, Zhang Y (2009) Building on traditional shifting cultivation for rotational agroforestry: experiences from Yunnan, China. For Ecol Manag 257(10):1989–1994

Lin H, Webster D, Johnson JA, Gray CA (2015) Anti-mycobacterial triterpenes from the Canadian medicinal plant *Alnus incana*. J Ethnopharmacol 165:148–151

Liu N, Wang H, Liu M, Zheng W, Huang Y (2009) Streptomyces alni sp. nov., a daidzein-producing endophyte isolated from a root of *Alnus nepalensis* D. Don. Int J Syst Evol Microbiol 59(2):254–258

Lohmus K, Mander U, Tullus H, Keedus K (1996) Productivity, buffering capacity and resources of grey alder forests in Estonia. In: Short rotation willow coppice for renewable energy and improved environment. Institutionen för lövträdsodling, Uppsala, pp 95–105

Luetz J (2017) Climate change and migration in the Maldives: some lessons for policy makers. In: Climate change adaptation in pacific countries. Springer, Cham, pp 35–69

Malézieux E (2012) Designing cropping systems from nature. Agron Sust Dev 32(1):15–29

Mayle FE, Levesque AJ, Cwynar LC (1993) *Alnus* as an indicator taxon of the younger dryas cooling in eastern North America. Quaternary Sci Rev 12:295–305

Meena RS, Lal R (2018) Legumes for soil health and sustainable management. Springer, Singapore, p 541. https://doi.org/10.1007/978-981-13-0253-4_10

Meena RS, Kumar V, Yadav GS, Mitran T (2018) Response and interaction of *Bradyrhizobium japonicum* and Arbuscular mycorrhizal fungi in the soybean rhizosphere. Rev Plant Growth Regul 84:207–223

Meena RS, Kumar S, Datta R, Lal R, Vijaykumar V, Brtnicky M, Sharma MP, Yadav GS, Jhariya MK, Jangir CK, Pathan SI, Dokulilova T, Pecina V, Marfo TD (2020a) Impact of agrochemicals on soil microbiota and management: a review. Land (MDPI) 9(2):34. https://doi.org/10.3390/land9020034

Meena RS, Lal R, Yadav GS (2020b) Long term impacts of topsoil depth and amendments on soil physical and hydrological properties of an Alfisol in Central Ohio, USA. Geoderma 363:1141164

Meena RS, Lal R, Yadav GS (2020c) Long-term impact of topsoil depth and amendments on carbon and nitrogen budgets in the surface layer of an Alfisol in Central Ohio. Catena 194:104752

Metcalfe CR, Chalk L (1950) Anatomy of the dicotyledons, vol 2. Clarendon Press, Oxford, pp 1302–1309

Navarro E, Bousquet J, Moiroud A, Munive A, Piou D, Normand P (2003) Molecular phylogeny of *Alnus* (Betulaceae), inferred from nuclear ribosomal DNA ITS sequences. In: Frankia Symbiosis. Springer, Dordrecht, pp 207–217

Neilson AH, Doudoroff M (1973) Ammonia assimilation in blue-green algae. Arch Mikrobiol 89 (1):15–22

Nemli G, Kırcı H, Serdar B, Ay N (2003) Suitability of kiwi (*Actinidia sinensis* Planch.) prunings for particleboard manufacturing. Ind Crop Prod 17(1):39–46

Normand P, Nouioui I, Pujic P, Fournier P, Dubost A, Schwob G, Klenk HP, Nguyen A, Abrouk D, Herrera-Belaroussi A, Pothier JF (2018) Frankia canadensis sp. nov., isolated from root nodules of Alnus incana subspecies rugosa. Int J Syst Evol Microbiol 68(9):3001–3011

Packer L, Rimbach G, Virgili F (1999) Antioxidant activity and biologic properties of a procyanidin-rich extract from pine (*Pinus maritima*) bark, pycnogenol. Free Radic Biol Med 27(5–6):704–724

Pande PC, Tiwari L, Pande HC (2006) Folk-medicine and aromatic plants of Uttaranchal, Bishen Singh, Mahendra Pal Singh Publishers, Dehradun

Parrotta JA, Turnbull JW, Jones N (1997) Catalyzing native forest regeneration on degraded tropical lands. For Ecol Manag 99:1–7

Perveen ANJ, Qaiser M (1999) Pollen flora of Pakistan–XXXI. Betulaceae. Pak J Bot 31:243–246
Raj A, Jhariya MK, Harne SS (2018) Threats to biodiversity and conservation strategies. In: Sood KK, Mahajan V (eds) Forests, climate change and biodiversity. Kalyani Publisher, New Delhi, pp 304–320
Raj A, Jhariya MK, Yadav DK, Banerjee A, Meena RS (2019a) Agroforestry: a holistic approach for agricultural sustainability. In: Jhariya MK, Banerjee A, Meena RS, Yadav DK (eds) Sustainable agriculture, forest and environmental management. Springer, Singapore, pp 101–131. https://doi.org/10.1007/978-981-13-6830-1
Raj A, Jhariya MK, Banerjee A, Yadav DK, Meena RS (2019b) Soil for sustainable environment and ecosystems management. In: Jhariya MK, Banerjee A, Meena RS, Yadav DK (eds) Sustainable agriculture, forest and environmental management. Springer, Singapore, pp 189–221. https://doi.org/10.1007/978-981-13-6830-1
Raj A, Jhariya MK, Yadav DK, Banerjee A (2020) Climate change and agroforestry systems: adaptation and mitigation strategies. Apple Academic Press, Palm Bay, p 383. https://doi.org/10.1201/9780429286759
Rascio N, Navari-Izzo F (2011) Heavy metal hyperaccumulating plants: how and why do they do it? And what makes them so interesting? Plant Sci 180(2):169–181
Rathore SS, Karunakaran K, Prakash B (2010) Alder based farming system a traditional farming practices in Nagaland for amelioration of jhum land. Indian J Trad Knowl 4:677–680
Richard L, Tonnel A (1987) Contribution a` l'e'tude des valle´ es internes des Alpesoccidentales. Documents de cartographie e'cologique, Grenoble XXX:113–136
Romero P, Botia P, Garcia F (2004) Effects of regulated deficit irrigation under subsurface drip irrigation conditions on vegetative development and yield of mature almond trees. Plant Soil 260(1–2):169–181
Saiz G, Byrne KA, Butterbach-bahl KLA, Kiese R, Blujdea V, Farrell EP (2006) Stand age-related effects on soil respiration in a first rotation Sitka spruce chronosequence in Central Ireland. Glob Change Biol 12(6):1007–1020
Sajid M, Khan MR, Shah NA, Shah SA, Ismail H, Younis T, Zahra Z (2016) Phytochemical, antioxidant and hepatoprotective effects of *Alnus* nitida bark in carbon tetrachloride challenged Sprague Dawley rats. BMC Complement Altern Med 16(1):268
Sarma B (2017) Study on dynamics of carbon Sequestration and its enhancement from upland agro-ecosystems of Assam. Doctoral dissertation, Tezpur University
Sharma R, Sharma E, Purohit AN (1995) Dry matter production and nutrient cycling in agroforestry systems of mandarin grown under *Albizia* and mixed tree species. Agrofor Syst 29(1):160–165
Sharma E, Sharma R, Sharma G, Rai SC, Sharma P, Chettri N (2008) Values and services of nitrogen-fixing alder based cardamom agroforestry systems in the eastern Himalayas. In: Smallholder tree growing for rural development and environmental services. Springer, Dordrecht, pp 393–409
Shaw K, Stritch L, Rivers M, Roy S, Wilson B, Govaerts R (2014) The red list of Betulaceae. Botanic Gardens Conservation International, Richmond
Smith WH (1999) Influence of heavy metal leaf contaminants on the in vitro growth of urban-tree phylloplane fungi. Microb Ecol 3:231–239
Socha J, Ochał W (2017) Dynamic site index model and trends in changes of site productivity for *Alnus glutinosa* (L.) Gaertn. in southern Poland. Dendrobiology 77:45–57
Sohail A, Haq ZU, Ullah F, Ullah A (2020) Timber, fuel wood and medicinal plants of Maidan Valley district Dir (L) Khyber Pakhtunkhwa, Pakistan. Pak J Weed Sci Res 5(25):11–14
Sun XJ (2016) Late cretaceous—paleocene palynological flora in China. Acta Phytotaxon Sin 17 (3):8–23 (in Chinese with English summary)
Takhtajan A (1980) Outline of the classification of flowering plants (Magnoliophyta). Bot Rev 46:225–359
Uri V, Lõhmus K, Mander U, Ostonen I, Aosaar J, Maddison M, Augustin J (2011) Long-term effects on the nitrogen budget of a short-rotation grey alder (*Alnusincana* (L.) Moench) forest on abandoned agricultural land. Ecol Eng 37(6):920–930

Uri V, Aosaar J, Varik M, Becker H, Ligi K, Padari A, Lõhmus K (2014) The dynamics of biomass production, carbon and nitrogen accumulation in grey alder (*Alnusincana* (L.) Moench) chronosequence stands in Estonia. For Ecol Manag 327:106–117

van Wieren SE (1996) Digestive strategies in ruminants and nonruminants. Thesis (doctoral), Landbouwuniversiteit Wageningen

Vinthert E (1983) Invasion of *Alnusglutinosa* (L.) Gaertn. in a former grazed meadow in relation to different grazing intensities. Biol Conserv 25:75–89

Wada H, Tachibana H, Fuchino H, Tanaka N (1998) Three new diarylheptanoid glycosides from Alnus japonica. Chem Pharm Bull 46:1054–1055

Wiedmer E, Senn-Irlet B (2006) Biomass and primary productivity of an *Alnusviridis* stand – a case study from the Schächental valley, Switzerland. Bot Helv 116:55–64

Wolters DJ, Akkermans ADL, Van Dijk C (1997) Ineffective *Frankia* strains in wet stands of *Alnus glutinosa* L. Gaertn. in the Netherlands. Soil Biol Biochem 29(11–12):1707–1712

Wopara I, Mobisson SK, Pius EA, Uwakwe AA, Wegwu MO (2020) Inhibition of phosphodiesterase 5 enzyme by Pterine-6 carboxylic acid from Baphia nitida–related to erectile dysfunction: computational kinetic. Eur J Med Plants 31:49–55

Yadav D, Singh SC, Verma RK, Saxena K, Verma R, Murthy PK, Gupta MM (2013) Antifilarial diarylheptanoids from *Alnus nepalensis* leaves growing in high altitude areas of Uttarakhand, India. Phytomedicine 20(2):124–132

Yano K, Lanusosang T (2013) Globalisation and its impact on agriculture: an overview of Kohima District, Nagaland, India. Int J Biores Stress Manag 4:651–654

Yu YB, Miyashiro H, Nakamura N, Hattori M, Park JC (2007) Effects of triterpenoids and flavonoids isolated from Alnus firma on HIV-1 viral enzymes. Arch Pharm Res 30(7):820–826

Zare H, Amini T (2012) A review of the genus alnus gaertn. In Iran, new records and new species. Iran J Bot 18(1):10–21

Zhang ZY, Chen ZD (1991) Embryology of *Carpinus* L. and its systematic significance. Cathaya 5:59–68

Zhao C, Sander HA (2015) Quantifying and mapping the supply of and demand for carbon storage and sequestration service from urban trees. PLoS One 10(8):12–18

Soil Carbon Stock and Sequestration: Implications for Climate Change Adaptation and Mitigation

13

Nahid Khan, Manoj Kumar Jhariya, Abhishek Raj, Arnab Banerjee, and Ram Swaroop Meena

Abstract

The land resource and other natural resources are degrading day by day due to human greed of development and unsustainable management. These will not only affect the ecosystem structure and related services but also disturb environmental sustainability and overall ecological stability at global scale. Today, climate change becomes most highlighted and burning issue among policy makers, stakeholders, scientists, and academicians across various national and international platforms. However, the climate change and other perturbation have altered the natural balance of different ecosystems resulting into poor ecosystem services. This will not only affect yield and productivity but also affect ecosystem health in many dimensions. In this context, capturing of carbon (C) through the process of C sequestration will increase C values in vegetation and soil as soil organic carbon (SOC) pools that directly or indirectly link with food-soil-climate security. Soil organic matter (SOM) and C are the key management strategies for managing land resources wisely. Updated and advance technologies of soil C-friendly management are the major mitigatory strategy for different ecosystems. Soil C management requires the practices which add C inputs in

N. Khan · M. K. Jhariya (✉)
Department of Farm Forestry, Sant Gahira Guru Vishwavidyalaya, Sarguja, Ambikapur, Chhattisgarh, India

A. Raj
School of Agriculture, Lovely Professional University, Phagwara, Punjab, India

A. Banerjee
Department of Environment Science, Sant Gahira Guru Vishwavidyalaya, Sarguja, Ambikapur, Chhattisgarh, India

R. S. Meena
Department of Agronomy, Institute of Agricultural Sciences, Banaras Hindu University, Varanasi, Uttar Pradesh, India

M. K. Jhariya et al. (eds.), *Ecological Intensification of Natural Resources for Sustainable Agriculture*, https://doi.org/10.1007/978-981-33-4203-3_13

soil instead removing the soil C and nutrients reserve. The land-use systems must be eco-friendly and sustainable one to stop the land degradation and deterioration. Sustained research and developmental activities are needed to generate C dynamics knowledge base which subsequently helps to visualize the changes in soil C quantity and impact on the atmospheric C. Moreover, this information supports for terrestrial C management and climate change adaptation and mitigation. In the view of the above, a rigorous and comprehensive discussion has been made on soil C sequestrations in varying land use practices (forest, agroforestry, and fruits based land use system, etc.) and its role in climate change mitigation to achieve the goal of sustainable environment and maintaining overall ecological stability.

Keywords

Adaptation · Carbon sequestration · Climate change · Carbon management · Mitigation · Soil ecosystem · Sustainability

Abbreviations

C	Carbon
GHG	Greenhouse gases
SOC	Soil organic carbon
SOM	Soil organic matter

13.1 Introduction

The soil ecosystem of the world is facing the various problems of degradation. These are due to various biotic and abiotic perturbations and alteration in the different segments of the environment and ecosystems. This change in the environment can alter the ecological process as well as natural balance and resource base (Kumar et al. 2020). Therefore, the better and efficient management of global soil resources is needed on urgent basis. Soil quality and health are backbone for agricultural production, plant development, and source of various resources. Soil ecosystem also contributes significantly in the C (carbon) sink across the globe especially in the tropics (Jhariya et al. 2019a; Meena et al. 2020).

In the Indian tropics the soils are mostly nutrient deprived and mostly utilized in unsustainable manner which gradually results into soil degradation after certain time interval (Jhariya et al. 2018a, b). The tropical soils possess 1/3rd of global soil organic carbon (SOC) which includes 128 Pg in tropical wet, 151 Pg in moist, 136 Pg in dry, and 56 Pg C in mountainous areas (Hiederer and Köchy 2011). The maintenance of SOC in the tropics is difficult one due to fast oxidation of organic material under high temperature regimes which lead to soil degradation (Lal et al. 2003; Meena et al. 2018; Meena and Lal 2018).

The soil C management is an important aspect for agricultural sustainability and from earth's C cycle point of view. The fluctuation in the vegetation C pool or in the soil C pool can enhance or reduce the C source in the atmosphere (Chen et al. 2004). Therefore, judicious and scientific management of C both in vegetation and soil system is essential for climate change mitigation and reducing the earth's warming (Raj et al. 2020; Banerjee et al. 2020). Therefore, in the present time C cycle and C dynamics trigger the scientific communities and people concern across the world. Lal (2004a) mentioned that the major atmospheric C pools contributed by the terrestrial ecosystem in which soil pool share significantly. The soil organic matter (SOM), its allocation and turnover potential (time and rate) is key process which affects the fertility of soil and its function. This phenomenon decides that the soil system becomes a C source or sink of in the C cycle of the earth which alters the climatic segments at a great extent (Post and Kwon 2000; Meena et al. 2020a, b). The human interferences substantially influence the C flux, pools, and C dynamics through human-built environmental process (Bolstad and Vose 2005). Soil C is essential pools which facilitates various functions and services and offer various direct and indirect benefits to the human being and biota (Fig. 13.1).

The present title deals with the soil ecosystem as an effective tool for C sink for adapting and mitigating the climate change through best management practices which enhance the C sink in soil ecosystem as well as management aspects towards sustainability.

13.2 Soil and Forest Ecosystems

Soil provides key ecosystem services to various ecosystems. The plant growth and development is affected by soil quality and health. Soils support the biota as habitat and the microorganisms involved in the mineralization process. The growth and development of vegetation depends upon the soil (Fig. 13.2). Therefore, the knowledge regarding soil resources and properties are key concern for the reforestation and afforestation program (Raj et al. 2019a).

However, rising temperature exert negative impacts on forest ecosystem by disturbing various biological processes, productivity, phenology, and morphological characteristics that cause death and heavy mortality of tree species. In this context, Table 13.1 represents extreme weather impacts on tree species in various regions in the world.

Relative abundance of C value in soil versus vegetations in different parts of the world is depicted in Fig. 13.3 (Dixon et al. 1994). From this figure, Russia contributed maximum value of the ratio of soil C to vegetation C as 3.38 which are followed by 1.84, 1.74, 1.21, 1.19, and 1.05 in Australia, USA, Africa, China, and Asia, respectively. Although these are very old report but would be helpful in synthesizing more new data based on these old report. The soil quality in a forest stand is influenced by the vegetal cover, composition of minerals, species mix, type of species, site condition and litter or organic matter addition in to soil systems. Therefore, these should be properly considered for the conservation and

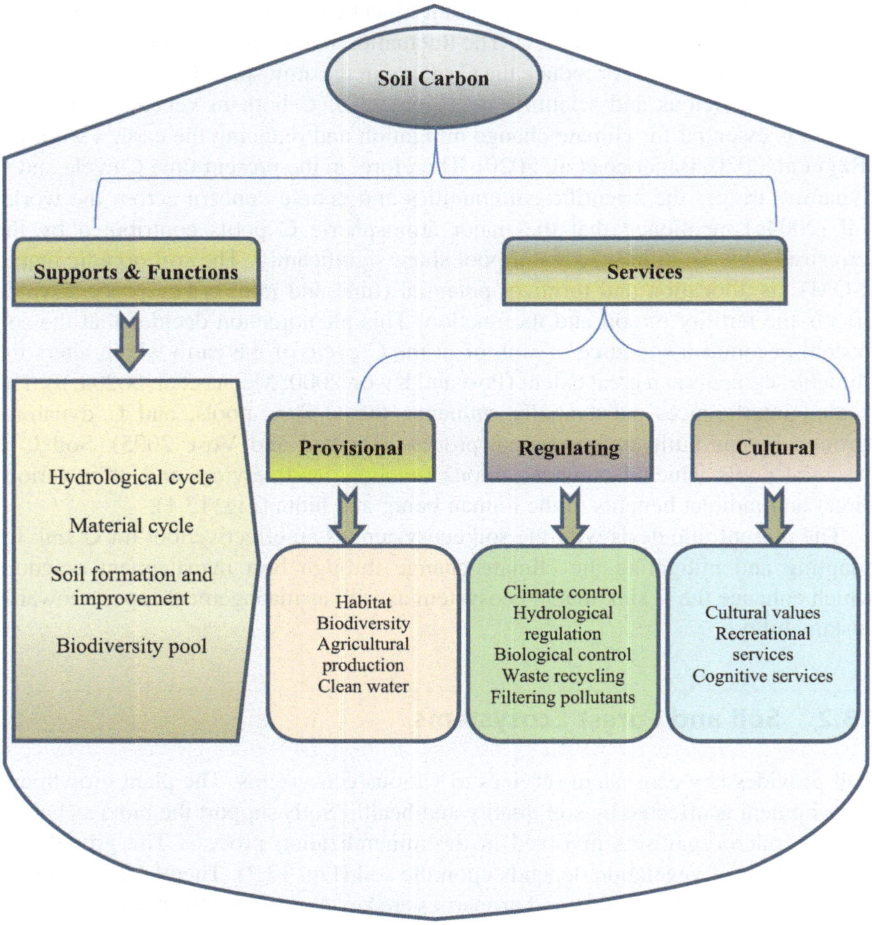

Fig. 13.1 Services provided by soils ecosystem (Compiled from Laban et al. 2018)

management of natural stands as well as regional biodiversity (Paustian et al. 2016). However, C sinks value in different forest biomes and its characteristics are depicted in Table 13.2 (Prentice 2001).

13.3 Soil and Sustainability

Soil, itself represents a largest natural resource that supports a huge biodiversity, upholds variety of flora and fauna, maintains livelihood for forest fringe peoples, and takes part in conservation and management of other important resources for maintaining better ecosystem and environmental sustainability (Raj et al. 2018). Although we cannot overlook the present and ongoing problems of land degradation,

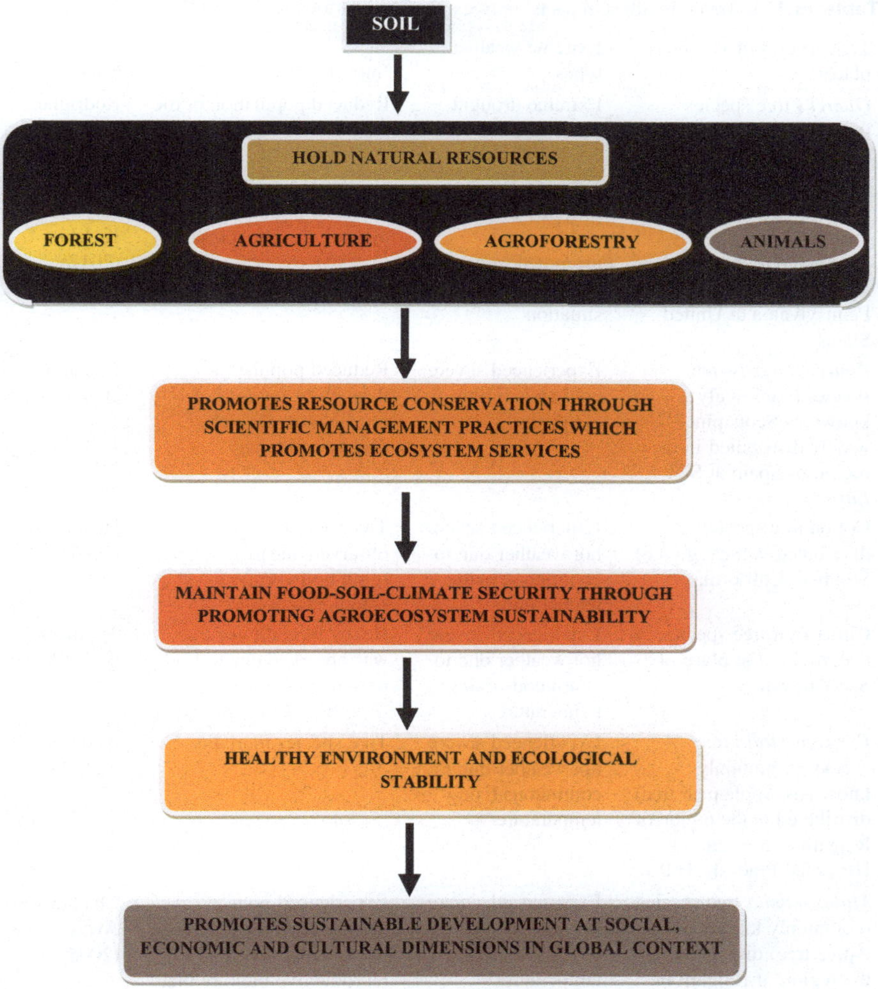

Fig. 13.2 The journey from soil to sustainable development through resource conservation strategy

deprive soil health and quality, and related poor ecosystem service due to deleterious and unscientific practices of agricultural and horticultural land use systems, agricultural intensification, high synthetic inputs, deforestation activity, illicit and overexploitation, etc. which not only affects other governing resources (forest, agriculture, agroforestry, soil nutrients, animals, etc.) but also affects overall food-soil-climate security at global level (Lal 2009).

Therefore, resource conservation is important for sustainable soil ecosystem. Healthy soil reserve pool for longer period without affecting present and future

Table 13.1 Extreme weather impacts on tree species in various regions in the world

Tree species of varying places	Extreme weather types	Impacts	Source
Quercus tree species (commonly known as Oak tree) in the Oklahoma region of United States	Extreme drought condition	Reduced population of the Oak	Rodríguez-Calcerrada et al. (2017)
Fraxinus tree (commonly known as Ash tree) distributed in the region of North western Pennsylvania at United States	Extreme drought and freezing promoted insect pest outbreak situation	Reduced population by dieback disease mortality	Royo and Knight (2012)
Pinus sylvestris tree species (commonly known as Scots pine) widely distributed in the region of Spain at S-W *Europe*	Experienced severe cold and chilling temperature	Reduced population through heavy loss of needles and dieback disease mortality	Camarero et al. (2015)
Cotton tree species distributed in the region of Southern California	Experienced severe hot weather due to continuous rising temperature	Tree mortality was observed due to insect pest (pink bollworm) emergence	Henneberry (2007)
Citrus fruit tree species occurred in the place of S-E Australia	Experienced severe hot weather due to continuous rising temperature	Heavy mortality of tree was observed due to Light brown apple moth (*Epiphyas postvittana*)	Thomson et al. (2010)
Pyrus pyrifolia tree species (commonly known as Apple pear tree) distributed in the region of Rajgarh in the state of Himachal Pradesh (H.P.)	Experienced severe hot weather due to continuous rising temperature	Tree species mortality	Anonymous (2008)
Malus pumila tree species (commonly known as Apple tree) distributed in the region of Rajgarh in the state of Himachal Pradesh (H.P.)	Experienced severe hot weather due to continuous rising temperature	Experienced poor in growth and development due to disturbed reproductive biology that resulted lowering in the fruit bud formation	Chadha and Awasthi (2005)
Malus pumila tree species (commonly known as Apple tree) widely distributed in the region of Kullu and Shimla in the state of Himachal Pradesh (H.P.)	Frequent changing temperature	Species diversions from apple (*Malus pumila*) to Kiwi was observed	Gulati (2009)
Distribution of both *Shorea robusta* (Sal tree species) and *Dipterocarpus turbinatus*	Experienced climate change phenomenon along with deforestation activity	Both tree species are threatened badly and loss of species are occurred	Deb et al. (2017)

(continued)

Table 13.1 (continued)

Tree species of varying places	Extreme weather types	Impacts	Source
(Gurjan tree species) in the regions of S.E. Asia			
Garcinia indica tree species (commonly known as Mangosteen which is distributed in the region of Northern-Western Ghats	Wide and frequent changing temperature due to extreme weather events	Tree species are threatened badly and loss of species are occurred	Pramanika et al. (2018)

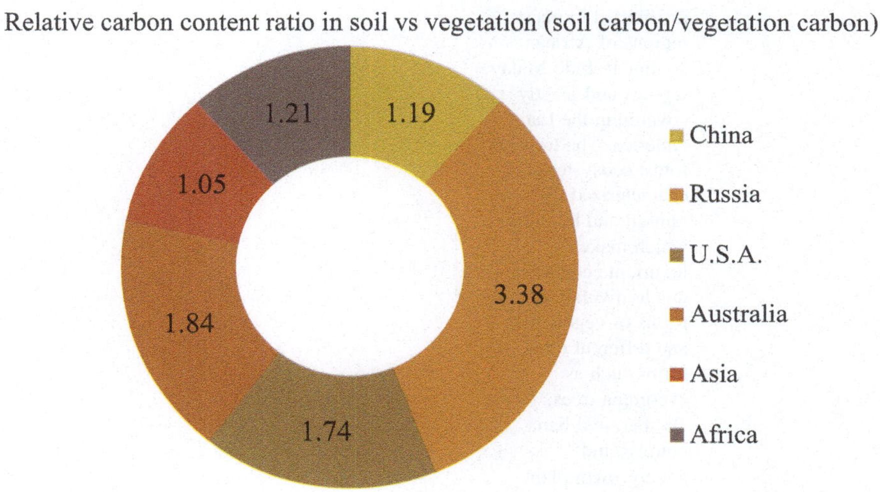

Fig. 13.3 Relative abundance of carbon value in soil versus vegetations in different parts of the world (Dixon et al. 1994)

resources shows a promise for moving towards sustainable development. Soil makes resources conservation in various ways such as enhancing soil fertility, efficient nutrient and water cycling, maximizing water availability and holding capacity, improving nutrient use efficiency, better nutrient availability, minimizing nutrient losses through leaching, optimizing the infiltration rates, improving the microbial populations and its health, better rhizosphere biology, greater climate resilience through C sequestration, etc. which maintains overall soil-food-climate security and ecological stability which is a true pillar of sustainable development at social, economic, and environmental dimensions (Lal 2009; Pimentel and Burgess 2013).

As we know soil is a global treasure and very important natural resources that uphold other resources such as forest, agriculture, agroforestry, and animals and stores various soils inhabiting microorganism which play an important role in ecosystem processes and nutrient cycling. However, practicing resource conservations practices through scientific management practices promotes soil

Table 13.2 Carbon sink value in different forest biomes and its characteristics (Modified: Prentice 2001)

World forests	Shorts characteristics	Total carbon sink value in Pg (pictogram)	Relative percentage contribution
Tropical forest biomes (covered approx. 1.76 million hectare of area)	This is largest forest biome and also known as "lung of the world" comes into the mark in between tropic of cancer and Capricorn which is widely distributed in many regions of African continent, Indo Malaya regions and mostly covered in the Latin America. The tropical forest ecosystem is characterized by high rainfall and humidity which makes a luxuriant ecosystem due to diversified forms of vegetations and different other life forms such as evergreen trees, climber, and liana, buttress and undergrowth plant species exit in the regions. The nature of complexity and diversity represents a better ecosystem services in this tropics	A total 553 Pg C pools value was observed of which vegetations contributed maximum value (340) as compared to soil having 213 Pg C	Overall vegetation portions contributed around 61.48% C whereas soil recorded 38.52%, respectively
Temperate forest biomes (covered approx. 1.04 million hectare of area)	This biomes covered least areas and particularly distributed in regions of northern and southern hemisphere, having moderate rainfall and temperature, less diverse, and are deciduous in nature. Tree like ash, redwood, oak, and beech, etc. are peculiar species of this biome	A total 292 Pg C pools value was observed of which soil contributed maximum value (153) as compared to vegetations having 139 Pg C	Overall vegetation portions contributed around 47.60% C whereas soil recorded 52.40%, respectively

(continued)

Table 13.2 (continued)

World forests	Shorts characteristics	Total carbon sink value in Pg (pictogram)	Relative percentage contribution
Boreal forest biomes (covered approx. 1.37 million hectare of area)	This biome covered the areas in between tropical and temperate forests which are distributed in subarctic and cold climate comprising the regions of Indonesia, China, Alaska, Finland, Canada, and the USA, respectively. This ecosystem is experiencing low temperature winter climate along with very short summer. Vegetations are generally evergreen in nature having pine and larch forest which having closed canopy and needle shaped leaves	A total 395 Pg C pools value was observed of which soil contributed maximum value (318) as compared to vegetations having 57 Pg C	Vegetation portions contributed around 14.43% C whereas soil recorded 85.87%, respectively

ecosystem services that maintain food-soil-climate security along with agroecosystem sustainability. It is the basis for healthy environment and ecological stability which helps in achieving sustainable development goals at social, economic, and cultural dimensions in global context.

13.4 Climate Change: A Burning Issue

Today, we cannot deny the issues and impacts of climate change on our ecosystem structure and related uncountable services for welfare of mother earth. Changing weather pattern, uncertain rainfall, extreme temperature, and other unusual climatic regimes are surely affecting our lifestyle, health, and productivity of various land use systems (agriculture, forestry, agroforestry, horticulture, etc.). The CO_2 is the most potent GHGs that contribute major part in overall atmospheric emissions. Climate change is directly or indirectly linked with carbon balance and related footprints which determines "*How much C is released (through source) and absorbed (by sinks) to make C balance into the atmosphere?*"

US EPA (2019) has reported GHGs emissions value as 26, 19, 17, 14, and 13% from different sectors such as agriculture, transportation, industry, electricity, and commercial and residential activities, respectively (Fig. 13.4). Of these sources,

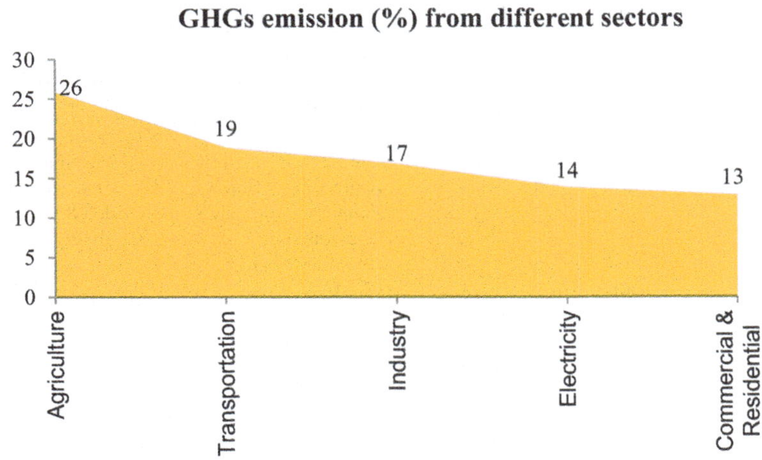

Fig. 13.4 Emission percentage of GHGs from different sectors (US EPA 2019)

agriculture practices represent higher position due to unscientific way of practices, intensive agricultural practices, high synthetic inputs, unsustainable land management practices, and overall we can say the farming practices which stand for removal and emissions of C into the atmosphere that leads to global warming and climate change issues. In turn, these unstoppable emissions definitely affect overall yield, health, and production systems due to deleterious impacts on environment sustainability and ecological stability.

13.5 Nexus Between Climate Change and Soil Ecosystem

Climate change is a mega event and therefore it is directly linked with various components of environment including air, water, soil, and biota. The nexus of soil ecosystem with the event of climate change is critical on various aspects. Some are direct and some are indirect. For example, the direct impact includes alterations in the OC (organic carbon) transformation, biogeochemical cycling on various moisture regimes. Further, climatic alteration may promote soil erosion due to higher level of precipitation. In this way climate change has its important influence over soil ecosystem services. Further, advancement in research has revealed the occurrence of drought or flood incidences within soil ecosystem is linked with the temperature and moisture regime of the climate (Coyle et al. 2016).

Climate change imposes significant alteration in the soil physical structure through changes in the soil mineralogy. This leads to loss of soil fertility and increases the demand for synthetic fertilizer. Climate change is associated with alteration in the precipitation pattern, changes in temperature and elevated level of CO_2. These are associated with the climatic elements. Irregularities in such process may lead to frequent occurrence of flood and drought. Climatic variables such as the moisture, temperature, influences the process of soil formation, weathering, and hence the development of the soil (Pareek 2017).

The indirect nexus rely on the mitigatory strategies for climate change adopted for a particular area. In order to mitigate climate change various technologies in the form of organic farming, zero tillage agriculture, climate resilient agriculture practices have been adopted across the globe. As a consequence of that modification has taken place in the agriculture practice. This may influence the soil ecosystem as reflected from the quality and quantity of the yield (Mandryk et al. 2017).

Therefore, there is close nexus between the various climatic elements and soil ecosystem which needs to be maintained properly for sustainability of agroecosystem. Further, for maintaining ecosystem homeostasis proper balance needs to be maintained by adapting to suitable strategies and policy formulation to combat climate change.

13.6 Soil Carbon: Fact and Figures

SOC is an important key for soil health and quality that determines the fertility status and productivity. Addition, decaying and decomposition of leaf litter and other residues from nitrogen fixing leguminous woody perennial trees would help in enhancing organic pools into the soil that promotes microbial populations and rhizosphere biology which in turn improves soil fertility and overall productivity level. This will surely improve health and quality of soil and healthy soil having greatest potential to capture C through C sequestration that also maintains C balance in the atmosphere and mitigate the ongoing burning issue of changing climate to make more sustainable environment. However, the SOC pools varies as per varying topography, soil class, types, orders, and can be influenced by various biotic and abiotic factors. Meanwhile, soil depth and its variations also affect the organic C content. For example, 1 m of top soil layers comprised 1500 pg C of SOC stock value around the world (FAO and ITPS 2015). Moreover, the regions of peatlands and wetlands contributed higher value of SOC pools, whereas the tropical and permafrost regions also comprise maximum C pools (Gougoulias et al. 2014; Köchy et al. 2015). Therefore, climatic situations are also key factors which determine the status of SOC pools. However, the main question is "How soil order affects SOC pools?" Thus, we need to explore and quantify the value of SOC pools as per varying soil order, types, depth, etc. which can be modified by varying practices of soil management. In this context, the value of SOC as per varying soil order in the world is depicted in Fig. 13.5.

From this figure, the maximum (316 billion tons) value of SOC pools was contributed by the soil order Gelisols followed by 190 billion tons in Inceptisols and least value (20 billion tons) was observed in Andisols (Eswaran et al. 2000).

As per one estimates, a total 4.03 billion ha of earth area was covered by forest ecosystem of which area both tropical and boreal forest contributed highest value of SOC pools (Pan et al. 2013). Deforestation and other anthropogenic factors affect the overall SOC pools by releasing 25% of soil C into the atmosphere (FAO and ITPS

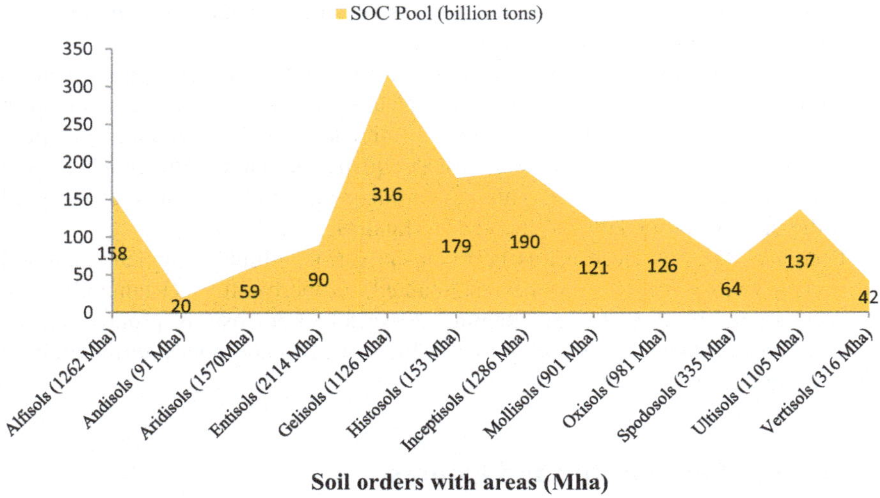

Fig. 13.5 SOC Pools in different orders of the world soils (Adapted: Eswaran et al. 2000)

2015). It is well understood that the extent of SOC pools is directly or indirectly connected with climate change mitigation (IPCC 2007; FAO 2015). Moreover, these SOC values also represent the extent of available essential soil nutrients that will affect the quantity and quality of food grains and fruits which maintains overall food and nutritional security. Thus, SOC is the interesting determinant of soil-water availability, soil fertility, efficient nutrient cycling, and nutrient use efficiency, and overall soil physico-chemical properties that improve yield and productivity by promoting plant growth and its development (Zdruli et al. 2017).

13.7 Soil Carbon Under Changing Climate

Climate change is the biggest issue in the forefront of the scientific communities of the world. The alarming climatic situation leads the fluctuations and rate of existing emission pattern throughout the globe. The land-use change, urbanization, industrialization, modernization, and laxative consumptive pattern and lifestyles change the global scenario of the GHGs emission (IPCC 2014). These changes create the downfall of various sectors and segments of the ecosystem. In the context of GHGs emission from various sectors the leading one is agriculture, forestry, and other land uses contribute about 1/4th of the total global emission as presented in Fig. 13.6.

Soil C pool is essential from global climate and agroecosystem sustainability perspectives. The change in climatic segments alters the cropping pattern, temperature regimes, production schedules, and shift in land use, adaptive traits of flora and fauna and their associated role and performance in various ecological processes. As

Fig. 13.6 Sectoral scenario of worldwide greenhouse gas emission (Adapted: IPCC 2014)

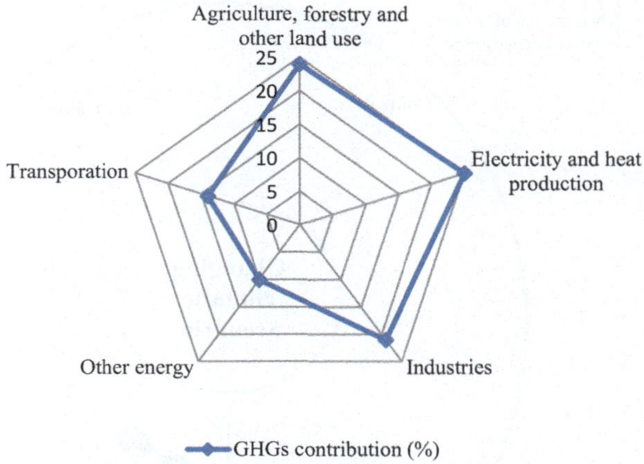

per Wiesmeier et al. (2016), SOC is reduced about 11–16% with the increment of mean annual temperature of 3.3 °C. Beside the temperature, higher rate of rainfall can accelerate the accumulation of SOC than the region having low rainfall due to higher biomass accumulation by plants in the presence of water (Trost et al. 2013).

The soil moisture and temperature are important aspect which decides the flux of C in various natural and managed ecosystems (Schlesinger and Andrews 2000). The bio-activation process (respiration and function) both needed specific temperature and available moisture regime and control the soil biology (Poll et al. 2013). Soil wetness and dryness is regulated by rainfall and irrigation which accelerates activity of soil biota and turnover of nitrogen and C in the soil (Unger et al. 2010). The level of atmospheric CO_2 can increase the productivity of plants in those areas where nutrient and moisture are not limiting factors (Runion et al. 2009). Further the impact on various ecological processes due to climate change at site specific level needs to be monitored and determined with their suitable management options.

13.8 Adapting and Managing the Impact of Climate Change

Mitigation of climate change through SOC stock is acknowledged in various researches. The increasing C sink in the agroecosystem is challenging and the capability of soil as source of sink towards combating climate change is uncertain. Soil, biophysical environment, socio-economic system, and the climate are intricately linked with each other. The GHGs reduction and its stabilization is the biggest environmental challenge which alters the environmental chemistry and causes the various undesirable events like global warming and climate change (Khan et al. 2020a, b). In this context, soil and plant systems seem to be effective part towards C sinks. Emission reduction is fundamental aspects for mitigating the negative consequence of climate change on the earth's biota. Therefore, C sinks in the form of storage and sequestration has become an integral part for adaptive

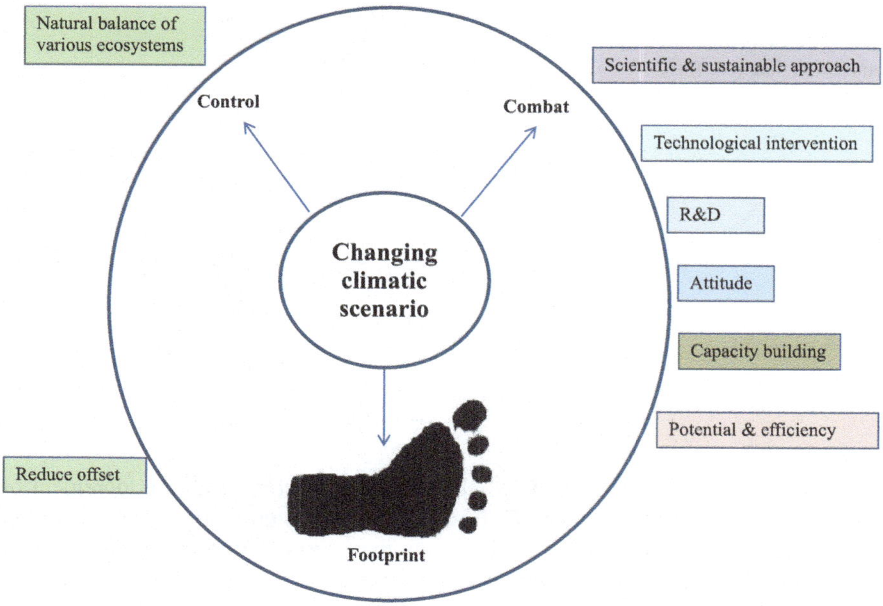

Fig. 13.7 Various adaptive and management approaches for combating climate change (Modified: Jhariya et al. 2019a, b; Raj et al. 2020; Banerjee et al. 2020)

mechanism and C management (Jhariya 2017). Presently the management of soil resources are not properly aligned to cope up with climate change. Therefore, effective soil conservation, protection, restoration, utilization, and management are the essential component to adapt and mitigate the current climatic scenario in coming future. In this context, various adaptive and management approaches are used for combating climate change (Fig. 13.7) (Jhariya et al. 2019a, b; Raj et al. 2020; Banerjee et al. 2020).

13.9 Managing the Soil Carbon

Managing and conservating the soil C is more effective way towards climate change adaptation and mitigation strategies which reduces the concentration of atmospheric CO_2 (Mayer et al. 2018). Therefore, the information regarding the SOC and associated C sequestration potential in the soil ecosystem is essential step to determine the source and sink of C altered by various biotic influences. There is several segments where the sink of C can be improved and increase for long term pool through best management and alternative practices in agroecosystem and other land-use. In farming system, soils are the potential means of C sink which is regulated by cropping system and management regimes. Thus, implementing proper practices like conservation agriculture, agroforestry, nutrient and soil-water management, diversification, etc. serves increment in the C sink and sequestration.

An estimate revealed the soil C pool, biotic pool, and atmosphere possess 2500 Pg, 650 Pg, and 560 Pg, respectively. Soil conservation and protection practices especially erosion control measure enhance the soil C reserve. The degradation of soil, its health and quality depletes the organic source and reduces the soil fertility and productivity up to 50% due to erosion and desertification. Therefore, SOC is the key feature of soil quality and sustainability of agroecosystem and C cycle. The C sink rate is regulated by C accumulation, stabilization, C stock equilibrium, and equilibrium phases or saturation point, etc. Among the abiotic component precipitation and temperature regime of the region exert key impact on quantity and quality of organic materials. Subsequently its decomposition and nutrient release in the system determines the storage capacity of the soil (Mandal 2020). Therefore, proper management of soil for increasing the C concentration in the tropical region from 0.1–0.2% to 1.10% is extremely difficult and need more attention under the era of changing climate and for achieving the agroecosystem sustainability and sustainable development goals (Lal 2004b).

Soil C in forest ecosystem varied significantly as per the forest types, vegetation mix, climatic regimes, and scale of interferences. As per Sreekanth et al. (2013) soil C of different forests was 165.24 mg ha^{-1} in temperate forest, 138.64 mg ha^{-1} in deciduous forest, 135.42 mg ha^{-1} in tropical thorn forest, and 104.94 mg ha^{-1} in riparian forest. Among these tropical thorn forest and riparian forest possess higher labile C pool, while the temperate forest has highest non-labile C pool. These changes are also observed in rangeland, pasture, and grassland ecosystems. The soil C in the global soil pools varies from 1500 to 2000 Gt (giga tons) and 600 to 700 Gt C in the vegetation pools (Mandal 2020). The soil ecosystem is becoming fragile due to its continuous degradation in an unsustainable manner. The increasing demand of the commodities is putting the pressures on global soil ecosystem. Nearly 1/3rd of the land resource is under moderate to severely degraded condition due to toxic impurities, salinization, acidification, compaction, and soil erosion as well as chemical pollutants. Therefore, soil health is the prime concern of today to serve the food to global population. Thus, it is included in the sustainable development goals under the target of 15.3, i.e., *Land Degradation Neutrality*, and UN Convention to *Combat Desertification* (Laban et al. 2018). Further, as per international forum (COP-21 in Paris) managing of GHGs and climate change *French 4 per mille initiative* needs to be addressed for effective management and balancing the source and sink of atmospheric CO_2. According to this, one should increase the C contained in the soil by 4% or 0.4% a year, then it will be effective measure to halt the CO_2 increment in the atmosphere (Lal 2016; Amundson and Leopold 2018). Thus, there are various management practices through different innovation aspects which are used for better soil C dynamics and are depicted in Fig. 13.8 (Funk et al. 2015).

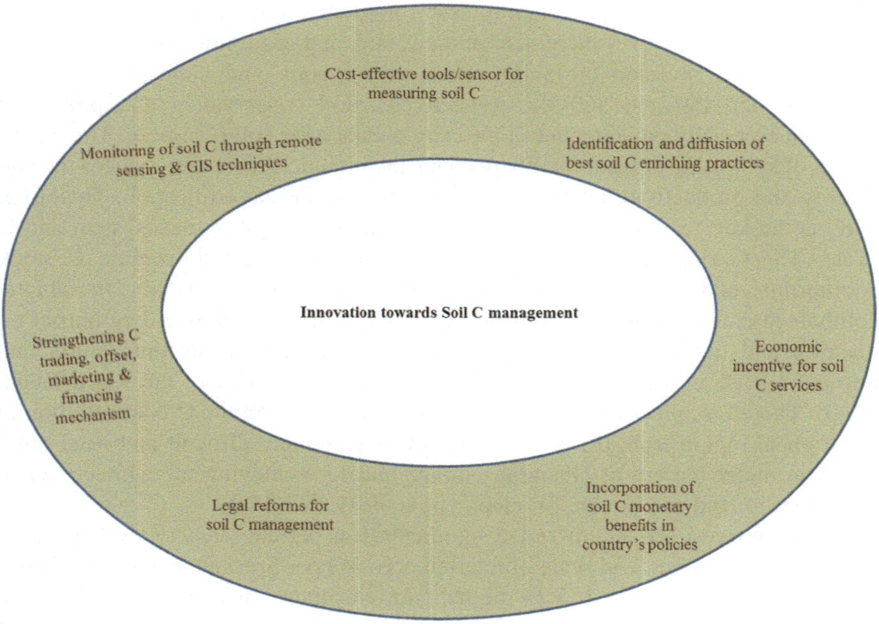

Fig. 13.8 Management and innovation aspects in soil carbon dynamics (Modified: Funk et al. 2015)

13.10 Agroforestry for Soil Carbon Improvement and Management

The current human population growth and demand of the resources for wealth and social development creating the serious concern related to soil-water-environment. Further the food security and crisis along with changing climatic pattern put pressure to intensify the land use, agroecosystem, and other allied sectors which leads to natural disruption and depletion of the various resources (Jhariya et al. 2015). The world's soil is under the threat of desertification due to erosion, land-use change, deforestation, habitat degradation, agricultural intensification, urbanization, industrialization, loss of soil C, and mismanagement. Further, climate change also exerts the significant challenge across the globe. All these demands sustainble approaches, management and planning for improvement of soil, water, ecology and environment to adapt and mitigate the various sort of risks associated with global change. In this context, agroforestry seems to be very promising to facilitate the climate change risk adaptation, mitigation, and improving the overall resilience besides the various social, economic, ecological, and environmental benefits (Raj et al. 2019b; Jhariya et al. 2019b).

Agroforestry is an age old practice but became modernized science in recent few decades. Agroforestry is simply the application of forestry with agricultural production in the same land resources and management regimes to diversify and maximize the output (Singh and Jhariya 2016; Painkra et al. 2016). The multilayer species combination in this practice judiciously utilizing the resources and protect the soil and environment through various functions. The choice of species, correct crop combination, crop rotation, planting legumes, conservation practices, and management aspect builds the healthy soil environment without deteriorating the quality of the soil and maximizes the production in a sustainable manner (Jhariya et al. 2019a; Raj et al. 2020; Banerjee et al. 2020). The species having fast growth rate, biomass accumulation, and vigorous root architecture are the key to restore the soil C. Moreover, the herbaceous species sustained the belowground biomass which builds the C pools (Lemus and Lal 2005). These C pools collectively permit to implement sustainable soil management design to enhance the possible soil C in the system. However, many authors have reported the C sink value in soil ecosystem of varying agroforestry systems in the most parts of the world that represents a diverse form of C sequestrations as per diversifying soil ecosystem (Table 13.3).

13.11 Fruit Based Agroforestry for Carbon Sequestration

Perennial fruit trees are good components of C farming systems due to a great potential of C capture and store as biomass into the vegetations and soils as SOC pools through soil C sequestration. However, C capture potential of different perennial tree species in the world is reported in Table 13.4. No doubt, agroforestry is sustainable land-use farming practices and integration of leguminous N_2 fixing perennial fruit trees can effectively enhance the C sink capacity that confirm C balance in our ecosystem and promising higher yield and productivity.

Soil enrichment through fruit based agroforestry system is understood by studying and exploring the value of SOM and SOC pools into the soil. It helps in enhancing soil fertility by efficient nutrient cycling, promotes microbial populations by improving rhizosphere biology, and enhances nutrient use efficiency as well as its capture by perennial and annual plants in agroforestry system that maintains overall food-soil-climate security at global scale. However, many authors have reported the comparative value of C sink under sole vs. combined agroforestry system in different agroclimatic zones of the country. For example, Singh and Singh (2015) have calculated the C sink capacity in the form of biomass C, soil C, and total C sink values from sole plant system vs. various fruit tree combinations in horti-silviculture based land use systems in dry regions of Rajasthan. This work having more emphasis on comparative studies of C sinks value among varying fruits based agroforestry systems and according to them, fruit tree based agroforestry systems (horti-silviculture system) contributed more C sink value as compared to sole based cropping systems due to complexity and higher diversity exist in horti-silviculture system. Similarly, overall C sink value is higher into the soil ecosystem than biomass C. Obviously, nutrient losses through leaching is minimized under horti-silviculture

Table 13.3 Soil carbon value of varying agroforestry systems in different parts of the world

Agroforestry systems in the world	Soil carbon values	Source
Gmelina arborea tree based Agri-silviculture model practiced in the region of Chhattisgarh of Indian subcontinents	30.20 US t/ha	Swamy and Puri (2005)
The practices of Home garden system in both central and S. America	Varies from 49.6 to 2.5 US t/ha	Kirby and Potvin (2007)
The practices of Home garden system in the region of Africa	220.5 US t/ha	Nair (2012)
The practices of Silvopastoral systems comprised varying tree-pasture combinations in most parts of Africa	Varies from 1.65 to 3.85 US t/ha	
Tree-crop combinations of poplar + maize, wheat and soybean under hedgerow intercropping systems prevalent in the regions of Canada	1.40 US t/ha	Oelbermann et al. (2006)
The practices of Silvopastoral systems comprised *Pinus Elliottii* (tree component) and grass like *Paspalum notatum* in the region of the USA	Varies from 7.60 to 27.0 US t/ha	Haile et al. (2008)
Multiple tree combinations comprised both *Eucalyptus* and *Casuarina* tree species which is prevalent in Puerto Rico (USA)	68.23 US t/ha	Parrotta (1999)
Multiple tree combinations comprised both *Leucaena* (subabul) and *Casuarina* tree species which is prevalent in Puerto Rico (USA)	62.40 US t/ha	
Multiple tree combinations comprised both *Leucaena* (subabul) and *Eucalyptus* tree species which is prevalent in Puerto Rico (USA)	68.01 US t/ha	
The practices of silvopastoral systems comprised cork oak tree (*Quercus suber*) along with valuable pastures in the region of Spain	Varies from 29.21 to 55.3 US t/ha	Howlett (2009)
The practices of silvopastoral systems comprised *Betula pendula* tree species along with valuable pastures in the region of Spain	Varies from 147.0 to 165.3 US t/ha	Howlett et al. (2011)
The practices of silvopastoral systems comprised *Eucalyptus* tree species which is combined with *Brachiaria* grass species in the region of Brazil	389.1 US t/ha	Tonucci et al. (2011)
The practices of silvopastoral systems comprised various combination of tree and grasses/pastures in the varying parts of the USA	564.4 US t/ha	Haile et al. (2010)
Model of hedgerow intercropping comprised poplar tree species (*Populus deltoids*) practiced in the region of Canada	63.0 US t/ha	Bambrick et al. (2010)
Poplar tree based Agri-silviculture system practiced in the region of Punjab of Indian subcontinents	10.4 US t/ha/year	Chauhan et al. (2010)
Subabul tree based Agri-silviculture system practiced in the region of Andhra Pradesh of Indian subcontinents	3.05 US t/ha/year	Rao et al. (1991)
The practices of Silvopastoral systems comprised babul (*Acacia nilotica*) with various important grasses/ pastures species in Haryana region of India	3.10 US t/ha/year	Kaur et al. (2002)

(continued)

Table 13.3 (continued)

Agroforestry systems in the world	Soil carbon values	Source
Model of home garden practiced in the state Kerala in Indian sub-continent	1.80 US t/ha/year	Saha et al. (2009)
Casuarina equisetifolia tree based Agri-silviculture model practiced in the region of Tamil Nadu of Indian subcontinents	1.73 US t/ha/year	Viswanath et al. (2004)
The practices of Silvopastoral systems comprised fodder species like bread grass (*Brachiaria brizantha*) combined with bay cedar (*Guazuma ulmifolia*) and salmwood (*Cordia alliodora*) in the region of Costa Rica	145.5 US t/ha	Amezquita et al. (2005)
A model of Silvopastoral system comprised *Acacia mangium* (tree species) + *Arachis pintoi* (fodder species) in the region of Costa Rica	191.0 US t/ha	
Poplar (*Populus deltoids*) + barley (*Hordeum vulgare*) based Agri-silviculture model practiced in the region of Canada	86.5 US t/ha	Peichl et al. (2006)
Douglas-fir (*Pseudotsuga menziesii*) + *Trifolium subterraneum* based Agri-silviculture model which is mostly practiced in the region of the USA	106.0 US t/ha	Sharrow and Ismail (2004)
Subabul (*Leucaena leucocephala*) based hedgerow intercropping system prevalent in Nigeria, Africa	15.0 US t/ha	Lal (2005)
A ecological sustainable model of fodder or protein bank system comprised the tree species of *Pterocarpus* and *Gliricidia* (a good N_2 fixing leguminous tree) in the region Mali	37.0 US t/ha	Takimoto et al. (2008)
Gliricidia (N_2 fixing leguminous tree) + maize (*Zea mays*) based Agri-silviculture system practiced in the region of Malawi	135.6 US t/ha	Makumba et al. (2007)

system due to process of closed and efficient nutrient cycling as compared to higher nutrient loss in sole cropping system due to open (leaky) nutrient cycling. Also, a maximum total C value was observed in the combination of *Cordia myxa* + *P. cineraria* based AFs due to greater potential of C sequestration which reflects the importance of tree-crop combination, types and its nature.

In this context a figure has been developed on comparative studies of biomass C and total C sink (Fig. 13.9) along with biomass C, soil C sink (Fig. 13.10) in horti-silviculture vs. sole tree systems in dry regions of Rajasthan (India) that is self-representative which explains that "*How the changing land use patterns i.e. sole vs. combined fruit based agroforestry affects C captures and biomass*?" Thus, these figures and reports are enough to represent the significance of perennial fruits based agroforestry systems in C sequestration and biomass production to make overall ecosystem stability for environmental sustainability.

Table 13.4 Carbon capture potential of different perennial tree species in the world

Perennial tree species	Botanical name	Carbon capture potential	Source
Australian wattle tree	*Acacia auriculiformis*	7.7 Mt C $year^{-1}$	Raizada et al. (2003)
North Indian rosewood	*Dalbergia sissoo*	3.6 Mt C $year^{-1}$	
Coast she oak	*Casuarina equisetifolia*	1.9 Mt C $year^{-1}$	
Gamhar	*Gmelina arborea*	1.4 Mt C $year^{-1}$	
California redwood	*Sequoia sempervirens*	5000 t C ha^{-1}	Runyon et al. (1994)
Douglas-fir	*Pseudotsuga menziesii*	1000 t C ha^{-1}	
Deodar	*Cedrus deodara*	469.1 t C ha^{-1}	Sharma et al. (2011)
Bahera	*Terminalia bellirica*	327.78 t C ha^{-1}	Hangarge et al. (2012)
Eucalyptus spp.	*Eucalyptus globulus*	320.67 t C ha^{-1}	Chavan and Rasal (2011)
Black wattle	*Acacia mangium* Willd.	292.02 t C ha^{-1}	Ilyas (2013)
Tropical clumping bamboo	*Bambusa balcooa*	234.17 t C ha^{-1}	Borah and Chandra (2010)
Indian bat tree	*Ficus amplissima*	221 t C ha^{-1}	Hangarge et al. (2012)
Teak	*Tectona grandis*	181 t C ha^{-1}	Sreejesh et al. (2013)
Rubber tree	*Hevea brasiliensis*	136 t C ha^{-1}	Dey (2005)
Poplar	*Populus deltoids*	115 t C ha^{-1}	Gera et al. (2006)
Mango	*Mangifera indica*	104.41 t C ha^{-1}	Chavan and Rasal (2012)
Ban oak	*Quercus leucotrichophora*	77.3 t C ha^{-1}	Sharma et al. (2011)
Siris tree	*Albizia lebbeck*	11.97 t C ha^{-1}	Jana et al. (2009)
Sal	*Shorea robusta*	8.97 t C ha^{-1}	
Mango (10 years old)	*Mangifera indica*	58.1 kg/tree	Shinde et al. (2015)
Mango (15 years old)	*Mangifera indica*	115.4 kg/tree	
Coconut (10 years old)	*Cocos nucifera*	56.6 kg/tree	
Coconut (15 years old)	*Cocos nucifera*	126.3 kg/tree	
Jamun (10 years old)	*Syzygium cumini*	38.7 kg/tree	
Jamun (15 years old)	*Syzygium cumini*	78.8 kg/tree	
Guava (10 years old)	*Psidium guajava*)	32.9 kg/tree	
Guava (15 years old)	*Psidium guajava*	54.3 kg/tree	
Mango based orchards throughout India	*Mangifera indica*	285.0 MT C	Ganeshamurthy et al. (2019)

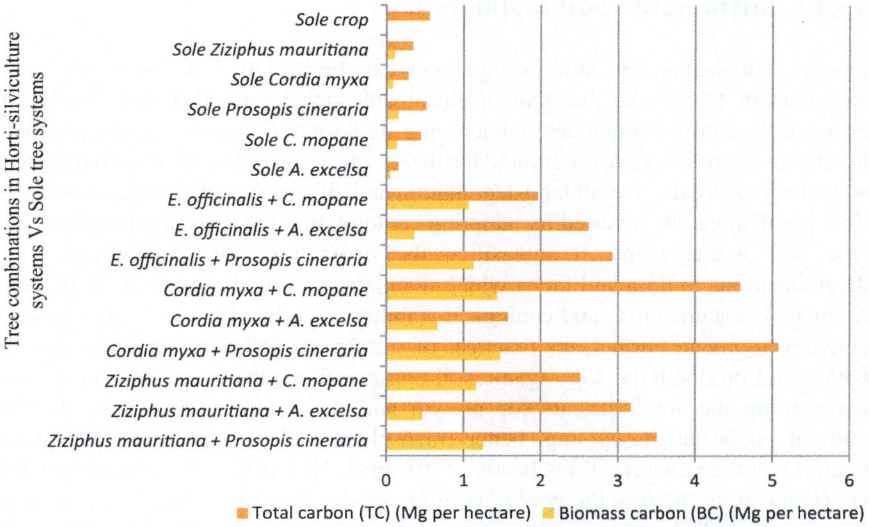

Fig. 13.9 Comparative studies of biomass carbon and total carbon sink in horti-silviculture vs. sole tree systems in dry regions of Rajasthan (India) (Singh and Singh 2015)

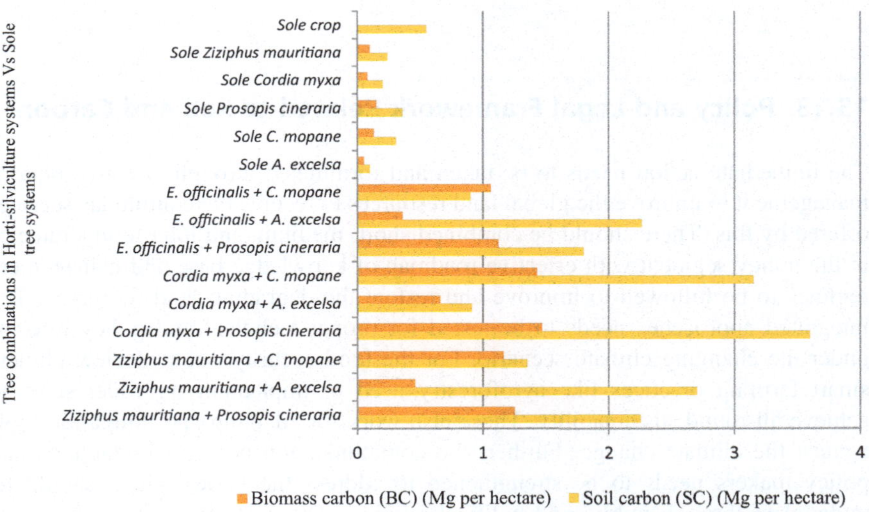

Fig. 13.10 Comparative studies of biomass carbon and soil carbon sink in horti-silviculture vs. sole tree systems in dry regions of Rajasthan (India) (Singh and Singh 2015)

13.12 Sustainable Soil Management

Practices of sustainable soil management can boost the SOC pools for better environment to achieve the goal of sustainable development. Today, faulty and unscientific land-use practices, intensifying agricultural systems, higher synthetic inputs, uncontrolled use of inorganic fertilizers, etc. deprive the health and quality of soil which insure the overall land degradation and affects overall ecosystem services. This problem can be reverted by adopting ecology oriented scientific based sustainable soil management. It intensifies the ecosystem services through more diversifications in flora and fauna which maintains soil-food-climate security, environmental sustainability, and ecological stability for achieving the goal of sustainable development. Further, the practices of sustainable soil management ensure the better land productivity, top organic soil conservations, minimize the soil erosion, maintaining the soil nutrients for growth and development of plants, effective nutrient pools and its cycling, better rhizosphere biology and as a consequence overall resource conservations becomes effective. However, the common question is "*How can we achieve the goal of sustainable development through soil management practices*?" A better soil management practices will surely improve SOC pools which is a good indicator of fertility status into the soil that directly or indirectly is connected with environmental sustainability, ecological stability which can touch the mark of sustainable development at global scale.

13.13 Policy and Legal Framework Related to Soil and Carbon

The immediate action needs to be taken and formulated through effective plan of management to improve the global land resources to receive the continuous services offered by this. There should be combined short, medium, and long-term strategies in the policy segment with effective roadmap of knowledge base and management regimes to be followed to improve and restore the degrading land resources. The integrated approaches needs to be tested for proper technical and policy reforms under the changing climate scenario. For the food security point of view climate smart farming practices like agroforestry need to implement on wider scale to achieve all-round sustainability. These also work for adapting and mitigation tools against the climate change. Further, the communication between researchers and policy makers needs to be strengthened to address the issues which should be undertaken through public and political opinion. The cost-effective sensors of C monitoring, site specific smart practices, C trading, C marketing and C payments services, monetization of soil C welfare in policies, legal framework to regulate the C, national and international and certification strategies for better land resource management, etc. should be addressed on priority basis to cut-off the C emission and improve C sink.

13.14 Conclusion

The soil management, combating climate change, food security, and improving the agricultural production and technology needs to be transformed through best practices having eco-friendly approach. However, better management practices of different land use systems would be helpful in enhancing the potential of C sequestration into the soils that not only build soil fertility, microbial populations, and rhizosphere biology but also to solve the problem of climate change and maintain overall soil-food-climate security for making environmental sustainability. It is required to create awareness among the farming community to conserve the SOM and soil resources. Farming communities needs to be communicated strictly regarding avoiding the residue burning, clean cultivation, intense tillage, and other traditional practices which are not eco-friendly or efficient. Further, conservation of land resources and organic material is essential from soil quality and health point of view.

13.15 Future Thrust

Resource conservation, reforestation, and rehabilitating of forest lands are the devoted program and schemes which are continuously implemented in various countries of world especially in developing countries where deforestation rate is higher and still continuing at an alarming rate causing severe environmental problems to a greater extent. Therefore, management of the natural resources with effective plans, policy, and strict legal implementation at ground level is the need of the present scenario. The focus must be on restoring the soil, water, and the environment for better performance of agroecosystem and human survival at comfort zone within the carrying capacity of the various ecosystems.

The C management in terrestrial ecosystem should be well placed within the planned framework of eco-friendly and green techniques to reduce the ecological and environmental footprints by improving the existing agricultural practices, resource use efficiency, organic and conservative farming, reducing the residue burning, proper environmental clearance monitoring as well as educating the people for go green concept to minimize the pollutant load and increasing the sink. The SOM in the earth's pedosphere needs to be improved and increased. In this perspective every country should plan the systematic monitoring, identification and implementation of best management practice in local and regional level to manage the soil ecosystems. The present farming system is needed to be shifted into *climate smart soils system* to make the natural balance between sources and sink under changing climate. But the community engages in the farming system frame the land resources as *weather proofing soil system* which has more potential to perform productive traits under varying climatic scenario. Thus, future research and developmental activities needs to be interlinked with these two aspects towards win-win strategy for the farming community as well as for global wellbeing.

References

Amezquita MC, Ibrahim M, Llanderal T, Buurman P, Amezquita E (2005) Carbon sequestration in pastures, silvopastoral systems and forests in four regions of the Latin American tropics. J Sustain For 21:31–49

Amundson R, Leopold B (2018) Soil carbon sequestration is an elusive climate mitigation tool. PNAS 115(46):11652–11656

Anonymous (2008) ENVIS newsletter Jul.–Dec., 2008, volume II. http://www.hpenvis.nic.in

Bambrick AD, Whalen JK, Bradley RL, Cogliastro A, Gordon AM, Olivier A, Thevathasan NV (2010) Spatial heterogeneity of soil organic carbon in tree-based intercropping systems in Quebec and Ontario, Canada. Agrofor Syst 79:343–353

Banerjee A, Jhariya MK, Yadav DK, Raj A (2020) Environmental and sustainable development through forestry and other resources. Apple Academic Press, Palm Bay, p 400. https://doi.org/10.1201/9780429276026

Bolstad PV, Vose JM (2005) Forest and pasture carbon pools and soil respiration in the southern Appalachian Mountains. For Sci 51:372–383

Borah RP, Chandra A (2010) Carbon sequestration potential of selected bamboo species of Northeast India. Ann For 18(2):171–180

Camarero JJ, Gazol A, Sancho-Benages S, Sangüesa-Barreda G (2015) Know your limits? Climate extremes impact the range of Scots pine in unexpected places. Ann Bot 116:917–927

Chadha KL, Awasthi RP (2005) The apple improvement: production and post harvest management. Malhotra Publishing House, New Delhi, pp 16–23

Chauhan SK, Sharma SC, Chauhan R, Gupta N, Srivastava R (2010) Accounting poplar and wheat productivity for carbon sequestration in agrisilviculture system. Indian For 136(9):1174–1182

Chavan BL, Rasal GB (2011) Sequestered carbon potential and status of Eucalyptus tree. Int J Appl Eng Technol 1(1):41–47

Chavan B, Rasal G (2012) Total sequestered carbon stock of Mangifera indica. J Environ Earth Sci 2:37–48

Chen CR, Xu ZH, Mathers NJ (2004) Soil carbon pools in adjacent natural and plantation forests of subtropical Australia. Soil Sci Soc Am J 68:282–291

Coyle C, Creamer RE, Schulte RPO, O'Sullivan L, Jordan P (2016) A functional land management conceptual framework under soil drainage and land use scenarios. Environ Sci Policy 56:39–48. https://doi.org/10.1016/j.envsci.2015.10.012

Deb JC, Phinn S, Butt N, McAlpine CA (2017) The impact of climate change on the distribution of two threatened Dipterocarp trees. Ecol Evol 7:2238–2248. https://doi.org/10.1002/ece3.2846

Dey SK (2005) A preliminary estimation of carbon stock sequestrated through rubber (*Hevea brasiliensis*) plantation in north eastern region of India. Indian For 131:1429–1436

Dixon RK, Brown S, Houghton RA, Solomon AM, Trexler MC, Wisniewski J (1994) Carbon pools and flux of global Forest ecosystems. Science, New Series 263(5144):185–190. http://www.jstor.org/stable/2882371

Eswaran H, Reich FP, Kimble JM, Beinroth FH, Padamnabhan E, Moncharoen P (2000) Global carbon stocks. In: Lal R, Kimble JM, Eswaran H, Stewart BA (eds) Global climate change and pedogenic carbonates. CRC/Lewis, Boca Raton, FL

FAO (2015) Learning tool on nationally appropriate mitigation actions (NAMAs) in the agriculture, forestry and other land use (AFOLU) sector. FAO, Rome

FAO, ITPS (2015) Status of the world's soil resources. FAO, Rome

Funk R, Pascual U, Joosten H, Duffy C, Pan G, la Scala N, Gottschalk P, Banwart SA, Batjes N, Cai Z, Six J, Noellemeyer E (2015) From potential to implementation: an innovation framework to realize the benefits of soil carbon. In: Banwart SA, Noellemeyer E, Milne E (eds) Soil carbon: science, management and policy for multiple benefits. CABI, Wallingford, pp 47–59

Ganeshamurthy AN, Ravindra V, Rupa TR (2019) Carbon sequestration potential of mango orchards in India. Curr Sci 117(12):2006–2013

Gera M, Mohan G, Bisht NS, Gera N (2006) Carbon sequestration potential under agroforestry in Rupnagardistrict of Punjab. Indian For 132(5):543–555

Gougoulias C, Clark JM, Shaw LJ (2014) The role of soil microbes in the global carbon cycle: tracking the below-ground microbial processing of plant-derived carbon for manipulating carbon dynamics in agricultural systems. J Sci Food Agric 94:2362–2371

Gulati V (2009) From apple to kiwi, a journey of returns. http://www.commodityonline.com/news/From-apple-to-kiwi-%96-a-journey-of-returns-14070-3-1.html

Haile SG, Nair PKR, Nair VD (2008) Carbon storage of different soil-size fractions in Florida silvopastoral systems. J Environ Qual 37:1789–1797

Haile SG, Nair VD, Nair PKR (2010) Contribution of trees to carbon storage in soils of silvopastoral systems in Florida, USA. Glob Change Biol 16:427–438

Hangarge LM, Kulkarni DK, Gaikwad VB, Mahajan DM, Chaudhari N (2012) Carbon sequestration potential of tree species in Somjaichirai (sacred grove) at Nandghur village, in Bihar region of Pune district, Maharashtra state, India. Ann Biol Res 3(7):3426–3429

Henneberry TJ (2007) Integrated Systems for Control of the pink bollworm *Pectinophora gossypiella* in cotton. In: Vreysen MJB, Robinson AS, Hendrichs J (eds) Area-wide control of insect pests. Springer, Dordrecht

Hiederer R, Köchy M (2011) Global soil organic carbon estimates and the harmonized world soil database, EUR 25225 EN. Publications Office of the European Union, Luxembourg, p 79

Howlett D (2009) Environmental amelioration potential of silvopastoral agroforestry systems in Spain: soil carbon sequestration and phosphorus retention. Ph.D. Dissertation, University of Florida, Gainesville

Howlett DS, Mosquera-Losada MR, Nair PKR, Nair VD, Rigueiro-Rodríguez A (2011) Soil C storage in silvopastoral systems and a treeless pasture in northwestern Spain. J Environ Qual 40:784–790

Ilyas S (2013) Allometric equation and carbon sequestration of Acacia mangium Willd. in coal mining reclamation areas. Civil Environ Res 3(1):8–16

IPCC (2007) Climate change 2007: mitigation. Contribution of working group III to the fourth assessment report of the intergovernmental panel on climate change. Cambridge University Press, Cambridge, New York

IPCC (2014) Climate change 2014: impacts, adaptation and vulnerability. synthesis report, Contribution of Working Groups I, II and III to the Fifth Assessment Report of the Intergovernmental Panel on Climate Change. IPCC, Geneva, Switzerland

Jana BK, Biswas S, Majumder M, Roy PK, Mazumdar A (2009) Comparative assessment of carbon sequestration rate and biomass carbon potential of young *Shorea robusta* and *Albizzia lebbek*. Inter J Hydro-Clim Eng Assoc Water Environ-Model 1(2):1–15

Jhariya MK (2017) Vegetation ecology and carbon sequestration potential of shrubs in tropics of Chhattisgarh, India. Environ Monit Assess 189(10):518. https://doi.org/10.1007/s10661-017-6246-2

Jhariya MK, Bargali SS, Raj A (2015) Possibilities and perspectives of agroforestry in Chhattisgarh. In: Zlatic M (ed) Precious forests-precious earth. InTech, Rijeka, pp 237–257. https://doi.org/10.5772/60841

Jhariya MK, Yadav DK, Banerjee A (2018a) Plant mediated transformation and habitat restoration: phytoremediation an eco-friendly approach. In: Gautam A, Pathak C (eds) Metallic contamination and its toxicity. Daya Publishing House, A Division of Astral International Pvt Ltd, New Delhi, pp 231–247

Jhariya MK, Banerjee A, Yadav DK, Raj A (2018b) Leguminous trees an innovative tool for soil sustainability. In: Meena RS, Das A, Yadav GS, Lal R (eds) Legumes for soil health and sustainable management. Springer, Singapore, pp 315–345. https://doi.org/10.1007/978-981-13-0253-4_10

Jhariya MK, Banerjee A, Meena RS, Yadav DK (2019a) Sustainable agriculture, forest and environmental management. Springer, Singapore, p 606. https://doi.org/10.1007/978-981-13-6830-1

Jhariya MK, Yadav DK, Banerjee A (2019b) Agroforestry and climate change: issues and challenges. Apple Academic Press, Palm Bay, p 335. https://doi.org/10.1201/9780429057274

Kaur B, Gupta SR, Singh G (2002) Carbon storage and nitrogen cycling in silvi-pastoral systems on a sodic soil in northwestern India. Agrofor Syst 54:21–29

Khan N, Jhariya MK, Yadav DK, Banerjee A (2020a) Herbaceous dynamics and CO_2 mitigation in an urban setup- a case study from Chhattisgarh, India. Environ Sci Pollut Res 27(3):2881–2897. https://doi.org/10.1007/s11356-019-07182-8

Khan N, Jhariya MK, Yadav DK, Banerjee A (2020b) Structure, diversity and ecological function of shrub species in an urban setup of Sarguja, Chhattisgarh, India. Environ Sci Pollut Res 27 (5):5418–5432. https://doi.org/10.1007/s11356-019-07172-w

Kirby KR, Potvin C (2007) Variation in carbon storage among tree species: implications for the management of a small-scale carbon sink project. For Ecol Manag 246:208–222

Köchy M, Hiederer R, Freibauer A (2015) Global distribution of soil organic carbon - part 1: masses and frequency distributions of SOC stocks for the tropics, permafrost regions, wetlands, and the world. Soil 1:351–365

Kumar S, Meena RS, Jhariya MK (2020) Resources use efficiency in agriculture. Springer, Singapore, p 760. https://doi.org/10.1007/978-981-15-6953-1

Laban P, Metternicht G, Davies J (2018) Soil biodiversity and soil organic carbon: keeping drylands alive. IUCN, Gland, viii + 24 p. https://doi.org/10.2305/IUCN.CH.2018.03.en

Lal R (2004a) Soil carbon sequestration impacts on global climate change and food security. Science 304:1623–1627

Lal R (2004b) Soil carbon sequestration to mitigate climate change. Geoderma 123:1–22

Lal R (2005) Forest soils and carbon sequestration. For Ecol Manag 220:242–258

Lal R (2009) Soil degradation as a reason for inadequate human nutrition. Food Secur 1:45–57

Lal R (2016) Beyond COP 21: potential and challenges of the "4 per thousand" initiative. J Soil Water Conserv 71:20A–25A

Lal R, Follett RF, Kimble JM (2003) Achieving soil carbon sequestration in the U.S: a challenge to the policy makers. Soil Sci 168:827–845

Lemus R, Lal R (2005) Bioenergy crops and carbon sequestration. Crit Rev Plant Sci 24:1–21

Makumba W, Akinnifesi FK, Janssen B, Oenema O (2007) Long-term impact of a Gliricidia-maize intercropping system on carbon sequestration in southern Malawi. Agric Ecosyst Environ 118:237–243

Mandal D (2020) Soil management for regulating C pools: perspective in tropical and subtropical soils. In: Ghosh PK, Mahanta SK, Mandal D, Mandal B, Ramakrishnan S (eds) Carbon management in tropical and sub-tropical terrestrial systems. Springer, Singapore, pp 57–70. https://doi.org/10.1007/978-981-13-9628-1_4

Mandryk M, Reidsma P, van Ittersum MK (2017) Crop and farm level adaptation under future climate challenges: an exploratory study considering multiple objectives for Flevoland, the Netherlands. Agric Syst 152:154–164. https://doi.org/10.1016/j.agsy.2016.12.016

Mayer A, Hausfather Z, Jones AD, Silver WL (2018) The potential of agricultural land managementto contribute to lower global surface temperatures. Sci Adv 4:eaaq0932

Meena RS, Lal R (2018) Legumes for soil health and sustainable management. Springer, Singapore, p 541. https://doi.org/10.1007/978-981-13-0253-4_10

Meena RS, Kumar V, Yadav GS, Mitran T (2018) Response and interaction of *Bradyrhizobium japonicum* and Arbuscular mycorrhizal fungi in the soybean rhizosphere: a review. Plant Growth Regul 84:207–223

Meena RS, Kumar S, Datta R, Lal R, Vijaykumar V, Brtnicky M, Sharma MP, Yadav GS, Jhariya MK, Jangir CK, Pathan SI, Dokulilova T, Pecina V, Marfo TD (2020) Impact of agrochemicals on soil microbiota and management: a review. Land (MDPI) 9(2):34. https://doi.org/10.3390/land9020034

Meena RS, Lal R, Yadav GS (2020a) Long term impacts of topsoil depth and amendments on soil physical and hydrological properties of an Alfisol in Central Ohio, USA. Geoderma 363:1141164

Meena RS, Lal R, Yadav GS (2020b) Long-term impact of topsoil depth and amendments on carbon and nitrogen budgets in the surface layer of an Alfisol in Central Ohio. Catena 194:104752

Nair PKR (2012) Climate change mitigation and adaptation: a low hanging fruit of agroforestry. In: PKR N, Garrity DP (eds) Agroforestry: the future of global land use. Springer, Dordrecht, pp 31–67

Oelbermann M, Voroney RP, Gordon AM, Kass DCL, Schlnvoigt AM, Thevathasan NV (2006) Carbon input, soil carbon pools, turnover and residue stabilization efficiency in tropical and temperate agroforestry systems. Agrofor Syst 68:27–36

Painkra GP, Bhagat PK, Jhariya MK, Yadav DK (2016) Beekeeping for poverty alleviation and livelihood security in Chhattisgarh, India. In: Narain S, Rawat SK (eds) Innovative technology for sustainable agriculture development. Biotech Books, New Delhi, pp 429–453

Pan Y, Birdsey RA, Phillips OL, Jackson RB (2013) The structure, distribution, and biomass of the world's forests. Annu Rev Ecol Evol Syst 44:593–622

Pareek N (2017) Climate change impact on soils: adaptation and mitigation. MOJ Eco Environ Sci 2 (3):136–139

Parrotta JA (1999) Productivity, nutrient cycling and succession in single- and mixed-species stands of *Casuarina equisetifolia*, *Eucalyptus robusta* and *Leucaena leucocephala* in Puerto Rico. For Manag 124:45–77

Paustian K, Lehmann J, Ogle S, Reay D, Robertson GP, Smith P (2016) Climate-smart soils. Nature 532:49

Peichl M, Thevathasan NV, Gordon AM, Huss J, Abohassan RA (2006) Carbon sequestration potentials in temperate tree based intercropping systems, southern Ontario, Canada. Agrofor Syst 66:243–257

Pimentel D, Burgess M (2013) Soil erosion threatens food production. Agriculture 3:443–463

Poll C, Marhan S, Back F, Niklaus PA, Kandeler E (2013) Field-scale manipulation of soil temperatureand precipitation change soil CO_2 flux in a temperate agricultural ecosystem. Agric Ecosyst Environ 165:88–97

Post WM, Kwon KC (2000) Soil carbon sequestration and land-use change: processes and potential. Glob Chang Biol 6:317–327

Pramanika M, Paudel U, Mondal B, Chakraborti S, Debd P (2018) Predicting climate change impacts on the distribution of the threatened Garcinia indica in the Western Ghats, India. Clim Risk Manag 19:94–105

Prentice IC (2001) The carbon cycle and atmospheric carbon dioxide. Climate change 2001: the scientific basis IPCC. Cambridge University Press, Cambridge, pp 183–237

Raizada A, Parandiyal AK, Ghosh BN (2003) Estimation of carbon flux through litter fall in forest plantations of India. Indian For 129(7):881–894

Raj A, Jhariya MK, Harne SS (2018) Threats to biodiversity and conservation strategies. In: Sood KK, Mahajan V (eds) Forests, climate change and biodiversity. Kalyani Publisher, New Delhi, pp 304–320

Raj A, Jhariya MK, Banerjee A, Yadav DK, Meena RS (2019a) Soil for sustainable environment and ecosystems management. In: Jhariya MK, Banerjee A, Meena RS, Yadav DK (eds) Sustainable agriculture, forest and environmental management. Springer, Singapore, pp 189–221. https://doi.org/10.1007/978-981-13-6830-1

Raj A, Jhariya MK, Yadav DK, Banerjee A, Meena RS (2019b) Agroforestry: a holistic approach for agricultural sustainability. In: Jhariya MK, Banerjee A, Meena RS, Yadav DK (eds) Sustainable agriculture, forest and environmental management. Springer, Singapore, pp 101–131. https://doi.org/10.1007/978-981-13-6830-1

Raj A, Jhariya MK, Yadav DK, Banerjee A (2020) Climate change and agroforestry systems: adaptation and mitigation strategies. Apple Academic Press, Palm Bay, p 383. https://doi.org/10.1201/9780429286759

Rao MR, Ong CK, Pathak P, Sharma MM (1991) Productivity of annual cropping and agroforestry systems on a shallow Alfisol in semi-arid India. Agrofor Syst 15:51–63

Rodríguez-Calcerrada J, Sancho-Knapik D, Martin-StPaul NK, Limousin JM, McDowell NG, Gil-Pelegrín E (2017) Drought-induced oak decline—factors involved, physiological dysfunctions, and potential attenuation by forestry practices. In: Gil-Pelegrín E, Peguero-Pina J, Sancho-Knapik D (eds) Oaks physiological ecology. Exploring the functional diversity of genus Quercus L. tree physiology, vol 7. Springer, Cham

Royo AA, Knight KS (2012) White ash (*Fraxinus americana*) decline and mortality: the role of site nutrition and stress history. For Ecol Manag 286:8–15

Runion GB, Torbert HA, Prior SA, Rogers HH (2009) Effects of elevated atmospheric carbon dioxide on soil carbon in terrestrial ecosystems of the Southeastern United States. In: Lal R, Follett R (eds) Soil carbon sequestration and the greenhouse effect, 2nd edn. Soil Science Society of America, Madison, WI, pp 233–262

Runyon J, Waring RH, Goward SN, Welles JM (1994) Environmental limits on net primary production and light-use efficiency across the Oregon transect. Ecol Appl 4:226–237

Saha S, Nair PKR, Nair VD, Kumar BM (2009) Soil carbon stocks in relation to plant diversity of home gardens in Kerala, India. Agrofor Syst 76:53–65

Schlesinger WH, Andrews JA (2000) Soil respiration and the global carbon cycle. Biogeochemistry 48:7–20

Sharma CM, Gairola S, Baduni NP, Ghildiyal SK, Sarvesh S (2011) Variation in carbon stocks on different slope aspects in seven major types of temperate region of Garhwal Himalaya, India. J Biol Sci 36(4):701–708

Sharrow SH, Ismail S (2004) Carbon and nitrogen storage in agroforests, tree plantations, and pastures in western Oregon, USA. Agrofor Syst 60:123–130

Shinde SM, Turkhade PD, Deshmukh SB, Narkhede GW (2015) Carbon sequestration potential of some fruit trees in Satara district of Maharashtra India. Ecol Environ Conserv Paper 21 (1):359–362

Singh NR, Jhariya MK (2016) Agroforestry and agrihorticulture for higher income and resource conservation. In: Narain S, Rawat SK (eds) Innovative technology for sustainable agriculture development. Biotech Books, New Delhi, pp 125–145

Singh B, Singh G (2015) Biomass production and carbon stock in a Silvi-Horti based agroforestry system in arid region of Rajasthan. Indian Forester 141(12):1237–1243

Sreejesh KK, Thomas TP, Rugmini P, Prasanth KM, Kripa PA (2013) Carbon sequestration potential of Teak (*Tectona grandis*) plantations in Kerala. Res J Recent Sci 2(ISC 2012):167–170

Sreekanth NP, Santhi Prabha V, Babu P, Thomas AP (2013) Soil carbon alteration of selected forest types as an environmental feedback to climate change. Int J Environ Sci 3:1516–1530

Swamy SL, Puri S (2005) Biomass production and C-sequestration of *Gmelina arborea* in plantation and agroforestry system in India. Agrofor Syst 64:181–195

Takimoto A, Nair PKR, Nair VD (2008) Carbon stock and sequestration potential of traditional and improved agroforestry systems in the West African Sahel. Agric Ecosyst Environ 125:159–166

Thomson LJ, Macfadyen S, Hoffmann AA (2010) Predicting the effects of climate change on natural enemies of agricultural pests. Biol Control 52(3):296–306

Tonucci RG, Nair PKR, Nair VD, Garcia R, Bernardino FS (2011) Soil carbon storage in silvopasture and related land use systems in the Brazilian Cerrado. J Environ Qual 40 (3):833–841. https://doi.org/10.2134/jeq2010.0162

Trost B, Prochnow A, Drastig K, Meyer-Aurich A, Ellmer F, Baumecker M (2013) Irrigation, soilorganic carbon and N_2O emissions. A review. Agron Sustain Dev 33(4):733–749

Unger S, Máguas C, Pereira JS, David TS, Werner C (2010) The influence of precipitation pulses on soil respiration–assessing the 'birch effect' by stable carbon isotopes. Soil Biol Biochem 42 (10):1800–1810

US EPA (2019) U. S. Environmental Protection Agency, Inventory of U.S. Greenhouse Gas Emissions and Sinks: 1990–2017 released on 2019. https://www.epa.gov/sites/production/files/2019-04/documents/us-ghg-inventory-2019-main-text.pdf

Viswanath S, Peddappaiah RS, Subramoniam V, Manivachakam P, George M (2004) Management of *Casuarina equisetifolia* in wide-row intercropping systems for enhanced productivity. Indian J Agrofor 6(2):19–25

Wiesmeier M, Poeplau C, Sierra CA, Maier H, Frühauf C, Hübner R, Kühnel A, Spörlein P, Geuß U, Hangen E, Schilling B (2016) Projected loss of soil organic carbon in temperate agriculturalsoils in the 21st century: effects of climate change and carbon input trends. Sci Report 6:32525

Zdruli P, Lal R, Cherlet M, Kapur S (2017) New world atlas of desertification and issues of carbon sequestration, organic carbon stocks, nutrient depletion and implications for food security. In: Carbon management, technologies, and trends in Mediterranean ecosystems. Springer, Cham, pp 13–25

Ecomodelling Towards Natural Resource Management and Sustainability

14

Arnab Banerjee, Manoj Kumar Jhariya, Nahid Khan, Abhishek Raj, and Ram Swaroop Meena

Abstract

Resource depletion is a mega event that is hampering the prosperity and well-being of human kind. Technological growth has increased the demand of resources considerably and as a result there is overuse and abuse of resources. Natural resource management is the major aspect in terms of adressing environmental sustainibility. In this direction ecomodelling is a tool that is helpful for proper decision making and screen the sustainable practices of resource excavation. Various models were used to assess the situation of various forms of natural resources and these models were used as decision making tool for sustainable management of resources. The present chapter has attempted to explore the role of ecomodelling towards natural resource management and addressing sustainability. It has also critically analysed the future perspectives of ecomodelling towards resource conservation and management.

Keywords

Ecomodelling · Environment · Management · Natural resources · Sustainability

A. Banerjee (✉)
Department of Environmental Science, Sant Gahira Guru Vishwavidyalaya, Ambikapur, Chhattisgarh, India

M. K. Jhariya · N. Khan
Department of Farm Forestry, Sant Gahira Guru Vishwavidyalaya, Ambikapur, Chhattisgarh, India

A. Raj
School of Agriculture, Lovely Professional University, Phagwara, Punjab, India

R. S. Meena
Department of Agronomy, Institute of Agricultural Sciences, Banaras Hindu University, Varanasi, Uttar Pradesh, India
e-mail: meenars@bhu.ac.in

M. K. Jhariya et al. (eds.), *Ecological Intensification of Natural Resources for Sustainable Agriculture*, https://doi.org/10.1007/978-981-33-4203-3_14

Abbreviations

C	Carbon
CAEDYM	Computational aquatic ecosystem dynamics model
CASM	Comprehensive aquatic system model
CATS	Contaminant in aquatic and terrestrial ecosystems
CO_2	Carbon dioxide
ERP	Enterprise resource planning
GHGs	Greenhouse gases
GIS	Geographic information systems
IAM	Integrated assessment models
IPCC	Intergovernmental Panel on Climate Change
LCA	Life cycle assessment
LUC	Land use change
NRM	Natural resource management
PEM	Partial economic sector models
SBP	Scenario building and planning
SDMS	Structurally dynamic models
USA	United States of America
WBM	Water balance model

14.1 Introduction

Simulation modelling has become an important aspect of multidisciplinary studies as well as various aspects of environmental management. It clearly reflects the interrelationship between society and environment. The increasing potential of the computers and their applications have increased the capacity to deal with a problem of multidimensional nature. In the process of decision making and policy formulation simulation modelling may aid in proper management of natural resources and address the issue of sustainability at various spheres.

The present era is undergoing a rapid change in terms of its environmental scenario due to occurrences of mega events such as climate change, biodiversity loss, environmental pollution which is exerting their effect on the natural resources of the environment (Slocombe 1993; Raj et al. 2020; Banerjee et al. 2020; Jhariya et al. 2019a, b). Huge loss of biodiversity along with ecological services across various ecosystems of the globe is prevalent under the present times of climate change (Keane et al. 2015; Raj et al. 2018). This particular aspect of change in environmental scenario is yet to be explored properly (Gustafson 2013). Knowledge regarding the appropriate aspect of ecosystem services is also not well known (Creutzburg et al. 2015a, b; Meena et al. 2018). It is therefore need of the society and human civilization to formulate proper strategies, go for proper decision making which would help to achieve the sustainable development (Khan et al. 2020a, b). It would also help to fight against the mega event of climate change and ecological consequences (Cuddington et al. 2013; Keane 2019).

From climatological perspective climatologists tend to predict future climate from past experiences (Allen 2007; Meena and Lal 2018). Later on problems related to misappropriation of the carbon dioxide (CO_2) equilibrium in the climatological aspects created the need of mathematical based analysis for specific results (Green and Sadedin 2005). As a consequence of this nowadays climatologists are adopting simulation modelling in order to predict the future climate. Such approaches helped to analyse the modification of the climate as well as future prediction. This was represented by the IPCC (Intergovernmental Panel on Climate Change) in the form of reports which have helped us to identify the likely consequences of climate change as well as future trends through ecological modelling approach (Loehman et al. 2017). On the basis of such climatological modelling approach it is predicted that in upcoming times there would be a rise of 4 °C rise in earth surface temperature which would have its impact on variation in rainfall and sea level rise (Loehman et al. 2017). Further all these findings are transformed in to policy matters by the IPCC (Keane et al. 2018).

The proper management of natural resource is another biggest challenge for the twenty-first century. Individual resources are suffering from the rapid problem of resource depletion (Meena et al. 2020; Kumar et al. 2020). It is the unprecedented growth of the population, their greed, growth and development of science and technology that is increasing the pressure on natural resources day by day. For instance, in many areas across the globe has become water starved area and many more to come under this category in near future. The soil quality is depleting at a faster rate under the pressure of over production (Raj et al. 2019a, b). The bioresources in the form of biodiversity are also suffering from the climatic extremes. This therefore necessitates the formulation of proper ecological modelling approach in order to understand and critically analyse the ecological consequences and to also have a proper future prediction. This would have important implications in the policy formulation network at the global level forum on biodiversity and ecosystem services (Keane 2019; Meena et al. 2020a, b). Therefore, from future stake it is necessary to have ecological modelling for proper future prediction about the ecological system and associated various ecosystem services (Botkin 1993). In this connection, the present title deals with the ecomodelling and their role in natural resources management, sustainability and sustainable development.

14.2 Concept of Modelling

Models can be defined as an abstract form of the events that is taking place around us (Bugmann and Cramer 1998). From ecological perspective models are considered as objects of mathematics in ecology (Heppell et al. 1994). A model is a collection of various elements and components and it is the responsibility of the modeller to incorporate appropriate components and elements in the simulation to get better result and output. Overall the attribute of a model is general, realistic and precise in nature. The generalistic nature implies the applicability of the views of the model in various spheres, and the realistic viewpoint highlights the similarity of the results of the model with real life situation followed by the precision based approach of the model.

Modelling has got a long history on its back. Levins in 1966 worked on ecological modelling with the help of ecological models based on ecological phenomena based on meta-population, population dynamics and community ecology. Initially modelling was referred to as formulation of simple mathematical equations to get precise answer about the research hypothesis (Miller and Dean 2000). However, Voltera at the early twentieth century initiated ecomodelling based on his famous prey–predator model on fish population (Miller and Dean 2000). Similar applications were used by Anderson and May for five decades regarding ecological modelling of disease biology of fishes (He 2008). Simple models give a generalized response in comparison to specific models.

Modelling is a systematic approach that aids in proper logical reasoning with the help of statistical and advance computational methods. Models help in proper valuation, computation of the events based on qualitative and quantitative data as well as an overall integration of the events. Models of the earth system comprise various components. Mathematical models are based on some variables along with the interaction between them which ultimately leads to generation of some mathematical equations. Variables may be an object, an event, may be part of mathematical equations in the form of input variables as well as land use type in the form of output variables. Further the nature of model may be of various types of mathematical expression (Fig. 14.1).

Models used to perform the task of scoping for a particular issue, object, and may involve higher participation of the community stakeholders along with proper adjustment through hypothesis testing or calibrations of the variables. Further management models provide suitable options of management. Proper selection of attributes/variable as well model ultimately determines the fate of the model output.

Models are concerned with future prediction considering the changes associated with ecology and environmental aspect within an ecosystem (Scheller 2018). Phenomenological description of data is concerned with projection of environmental change for the area of which the data has been collected (Hansen et al. 1995). Models are built for climatological perspective in the form of process-based models in order to understand the mechanism of climatic events. In modelling proper understanding of model is vested upon the clear definition of system as it is target point to study the systems ecology. A system is basically an interacting unit working altogether to function efficiently. Alteration in the properties of a system makes a total change in the system. If we consider the matter from ecosystem perspective, then predicting the continuous change that is taking place within the ecosystem is considered as system ecology. It would help to precisely predict the changes that take place in the environment which ultimately goes to biological world up to the organisms (Lucash et al. 2018). Further, higher datasets have led to emergence of a new discipline known as systems biology (He and Mladenoff 1999). This is combined with the recognition of the behavioural complexicity. Systems biology attempts to

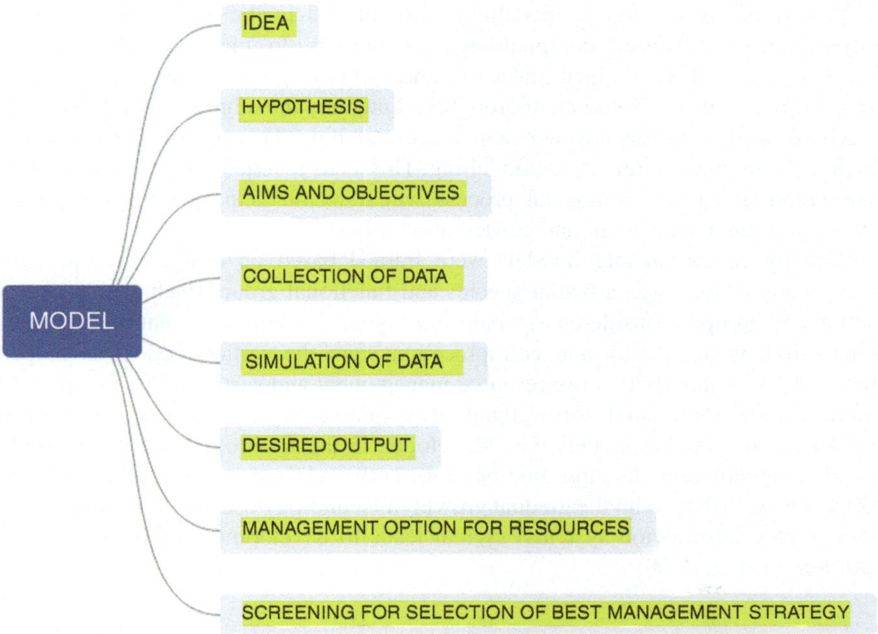

Fig. 14.1 Concept of modelling and output

understand the interaction between the components of the system which reflects the behaviour of the system. Such types of modelling approach are systematic method of collection of information and its proper interpretation. Subsequently these interpretation gives the outcome of generate prediction. The systems ecology in 1960s was considered as a tool for effective management of natural resources (Keane and Finney 2003).

14.3 Development of Ecological Modelling

The approach of ecological modelling dates back with the earlier attempts of formulating mathematical equations for assessment of ecological process (Streeter and Phelps 1925). Further improvement in the models took place through algae inclusion to address the issue of eutrophication (Patten 1968). Later on the feedback mechanisms as well as various interrelationships were incorporated on time to time basis (Spofford 1975). Later on Lotka–Volterra equations were framed based on prey–predator model. Later on it was found that the model proposed by Lotka Volterra was very simpler in nature and therefore cannot quantify the interaction that is exactly taking place within the ecosystem. Later on further advancements in the approach of ecological modelling took place through the development of systems ecology and cybernetics (Odum 1953).

A new era of ecological modelling was initiated in the early 1960s through application of advanced computational technology. In the late 1960s in San Francisco of the USA (United States of America) issues of ecomodelling related to zero emission were discussed thoroughly. Later on the impracticability of zero discharge leads to further advancement of ecomodelling. This necessitated the access to digital computer system in ecomodelling. This leads to requirement of quantitative estimation of various ecological process which would help to formulate proper strategy in the management and conservation process.

Initially compartmental' models were framed based on differential equations specifically addressing particular species and functional group. Each of the species action and group is considered as single compound. Modelling in relation to species interaction, water quality and eutrophication were the earlier models developed before 1970. After 1970s from resource management and sustainability perspective models were formulated for wetland ecosystem followed by eco-toxicological models. Later flexible models were developed in the form of SDMS (structurally dynamic models) by allowing modifications in the selected parameters in the study (Zhang et al. 2010). With the gradual growth of technology use of application of GIS (geographic information systems) system leads to development of spatial models (Jørgensen et al. 2009).

14.4 Need of Ecological Modelling

Model itself has got some specific applications to be used as management tools. Therefore, models serve the purpose of solving various problems and issues of wide dimension. From resource management perspective, model acts as synthesizer of theory as well practical knowledge and provides information for individual components of a system. For example, if we consider water resource, then we need to know about its reserve, stock volume, the specific flow rates and its future prediction. They also represent the concept of the system in a formal way. Models also pave the way of new scientific research while exploring through the existing knowledge. Models also serve the important function of hypothesis testing to answer the research problem and question. It also allows proper understanding of the system components.

14.5 Spheres of Modelling

14.5.1 Socio-economic Modelling

It covers the various aspects of socio-economic condition and associated problems to reflect the clear picture of the society, economy system prevailing over an area. Economic models are the abstract form of the prevailing economic system for a particular area. They entirely reflect the prevailing economy of a particular area. Env-linkages, GTAP are examples of some of these models. Demographic models give the long term prediction about the human population growth as well their future projections. Some example includes the Phoenix. For highlighting the specific sector under economic system, there is PEM (partial economic sector models).

14.5.2 Biophysical Modelling

Biophysical models deal with the specific natural processes and specific components. For example, hydrological models deal with the hydro-geochemistry of the various aquatic systems prevailing over the earth surface without considering the ocean and the atmospheric environment. WBM (water balance model) is the typical example of such hydrological model. BGC (biogeochemistry) model includes various concepts such as ecological processes of the vegetation, material and energy exchanges between the various components of the environment. The concept is also applicable for other disciplines such as agriculture as well as in the forestry sector.

14.5.3 Integrated Modelling

Under this concept there is an integrated approach to include the societal and environmental aspect of human environment along with interlinked sectors such as transport, agriculture, livestock population, forestry practices, etc. This type of modelling approach is closely associated with various forms of land uses (Pontius et al. 2007). One of the types includes the LUC (land use change) models. Another form of integrated models includes the IAM (Integrated Assessment Models) which has holistic approach to include the entire socio-economic aspects human environment. SBP (Scenario Building and Planning) models also work on the principle of integration of various dimensions of sustainability of the earth ecosystem.

For resource management, the integration of the modelling system has a huge role to play in proper decision making towards sustainable development. Simultaneously it helps to frame the policy and strategies towards conservation of socio-cultural and other ecological resources. From sustainability point of view the socio-economic dimension is the most critical one as it involves the solid interactions between the components. Such type of approach helped to predict the changes and its subsequent management in Baltic Sea region (Baltic 21 Report 2006). The integrated models therefore consist of three key components which include evaluation of model, proper analytical framework as well as optimum management of knowledge (US EPA 2008). It requires collaboration between the academicians, researchers and community stakeholders to percolate the utility of modelling and its application.

At the global scale integrated model considering the ecological risk and adaptive strategies was framed by various workers to make proper policy decision (Gentile et al. 2001). It requires a comprehensive approach to screen out the factors causing the changes followed by proper identification of societal needs and preferences. The model is very much capable to reflect the nexus between management option, societal paradigm and environmental domain. A work in river basin in Germany reported the integration of socio-economic and ecological factors in decision making process towards sustainable management (Volk et al. 2008). The results of the analysis revealed that proper land management is required for sustainability of the river basin system. Similarly, Turner et al. (2000) reported a work on wetland that represents a comprehensive integration between socio-economic valuation and conservation perspective.

Another example of integrated model was given by Chang et al. (2008) on coral reef ecosystem. Coral reef is an important and attractive ecosystem from its various aspects. The model developed for sustainable management of coral reef ecosystem integrates maintaining ecological health, proper land use, wastewater treatment as well as regulation of tourist activity. Integration of socio-economic and ecological condition through modelling approach was further designed by Nobre et al. (2009) for East China Sea. The model helped to screen the effective areas for cultivation along with associated constraints. Further, the model proposed the reduction of cultivable area for proper management of the aquatic ecosystem.

14.5.4 Biodiversity Modelling

Assessment of bioresources in the biodiversity can be done through the use of biodiversity modelling. They indirectly work for assessing the ecosystem services of the various ecosystems (Braat et al. 2008). Biodiversity models have two approaches which include indicator-based models and species distribution/climate envelope models. In the first approach indicator estimation is done without giving emphasis to particular species and the second approach deals with the issue of species distribution under various climatic niche.

14.6 Systems Approaches and Ecological Modelling

Systems approach in biological and ecological research is an important aspect which involves significant level of modelling approach. Even the concepts of community succession and the ecological role of forest were also studied by using suitable ecological models (Purves et al. 2008; Moorcroft et al. 2001). Predictions of secondary growth of tree species in the USA were also studied through ecological modelling (Pacala et al. 1996). Forest dynamics of tree species for 100 years were also studied by using ecological modelling (Purves et al. 2008). Similar types of system approaches and modelling were used to assess the ecosystem functionality in tropical part of South America (Moorcroft et al. 2001). In Austria simulation modelling was used to study the impacts of climate change on the structure and functional aspect of forests (Moorcroft et al. 2001). Systems approach and ecological modelling are also used in ecosystem modelling that incorporates all the trophic level helps to predict the future state of the ecosystem (Bithell and Brasington 2009).

From bioresource conservation point of view there are various critical issues which simple modelling approach cannot resolve. For example, events such as species conservation, productivity of a particular species within an ecosystem require complex modelling approaches in order to use the modelling approach as a decision making tool. Therefore, these questions need to be resolved through proper systems approaches under the AEGIS of ecological modelling.

14.7 Application of Modelling Towards Sustainability

The modelling approach has a wide dimensional effect on sustainability. It has become a tool for addressing various issues of environment, resource management, sustainable development, ecology and many other allied fields.

14.7.1 Application of Modelling in Agriculture Sector

In the various sectors of agriculture people engaged in can become critical factors for various forms of sustainability issues. There is the possibility of overexploitation of natural resources in terms of water resources, loss of soil fertilizer, use of chemical fertilizer for over production. Therefore, selecting appropriate parameters and variables in the modelling approaches is the key to get proper decision making towards sustainable development (Rebaudo and Dangles 2013; Martins et al. 2014).

14.7.2 Ecosystem and Climate

Various modelling approaches were implied in order to study, analyse as well as make proper decision making. Through the modelling approach various forest issues and their sustainable management have been addressed. Policy investigation on both national and international level on the issue of climate change and its future prediction has also been studied. The issue of trading GHGs (greenhouse gases) has also being studied through modelling approaches (Aubert et al. 2015; Gerst et al. 2013).

14.7.3 Energy

Energy has a very important value from resource point of view and there is very critical to achieve sustainability in various dimensions. Simulation modelling was used to analyse the impact of various energy policies at various levels (Barisa et al. 2015; Franco et al. 2015). Modelling approaches were also used to assess the market economy of energy resources, policies related to renewable energy resources, energy management systems (Franco et al. 2015). Suitable alternatives were framed in place of LCA (life cycle assessment) through simulating modelling (Miller et al. 2012; Reddi et al. 2013).

14.7.4 Human Health

Human health is a big issue of the present times. There are various problems such as environmental pollution in terms of water and soil which has got significant impacts on health (Petering et al. 2015). Various healthcare systems are there whose efficiency were also assessed through simulation modelling and screening of the best system, their optimization was done through modelling approach (Petering et al. 2015).

14.7.5 Land Use

Simulation modelling was used to assess the alterations in the land use pattern, as well as integration of the social and environmental issues with the land use pattern. Use of simulation modelling in these aspects provides a deeper insight on these particular issues. Particular issues of desertification, deforestation were also addressed through simulation modelling (Chen et al. 2014).

14.7.6 Resource Use and Consumption

It is the most important aspect on the context of sustainability and sustainable development. This needs to be addressed properly through the modelling approaches. However, some attempts were made in this regard with application of simulation modelling in the sector of policy formulation in relation to reducing GHGs emissions, energy consumption and use of materials (Lindskog et al. 2011; Sproedt et al. 2015).

14.7.7 Mining

Mining is the process of extracting earth resources causing several negative consequences in the form of land degradation, loss of biodiversity, deforestation, soil erosion, loss of top soil cover, etc. On the other hand, it is the pillar of economy for a particular area or country. Different policies regarding these issues on mining and their interactions were addressed through simulation modelling approaches (Maluleke and Pretorius 2013). The application of simulation modelling in the form of LCA approach of the mining process has been done by Nageshwaraniyier et al. (2011) and integrated with ERP (Enterprise Resource Planning) system.

14.7.8 Sustainable Development

Sustainable development is a broad concept which works in wider dimension of social, economic and environmental aspects. Simulation approaches were used and designed to formulate specific combinations to achieve sustainable developmental goals (Nikolaou et al. 2015). From urban and community planning, the United Nations mentioned that >50% of the global population is residing in the urban area and this trend is gradually increasing day by day. As a consequence of that there is improper planning in development of urban planning. Therefore, modelling approaches were used to know about the future projection in relation to sustainability issues (Gaube and Remesch 2013; Haase et al. 2010). Various perspectives of residential development and social structure have also been studied through the modelling approach (Steinhoffel et al. 2012; Xu and Coors 2011).

14.7.9 Waste-Recycling and Reuse

From environmental perspectives waste and its proper management have become a serious concern. If not properly managed, it creates the problem of environmental pollution. The most advanced approach includes conversion of waste into resources through recycling and reuse as much as possible. This may have contribution towards sustainable development. The process of reuse and recycling involves various stakeholders in terms of various forms of resources, technologies, funding, involvement of the government, public participation as well as proper policy formulation. Therefore the effectivity of the best recycling and reuse process needs to be screened precisely. Simulation modelling was done for the various types of waste management process by various workers (Blumberga et al. 2015; Shokohyar and Mansour 2013).

14.7.10 Water Resources

Among the various natural resource water resource is a precious one as its scarcity across the globe is increasing day by day. Many people across the globe are suffering the crisis of water stress condition. Therefore, proper management in terms of sustainable use of water resources is the need of the hour. Various workers have used different modelling approaches such as watershed modelling, modelling the effective way of management of water resources as well as interactions between the policymakers, scientists and community stakeholders has been done for proper management of water resources (Dai et al. 2013; Ali et al. 2014; Sahin et al. 2015) .

14.8 Natural Resources and Their Management

Natural resource can be considered as natural assets which are the gift of nature to mankind for its social well-being and development. This purpose of natural resources has been misinterpreted by human beings through their growth in science and technology. Natural resources comprise every environmental components that are used for the benefits of the mankind (Fig. 14.2).

The major problem associated with natural resource is its depletion rate through over extraction and exploitation. For instance, there is urgent need to increase the food production to feed approximately 10 billion of people across the world. Many people in the Asian and African subcontinent are under the stress of food crisis (FAO 2011).

From forest resource perspective UN has already declared 2011 as International Year of Forests and has emphasized for sustainable management of forest globally. As per the reports of FAO (2010) forest area of up to about 13 million ha has been converted in to other land uses. Further, FAO reports that such deforestation activities have led to 20% emission of GHGs. Some earlier research reports reveal that 13% of the cultivable land across the globe are suffering from loss of productivity and 84% of these areas have been affected by the erosive forces of wind and water. Grazing activity, deforestation, agricultural expansion have committed such overuse and high depletion of natural resources (Pimentel 2006). Human mediated land degradation approximates up to 1.9 billion hectares of land (Lal 2001).

Further as per the reports of UN environment programme more than 10% of the vegetation covered area along with its productivity has been degraded. The nature of degradation was found to be higher for the African continent (3.2 million km). Further, 210,000 km^2 area of cropland becomes degraded up to 1998 (Von Baratta 1998).

As per the reports of FAO (2010) within a span of 5 years (2005–2010) the forest area has depleted up to 30% of the global land area. The deforestation activity further

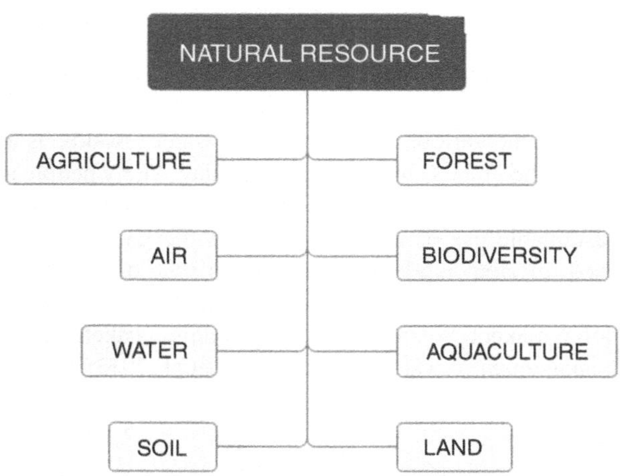

Fig. 14.2 Types of natural resources

continues even various conservation planning for forest conservation is going on. Nearly 5.7 million ha of forest area is lost through deforestation activities in the last two decades. Therefore, resource depletion has become an inevitable truth for the mankind as the lifestyle of human beings as well as the decision making has changed gradually with growth of technological knowledge. Sustainable approaches and proper decision making are required to address the issue of resource depletion worldwide (Kumar et al. 2020; Jhariya et al. 2015; Singh and Jhariya 2016).

14.9 Ecomodelling and Natural Resource Management

14.9.1 Modelling of Water Quality for Water Resource Management

Water quality is a very important aspect as many people across the globe are devoid of safe drinking water day by day. This particular issue needs to be addressed and solved with utmost priority. Very early practices in water modelling include prediction of water qualities of surface water bodies. The study comprised of studying the events of eutrophication (Jørgensen et al. 1978). Proper modelling gave clear cut results of exchange of nutrients between the sediment and water playing the significant role in the process of eutrophication. The results revealed that after third year reduction of the nutrient inflow tends to change the transparency of the aquatic medium. This therefore helped for realization of the fact that reduction of eutrophication may take place through reduction of phosphorous flow (Jørgensen 1986). The success of such systems approach of modelling is based on model's complexity as well as proper collection of data. Table 14.1 represents the various ecological models used in various times to assess the level of water quality along with their proper management strategy.

Table 14.1 Various water quality models for water resource management

Type of modelling approach	References
Dynamic reservoir simulation model (DYRESM)	Hamilton and Schladow (1997)
Contaminant in aquatic and terrestrial ecosystems (CATS)	Traas et al. (1998)
Comprehensive aquatic system model (CASM)	Bartell et al. (1999)
System dynamics approach—ErhaiSD	Guo et al. (2001)
CASM–SUWA	Naito et al. (2002)
Bayesian networks	Borsuk et al. (2004)
Eco-hydrodynamic model	Lee et al. (2005)
Structurally dynamic models	Gurkan et al. (2006)
Bayesian calibration of process-based models	Arhonditsis et al. (2007)
AQUATOX	Park et al. (2008)
Computational aquatic ecosystem dynamics model (CAEDYM)	Gal et al. (2009)

There are various approaches in various hydrological models. The DYRESM model was a one-dimensional model using multiple variables to find out the interaction between the various variables. CAEDYM model was used to screen the potential parameter affecting the water quality of surface water body. AQUATOX MODEL was used for assessing the ecological risk and it includes various trophic levels of the aquatic ecosystem. The concerned model can perform their task for various contaminants and toxicants present in the ecosystem. The model is cable of tracking alterations in the biomass also. It helps to assess the impact of chemicals on organisms as well as on the food web and other organisms. The CATS model is based on bioenergetics which reveals the interaction between the biotic and abiotic components of the ecosystem. In this perspective modelling of bioaccumulation event is also done for aquatic ecosystem. Later on eco-hydrodynamic model was proposed for pollution load estimation for their further reduction for water quality betterment. Ecological risks of surface water bodies were also assessed for various surface water bodies by using CASM-SUA model. Such models acted as effective tool for proper management of water quality and sustainability of water resources. Model ErhaiSD was used as a tool for water resource management on regional basis which was found to be related with socio-economic feedback of the local people. Under this model proper planning for water resource management along with suitable alternatives was formulated. Process-based models were used for effective management of eutrophication process Borsuk et al. (2004). Structurally dynamic models were used to find out the effectivity of various forms of restoration practices for restoration of water quality.

14.9.2 Modelling for Pisciculture Resource Management

One of the most significant economic activities of humankind is the aquaculture practice which results into significant economic gain for the rural stakeholders as well as fishermen community. It is also a well-known fact that aquaculture practices also gave rises to various forms of environmental degradation and pollution scenario. Therefore, proper management of aquaculture practices along with the aquatic system involved in it is very much essential. Various modelling approaches were attempted for proper management of aquaculture practices which are presented in Table 14.2.

EcoPath model was used in fisheries management which was used to understand the various trophic relationships on the basis of their efficiencies. Assessing the environmental impact of fishing and other aquaculture practices was done through Ecosim modelling approach. Management in terms of banning of fish and its after effect of improvement of water quality were reflected through Ecosim modelling approach. The combined application of EcoPath with EcoSim (EwE) modelling approach was used for biomass and productivity data in terms of fish catch in the food web of Baltic sea Harvey et al. (2003). System dynamic model was applied for screening of best management practices for sustainable harvesting. FORMAN model was used for ecological restoration of mangrove ecosystem in South America. Such

Table 14.2 Various aquaculture models for managing fisheries and aquaculture

Type of modelling approach	References
EcoPath—mass balanced model	Christensen and Walters (2004)
EcoSim—ecosystem simulator	Panikkar and Khan (2008)
EwE—combined application of EcoPath and EcoSim	Harvey et al. (2003)
Dynamic model for shellfish production	Bald et al. (2009)
Bivalve production models	McKindsey et al. (2006)
FORMAN—restoration simulation modelling	Twilley et al. (1998)

Table 14.3 Various models for managing forest resource

Type of modelling approach	References
Growth-yield models and gap models	Liu and Ashton (1995)
Mixed forest growth models	Porte and Bartelink (2002)
Lumped-parameter process-based models and hybrid models	Battaglia and Sands (1998)
Ecological process model	Zeide (2001)
Individual-based models (IBMs)	Shugart (1984)
Individual tree stand models	Shugart and West (1977)

models help in proper understanding of the spatio-temporal dimension of conservation of mangrove ecosystem.

14.9.3 Management of Forest Resources Through Ecomodelling

Forest is an important resource for the mankind as we get numerous ecological services from it. In this regard forest tend to reduce soil erosion, regulate the climate of an area, maintain the hydrological cycle, act as largest terrestrial C (Carbon) sink through C sequestration and gradual development of C stock (Jhariya et al. 2018a, b). It also acts as habitat for diverse form of living organisms. Apart from these services, forest help to maintain the rural livelihoods. Therefore, considering these services proper management of forest resource needs to be done accordingly to maintain the very existence of the human beings on earth. Modelling approaches have been used to understand, analyse the critical situation of forest as well as its sustainable management. Some approaches of modelling of forest resource have been represented in Table 14.3.

Forest model also has its application for the coastal area. The concept of using ecological modelling has its origin long back with some diverse problems in application of these models for better management of the resource. The major problem associated with forest modelling is to find out the fixed approach to apply on case to case basis. In this connection, modelling for commercial species and study of structure and function of forest ecosystem were studied using specific models. However, such approaches are applicable for monoculture practice. For mixed culture separate models were designed for various purposes. The major objectives include identifying and analysing the forest dynamics and give better prediction for its future pathway. Models were also framed for understanding the ecological processes and their role towards forest growth and development. Modern models of forestry also include the scenario of changing environment and its subsequent management. Specific events such as thinning were also addressed through specific models. Individual models helped to study the various attributes of species in terms of growth and development. The successional changes were also included under this model.

14.10 Ecomodelling and Landscape Management

NRM (Natural resource management) is a broad issue from future perspective as all the non-renewable resources are depleting at a faster rate over the earth surface. The success of it lies between specific projections that would help to develop management plan fulfilling the desired objective (Herrick et al. 2006). Initially the field of landscape management is based upon the experience and traditional knowledge regarding the past trends and accordingly management activities were planned. But since last two decades the scenario of earth has gone beyond prediction based on the traditional knowledge due to the mega events of climate change, greenhouse effect, ozone hole, etc. (Marlon et al. 2009).

Therefore, the purview of past experiences and traditional knowledge would not work perfectly for proper management of the landscape resources of today (Keane et al. 2015). The problems across the globe have further been aggravated through ecological invasion, biodiversity loss, rapid pest and disease outbreak in the agriculture sector leading overall rise in the footprint value of human beings (Creutzburg et al. 2015a, b). Under this condition use of ecomodelling of the landscape for proper decision making and prediction is the need of the hour. The models for landscape are usually developed through integration of various environmental components along with historical database (Cuddington et al. 2013).

Simulation modelling approach in this perspective helps to integrate the current scenario for better understanding of the dynamics of the landscape. Landscape models are very much useful to collect data for a larger area and for larger time scale to take proper decision for effective management of landscape resources (Keane 2012). Such models tend to help in collecting data which are not easily available in other ways (Bugmann and Cramer 1998). The major aspect of ecomodelling in the landscape management sector lies with assessing the impacts with altered management practices (Miller and Dean 2000). The modelling approach in landscape sector can be effectively utilized for proper planning and decision making towards effective management of the landscape (Keane and Karau 2010).

14.11 Ecosystem Modelling for Mapping Ecosystem Services

Ecosystems can be considered as a main pillar behind the existence of life on earth surface. It is a self-sustaining unit that supports all the living biota over the earth surface. Therefore ecosystem has a wide range of functionality in terms of ecosystem services (de Groot et al. 2012). Therefore, proper valuation of ecosystem service is required in order to manage the ecosystem in a sustainable way. The approach of attaining sustainability can be further improved through application of ecomodelling approach (McKenzie et al. 2011).

The supply and demand management within the ecosystem is the key for effective management of ecosystem services. However, it may vary on case to case basis (Bastian et al. 2012a). Various ecomodelling studies have been conducted by the various workers at various geographical scales (Schulp et al. 2012; Bateman et al. 2011). Mapping of the ecosystem services was also done by various workers (Nedkov and Burkhard 2012). Food production was also mapped properly in terms of ecosystem services. The source of food may be from agricultural crops as well as animal husbandry (Bryan et al. 2011). Such types of modelling involve the data integration between soil, land use and climatic attributes. Water is a key component in the food production process. In this connection determining the availability of water for consumptive purpose is a key process that needs to be addressed through modelling approach. The models used in such mapping process tend to calculate the volume of water available from various sources and water budget of the components of water cycle existing in the nature. The storage potential and extraction rate of water from ecosystem components were also calculated through some complex modelling approach (Mendoza et al. 2011).

Typical models such as basin scale models, rainfall runoff models were some of the examples of quantification of the available water resources for food production in the form of ecosystem services. Production of goods for human consumption is another potential service of the ecosystem offered by nature. In this connection, the amount of goods produced was mapped through modelling approach (Maes et al. 2012). Integration of data regarding harvesting of forest produce along with forest types, amount of product available followed by resource dependency of local community stakeholders were also evaluated through complex modelling approach (van Jaarsveld et al. 2005). Various other forms of resources in the form of non-timber forest products, genetic resources were also evaluated through mapping and modelling approach (Fisher et al. 2011; Vihervaara et al. 2010). Modelling of removal of air pollutants by tree species has been attempted considering the physiognomic and phytosociological attributes of the vegetation by various workers (Maes et al. 2012). Combating and mitigating climate change is another most important ecosystem service to develop ecosystem resistance which was addressed through the modelling approach by mapping the C pool and stock in various abiotic and biotic components of ecosystems (Bastian et al. 2012b). Various complex models can be used to map the C flow and establish the relationship between the biotic and abiotic components of the ecosystem. Further, simulation approaches can be used to predict the C flow for a particular land use or its modification (Crossman et al. 2011).

Reducing the risk and calamities of various natural hazards is also one of the prime concerns in terms of ecosystem services. Modelling the different land use and screening of the best practices are one of the major approaches in reducing the negative consequences of natural calamities. For example using simulation studies for flood mapping were done by the various workers to estimate the volume of water to be retained by the vegetation and soil to reduce the impact of flood (Schulp et al. 2012). Same studies of reducing coastal surges by the coral reef as well as mangrove ecosystem were studied by several workers (Costanza et al. 2008). Another study was conducted to establish relationship between riparian zone and existing land uses were done in order to regulate water flow (Pert et al. 2010). Modelling approach was also used to assess the waste management potential of fresh water system and vegetation (Simonit and Perrings 2011). Mapping of nutrient flow from agricultural land to wetlands and floodplain area was also mapped through ecomodelling (Posthumus et al. 2010). Such approaches of modelling used to assess the sediment flow along with associated soil erosion (Simonit and Perrings 2011). Other studies were conducted to assess the waste accumulation of humans and other living biota (Bryan and Kandulu 2009).

Soil is an important ecosystem as it helps in crop production as well as it acts habitat for many living soil biota. In the present time the human population explosion has increased the demand for food production creating the problem of food security and crisis. Under this purview, maintenance of soil fertility in order to maximize production is the biggest challenge for mankind. The growth of industrial based agriculture having high synthetic outputs has decreased the soil fertility all over the world. Under these circumstances modelling the soil fertility is the most essential approach. Some works have been done considering the existing land use practices and available soil data (Maes et al. 2012). Egoh et al. (2008) reported modelling approach considering soil depth and litter cover as the indicator for soil quality. Sandhu et al. (2008) modelled the soil fertility on the basis of soil earthworm population.

Pollination is one of the most important ecosystem services which helps to maintain the population of pollinators. In doing so, biodiversity of an area is maintained. Using land use and cover types, habitat condition followed by the yield of crops data was used to assess the pollination process in terms of ecosystem services (Schulp et al. 2012). Complex modelling was done in pollination services by Lonsdorf et al. (2011) through inclusion of distance and climatic factors along with previously said dataset. Land use, tree density and data on pest population were used to model the biological control of pest population (Brenner et al. 2010; Petz and van Oudenhoven 2012).

TEEB (2010) mentioned about the various factors to maintain the life cycle of a species or ecosystem. It helps to maintain diversification of living organisms in a particular habitat. Modelling life cycle maintenance is based upon determining habitat suitability of species, distribution of biodiversity and factors regulating distribution of species. Some of the work related to this aspect has been reported by the various workers (Summers et al. 2012; Guisan and Zimmermann 2000). Genetic diversity is an important aspect in order to maintain the diversity at the species level as well as promote germplasm conservation. Under natural condition maintaining the species endemism serves this purpose (TEEB 2010). From biodiversity point of view, this was done through mapping of biodiversity hotspots through modelling approach.

The socio-cultural services of ecosystem in terms of recreation, amenity functions and cultural heritage were also studied through various modelling approach. Modelling was done through survey questionnaires followed by personal interviews to identify the most suitable places for recreation in the nature (de Vries et al. 2007; Crossman et al. 2010). Such models involve the criteria of accessibility of the area and existing land cover over that area for structure development of the model. The social and spiritual value of a site was modelled by some workers who identified some locations (Bryan et al. 2010, 2011).

14.12 Research and Development in Ecomodelling for Resource Management

The concept of using mathematical model for natural resource management was initiated since 1970 (Dasgupta and Heal 1979). Mostly these models were used in neoclassical economics. The major objective of such models was maximization of benefits with sustainable use of resources. Initially simple model for various species population was framed but later on complexity aroused for both aquatic as well as terrestrial ecosystems (Holdo et al. 2009). Species conservation along with ecosystem conservation was highlighted by Coulson et al. (2001). New models were framed in order to promote resource management and conservation of wildlife resources, aquaculture resources followed by rangeland management. In case of fishery resources prime objective was to reduce fish mortality rate followed by increase in the yield. Initial modelling approaches were biomass based but later on it shifted towards management implications (Tahvonen 2010). Further, the concepts of applying modelling were shifted to multiple species in fisheries from monoculture species perspective (Pikitch et al. 2004). In this connection research and developement activities lead to formulation of SES (socio-ecological systems modelling) approach (Pauly et al. 2000). Further, development took place in the form of integration of computation of the database involving larger database of species interaction and age class structure to form complex models (Voss et al. 2011). Later on human involvement in fishery practices was also incorporated (Fulton et al. 2011).

Grazing activity is a phenomenal activity as it triggers massive negative changes within the ecosystem leading to events such as soil erosion, desertification which is usually irreplaceable in nature. This is very much true under arid and semiarid conditions of habitat. Research and developmental activities have significantly contributed towards the ecological modelling approach in proper management of this grazing land or rangelands. Initially the conversion of one ecosystems type to another was assessed (Bestelmeyer et al. 2004). Further the human component was added into it through different modelling approach. Some major drawbacks associated with such type of modelling in grazing land include tracking the ecological changes, variability of rainfall as well as heterogeneous nature of the vegetation (Illius and O'Connor 1999). Later on all these components were added to the modelling approaches incorporating various climatic elements (Pickup 1996). Further, multiuser model was developed in order to formulate proper management policy for the mankind.

Wildlife is the central theme of biodiversity and bioresources. It is an integral component of forest ecosystem. Initial modelling of wildlife was done on the basis of wild life growth characteristics and attributes (Sinclair et al. 2005). Later on the conflict between wildlife management and users of bioresources, Barrett and Arcese (1998) developed bioeconomic model in order to provide suitable alternative for the end users of bioresources (Winkler 2011). For sustainable management of wild life resources including the decision making process within the modelling system was developed which requires further development in various countries across the globe (Bunnefeld et al. 2011).

14.13 Conclusion

Ecological modelling is an important aspect for sustainable management of resources. The approach of applying models was initiated from early 1960s with the initial compartment model. New developments in the field of modelling generated more complex models which supplemented the traditional models in due course of time for natural resource management. The complexity in the modelling approach has taken place due to inclusion of human dimension and decision making process in the modelling approach followed by resource specific approach towards the sustainability. The recent models tend to very much upgraded in the form of simulation modelling including balanced parameters and values with multifacets of applications in various spheres of natural resource management. The eco models of today are well equipped to analyse various types of data or combination of them. Future research towards understanding the complexity of the models would help to reach the sustainability of natural resources to a considerable extent.

14.14 Future Research in Ecomodelling towards Natural Resource Management and Sustainability

Future perspective of ecomodelling seems to be having significant role to play in upcoming times. Hybrid type of models needs to design for effective management of resources. There are some basic conceptual problems in modelling approach that is creating the problem in its application in NRM. The integration of ecology along with other disciplines needs to be integrated and models should be developed on this basis in upcoming times. Future research in this field should be aimed towards this direction. Further a more comprehensive knowledge in terms of theoretical knowledge needs to be build up for effective modelling.

Within the modelling approach nonrandomized database needs to be used for formulation of model. This would lead to multi-parameter analyses. Various models have been build up from time to time. But the most important aspect is the model validation and calibration. Improper validation and calibration may lead to wrong decision making in the management process. Better validation and calibration of models would help to increase the feasibility of application of the models. Further the models should be flexible enough to find out the suitable alternatives in the decision making process. Heterogeneity of models should be accounted during NRM and addressing resource sustainability. Specific solution for specific problems in terms of resource depletion needs to be formulated. Future research in ecomodelling should be aimed towards specific use and assessment of impact of the model to give the positive outcome. From resource conservation point of view the output communication in terms of resource management option should be evaluated in the future research in this field.

Policy makers have a significant role to play in the process of modelling. Future research and development should attempt to minimize the error of the modelling process, provide best models for effective management of resource as well as address the issue of sustainability, develop multidimensional models to deal with heterogeneous problems of natural resource depletion.

References

Ali AM, Shafiee ME, Berglund EZ, Arumugam S (2014) A complex adaptive systems approach to simulate interactions among water resources, decision-makers and consumers and assess urban water sustainability. In: 11th international conference on hydroinformatics HIC 2014, New York City, pp 1–9

Allen CD (2007) Interactions across spatial scales among Forest dieback, fire, and erosion in northern New Mexico landscapes. Ecosystems 10(5):797–808

Arhonditsis GB, Qian SS, Stow CA, Lamon EC, Reckhow KH (2007) Eutrophication risk assessment using Bayesian calibration of process-based models: application to a mesotrophic lake. Ecol Model 208:215–229

Aubert S, Muller JP, Ralihalizara J (2015) MIRANA: a socio-ecological model for assessing sustainability of community-based regulations. In: DA Swayne et al. eds. International congress on environmental modelling and software. Ottawa, pp 1–9

Bald J, Sinquin A, Borja A, Caill-Milly N, Duclercq B, Dang C, de Montaudouin X (2009) A system dynamics model for the management of the Manila clam, Ruditapes philippinarum in the Bay of Arcachon (France). Ecol Model 220:2828–2837

Baltic 21 Report (2006) Realizing a common vision of a Baltic Sea eco-region. Baltic 21 Series No. 1/2006

Banerjee A, Jhariya MK, Yadav DK, Raj A (2020) Environmental and sustainable development through forestry and other resources. Apple Academic Press, New York. ISBN: 9781771888110, p 400. https://doi.org/10.1201/9780429276026

Barisa A, Romagnoli F, Blumberga A, Blumberga D (2015) Future biodiesel policy designs and consumption pattern in Latvia: a system dynamics model. J Clean Prod 88:71–82. https://doi.org/10.1016/j.jclepro.2014.05.067

Barrett CB, Arcese P (1998) Wildlife harvest in integrated conservation and development projects: linking harvest to household demand, agricultural production, and environmental shocks in the Serengeti, land. Econometrica 74:449–465

Bartell SM, Lefebvre G, Kaminski G, Carreau M, Campbell KR (1999) An ecosystem model for assessing ecological risks in Quebec rivers, lakes, and reservoirs. Ecol Model 124:43–67

Bastian O, Grunewald K, Syrbe RU (2012a) Space and time aspects of ecosystem services, using the example of the EU water framework directive. Int J Biodivers Sci Ecosyst Services Manage 8:1–12

Bastian O, Haase D, Grunewald K (2012b) Ecosystem properties, potentials and services – the EPPS conceptual framework and an urban application example. Ecol Indic 21:7–16

Bateman IJ, Mace GM, Fezzi C, Atkinson G, Turner K (2011) Economic analysis for ecosystem service assessments. Environ Resource Econ 48:177–218

Battaglia M, Sands PJ (1998) Process-based forest productivity models and their application in forest management. For Ecol Manage 102:13–32

Bestelmeyer BT, Herrick JE, Brown JR, Trujillo DA, Havstad KM (2004) Land Management in the American Southwest: a state-and-transition approach to ecosystem complexity. Environ Manag 34:38–51

Bithell M, Brasington J (2009) Coupling agent-based models of subsistence farming with individual-based forest models and dynamic models of water distribution. Environ Model Software 24:173–190. https://doi.org/10.1016/j.envsoft.2008.06.016

Blumberga A, Timma L, Romagnoli F, Blumberga D (2015) Dynamic modelling of a collection scheme of waste portable batteries for ecological and economic sustainability. J Clean Prod 88:224–233. https://doi.org/10.1016/j.jclepro.2014.06.063

Borsuk ME, Stow CA, Reckhow KH (2004) A Bayesian network of eutrophication models for synthesis, prediction, and uncertainty analysis. Ecol Model 173:219–239

Botkin DB (1993) Forest dynamics: an ecological model. Oxford University Press, New York, p 309

Braat L, ten Brink P, Bakkes J, Bolt K, Braeuer I, ten Brink B, Chiabai A, Ding H, Gerdes H, Jeuken M, Kettunen M, Kirchholtes U, Klok C, Markandya A, Nunes P, van Oorschot M, Peralta-Bezerra N, Rayment M, Travisi C, Walpole M (2008) The cost of policy inaction. The cost of not halting biodiversity loss. Report to the European Commission. Alterra Report 1718. Wageningen, p 314. https://www.cbd.int/financial/doc/copi-2008.pdf

Brenner J, Jimenez JA, Sarda R, Garola A (2010) An assessment of the nonmarket value of the ecosystem services provided by the Catalan coastal zone, Spain. Ocean Coast Manag 53:27–38

Bryan BA, Kandulu JM (2009) Cost-effective alternatives for mitigating cryptosporidium risk in drinking water and enhancing ecosystem services. Water Resour Res 45:8

Bryan BA, Raymond CM, Crossman ND, Macdonald DH (2010) Targeting the management of ecosystem services based on social values: where, what, and how? Landsc Urban Plan 97:111–122

Bryan BA, King D, Ward JR (2011) Modelling and mapping agricultural opportunity costs to guide landscape planning for natural resource management. Ecol Indic 11:199–208

Bugmann H, Cramer W (1998) Improving the behavior of forest gap models along drought gradients. For Ecol Manage 103(2/3):247–263

Bunnefeld N, Hoshino E, Milner-Gulland EJ (2011) Management strategy evaluation: a powerful tool for conservation? Trends Ecol Evol 26:441–447

Chang YC, Hong FW, Lee MT (2008) A system dynamic based DSS for sustainable coral reef management in Kenting coastal zone, Taiwan. Ecol Model 211:153–168

Chen H, Chang YC, Chen KC (2014) Integrated wetland management: an analysis with group model building based on system dynamics model. J Environ Manage 146(C):309–319

Christensen V, Walters CJ (2004) Ecopath with Ecosim: methods, capabilities and limitations. Ecol Model 172:109–139

Costanza R, Perez-Maqueo O, Martinez ML, Sutton P, Anderson SJ, Mulder K (2008) The value of coastal wetlands for hurricane protection. Ambio 37:241–248

Coulson T, Mace GM, Hudson E, Possingham H (2001) The use and abuse of population viability analysis. Trends Ecol Evol 16:219–221

Creutzburg MK, Halofsky JE, Halofsky JS, Christopher TA (2015a) Climate change and land management in the rangelands of Central Oregon. Environ Manag 55(1):43–55

Creutzburg MK, Halofsky JE, Halofsky JS, Christopher TA (2015b) Climate change and land management in the rangelands of Central Oregon. Environ Manag 55(1):43–55

Crossman ND, Connor JD, Bryan BA, Summers DM, Ginnivan J (2010) Reconfiguring an irrigation landscape to improve provision of ecosystem services. Ecol Econ 69:1031–1042

Crossman ND, Bryan BA, Summers DM (2011) Carbon payments and low-cost conservation. Conserv Biol 25:835–845

Cuddington K, Fortin M-J, Gerber L, Hastings A, Liebhold A, O'connor M, Ray C (2013) Process-based models are required to manage ecological systems in a changing world. Ecosphere 4 (2):1–12

Dai S, Li L, Xu H, Pan X, Li X (2013) A system dynamics approach for water resources policy analysis in arid land: a model for Manas River basin. J Arid Land 5(1):118–131. https://doi.org/10.1007/s40333-013-0147-1

Dasgupta PS, Heal GM (1979) Economic theory and exhaustible resources. Cambridge University Press, Cambridge

de Groot R, Brander L, van der Ploeg S, Costanza R, Bernard F, Braat L, Christie M, Crossman N, Ghermandi A, Hein L, Hussain S, Kumar P, McVittie A, Portela R, Rodriguez LC, ten Brink P, van Beukering P (2012) Global estimates of the value of ecosystems and their services in monetary units. Ecosyst Serv 1:50–61

de Vries S, Lankhorst JRK, Buijs AE (2007) Mapping the attractiveness of the Dutch countryside: a GIS-based landscape appreciation model. For Snow Landscape Res 81:43–58

Egoh B, Reyers B, Rouget M, Richardson DM, Le Maitre DC, van Jaarsveld AS (2008) Mapping ecosystem services for planning and management. Agric Ecosyst Environ 127:135–140

FAO (2010) Global Forest resources assessment 2010, main report. Forestry Department, Food and Agriculture Organization of the United Nations, Rome

FAO (2011) World food day focuses on swinging food prices, the world food day ceremony 2011. FAO, Rome. http://www.fao.org/news/story/en/item/93080/icode/. Accessed 17 Oct 2011

Fisher B, Turner RK, Burgess ND, Swetnam RD, Green J, Green RE, Kajembe G, Kulindwa K, Lewis SL, Marchant R, Marshall AR, Madoffe S, Munishi PKT, Morse-Jones S, Mwakalila S, Paavola J, Naidoo R, Ricketts T, Rouget M, Willcock S, White S, Balmford A (2011) Measuring, modeling and mapping ecosystem services in the eastern Arc Mountains of Tanzania. Prog Phys Geogr 35:595–611

Franco CJ, Castaneda M, Dyner I (2015) Simulating the new British electricity-market reform. European J Operational Res 245(1):273–285

Fulton EA, Smith ADM, Smith DC, van Putten IE (2011) Human behaviour: the key source of uncertainty in fisheries management. Fish 12:2–17

Gal G, Hipsey MR, Parparov A, Wagner U, Makler V, Zohary T (2009) Implementation of ecological modeling as an effective management and investigation tool: Lake Kinneret as a case study. Ecol Model 220:1697–1718

Gaube V, Remesch A (2013) Impact of urban planning on household's residential decisions: an agent-based simulation model for Vienna. Environ Model Software 45:92–103

Gentile JH, Harwell MA, Cropper W, Jr Harwell CC, DeAngelis D, Davis S, Ogden JC, Lirman D (2001) Ecological conceptual models: a framework and case study on ecosystem management for South Florida sustainability. Sci Total Environ 274:231–253

Gerst MD, Wang P, Roventini A, Fagiolo G, Dosi G, Howarth RB, Borsuk ME (2013) Agent-based modeling of climate policy: an introduction to the ENGAGE multi-level model framework. Environ Model Software 44:62–75. https://doi.org/10.1016/j.envsoft.2012.09.002

Green DG, Sadedin S (2005) Interactions matter—complexity in landscapes and ecosystems. Ecol Complex 2(2):117–130

Guisan A, Zimmermann NE (2000) Predictive habitat distribution models in ecology. Ecol Model 135:147–186

Guo HC, Liu L, Huang GH, Fuller GA, Zou R, Yin YY (2001) A system dynamics approach for regional environmental planning and management: a study for the Lake Erhai Basin. J Environ Manage 61:93–111

Gurkan Z, Zhang J, Jørgensen SE (2006) Development of a structurally dynamic model for forecasting the effects of restoration of Lake Fure, Denmark. Ecol Model 197:89–102

Gustafson E (2013) When relationships estimated in the past cannot be used to predict the future: using mechanistic models to predict landscape ecological dynamics in a changing world. Landsc Ecol 28(8):1429–1437

Haase D, Lautenbach S, Seppelt R (2010) Modeling and simulating residential mobility in a shrinking city using an agent-based approach. Environ Model Software 25(10):1225–1240

Hamilton DP, Schladow SG (1997) Prediction of water quality in lakes and reservoirs. Part I—model description. Ecol Model 96:91–110

Hansen AJ, Garman SL, Weigand JF, Urban DL, McComb WC, Raphael MG (1995) Alternative silvicultural regimes in the Pacific northwest: simulations of ecological and economic effects. Ecol Appl 5(3):535–554

Harvey CJ, Cox SP, Essington TE, Hansson S, Kitchell JF (2003) An ecosystem model of food web and fisheries interactions in the Baltic Sea. ICES J Marine Sci 60:939–950

He HS (2008) Forest landscape models, definition, characterization, and classification. For Ecol Manage 254:484–498

He HS, Mladenoff DJ (1999) Spatially explicit and stochastic simulation of forest-landscape fire disturbance and succession. Ecology 80(1):81–99

Heppell SS, Walters JR, Crowder LB (1994) Evaluating management alternatives for red-cockaded woodpeckers: a modeling approach. J Wild Manage 58(3):479–487

Herrick J, Bestelmeyer B, Archer S, Tugel A, Brown J (2006) An integrated framework for science-based arid land management. J Arid Environ 65(2):319–335

Holdo RM, Holt RD, Fryxell JM (2009) Grazers, browsers, and fire influence the extent and spatial pattern of tree cover in the Serengeti. Ecol Appl 19:95–109

Illius AW, O'Connor TG (1999) On the relevance of nonequilibrium concepts to arid and semiarid grazing systems. Ecol Appl 9:798–813

Jhariya MK, Bargali SS, Raj A (2015) Possibilities and perspectives of agroforestry in Chhattisgarh. In: Zlatic M (ed) Precious forests-precious earth. InTech, Croatia, pp 237–257. https://doi.org/10.5772/60841ISBN: 978–953–51-2175-6

Jhariya MK, Yadav DK, Banerjee A (2018a) Plant mediated transformation and habitat restoration: phytoremediation an eco-friendly approach. In: Gautam A, Pathak C (eds) Metallic contamination and its toxicity. Daya Publishing House, New Delhi, pp 231–247

Jhariya MK, Banerjee A, Yadav DK, Raj A (2018b) Leguminous trees an innovative tool for soil sustainability. In: Meena RS, Das A, Yadav GS, Lal R (eds) Legumes for soil health and sustainable management. Springer, Singapore., ISBN 978-981-13-0253-4 (eBook), ISBN: 978-981-13-0252-7 (hardcover), pp 315–345. https://doi.org/10.1007/978-981-13-0253-4_10

Jhariya MK, Banerjee A, Meena RS, Yadav DK (2019a) Sustainable agriculture. In: Forest and environmental management. Springer, Singapore, p 606. https://doi.org/10.1007/978-981-13-6830-1

Jhariya MK, Yadav DK, Banerjee A (2019b) Agroforestry and climate change: issues and challenges. Apple Academic Press, New York. ISBN: 978-1-77188-790-8 (hardcover), 978-0-42957-274-8 (E-book), p 335. https://doi.org/10.1201/9780429057274

Jørgensen SE (1986) Validation of "prognosis based upon a eutrophication model". Ecol Model 32:165–182

Jørgensen SE, Mejer HF, Friis M (1978) Examination of a lake model. Ecol Model 4:253–278

Jørgensen SE, Chon T, Recknagel F (2009) Handbook of ecological modelling and informatics. WIT Press, Southampton, p 432

Keane RE (2012) Creating historical range of variation (HRV) time series using landscape modeling: overview and issues. In: Wiens JA, Hayward GD, Stafford HS, Giffen C (eds) Historical environ- mental variation in conservation and natural resource management. Wiley, Hoboken, pp 113–128

Keane RE (2019) Using ecosystem and landscape models in natural resource management applications. Wiley, Hoboken

Keane RE, Finney MA (2003) The simulation of landscape fire, climate, and ecosystem dynamics. In: Veblen TT, Baker WL, Montenegro G, Swetnam TW (eds) Fire and global change in temperate ecosystems of the Western Americas. Ecological studies, vol 160. Springer, New York, pp 32–68

Keane RE, Karau EC (2010) Evaluating the ecological benefits of wildfire by integrating fire and ecosystem models. Ecol Model 221:1162–1172

Keane RE, McKenzie D, Falk DA, Smithwick EAH, Miller C, Kellogg L-KB (2015) Representing climate, disturbance, and vegetation interactions in landscape models. Ecol Model 309–310:33–47

Keane RE, Loehman RA, Holsinger LM, Falk DA, Higuera P, Hood SM, Hessburg PF (2018) Use of landscape simulation modeling to quantify resilience for ecological applications. Ecosphere 9 (9):e02414. https://doi.org/10.1002/ecs2.2414

Khan N, Jhariya MK, Yadav DK, Banerjee A (2020a) Herbaceous dynamics and CO_2 mitigation in an urban setup- a case study from Chhattisgarh, India. Environ Sci Pollut Res 27(3):2881–2897. https://doi.org/10.1007/s11356-019-07182-8

Khan N, Jhariya MK, Yadav DK, Banerjee A (2020b) Structure, diversity and ecological function of shrub species in an urban setup of Sarguja, Chhattisgarh, India. Environ Sci Pollut Res 27 (5):5418–5432. https://doi.org/10.1007/s11356-019-07172-w

Kumar S, Meena RS, Jhariya MK (2020) Resources use efficiency in agriculture. Springer, Singapore, p 760. https://doi.org/10.1007/978-981-15-6953-1

Lal R (2001) Soil degradation by erosion. Land Degradation Dev 12:519–539

Lee DI, Park CK, Cho HS (2005) Ecological modeling for water quality management of Kwangyang Bay, Korea. J Environ Manage 74:327–337

Lindskog E, Lund L, Berglund J, Lee T, Skoogh A, Johansson B (2011) A method for determining the environmental footprint of industrial products using simulation. Winter Simulation Conf 43:2136–2147. https://doi.org/10.1109/WSC.2011.6147926

Liu J, Ashton PS (1995) Individual-based simulation models for forest succession and management. For Ecol Manage 73:157–175

Loehman RA, Keane RE, Holsinger LM, Wu Z (2017) Interactions of landscape disturbances and climate change dictate ecological pattern and process: spatial modeling of wildfire, insect, and disease dynamics under future climates. Landsc Ecol 32(7):1447–1459

Lonsdorf E, Ricketts TH, Kremen C, Winfree R, Greenleaf S, Williams N (2011) Crop pollination services. In: Kareiva P, Tallis H, Ricketts TH, Daily G, Polasky S (eds) Natural capital: theory and practice of mapping ecosystem services. Oxford University Press, Oxford, pp 168–187

Lucash MS, Scheller RM, Sturtevant BR, Gustafson EJ, Kretchun AM, Foster JR (2018) More than the sum of its parts: how disturbance interactions shape forest dynamics under climate change. Ecosphere 9(6):e02293

Maes J, Paracchini ML, Zulian G, Dunbar MB, Alkemade R (2012) Synergies and trade-offs between ecosystem service supply, biodiversity, and habitat conservation status in Europe. Biol Conserv 155:1–12

Maluleke G, Pretorius L (2013) A systems thinking approach to sustainable manganese mining: a case for system dynamics modelling. In: IEEE international conference on industrial technology. Cape Town, South America, pp 1–6

Marlon JR, Bartlein PJ, Walsh MK, Harrison SP, Brown KJ, Edwards ME, Higuerag E, Powerch MJ, Andersoni RS, Brilesg C, Brunelleh A, Carcailletj C, Danielsk M, Hul FS, Lavoiem M, Longn C, Minckleyo T, Richardp PJH, Scottq AC, Shaferr DS, Tinners WJ, Umbanhowar CE,

Whitlockg C (2009) Wildfire responses to abrupt climate change in North America. Proc Natl Acad Sci U S A 106(8):2519–2524
Martins JH, Camanho AS, Oliveira MM, Gaspar MB (2014) A system dynamics model to support the management of artisanal dredge fisheries in the south coast of Portugal. Int Transac Operational Res 22(4):611–634. https://doi.org/10.1111/itor.12090
McKenzie E, Irwin F, Ranganathan J, Hanson CE, Kousky C, Bennett K, Ruffo S, Conte M, Salzman J, Paavola J (2011) Incorporating ecosystem services in decisions. In: Kareiva P, Tallis H, Ricketts TH, Daily GC, Polasky S (eds) Natural capital: theory and practice of mapping ecosystem services. Oxford University Press, Oxford, pp 339–355
McKindsey CW, Thetmeyer H, Landry T, Silvert W (2006) Review of recent carrying capacity models for bivalve culture and recommendations for research and management. Aquaculture 261:451–462
Meena RS, Lal R (2018) Legumes for soil health and sustainable management. Springer, Singapore, p 541. ISBN 978-981-13-0253-4 (eBook), ISBN: 978-981-13-0252-7(hardcover). https://doi.org/10.1007/978-981-13-0253-4_10
Meena RS, Kumar V, Yadav GS, Mitran T (2018) Response and interaction of *Bradyrhizobium japonicum* and Arbuscular mycorrhizal fungi in the soybean rhizosphere: a review. Plant Growth Regul 84:207–223
Meena RS, Kumar S, Datta R, Lal R, Vijaykumar V, Brtnicky M, Sharma MP, Yadav GS, Jhariya MK, Jangir CK, Pathan SI, Dokulilova T, Pecina V, Marfo TD (2020) Impact of agrochemicals on soil microbiota and management: a review. Land (MDPI) 9(2):34. https://doi.org/10.3390/land9020034
Meena RS, Lal R, Yadav GS (2020a) Long term impacts of topsoil depth and amendments on soil physical and hydrological properties of an Alfisol in Central Ohio, USA. Geoderma 363:1141164
Meena RS, Lal R, Yadav GS (2020b) Long-term impact of topsoil depth and amendments on carbon and nitrogen budgets in the surface layer of an Alfisol in Central Ohio. Catena 194:104752
Mendoza G, Ennaanay D, Conte M, Walter MT, Freyberg D, Wolny S, Hay L, White S, Nelson E, Solorzano L (2011) Water supply as an ecosystem services for hydropower and irrigation. In: Kareiva P, Tallis H, Ricketts TH, Daily GC, Polasky S (eds) Natural capital: theory and practice of mapping ecosystem services. Oxford University Press, Oxford, pp 53–72
Miller CU, Dean L (2000) Modeling the effects of fire management alternatives on Sierra Nevada mixed-conifer forests. Ecol Appl 10(1):85–94
Miller SA, Moysey S, Sharp B, Alfaro J (2012) A stochastic approach to model dynamic systems in life cycle assessment. J Ind Ecol 17(3):352–362. https://doi.org/10.1111/j.1530-9290.2012.00531.x
Moorcroft PR, Hurtt GC, Pacala SW (2001) A method for scaling vegetation dynamics: the ecosystem demography model (ED). Ecol Monogr 71:557–586
Nageshwaraniyer SS, Meng C, Son Y, Dessureault S (2011) Simulation-based utility assessment of real-time information for sustainable mining operations. In: Proceedings of the 2011 winter simulation conference (WSC), Phoenix, AZ, pp 871–882. https://doi.org/10.1109/WSC.2011.6147813
Naito W, Miyamoto K, Nakanishi J, Masunaga S, Bartell SM (2002) Application of an ecosystem model for aquatic ecological risk assessment of chemicals for a Japanese lake. Water Res 36:1–14
Nedkov S, Burkhard B (2012) Flood regulating ecosystem services—mapping supply and demand, in the Etropole municipality, Bulgaria. Ecol Indic 21:67–79
Nikolaou I, Evangelinos K, Filho WL (2015) A system dynamic approach for exploring the effects of climate change risks on firms' economic performance. J Clean Prod 103:499–506
Nobre AM, Musango JK, de Wit MP, Ferreira JG (2009) A dynamic ecological- economic modeling approach for aquaculture management. Ecol Econ 68:3007–3017
Odum EP (1953) Fundamental of ecology. W.B. Saunders, Philadelphia, p 384

Pacala SW, Canham CD, Saponara J, Silander JA, Kobe RK, Ribbens E (1996) Forest models defined by field measurements: estimation, error analysis and dynamics. Ecol Monogr 66:1–43. https://doi.org/10.2307/2963479

Panikkar P, Khan MF (2008) Comparative mass-balanced trophic models to assess the impact of environmental management measures in a tropical reservoir ecosystem. Ecol Model 212:280–291

Park RA, Clough JS, Wellman MC (2008) AQUATOX: modeling environmental fate and ecological effects in aquatic ecosystem. Ecol Model 213:1–15

Patten BC (1968) Mathematical models of plankton production. Int Revue Gesamten Hydrobiologie 53(3):357–408

Pauly D, Christensen V, Walters C (2000) Ecopath, Ecosim, and eco space as tools for evaluating ecosystem impacts of fisheries. ICESJ Mar Sci 57:697–706

Pert PL, Butler JRA, Brodie JE, Bruce C, Honzak M, Kroon FJ, Metcalfe D, Mitchell D, Wong G (2010) A catchment-based approach to mapping hydrological ecosystem services using riparian habitat: a case study from the wet tropics, Australia. Ecol Complexity 7:378–388

Petering MEH, Aydas OT, Kuzu K, Ross A (2015) Simulation analysis of hospital intensive care unit reimbursement policies from the triple bottom line perspective. J Simulation 9(2):86–98. https://doi.org/10.1057/jos.2014.24

Petz K, van Oudenhoven APE (2012) Modelling land management effect on ecosystem functions and services: a study in the Netherlands. Int J Biodivers Sci Ecosyst Services Manage 8:1–21

Pickup G (1996) Estimating the effects of land degradation and rainfall variation on productivity in rangelands: an approach using remote sensing and models of grazing and herbage dynamics. J Appl Ecol 33:819–832

Pikitch EK, Santora C, Babcock EA, Bakun A, Bonfil R, Conover DO, Dayton P, Doukakis P, Fluharty D, Heneman B, Houde ED, Link J, Livingston PA, Mangel M, McAllister MK, Pope J, Sainsbury KJ (2004) Ecosystem-based fishery management. Science 305:346–347

Pimentel D (2006) Soil erosion: a food and environmental threat. J Environ Dev Sustain 8 (1):119–137

Pontius RG Jr, Boersma W, Castella JC, Clarke K, de Nijs T, Ditzel C, Duan Z, Fotsing E, Goldstein N, Kok K, Koomen E, Lippitt CD, McConnell W, Mohd Sood A, Pijanowski B, Pithadia S, Sweeney S, Ngoc Trung T, Veldkamp AT, Verburg PH (2007) Comparing the input, output, and validation maps for several models of land change. Annals Regional Sci 42:11–37

Porte A, Bartelink HH (2002) Modelling mixed forest growth: a review of models for forest management. Ecol Model 150:141–188

Posthumus H, Rouquette JR, Morris J, Cowing DJG, Hess TM (2010) A framework for the assessment of ecosystem goods and services; a case study on lowland floodplains in England. Ecol Econ 69:1510–1523

Purves DW, Lichstein JW, Strigul N, Pacala SW (2008) Predicting and understanding forest dynamics using a simple tractable model. Proc Natl Acad Sci U S A 105:17018–17022. https://doi.org/10.1073/pnas.0807754105

Raj A, Jhariya MK, Harne SS (2018) Threats to biodiversity and conservation strategies. In: Sood KK, Mahajan V (eds) Forests, climate change and biodiversity. Kalyani Publisher, New Delhi, pp 304–320

Raj A, Jhariya MK, Yadav DK, Banerjee A, Meena RS (2019a) Agroforestry: a holistic approach for agricultural sustainability. In: Jhariya MK, Banerjee A, Meena RS, Yadav DK (eds) Sustainable agriculture, forest and environmental management. Springer, Singapore. eISBN: 978-981-13-6830-1, hardcover ISBN: 978-981-13-6829-5, pp 101–131. https://doi.org/10.1007/978-981-13-6830-1

Raj A, Jhariya MK, Banerjee A, Yadav DK, Meena RS (2019b) Soil for sustainable environment and ecosystems management. In: Jhariya MK, Banerjee A, Meena RS, Yadav DK (eds) Sustainable agriculture, forest and environmental management. Springer, Singapore. eISBN: 978-981-13-6830-1, hardcover ISBN: 978-981-13-6829-5, pp 189–221. https://doi.org/10.1007/978-981-13-6830-1

Raj A, Jhariya MK, Yadav DK, Banerjee A (2020) Climate change and agroforestry systems: adaptation and mitigation strategies. Apple Academic Press, New York. ISBN: 9781771888226, p 383. https://doi.org/10.1201/9780429286759

Rebaudo F, Dangles O (2013) An agent-based modeling framework for integrated pest management dissemination programs. Environ Model Software 45:141–149

Reddi KR, Li W, Wang B, Moon Y (2013) System dynamics modelling of hybrid renewable energy systems and combined heating and power generator. Int J Sustain Eng 6(1):31–47. https://doi.org/10.1080/19397038.2012.689781

Sahin O, Stewart RA, Porter MG (2015) Water security through scarcity pricing and reverse osmosis: a system dynamics approach. J Clean Prod 88:160–171

Sandhu HS, Wratten SD, Cullen R, Case B (2008) The future of farming: the value of ecosystem services in conventional and organic arable land- an experimental approach. Ecol Econ 64:835–848

Scheller RM (2018) The challenges of forest modeling given climate change. Landsc Ecol 33 (9):1481–1488

Schulp CJE, Alkemade R, Klein Goldewijk K, Petz K (2012) Mapping ecosystem functions and services in Eastern Europe using global-scale data sets. Int J Biodivers Sci Ecosyst Services Manage 8:1–13

Shokohyar S, Mansour S (2013) Simulation-based optimisation of a sustainable recovery network for waste from Electrical and electronic equipment (WEEE). Int J Computer Integr Manufact 26 (6):487–503

Shugart HH (1984) A theory of Forest dynamics: the ecological implications of Forest succession models. Springer, New York, p 278

Shugart HH, West DC (1977) Development of an Appalachian deciduous forest succession model and its application to assessment of the impact of the chestnut blight. J Environ Manage 5:161–179

Simonit S, Perrings C (2011) Sustainability and the value of the 'regulating' services: wetlands and water quality in Lake Victoria. Ecol Econ 70:1189–1199

Sinclair ARE, Fryxell JM, Caughley G (2005) Wildlife ecology, conservation and management. Blackwell, Oxford

Singh NR, Jhariya MK (2016) Agroforestry and Agrihorticulture for higher income and resource conservation. In: Narain S, Rawat SK (eds) Innovative technology for sustainable agriculture development. Biotech Books, New Delhi, pp 125–145ISBN: 978–81–7622-375-1

Slocombe DS (1993) Implementing ecosystem-based management. Bioscience 43(9):612–622

Spofford WO (1975) Ecological modelling in a resource management framework: an introduction. In: Russell CS (ed) Ecological Modelling: in a resource management framework. National Oceanic and Atmospheric Administration and Resources for the Future. Resources for the Future, Washington

Sproedt A, Plehn J, Schonsleben P, Herrmann C (2015) A simulation-based decision support for eco-efficiency improvements in production systems. J Clean Prod 105:389–405. https://doi.org/10.1016/j.jclepro.2014.12.082

Steinhoefel J, Anders F, Kalisch DPH, Kohler H, Koenig R (2012) Investigation of cognitive neighbourhood size by agent-based simulation. In: European conference on modelling and simulation, pp 1–7. https://doi.org/10.7148/2012-0669-0675

Streeter HW, Phelps EB (1925) A study of the pollution and natural purification of the Ohio River. Public health bulletin no. 146. U.S. Public Health Service, Washington, DC

Summers DM, Bryan BA, Crossman ND, Meyer WS (2012) Species vulnerability to climate change: impacts on spatial conservation priorities and species representation. Glob Chang Biol 18:2335–2348

Tahvonen O (2010) Age-structured optimization models in fisheries bioeconomics: a survey. In: Boucekkine R, Hritonenko N, Yatsenko Y (eds) Optimal control of age-structured populations in economy, demography, and the environment. Routledge, New York, pp 140–173

TEEB (2010) In: Kumar P (ed) The economics of ecosystems and biodiversity: ecological and economic foundations. Earthscan, London and Washington

Traas TP, Janse JH, Aldenberg T, Brock TCM (1998) A food web model for fate and direct and indirect effects of Dursban 4E (active ingredient chlorpyrifos) in freshwater microcosms. Aquatic Ecol 32:179–180

Turner RK, van den Bergh JCJM, Söderqvist T, Barendregt A, van der Straaten J, Maltby E, van Ierland EC (2000) Ecological–economic analysis of wetlands: scientific integration for management and policy. Ecol Econ 35:7–23

Twilley RR, Rivera-Monroy VH, Chen R, Botero L (1998) Adapting an ecological mangrove model to simulate trajectories in restoration ecology. Mar Pollut Bull 37:404–419

US EPA (2008) Integrated modeling for integrated environmental decision making. White Paper 100/R-08/010. http://www.epa.gov/crem/library/

van Jaarsveld AS, Biggs R, Scholes RJ, Bohensky E, Reyers B, Lynam T, Musvoto C, Fabricius C (2005) Measuring conditions and trends in ecosystem services at multiple scales: the southern African millennium ecosystem assessment (SAfMA) experience. Philos Transac Royal Soc B: Biol Sci 360:425–441

Vihervaara P, Kumpula T, Tanskanen A, Burkhard B (2010) Ecosystem services– a tool for sustainable management of human–environment systems. Case study Finnish Forest Lapland. Ecol Complexity 7:410–420

Volk M, Hirschfeld J, Dehnhardt A, Schmidt G, Bohn C, Liersch S, Gassman PW (2008) Integrated ecological–economic modelling of water pollution abatement management options in the upper Ems River basin. Ecol Econ 66:66–76

Von Baratta M (ed) (1998) Der Fischer Weltalmanach. Frankfurt am Main. Fischer Taschenbuchverlag, Berlin

Voss R, Hinrichsen HH, Quaas MF, Schmidt JO, Tahvonen O (2011) Temperature change and Baltic sprat: from observations to ecological-economic modeling. ICES J Mar Sci 68:1244–1256

Winkler R (2011) Why do ICDPs fail? The relationship between agriculture, hunting and ecotourism in wildlife conservation. Resour Energy Econ 33:55–78

Xu Z, Coors V (2011) Application of system dynamics, GIS and 3D visualization in a study of residential sustainability. In: Computational science and its applications—ICCSA. Springer, Berlin, pp 300–314

Zhang JZ, Gurkan JSE (2010) Application of eco-energy for assessment of ecosystem health and development of structurally dynamic models. Ecol Model 221:693–702

Zeide B (2001) Natural thinning and environmental change: an ecological process model. For Ecol Manage 154:165–177

Ecological Intensification for Sustainable Agriculture: The Nigerian Perspective

15

L. N. Muoghalu and A. O. Akanwa

Abstract

Nigeria has vast potential resources such as land, water and large population that can be utilized for maximum agricultural production to combat its food insecurity and malnutrition challenges. It is estimated that over 70% of the total land area covering 68 Mha (million hectares) can be employed for agricultural production, while only 33 Mha is used for cultivation. Also, of the estimated 3.14 Mha irrigable land, only about 220,000 ha or 7% is engaged coupled with the ecological diversity that can support a large community of livestock. It also has a surface and underground water of about 267.7 billion m^3 and 57.9 billion m^3, respectively, and over 70% of the population of 200.9 million people are participants in the agricultural sector. All these have contributed to the Nigerian economy in terms of food supply, a national food market, employment, national income generation, livelihoods and industrialization. Agriculture in Nigeria employs approximately 26 million people; more than 80% of the population are involved in small scaled farms that represent half of all jobs accounting for 20% of Nigeria's gross domestic products (GDP). In Nigeria, sustainable agricultural production has been a priority in the policy implementation all in a bid to overcome the general persistent problems of food shortages, insecurity and malnutrition. However, agricultural productivity has been constrained by a range of setbacks in Nigeria, which include but not limited to climate change, youth unemployment and poor policy implementation and consistency, lack of education, inadequate agri-digitalization, poor farming practices, finance, lack of mechanized agriculture among others. The agricultural sector has been grossly affected by climate change as seen in its glaring consequences such as droughts, floods, invasion of pests and diseases. It is on account of these recognizable huge

L. N. Muoghalu · A. O. Akanwa (✉)
Department of Environmental Management, Faculty of Environmental Sciences, Chukwuemeka Odumegwu Ojukwu University (COOU), Uli, Anambra, Nigeria

M. K. Jhariya et al. (eds.), *Ecological Intensification of Natural Resources for Sustainable Agriculture*, https://doi.org/10.1007/978-981-33-4203-3_15

potentials but surrounded by myriads of challenges in the Nigerian agricultural sector that this review study seeks to determine the best practices that would ecologically intensify sustainable agricultural production in Nigeria. This study used a qualitative and descriptive appraisal of agricultural practices within the context of ecological and agricultural intensification with a view to identify the factors that are responsible for the poor agricultural productivity. Against the background of the concept of sustainable agriculture and agricultural intensification, the paper assesses the levels of agricultural productivity and its challenges in Nigeria. This present chapter mainly covers and delineates the theoretical base for this assessment. It focuses on agricultural practices in Nigeria and interrogates their sustainability. It further examines the policy implications of our assessment. Finally, it recommends appropriate policies that will lead to sustainable intensification of agriculture while maintaining or enhancing environmental health and services.

Keywords

Agriculture · Eco-intensification · Food security · Nigeria · Policy · Planning · Sustainability

Abbreviations

ADP	Agricultural Food Development Programme
APPEALS	Agro-processing productivity enhancement and livelihood support
ATA	Agricultural transformation agenda
DFRRI	Directorate for Foods, Roads and Rural Infrastructure
GDP	Gross domestic products
GHG	Greenhouse gases
IRDP	Integrated rural development programme
Mha	Million hectares
NAFSS	National Agricultural and Food Security Strategy
NALDA	National Agricultural Land Development Agency
NPFS	National programme for food security
OFN	Operation feed the nation
PIA	Presidential initiative on agriculture
RBRDA	River Basin and Rural Development Authorities
SI	Sustainable intensification
SSA	Sub-Saharan Africa
TVA	Tennessee Valley Authority
USAID	United States Agency for International Development

15.1 Introduction

Agriculture remains the dominant economic sector to Africa's future. It has established an extended social, economic and environmental footprint in Africa. Over 60% of the populations of SSA are small-scale agriculturists, while about 23% of SSA GDP's flows from agriculture. However, the continent is faced with the pressing need of providing food for about 1.5 billion people by 2030 and 2 billion by 2050. Practical solutions are expedient in order to attain food security for the fast growing population and highly urbanizing African cities with mostly scattered smallholder farms. This is a present problem for the African continent, since, it currently imports 15 billion (USD) on food imports (grains, edible oils and sugar), primarily from regions in Asia and South America (FAO 2020).

Achieving Africa's agricultural potential is dependent on its investment in the sector. You et al. (2010) reported that Sub-Saharan Africa (SSA) needs a minimum of 8 billion (USD) of investment in providing basic storage (without the inclusion of cold-chain investments for horticulture/animal products) and about 65 billion (USD) in irrigation to actualize its agricultural prospects. World Bank (2009a) report on Africa estimates that about a range of 480 to 840 million hectares of agricultural lands are still untapped and can be employed to increase production. Unfortunately, most of these lands are inaccessible due to poor transport infrastructures; most of them are under conflict zones or under forest cover or part of a conservation zone. Observation of issues such as market access, population density and agro-ecological conditions indicates that only about 20–30 million hectares of additional cropland in SSA, essentially in nine countries can be cultivated (Chamberlain et al. 2014; Meena et al. 2018).

This represents only a 10% fraction of the potential increase in African's cultivated land. This number can increase only if new infrastructure and investments are provided across inaccessible areas. Moreover, Africa continues to be a target for increased agric-production if more than 420 projects comprising ten million hectares have been completed between 2000 and 2016. Unfortunately, only a few of these projects have been implemented (Chamberlain et al. 2016; Meena and Lal 2018). This few implemented projects suggest that land expansion will not hinder increased production. Additionally, African countries and regions are at different levels of agricultural production. Some regions experience large agric-productions of about 100 hectare farm sizes that give large farm outputs and others use less than that. For example, Nigeria is mostly dominated by about 50 hectare farm sizes (Jayne et al. 2016).

Agriculture employs over half of the African population and is the largest contributor to the total GDP. Table 15.1 shows that agriculture contributes immensely to Africa's GDP. About 30 countries represented in Table 15.1 have agriculture as the dominant sector. Agriculture has created most of the jobs and livelihoods in Africa (Akanwa and Ezeomedo 2018). Also, the extractive industry has had a great impact on the GDP of 15 of these countries which happens to be either equal to or greater than that of agriculture. However, some countries have

Table 15.1 Gross domestic product of African countries (Source: NEPAD 2010)

Countries	GDP (USD)	Countries	GDP (USD)
Libya	11,314	Lesotho	836
Equatorial Guinea	11,033	Kenya	809
Seychelles	10,681	Comoros	802
Gabon	8724	Chad	767
Botswana	7627	Mali	691
Mauritius	7593	Benin	689
South Africa	7157	Gambia	616
Namibia	5651	Burkina Faso	597
Angola	4477	Zimbabwe	594
Algeria	4435	Rwanda	562
Tunisia	4200	Tanzania	548
Morocco	3248	Guinea-Bissau	508
Cape-Verde	3156	Uganda	500
Swaziland	3061	Mozambique	458
Congo	2983	Togo	458
Egypt	2788	Guinea	448
Sudan Republic	1705	Central African	435
Nigeria	1389	Eritrea	397
Djibouti	1383	Madagascar	391
Ghana	1311	Niger	381
Zambia	1221	Ethiopia	350
Mauritania	1194	Sierra Leone	325
São Tomé and Príncipe	1183	Malawi	321
Cameroon	1100	Liberia	226
Côte d'Ivoire	1036	Dem. Rep. of the Congo	186
Senegal	980	Burundi	180

succeeded in achieving industrialization. For example, South Africa operates a diversified and integrated economy and it generates 30% of the continent's GDP, even though it is home to barely 5% of its population. Additionally, North Africa is exceptional with its diverse secondary and tertiary sectors targeting the European market, with only 35% of GDP and 20% of the African population. Aside from North and Southern African countries, in some oil-producing countries in the Gulf of Guinea, with the exception of Nigeria, Cote d'Ivoire and Cameroon, agriculture employs half or more of the working population. The agricultural population in Africa stands at 530 million people, and is expected to exceed 580 million by 2020 (AGRA 2017a). The dependent population relying on agriculture accounts for 48% of the total population in Africa with a high of 70% is in East Africa (AGRA 2017b). A special feature of African agriculture in comparison to the rest of the world over the last 30 years is that the sector has continued to absorb a large proportion of the working population; half of all new entrants to Africa's working population have turned to agriculture, whereas in Asia, this statistic is only 30% (Mellor 2017).

Notably, Africa's food production is still insufficient and since the 1980s it has experienced stagnant growth due to inadequate infrastructure, sci-tech, policy implementation, increasing population growth and the impact of climate change among other challenges that Africa must overcome. Certain African countries such as Botswana, Morocco, Cote'd Ivoire, Ethiopia, Ghana, Kenya, Rwanda and Senegal have nevertheless made huge reforms in their agricultural sector and gained strides in poverty reduction, while still harnessing the potentials of the sector (AGRA 2017c). This is because poverty is the greatest threat to food security or the ability to obtain sufficient food (Cunningham and Cunningham 2006). Central Africa has the highest percentage of hungry people in the world.

From the foregoing, the goal of this chapter is to appraise the agricultural practices within the theoretical context of ecological and agricultural intensification in order to identify the factors responsible for the poor agricultural productivity and to proffer solutions using Nigeria as a case study. This can be achieved by increasing agricultural production through sustainable agricultural intensification. This means fostering access to inputs—including the use of "smart" subsidy policies, encouraging the adoption of innovations and securing access to resources for women and young people in particular, possibly by law. Support will be offered as a matter of priority to family farms that make optimal use of land and labour on small surface areas. Recognizing the role of women in food production is, according to Cunningham and Cunningham (2006) a sure step to food security for all. All over the developing world they assert that women do 50–70% of farm work, but control only a tiny proportion of land, scarcely have access to capital or developmental assistance. Citing example of Nigeria they observe that home garden (cultivated by women) makes up on the 2% of all crop land but provides 50% of family food. They conclude that making land, credit, education and access to market available to women could contribute greatly to family nutrition.

This chapter discusses nine topical areas which include the concept of sustainable agriculture and the theory of agricultural intensification as put forward by Boserup (1965), and Ruthenberg (1980). Also, agricultural systems and practices in Nigeria, the ecological and agricultural intensification content of the Nigerian agricultural systems and practices were highlighted with a view to highlighting their success or failure.

Further areas reviewed include the sustainability of Nigerian agricultural systems and practices using the indicators of sustainability as outlined in the paper, the challenges and strategies confronting the success of ecological/agricultural intensification in Nigeria. Desired perspectives that will ensure successful and sustainable agricultural intensification in the country were identified. Policy implications for achieving the objectives of agricultural/ecological intensification were pointed out and finally, the conclusion and further areas for introspection were discussed.

15.2 Agriculture and Livelihoods in Sub-Saharan Africa

One striking phenomena globally is the continuous population increase which seems to have stabilized in global North, while in the global South; population growth is occurring at an astronomical rate much higher than in the heydays of growth in the developed world. For example, world population distribution and projections from

2012 to 2050 show that as at 2012 world population stood at 7.058 billion, while the projected population figures for 2025 and 2050 stand at 8.082 billion and 9.624 billion, respectively. The slices for the more developed countries for the three years are 1.243 billion, 1.292 billion and 1.338 billion, respectively. For the developing countries, their proportions for the three years are 5.815 billion, 6.790 billion and 8.286 billion, respectively. Africa's shares for the three years are 1.072 billion, 1.466 billion and 2.339 billion (Chandra 2014). In Africa Nigeria's shares of the population are 170.1 million, 234.4 million and 402.4 million, respectively for 2012, 2025 and 2050. In terms of density there were 184 persons/km^2 in Nigeria during 2012.

Another disturbing demographic aspect is urbanization, a process by which population movement is from rural to urban centres in addition to natural increase in cities. As at 1950 the more developed regions around globe had 58.3% of total global urban population, while the less developed regions had 41.7%. By 1975 the respective ratios were 46.4% and 53.6%, in 2000 they are 30.9% and 69.1%, respectively, and by 2030 projections give 20.5% and 79.5%, respectively (Cohen 2006).

The above data have implications for resources and environment especially for food production. In the recent past, farming activities have spread to rural communities and cities providing livelihoods and increased economic output globally except in Africa where agricultural performance has been slowed down (Pretty et al. 2011). The African continent has vast resources in its abundant land, water, markets and human resources that have contributed to the global food market, while boosting its continental agricultural sector (GSMA 2015). These resources to a large extent have been able to provide for its growing population through its numerous smallholder farms though it has faced shortcomings in the process of combating food insecurity, hunger and malnutrition (Lowder et al. 2016; World Bank 2009b, 2019). SSA is fundamentally agrarian though there are other non-farming activities that have contributed to the economies of the countries in SSA (Rapsomanikis 2015). Agriculture has transformed the economies through increased food production and diversified products, intra-Africa trade, boosted the flow of products to various local markets and introduced the use of improved seeds. It has encouraged greater trade with nations outside Africa and more importantly it has created jobs, generated income and sustained livelihoods (McArthur and McCord 2014; Meena et al. 2020a, b).

In 2050, Africa's population is estimated to be dominated by women and youths totaling about two billion people (World Bank 2019). Africa's population has increased tremendously in the last couple of decades both in villages and cities. It is important to note that population increase has placed severe pressure on food supply and production in Africa (Obonyo 2018). Moreover, agricultural production in Africa has not increased as much as South America, but is comparable to agricultural growth in Asia. This estimation is a call to the need for food security, intensification and transformation of the agricultural sector whereby climate volatility can be minimized (Wiebe et al. 2017; Siba 2018; Siba and Signé 2017; Akanwa et al. 2019, 2020; Akanwa and Joe-Ikechebelu 2020).

However, agricultural growth has not matched unprecedented demographic growth especially in rural communities where population has increased tremendously. Generally, this has affected food production levels in Nigeria where the growing population has by far bypassed food supplies making agricultural practice different from the rest of the world. Also, agriculture has provided job opportunities for majority of the young working population in Nigeria and there are estimates that about 330 million young Africans by 2025 will require employment. Unfortunately, with high unemployment rates in other sectors of the economy only agriculture which provides various means of generating income for young people (World Economic Forum 2017).

However, in Nigeria the agricultural sector employs approximately 75% of the country's labour force. The sector has diverse potentials for socio-economic production and growth covering multi-faceted approaches that can sponsor food supplies that can feed both West and Central Africa regions. This is important since there is a relationship between agriculture and food security; moreover, Africa is largely besieged by hunger and undernourished people. It is of great concern since the numbers of undernourished people in Africa have increased in the past 30 years with observesations showing that 20% of total SSA received less than the minimum amount of food to sustain health and in 1995, 100 million people were malnourished. By the end of the 1970s Africa's marketed volumes of key agricultural exports were at the level of the 1950s, a serious defect in the concept of sustainable development.

Food insecurity in Africa is a persistent problem today especially in remote rural settings. The villages are mostly affected by inadequate farm products since all farming operations are basically carried out at the subsistence levels where majority can barely feed their families. Also, they lack primary facilities such as land, fertilizers, labour and machinery; in addition, rural farmers are poor and unable to boost their production levels thereby increasing food scarcity (Siba and Signé 2017). There is need for constant flow and access to food products to ensure that every trace of food insecurities or shortages is minimized. Agriculture in developed countries has transformed over the years, yet, Africa still practices smallholder farming or family farming which depends on family labour (Sustainable Food Lab 2017). This is responsible for Africa's 33 million farms which are less than 2 hectares amounting to 80% of all farms. Africa's farms workers are largely comprised of women, however, land tenure systems covering use and transfer of land right of ownership do not consider women as heirs unlike women domiciled in Asia and Latin America (Christine 2019).

Before now, SSA has experienced severe stress in agricultural production since it is a rainfall-dependent farming region making its production levels quite poor. This makes farmers work within poor budgets, scarce resources and labour even worse on degraded and infertile lands that have been farmed continuously over the years added to the erratic temperatures and rainfall patterns (Muoghalu 2019). In certain areas where drought and high temperatures are prevalent farmers have employed the system of intercropping shallow and deep rooted crops. For example, farming maize and beans, cowpea and sorghum, millet and groundnut interchangeably all targeted towards sustainable agriculture and livelihoods.

Hadley Centre for Climate Change has projected that regions showing threats of drought will experience more aridity to the degree of about 60–90 million hectares and by 2050–2090 some areas will become endangered. Fisher et al. (2005) argue that Southern African rain-fed agricultural productivity would be reduced by up to 60% between 2000 and 2020 and that even in areas in Africa where rainfall is adequate, the threats of flooding, deforestation and soil loss through sheet and gully erosion will pose serious challenges. In regions that produce minerals and fossil fuels agricultural productivity will face untold stress as the ecological devastation by the Niger Delta region of Nigeria shows. Before the oil boom in Nigeria in the 1970s, the nation was heavily dependent on agricultural sector where agricultural practice and crop and tree specialization and method of livestock breeding were configured on the basis of ecological conformities of soil, climate, altitude and biodiversity.

Shifting cultivation and crop rotation characterized agricultural practices dictated by land tenure system with lack of intensification knowledge and use. Farming implements were rudimentary, mainly digging sticks, hoe and cutlass and sickles which reduced the level of agricultural output. Relevance was placed on mainly food crop production: yams (*Dioscorea alata*), cocoyam (*Colocasia esculenta*), okra (*Abelmoschus esculentus*), vegetables (*Brassica oleracea*), maize (*Zea mays*), cassava (*Manihot esculenta*), plantain (*Musa paradisiaca*), bananas (*Musa acuminata*), kola nuts (*Genus cola*), cocoa (*Theobroma cacao*), oil palm (*Elaeis guineensis*), millets (*Pennisetum glaucum*), guinea corn (*Sorghum bicolor*) and tomatoes (*Solanum lycopersicum*) and were spatially distinguished (Adeniran et al. 2019; Fasimirin and Braga 2009).

British colonial intervention brought about plantation agriculture mainly for cocoa (South Western Nigeria), oil palm (South Eastern Nigeria) groundnut and cotton (Northern Nigeria) and rubber (Edo and Delta States) for export to metropolitan country. Population growth consequent on low morbidity and mortality and improved medical health and access, advent of oil and gas economy and urbanization rewrote the attention given to agriculture. Participation of able-bodied young men in agriculture decreased visibly following the large-scale migration of these men from their local communities to urban areas where they can get higher paying jobs.

Uncontrolled land tenure, alternative land use competition from other sectors and mode of land inheritance among some ethnic nationalities put severe pressure on land, often resulting in fragmentation of landholding. The net effect is that aggressive intensification of agriculture was not possible. Other countries have intervened to make large-scale agriculture through land collectivization and pooling of small and scattered small farms possible. The attempt to do so through the Land Use Act of 1978 failed woefully because of lack of courage and political commitment. The result has been precipitous decline in food production and food insecurity.

Despite the apparently laudable agricultural programmes introduced in Nigeria: the Green Revolution, Operation Feed the Nation, National Agricultural Land Development Agency (NALDA), the Directorate for Foods, Roads and Rural Infrastructure (DFRRI), the River Basin and Rural Development Authorities (initially 11, later increased to 18 and finally brought down to 11), Agricultural Food

Development Programme (ADP), later transformed into the Integrated Rural Development Programme (IRDP), Presidential Initiative on Agriculture (PIA) (2002–2007), National Agricultural and Food Security Strategy (NAFSS, 2008–2011) and Agricultural Transformation Agenda (ATA, 2011–2015), food production is not yet where it should be in terms of productivity, putting more money in the pockets of farmers, self-sufficiency and operation within ecological limits.

In an assessment by Biswanger et al. (2017) indicators of agricultural intensification are weak—only 41% of households use inorganic fertilizer, 34% use agrochemicals, while use of organic manure is low—only 3%. Terrorism and Fulani cattle herdsmen take a toll on agricultural productivity. To worsen all these, agricultural intensification in West Africa is barely being introduced through diverse means (Okike et al. 2005).

The result is a higher incidence of hunger where about 80% of the people still live under poor conditions, presently. Nigeria happens to be the poverty capital of the world. The resultant effect is malnutrition. As research has demonstrated the three latest government initiatives focused on rice and cassava production which did not provide the desired increased crop output in yields but only succeeded in destroying forested areas thereby interfering with environmental services forests provide (Udondian and Zimilia 2018). In addition, all Nigerian food production programmes have not stated "sustainable intensification" (SI) in clear terms as targets, but contain some elements of the concept, including supply of chemical fertilizers, pesticides, provision of irrigation technology all leading to higher yields. Regrettably Nigeria falls short of FAO suggested deployment of 10% annual budget to agriculture.

15.3 Conceptual and Theoretical Frameworks

Before we define agricultural intensification, it will be proper to premise that with an overview of sustainable agriculture (SA).

SA has been defined by the Brundtland Report (1987) as "agricultural and agri-food systems that are economically viable and meet society's need for safe and nutritious food, while conserving or enhancing natural resources and the environment for future generations". Apart from this perspective, Sulphey (2013) contends that SA is viewed differently by various entities. For example, industry views it as the responsible use of available resources to meet energy, food and fibre needs of the population.

Farmers see it as production that ensures environmental, economic and social harmony with surrounding areas. Governments view it as agricultural practices that are sustainable over time, while yet others see it as agricultural practices that are socially just and environmentally and culturally sound. In the midst of this plethora, SA was defined to cover four major areas, namely: the biophysical environment, institutional and policy environment, social and cultural concerns and economic viability (Sala et al. 2015). According to Sala et al. (2015) it presents the foundation for a solid and operational framework for sustainability assessment. Sustainability assessment principles are presented in an integrated way so as to design a family of assessment tools. Hence, these assessment tools would enable the implementation of agricultural development goals and therefore reducing the colossal ecological and social damages. Moreover, these goals are aligned with the germane issues emphasized on agricultural sustainability by Smith and Donald (1995).

Firstly, a sustainable agriculture requires much more than just maintaining high crop production levels, but there must be consistent increase in crop yields especially for major crops that are on high demand. Without increase in yields, there will be challenges over food supplies and income generation for the working populace and even future generations. Secondly, intensifying crop production on deforested lands provides more options for conservation of other green areas. Thirdly, SA does not refer to shortages in food production levels; obviously, poor production levels are not sustainable from eco-social vintage points. On the other hand, high agricultural productivity approaches can cause immense harm to the environment. They also argue that sustainability in tropical agriculture includes the challenge of degraded lands, invasion of pest and diseases, coupled with the various impacts of weeds, the eco-social contribution of tree crops among others and the need to employ a combination of workable natural and modern agric-techniques.

One of the most prominent theories on the impact of increased population on agricultural production was proposed by Thomas Malthus in 1798. Malthus (1998) argued that the changes in quality of life are dependent on variations between birth and death levels. He further implied that the income accrued from labour is eventually reduced since agricultural lands were naturally fixed and hence, could not be increased. He contends that as humans continue to increase, food production rates, however, does not grow in tandem, making income levels of workers involved in agricultural production to reduce and bringing with it a decrease in birth rate and increase in death rate. As a result, agricultural production determines the extent of population increase. In addition Malthus argued that population growth impacts negatively on the ecology by destroying all forest resources and so resulting in environmental degradation and famine, as well as sabotaging agricultural transformation and inducing massive land destruction beyond recovery. Malthus' depositions were confirmed by Chu and Karr (2017) who discovered that where high populations have been steady over a long time that all human injustice on natural resources ranging from deforestation, biodiversity loss, stagnation and reverse productivity would prevail.

Boserup (1965) in contrast to Malthus' depositions argues that agriculture would intensify in response to increasing population pressure. That is, that increased demand on resources consequent on increased human growth can cause positive agricultural prospects and transformation. Boserup (1965), therefore, conceptualizes that growth in population levels (or land scarcity) is an independent variable that can stimulate agricultural intensification which follows agricultural extensification. In this way Boserup (1965) and Ruthenberg (1980) provided a framework for the analysis of the effect of increased population levels and market access on the intensification of farming systems. Over the last couple of years rapid rise in population has placed farming systems under pressure, while swift urban expansion and economic growth have provided new markets opportunities for the distribution of agricultural products.

The evidential effect of rapid population growth on agriculture or decrease in land on agricultural production is that the length of fallow period has shortened or even been disappearing. Hence, this has increased low soil fertility levels where crop outputs are reduced even with the application of organic and chemical fertilizers. To further aggravate the situation there are inadequate investments targeted towards irrigation. Process of intensification is weak. Farm sizes are reduced as land inheritance results in land fragmentation. This diminution would have led to intensification and production.

According to the intensification of farming systems model known as Boserup and Ruthenberg (BR) model, rising population levels and market opportunities can lead to a continuous cycle of agricultural intensification. The factors behind the soaring population density, market access, growth of cities and economic growth, which led to reduced farm size, would stimulate increased application of organic manure and fertilizer to alleviate the consequences of threatened soil fertility, investment in mechanization, land and irrigation. All these will potentially offset negative impacts of an increasing population on available agricultural lands, maintain or increase per capital crop production and even increase income levels generated by farmers. Rising population levels thus orchestrates the need for intensification, while access to larger market provides the opportunity. Increase in output depends on increased labour and other inputs per unit area of cultivated land.

It is however argued that even though agricultural intensification is dependent on population growth and opportunities to market access, the agro-ecological potential of a place is a critical factor, which is dependent on human migration to areas with high potential such as the tropical highlands in East and Central Africa. However, while governments may choose to invest on public facilities such as roads and a market which enables them to maximize food productivity levels that would provide for the high population.

This was the philosophy behind roads and rural infrastructure development in the General Babaginda institutional infrastructure of the Directorate for Foods, Roads and Rural infrastructure in Nigeria in the mid-eighties (1987–1990). Investments in roads and markets are traceable to the capacity of cities demand levels on food products and distances from rural communities. However, market access is dependent on two factors, namely the demands from cities and those for export markets. Here, the developed roads make accessibility easier for the farmers.

With rapid population growth and expansion coupled with ineffective policy implementation, it is likely that institutional or agro-ecological environment intensification would lead to involution and diminution of eco-social health and ecological damages. Gerber et al. (2010) defined involution as the situation where high demand for food products merges with high labour referred to as intensive intensification, though this happens at small decreasing and average returns to inputs. Moreover, traditional agricultural intensification is carried out in three major ways (1) increasing yields per unit area of land, (2) increasing intercropping intensity (e.g. two or more crops per unit area of land or other inputs (such as water) or livestock intensity (such as faster maturing breeds); (3) changing land use from low value crop or commodities to those getting higher market prices or have better nutritional content (Pretty and Stella 2011).

On the other hand, ecological intensification implies green concept which involves diverse use and application of natural ecological functions. It pursues a functional design that can serve several agro-ecosystems, while it obtains natural sustainability and applies regulatory natural functions for its agro-economic design (Kumar et al. 2020; Banerjee et al. 2020; Raj et al. 2020). Since agricultural systems are amended ecosystems, sustainable agro-ecosystems seek to move most of the recovered properties towards natural systems without necessarily jeopardizing its productivity. This means that agro-ecological design of systems should produce both crops and ecological products (Jhariya et al. 2019a, b). Sustainable agro-ecological systems refer to all contributions made on the natural, social and human segments of the economy, but when it is unsustainable it degrades both the present and future resources (Pretty and Stella 2011). Tittonell (2014) has argued that the present approach to agricultural intensification cannot be sustained both socially and thermodynamically since it is neither ecological nor eco-efficient. Hence, it is not appropriate in terms of global food security as it also endangers the environment by biodiversity destruction. From the foregoing, all these emphasize the necessity for substitutes in agricultural intensification practice.

Clearly, continuous agricultural activities on a particular piece of land can degrade the natural/forest resources and productivity rates thereby producing high levels of waste and pollutants (Pretty and Stella 2011; Jhariya et al. 2018a, b). For example, an important index of intensification is seen in the influential capacity of externalities to negatively affect the productivity abilities of farm land thereby destroying the environment and people who live in them downstream of farming areas. For instance, fertilizers/pesticides not used up by plants are flushed during run-off draining into public water supplies carrying pollutants that can threaten the survival of soil organism and aquatic life such as fishes and turtle species (Meena et al. 2020).

Highly intensive farming practices will consume greater energy such that consumption of fossil fuels in the short term will increase in rural areas causing increase in Greenhouse Gases (GHGs). Intensive agriculture, such as paddy rice, emits large quantities of methane. Because of this, intensive agriculture calls for monitoring of pollutants/GHG emissions that can trigger local or global warming that threaten climate change, monitoring the health of soils (such as soil erosion), organic manure, acidity/alkaline levels of trace metals and compaction (Akanwa and Joe-Ikechebelu 2020; Khan et al. 2020a, b).

These considerations gave rise to the concept of SI which is now popular with global research and international bodies on policy-making and practice such as FAO and the World Economic Forum agribusiness world and large-scale international donor organizations. The motto of this concept encourages eco-efficiency, which implies maximum production at minimal waste and pollutants emission on the environment. Pretty et al. (2011) define SI and made specific references to the Africa continent where SI refers to the process of producing high outputs within a particular land area. It exploits natural resources to provide capital and effective environmental performance, while the negative effects on the environment are adequately minimized.

Dore et al. (2011) defined it as a unique means or process of generation of new products and services at higher quality and output levels using different technology. Environmental bodies at the grassroots level and global organizations have expressed displeasure at the use of SI, which they regard as an incomprehensive approach to best describe the complex strategy of intensification. In 2008, the international organization—FAO called for a global attention over the need to increase food supply levels for a fast growing world of 9 billion people in 2050 and in doing so every sci-tech strategy should be put in place to double food production levels (Graziano da Silva 2012). With this, biotechnology claimed that the world cannot feed itself without genetically modified crop cultivated. The fertilizer industry followed such that at a gathering of African Ministers of Agriculture in Nigeria in the presence of fertilizer industry executives in 2006, they agreed to encourage higher use of fertilizers from the then 8kg/ha to an average of 50kg/ha by 2015 to achieve African Green Revolution.

However, there are four basic limitations to food production and sustainable development of the ecosystems comprising water, soil, biodiversity and land. Pretty (2008) contends that a defined sustainable approach should include the following valuable conditions during and at the end of the process of food production:

1. Farming systems should apply the use of crop and livestock varieties that have inherent high quality and productivity abilities to provide high output levels;
2. The overuse of external means can be avoided except when it is expedient;
3. The application of natural agro-ecological processes that aid in high food production is encouraged such as nutrient-cycling among others;
4. The application of excessive high technology is reduced in order to avoid their negative effects on health and environmental risks;
5. The use of human and social capital in the area of innovative ideas, knowledge of proven ideas and concepts that can be applied to contend with common problems such as pests, diseases, soil management among others and
6. The application of proven means to manage persistent problems that have huge impacts on the environment should be practiced as they affect global warming, GHGs emission, plants and animal extinction, spread of pests and weeds among others.

The Royal Society (2009) defines SI to refer to an effective means that considers the maximum use of land space, high food productivity and low environmental degradation. It encourages the end product of high food supplies rather the process which can incorporate technologies such as new breeds, varieties and other approaches. It is distinct from other applied concepts of agricultural intensification because of its unusual combination of factors that pursues peculiar goals targeted at productive enhancement. SI combines various approaches, methods and technologies aimed at providing higher food supplies within a given area while exploiting natural resource. It also provides policies that allow natural resource expliotation and the production of higher food supplies while environmental protection is guaranteed.

However, problems have been pointed out as to the workability of SI especially at the early process where food productivity and other important products along the food chain are high while upgrading the state of natural resources. Pretty and Zarean (2014) have argued that SI concept is complex and yet every aspect needs to be included in the process; hence, this makes it a herculean task to capture all the process knowing that agro-ecological approaches are multi-faceted and can only be practiced based on location, community needs and farmers priorities. Similarly, Milder et al. (2012) argued that SI approach has no distinctive methods or processes that clearly separate an alternative from a conventional practice that a farmer can apply; it all depends on the need and capacity to apply agro-ecological principles to industrial farms. Elliot et al. (2013) posit that where researches are targeted at achieving increased crop outputs and environmental outcomes, the end products will be based on applied conditions programmed to give environmental improvements, timing and weightings employed.

Jennings (2007) has called for an "agronomic revolution" to achieve sustainable agriculture because crop yield gaps in rice, for example, result from agronomic failings. Agronomy refers to proper care of plants and livestock under peculiar conditions that coincides with the emergence of the concept of agro-ecology which shows the relevance for investing in science and practice. This confronts farmers with various best options of seeds and breeds and their care under local ecological contexts. Since SI is multi-faceted, it requires appropriate skills, knowledge and managerial technology for farmers to know that farm inputs can complement or contradict biological processes and ecosystem services that inherently support agriculture (Royal Society 2009). Farmers require the assurance that increased productivity levels generate high income through sales. Additionally, this creates social trust, connectivity and ways of doing things which are valuable to agric learning of other areas such as soil management and maximizing input efficiencies.

15.4 Agricultural Systems and Practices in Nigeria

The agricultural systems in Nigeria include shifting cultivation, terrace agriculture, sedentary cultivation or permanent farming, mixed cultivation and intensive irrigated cultivation (Onakuse 2012; Asadu and Asadu 2015). Shifting cultivation or bush fallowing (in which mixed cropping is practiced) is a system in which farm lands are rotated involving short periods of land occupancy and extended periods of fallow.

The main determinant of this system is the ratio between the length of the time the soil will sustain cultivation with satisfactory results and the periods required for restoration of fertility. The fallow period ranges from 5 to 15 years depending on the availability of land and the fertility of the soil. Shifting cultivation has been dubbed an inefficient system in terms of agricultural productivity. It requires about 60.75 ha of land per farm family in the Savanna zone and 40.5 ha in the rainforest region per farm family. These figures are dream figures given conditions of food arable land area, individualized land tenure in the south, increasing population density in Nigeria today and peasant agriculture with traditional hoe and knife technology. In Africa crop parcels are very small (60–86%) had sizes ranging from 0.25 to 2 acres compared to average farm size in Africa of 2.5 ha; North America (121 ha), Latin America (67 ha) and Europe (27 ha) (Kanu et al. 2014; OSMARD (Ondo State Ministry of Agriculture and Rural Development) 2004).

Nielsen and Calderón (2011) identified four main problems of soil fertility emanating from bush fallow systems: maintenance of soil fertility vis-à-vis rapid population growth and urbanization; difficulties of getting farmers to adopt innovation; inability of the system to keep pace with rapid population increase and consequent increases in food requirements and finally the shortening of fallow periods (an inevitable response to rapid population growth) with its grave consequences of soil erosion and soil ruination.

Terrace agriculture is found in Benue, Plateau, Borno, Adamawa, Taraba and Enugu States. It represents a community located on hill-top and this defence regime requires that farmers will adjust to the rigorous movement around the upland environment. The system is based on four principles—prevention of erosion by fallowing; heavy use of animal and human waste to build up the nutrient status of the soil; a simple rotation of the crops to minimize depletion of soil fertility; and the planting and protection of trees to provide supplementary animal and human food (Durán and Pleguezuelo 2008; Raj et al. 2019a, b).

Sedentary or permanent or continuous cultivation, cushioned on simple crop rotation and intensive use of manure on permanent well-defined land holdings, is practiced in the densely populated areas of Zaria, Katsina, Kano and parts of Anambra and Imo States. It is sustained by the application of household organic manure and crop rotation (Muoghalu 1995).

Mixed farming, an innovation introduced by the various State ministries of Agriculture, is a rudimentary system involving a balance of crops and livestock on individual holdings and the use of animal manure to reduce the need for bush fallowing (Onakuse 2012). Although traditional form of this system existed in some parts of today's north-eastern geopolitical zone (Babalobi and Akinwum 2000), obstacles to this system include the nomadic culture of cattle Fulani, who have no interest in crop production (Ducrotoy et al. 2016).

Finally is the intensive irrigated cultivation, a development occasioned by the seasonality of rainfall in Nigeria and its unreliability especially in the Sudan Savanna zone of northern Nigeria, plagued by meteorological, agricultural, hydrological and environmental droughts. Adelodum and Choi (2018) identify four principal methods of irrigation—flood water irrigation often called the Fadama schemes, the shadoofs, canal irrigation and overhead irrigation. The Fadama or naturally/seasonally flooded swamps are widespread in Nigeria around Kano, Bida (Niger State), Zaria (Northern Kaduna State) and Sokoto State where sugarcane, rice and various forms of vegetables are grown. This system is also practiced in riverine areas in Anambra State in which rice is grown, as well as in Ebonyi and Enugu States. The shadoof, used to supplement the water supplied by the seasonal floods, produces vegetables in the drier areas of Oyo and Kwara States.

Development of large-scale integrated hydroelectric dams for purposes of agriculture is a development that started in the 1960s. The major dams are all concentrated in the north and include Kainji Dam, Tiga Dam and Bagauda (in Kano State), the Dadinkowa tomato irrigation scheme, Jebba, Bacita Sugar Estate and the Chad Basin. Other comprehensive multipurpose irrigation schemes are the River Basin and Rural Development Authorities (RBRDAs) aimed at all round socio-economic development of all the resources—water, land, plant, animal, etc.,—of all the major drainage basins in the country.

Originally 11, these were increased to 18 and later reduced to 11. Some of the powers/functions of RBRDAs germane to this chapter include—providing an expanded development of ground water supplies that can serve several purposes; providing watershed management schemes for flood and erosion control, building and managing dams, dykes, drainage systems, wells/boreholes, irrigation and drainage systems; providing irrigation schemes for the production of crops and livestock; and monitoring pollution water bodies within the authority's area in accordance with national standards.

According to Phelps and Kaplan (2017), the system of livestock production is governed by three major factors: the degree of specialization in animal husbandry, the degree of mobility and the type of animals. The factors produced five specialized systems of livestock farming: nomadic cattle breeding, settled montane cattle breeding, controlled cattle breeding, cattle ranching and intensive stock breeding, fattening and milking.

Since almost all parts of Nigeria receive fairly enough rainfall for agriculture, two major ecological/environmental factors affect livestock specialization. The long dry season and the presence or absence of tsetse fly is the vector of trypanosomiasis. The first restricts the use made of the northern Savanna during period of low humidity, while the second restricts the use which can be made of the southern guinea Savanna during periods of high humidity (King 2006; Enete and Amusa 2010). Scarcity of water and tsetse fly prevalence localize livestock to a few convenient places in the northern grasslands, some of which are experiencing the tragedy of the commons due to over stocking of livestock. This is made worse by periodic droughts such as those of 1968 and 1972–1973. Some northern states are improving rural water supplies, especially in Borno and Yobe states, through irrigation by exploiting artesian wells in the southern and western parts of Lake Chad (Fig. 15.1).

In terms of livestock population cattle is by far the most significant numbering over 8 million by 1982, followed by sheep over 7 million. Goats numbering over 20 million are widespread in the north and south. Pigs numbered about 1.4 million and are dominant in the south and non-Muslim Middle Belt, while camels, horses and donkeys numbering about 3.1 million are beasts of burden are mostly found in the north (Gentry et al. 2004).

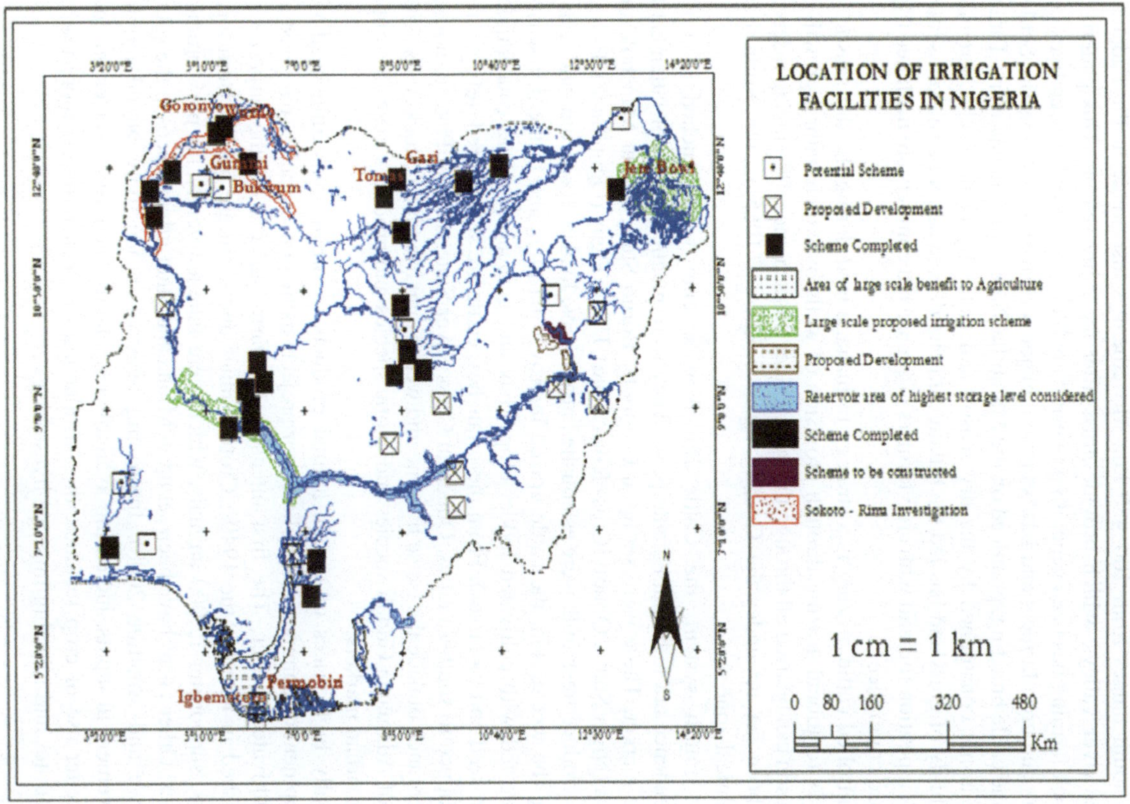

Fig. 15.1 Showing the location of irrigation facilities in Nigeria

Nomadic cattle rearing involves a form of transhumance, a latitudinal movement from the north to the south in the dry sea and a vertical movement from upland to lowland and back again in the wet season. This system found especially among the Fulani herdsmen combines the rearing of livestock. Distances moved from the north to the south ranges from 100–480 km. Movement is also made to the Fadamas, flooded river valleys, which provide grass after the rainy season. This pattern of livestock farming has been generating intermittent bloody clashes between livestock farmers and crop farmers and has become a major security issue in Nigeria. Settled montane cattle breeding, practiced on the Mambilla Plateau in Adamawa and Taraba States is not constrained by the dry season and tsetse fly factors. The perennial availability of grassland has led the Fulani cattle herders to adopt permanent settled life. Competition for land with crop farmers is intense but conflicts are mediated by a land use committee.

Controlled cattle grazing is a strategy to manage land during the dry season. It employs rotational grazing through provision of wells by which natural growth of the season can be used all through the year. Under this scheme, 109 grazing reserves have been set up under the Grazing Reserve Law of 1965 in the North East geopolitical zone.

The fourth system, the Cattle Ranching scheme was launched in post-independence era to be public enterprise. They were meant to breed or fatten cattle for the market. These were set up at Gombole, Borno State, and at Manchok and Mokwa (Niger State), Obudu (Cross River State), Upper Ogun (Ogun State). Dairies which produce fresh milk were established near Lagos, Ibadan, Jos (Vom) and Kano. During the wet season, the cattle grazed improved pastures and in the dry season they are fed with cultivated fodder crops (hay and guinea corn) and concentrates (cotton seed and molasses). From these ranches, cattle are slaughtered, refrigerated and shipped to southern cities. Obudu in Cross River State, another Plateau, houses another ranch outside the tsetse fly zone and wetter than Mambilla, stocks over 3500 cattle of Adamawa Gudali species and exotic species and produces meat for south-eastern urban markets.

Finally high prices of certain animal products in urban areas influenced the development of commercial public sector projects using concentrates, sown pasture and cultivated fodder. The first intensive pig farm and one of the largest was established in Kano in the 1940s. Crop residues and grain are bought from local farmers, supporting 12,000 animals, which were fattened yearly for pork product factories. Other farms have been set up in Minna and other urban places in the south (Ironkwe and Amefule 2008). Improving livestock productivity would require improvement in inputs, increasing available rangelands; increasing fodder production, better use of crop residues, the introduction of sown pastures, more use of concentrates and injecting improve livestock.

Fishery

Nigeria has vast inland water resources estimated at about 12.5 million hectares, comprising the rivers Niger, Benue, Gogola, Argungu, Komadugu, Yobe, Anambra, Cross River, Ogun, Osun and numerous other smaller rivers in addition to the Lake

Chad, Kainji Lake, and Oguta Lake. All these produce over 5 million metric tons of fish annually. Creeks and lagoons are important fishing grounds in Nigeria (Fig. 15.2).

We can divide the fishing industry in Nigeria into three: Artisanal (producing 80% of annual fish output), industrial and land-based aquaculture. Lake Chad dominates the industry with a total production of 25%. Most of the fish caught in the first category are smoked in absence of large-scale refrigeration facilities. The Argungu fishing waters have given rise to a growing annual tourist festival based on River Rima. Large scale fishing has expanded since 1960 and is based on distant Marin and inshore and costal grounds, operated by foreign trawlers from Russia, Japan and Poland in charter to Nigerian companies. These companies have established the efficient and far-flung distribution and marketing network furnished with cold storage facilities. Ogbonna (2001) stated that shrimp catch in Nigeria was 10,807 tons in 1997, pink shrimp (*Penaeus notialis*) being the most common shrimp species with the increase in demand, shallow water brown shrimp is now caught in larger quantities. The small-sized fish caught by the shrimping vessels are sold at local markets, fresh, smoked or dried by women. In 1999, 187 vessels were licensed for inshore shrimp fishing. The major companies with large fleets are joint-ventures. All are grouped within a "Nigerian Trawler Owners' Association" (NITOA). However, Cross River and Rivers states have entered partnerships with foreign countries to enhance the fishing industry. Inland aquaculture is expanding rapidly not only to provide employment, but also to boost fish production. These are based mostly in the burgeoning urban centres especially Lagos, Ibadan, Port-Harcourt, Onitsha, Kano, etc.

15.5 Ecological Intensification in Agricultural Practice in Nigeria

Here, we look at agricultural practice in terms of ecological zones and their crops, agricultural inputs, proportion of land under different crops, agricultural output and productivity, crop yields, farming methods, state of forests and biodiversity.

Developing countries including Nigeria experience food shortages and acute hunger. African Development Bank Group (2018) pointed a horrendous picture of African agriculture. Since 1970 many African countries face acute hunger and starvation. Foods produced locally have been inadequate to satisfy the fast growing population; while per capital earnings cannot support enough food importation. For example, Nigeria food production index was 69 from 1990–2002 and 105 from 2003–2005. For crop production, corresponding indices were 68.7 and 105. For livestock production corresponding indices were 76.8 and 107.0, respectively. Cereal yield for 1990–1992 was 1.135 kg/ha and 1.460 kg/ha for 2003–2005, a slight rise. Total catch of fish by artisanal, industry and aquaculture was 360, 219.90; 27, 701.0 and 25, 204.6 metric tons, respectively, in 1997 (World Bank 2009c). If we examine the data for the production of individual food crops, the horrendous picture painted by hutchie becomes clear. Table 15.2 shows the production of food crops

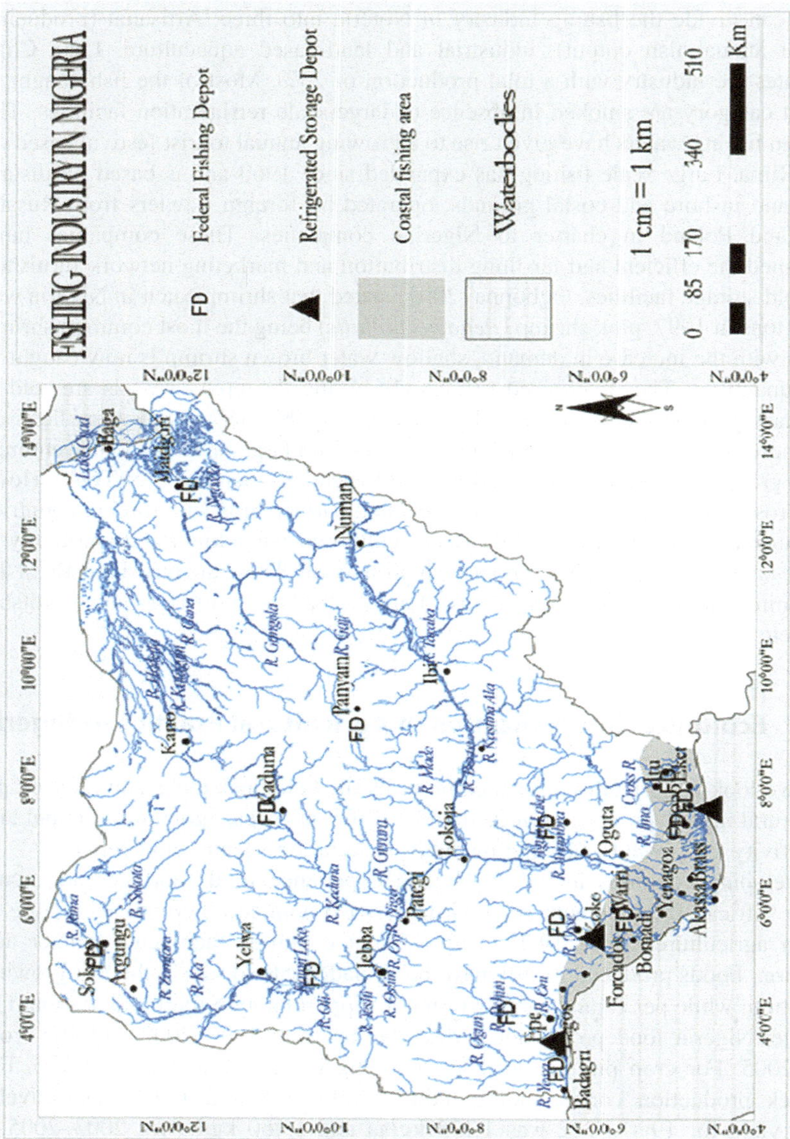

Fig. 15.2 Showing the fishing industry in Nigeria

Table 15.2 National production of food crops in Nigeria (Source: National Bureau of Statistics 2007)

	1994/95	2000/01	2005/06	%
Food crops	In 1000 metric tons			
Millet (*Pennisetum glaucum*)	7271.23	4158.86	4323.86	−59.46
Guinea corn (*Sorghum bicolor*)	6329.91	4963.48	5039.20	−79.62
Beans (*Phaseolus* spp)	2338.15	1581.90	1650.09	−70.57
Yam (*Dioscorea* spp)	23,395.75	24,654.74	25,707.45	+9.88
Maize (*Zea mays*)	5120.70	4719.37	5768.94	+12.66
Cassava (*Manihot esculenta*)	23,831.39	27,702.93	35,614.05	+49.49
Rice (*Oryza sativa*)	1994.02	3159.65	3286.50	+64.79
Melon (*Cucumis melo*)	287.25	231. 91	357.65	+24.39
Cocoyam (*Colocasia esculenta*)	1560.45	1700.13	2149.32	+37.76

from 1994/95 to 2005/06 for millet, guinea corn, beans, yam, maize, rice, melon and cocoyam.

In terms of agricultural input, we look at percentage of total arable land devoted to agriculture, land devoted to cereals, fertilizer application and the proportion of land under irrigation. In terms of proportion of total land under agriculture, from 1990–1992, 79.2% of total land was given over to agriculture, while for 2003–2005, it was 80.4%, a slight increase. Total slice of land cultivated with cereals was 66,416,700 ha in 1990–1992 and for 2003–2005, it was 199,572,300 ha, an increase of 1105.16% due mainly to various rural development schemes introduced. The total land under irrigation was 0.7% of all the land under agricultural production in the period 1990–1992 and 0.8% in 2003–2005 (World Bank 2009b). In 2002, while irrigation contributed 50% to the world's food supply, in Nigeria it was 8.5%. Studies and proposals showed that 2.7 Mha out of 72 Mha have been identified for irrigation. In recent times, about 1 Mha of lands are under irrigation (see Fig. 15.2) (Arav and Uza 2002). The World Bank source shows that fertilizer use in Nigeria was 142.2 g/ha in 1990–1992 and a reduced 63.9 g/ha in 2003–2005, a decrease of 55.06%.

It was observed from Table 15.2, that food products such as yam, rice, cassava among others exhibited a decrease in their levels of production. Yam, a major staple food item throughout the country showed a marginal increase of 5.38% as at 2000/2001 and an increase in productivity of 9.37% in 2005/06 on the output in 2000/01; cassava showed a better performance with 16.24% growth rate in 2000/01 and a 28.56% increase in 2005/06 over that of 2000/01. The picture for rice is dismal with an improvement of 58.43% in production in 2000/01 vis-à-vis 1994/95 figure and only sliding down to a mere 4.02% increase in 2005/06 over the figure for 2000/01 production year. Considering that rice is a major staple food for all Nigerians, it is a poor showing. In effect, apart from a few crops on Table 15.1, the rest have violated the norm of agricultural sustainability that yields should continuously rise, without which it would be impossible to improve incomes and compensate for population increases. As at 2006 population census annual population increase was put at 3.18% between 1991 and 2005 (National Population Commission 2006). Fertilizer application fell by 55.06% in 2003–2005. Taken in totality, it is, therefore, clear why Nigeria has been ranked as the second largest importer of rice in the world (Index Mundi 2019) and that she expended between $6 and $8 billion to import food annually (Stephen 2019).

Since agriculture includes forestry a comment on the condition of national forest resources is necessary. It is notable that as at 1990 total forested area was 172,000 km^2, while in 2005 it had gone down to 111,000 km^2 giving a loss of 64.43%. Average annual rate of deforestation stood at 2.7% in 1990–2000 and 3.3% as at 2003–2005. As of 2000 the National Coordinator of the National Forestry Commission observed that Nigeria was losing between 100–400,000 ha per year (USAID 2008). As at 2004 total known species of higher plants numbered 4715, while in 2008 a total of 171 species were already threatened. For animal species total known species stood at 1189 in 2004 and by 2008 it was 79. Nationally protected area was 56,000 km^2 representing 6.2% of total land area.

The inability to consume quantitative and qualitative food has practical visible health effects. Starving children become visible symptoms of deepening economic deterioration. The World Development Report (World Bank 2009b) reports that as at 2009, prevalence of malnutrition in Nigeria showed 27.2% of under 5 age children being underweight and 43.0% suffering from stunted growth. Underweight is the most common malnutrition indicator. It must be noted that mild underweight increases the risk of death among under 5 age children, inhibits their cognitive development and perpetuates the problem across generations. In the same vein women lacking vital nutrients are more likely to have low birth weight babies. Stunting is used as a proxy indicator of long-term changes in malnutrition.

Taking an example with cotton and groundnut, the key export crops from the Sudan ecological region of Nigeria, cotton production fluctuated from 387,940 metric tons in 1994/95 to 380,080 metric tons in 2000/01 to 481,180 in 2005/06. For groundnut, corresponding statistics were 3,092,350, 2,240,113 and 2,752,730 metric tons, respectively, a continuous slide (World Bank 2009b).

A sliding productivity of key agricultural export commodities and equally sliding performance in food production create dilemma in terms of food importation. The unfortunate development is that shortage of money to finance food importation makes political leadership cut back on money for social services, especially health and education, two sectors that should really drive sustainable development.

There is need to emphasize a few issues before drawing the curtain on sustainable agricultural development in Africa, particularly in Nigeria. Firstly, Nigerian agricultural enterprise is prosecuted in definite ecological zones in conformity with climate, pedologic and vegetation zones of the country. Three ecological zones are generally recognized as the Southern tree and root crop zone, which occupies the rainforest belt; the mixed crop and tree zone of the Guinea Savanna belts (Middle Belt) in which climate conditions allow the growth of both root and grain crops; and the northern Savanna Zone, which is dominated by grain economy and covers the Sudan Savanna (Fig. 15.1) (Muoghalu 1995).

Conspicuous crops of the southern tree and crop zone include maize, yams, water yam, kola, cocoa, cassava, oil palm, rubber, and plantain and raffia palm. The Guinea

Savanna zone produces rice, yams, vegetables, guinea corn, cassava, livestock, cow pea, seed and nuts, soya beans, Irish potato, while the Sudan Savanna Zone produces rice, groundnuts, guinea corn, soya beans, millet, cow peas, wheat, vegetables, seeds, nut and livestock. The ecological zones and their agricultural crops cultivated in Nigeria are given in Fig. 15.3.

Secondly, it appeals that from 2007 crop production increased as shown on Table 15.3. Of particular note is that rice, maize, beans, potatoes, cassava and palm oil grew by over 7.0%, bolstered by the implementation of the National Programme for Food Security (NPFS). The major factors in the increase are the visible increase in the quantity of assorted fertilizers procured and distributed nationwide, rehabilitation and expansion of existing irrigation schemes and the abolition of tariffs on imported agro-chemicals. Thirdly, the Federal Government targeted intervention on 13 strategic crops as shown on Table 15.3 in addition to the use of improved cassava cuttings and expansion of processing facilities is also a major factor. The production of paddy rice had a boost of 7.3% increase in 2008 because of the adoption of the high yielding NERICA rice type and Rice Box Technology by farmers (Central Bank of Nigeria 2008). As of 2019–2020 private investors are investing on extensive mechanized rice production. In Anambra State, Coscharis started producing on a 4,000 hectare farm in Ayamelum LGA backed up by processing facility located at Igbariam in Anambra East LGA.

Fourthly, it bears stressing that crop production in Nigeria must battle continually with low input, land degradation, lack of water, storage facilities, poor infrastructure (especially road and railway rehabilitation and new construction), climate change and population growth (AFDB 2013). Some scholars have isolated a fourth ecological zone known as the high attitude zone, specializing in the production of temperate and subtropical crops. The Jos, Adamawa and Obudu Plateau readily come to mind with market garden products, carrot, tomatoes, onions, pepper, potatoes and livestock.

15.6 Sustainability of Nigerian Agricultural Systems

How sustainable are the agricultural practices in Nigeria vis-a-vis the maintenance of ecological balance? This discussion will be conducted within the confines of the definition of sustainable agriculture and sustainable development advanced in the earlier part of this chapter. Specifically, we will isolate the Brundtland Report (1987) definition and Smith and Donald (1995) definition and ecological intensification as the framework of analysis.

In terms of ecological constraints, the following deserve special mention. Deforestation constrains agricultural sustainability because its reduction causes a reduction in the quantity of rainfall, exposes the soil to desiccation and engenders slower plant/crop growth. Deforestation has been ahead of human population growth rate in negative decline. The enormity of deforestation is gleaned from the fact that Nigeria had the highest deforestation rate as published by the Food and Agriculture Organization from 2000–2005 having lost 55.7% of its primary forests (Food and

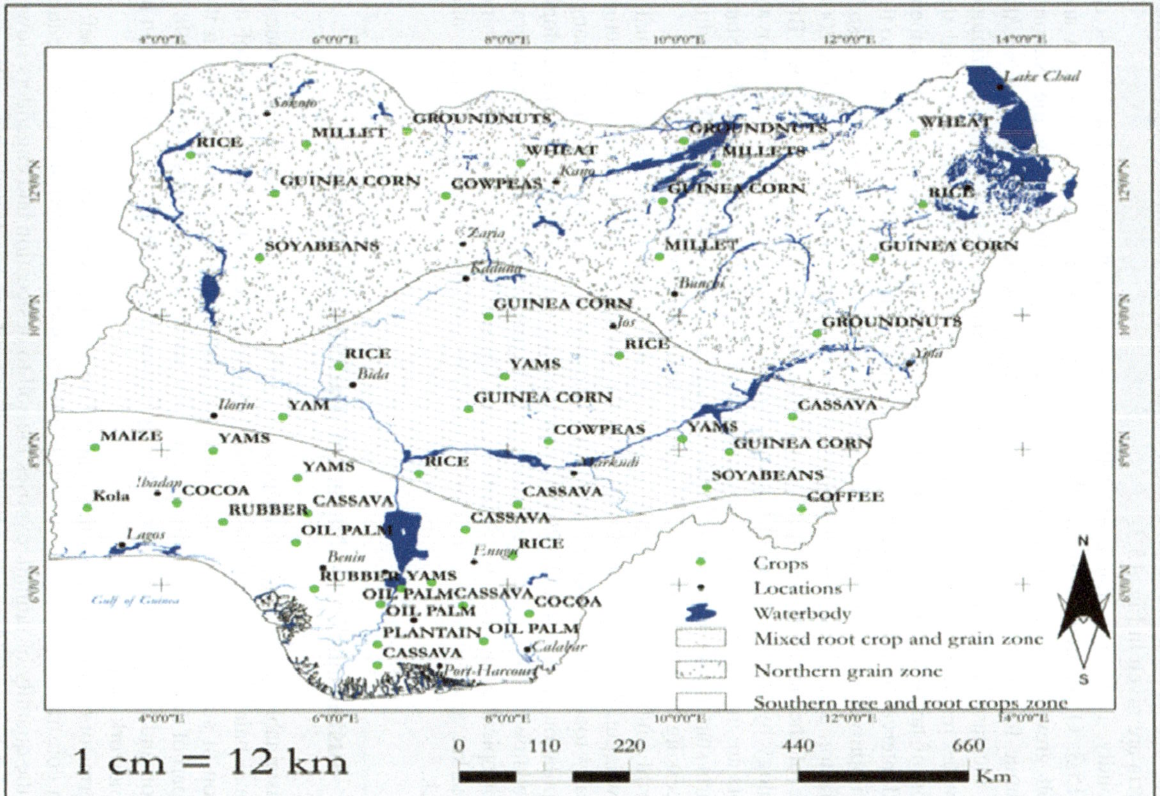

Fig. 15.3 Showed the ecological zones in Nigeria and the distribution of agricultural products

Table 15.3 Growth in major crop production (%) (Central Bank of Nigeria 2008)

Crop	2007	2008	Crop	2007	2008
Wheat	6.6	6.3	Plantain	6.6	6.0
Sorghum	5.9	6.0	Potatoes	7.3	6.4
Rice	7.7	7.3	Yam	5.4	5.9
Maize	7.1	7.0	Beans	7.0	7.2
Millets	6.3	6.6	Cassava	7.4	9.1
Soya beans	5.7	5.7	Palm oil	11.4	9.0
Rubber	6.8	6.4	Cocoa	5.5	5.6

Agricultural Organization 2005; Rhett 2005). Every year more than 400,000 ha of forests are devastated. The high rate of deforestation is attributed to rising human population, lack of political will on the part of governments (State, Federal and Local Governments) to enforce regulations against deforestation and re-afforestation, poverty, agricultural practices of shifting and bush fallow systems of cultivation, mining of solid minerals, exploitation of oil and gas deposit, bush fire, livestock over-grazing, infrastructure development, desiccation and desertification, urbanization, fuel wood exploitation, wars and terrorism, hydroelectric projects, pollution, hunting and poaching.

There are constraints arising from soil degradation, especially erosion. Soil erosion affects agriculture in marginal areas because of a combination of steep slopes and poor vegetation, which aid fast run-off and wind erosion, in addition to diminishing infiltration potentials. Intense and heavy rainfall in southern Nigeria is increasing the range of gully erosion because of increase in overland flows in fragile sandstone-dominated environment. Soil and coastal erosion adversely affect over 80% of the land of Nigeria. Africa's largest single erosion complex exists in Nigeria. While sheet erosion is fairly widespread, gully erosion presents spectacular catastrophe. Although gully erosion affects only 0.1% of total land area of Nigeria, the number of gullies is dauntingly large and sizes of some of them are astonishingly large (Akanwa and Ezeomedo 2018).

Figure 15.4 showed the spatial incidence of gully erosion covering five states in the South East geopolitical zone, Cross River and Edo States in the South-South geopolitical zone, Ondo State in the South-West geopolitical zone, Benue State (North Central) and Gombe in the North-East geopolitical zone.

In the Niger Delta Region coastal areas salt water intrusion from the Atlantic Ocean and inundation affect water supply, while intricate construction of roads and pipelines distort the Niger distributaries leading to depletion of mangrove trees. This affects breeding grounds for fish.

Drought, acidification and desertification constrain agricultural sustainability. The impact of drought on the availability of nitrogen and phosphorus is generally high after drought. Understanding nitrogen levels can lead farmers to sow inappropriate crops. The calculation of nitrogen should deal with the residual nitrogen from fertilizer application in the previous year as well as sub-soil and top soil nitrogen mineralization.

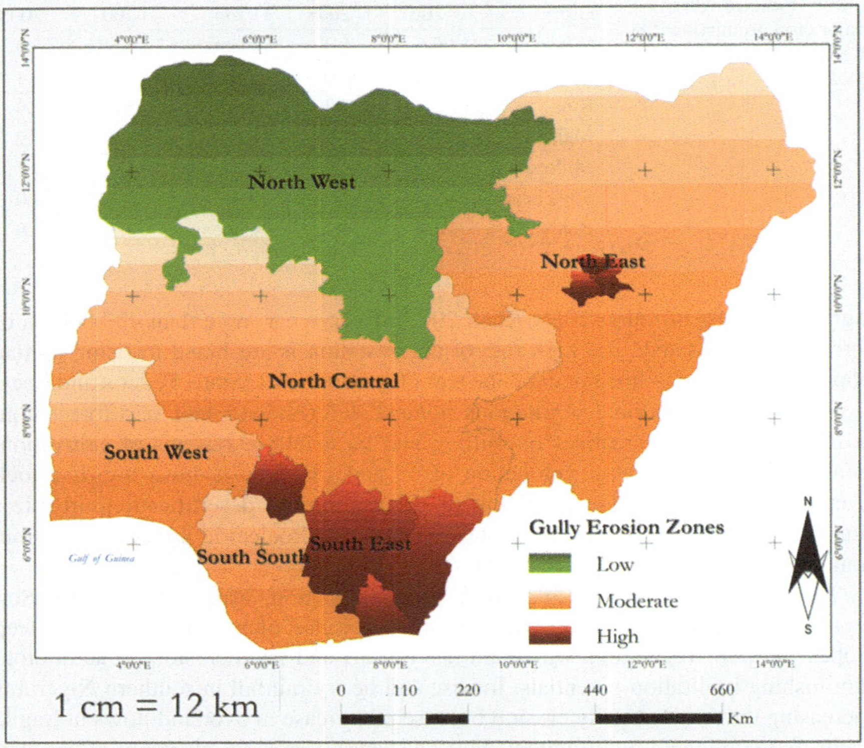

Fig. 15.4 The spatial incidence of gully erosion spread in Nigeria

From 10° North latitude in Nigeria is classed as the most desertification-prone area. Figure 15.5 showed the Desert-Prone Zones in Northern Nigeria covering Adamawa, Bauchi, Borno, Gombe, Jigawa, Kano, Katsina, Kebbi, Sokoto, Yobe and Zamfara States. These are characterized by a gradual shift in vegetation from grasses, bushes and dotted trees to expansive areas of desert-like sand/sand dunes. At the moment Nigeria loses 351,000 ha per annum of its landmass to desertification, or an annual loss rate of 0.6 km (Federal Ministry of Environment 2012). As of 2012, 4830 km^2 were covered by sand dunes. In terms of livestock farming, the migration of human and livestock from the 11 frontline states southward has been the cause of life-threatening security issue between sedentary crop farmers of the North Central States, South-East and South-West and the Fulani nomadic livestock rearers. In addition, this migration is resulting to intensive use and degradation of marginal ecosystems of these areas.

As dry seasons decline they do not stop the growth of volunteer plants in fields, pastures and plantations. As they argue, weeds constitute the rationale why farmers abandon their swidden fields to clear new forests areas. Often annual burning to

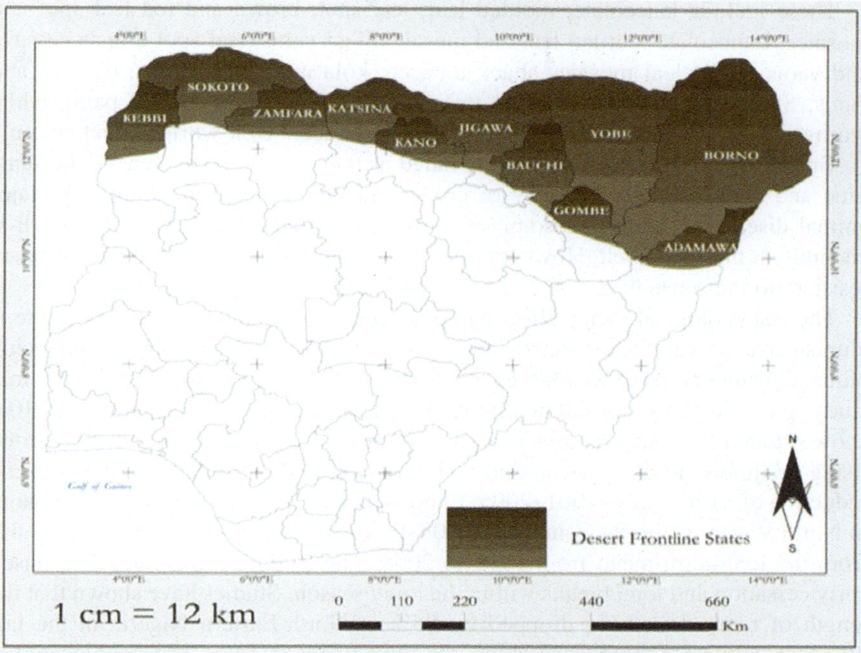

Fig. 15.5 Showed the desert-prone zones in northern Nigeria

control weeds leads to a buildup of perennial grasses, as is the case with many invasive species in Southern Nigeria.

The negative impact of crop diseases and pests in detracting from agricultural sustainability needs not belabouring. The hot and humid climate of Nigeria provides a favourable environment for the wide variety of pests and diseases afflicting livestock and crops. The effects of climate on pests and diseases are shown in their localization and occurrence as well as the geographical limits of the range of crop cultivation, variations in yield due to differences in the seasons or date of planting and annual variations in yield corresponding to changes in weather (Akanwa and Joe-ikechebelu 2020).

The disease of rosette which attacks groundnuts increases in intensity southward as the volume of rainfall increases. Ootheca afflicts cowpeas and causes yellow mosaic disease which thrives more in areas of heavy rainfall. Diseases of crops in Nigeria are of three types; virus, fungal and bacterial. Fungal diseases affect a great number of crops than viral diseases (Williams et al. 2017). The swollen root disease thrives in areas of high temperature and humidity. Cocoa is the major sufferer, leading to poor tree development and reduced yield. Most food crops are affected by a variety of viral diseases.

These include lanceolate, mottled leaf, leaf spot, brown and red leaf spot and mosaics. Lanceolate mottled leaf and mosaic affect yams; leaf spot affects cassava and yams, while leaf mosaics attack tobacco, kola nuts, groundnuts, cassava and yams. Stem borers damage maize, rice, millet, kola, coffee and oil palm, while worms destroy large fields of maize, yams and tobacco. Bolt worms affect cotton.

Grasshoppers such as locusts are dreaded as they invade large areas at the same time and damage a great variety of crops, causing widespread famine. A major animal disease is the trypanosomiasis, which is carried by the tsetse fly which is endemic in the forest belt. However, the N'dama and Muturu varieties of cattle are resistant to the tsetse fly.

The last ecological factor affecting agricultural sustainability we discuss here is climate change. Ample research evidence shows that Nigeria is already experiencing climate change extreme weather events (Odjugo 2010a; Ekpo and Nsa 2011). In a study spanning 105 years and involving 30 meteorological stations, Odjugo (2010a) showed that mean temperature has been constant until the late 1960s, when it started rising gradually till the year of study. He discovered that rainfall had experienced reduction of 81 mm from 1901–2005. Ekpo and Nsa (2011) in their study of Sokoto in North Western Nigeria from 1968–2008 discovered that rainfall reduced by 8.8% from the long-term mean from 1915 to 2008. The study also revealed late onset, early cessation and long breaks within the rainy season. Studies have shown that the length of rainy days have dropped by 55% in North-Eastern Nigeria in the last 30 years, while for the coastal areas, the drop is put at 14%. The double rainfall maxima have shifted farther south, while the "little dry season" usually experienced in August is now experienced in July (Odjugo 2010b; Ekpo 2009).

Crop specific studies and climate change effects have been conducted extensively in Nigeria. Odusina and Kassim (2004) discovered that rainfall variability affected rice production in the period of study from 1996 to 2011 in Ogun State. They showed that rainfall boosted rice production in the absence of flooding, while with flooding the reverse is the case. Their study however showed that the expected temperature at which rice production is constrained as shown by IPCC (2001) study has not been reached.

A study by Ufoegbune et al. (2014) showed that water melon production is better in the dry season under irrigation technology regime than in the wet season with erratic rainfall regime. The former ensures longer life cycle for water melon, as well as relative safety from pest attack. A trend study by Oguntunde et al. (2014) on the relationship between cocoa production and climate change variables from 1976 to 2009 in Ondo State (which produces between 45 and 65% of all cocoa in Nigeria) showed a reduction of 910.44 metric tons per year or a total reduction of 30,044.52 metric tons for the period of study. This is attributable to rainfall, maximum temperature, minimum temperature, mean temperature, potential evapotranspiration and water vapour deficits. There was a rise in diurnal temperature range above average measurements in the past four decades.

Ukhurebor and Abiodun studied the annual rainfall data of forty years (1978–2017) for South-South, Nigeria. The results show that the differences between the two means of the equal-length time scales revealed variability of

7.00 mm. Similarly, the CV of rainfall was 0.145 signifying low variability. However, the anomaly results revealed that 21 years (52.5%) recorded less rainfall; while 19 years (47.5%) recorded much rainfall. The Sen's estimator slope revealed downward trend of 94 mm/year in 1978–1987 decades; while it recorded upward trends of; 90 mm/year, 30 mm/year and 118 mm/year during the 1988–1997, 1998–2007 and 2008–2017 decades, respectively.

The implication of this study is that there are variations in the rainfall. Consequently, there is an optimum need to sensitize the general public about its existence in order to take the necessary measures and adaptation options for its mollification and management (Olaniran and Sumner 2006). Shortened growing season leads to reduced crop performance as in the case of Bida, Ilorin, Zaria and Yelwa. The authors attributed these variabilities to the patterns of the northward and southward march of the Inter Tropical Discontinuity (ITD). Except for Makurdi, the authors recommended irrigation farming for the other 15 stations and their farming regions. In the alternative, genetically modified crops of shorter growing periods are recommended.

Odjugo (2010b) studied shift in crop production as an adaptation to climate change in semi-arid region of Sokoto and Zamfara from 1975 to 2007 and discovered increase in temperature and decrease in rainfall from 783 mm in 1975–1977 to 618 mm in 2005–2007 (a decrease of 146 mm). As a result of this, farmers switched to millet instead of guinea corn, followed by maize production (2–3 months production period) and beans and groundnuts, requiring only 3 months for production.

Apparently the most demonstrable evidence of climate change in Nigeria is to be found in the Lake Chad Basin, situated in an intense region of evapotranspiration in the semi-arid region of Nigeria's north eastern region. The basin has witnessed a diminution of its size. Rainfall has decreased from 800 mm to 400 mm (50% decreases) in the last 40 years, resulting in incessant drought. Decreased water in the region derives from dams constructed on the upper reaches of rivers draining into the basin, reduced rainfall, high evapotranspiration, scarcity of cloud cover and poor land management. Lake Chad has dwindled in size from a maximum surface area of 25,000 km^2 (9650 $miles^2$) in 1963 to just 1350 km^2 or 521 $miles^2$ today. In the 1960s the theoretical lake basin covered 2,397,423 km^2 straddling the borders of Nigeria, Chad, Niger and Cameroon. The US NASA forecasts show that the Lake could disappear by 2030.

Initially the Lake served as a source of fresh water, fisheries (120 species), pastoral and agricultural land for some 20–30 million people. Today it is at the brink of a humanitarian and environmental disaster, drying up. A reed *Typha australis* (known locally as *kachalla*) has been covering the River Komadugu-Yobe fertile plain since the completion of the Tiga Dam in Kano State. *Kachalla* has also been an ideal habitat for the destructive quelea bird, which ravage crops. As a result of the current ecological condition of Lake Chad, there has been 60% decline in fish production. Degradation of pasture lands had resulted to 46.5% shortage of dry matter as at 2006, a reduction in livestock population and threat to biodiversity (Nation Newspaper 2019).

Economic estimate of fish sold in the region is put at N 350 million per week or N 1.4 billion a month. Estimated 50,000 metric tons annual loss is sustained in fresh water fish supply due to desiccation, 55% estimated loss in pasture land due to desertification and 95% drop in quantity of milk product per cow as a result of poor cattle nutrition and massive reduction in crop yield in the past 20 years, 45% in tomato and 55% in wheat production (Kenechukwu 2016).

In Kano State only 42,000 metric tons of groundnut was produced in 1978 as against 500,000 metric tons in earlier years; in Borno State 400,000 cattle, 7 million animals and 3 million goats were affected by desertification and drought (Federal Ministry of Environment 2012). About 65% of the landmass was affected by desertification in Sokoto and Kebbi States, while 55% of land in Borno and Yobe States was affected.

Nigerian agricultural practice, despite ecological intensification, does not make for survival. The declining agricultural production does not guarantee adequate nutrition or even the supply of raw materials to local agriculture-based industries. A combination of climate change, desertification, deforestation, slash and burn practice, soil and gully erosion, poor land management and drought has affected productivity.

An evidence of this is that for 2004–2005 the values of agricultural export and import amounted to $623 million and $2285 million, respectively. For food, respective figures were $548 and $2024 million. A total of $1bn was posted as annual loss in non-timber forest products due to rapid deforestation, while 90% permanent loss is sustained in the natural habitat of pollinators, critical to agricultural productivity. About 1.5 million trees are being felled daily (Kenechukwu 2016).

Despite the fact that Nigeria has enormous water resources potential of about 319 billion m^3 and with surface water accounting for 267 billion m^3 and ground water of 52 billion m^3, it has only 220 dams with a combined capacity of only 33 billion m^3 (Federal Ministry of Environment 2012). With 11 River Basins and Rural Development Authorities, Nigeria has not tapped into its water resources for sustainable intensive agriculture.

In terms of soil, Nigeria's soil types are fluvisols, regosols, acrisols, ferrasols, alfisols, lixisols, luvisols, nitosols, arenosols and vertisols, all varying in their potentials for agriculture. Nearly 48% of Nigeria's soils, especially vertisols, alfisols, acrisols and arenosols are found in the dry land area of the country and these falls into low classes of productivity. The poor quality of soils is accentuated by deforestation, desiccation, desertification, sand dunes (barchans), soil and gully erosion, and increase in acidity and decrease in soil fertility that farmers seek out adaptation such that cassava has virtually supplanted yam cultivation in southern Nigeria (Onakuse 2012). The figures of cassava production were 23.83 m and 35.61 m metric tons for 1994/95 and 2005/06, respectively (National Bureau of Statistics 2007).

Inconsistencies in policy implementation and introduction of agricultural development programmes/systems have bedeviled agricultural productivity (Muoghalu 1992a). Each military or civilian administration introduced its own agricultural programme and dropped the one it met. The regional governments in the western and eastern regions before independence in 1960 operated the Israeli-based farm settlement scheme. The military regime in 1972 introduced the Agricultural Development Programme (ADP) sponsored by the World Bank (later renamed the Integrated Rural Development Programme (IRDP) in the third and fourth plan (1975–1980 and 1981–1985)). The military introduced the 11 River Basin and Rural Development Authorities in 1976, modelled on the Tennessee Valley Authority (TVA) in the USA. By 1982, these have gulped N 1.56tr. In the main, corruption crippled them. While Obasanjo introduced "Operation Feed the Nation", Shehu Shagari came up with the Green Revolution, Babangida introduced the Directorate of Foods, Roads and Rural Infrastructure (Muoghalu 1992b).

A major constraint to agricultural productivity is agricultural landholding. As at 1974, landholding ranged from 0.405 ha to 3.04 ha for peasant farmers (Federal Ministry of Agriculture and Natural Resources (FMANR (Federal Ministry of Agriculture and Natural Resources) 1974). This has been on the decline. For example, in 1994/95 total hectare devoted to millet cultivation by peasant farmers was 7.53 Mha. This declined to 3.91 Mha in 2005/06, a decline of 48.07%. For guinea corn, the story is the same: 6.00 Mha in 1994/95 and 3.95 Mha in 2005/06 and for beans it was 3.467 Mha in 1994/95 and 2.313 Mha in 2005/06.

The categories of peasant farmer's access to land consisted of owner's land, family land, rented, squatter and others. For 1994/95 and 1995/96 the proportion of land farmed by these were 69.44%, 25.97%, 6.59%, 0.58% and 1.41%, respectively. For 2004/05 and 2005/06 corresponding proportions were 58.96%, 23.60%, 14.03%, 1.00% and 2.41% (National Bureau of Statistics 2007). This shows that land owners and family land cultivators decreased, while the rest increased. The extent of land farmed is a function of the failure of the Land Use Act of 1978, which was meant to give developers and investors access to land. Today most land many kilometres away from cities have been bought from real owners by land speculators. Even in villages poverty is forcing poor people to sell off their land to the rich capitalist few.

In terms of farm inputs, fertilizers come readily to mind. The application of fertilizers has witnessed sharp decreases. In 1994/95, 112,130 metric tons of fertilizer was used, while 63,262 metric tons was used in 2005/2006. Use showed spatial variation. For example, Benue, Borno, Jigawa, Kano, Katsina, Nasarawa, Niger, Plateau, Sokoto and Yobe States each used more than 3000 metric tons per annum, while southern States (Abia, Anambra, Bayelsa, Ekiti) used less than 300 metric tons (National Bureau of Statistics 2007). It must be observed that the intensity of fertilizer use is related to the landmass cultivated, the extent to which agriculture is the major economic engagement and the dominance of food crop cultivation over tree crops, relative affluence of farmers and readiness of farmers to adopt innovation.

15.7 Challenges and Strategies for Agricultural/Ecological Intensification in Nigeria

Federal Ministry of Agriculture and Rural Development (2000) stated some of the major challenges of agricultural sector in Nigeria to include lack of experience of improved farming systems, land tenure insecurity, low and unstable investment in agricultural research, financial constraints where there are high interest rates on loans, inaccessible credit due to tough conditions and expensive manually irrigated and controlled agriculture. Also, there are storage constraints where majority of the smallholder farmers practice poor storage methods, lack of post-harvest preservation skills and consistent theft of farm products accentuate the challenges. In addition, there are the issues of farm input constraints where prices of farm inputs are outrageously high, confusion on choice of variety of input to use, infrastructural constraints such as inaccessible roads, poor market facilities, marketing constraints where prices of goods fluctuate, presence of middlemen who make more financial gains at the expense of the rural farmers and intense competition. From the forego-ing, increasing agricultural production can only be achieved through sustainable agricultural intensification. This requires the application of smart policies that would adopt innovations while granting access to local farmers (Idachaba 2000).

This involves improving the market structure to organize and expand the rural market services beyond the sale of local farm products to highly industrialized global market levels where it can enjoy organized international negotiations (Punch 2017; AGRA 2017a, c, d; AgroNature Nigeria 2018). Nigerian agricultural sector should promote investment in the area of providing local farmers with credit facilities to boost food supply and reduce risks (AFI 2017). This will involve the government at the federal, state and local government levels to support local farmers, presently; the Nigerian Federal Government has launched the Agro Processing Productivity Enhancement and Livelihood Improvement Support (APPEALS) project.

It is targeted to improve farmers' productivity and boost the agricultural sector by providing a loan of six hundred billion dollars to support 2.4 million farmers with zero interest across the nation (ThisDay 2020). APPEALS project in Nigeria is for small and medium farmers in Kano, Kaduna, Cross Rivers, Enugu, Lagos and Kogi. It is vital to revisit our nation's legal rights and access to lands so that local farmers can better enjoy access to large farm lands for farming either within the customary framework or a formalized modern law (Eboh 2004; IFDC/IITA/WARDA Report 2000; Kormawa et al. 2003).

In addition, climate-smart agriculture is vital for transforming and reorienting agricultural production systems and food value chains so that they support sustain-able development and ensure food security under climate change. This will prevent climate-related disasters and crises and aid in rapid recovery in a sustainable manner. It includes protecting, restoring and improving food and agricultural systems under climate threats that impact food and nutrition security, agriculture, and food safety/ public health. Further, agri-businesses should collectively be associated with the production, processing and distribution of agricultural products, including business entities involved in the production and distribution of agricultural inputs and machinery to farmers and those involved in purchasing, aggregating, processing and distributing farm produce (CGAP 2018).

15.8 Research and Development in Ecological Intensification of Agricultural Practices in Nigeria

For ecological intensification in the agricultural sector in Nigeria to be achieved, there is need for digitally delivered information on areas such as agronomic best practices, pests and diseases, weather, and market prices, as well as more sophisticated digital services and farm management software tailored to the specific farmer, farm or field that enable smallholder farmers to make decisions that maximize output from their land, improve the quality of agricultural production and maximize farm revenues and profits via lower costs of production, improved ability to identify markets and/or better price realization (Addom and Enghild 2018; Bobbi 2018; Ehui 2018).

Businesses collectively associated with the production, processing and distribution of agricultural products, including business entities involved in the production and distribution of agricultural inputs and machinery to farmers for those involved in purchasing, aggregating, processing and distributing farm produce should be developed. The area of agricultural transformation should be harnessed since it is a necessity in Nigerian agric-business sector where agriculture is a vibrant, modern and sustainable business that creates value for farmers, entrepreneurs, youth and women, and produces affordable, nutritious and healthy food for all (African Center for Economic Transformation 2017).

In addition, the area covering pest and disease surveillance and monitoring at regional, national or even farm and field levels to record the prevalence and severity of pests and plant disease typically goes beyond simple monitoring but should include early warning and advice on pest and disease management.

The study of large, diverse, complex data sets generated from instruments, sensors, financial transactions, social media and other digital means should be given attention (Africa News 2018). Finally, women farmers should be given more attention through research studies, since they make huge contribution to the agricultural sector in Africa and Nigeria where over 64% of the labour force working in the agricultural sector are women who produce 80% of food resources.

15.9 Policy Implications for Sustainable Agricultural Intensification in Nigeria

Our discussion above has obvious policy implications for agricultural sustainability and agricultural and ecological intensification. The first is that Nigerian agriculture does meet the energy, food and fibre needs of the population. It does not create a balance of environmental, economic and social criteria of sustainable agriculture. It is not sustainable over time. Yields do not meet current needs as shown in variability in annual productivity. Yields do not rise overtime to meet added consumption demand in line with population rise. Given increasing soil degradation and decreasing total annual yield, the system is incurring inter-generational debt for the future. Intensification of production on cleared existing space is no guarantee for uncleared portions of land, because deforestation, shifting cultivation, desertification and poor land management are consuming cleared and uncleared spaces.

Nigerian agriculture is characterized by low input investment. Fertilizers do not reach the real farmers; it is often corruptly diverted and rarely sold at government officially announced prices. Introduction of machinery for land preparation and other activities is beyond the reach of peasant farmers, while agricultural extension services are no longer there. There is the challenge of marginal lands. The practice of agroforestry is not widespread. Access to land is inequitable. The practice as at present does not guarantee long farm carrying capacity of the land as there is no soil evaluation to understand deficient soil nutrients. Wetlands along the coast are threatened by urban development, oil and gas industry, gas flaring, pollution from industry and intrusion of salt sea water.

An important question relates to whether sustainable agricultural development can be achieved within existing social, economic and political structures of the country. These relate to land availability and access. There is the crying need to integrate the conservation of ecological resources and development. This is because development projects often are sources of ecological degradation rather than solutions to problems. For example, sand stone and laterite quarrying, transport development (and its accompanying pollution) impact negatively on agriculture. These are important because desirable economic, social and political structures belong to the development aspect of SD, while sustainable aspects focus on ecological sustainability (Kates et al. 2005; Ndondiana and Elizabeth 2018).

Economic inputs in Nigeria's agriculture consist of access to productive assets-land, water, fertilizers, germicides and insecticides. These evoke social and cultural distribution of incomes that determine access. Not compromising the ability of future generations in meeting their own needs touches on water resources, land and soil degradation or even the displacement of agricultural or potential agricultural land by investments other than agriculture and keeping to sustainable ecological footprints in terms of land area on which farmers and consumers draw agricultural crops and livestock. It is necessary to ensure that agricultural wastes and those of other sectors are kept within the absorptive capacities of agricultural land, especially rivers and coastal areas for fishing and maintaining the ability of rivers to break down biodegradable wastes without ecological degradation.

Strategies for achieving social equity, social integration and social stability are essential attributes of SA. In this area terrorism and the creation of cattle colonies or the so-called RUGA or forced settlement for Fulani cattle rearers outside their States of origin cast ominous shadows on the sustainability of Nigerian agriculture. It is against the above observations that we present a few policy implications for the future of Nigeria's agriculture.

One, since the current land tenurial system stifles land utilization and modernization of traditional agriculture, in terms of land ownership, land use rights and land acquisition, there is need for the enforcement of agrarian land use laws aimed at the unification of land use patterns and procedure with regard to freehold, purchase, leasehold, use rights, excessive fragmentation (which leads to scattering of tiny farm plots), inequality in the distribution of farmlands and excessive land use costs. There is an urgent need to re-examine the 1978 Land Use Act, to remove bottlenecks in its implementation, especially as few wealthy individuals are buying up available rural and rural–urban land.

The issue of farmers cultivating marginal land is related to the problem created by the imprecision and ambiguity of current land tenure system. It would have made sense to concentrate farming on optimal lands, but the factors of population pressure, areas receiving excessive rainfall, poor soils, steep slopes and inadequate drainage force the poor to farm marginal lands. The way out of the cultivation of marginal lands may lie in Tittonell (2014) strategy which requires the deployment of different crops and several varieties of each crop that requires different nutrients and have different abilities to tolerate environmental stress.

This is the platform in which research strategies will develop crop varieties that are resilient to moisture stress and to poor soils. A typical example already in operation is the case of cassava and mango displacing yam. Coming up with crop varieties that can withstand diseases and pests is necessary. Allied to this is the necessity for land evaluation which is the process of estimating the potential of land for a specific use or for a multiplicity or alternative uses. The uses may include the productive potentials of land for arable crops, livestock production, and forestry yield or water catchment areas. Land evaluation requires information from land, land use, economics, agronomy, forestry and geographic information system, thus emphasizing its multi-disciplinary nature.

The issue of soil conservation techniques is a priority in hilly and mountainous areas or areas of severe soil and gully erosion, as well as in the arid and semi-arid areas where wind erosion is aggressive. It must be borne in mind that fertilizers are not compensatory for soil fertility lost during erosion as sub-soils have different physical and chemical characteristics than the more friable and fertile top soils and these can interfere with proper root development. One strategy for soil conservation is to irrigate the land to conserve the top soil. Another strategy is to adopt grass and tree strips strategy to trap soils, leaving bands of natural vegetation along contours. Surplus water can be removed by grass water ways. In the bare surface areas netting can be used after seeds have been sown to achieve temporary erosion control until plants have grown to bind soil together.

In the extreme north where wind erosion is ferocious, the Great Green Belt project should be continued. But to counteract desertification the use of *Cactus opuntia* is highly recommended. This plant is recommended because it is well adapted to desert and desert-like and infertile soils, in addition to tolerating temperatures of up to 50–55 °C. It is drought resistant and can survive long periods of drought. Its preference for ecological tolerance is boosted by its vast economic utility providing highly nutritional fodder for animals, raw materials for production of drugs, cosmetics, confectionaries, while its plantations provide employment, income for farmers and better atmospheric conditions for the people through reduction of CO_2.

Much more importantly, the establishment of *Cactus opuntia* plantation in Local Government Areas of the 11 States of the shelter belt can make cattle Fulani herders adopt permanent settlement and embrace ranching as a new and modern way to herd cattle. Because of its fast growth, other advantages of *Cactus opuntia* can be realized through improved quality of meat, milk, leather and wool. This will remove one destabilization problem in the Nigerian nation-armed violence of cattle Fulani herders.

The use of agro-chemical, such as chemical inorganic fertilizer and synthetic pesticides should be minimized, as they pollute surface water bodies and encourage the proliferation of biotic plants in rivers and river plants, such as *water hyacinth* and *Typha australis* through eutrophication. Also pesticides can often be too costly for most crops in the humid tropics, but can be economically viable for certain cash crops to some extent. A better alternative as has been observed above, to deal with pests and diseases is to generate resistant crop varieties and build on traditional knowledge and modern methods of science, through poly-culture of several cultivars for each crop through biotechnological science. Genetic resistance is cheaper for farmers, consumers and the environment in the long run.

The use of herbicides can be encouraged because they have minimal environmental impact and can be cheaper than labour. But the use of inter crops that out-compete the weeds is preferable because they provide fodder, mulch or nitrogen fixation (Mobasser et al. 2014). The use of nitrogen fixing plants (e.g. *Centrosema*) can withstand trampling and heavy grazing. Weeds can also be shaded out by using various cropping systems, such as agroforestry.

Finally is the place of the individual farmers in agricultural intensification. How much does the individual farmer know of the intricate and delicate relationship between the ecosystem and his use of the same? Since the individual producers and consumers wish to maximize their utility of natural capital, there is every need to restrain their anthropocentric propensities. This emphasizes the need for information and knowledge sharing. The need for agricultural extension officers is undeniable, as well as the efforts of NGOs, CBOs, cooperative societies and agricultural research institutes in producing and disseminating knowledge. Farmers should be led into appreciating the need for responsible stewardship. Innovations that improve on traditional methods of production and land management should be introduced.

15.10 Conclusion

The chapter stated agricultural systems and practices in Nigeria in terms of addressing the tenets of agricultural sustainability, ecological and agricultural intensification. Nigeria scored low due to a multitude of factors, including poor land management practices, inadequate farm inputs, inaccessible land tenure system, poverty, corruption, inadequate and degraded infrastructure, inconsistency in policy and programme formulation among others. However, there are redeemable means for agricultural intensification in Africa and Nigeria. This should include the need for maximum budgetary allocation to the agricultural sector. Also, the Federal

Government should concern itself with general policy and programme formulation, the implementation realm should devolve on the states and local governments that closely interact with the real farmers. Further, research and development should address local ecological matters, not national global issues because the problems of agricultural productivity in Nigeria are primarily at the local level. This emphasizes the need for intensive research and development and cooperation between tertiary institutions in confronting the challenges of the agricultural sector. Unfortunately, these institutions are operationally distant from the real farmers and should be supplied with the facilitation for research.

Is there a real link between agricultural extension staff and those who churn out research? How relevant is the research to pressing agricultural challenges? What is the size of our multi-disciplinary strategy in addressing our ecological and agricultural challenges? What is the role of the digital economy in solving our problems? What is the impact of urbanization on the effort to increase agricultural productivity? Above all, how much do the policy formulators know about the inter-twining relationship between the biophysical environment and agricultural endeavours? What do we do with the geometrical rise in our population?

The perspectives are multitude, but suffice it to end by saying that for Africa to come out of her agricultural quagmire there is a crying need for agricultural revolution involving political, structural, financial, knowledge, communication, transparency and accountability in this re-engineering.

15.11 Future Perspective

From the foregoing, it is obvious that for agricultural production in Nigeria to attain sustainability it requires a radical revolution. This revolution should include a highly mechanized, digitalized, use of improved seeds resistant to climate change and vertical farming in urban areas all aimed at achieving a broad national agri-food supply beyond the subsistence smallholder farmers and pastoralism practiced in Nigeria. In addition, the consistent application of the right policies, innovation and investment in Nigeria's agriculture could be ecologically expanded for agri-production and industrialization through processed by-products that can meet international standards for exportation and thus yielding foreign exchange for the nation.

Government can apply the knowledge of ecological zones to improve macro-decisions on policy-making, as well as the design and implementation of their programmes. When this is fully implemented, it will provide a highly connected agricultural ecosystem that is efficiently more productive and transparent than ever before. The growing quantity and quality of agricultural data and digital agricultural solutions significantly can reduce the cost of service, inputs and information delivery to farmers and other value chain intermediaries. It must be a mass movement in which everybody is involved—the real farmer/producers, consumers, service providers, politicians, academics, civil societies, community organizations, men and women and youth, researchers, implementers and traders.

References

Addom B, Enghild A (2018) Attracting private sector investment to digitalize African farming. Technical centre for agricultural and rural cooperation. CTA blog post. https://www.cta.int/en/blog/article/attracting-private-sector-investmentto-digitalise-african-farming-sid053a6cbed-475f-46fe-bc5da5feb137495c

Adelodum B, Choi KS (2018) A review of the evaluation of irrigation practice in Nigeria: Past, present and future prospects. Afr J Agric Res 13(40):2087–2097

Adeniran A, Olaniyi A, Olawale M, Sidiq OB (2019) Strategy for agricultural intensification in Nigeria: emphasis for agricultural aviation. Int Educ Sci Res J 5(6):32–36

AFDB (African Development Bank) (2013) Special teacher on: facilitating green growth in Africa: perspectives from the African development bank in gender, poverty and environmental indicators on African countries. Economic and Social Statistics Division, Tunis

AFI (2017) Digitally delivered credit: consumer protection issues and policy responses to new models of digital lending. https://www.afi-global.org/publications/2633/Digitally-Delivered-Credit-Consumer-Protection-Issues-and-Policy-Responses-to-New-Models-of-Digital-Lending

Africa News (2018) Digital: Africa's internet users increase by 20%'. http://www.africanews.com/2018/02/06/digital-in-2018-africa-s-internet-users-increase-by-20-percent/

African Center for Economic Transformation (2017) African transformation report-agriculture powering Africa's economic transformation. http://acetforafrica.org/acet/wp-content/uploads/publications/2017/10/ATR17-full-report.pdf

African Development Bank Group (2018) Unlocking Africa's agricultural potential to create wealth. https://www.afdb.org/en/news-and-events/unlocking-africas-agricultural-potential-to-create-wealth-18437

AGRA (2017a) Africa agriculture status report: the business of smallholder agriculture in Sub-Saharan Africa. Alliance for a green revolution in Africa (AGRA), Nairobi

AGRA (2017b) Strategy roll-out report: country support and policy engagement. Alliance for a green revolution in Africa (AGRA), Nairobi

AGRA (2017c) AGRA strategy and business plan (2017-2021). Alliance for a green revolution in Africa (AGRA), Nairobi

AGRA (2017d) Doubling productivity and incomes of smallholder farmers in Africa. https://agra.org/doubling-productivity-andincomes-of-smallholder-farmers-in-africa/

AgroNature Nigeria (2018) FG, IFAD flags-off construction of 22 rural roads for farmers in 7 states. https://www.agronaturenigeria.com/fg-ifad-flags-off-construction-of-22-ruralroads-for-farmers-in-7-states/

Akanwa AO, Ezeomedo (2018) Changing climate and the effect of gully erosion on Akpo community farmers. J Ecol Nat Res 2(6):2–12

Akanwa AO, Joe-ikechebelu N (2020) The developing world's contribution to global warming and the resulting consequences of climate change in these regions: a Nigerian case. In: Tiefenbacher J (ed) Global warming and climate change. InTech Open, London

Akanwa AO, Mba HC, Jiburum U, Ogboi KC (2019) Strategies for combating climate change. In: Jhariya MK, Banerjee A, Meena RS, Yadav DK (eds) Sustainable agriculture forest and environmental management. Springer, Singapore

Akanwa AO, Mba HC, Ogbuene EB, Nwachukwu MU, Anukwonke CC (2020) Potential of agroforestry and environmental greening for climate change minimization. In: Raj A, Jhariya MK, Dhiraj DK, Banerjee A (eds) Climate change and agroforestry system. Apple Academic Press, London

Arav T, Uza DV (2002) Agriculture in les Éditions. Africa Atlas: Nigeria, Paris

Asadu CLA, Asadu AN (2015) Analytical overview of agricultural conditions in Nigeria. J Trop Agric Food Environ Extens 14(1):1–17

Babalobi OO, Akinwum JA (2000) Mixed farming in a grazing reserve in Northern Nigeria. Nigerian J Anim Sci 3(2):125–130

Banerjee A, Jhariya MK, Yadav DK, Raj A (2020) Environmental and sustainable development through forestry and other resources. Apple Academic Press Inc., New York, p 400
Biswanger M, Hons P, Savastano S (2017) Agricultural intensification. The status in six African countries. J Food Policy 67:26–40
Bobbi G (2018) Digital farmer profiles: re-imagining smallholder agriculture. Grameen Foundation. https://grameenfoundation.org/resource/digital-farmer-profilesreimagining-smallholder-agriculture
Boserup E (1965) The conditions of agricultural growth: the economics of agrarian change under population pressure. Aldine Publishers, New York
Brundtland Report (1987) Our common future. Oxford University Press, Oxford
Central Bank of Nigeria (2008) Annual Report and Financial Statements, 31st December, 2008, p 91
CGAP (2018) Super platforms: connecting farmers to markets in Africa. https://www.cgap.org/blog/super-platformsconnecting-farmers-markets-africa)
Chamberlain J, Headey D, Jayne TS (2014) Scarcity amidst abundance? Reassessing the potential cropland expansion in Africa. Food Policy 48:51–65
Chamberlain W, Giger M, Nolte K (2016) International deals for agriculture: fresh insights from the land Matrix: Analytical report 11, land Matrix. Landmatrix.org
Chandra RC (2014) Geography of population: concept, determinants and patterns, 11th edn. Kalyani Publishers, New Delhi
Christine V (2019) Revue internationale de politique de développement (11):152–176
Chu EW, Karr JR (2017) Environmental impact: concept, consequences, measurement. Module Life Sci. https://doi.org/10.1016/B978-0-12-809633-8.02380-3
Cohen B (2006) Urbanization in developing countries: current trends, future projections and key challenges for sustainability. J Technol Soc 28:63–80
Cunningham W, Cunningham MA (2006) Principles of environmental science: inquiry and applications, 3rd edn. McGraw Hill Higher Education, Boston, p 464
Dore T, Makowski D, Malezieux E, Munier-Jolain N, Tchamitchian M, Tittonell P (2011) Facing up to the paradigms of ecological intensification in agronomy: revising methods, concepts and knowledge. Eur J Agron 34:197–210
Ducrotoy MJ, Majekodunmi AO, Shaw APM, Baba Musa HBU, Bertu WJ, Gusi AM, Ocholi RA, Bryssinckx W, Welburn SC (2016) Fulani cattle productivity and management in the Kachia Grazing Reserve, Nigeria. Pastoralism 6(25):1–19
Durán ZHV, Pleguezuelo RRC (2008) Soil-erosion and runoff prevention by plant covers. A review. Agron Sustain Dev 28(1):65–86
Eboh EC (2004) Large Scale Arable crops farming development in Nigeria: Policy Questions and current challenges in Nigerian Agriculture: Transiting to Large scale Agricultural Competitiveness, Report of 3rd NESG Agriculture Summit
Ehui S (2018) How can digital technology help transform Africa's food system? World Bank. Blog post. http://blogs.worldbank.org/ic4d/how-can-digital-technology-help-transformafrica-food-system
Ekpo IJ (2009) Climate, society and environment. St. Paul's Publishing Company, Calabar
Ekpo IJ, Nsa E (2011) Extreme climate variability in North Western Nigeria: an analysis of rainfall trends and patterns. J Geogr Geol 3(1):51–62
Elliot J, Firbank LG, Drake B, Cao Y, Gooday R (2013) Exploring the concept of sustainable intensification. http://www.snh.gov.nd/docs/A931058
Enete AA, Amusa TA (2010) Challenges of agricultural adaptation to climate change in Nigeria: a synthesis from the Literature. J Field Sci Action Rep 4:1–11
FAO (2020) World's food systems rely on biodiversity. Food and Agriculture Organization, Rome. http://www.fao.org/news/story/en/item/1263301/icode/
Fasimirin JT, Braga F (2009) Agriculture for sustainable food, energy and industrial development in the Sub-Saharan Africa: the case of Nigeria. Afr J Food Sci 3(13):429–433

Federal Ministry of Agriculture and Rural Development (2000) Report of the standing committee on the problems of farmers in Nigeria (Implementation Modalities)
Federal Ministry of Environment (2012) The great green wall for the Sahara and Sahel initiative. the national strategic action plan. Abuja, Nigeria
Fisher GM, Shah NT, Van VN (2005) Social-economic and climate change impact on agriculture: an integrated assessment, 1990–2000. Phil Trans R Soc A 360:2067–2083
FMANR (Federal Ministry of Agriculture and Natural Resources) (1974) Agricultural development in Nigeria (1973–1985). Federal Ministry of Agriculture, Nigeria, Lagos
Food and Agricultural Organization (2005) Food and Agricultural Organization Annual Statistical Report. 56, pp 16
Gentry A, Clutton-Brock J, Groves CP (2004) The naming of wild animal species and their domestic derivatives. J Archaeol Sci 31(5):645–651
Gerber P, Mooney HA, Dijkan J, Tarawali S, de Haan C (2010) Livestock in a changing landscape: experiences and regional perspective. www.Islandpress.org/bookstore/details/5173.html?prod-id=1950
Graziano da Silva J (2012) Feeding the world in 2050, keynote address at the economist conference. Switzerland, Geneva
GSMA (2015) Market size and market opportunity for agricultural value-added-services. https://www.gsma.com/mobilefordevelopment/wpcontent/uploads/2015/02/Marketsize-and-t-ity-for-agricultural-value-added-services
Idachaba FS (2000) Desirable and workable agricultural policies for Nigeria. Topical issues in Nigerian agriculture, Dept. of Agric.Eco. Uni. Ibadan, Nigeria
IFDC/IITA/WARDA Report (2000) Agricultural inputs markets in Nigeria: an assessment and a strategy for development report
Index Mundi (2019). Available https://www.legit.nig11149140-nigeriarice-production-statistics.html
IPCC (2001) Climate 2001: the impacts, adaptation and vulnerability. Contribution of Working Group II to the Third Assessment Report of the IPCC
Ironkwe MO, Amefule KU (2008) Appraisal of indigenous pig production and management practices in rivers state, Nigeria. J Agric Soc Res 8(1):1–7
Jayne TS, Chamberlain J, Traub L, Sitko N, Muyanga M, Yeboah FK, Anseeuw W, Chapoto A, Wineman A, Nkonde C, Kachule R (2016) Africa changing farm size distribution patterns: the rise of medium scale farms. Agric Econ 47(1):197–214
Jennings P (2007) Luck in the residue of design: the IRRI pioneer interviewed. IRRI, Philippines
Jhariya MK, Yadav DK, Banerjee A (2018a) Plant mediated transformation and habitat restoration: phytoremediation an eco-friendly approach. In: Gautam A, Pathak C (eds) Metallic contamination and its toxicity. Daya Publishing House, New Delhi, pp 231–247
Jhariya MK, Banerjee A, Yadav DK, Raj A (2018b) Leguminous trees an innovative tool for soil sustainability. In: Meena RS, Das A, Yadav GS, Lal R (eds) Legumes for soil health and sustainable management. Springer, New York, pp 315–345
Jhariya MK, Banerjee A, Meena RS, Yadav DK (2019a) Sustainable agriculture, forest and environmental management. Springer, Singapore, p 606
Jhariya MK, Yadav DK, Banerjee A (2019b) Agroforestry and climate change: issues and challenges. Apple Academic Press Inc., New York, p 335
Kanu BS, Salami AO, Numasawa K (2014) Inclusive growth: an imperative for African agriculture. Agriculture and agro-industry department (OSAN). Development Research Department (EDRE). African Development Bank Group. http://www.afdb.org/fileadmin/uploads/afdb/Documents/Publications/Inclusive_Growth_-_An_imperative_for_African_Agriculture.pdf
Kates RW, Parris TM, Leiserowitz AA (2005) What is sustainable development? Environ Sci Policy Sustain Dev 47(3):8–21
Kenechukwu O (2016) Why we should care for the environment, in Daily Sun Newspaper, p 14

Khan N, Jhariya MK, Yadav DK, Banerjee A (2020a) Herbaceous dynamics and CO_2 mitigation in an urban setup- a case study from Chhattisgarh, India. Environ Sci Pollut Res 27(3):2881–2897. https://doi.org/10.1007/s11356-019-07182-8

Khan N, Jhariya MK, Yadav DK, Banerjee A (2020b) Structure, diversity and ecological function of shrub species in an urban setup of Surguja, Chhattisgarh, India. Environ Sci Pollut Res 27 (5):5418–5432. https://doi.org/10.1007/s11356-019-07172-w

King G (2006) Factors influencing livestock production department of animal & poultry science, University of Guelph. http://animalbiosciences.uoguelph.ca/~gking/Ag_2350/factors.htm

Kormawa PM, Ezedinma CI, Adekunle AA, Makinde KO (2003) Promoting market-led agricultural technology transfer and commercialization in Nigeria. Rural Sector Enhancement Program, IITA

Kumar S, Meena RS, Jhariya MK (2020) Resources use efficiency in agriculture. Springer, Singapore, p 760

Lowder SK, Skoet J, Raney T (2016) The number, size, and distribution of farms, smallholder farms, and family farms worldwide. World Dev 87:16–29

Malthus T (1998) An essay on principle of population as it affects the future improvement of society with remarks on the speculations of Mr. Godwin M. Condorcet and other writers

McArthur JM, McCord GC (2014) Fertilizing growth: agricultural inputs and their effects in economic development. Global Economy and Development (Working Paper 77). Brookings. https://www.brookings.edu/wp-content/uploads/2014/09/fertilizing-growth-final-v3.pdf

Meena RS, Lal R (2018) Legumes for soil health and sustainable management. Springer, Singapore

Meena RS, Kumar V, Yadav GS, Mitran T (2018) Response and interaction of *Bradyrhizobium japonicum* and Arbuscular mycorrhizal fungi in the soybean rhizosphere: a review. Plant Growth Regul 84:207–223

Meena RS, Kumar S, Datta R, Lal R, Vijaykumar V, Brtnicky M, Sharma MP, Yadav GS, Jhariya MK, Jangir CK, Pathan SI, Dokulilova T, Pecina V, Marfo TD (2020) Impact of agrochemicals on soil microbiota and management: a review. Landscape 9(2):34. https://doi.org/10.3390/land9020034

Meena RS, Lal R, Yadav GS (2020a) Long term impacts of topsoil depth and amendments on soil physical and hydrological properties of an Alfisol in Central Ohio, USA. Geoderma 363:1141164

Meena RS, Lal R, Yadav GS (2020b) Long-term impact of topsoil depth and amendments on carbon and nitrogen budgets in the surface layer of an Alfisol in Central Ohio. Catena 194:104752

Mellor JW (2017) Agricultural development and economic transformation: promoting growth with poverty reduction. Palgrave/Macmillan, Cham

Milder JC, Garbacti K, Declerck FAJ, Driscoll L, Montenegro M (2012) An assessment of the multi-functionality of agro-biological intensification. A report presented for the Bill and Melinda Gates Foundation

Mobasser HR, Vazirimehr MR, Rigi K (2014) Effect of intercropping on resources use, weed management and forage quality. Int J Plant Anim Environ Sci 4(2):706–713

Muoghalu LN (1992a) Rural development in Nigeria: review of previous initiatives. In: Michael SO, Johnny O, Obiukwu I (eds) Rural development in Nigeria: dynamics and strategies. MEKSLINK Publishers, Awka, pp 77–104

Muoghalu LN (1992b) The task force approach to rural development in Nigeria: an evaluation of the directorate for food roads and rural infrastructure. In: Michael SO, Jonny O, Obiukwu I (eds) Rural development in Nigeria: dynamics and strategies. MEKSLINK Publishers, Awka, pp 297–312

Muoghalu LN (1995) The interplay of ecological and economic theories in relation to spatial distribution of agricultural emplacement in rural settings. In: Niba HC, Ogbazi JU (eds) Introduction to rural development planning. Amazing Printing and Publishing Company, Enugu, pp 30–52

Muoghalu LN (2019) Climate change and food security in Nigeria, African. J Environ Res 2 (1):146–166
Nation Newspaper (2019) Saturday, June 1, 2019, pp 13-16
National Bureau of Statistics (2007) Agricultural survey report 1994/95–2005/06. Federal Republic of Nigeria
National Population Commission (2006) Population and housing census of the Federal Republic of Nigeria, Priority Tables (Volume 1). Federal Government Printers
Ndondiana NS, Elizabeth JZR (2018) Exploring agricultural intensification: a case Study of Nigerian government rice and cassava initiatives. Int J Agric Econ 3(5):118–128
NEPAD (2010) Statistics and analysis for GDP and the structure of African Economies. NEPAD-OECD Africa Investment Initiative. www.un.org/africa/osaa and atwww.oecd.org/daf/investment/africa
Nielsen DC, Calderón FJ (2011) Fallow effects on soil. Publications from USDA-ARS/UNL Faculty U.S. Department of agriculture: agricultural research service, Lincoln, Nebraska, pp 2–19. https://digitalcommons.unl.edu/usdaarsfacpub
Obonyo R (2018) Towards a food-secure Africa. UN Africa Renewal Magazine. UN. https://www.un.org/africarenewal/magazine/december-2017-march-2018/towards-food-secureafrica
Odjugo PAO (2010a) Shift in crop production as a means of adaptation to climate change in the semi-arid region in Nigeria. J Meteorol Clim Sci 8(1):1–6
Odjugo PAO (2010b) General overview of climate change in Nigeria. J Hum Ecol 290:47–55
Odusina OA, Kassim HG (2004) Analysis of impact of climate variability on rice production in Ogun State, Nigeria. J Meteorol Clim Sci 12(1):1–6
Ogbonna JC (2001) Reducing the impact of tropical shrimp trawling fisheries on the living marine resources through the adoption of environmentally friendly techniques and practices in Nigeria. Federal Ministry of Agriculture. Water Resources and Rural Development Depart. of Fisheries; Wuse zone 1, Abuja, Nigeria. http://www.fao.org/3/y2859e/y2859e07.htm
Oguntunde SO, Ajibefun IA, Oguntunde PG, Anugwo SC (2014) Analysis of Trends and Association between Climate Variables and Cocoa Yield in Ondo State, 1976–2009. J Meterol Clim Sci 12(1):42–50
Okike IMA, Jabba VMM, Smith JW (2005) Ecological and socio-economic factors affecting agricultural intensification in the West African Savannas evidence from Northern Nigeria. J Sustain Agric 27(2):5–37
Olaniran OJ, Sumner GN (2006) A study of climate variability in nigeria based on the onset, retreat and length of the rainy season in West Africa, using only surface data. Int J Climatol 9:865–880
Onakuse S (2012) The future of subsistence agriculture in the rural community of Uzanu, Edo state, Nigeria. J Agric Food Syst Commun Dev 3(1):61–71
OSMARD (Ondo State Ministry of Agriculture and Rural Development) (2004) Ondo state ministry of agriculture and rural development, Annual Report, Akure, Nigeria
Phelps LN, Kaplan JO (2017) Land use for animal production in global change studies: defining and characterizing a framework. Glob Chang Biol 23(11):4457–4471
Pretty J (2008) Agricultural Sustainability: concept principles and evidence. Philos Trans R Soc 363:447–466
Pretty JCT, Stella W (2011) Sustainable Intensification in African agriculture. Int J Agric Sustain 9 (1):5–24
Pretty J, Zarean PB (2014) Sustainable intensification in agricultural systems. J Anim Botany 114:1571–1596
Punch (2017) Nigeria needs more investment in rural infrastructure to boost agriculture. https://punchng.com/nigeria-needs-more-investment-in-rural-infrastructure-toboost
Raj A, Jhariya MK, Yadav DK, Banerjee A, Meena RS (2019a) Agroforestry: a holistic approach for agricultural sustainability. In: Jhariya MK, Banerjee A, Meena RS, Yadav DK (eds) Sustainable agriculture, forest and environmental management. Springer, Singapore, p 606

Raj A, Jhariya MK, Banerjee A, Yadav DK, Meena RS (2019b) Soil for sustainable environment and ecosystems management. In: Jhariya MK, Banerjee A, Meena RS, Yadav DK (eds) Sustainable agriculture, forest and environmental management. Springer, Singapore, p 606

Raj A, Jhariya MK, Yadav DK, Banerjee A (2020) Climate change and agroforestry systems: adaptation and mitigation strategies. Apple Academic Press Inc., New York, p 383

Rapsomanikis G (2015) The economic lives of smallholder farmers: an analysis based on household data from nine countries. Food and Agriculture Organization of the United Nations

Rhett AB (2005) World forest situational report. http://news.mongabay.com/2005/11Tforests.html

Royal Society (2009) Reaping the benefits: science and the sustainable intensification of global agriculture. The Royal Society, London

Ruthenberg HP (1980) Farming systems in the tropics. Clarendon, Oxford

Sala S, Ciuffo B, Nijkamp P (2015) A systemic framework for sustainability assessment. Ecol Econ 119:314–325

Siba E (2018) Foresight Africa viewpoint – Nutrition security: The last mile of Africa's food security agenda. Brookings blog post. https://www.brookings.edu/blog/africa-nfocus/2018/01/26/foresight-africa-viewpoint-nutrition-securitythe

Siba E, Signé L (2017) Four things you should know about food security in Africa. The Washington Post. https://www.washingtonpost.com/news/monkey-age/wp/2017/10/30/four-things-you-should-know-about-foodsecurity-in-africa/?utm_term=.b245eb26bfe0

Smith NJH, Donald P (1995) Sustainable agriculture in the tropics: issues, indicators and measurement. In: Munasinghe M, Shearer W (eds) Defining and measuring sustainability: the biogeophysical foundations. IBRD/The World Bank, Washington DC, pp 237–247

Stephen (2019) Confronting the unemployment crisis, The Nation, Sept. 13, 2019, p 15

Sulphey MM (2013) Introduction to environment management. PHI Learning Limited, Delhi

Sustainable Food Lab (2017) Enabling smallholder farmers to improve their incomes. Improving incomes challenge white paper. http://sustainablefoodlab.org/wp-content/uploads/2017/09/BFP-Improving-Incomes-WEB.pdf

The World Bank (2009) Africa's development report. The World Bank, New York

ThisDay (2020) 2.4 million farmers to benefit from FG's 600 billion dollars loan. From thisdaylive.com.cdn.ampproject.org

Tittonell P (2014) Ecological intensification of agriculture – sustainable by nature. Curr Opin Environ Sustain 8:53–61

Udondian NS, Zimilia REJ (2018) Exploring agricultural intensification: a case study of Nigerian government rice and cassava initiatives. Int J Agric Econ 3(5):118–128. https://doi.org/10.11648/j.ijae.20180305.14

Ufoegbune GC, Fadipe OA, Bello NJ, Eruola AO, Makinde AA, Oyekunle OZ (2014) Growth and development of watermelon in response to seasonal variation of rainfall. J Meterol Clim Sci 12 (1):31–41

USAID (2008) United States Agency for International Development. Nigeria Biodiversity and Tropical Forestry Assessment

Wiebe KD, Sulser TB, Rosegrant MW (2017) The effects of climate change on agriculture and food security in Africa. In: Alessandro De Pinto, Ulimwengu JM (eds) In a thriving agricultural sector in a changing climate. Meeting Malabo Declaration goals through climate-smart agriculture, International Food Policy Research Institute (IFPRI), Washington DC. http://www.ifpri.org/publication/effects-climate-change-agriculture-and-food-security-africa

Williams SD, Boehm MJ, Mitchell TK (2017) Fungal and Fungal-like Diseases of Plants. The Ohio State University, College of Food, Agricultural, and Environmental Sciences, Ohio State ATI. www.osu.edu

World Bank (2009a) Africa's development report. The World Bank, New York

World Bank (2009b) World development report. Washington. World Bank Group (2018) Working with Smallholders'. A Handbook for Firms Building Sustainable Supply Chains. http://documents.worldbank.org/curated/en/284771480330980968/

World Bank (2009c) Awakening the sleeping giant. Prospects for commercial agriculture in the Guinea Savannah Zone and beyond. World Bank, January 1, 2009. Documents.worldbank.org

World Economic Forum (2017) The future of jobs and skills in Africa: preparing the region for the fourth industrial revolution. https://www.weforum.org/reports/the-future-of-jobs-and-skills-in-africa-preparing-the-region-for-the-fourth-industrial-revolution

You L, Ringler C, Wood-Sic U, Robertson R, Wood S, Zhu T, Charles G, Guo Z, Sun Y (2010) What is the irrigation potential of Africa? International Food Policy Research Institute

Eco-Designing for Sustainability

16

Nahid Khan, Manoj Kumar Jhariya, Abhishek Raj, Arnab Banerjee, and Ram Swaroop Meena

Abstract

Concept of sustainable development (SD) has forced the society and many industries to rethink about the way of development as environmental degradation is the global problem. Higher environmental degradation leads to depletion of resources, causes environmental pollutions, reduces the corporate social responsibility (CSR) and overall has its impact on sustainability. In every sector green approach is the requirement for sustenance of human civilization. Green designing, eco-labelling, green marketing, green consumerism are the essential requirement for addressing sustainability through eco-designing. Adopting eco-designing would generate CSR, green consumerism, energy intensive behaviour, green growth and would lead to formulation of suitable policies for SD. It would also help to reduce environmental footprint, address social and economical aspects of sustainability, promote sustainable management policies in various developmental sectors as well as combat the mega event of climate change.

N. Khan · M. K. Jhariya (✉)
Department of Farm Forestry, Sant Gahira Guru Vishwavidyalaya, Sarguja, Ambikapur, Chhattisgarh, India

A. Raj
School of Agriculture, Lovely Professional University, Phagwara, Punjab, India

A. Banerjee
Department of Environment Science, Sant Gahira Guru Vishwavidyalaya, Sarguja, Ambikapur, Chhattisgarh, India

R. S. Meena
Department of Agronomy, Institute of Agricultural Sciences, Banaras Hindu University, Varanasi, Uttar Pradesh, India

M. K. Jhariya et al. (eds.), *Ecological Intensification of Natural Resources for Sustainable Agriculture*, https://doi.org/10.1007/978-981-33-4203-3_16

Keywords

Ecological sustainability · Eco-design · Green technology · Sustainability · Natural resource management

Abbreviation

CSR	Corporate social responsibility
GHG	Greenhouse gases
GT	Green technology
R&D	Research and development
SD	Sustainable development
SM	Sustainable manufacturing

16.1 Introduction

Eco-design or environmentally sustainable designs are made to reduce the negative impact on environment. The main principle of the design is ecological sustainability. Eco-design also includes the sustainable architect, eco-friendly products, growing green marketing for natural resource sustainability, as well as reducing the overuse and destruction of resources. In eco-design life cycle assessment of product is done for environmental protection over product service. Concept of sustainable development (SD) has forced the society and many industries to rethink about the way of development as environmental degradation is the global problem. To attain the SD, many new eco-friendly technologies have emerged. For maintaining and improving the environment with resource efficient approaches, various green initiatives are being taken for sustainable society (Yilmaz et al. 2019; Jhariya et al. 2019a, 2019b).

For concerning the environment and to overcome the current environment problem, applications of science and technology with green technologies (GT) have assessed and have great role in responding to mitigate the environmental threats. Environment and sustainability are important consideration in making and adapting new technologies (Raj et al. 2020; Banerjee et al. 2020). For enhancing the designing and manufacturing process, there is need of utilizing suitable tools and technique and focuses on sustainability, life cycle assessment and other practices that are aware of the entire life cycle which does not affect the environment (Shafiei and Abadi 2017).

Agricultural industries and other industrial sector are becoming one of the central issues for environmental problem including many other aspects. For this, green space are not only way to reduce the impact but also green production process and eco-friendly products, green marketing are applied in marketing methods (Dzulkarnain et al. 2019).

Understanding of human and nature interaction causing severe damage to environment by scientific communities which uses non-renewable source of energy.

Development in industrial sectors consumes more raw material and energy unsustainably which leads to degradation of resources and energy through deep transformations (Marques and Loureiro 2013). To overcome the negative effects on environment construction sector is working to make their work more sustainable by using eco-design and trying to use materials which have less effect on environment. Construction and demolition of building due to urbanization cause many difficulties and generate high amount of waste material causing bad impact on environment. Due to high cost of raw material, humans go for cost-effective resources and use the resource without concerning about its management and have always taken advantage of free resources. Population expansion also increases the need of raw material, with knowledge about the resource management and life cycle assessment of resources leads to decline in environment which inversely affects the environment. Many environmental impacts occur due to production process and products such as water pollution, air pollution, carbon and amount of energy during life cycle of product. Eco-friendly products and material should be considered for promoting the sustainability and maintaining environment sustainability (Allione et al. 2012). Over extraction of resources leads to damage to the planet earth which severely reduces the biodiversity and landscape alteration occurs due to mining process and many other anthropogenic activities (Khan et al. 2020a, 2020b; Raj et al. 2018).

In age of globalization it is difficult to fulfil the need as well as keeping our environment safe has become the major challenge to manufacturers and producer nowadays (Reddy 2017). Sustainability is a challenge for consumers and it is linked with the consumption of material which causes environmental stress (Hojnik et al. 2019). To maintain the social responsibility and SD companies are practicing green marketing, which include the marketing of eco-friendly products which causes less harm to the environment (Green Business Bureau 2020).

Improvement of new technologies and eco-friendly products are focused on eco-design with sustainability of products and for economy development concept. The motto behind the eco-designing is to reduce the environmental risk without altering or compromising with commodity quality and other parameters and consumer satisfaction (Haase et al. 2017; Stal and Corvellec 2018).

Using eco-design for sustainability can reduce the impact of urbanization, climate change and other factors on environment and natural resources. Sustainable architecture, green marketing, using eco-friendly product from different sector and in day today life can reduce the impact on environment and resources and this can also improve human health and natural environment. This chapter discusses the concept of eco-designing for sustainability of natural resources and reducing the impact of anthropogenic activities and impact of over use of resources by human beings.

16.2 Eco-Design and Sustainability

In the corporate sector proper designing of products is a crucial step for effective marketing process. In this context, incorporation of environmental aspects leads to eco-designing which focuses on reducing environmental impact and eco-friendly

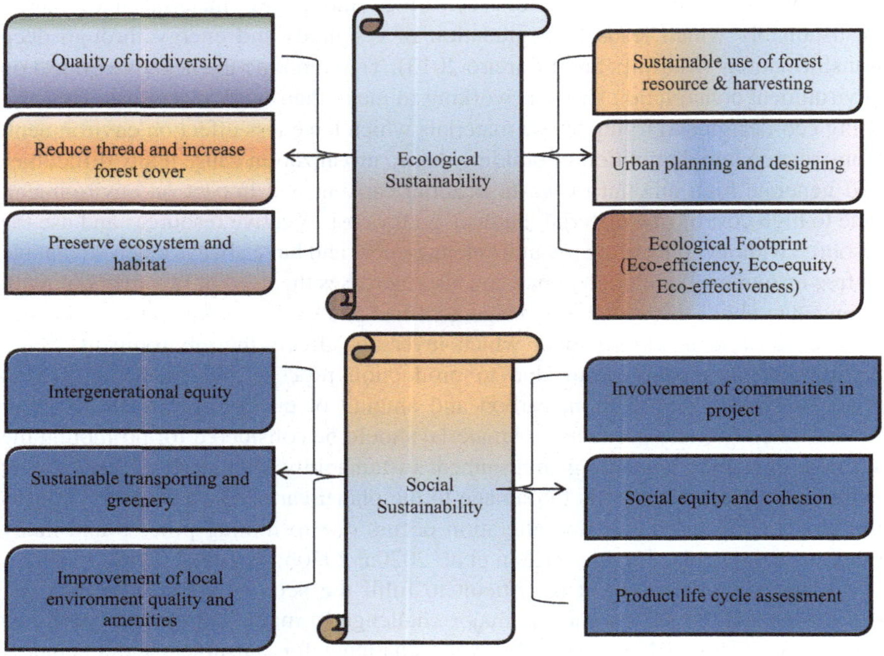

Fig. 16.1 Dimensions of sustainability (Modified: Pereira et al. 2018; Degato 2017)

process formulations. From sustainability perspectives eco-friendly product designing is the key concept for various countries across the world (Pigosso et al. 2013).

Sustainability is a big issue which requires a comprehensive approach for attaining the ecological and social sustainability (Fig. 16.1). From social perspective equity in terms of resource utilization, distribution followed by green designing to bring aesthetic pleasure is a major issue. Further, to generate environmental consciousness both in consumers as well as in the production cell, green consumerism through purchasing eco-friendly product is the main stay. Further, schemes such as eco-labelling, maintaining environmental standards at the societal level are very important to promote social and ecological sustainability. In ecological sustainability point of view various processes and marketing mechanism should have green approaches in order to promote conservation as well as maintain ecological integrity.

Developing eco-friendly products has also lead to the development of life cycle assessment approach which reduces the impact on environment. Eco-designing has a compliance with SD. Sustainability is big issue for human society which includes better life, healthy environment and overall improvement of human civilization (Ortiz et al. 2009). Every eco-designing project should aim towards efficiency in resource and energy sector, reduction of GHGs emission, and prevention of pollution, maintain the quality of life and harmony with the environment (Ortiz et al. 2010).

16.3 Natural Resource, Sustainability and Eco-Designing under Climate Change

Natural resources are global treasure that sustains lives of diverse flora and fauna by providing uncountable and multifarious ecosystem services. These are categorized into renewable (agriculture, forest and animals) and non-renewable resources (soil and minerals). These resources are quite important due to variety of ecosystem services that not only sustain human lives but also maintain environmental sustainability and ecological stability at global scale. Agriculture sustains lives of billions of peoples by providing quality foods and fruits. Forests are the largest natural resource, having greater potential of C sink and hold diversified plants, animals and provide timber and non-timber forest products for sustaining lives of forest fringe peoples. Animals are integral parts of both agriculture and forestry which provides milk, meat, wood and recreations, etc. Similarly, soil is the pillar of life that holds forest, agriculture, animals and other resources. Anchoring root system of plants provides essential nutrients for proper growth and development of crop plants along with climate change mitigation through soil C sequestration. Whereas minerals are source of life and maintain efficient mineral cycling. Thus, conservation and management of these resources are utmost important that contributes in making foundation for SD (Fig. 16.2) (Wellmer et al. 2019).

From sustainability perspective natural resource is a big issue as it challenges the supportive and assimilative capacity of the earth under changing climate. Climatic perturbations have increased the dependency on natural resource across the globe. Under changing climate humans are going for more production, adopting consumptive lifestyle and maximum utilization of natural resource. Therefore, for various sectors such as agriculture, corporate, business, marketing, ecology, environment requires proper planning and designing to achieve the goal of SD.

16.3.1 Sustainable Design for Natural Resource Management

Degradation of natural resources due to bad management practices, unsustainable and unscientific technology is major concern today and its management is utmost important. The goal of sustainability can be achieved by managing natural resources through application of eco-designing such as green designing, eco-labelling, eco-marketing and green consumerism. These will not only manage natural treasure but also maintain environmental sustainability and ecological stability at global scale. In this context, a figure is drawn that represents "how can we achieve the goal of sustainability through eco-designing for natural resource management?" Thus, we can say a great nexus exist among natural resource management, eco-designing and SD (Fig. 16.3).

SD involves social, ecological and economical objectives and improves the sustainability of resource exploitation, technological development and change in institutional development (Fig. 16.4). SD is said to help in developing social, environmental and economic goals (Koltun 2010). For sustainable designing, it is

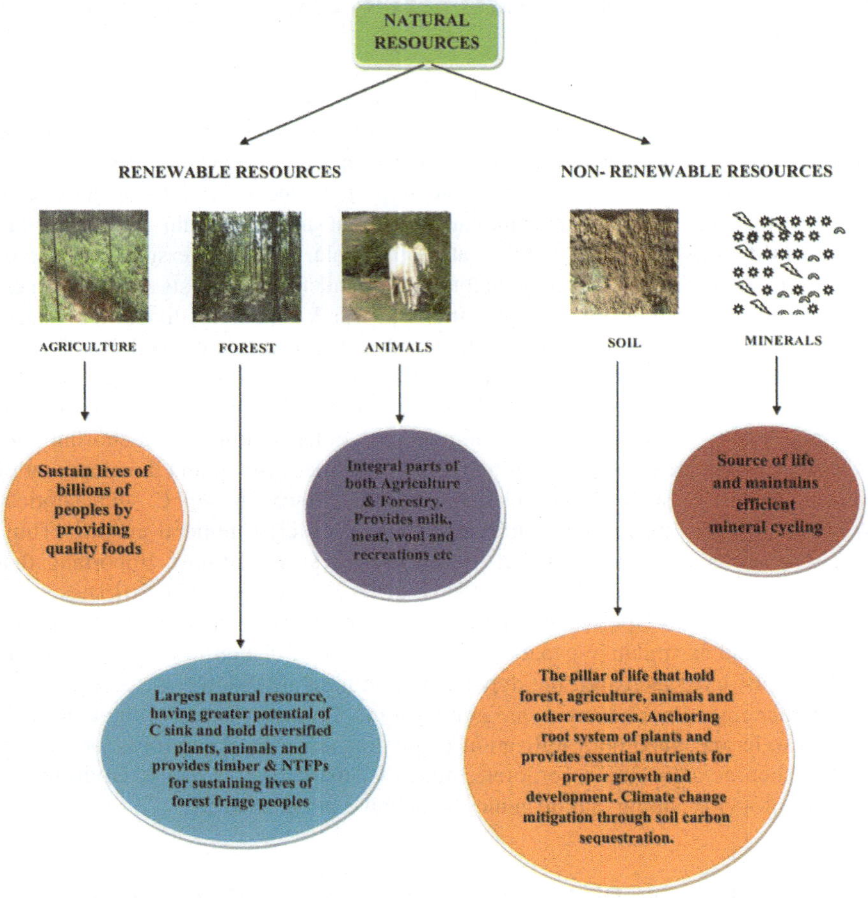

Fig. 16.2 Natural resource classification and its performance (Modified: Wellmer et al. 2019)

very difficult to design the product and systems. Sustainable design is the instrument for the SD and for protecting the environment (Skerlos 2015). SD should contribute to meet the need of today without affecting the ability of future generation. SD defines the problem and choses the approaches and makes solution toward environmental problem (Tomasowa 2018).

Comprehensive goal of education and implementing the sustainable design in field of architecture, engineering, construction and facility management is very important and necessary for making sustainable environment with less damage to the environment. Excessive extraction and use of natural resources are the reason for the degradation of natural resources and causing environmental problem across the world (Tomasowa 2018; Meena et al. 2018).

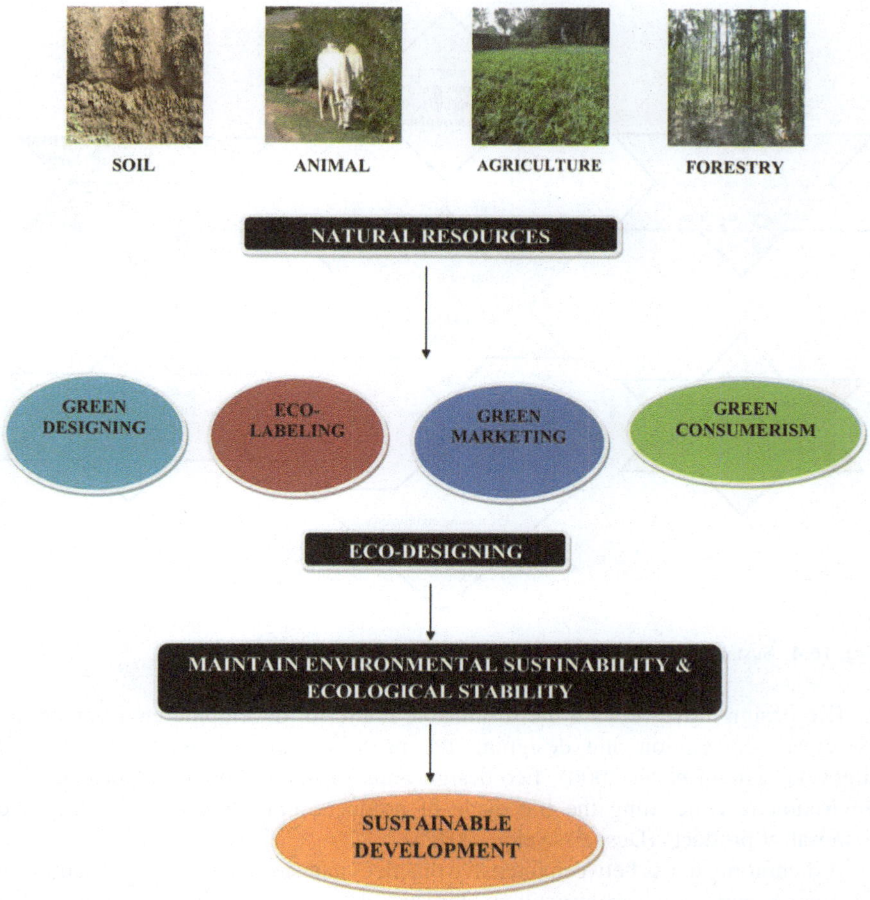

Fig. 16.3 Nexus among natural resource management, eco-designing and sustainable development

Integration of sustainability with manufacturing is termed as sustainable manufacturing (SM) and it deals with design objectives, function, profits and productivity of the system. It influences the advances and priorities for manufacturing technologies and operations. There is need to practice the sustainability in industries also, manufacturers and decision maker have to practice to establish the sustainable culture for fulfilling the need and demand of today and future (Yadav and Pathak 2017; Meena and Lal 2018). SM deals with all three component of sustainability such as environment, economy and society. It is broader than eco-manufacturing, eco-machining and clean production. Material toxicity, GHG emission and pollution all the environmental concern are covered by SM. It uses both non-technological and technological solution, from material selection to organizational mission and performance of reports.

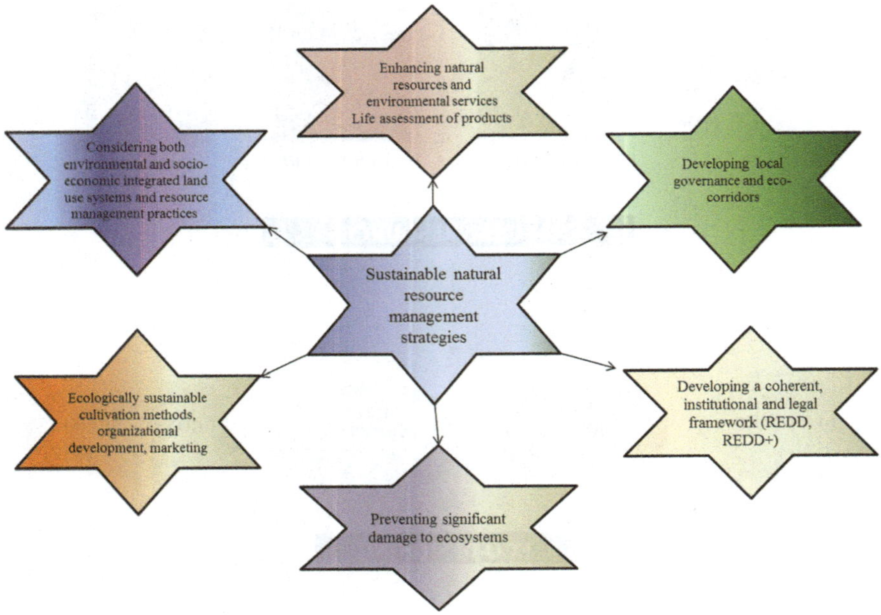

Fig. 16.4 Sustainable natural resource management strategies (FAO 2020)

Eco-design moves toward the sustainable future for developing environment by systematic integration and designing the process across assessing the product lifecycle (Simon et al. 2000). Eco-design aims to reduce impact of products on environment, concerning the life cycle of products and using the recycling and disposal of products (Despeisse et al. 2012).

A dichotomy exists between intensive practices and sustainable intensifications in land use systems such as agriculture, forestry, agroforestry, horticulture and other farming practices. Both are contrary just like a two faces of one coin. Intensive practices need high synthetic inputs with heavy mechanizations that will affect land quality and cause resource depletions (Meena et al. 2020; Kumar et al. 2020). In this context, an eco-designing must be practiced for managing land and other resources. For example, the practices of eco- and sustainable based intensifications (that are assuming to be as eco-designing) would be helpful in managing resources, land quality, soil fertility and enhance biodiversity that intensify ecosystem services. The characteristics, principles and practices for sustainability are very clear which can be utilized for resource conservation along with maintaining environmental sustainability and ecological stability. Thus, a sustainable design including sustainable intensification is used in different land management practices for betterment of environment and would be helpful in achieving the goal of SD (Table 16.1).

Table 16.1 Sustainable design in different land use practices for resource conservation

Feature	Sustainable intensification in different land use practices	Source
Characteristics	Maximize yield and productivity along with natural resource conservation	Pretty (1997)
	Improve nutrient availability, its efficient utilizations and maintain overall soil fertility and SOC pools that promote return from different land use practices such as agriculture, horticulture, forestry, agroforestry, etc.	Ruerd and Lee (2000)
	Maintain a better tree–crop interaction with less competitions among them for resources and improve overall productions without destroying our natural ecosystem and environment	Baulcombe et al. (2009), Pretty and Bharucha (2014)
	Enhance resource use efficiency and judicious utilization of resources and less synthetic inputs are other important characteristics of sustainable intensification that maintains environmental sustainability and ecological stability at global scale	Pretty (2008)
	Strengthen the productivity along with input-output balance and sustainable livestock's productions without resource and environmental degradations are other peculiar features that signify the ecosystem services	Gibon et al. (1999)
Principles	This is based on the principle of less inputs and proper utilizations of renewable resources like land, water, light, space, etc. for maintaining efficient productions at farm level	Godfray et al. (2010), Firbank et al. (2013)
	Efficient utility of tree-crops varieties along with essential breeds of cattle and animals	Ruerd and Lee (2000), Pretty (2008)
	Optimization of outside inputs, better resource use efficiency, improves food production systems and reduces its impact on our environment	Pretty (1997), Matson et al. (1997)
	Reducing food wastage with higher productivity is important principle on which sustainable intensification will depend	Garnett et al. (2013)
Practices	Conservation tillage and mulching practices are very important for better land quality, soil fertility and overall conservation of natural resources	Wezel et al. (2015)
	IPM (integrated pest management) practices which is very viable for managing emergence of insect pest and related diseases	Pretty (1997)
	Practices of crop rotations system along with integration of cash and cover crops including beans and their proper harvesting maintained existing resources by minimizing its depletion	Tilman et al. (2011a)

(continued)

Table 16.1 (continued)

Feature	Sustainable intensification in different land use practices	Source
	Incorporation of improved varieties of woody perennial trees, agricultural crops and livestock's along with protection and conservation of important plant genetic resources that are economically and ecologically viable	FAO (2004)
	Applying of soil and water resource based conservations practices that improve the resource availability and utilization	FAO (2004), Wezel et al. (2015)
	Fertigation and irrigation based water management practices are involved	FAO (2004)

16.3.2 Climate Change and Natural Resource

Globally, changing climate as well as degradation of natural resources is severe problems that people experiencing to a great extent. The changing climatic condition also puts pressure on the food availability, quality, food security and poverty for achieving the global SD. Natural resource conservation and management along with changing climate adaptation and mitigation are key sustainability issues in the sector of food, economic growth and good governance globally (Singh and Jhariya 2016; Jhariya et al. 2018a, 2018b). Eco-friendly product development, promotion and utilization can meet out the environmental resilience and can improve the local and national economy. Climatic perturbations have influenced the human and natural ecosystem worldwide. The continuous rise in the global temperature leads various major changes in various ecosystems (Raj et al. 2020; Banerjee et al. 2020; Jhariya et al. 2019a). Thus, conserving natural resources, it's efficient and sustainable utilization is the need of the hour to meet increasing global needs.

As per prediction, the world's food and water requirement is expected to rise twice in coming three decades. Moreover, the changing climate imposes the problems directly and indirectly on the productivity in agricultural sectors through climatic irregularities. The reduction in the natural resource base also creates conflicts and competition at local, national and international level if the proper conservation, management and sustainable utilization of resources are not given proper consideration (Raj et al. 2019a, 2019b). Increasing human population in addition to changing climates puts the pressure on natural resources. Further the over use of natural resources has multiple issues like mass extinction of species in the one hand and threatens the environment and ecosystem on the other. This affects the energy and food supply system in third world nations and challenging the social, economic and environmental development (Meena et al. 2020a, b). Therefore, judicious utilization of these resources is pre-requisite for human health and environment sustainability. Climatic alteration influence on the natural resource is very complex and regulated by direct and indirect ways. SD is the key to balance bridge between environment and economic developments of community at present and future context. It infers the equitable management of resources at sectoral level to satisfy the people needs as per space and time (Cruz et al. 2007).

16.3.3 Social and Economic Perspective on Sustainability

As we know, social, economic and environment are the three pillars for the sustainability. Society welfare, social cohesion, gender discrimination, maintenance of people capital, etc. are studied under social aspects whereas the value of production/consumption, economic growth, efficiency and competitiveness, foreign trade system, stability/flexibility and employment generations are recognized under economic aspects. The maintenance of these two pillars decides our environmental quality and base for the SD.

It has been found that consumeric lifestyle is the root cause of modern day problems. It is causing pollution, depletion of natural resource and loss of environmental balance. Adopting green practices and behavioural approach one can change from consumeric to conservative lifestyle (Culiberg and Elgaaied-Gambier 2016). This is very much important for the finite resources of the earth such as fossil fuels (Maidment 2015). It is therefore urgent need to adopt such mechanism of green practices very quickly otherwise we would be devoid of our essential resources for existence of life on the earth.

16.4 Green Technologies and Sustainable Development

The changing climate, earth's warming, resource depletion and other environmental concern have triggered the scientific communities towards GT for sustainable development. It is noted by scientific research that increment in the SD level creates sustainable society and economy (Klimova et al. 2016). From sustainability point of view, technological transformation and upgradation through innovation, research and creativeness are the need of modern society from the one hand and it should be eco-friendly on the other. The technologies have negatives consequences on environment and ecology from regional to global scale. In this context, GT seems to be promising to avail the economic sustainability with bridging the balance between society and the environment. Moreover, the technological effectiveness, efficiency, economical and environmental impact must be tested and evaluated prior to its implementation (Shaikh 2018). GT protects and conserves the environment and therefore causes least harm to the ecosystem by biotic interference (Huesemann and Huesemann 2011). The GT gives environmentally safe produce and cuts the GHGs production, waste generation and ensures the human's live better today and tomorrow (Williams and Helm 2011). Thus, SD is comprised of all round social, economic and environment development (Ahmed et al. 2016).

16.5 Marketing and Environmental Issues

Marketing is a major aspect in terms of customer satisfaction for particular firms. It is the responsibility of the firm to properly communicate and generate value for its customers. Better performance and economic output become a firm capital and develop competency in market mechanisms. In this mechanism after competency development proper use of resources helps the firms to achieve the aims and objectives of SD. This is the area of success for a firm to effectively produce own resources through market competency (Arnett and Wittmann 2014). Development process is going on in the direction of developing interest in environmental issue related to eco-design and marketing (Polonsky 2011).

According to Fisk (1974) marketing tends to have detrimental effects upon the environment due to consumeric lifestyle of the consumers. However the opinion varies on case to case basis. The social and environmental aspects can also be addressed under marketing mechanism (Sheth and Sisodia 2015). However, this particular aspect needs to be scientifically explored properly for policy framing in the corporate sector towards SD. Environmental issues would be an aided advantage in the marketing mechanism to increase the competitiveness (Arnett and Wittmann 2014).

Besides their benefits GT implementation is a hard task and is associated with some challenges. The first and foremost challenge includes adequate funding for research and development (R&D) activities. For development of new innovative technologies sufficient funds should be available for R&D activities. Environmental impact assessment study is often prove to be non-fruitful for adoption of GT. Lack of extensive support system often hampers to maximize the potential of GT. Due to the costing GT sometimes becomes a luxury approach. Conservative approach in our culture often hampers the implementation of GT.

16.6 Green Market and Marketing

In modernized world, it is challenging task to keep user aware about the environment and its related concern. The awareness and human interactions with surrounding environment in safe manner are the key factors governing the success of green marketing. In the present context, green marketing become important phenomenon in India and other developing countries of the world to address the SD because in these regions the environmental pollution is the biggest issues (Sasikala 2017). Therefore, by the firms it is needed to create the environmental friendly products with communicating the people to adopt eco-friendly approaches.

Incorporation of environmental issues in marketing develops the concept of green marketing. Under this concept marketing strategies are oriented towards eco-friendly direction to get positive results towards sustainability.

From the marketing mechanism it is evident that unsustainable consumption pattern causes environmental degradation (Christensen et al. 2007). Therefore, adopting sustainable practices in the marketing process would create a positive

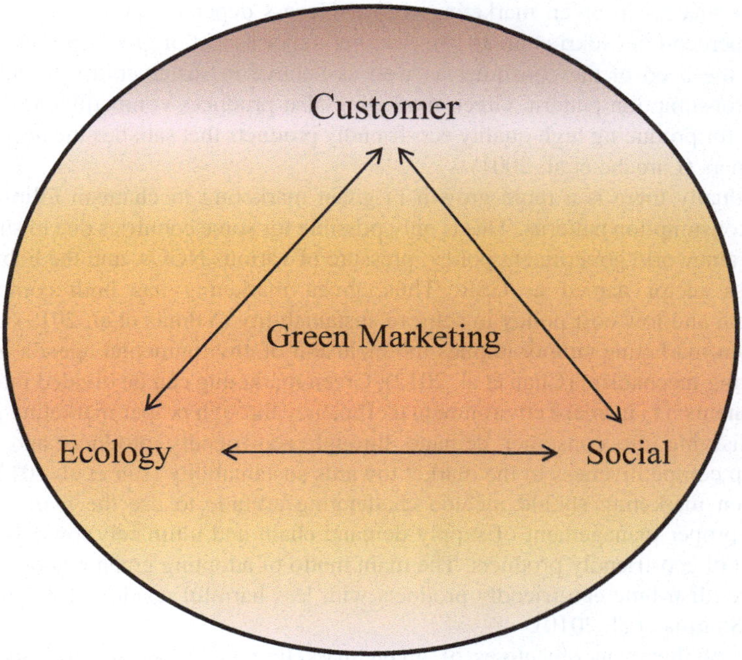

Fig. 16.5 Sphere of green marketing towards sustainability (Folasayo 2019)

image in the society and development of eco-friendly products (Fig. 16.5). It may lead to bring the products under green category but not at the cost of expectations of the stakeholder stand points. For effective implementation of green marketing, one needs to understand the concept of SD, alter the behavioural attitude of the consumers to become green consumer. Changing consumption pattern is a part of green marketing mechanism which indicates the corporate responsibilities towards the society in reducing environmental degradation. Thus, green marketing has become the thrust area of research in the present century (Alhamad et al. 2019).

In the present time the marketing trend has changed to a considerable extent focusing on development of eco-friendly products for effective use of consumers and safety of the environment (Schiffman and Wisenblit 2019). Climate change and global warming are posing significant threat to humankind as well as marketing sector. Therefore, production of eco-friendly products is the need of the hour to make the environment hostile and sustainable (Alnoor et al. 2018; Abdulsahib et al. 2019).

From global context, several countries are approaching towards motivation process among the consumer to become environmentally conscious in terms of purchasing decision. This may lead to change in behavioural attitude. As a consequence green labelling and eco-labelling have become prevalent for eco-friendly product in the market. This also motivates the consumer to purchase green products other than the products which are not eco-friendly. This is essential in order to achieve success in green marketing whose success depends upon change in consumer perception (Sukri et al. 2015). Another major aspect of green marketing is to satisfy the need of the consumer as well as believe in sustainability by adopting green consumption pattern. Green marketing also produces competitiveness in the market for producing high quality eco-friendly products that satisfies the need of the consumers (Laroche et al. 2001).

Gradually there is a rapid growth in green marketing mechanism followed by green consumption patterns. This is only possible for some countries due to stringent legal framework, government policy, pressure of various NGOs, and the liability of business sector named as CSR. Thus, green marketing has both competitive approach and low cost policy to achieve sustainability (Yılmaz et al. 2019).

Green marketing simply implies the inclusion of environmental agenda into the marketing mechanism (Chan et al. 2012). Green marketing can be divided into three components: (1) increase environmental efficiency through proper marketing policy, (2) satisfying the consumer demand through eco-friendly products and (3) to develop competitiveness in the market towards sustainability (Liu et al. 2012).

Green marketing should include challenging attitude to use the market effectively, proper management of supply-demand chain and ultimately towards development of eco-friendly products. The main motto of adopting green marketing is to produce, distribute eco-friendly products with less harmful residue in the environment (Sharma et al. 2010).

One of the main objectives of green marketing includes inventorization and identification of consumer demand and its successful fulfilment for societal well-being in a sustainable way (Chan 2014). With gradual growth of green marketing researches were focussed on identifying the impact of green marketing over a firm performance. Various assessments were done for the firms through their performance in producing various products with various qualities (Robins 2006). From the results it was observed that despite the variable approaches of green marketing focussing on environmental improvement most of the cases results were unsatisfactory in terms of its outputs (Crane 2000). Green marketing expands the dimension of the business sector by incorporating the environmental factors into marketing strategies. For instance, under green marketing, one needs to consider the societal and environmental value along with the economic value. This would help to understand the interaction between human-environment and marketing process (Robins 2006).

Suitable strategies for effective implementation of green marketing have two basic components which include the basic component and on the other applied component. Basic component strives for improvement in the performance irrespective of organization, individual as well as maximum benefit for people. The applied aspect includes green positioning, green designing and green pricing which would work in an integrated manner for sustainability. Further the component of green logistic and disposal is the supplementary part which would reflect the effectivity of the process (Fig. 16.6).

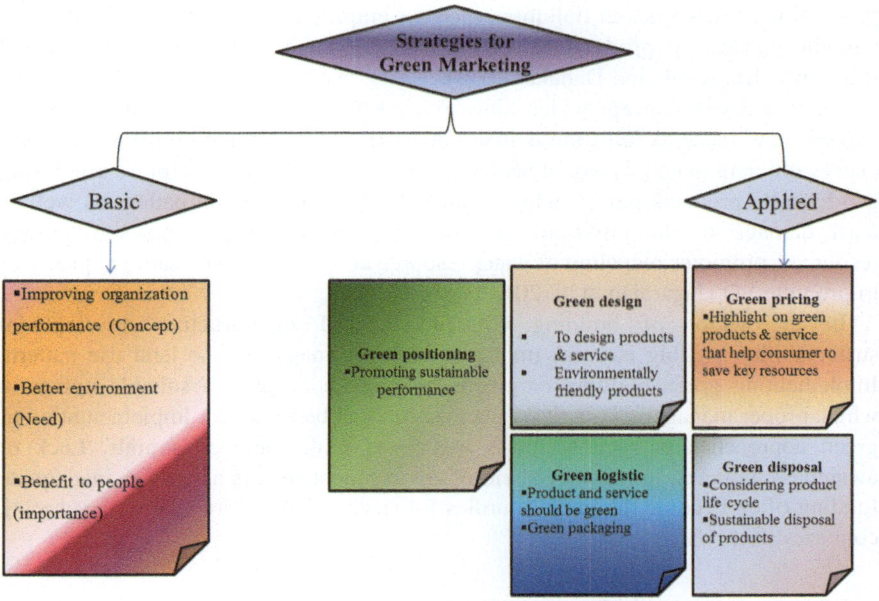

Fig. 16.6 Strategies for green marketing (Alhamad et al. 2019)

16.7 Eco-Friendly Product and Sustainable Architecture

Green designing is the need of the hour in order to avoid the negative consequences of global warming and climate change. Various countries across the globe are practicing green architecture in building constructions which is very much important for reducing GHGs emission in the atmosphere (Ghaffarian Hoseini et al. 2013). Due to such positive outputs the concept of green designing, green building is gaining worldwide acceptance and recognition (Aithal and Aithal 2016). Such approaches are gaining more importance in the industrial sector with gradual development of latest and new technology of building constructions. The major output takes place in the form of environmental sustainability through adapting green building practice (Jagarajan et al. 2017).

In the developing countries the green building practice is a suitable strategy for SD. It is little bit costlier than the normal and conventional building constructions due to modelling and designing approach but in long term it would surpass the benefits of normal buildings. Initially green constructions seem to be little bit costly but however by using green materials and technologies one can reduce the environmental cost and externalities for a better health of human civilization. GT leads to produce eco-friendly products which increase the property values, societal value and above all environmental values (Ghaffarian Hoseini et al. 2013; Paritosh et al. 2017). Such technologies focus on sustainable use of water and energy resources, lesser production of waste, lesser pollution, increase employment opportunities and maintain the health and productivity of human ecosystem with least environmental footprints (Bhowmik and Dahekar 2014).

GT is a newer concept which aims to serve the purpose of human beings and explore new ideas to fulfil the demand and comfort of human civilization. Using energy efficient materials would lead to produce energy source for future use. Using eco-friendly products would help to minimize the consumption pattern as well as least damage to the environment. GT therefore identifies renewable energy resources, promotes recycling of water resource and waste water treatment practices to promote SD (Jagarajan et al. 2017).

In the sector of building construction and infrastructure development sustainability is a big issue as improper practices may alter the land use pattern. Implementing green building or green designing concept is a suitable option in which proper management in the concern sector can be achieved. Implementation of green approaches in such sector is associated with some constraints. Lack of awareness, policies, non-involvement of environment sectors as well as inadequate funding often become the bigger hurdles for sustainable architecture in developing countries (Fig. 16.7).

16.8 Eco-Labelling

Sustainability approach in corporate sector is including environmental issues with gradual growth of science and technology. Production process is very much important for both the industries as well as the corporate world. Therefore, eco-friendly technologies, products are the need of the hour in order to improve competitiveness and economic output. It was observed that some of the firms tend to be green in their message and not in activity. Under the current time they would be obliterated from the market due to poor quality and lack of environmental consciousness. So, firms need to be innovative in their approach, go more towards designing eco-friendly products and environment friendly practices (Janßen and Langen 2016; Del Rio et al. 2016).

In these connections eco-labelling is a management policy to promote eco-friendly products into the market. This is also good for consumers from their health perspective as well as their awareness related to environmental issues in the corporate world. Therefore, for proper marketing of a product eco-labelling concept is being used by the companies both in developed and developing world. It also helps the consumer for screening of products and their quality within the market mechanisms (Rex and Baumann 2007).

For effective implementation of eco-labelling one needs to reduce the gap of understanding between consumers and sellers regarding various eco-labels. Consumers need to understand the various eco-labels to gain quality as well as fulfilling their own aspirations. In this way they would be reflecting their environmental responsiveness for safety and well-being (Nik Abdul Rashid 2009). On the other hand, eco-labelling is directly associated with consumer consciousness and

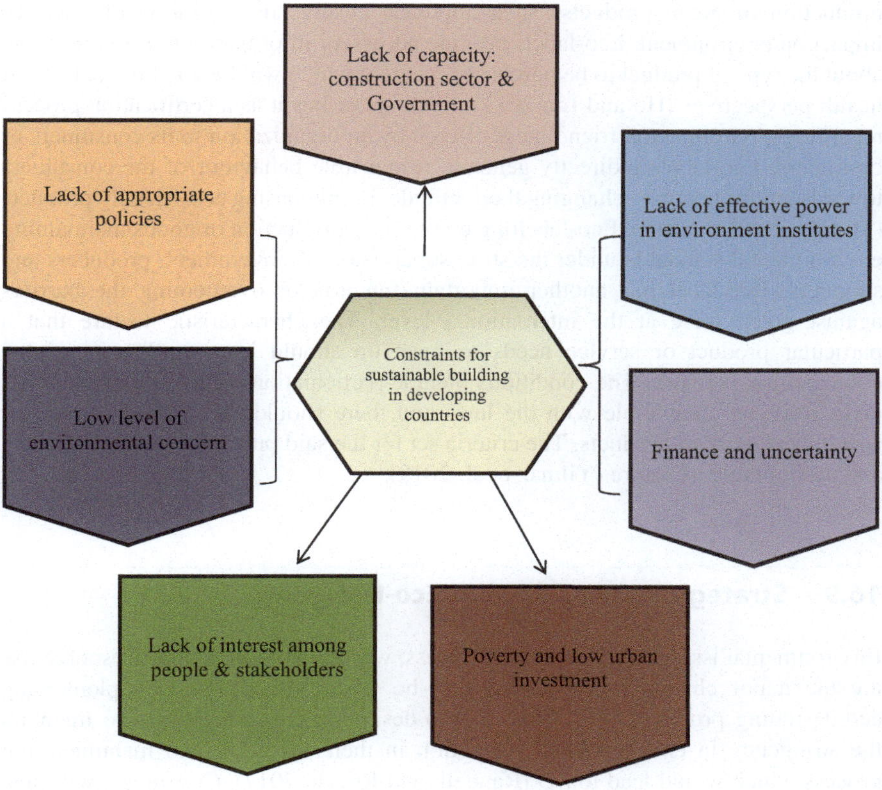

Fig. 16.7 Constraints for sustainable building in developing countries

green marketing. Therefore, consumer awareness should be promoted regarding adopting and purchasing of eco-friendly products and proper decision making (Alhamad et al. 2019).

Eco-labels are the most significant aspect in terms of environmental issues associated with a particular product or services. It performs to generate environmental consciousness among the consumers in relation to environmental quality. Eco-labelled product usually does not harm the environment and are eco-friendly (Buckley 2002). According to Lupu et al. (2013) it is an approach which stimulates environmentally responsive behaviour among the corporate sector to market such products and services which causes least damage to the environment. The associated problems in the life cycle of a product are usually not visible to a consumer but through proper eco-labelling consumer can be made well aware about the negative consequences of a product. Eco-labels highlight the purchase preference of a consumer for a particular product or service considering the environmental issues (Burgin and Hardiman 2010). Corporate firm and companies nowadays are under tremendous pressure due to CSR to maintain the eco-friendly practices and production of green products. Such approach ensures minimization of negative impact on environment. Eco-labels provide necessary information to the consumers about the type of product to be purchased by them which will be good in quality and health perspectives (Ho and Lin 2011). Some describes it as a certification process regarding environmental friendliness offered by an organization to its consumers or customers. Eco-labels indirectly generate responsible behaviour of the consumers towards environment by changing their attitude for purchasing eco-friendly products (Thorgersen et al. 2010). Eco-labelling emphasizes production of goods maintaining environmental standards under the strict supervision of government, producers and suppliers. Eco-label has another important function of overcoming the barriers against green trade at the international level. The characteristic feature that a particular product or service needs to meet up should be according to global eco-labelling network. The conditions include participation in the process from all ends, must be compatible with the laws and there should be suitable criteria for specific category of products. The criteria set for the said purpose should be flexible and accountable in nature (Yilmaz et al. 2019).

16.9 Strategic Approaches for Eco-Designing

Environmental issues such as pollution stress, waste accumulation and resource use are the major challenges which need to be taken care of while implementing eco-designing process (Witt 2011). In eco-designing companies need to maintain the stringency in environmental regulation in their operation and manufacturing process which would lead to SD (Randelli and Rocchi 2017). Consumers will play their active part by altering their purchasing behaviour towards eco-friendly products. Therefore, environmental component should be a key aspect of product development (De Medeiros and Ribeiro 2017). To achieve SD major emphasis should be given on green consumerism that would lead to environmental sustainability.

Under green marketing strategies the associated value of product may be increased by improving its quality followed by reducing the environmental risks. It would act as a fruitful strategy for the firm to win the competition in the market. The policies of green marketing tend to be effective by the firms approach to produce eco-friendly products along with consumer education for purchasing environmentally friendly products. As a marketing strategies firm may promote product development, proper pricing, improve services, and all other green practices to achieve the target of SD. Marketing policy and strategies also depend upon building up of an appropriate image of the firm for gaining confidence of consumers. This is very important for effective implementation of green marketing (Chen and Chang 2013).

Implementing green marketing helps the firm to explore new green opportunities, boost up the corporate sector, increase the value of products, and elevate the comparative benefits along with following environmental trends (Chen and Chang 2013).

16.10 Environmental Footprint and Eco-Design

Eco-design is flexible mechanisms which incorporated under the design of the product and reduce the environmental impact. However, it does not alter the basic structure and fundamentals of the production process but modifies them by including the environmental criteria in order to reduce various types of footprints and improve the quality and safety. Inclusion of environmental criteria for a product design should be planned at the time of designing of the product. This is very much important as once the product has be produced and sent to the market they therefore provides least opportunities for effective designing to reduce environmental impacts.

C footprint is a significant factor during production cycle of a particular product. It simply refers to amount of GHG emitted as a whole during the entire lifecycle of a product starting from production to end use, i.e., from cradle to grave approach. In order to reduce the C footprint the amount of GHG emission at various stages of the life cycle needs to be monitored. After having the detailed inventories and sound data base suitable designing considerations including the environmental aspects can be implemented in order to reduce the footprint.

In agricultural sector footprint is a serious issue due to emission of GHG in a significant amount. It also promotes the climate change event which is a serious issue nowadays for both developed and developing nations. In this sector, there are various processes associated in the production activity. Animal husbandry, aquaculture is the various activities which contribute at the significant level in GHG emission. Therefore, the footprint is increasing day by day (Vergé et al. 2012). Agro-products have variable amount of C footprint which is increasing day by day with increase in the production process. Therefore, minimizing the GHG emission and thereby reducing the C footprint through eco-friendly practices such as organic farming, eco-intensification, eco-designing and practices need to be implemented to fulfil the food requirement and reducing the C footprint (Tilman et al. 2011b).

Scientific investigation and explorations towards environmental footprint reduction and eco-designing are the need of the hour. The scientific findings should provide transparent database and also provide suitable policy formulation for GHG emission reduction and hence footprints.

16.11 Urban Green Space for Environment Protection

Urban greening is gaining importance nowadays at the event of climate change. For proper designing and structure of cities urban green spaces act as eco-friendly approach that performs multidimensional functions. For instance, it provides fresh oxygen to the ambient atmosphere improving the air quality of cities. It also acts as a site of aesthetic pleasure and recreation of the local people. It also tends to reduce the pollution load to a certain extent. However, the positive outcomes of urban green space seem to be under doubt due to faulty practices, improper land use, population explosion in cities, and over-exploitation of natural resources (Ebrahimpour et al. 2013).

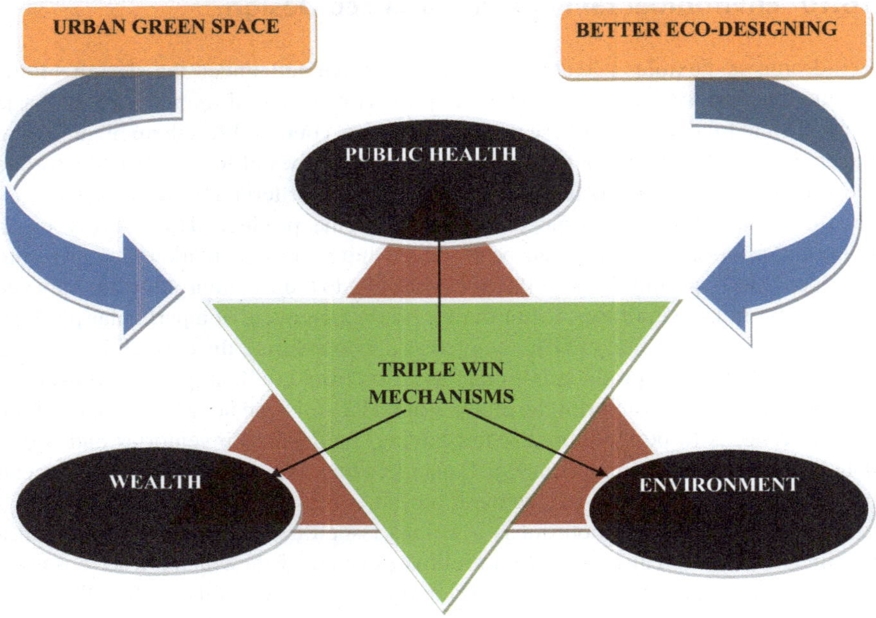

Fig. 16.8 Triple win mechanism through urban green space by adoption of eco-designing (Vargas-Hernández et al. 2018; Kruize et al. 2019)

Public health, neighbourhood stability and environmental sustainability are recognized significant through the proper designing of urban green space. Obviously, a proper designing of green space in urban areas would be highly significant in delivery of uncountable and multifarious ecosystem services. For example, adult work training, childhood educations, taxonomical practices and building social cohesion are certified services which can be delivered through proper designing of green space. Similarly, physical exercise, psychological treatment, clean air breathing and disease reduction (asthma curing) are significant from public health perspective. Pollution reductions, minimizing extreme noise and sound, environmental ameliorations and wildlife management and its protections are included in environmental remediation. Therefore, urban green space regulates the triple win mechanisms by promoting public health, wealth and environment. "How can eco-designing work for urban green space to achieve the triple wins?" However, this question is absolutely justified by above assumptions. Also, a figure represents promising of triple win mechanism through urban green space by adoption of a better eco-designing (Fig. 16.8). Thus, urban green spacing is a component of ecosystem that must be operationalized in such a way to keep better health and environment at sustainable basis (Vargas-Hernández et al. 2018; Kruize et al. 2019).

16.12 Business Strategies for Sustainable Management

In the modern world all the firms are promoting environmental friendly practices and products to minimize the degradation of environment and provide maximum consumer benefits. Some have transformed their technologies towards cleaner production fulfilling the demand of the consumers. It also improves their resource use efficiency of the products, process and other benefits and reduces the negative impacts on public health and environment (Bai et al. 2015).

16.13 Green Growth in Developing Countries

Sustainability is a big issue which integrates various aspects such as health safety and pollution free environment. Overall there should be integration between human beings with the nature. This requires a conservative approach to promote conservation of natural resources. Technologies reformation or reorientation needs to be done to achieve sustainable strategies for future. The concept of sustainable cities is based on harmonized growth and development which includes proper reuse and recycling, green designing of buildings in the form of roof top rainwater harvesting and zero energy buildings are the basic requirements. Further from infrastructure point of view proper drainage system, micro-irrigation facilities, use of renewable energy resource such as solar energy, wind power, hydrothermal energy systems may be adopted as GT. In the transportation sectors use of electric vehicles, compressed natural gas buses along with presence of bio-toilets can be implemented as GT to improve the overall performance on environmental issues (Aithal and Aithal 2016).

For promoting green growth three-dimensional approach of environment, economic and social perspective is necessary to address SD (Fig. 16.9). Green growth demands low C economy, resource efficiency, economically viable alternatives as well as connectivity with the society.

16.14 Eco-friendly Product and Sustainable Management in Different Sector

Eco-designing and eco-friendly technologies need to be implemented properly in the various sectors of agriculture, food production units and other sectors (Table 16.2). Agriculture is such an issue which involves both developing and developed nation. In order to maintain the production unit cleaner production activities are required for having resource saving approach as well as convert them into eco-efficient and feasible alternatives (Zhang et al. 2018).

For sustainable management of the agriculture sector, the problems need to be addressed on sectoral basis. The policies should be aimed to address the sustainability issues in terms of maintaining the sustainable yield and production under various biotic and abiotic stress conditions. Income generating agricultural practices should be identified in the agricultural sector in the form of agro-ecology,

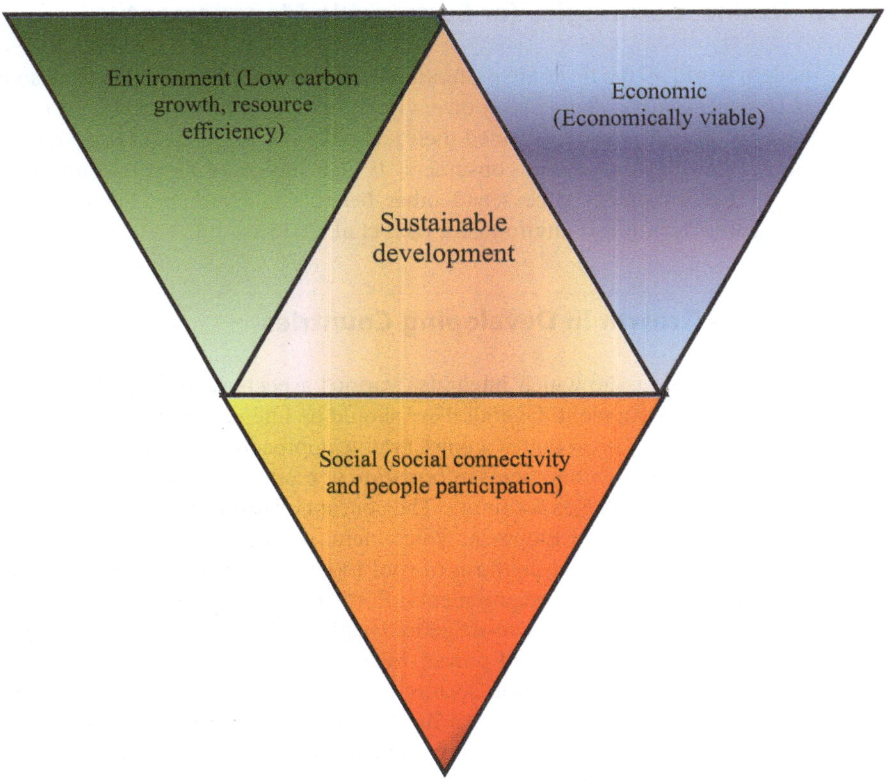

Fig. 16.9 Sustainable development approach (Jhariya et al. 2019a, 2019b; Raj et al. 2020; Banerjee et al. 2020)

ecology based agriculture, renewable energy production in rural sector, etc. Further, the national policies and its impact should be properly studied and on the basis of that extension activities of particular technologies should be implemented. The concept of sustainable agriculture addressed the issues of health, economy and social well-being in an integrated manner. This can also be considering as green agro-technology and designing for effective production of food. After production proper management and marketing are essential.

Water resource is a scarce resource due to its over use and abuse. As a consequence, it has been reported that approximately 1.1 billion people is devoid of safe access to drinking water, 2.4 billion people is without proper sanitation facilities, more than two/third of the developing world suffer from the water borne diseases (Aithal and Aithal 2016). Therefore, the demand for safe and clean water is increasing day by day. Green designing and process build up is necessary in this perspective in order to reduce water pollution and promote conservation of water. Nanotechnology is the most suitable example of that. It is such a technology which undergoes

Table 16.2 Green alternative for various sectors (Modified: Aithal and Aithal 2016)

S. N.	Sectors	Green Technological Perspective
1	Agriculture/farming/ cultivation	Adopt eco-friendly technology for reducing environmental degradation
2	Food processing and packaging	Elimination of contamination and hazardous impact of food to human health
3	Water	Water purification and filtration through bioremediation and nano-technological applications and other eco-friendly processes
4	Sustainable energy	Harvesting and assessment of non-conventional energy resources through green processes
5	Firm produce and green designing	Use of eco-labelling, development of green technology having least damage on environment. Green designing for building construction for energy efficiency and environment friendly structures
6	Automobiles, air navigation and industrial sectors	Energy efficient technologies in automobile sector having zero pollution, adoption of green materials and energy sources in air navigation. Development of industrial process having zero greenhouse emission, using recyclable items
7	Education, health and information technology	Go for green education, adoption of traditional medicinal treatment having no side effect, use of renewable energy sources for sustainable management of energy resource

filtration at molecular level followed by water conserving activities such as recycling of rainwater, desalinization of sea water, etc. It has been found that utilization of nanotechnology in purification of water can generate large amount of portable water by using wind and solar power.

Energy sector is also another important one which is suffering from the crisis situation due to over use and abuse of human beings. With the motto of comfortable lifestyle the energy consumption has increased considerably across the globe. As a consequence of that it is estimated that more than 1.3 billion people has no excess to electricity, 2.2 billion people depends on electricity from organic source. On the other hand, it is predicted that fossil fuel consumption would be double up to year 2025 (Aithal and Aithal 2016).

Green designing in the form of implementation of nanotechnology would address the energy crisis situation to a considerable extent. It may take place in the form of designing of solar photovoltaic cell, silicon cell, hydrogen fuel cell for energy production. Nanotechnology also promotes the solar and wind power as green technology.

In the infrastructure and building construction sectors GT has much role to play by using eco-friendly materials for infrastructure development. For example, construction materials such as cement can be modified up to nanometre scale leading to formation of novel super plasticizers which would help in cement bindings, improve the concrete stability and overall performance of building (Aithal and Aithal 2016).

Food and food processing industries is another big issue in order to feed the growing human population. Due to rising population of human beings GT needs to be implemented at the food production and processing sector (Boye and Arcand 2013). Various forms of technologies in the form of bio-preservation, adoption of electromagnetic forces for decontaminating food materials are the suitable example of GT in food processing sector.

16.15 Policy and Legal Framework Towards Eco-Designing and Sustainability

Sustainability has become a major issue from global context in recent century. Every sphere of human society such as economic, political, social, psychological aspects is going through transformations to achieve sustainability (Patterson et al. 2017). Policy formulation in this aspect is very important to evaluate the effectivity on green designing. For instance, community awareness towards purchasing eco-friendly products can be a key policy for effective implementation of green designing. Further various sectors of the society including the government, corporate organizations, local community stakeholders and policy makers should play their part of effective implementation of green designing policies. Policy should aim toward proper certification, environmental friendliness, green production, packaging, and green consumerism in order to maintain corporate social responsibility (Hojnik et al. 2019).

In this process proper communication between society and the corporate sector is the essential key process for moving towards sustainability (Fig. 16.10). In this, community participation and perception are very important to develop a conservative attitude for fulfilment of needs of future generation. This would ultimately bring a psychological change among the community people and lead to environmentally responsive behaviour (Hojnik et al. 2019). One example of green consumerism includes use of cotton cloth bags instead of plastic bags can be effectively implemented among the consumer during their purchasing activity. In this, consumers have to realize the benefit of using eco-friendly product and exhibit environmentally responsive behaviour. Mass media can serve the purpose of awareness generation in this direction to a considerable extent (Hojnik et al. 2019).

Policy should also be framed towards community awareness for better use of green designing, GT, eco-labelled products and therefore improving the motivation from consumer point of view (Liu et al. 2017). Researches revealed that with gradual improvement of consumer perception the inclination to purchase eco-friendly product gradually increases. In this perspectives consumer should be updated with latest information about eco-friendly products, their benefits that would lead to sustainable lifestyle (Hojnik et al. 2019).

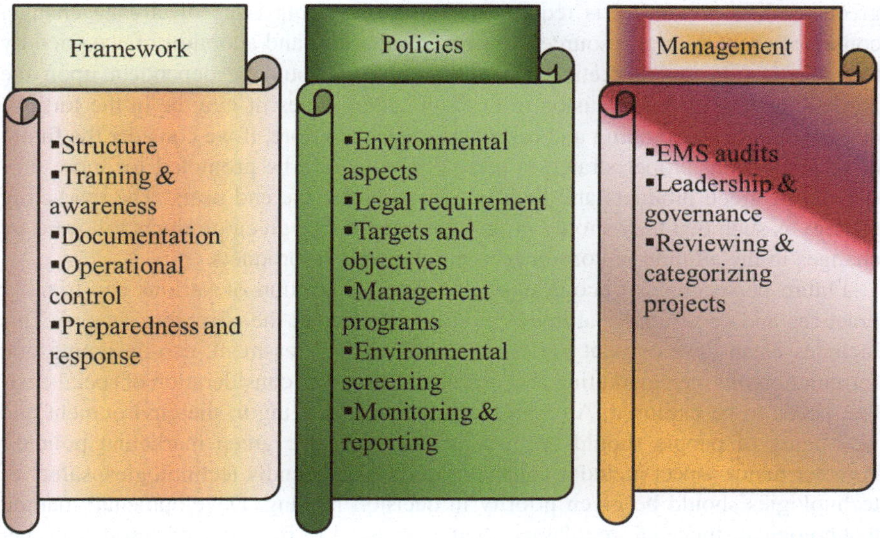

Fig. 16.10 Policies and legal framework towards sustainable management (Modified: Jabareen 2008; Baumgartner 2014)

16.16 Conclusion

Researches should be focussed on new developments in the issues of green marketing as well as developing community perception in these aspects. Positive outcomes should be assessed in terms of various green marketing strategies such as green consumerism, green mechanism, green economy, eco-labelling, etc. Contextual aspects also need to be considered in order to assess the firm performance in implementing green designing technologies. All such aspects need to be explored properly. On the other hand, various factors such as government policies, policies of corporate sectors for adopting green marketing and promotion of sustainable green practices should be the future perspective of green designing technologies. GT seems to have a promising future ahead in terms of social-economic-environmental benefits. Further, technological advancement is required in these aspects to implement the GT from local to regional to global platform towards SD.

16.17 Future of Eco-Design

In the production process for future perspective proper designing and planning are required in order to maintain the environmental sustainability. It is evident that due to technological growth and ever increasing human population one needs to arrest various forms of environmental degradation. For achieving this green designing and green product promotion is required to mitigate the big issue of climate change, conservation of natural resources, reuse and recycling and adoption of eco-friendly technologies. In the marketing sector the success would be dependent upon the marketing strategy for inclusion of environmental issues. It may be in the form of eco-labelling, eco-branding and eco-packaging. Therefore, if we consider the future of eco-designing proper strategies and policies should be promoted for more production of green products and their consumption by the end users. The marketing strategy is such that they should promote production of green products followed by changes in the attitude of consumer to purchase green products.

Future perspective of eco-designing implies promotion of various eco-friendly practices and eco-friendly attitude development among the corporate sectors. This includes clean development mechanism, life cycle assessment, eco-designing and formulation of green marketing policies. It also involves consideration of social costs that needs to be explored. Any approach that is damaging to the environment and well-being of people should be discouraged under the green marketing policies. Another major aspect includes consideration of eco-friendly technologies, safety of technologies should be given priority in decision making. Developmental market, technological, financial, regulatory challenges need to be managed through proper R&D activities followed by supportive and compatible policies for betterment of eco-designing and overall sustainability.

References

Abdulsahib JS, Eneizan B, Alabboodi AS (2019) Environmental concern, health consciousness and purchase intention of Green products: an application of extended theory of planned behavior. J Social Sci Res 5(4):1203–1215

Ahmed S, Ahmad M, Swami BL, Ikram S (2016) A review on plants extract mediated synthesis of silver nanoparticles for antimicrobial applications: a green expertise. J Adv Res 7(1):17–28

Aithal PS, Aithal S (2016) Opportunities & challenges for Green technology in 21st century. Int J Curr Res Modern Edu 1(1):818–828

Alhamad AM, Mat Junoh MZB, Tunku Ahmad SB, Eneizan B (2019) Green marketing strategies: theoretical approach. Am J Econ Business Mgt 2(2):77–94

Allione C, De Giorgi C, Lerma B, Petruccelli L (2012) From Ecodesign products guidelines to materials guidelines for a sustainable product, qualitative and quantitative multicriteria environmental profile of a material. Energy 39(1):90–99

Alnoor A, Eneizan B, Makhamreh HZ, Rahoma IA (2018) The effect of reverse logistics on sustainable manufacturing. Sciences 9(1):71–79

Arnett DB, Wittmann CM (2014) Improving marketing success: the role of tacit knowledge exchange between sales and marketing. J Business Res 67(3):324–331

Bai Y, Yin J, Yuan Y, Guo Y, Song D (2015) An innovative system for promoting cleaner production: mandatory cleaner production audits in China. J Clean Prod 108:883–890

Banerjee A, Jhariya MK, Yadav DK, Raj A (2020) Environmental and sustainable development through forestry and other resources. CRC Press, Boca Raton, FL, pp 1–400. https://doi.org/10.1201/9780429276026

Baulcombe D, Crute I, Davies B, Green N (2009) Reaping the benefits: science and the sustainable intensification of global agriculture. London, UK, Royal Society

Baumgartner RJ (2014) Managing corporate sustainability and CSR: a conceptual framework combining values, strategies and instruments contributing to sustainable development. Corp Soc Respon Env Mgt 21:258–271

Bhowmik A, Dahekar RM (2014) Green technology for sustainable urban life. Recent Res Sci Tech 6(1):4–8

Boye JI, Arcand Y (2013) Current trends in green technologies in food production and processing. Food Eng Rev 5(1):1–17. https://doi.org/10.1007/s12393-012-9062-z

Buckley R (2002) Tourism Ecocertification in the international year of ecotourism. J Ecotour 1:197–203

Green Business Bureau (2020). Financial benefits of an eco-friendly business. Accessed Jan 22, 2020

Burgin S, Hardiman N (2010) Ecoaccreditation: win-win for the environment and small business? Int J Bus Stud 18:23–38

Chan ESW (2014) Green marketing: hotel customers' perspective. J Travel Tour Marketing 31 (8):915–936

Chan HK, He H, William YC, Wang WYC (2012) Green marketing and its impact on supply chain management in industrial markets. Industrial Marketing Mgt 41(4):557–562

Chen YS, Chang CH (2013) Towards green trust: the influences of green perceived quality, green perceived risk, and green satisfaction. Manag Decis 51(1):63–82

Christensen TH, Godskesen M, Gram-Hassen K, Quitzau M, Ropke I (2007) Greening the Danes? Experience with consumption and environment policies. J Consumer Policy 30(2):91–116

Crane A (2000) Facing the backlash: green marketing and strategic reorientation in the 1990s. J Strategic Marketing 8(3):277–296

Cruz RV, Harasawa H, Lal M, Wu S, Anokhin Y, Punsalmaa B, Honda Y, Jafari M, Li C, Huu NN (2007) Climate change 2007: impacts, adaptation and vulnerability. In: Parry ML, Canziani OF, Palutikof JP, van der Linden PJ, Hanson CE (eds) Contribution of working group II to the fourth assessment report of the intergovernmental panel on climate change. Cambridge University Press, Cambridge, UK, pp 469–506

Culiberg B, Elgaaied-Gambier L (2016) Going Green to fit in—understanding the impact of social norms on pro-environmental behaviour, a cross-cultural approach. Int J Consum Stud 40:179–185

De Medeiros JF, Ribeiro JLD (2017) Environmentally sustainable innovation: expected attributes in the purchase of green products. J Clean Prod 142:240–248

Degato DD (2017) Innovation and paths to social-ecological sustainability. RISUS - J Innovation Sust 8(2):3–33

Del Rio P, Peñasco C, Romero-Jordán D (2016) What drives eco-innovators? A critical review of the empirical literature based on econometric methods. J Clean Prod 112:2158–2170

Despeisse M, Ball PD, Evans S (2012) Modelling and tactics for sustainable manufacturing: an improvement methodology. In: Sustainable manufacturing. Springer, Berlin, pp 9–16. https://doi.org/10.1007/978-3-642-27290-5_2

Dzulkarnain I, Santoso T, Maulida AN (2019) Green marketing strategy for local specialty agro-industry development to support creative agro-industry. IOP Conf Ser: Earth Environ Sci 230:1–9. https://doi.org/10.1088/1755-1315/230/1/012052

Ebrahimpour M, Saremi HR, Khakpour B (2013) Analyzing of urban green spaces development process with emphasis on sustainable principles (case study: Mashhad metropolitan). Am J Engg Res 2(4):113–119

FAO (2004) The ethics of sustainable agricultural intensification; ethics series. Food and Agriculture Organization of the United Nations, Rome, Italy, pp 3–5

FAO (2020) World's food systems rely on biodiversity. Food and Agriculture Organization, Rome. http://www.fao.org/news/story/en/item/1263301/icode/

Firbank LG, Elliott J, Drake B (2013) Evidence of sustainable intensification among British farms. Agric Ecosyst Environ 173:58–65

Fisk G (1974) Marketing and the ecological crisis. Harper & Row, New York

Folasayo AM (2019) Green marketing and perceived corporate image: a study of fast moving consumer goods in Lagos state Nigeria. Int JAcad Res Business Social Sci 9(7):202–224

Garnett T, Appleby MC, Balmford A (2013) Sustainable intensification in agriculture: premises and policies. Science 341:33–34

Ghaffarian Hoseini A, Dahlan ND, Berardi U, Ghaffarian Hoseinz A, Makaremi N, Ghaffarian Hoseini M (2013) Sustainable energy performances of green buildings: a review of current theories, implementations and challenges. Renew Sust Energ Rev 25:1–17

Gibon A, Sibbald AR, Flamant JC (1999) Livestock farming systems research in Europe and its potential contribution for managing towards sustainability in livestock farming. Lives Prod Sci 61:121–137

Godfray HCJ, Beddington JR, Crute IR (2010) Food security: the challenge of feeding 9 billion people. Science 327:812–818

Haase RP, Pigosso DC, McAloone TC (2017) Product/service-system origins and trajectories: a systematic literature review of PSS definitions and their characteristics. Procedia CIRP 64:157–162. https://doi.org/10.1016/j.procir.2017.03.053

Ho Y, Lin C (2011) An empirical study on Taiwanese logistics companies' attitudes toward environmental management practices. Adv Manag Appl Econ 2:223–241

Hojnik J, Ruzzier M, Ruzzier MK (2019) Transition towards sustainability: adoption of eco-products among consumers. Sustainability 11(16):4308. https://doi.org/10.3390/su11164308

Huesemann M, Huesemann J (2011) Techno-fix: why technology won't save us or the environment. New Society Publishers, Gabriola, BC, Canada

Jabareen Y (2008) A new conceptual framework for sustainable development. Environ Dev Sustain 10:179–192

Jagarajan R, Asmoni MNAM, Mohammed AH, Jaafar MN, Mei JLY, Baba M (2017) Green retrofitting–a review of current status, implementations and challenges. Renew Sust Energ Rev 67:1360–1368

Janßen D, Langen N (2016) The bunch of sustainability labels—do consumers differentiate? J Clean Prod 143:1233–1245

Jhariya MK, Yadav DK, Banerjee A (2018a) Plant mediated transformation and habitat restoration: phytoremediation an eco-friendly approach. In: Gautam A, Pathak C (eds) Metallic contamination and its toxicity. Daya Publishing, New Delhi, pp 231–247

Jhariya MK, Banerjee A, Yadav DK, Raj A (2018b) Leguminous trees an innovative tool for soil sustainability. In: Meena RS, Das A, Yadav GS, Lal R (eds) Legumes for soil health and sustainable management. Springer, Cham, pp 315–345. https://doi.org/10.1007/978-981-13-0253-4_10

Jhariya MK, Banerjee A, Meena RS, Yadav DK (2019a) Sustainable agriculture, forest and environmental management. Springer, Singapore, p 606. https://doi.org/10.1007/978-981-13-6830-1

Jhariya MK, Yadav DK, Banerjee A (2019b) Agroforestry and climate change: issues and challenges. CRC Press, Boca Raton, FL, p 335. https://doi.org/10.1201/9780429057274

Khan N, Jhariya MK, Yadav DK, Banerjee A (2020a) Herbaceous dynamics and CO_2 mitigation in an urban setup- a case study from Chhattisgarh, India. Environ Sci Pollut Res 27(3):2881–2897. https://doi.org/10.1007/s11356-019-07182-8

Khan N, Jhariya MK, Yadav DK, Banerjee A (2020b) Structure, diversity and ecological function of shrub species in an urban setup of Surguja, Chhattisgarh, India. Environ Sci Pollut Res 27 (5):5418–5432. https://doi.org/10.1007/s11356-019-07172-w

Klimova A, Rondeau E, Andersson K, Porras J, Rybin A, Zaslavsky A (2016) An international Master's program in green ICT as a contribution to sustainable development. J Clean Prod 135:223–239

Koltun P (2010) Materials and sustainable development. Progress in Nat Sci: Materials Int 20:16–29

Kruize H, van der Vliet N, Staatsen B, Bell R, Chiabai A, Muiños G, Higgins S, Quiroga S, Martinez-Juarez P, Aberg Yngwe M, Tsichlas F, Karnaki P, Lima ML, García de Jalón S, Khan M, Morris G, Stegeman I (2019) Urban Green space: creating a triple win for environmental sustainability, health, and health equity through behavior change. Int J Environ Res Public Health 16(22):4403. https://doi.org/10.3390/ijerph16224403

Kumar S, Meena RS, Jhariya MK (2020) Resources use efficiency in agriculture. Springer, Singapore, p 760. https://doi.org/10.1007/978-981-15-6953-1

Laroche M, Bergeron J, Barbaro-Forleo G (2001) Targeting consumers who are willing to pay more for environmentally friendly products. J Consum Marketing 18(6):503–520

Liu S, Kasturiratne D, Moizer J (2012) A hub-and-spoke model for multi-dimensional integration of green marketing and sustainable supply chain management. Ind Mark Manag 41(4):581–588

Liu Q, Yan Z, Zhou J (2017) Consumer choices and motives for eco-labeled products in China: an empirical analysis based on the choice experiment. Sustainability 9:331

Lupu N, Tanase MO, Remus-Alexandru TA (2013) Straightforward X-ray on applying the Ecolabel to the hotel business area. Amfiteatru Econ 15:634–644

Maidment A (2015) How big brands are using renewable energy to their advantage. Renew Energy Focus 16:84–86

Marques B, Loureiro CR (2013) Sustainable architecture: practices and methods to achieve sustainability in construction. Int J Engineering Tech 5(2):223–226

Matson PA, Parton WJ, Power AG (1997) Agricultural intensification and ecosystem properties. Science 277:504–509

Meena RS, Lal R (2018) Legumes for soil health and sustainable management. Springer, Singapore, p 541. https://doi.org/10.1007/978-981-13-0253-4_10

Meena RS, Kumar V, Yadav GS, Mitran T (2018) Response and interaction of *Bradyrhizobium japonicum* and Arbuscular mycorrhizal fungi in the soybean rhizosphere: a review. Plant Growth Regul 84:207–223

Meena RS, Kumar S, Datta R, Lal R, Vijaykumar V, Brtnicky M, Sharma MP, Yadav GS, Jhariya MK, Jangir CK, Pathan SI, Dokulilova T, Pecina V, Marfo TD (2020) Impact of agrochemicals on soil microbiota and management: a review. Land 9(2):34. https://doi.org/10.3390/land9020034

Meena RS, Lal R, Yadav GS (2020a) Long term impacts of topsoil depth and amendments on soil physical and hydrological properties of an Alfisol in Central Ohio, USA. Geoderma 363:1141164

Meena RS, Lal R, Yadav GS (2020b) Long-term impact of topsoil depth and amendments on carbon and nitrogen budgets in the surface layer of an Alfisol in Central Ohio. Catena 194:104752

Nik Abdul Rashid NR (2009) Awareness of eco-label in Malaysia's green marketing initiative. Int J Business Manage 4(8):132–141

Ortiz O, Castells F, Sonnemann G (2009) Sustainability in the construction industry: a review of recent developments based on LCA Constr. Build Mater 23:28–39

Ortiz O, Pasqualino JC, Castells F (2010) Environmental performance of construction waste: comparing three scenarios from a case study in Catalonia, Spain. Waste Manag 30:646–654

Paritosh K, Kushwaha SK, Yadav M, Pareek N, Chawade A, Vivekanand V (2017) Food waste to energy: an overview of sustainable approaches for food waste management and nutrient recycling. BioMed Res Int 2017:2370927. https://doi.org/10.1155/2017/2370927

Patterson J, Schulz K, Vervoort J, Van der Hel S, Widerberg O, Adler C, Hurlbert M, Anderton K, Sethi M, Barau A (2017) Exploring the governance and politics of transformations towards sustainability. Environ Innov Soc Transit 24:1–16

Pereira LM, Karpouzoglou T, Frantzeskaki N, Olsson P (2018) Designing transformative spaces for sustainability in social-ecological systems. Ecol Soc 23(4):32. https://doi.org/10.5751/ES-10607-230432

Pigosso DCA, Rozenfeld H, McAloone TC (2013) Ecodesign maturity model: a management framework to support ecodesign implementation into manufacturing companies. J Clean Prod 59:160–173. https://doi.org/10.1016/j.jclepro.2013.06.040

Polonsky MJ (2011) Transformative green marketing: impediments and opportunities. J Business Res 64(12):1311–1319

Pretty JN (1997) The sustainable intensification of agriculture. In: Natural resources forum. Blackwell Publishing, Oxford, UK

Pretty J (2008) Agricultural sustainability: concepts, principles and evidence. Philos Trans Biol Sci 363:447–465

Pretty J, Bharucha ZP (2014) Sustainable intensification in agricultural systems. Ann Bot 114:1571

Raj A, Jhariya MK, Harne SS (2018) Threats to biodiversity and conservation strategies. In: Sood KK, Mahajan V (eds) Forests, climate change and biodiversity. Kalyani Publisher, New Delhi, India, pp 304–320

Raj A, Jhariya MK, Yadav DK, Banerjee A, Meena RS (2019a) Agroforestry: a holistic approach for agricultural sustainability. In: Jhariya MK, Banerjee A, Meena RS, Yadav DK (eds) Sustainable agriculture, forest and environmental management. Springer, Singapore, pp 101–131. https://doi.org/10.1007/978-981-13-6830-1

Raj A, Jhariya MK, Banerjee A, Yadav DK, Meena RS (2019b) Soil for sustainable environment and ecosystems management. In: Jhariya MK, Banerjee A, Meena RS, Yadav DK (eds) Sustainable agriculture, forest and environmental management. Springer, Singapore, pp 189–221. https://doi.org/10.1007/978-981-13-6830-1

Raj A, Jhariya MK, Yadav DK, Banerjee A (2020) Climate change and agroforestry systems: adaptation and mitigation strategies. CRC Press, Boca Raton, FL, pp 1–383. https://doi.org/10.1201/9780429286759

Randelli F, Rocchi B (2017) Analysing the role of consumers within technological innovation systems: the case of alternative food networks. Environ Innov Soc Transit 25:94–106

Reddy DNVK (2017) A study on impact of Green marketing on sustainable development (with reference to Khammam District). National Conference on Marketing and Sustainable Development, pp 86–106

Rex E, Baumann H (2007) Beyond ecolabels: what green marketing can learn from conventional marketing. J Clean Prod 15:567–576

Robins F (2006) The challenge of TBL: a responsibility to whom? Business Soc Rev 111(1):1–14

Ruerd R, Lee D (2000) Combining internal and external inputs for sustainable intensification. Washington, DC, International Food Policy Research Institute (IFPRI), pp 1–2

Sasikala A (2017) Green marketing in India-a conceptual study. Int J Res Rew 1(4):45–57

Schiffman LG, Wisenblit JL (2019) Consumer behavior, 12th edn. Pearson, New York, p 477

Shafiei MWM, Abadi H (2017) The importance of Green technologies and energy efficiency for environmental protection. Int J Appl Environ Sci 12(5):937–951

Shaikh ZA (2018) Towards sustainable development: a review of Green technologies. Trends Renewab Energy 4(1):1–14. https://doi.org/10.17737/tre.2018.4.1.0044

Sharma A, Lyer GR, Mehrotra A, Krishnan R (2010) Sustainability and business to business marketing: a framework and implications. Ind Marketing Mgt 9:330–341

Sheth JN, Sisodia RS (2015) Does marketing need reform?: fresh perspectives on the future. Routledge, Abingdon, UK

Simon M, Poole S, Sweatman A, Evans S, Bhamra T, McAloone T (2000) Environmental priorities in strategic product development. Business Strategy Env 9(6):367–377

Singh NR, Jhariya MK (2016) Agroforestry and Agrihorticulture for higher income and resource conservation. In: Narain S, Rawat SK (eds) Innovative technology for sustainable agriculture development. Biotech Books, New Delhi, India, pp 125–145

Skerlos SJ (2015) Promoting effectiveness in sustainable design. Procedia CIRP 29:13–18

Stal HI, Corvellec H (2018) A decoupling perspective on circular business model implementation: illustrations from Swedish apparel. J Clean Prod 171:630–643. https://doi.org/10.1016/j.jclepro.2017.09.249

Sukri S, Meterang N, Waemustafa W (2015) Green marketing and purchasing decisions among teenagers: an empirical perspective. Aust J Basic Appl Sci 9(37):238–244
Thorgersen J, Haugaard P, Olesen A (2010) Consumer responses to Ecolabels. Eur J Mark 44:1787–1810
Tilman D, Balzer C, Hill J, Befort BL (2011a) Global food demand and the sustainable intensification of agriculture. Proc Natl Acad Sci U S A 108(50):20260–20264
Tilman D, Balzer C, Hill J (2011b) Global food demand and the sustainable intensification of agriculture. Proc Natl Acad Sci U S A 108:20260–20264
Tomasowa R (2018) Study on sustainable design awareness. IOP Conf Ser: Earth Environ Sci 195:012077. https://doi.org/10.1088/1755-1315/195/1/012077
Vargas-Hernández JG, Pallagst K, Zdunek-Wielgołaska J (2018) Urban Green spaces as a component of an ecosystem. In: Dhiman S, Marques J (eds) Handbook of engaged sustainability. Springer, Cham, pp 1–32
Vergé XPC, Dyer JA, Worth DE, Smith WN, Desjardins RL, McConkey BG (2012) A greenhouse gas and soil carbon model for estimating the carbon footprint of livestock production in Canada. Animals 2:437–454. https://doi.org/10.3390/ani2030437
Wellmer FW, Buchholz P, Gutzmer J, Hagelüken C, Herzig P, Littke R, Thauer RK (2019) Current status of natural resources-an overview. In: raw materials for future energy supply. Springer, Cham, pp 107–144
Wezel A, Soboksa G, Mcclelland S (2015) The blurred boundaries of ecological, sustainable, and agroecological intensification: a review. Agron Sustain Dev 35:1283–1295
Williams M, Helm A (2011) Waste-to-energy success factors in Sweden and the United States. Analyzing the Tranferability of the Swedish Waste-to-Energy Model to the United States
Witt U (2011) The dynamics of consumer behavior and the transition to sustainable consumption patterns. Environ Innov Soc Transit 1:109–114
Yadav R, Pathak GS (2017) Determinants of consumers' green purchase behavior in a developing nation: applying and extending the theory of planned behavior. Ecol Econ 134:114–122
Yilmaz Y, Unguren E, Kacmaz YY (2019) Determination of managers' attitudes towards eco-labeling applied in the context of sustainable tourism and evaluation of the effects of eco-labeling on accommodation enterprises. Sustainability 11:5069. https://doi.org/10.3390/su11185069
Zhang P, Duan N, Dan Z, Shi F, Wang H (2018) An understandable and practicable cleaner production assessment model. J Clean Prod 187:1094–1102

Ecological Intensification: Towards Food and Environmental Security in Sub-Saharan Africa

17

Stephen Syampungani, Ferdinand Handavu, Luckson Chama, Kennedy Ouma, Nalukui Matakala, Chabu Sumba, Stanford Siachoono, Kondwani Kapinga, and Paxie W. C. Chirwa

Abstract

Woodlands and forests in Sub-Saharan Africa (SSA) significantly contribute to environmental safety and economic growth of Africa. The provisions provided by these ecosystems are major sources of income in Africa's rural population's livelihood. Due to an increase in population in SSA coupled with unsustainable farming practices, woodland and forest ecosystem degradation has been on the rise. Addressing the current unsustainable utilisation of woodland resources requires the development and implementation of sustainable models and ecological intensification (EI) provides a basis upon which utilisation of woodland resources could be sustainable. EI includes various practices such as organic agriculture, some form of conservation agriculture, agroforestry (AF) and woodland silvilculture, among others. Literature reports the benefits of these practices ranging from, increasing crop yields, climate change mitigation through carbon (C) sequestration, water resources conservation, soil fertility preservation and biodiversity conservation, among others which promote food and environmental security. Climate change has greatly contributed to food insecurity in SSA; therefore, mitigating climate change would enhance food security in SSA. Climate change effects such as biodiversity loss, floods, droughts, fires, heat waves,

S. Syampungani (✉) · L. Chama · K. Ouma · N. Matakala · C. Sumba · S. Siachoono · K. Kapinga
Copperbelt University, ORTARCHi, Environment & Development, School of Natural Resources, Kitwe, Zambia
e-mail: stanford.siachoono@cbu.ac.zm

F. Handavu
Zambia Forestry College, Kitwe, Zambia

P. W. C. Chirwa
University of Pretoria, SAFCOL Forest, Forest Programme, Department of Plant & Soil Sciences, Pretoria, South Africa
e-mail: paxie.chirwa@up.ac.za

M. K. Jhariya et al. (eds.), *Ecological Intensification of Natural Resources for Sustainable Agriculture*, https://doi.org/10.1007/978-981-33-4203-3_17

pathogens and pest spread among others have all contributed to food insecurity in SSA. The projected rainfall variabilities, increased temperatures and reduced water availability will further increase the levels of food insecurity in SSA, hence the need for adaptation and mitigation measures. EI provides both a mitigation and adaption measure to climate change effects in the agriculture sector which is a key for food security attainment. EI systems in SSA include AF, improved silvicultural systems, entomoforestry and wildlife ranching. EI offers a practical pathway towards food security enhancement and environmental conservation in SSA. However, the implementation of EI systems in SSA is lagging due to inadequate policies. Therefore, there is a need to develop robust policies and strengthen rules and regulations governing EI and resources utilisation to promote food production and environmental stewardship for environmental sustainability in SSA. The current chapter presents EI potential contribution to food and environmental security attainment in SSA.

Keywords

Agroforestry · Ecological intensification · Environmental safeguard · Food security · Sub-Saharan Africa

Abbreviations

AF	Agroforestry
C	Carbon
EI	Ecological intensification
ESS	Ecosystem goods and services
FAO	Food and Agriculture Organization
LULC	Land use and land cover
NTFF	Non-timber forest foods
SOM	Soil organic matter
SSA	Sub-Saharan Africa
USD	United States dollar

17.1 Introduction

Sub-Saharan African (SSA) woodlands and forests significantly contribute to environmental safety and economic development in Africa. These ecosystems support various groups of people, namely herders, forest inhabitants, farmers and rural communities through provisioning of different important ecosystem services like watershed protection, employment, capital and constitute a major source of income. The increase in human population coupled with unsustainable farming practices in SSA produces environmental pressures that alter natural ecosystem processes.

According to FAO (Food and Agriculture Organization 2008), at least over 20% of farmland, 10% of grassland and 30% of forests are going through degradation. Land degradation is a critical environmental issue affecting SSA. SSA has the second highest rate of degradation at global level having accrued 24, 231,000 km^2 in the recent past (Daramola 2012). Additionally, SSA has been observed to have 95 million ha of land at risk of irreversible land degradation if unsustainable practices continue (Daramola 2012).

A number of factors have been attributed to land degradation in the SSA. However, agriculture expansion, population growth (Kamwi et al. 2015; Scholes et al. 2018; Handavu et al. 2019) and wood harvesting for energy (Handavu et al. 2019) are the most critical factors driving land use land cover (LULC) change and degradation across SSA. Given that most of SSA's poor people rely on agriculture for their livelihood, addressing land degradation in the region is critical for sustainable development. SSA is moving towards developing practices that provide for sustainable exploitation of natural resources for both the current and future generations (Geldenhuys 2010). However, the extent to which such practices have been undertaken in the region is not well documented. Therefore, the chapter examines how ecological intensification (EI) around the SSA can enhance environmental and food security. The chapter focuses on the innovation fostering EI in the SSA.

17.2 Concepts and Principles of Ecological Intensification

According to Bommarco et al. (2013) and Tittonell et al. (2016), EI encompasses the active management of farmland for increased ecological processes that promote production and sustainable ecological services. It is relatively new in scientific literature and has been shown to have functional links with ecosystem services (Tittonell et al. 2016). This is due to its proposition on landscape approaches that aim to utilise natural ecosystem functionalities that increase contributions to the flow of environmental services and natural capital. In order to increase beneficial biotic interactions, there is a need for a holistic approach to agroecosystems that intensifies ecological processes through redefining farming systems (Gaba et al. 2014).

The principles of EI emphasise an increase in ecological and biological processes and functions in agroecosystems for soil fertility management, increased biodiversity and nutrients and enhanced interactions among organisms (Agropolis 2013; Wezel et al. 2015). The focus is to design multifunctional agroecosystems that are sustainable. To achieve this, a transition from conventional to EI that aims at maintaining or increasing sustainable agricultural productivity while reducing dependence on artificial inputs through effective management of ecosystem services provided for by biodiversity (Bommarco et al. 2013) should be the main focus.

17.3 Food and Environmental Scenario in Sub-Saharan Africa

With over 80% of SSA's rural population classified as poor and depending on forests for their livelihood (Leakey et al. 2006), the reliance of people on the environment for food is highly evident. Ericksen (2008) suggested the complexity and multidimensional nature of the link between the environment and food security. This is because of food security's dependence on the environment while playing a role in environmental degradation. The current global call of attaining food security for all by 2030 entails adopting sustainable food production innovations that safeguard the environment (Meena et al. 2020a, b). In SSA, population growth is observed to cause environmental degradation as population results in increased need for agricultural land, infrastructure and energy and goods and services to meet modern demands resulting in land use and cover change (Poppy et al. 2014).

The unsustainable use of natural resources by humans has resulted in climate change, which has brought about more variability in rainfall patterns, droughts and fires, among others (Raj et al. 2020; Banerjee et al. 2020; Meena and Lal 2018). SSA has been reported as the most vulnerable region to climate change due to its majority population's dependence on rainwater for agriculture, thereby affecting food access and security (Ericksen 2008; World Bank 2009). Connolly-Boutin and Smit (2016) cited the biophysical environment, social and economic stresses of SSA as factors affecting the regions adaptability to climate change. Climate change has been observed to reduce food production in SSA (Poppy et al. 2014), thereby affecting the quest to attain food security. Attaining food security while protecting the environment is a challenge the region faces; however, with various sustainable agricultural production practices such as EI that provide an opportunity to attain food and environmental security, SSA could tap into the potential of these practices to enhance food and environmental security.

It is reported by NEPAD (2013) that 25% of the malnourished people across the globe reside in Africa. Undernourishment is said to have increased solely in African continent over the past three decades (Table 17.1). Insecurity in accessing food remains a vital rural phenomenon. Food insecurity impact on rural Africa is more than cities due to the fact that people producing food most often do not make enough food resources to sustain their families because they lack sufficient access to means

Table 17.1 Contribution to food security in African regions

	Contribution to African food production (%)		Contribution to total African food crop production (%)	
Region	1980	2010	1980	2010
Southern Africa	18	14	16	14
Central Africa	9	7	14	9
East Africa	28	23	26	21
West Africa	26	33	28	38
North Africa	19	23	16	18

of production. However, access to permanent economies to food has in the recent past become the decisive factor in food insecurity.

17.4 Food Security, Woodland Exploitation and Management

Food security in any part of the SSA can be met by addressing four pillars, namely: availability, access, utilisation and stability. Like other regions globally, SSA, especially rural areas faces critical food insecurity and to resolve this, local solutions are required (Tittonell et al. 2016). However, the forests and woodlands of SSA are increasingly at risk from human-induced pressures that remove woody species, deplete soil nutrients and other aspects that alter their ecological integrity and contribute to climate change. There are a wide variety of factors driving the degradation of Miombo woodlands (Gumbo et al. 2018).

The IPBES (2018) Africa regional summary assessment report identifies the major drivers of landscape transformation as land use change including agricultural expansion, climate change, growing population, changing consumption patterns, urbanisation and infrastructure and globalised demand for products (Scholes et al. 2018). For example, according to Leakey et al. (2006), approximately over 80% of SSA's rural population are classified to be poor and customarily depend on forests for their livelihoods. They practise slash and burn agriculture, timber and charcoal production for income generation. Fuelwood remains one of the main sources of energy for domestic and processing. It accounts for the highest ratio of the national energy budget in Southern African countries, for example, Malawi (88.5%) (Gamula et al. 2013), Mozambique (85%) (Brigham et al. 1996), Zimbabwe (52%) (Griffin 1999) and Zambia (76%) (Chidumayo 1997). Otsuka and Place (2014) found that arable land area in SSA has expanded mainly because of the conversion of forest and woodlands. They further observed that forest area accounted for about 30% of total land area in SSA in 2010 but has been decreasing rapidly over the last two decades.

17.5 Systems for Ecological Intensification in Sub-Saharan Africa

The EI is the foundation for addressing the current unsustainable harvesting practices across the region (Syampungani 2009) that have been practised for a long time.

A silvicultural system that promotes multifunctional landscapes requires both technological and institutional innovation. In doing so, it would be important to consider approaches of integrated food-forestry systems. However, due to increased human influence, there is a need for standardising the optimal levels of silvicultural intervention to guarantee future provisions for both the current and future generation (Schabel 2006). Developing models of sustainable natural resources management that provides for sustainable rural livelihood links well with ecosystem-based principles of EI, which among others include the practice of organic agriculture,

Table 17.2 Practices for ecological intensification in various African regions

	Ecological intensification practices	Source
1	Mixed cropping systems	Affholder et al. (2010), Agropolis (2013), Egger (1986), FAO (2009) and Bommarco et al. (2013)
2	Use of cover crops	
3	Direct seeding and mulch-based cropping systems	
4	Diversified crop rotation	
5	Conservation tillage, minimising soil detoxification and compaction	Agropolis (2013), Bommarco et al. (2013) and Cassman (2005)
6	Integrated pest management	Bommarco et al. (2013), Cassman (2005) and CIRAD (2008)
7	Improved fertiliser and nutrient management, regulation and monitoring of nutrient supply and fertigation	Agropolis (2013), Cassman (2005), CIRAD (2008) and Egger (1986)
8	Biodiversity preservation and promotion of positive allelopathic effects	Brussaard et al. (2010) and CIRAD (2008)

some forms of conservation agriculture, agroforestry (AF) and sustainable woodland silviculture.

Different practices are used in various African regions for EI. Some of the practices are shown in Table 17.2.

17.5.1 Agroforestry

AF could be defined as a common name for land use systems that grow agricultural crops and trees together or with livestock in a spatial arrangement (Young 2002). According to Leakey (1996), AF is an active natural resource management system that increases production for small-scale farmers, thereby enhancing the environmental, social and economic benefits through the combination of rangeland and trees on the farm. This is an old practice which farmers have been practising over years throughout the world. The potential for AF to mitigate climate change through carbon (C) sequestration, improving crop yields and enhancement of biodiversity on the same piece of land has been reported in the literature (Kirby and Potvin 2007; Akinnifesi et al. 2008; Murthy et al. 2013; Jhariya et al. 2015, 2019a, b). It is an efficient land management strategy that enhances soil quality and water resources conservation (Kumar 2006; Murthy et al. 2013; Raj et al. 2019a, b).

Agroforestry and Climate Change

African farmers, like those across the globe face the rising temperatures and increased rainfall variability associated with climate change (IPCC 2012; Meena et al. 2018). There is clear evidence that average temperatures have become warmer globally. For example, in Africa, trends in temperatures have been reported to be increasing at approximately 0.03 °C per annum since 1975 (NOAA 2018; Hartmann et al. 2013). There is strong evidence of an anthropogenic signal in continent-wide temperature increases in the twentieth century (Stott et al. 2010). Furthermore, there is a projected rise in temperature in Africa during the twenty-first century than global averages (James and Washington 2013). Based on General Circulation Models (GCMs) it is suggested that temperature increase for Africa, with the current emissions trajectory is poised to be 1.7 °C by the 2030s, 2.7 °C by the 2050s and 4.5 °C by the 2080s (Girvetz et al. 2019). Notably, historical patterns of precipitation show that much of Africa is drying (Fig. 17.1) (Hartmann et al. 2013; Niang et al. 2014). This situation is critical as many places are likely to have less available water both in streams and in the soil. These changes have serious implications for food production and security across the African continent (Niang et al. 2014; Rosenzweiga et al. 2014).

Much of Africa's vulnerability to climate change lies in the fact that agricultural systems are largely rain-fed with few technological inputs (Girvetz et al. 2019). However, the high levels of dependence on precipitation for agriculture in combination with observed crop sensitivity to higher temperatures during growing seasons indicate critical risk of the sector to climate change (Serdeczny et al. 2016). Additionally, serious gaps that exist are linked to insufficient historical weather data to help in assessing and projecting climate change trends (Girvetz et al. 2019).

Principally, most parts of Africa focus on reforestation and forest protection for the mitigation of climate change (Mbow et al. 2012). However, such efforts are at variance with the increasing need for more agricultural production to meet the rising demand of food driven by the population growth in Africa (Mbow et al. 2012). AF has potential to contribute to climate change mitigation, capacity to enhance soil organic matter (SOM) and store substantial quantities of C in the woody biomass (Syampungani et al. 2010). The amount of C sequestered in an AF system is dependent on the system, its age and the environment within which such an AF system is undertaken. For smaller AF systems in the tropics, for example, C sequestration rates could possibly range between 1.5 and 3.5 ton C ha^{-1} year^{-1} (Montagnini and Nair 2004). In Southern Africa, a non-cropping species was reported to sequester between 26 and 78 Mg ha^{-1} of C in a 2-year rotation, while a 120 Mg ha^{-1}of C was sequestered in a 4-year rotation (Makumba et al. 2007). Additionally, AF has the potential to reduce C emission as trees store and accumulate C in their biomass and in the soil, SOM and wood products by protecting existing forest ecosystems (Syampungani et al. 2010).

According to Sileshi et al. (2008) live biomass can store between 3 and 60 tons ha^{-1}, while wood product's capacity was reported to be between 1–100 ton ha^{-1}. Further SOM's capacity was between 10 and 50 ton ha^{-1} with forest ecosystems storing and up to 2000 ton ha^{-1} (existing forests), thereby counteracting greenhouse gas emissions through material substitution and energy. Mbow et al. (2012) reported varying ranges of C sequestration potential of different AF systems and practices across the SSA (Table 17.3).

Agroforestry and Crop Yields

AF systems have been observed to increase crop yields, thereby enhancing food security in the SSA. A number of technologies such as crop-traditional tree- and

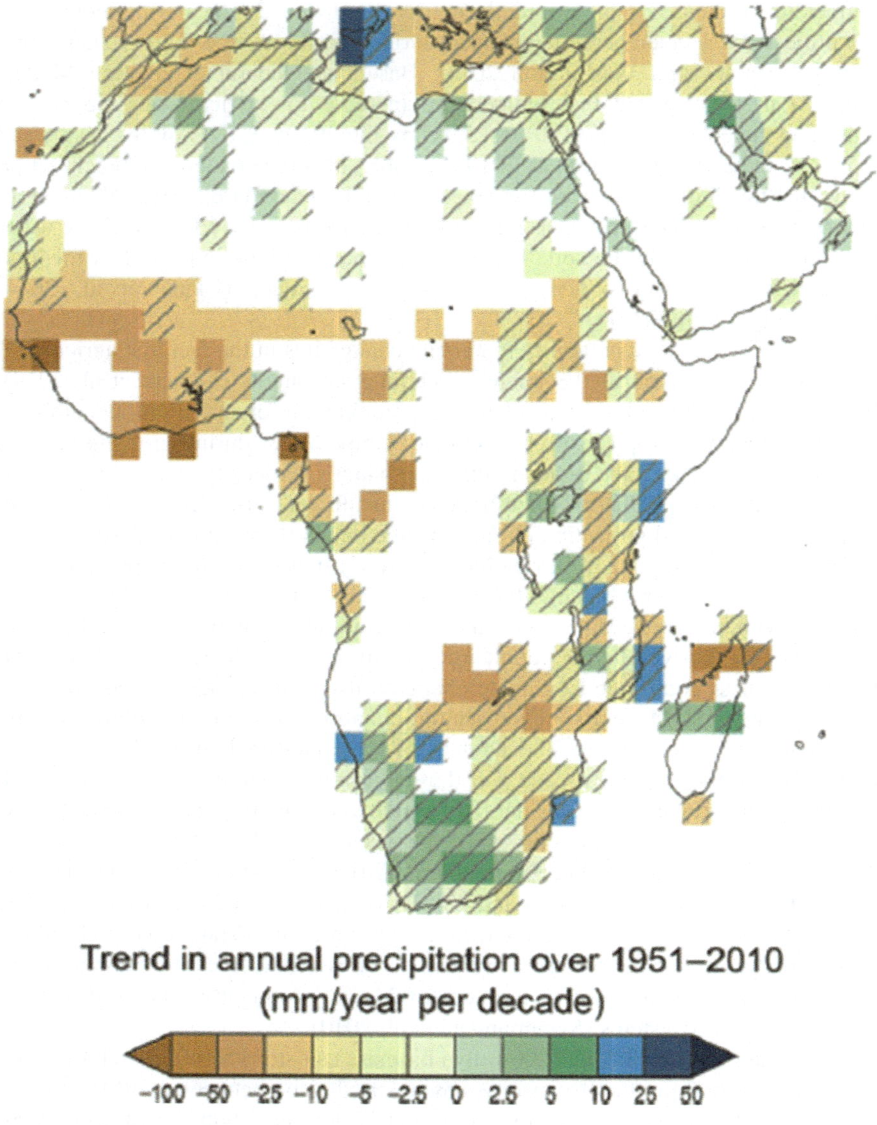

Fig. 17.1 Model projection of historical trends in precipitation from 1951 to 2010. Areas with insufficient data are marked white, while solid indicates statistically significant trends (at 10% level) and diagonal lines indicate areas with trends not statistically significant (Adopted from Niang et al. 2014)

Table 17.3 Potential C stock and C sequestration of some agroforestry systems in Africa (Source: Mbow et al. 2012)

Description	C sequestration (Mg C/ha/year) (range)	C stock (Mg C/ ha) (Range)	Maximum rotation period (year)
Parklands dominate AFS (*Faidherbia albida*)	0.5 (0.2–0.8)	33.4 (5.7–70.8)	50
Rotational woodlots	3.9 (2.2–5.8)	18.5 (11.6–25.5)	5
Tree planting windrows-home gardens	0.6 (0.4–0.8)	19.0 (ns)	25
Long terms fallows, regrowth of woodlands in abandoned farms	2.24 (1.0–6.7)	15.7 (ns)	25
AF system and integrated land use	3.12 (1.0–6.7)	77.9 (ns)	50
Soil C in agroforestry system	0.9 (0.25–1.6)	90.7 (13–300)	Ns

parkland system such as the *Faidherbia albida* (white acacia) based system improved fallow with short rotational species and improved fertiliser tree system (e.g. coppicing tree fallow) have been employed in improving soil fertility and therefore crop yields (Akinnifesi et al. 2008). These technologies have been observed to improve the crop yield up to double the amount. For example, in East Africa, Kenya in particular, *Leucaena leucocephala* (Lead tree/subabool) and *Gliricidium sepium* (quick stick) were observed to increase the crop yield by 53% and 42%, respectively (Akinnifesi et al. 2006).

In other parts of SSA, smallholder farms rotational fallows of *Sesbania sesban* (Egyptian riverhemp), *Cajanus cajan* (Pigeon pea), *Crotalaria* species and *Tephrosia vogelii* (Fish bean) yielded >4 t ha^{-1} of maize (*Zea mays*) compared to an unfertilised maize control with <1 t ha^{-1} (Akinnifesi et al. 2006). Several other studies on crop-traditional tree and parkland systems (Akinnifesi et al. 2006; Syampungani et al. 2010) have reported increased crop yields up to five times in many parts of the SSA. In Zambia, Chirwa et al. (2003) reported an increase in maize yield of up to five times in the Egyptian river hemp fallow, while Gama et al. (2004) and Haule et al. (2003) observed an increase of between 40 and 68% and more than four times in Tanzania and Malawi, respectively. Intercropping maize with coppicing legumes can enhance yields for more years continuously and seasons after establishment (Carsan et al. 2014) through

- Enhanced biological nitrogen fixation,
- Enhanced biological activity in the soil and nutrient turnover,
- Improvement in soil physico-chemical properties.

Agroforestry and Improved Food Security

SSA is mostly associated with persistent food insecurity and vulnerability to famine. Several examples of food insecurity exist in SSA, with the main cause of food insecurity being inadequate food production (Sasson 2012). Examples of food insecurity experience in the SSA from time to time are many; Harsch (1992) and Kandji et al. (2006) observed that up to about 30 million people in Southern Africa were at the brink of starvation mainly due to climate change. The major effect of climate change on the agriculture sector is water unavailability for a sector that is heavily dependent on rainfall which would exert pressure on irrigation systems to close the harvest gap (Tilman et al. 2011). According to Sileshi et al. (2008), AF systems using quick stick intercropped with maize showed enhanced water use efficiency and filtration in Malawi. This is because AF has supportive functions such as water recycling and soil fertility enhancement especially when management practices like conservation agriculture and mulching are applied (Bucagu et al. 2013). EI through AF systems is considered an improved farming practice for smallholder farmers for improved food security using management techniques that are both socially acceptable and cost-effective in the SSA. AF systems such as rotational fallow of *Cajanus cajan* can provide between 100 and 250 Kg Nha^{-1} after a 2–3 years rotation in addition to improving water use efficiency and filtration (Carsan et al. 2014). Furthermore, the system has the capacity to increase financial returns. For example, a 5-year cycle showed varying net profits, unfertilised maize had US$130 ha^{-1} profit, while maize intercropped with the quick stick had US$269 and a rotation with Egyptian river hemp yielded US$309 ha^{-1}, respectively (Carsan et al. 2014).

17.5.2 Improved Silvicultural Systems as an Ecological Intensification Approach

The wide range of currently available silvicultural systems has evolved over time, and the evolution has been due to various factors. In the past, the provision of material goods, such as harvesting tree resources for timber, was the main driver. However, the idea now is to use silvicultural systems in managed forests for a variety of ecosystem goods and services (ESS) within managed forests (Wagner et al. 2013). To increase the functions, biota and diversity of structures subsequently supporting a wide range of ESS, there is a need for diverse silvicultural systems and approaches across landscapes/region (Wagner et al. 2013).

Understanding the key attributes of African Savannas can inform and improve silvicultural practices to address unsustainable harvesting across the SSA woodlands and forests. The key attributes of the African Savannas and species have a direct implication on developing sustainable silvicultural systems for their management and improving local community's livelihoods across the SSA. Most of the dominant species of the African Savannas are light demanders and therefore opening up of their canopy would provide for increased temperature and light intensity on the woodland floor, thereby stimulating and enhancing germination of the soil seed reserves and the development of seedlings that otherwise would remain suppressed by the herbal layer. Currently, there is strong focus on developing silvicultural systems that integrate utilisation aspects such as charcoal production, slash and burn agriculture that opens up the woodland during timber harvesting so as to

Fig. 17.2 African Savanna woodland species persist through vegetative growth from rootstocks and cut stems. (**a, b**) Recovery from shoots, (**c**) Shoot from the base of the stump, (**d**) Shoots from underground stock). (Photos by Geldenhuys and Syampungani)

enhance the persistence of dominant species that would otherwise be under shade (Syampungani et al. 2016).

Many species sprout (regrow vegetative or coppice) from a persistent underground rootstock or from the stem base, as an adaptation to a variety of browsers (Fig. 17.2). This allows them to persist on site through vegetative regrowth from underground rootstocks and cut stems to produce fast-growing shoots after a fire, a clearing or when damaged or harvested (Syampungani 2009; Jhariya 2017; Jhariya and Singh 2020). The rootstocks or cut stems tend to resprout fast because of the availability of food reserves in either the stump or rootstock that supports resprouting

(Syampungani et al. 2017). Developing silvicultural systems for sustainable management of African Savanna requires incorporating such silviculture of the woodland species.

EI also incorporates climate change forest-based adaptation and mitigation projects. The contribution of such projects to rural livelihoods may be in form of 'cash incomes or through access to non-timber forest products generated' by enhanced forest management (Syampungani and Chirwa 2011). A number of C projects have been reported across SSA, such as in Mozambique and Western Sudan (Jindal et al. 2008), Tanzania (Scurrah-Ehrhart and Blomley 2006) and Kenya (Mutunga and Mwangi 2006). Chirwa et al. (2008) also highlighted a number of non-forest timber foods benefits associated with C projects, namely beeswax, edible insects, honey, fruits, mushrooms, fibres, traditional medicine and vegetables. Managing woodlands for C may also result in enhanced ecosystem services such as sedimentation, soil erosion control, steadier and higher quality water supplies (White et al. 2004). Jindal et al. (2008) and Mutunga and Mwangi (2006) observed that the two C projects in Western Sudan and Kenya generated improved local rangelands and the ecology of Lake Victoria Basin through control of soil erosion and watershed management activities.

17.5.3 Entomoforestry

Entomoforestry is a critical component which embraces insects of socioeconomic importance. Over the years, recognition of the contribution of insects to food security has grown considerably across the SSA. Apart from their fundamental contribution to disseminating pollen, aerating soils through burrowing and helping with dead plant and animal material decomposition, insects have significant contribution to rural and urban community livelihoods (Stack et al. 2003). Edible insects are also a popular food in many cultures across SSA and other parts of the globe. Edible insects supplement the diet of approximately two billion people (Afam et al. 2017) worldwide providing food and nutritional security. Of late, edible insects are gaining much attention for their high nutritional value and environmental advantages over meat production (van Huis et al. 2013a). According to van Huis and Vantomme (2014), insects have been rendered to be an excellent alternative to meat due to the following aspects; their short life cycle, low space requirement, efficient nutrient conversion rates and lower greenhouse gas production. In terms of nutritional status, crude protein content of insects has been reported to range from 40 to 75% on dry weight basis depending largely on species and stage in their life cycle.

The practice of eating insect is known as entomophagy (van Huis et al. 2013b). The use of insects as food by humans is an old African tradition which is intensely embedded in most African cultures (Mbata et al. 2002). The consumption of caterpillars in SSA is reported to constitute 30% of all edible species of insects and this is attributed to diversity of species. For example, based on intensive studies in the northern eastern plateau of Zambia and neighbouring areas of DRC and Zimbabwe, Malaisse (1997) identified 38 species of edible caterpillars.

Consumption of caterpillars (Lepidoptera) in SSA is estimated at 18% (van Huis et al. 2013b). Edible termites which belong to the macrotermites (Macrotermitinae) consist of the winged form which swarms from termite mounds or hills shortly after first rains. Stink bugs (*Encosternum delegorguei*) are considered a delicacy in parts of Southern Africa (van Huis 2003; Morris 2004). Edible grasshopper (*Ruspolia differens*) is a common food source in many parts of Eastern and Southern Africa (van Huis et al. 2013b).

As such, their potential needs more serious consideration in strategies for poverty alleviation and food security in SSA (DeFoliart 1992). The great diversity of forest and woodland habitats harbours large populations of edible insects across SSA. This provides an assortment of opportunities for managing food insects to concurrently contribute to preserving the ecological integrity and habitat diversity of other forms of life (Schabel 2006).

Given the overwhelming role of edible forest insects to human livelihood systems, there is a realisation for a rethink in developing silvicultural systems that incorporate management of forests to provide for edible insect production (such as caterpillars) (Schabel 2006). In addition to their role in the food basket, edible insects have been reported to improve rural economies. In Botswana, Zitzmann (1999) indicated that the sale of *Imbrasia belina*, a caterpillar species contributed 13% of the total household income. Edible insect farming and gathering has potential to contribute towards creating employment and enhancing cash income on both household and large-scale operations (Jideani and Netshiheni 2017).

Cultivation and Harvesting of Edible Insects

Edible insects, like other forest products have been customarily considered common goods taken freely either by specific collection during seasons of abundance or opportunistically, (Schabel 2006). At such a time, the harvesting methods used could have been socially and ecologically tolerable and possibly beneficial to the edible insects too. In addition, edible insect harvesting by cutting of branches and trees may have had silvicultural advantages as the felling of trees individually or in groups encouraged natural regeneration, enhanced the community structure of trees and provided vigorous saplings which are more palatable to certain food caterpillars (Schabel 2006).

Notwithstanding, Schabel (2006) further observed that human intrusion requires standardised silvicultural interventions at optimal levels to guarantee future supplies of forest resources and caterpillars. Additionally, drought, forest loss, loss of traditional authorities that regulate forest resource use, recurrent late-season fires and increased international demand for caterpillars have led to decline in caterpillar populations, thereby posing challenges to the fairly informal system (Balinga et al. 2004). Adverse effects of over collecting of caterpillars have been observed on common property elsewhere (Yhoung-Aree et al. 2005) and to mitigate tragedies of the commons like these efforts are being made to regulate collection seasons, issuance of caterpillar hunting permits. Developing silvicultural systems that support the development and establishment of tree species such as *Entandrophragma* spp., *Isoberlinia* and *Colophospermum* spp. that are associated with high value caterpillar is seen as one way of contributing to sustainable caterpillar production.

Bees and Ecological Integrity

Bees and trees are interdependent as they go side by side. Forests provide excellent forage resources (Bradbear 2009) and habitat for bees and therefore for the apicultural sector. Bees are a vital part of terrestrial forest ecosystems and the extensive losses in pollinator guilds and communities could disturb ecosystem integrity. Bees play an important role in influencing ecological relationships and also the provisioning of ecological services (Bradbear 2009; FAO 2009; McCracken et al. 2015; Painkra et al. 2016). FAO (2009) noted that approximately three quarters of all plants and plant products consumed by humans are pollinated by bees directly or indirectly. Cross pollination plays a critical role in enhancing genetic differentiation and survival of offspring (Agera 2011). In SSA, beekeepers depend on access to woodlands and forests for bee foraging, and therefore bee products and yields depend on the type and the condition of the forests (Shackleton et al. 2010). However, the control of other forest resources users such as charcoal producers is beyond beekeepers, smallholder farmers, timber harvester and has a limitation when it comes to safeguarding their interests. Legal authority of woodlands and dry forests in SSA ranges from solely state-owned land to several traditional customary tenures. The legislations in most SSA countries only marginally recognise beekeepers' access or management rights to forests. To promote sustainability in beekeeping, two major routes have been undertaken: (1) producer and forest-based initiatives whose main goal is reinforcement of traditional beekeepers rights to forest areas and (2) promoting the link between sustainable forest management and beekeeping.

Wildlife Ranching

The SSA region has observed increased economic growth in the previous decades. For instance, six of the fastest growing economies in the world between 2000 and 2010 emerged from SSA (The Economist 2011). However, economic growth must be sustainable in order to increase development and reduce poverty levels. In the SSA, sustainable economic growth and development can only be achieved when environmental safeguards are objectively designed and implemented to promote green economy. Green economy envisions the balance between improvement of the quality of human life and social equity and a significant reduction of environmental risks to promote provision of environmental services (UNEP 2011).

For many countries in SSA, important choices have to be made between conservation of wildlife areas and opening up of these natural systems either for human expansion or commercial activities such as agriculture and industrialisation to support economic development (Harrison et al. 2015). Conservation of natural capital assets such as wildlife and bio-diverse forest resources can provide income and new sources of livelihood at both local and national levels (GIZ 2012). Wildlife ranching can be a perfect example of EI for enhancing the sustainable management and utilisation of the natural environment if applied adequately. Several forms of game ranching exist, including (1) game farming; (2) mixed livestock–wildlife; (3) aquaculture–wildlife; (4) tourism–wildlife; (5) crocodile farming and (5) a combination of aquaculture, crocodile, livestock, tourism and wildlife.

Game Ranching

This kind of land use involves the subjection of land exclusively for farming and management of wild animals. Unlike intensive crop husbandry which often results in negative impacts on the land, the farming of wild animals reduces these impacts by conserving and restoring biodiversity (Pywell et al. 2012). Wildlife ranching is one of the emerging potentially and ecologically intensive contributors to the transition towards the development of the green economy among SSA countries. For example, in South Africa, wildlife ranching is recognised as a lawful and commercially viable venture with environmental and socioeconomic benefits (Taylor et al. 2016).

Several other SSA countries have commercialised wildlife ranches and sanctuaries which are actively managed for increased efficiency and productivity (e.g. Zambia, Lindsey et al. 2013a; Zimbabwe, Bond 2013; Namibia, Lindsey et al. 2013b; Kenya, Bos et al. 2000). While the farmer's primary objective may potentially centre on the production of live animals and game products such as meat, animal skins and trophies, the management of the ecosystems that supports the growth of these animals becomes equally critically important. Besides, the management of wild animals in game ranches acts as a buffer or back-up to the existing traditional conservation efforts that are largely centred on management of wildlife in areas designated as national parks. Therefore, game ranches can supplement local conservation efforts, especially for endangered species such as rhinos (*Rhinoceros* spp.), whose populations have widely been decimated across the continent over the years. The most indigenous wild animal species on the continent seem to have evolved to adapt to local ecological conditions such as inadequate rainfall or presence of certain disease organisms, they are often expected to naturally be productive and therefore easy to manage (Ntiamoa-Baidu 1997).

Mixed Livestock-Game Farming

Other than the utilisation of land for game farming only, here, farmers integrate the management of livestock with game on the same often private owned properties (Fig. 17.3). While livestock farming has long been part of the African tradition, there

Fig. 17.3 (**a**) Pastoralism in rangelands and (**b**) mixed livestock–wildlife ranching in marginal land areas in sub-Saharan Africa (Source: Liniger and Mekdaschi 2019)

has been a steady shift in the past few decades towards mixed livestock and game farming for maximised land utilisation.

Game animals are often managed on natural vegetation even though the habitat could be manipulated to enhance productivity. Therefore, the incorporation of game promotes not only the preservation of pastureland, but also the preservation of forest ecosystems existing within these farmlands since they provide habitat for wild animals. While livestock production has also been a good source of income, the farmer expects to benefit even more from the incorporation of wild animals on their farmland, as they are generally more profitable than livestock. For example, wild animals have been shown to require much lower daily water intake, faster breeding and growth, earlier maturity and ability to put on weight in grazing conditions under which livestock such as cattle can lose weight (Ntiamoa-Baidu 1997; Thresher 1980). In this case, mixing the two maximises not only the conservation value of the utilised land, but also financial returns for the farmer. For this reason, more and more private farmers in SSAs are converting their lands into incorporating wildlife on cattle ranches as mixed ranches for increased returns and environmental benefits (Campbell and Brigham 1993; Ntiamoa-Baidu 1997).

Mixed Aquaculture and Game Farming

Aquaculture, the production of fish for both protein and income generation is a highly exploited endeavour. The World Bank Brief (2014) projects would become the primary source of food globally by 2030, especially as demand grows globally from the middle class and wild capture fisheries approach their maximum take. As the global population is predicted to reach nine billion by 2050, the demand for food will undeniably increase too. If practised responsibly, fish farming has a huge potential to supplement other food sources by providing a means of livelihoods and feeding an ever-increasing global population. But for an aquaculture system to be truly sustainable, it must contribute to environmental and socioeconomic sustainability. In doing so, fish farming should not only reduce the exertion of significant disruption to natural ecosystems, or cause the loss of biodiversity or substantial pollution impact, but must also be a viable business with good long-term prospects that are socially responsible and support the overall well-being of local communities. To realise this potential, fish farming needs to be integrated into other ecological management models such as game farming. This approach provides an opportunity for the establishment of an effective system which imitates the ecological system which often occurs in the naturally undisturbed habitats and one that will potentially provide multiple benefits towards both the farmer and the natural environment. Further, this approach utilises inter-trophic transfers of resources for optimum resource use by using larger organism's waste as food sources for smaller organisms. This guarantees nutrient recycling, thereby reducing waste while increasing production. If properly executed, fish and game farming integrated on the same piece of land can help contribute towards resource utilisation sustainability in the context of food production, habitat restoration, enhancement of wild stock population and endangered species populations (The World Bank Brief 2014).

Although wild animals may cause challenges related to capture, translocation and taming, they are generally a very profitable and ecologically sound inclusion on the farm. Unlike wild animals, however, fish farming is easier to do, as it requires less maintenance. Instead, they only require food, good water and temperature conditions which often are not a major challenge in most SSAs countries as they are endowed with abundant naturally occurring freshwater resources and predominantly enjoy tropical to sub-tropical climatic conditions.

Tourism–Game Ranching

The tourism and game ranching system involve the promotion of tourism activities in areas managed as game ranchers. Here, the property owner's interest is more in revenue generation resulting from not only the sale of game in the form of live animals, game meat, trophies such as animal skins and horns, but also from tourism activities such as game drives, safari hunting and where possible, sport-fishing. Unlike in game farming, here, game is often allowed to naturally fend for itself in often real wild conditions like those found in traditional national parks. In this case, the primary goal of the property owner, where management is concerned, is to ensure that it is managed with little or no disturbance to the natural ecosystem that hosts these animals. Ultimately, this fosters environmental sustainability, especially in the face of changing climates and with increased calls for preservation of natural ecosystems (Parish et al. 2008).

Several game ranchers in SSAs seem to predominantly prefer the game ranching and tourism systems for economic and environmental sustainability reasons. For example, the number of game ranches incorporating ecotourism in Zambia is believed to have improved from 30 (occupying 1420 km^2) in 1997 to 177 (occupying approx. 6000 km^2) by 2012 (Lindsey et al. 2013a). Interestingly, these ranches generated an income of approximately United States Dollar (USD) 15.7 million per year, compared to US$16 million from the public game management areas which involve an area that is 29 times larger. Furthermore, at least 1200 people are employed by the ranching-ecotourism industry plus about 1000 individuals employed in associated industries (Lindsey et al. 2013a).

The growth in game ranches, especially those incorporating tourism activities also shows this enterprise's potential to contribute towards supplementing traditional conservation efforts centred on national parks. For example, in South Africa, 80% of nature conservation is taking place on privately owned game-ecotourism properties (Eloff 2000; Fox and Du Plessis 2000; Van der Merwe and Saayman 2003). The country has been recording an average increase of approximately 300,000 ha per year in land utilised for game ranching and tourism (Bothma 2002). In Zambia, wild ungulate populations on game-ecotourism ranches increased from 21,000 individuals in 1997 to about 91,000 in 2012. In contrast, those (ungulates) in state protected traditional national parks have declined sharply (Lindsey et al. 2013a). Therefore, the game-tourism system, if properly supported and managed, can contribute not only towards enhancing ecological integrity, but also sustained revenue gains for the property owner and indeed employment creation for the local communities.

A Combination of Aquaculture, Livestock, Tourism and Wildlife

Where local conditions allow, EI can opt for a land use system that incorporates fish, livestock, tourism and wildlife farming on the same property. In this case, the land should have naturally existing resources that are ideal for supporting all these systems uniquely or combined. For example, there should be freshwater resources to support fish farming; pastureland and forest for livestock and wild animals to graze and browse (Pretty and Bharucha 2014).

Nsobe game farm in Zambia sets a perfect example of this model. Having been established in 2001, Nsobe forms part of the over 1500 ha commercial Miengwe Farm Ltd., host to several species of ungulates (including rare species such as *Tragelaphus spekii* (Sitatunga), *Kobus leche* (Black Lechwe) and *Hippotragus niger* (Sable)), nocturnal animals, primates and over 300 species of birds. The property has taken advantage of the Kafulafuta stream which passes through its forests to establish two dams that have been stocked with a number of fish species to support sport-fishing activities. Some of the water from these dams is diverted towards supporting a large fish farm on the property. Further, the property has livestock including a beef herd and a piggery and a safari lodge to support ecotourism activities. Prior to becoming a private property, Nsobe was part of an existing national forest reserve (Genschick et al. 2017).

Zambia has over 400 national forest reserves, all of which are managed primarily for promoting plant protection with little or no benefits generated from these resources. In contrast, Nsobe generates revenue from ecotourism activities such as game viewing (birding, game drives, bush walks, swimming, sport-fishing and cycling), lodging and conference facilities. Additional revenue is generated from the fish farm and livestock production. The property has employed several members of the local community who are working in its various departments (game scouts, lodge managers, tour guides, fish farm managers, livestock production, management and beef marketing). As an additional cooperate socio-responsibility, the property constructed a primary and secondary school in 2009, which is accessed by children from the local communities. The fish farm, game and ecotourism activities critically depend on the sustainable utilisation and management of existing natural environment (forest and freshwater ecosystems). Nsobe has placed this as a priority in the management of their property. In doing so, they contribute to ecological sustainability. Besides, the value of these natural resources (forest and freshwater resources) have been enhanced, as they are not managed like the traditional national forest reserves but generate revenue which contributes towards both resource conservation and management as well as enhancing the well-being of local communities (Genschick et al. 2017).

17.6 Research and Developmental Activities Towards Food and Environmental Security

Food and environmental security have become global challenges that require urgent attention as the world moves towards sustainability. The attainment of food and environmental security has been met with various challenges globally ranging from population increase, climate change, land degradation and soil erosion, reduced water table, rising costs of energy and limited natural resources (Sage 2013; El Bilali 2019; Kumar et al. 2020; Meena et al. 2020c; Khan et al. 2020a, b). Environmental degradation has been reported as the main cause of food insecurity due to its bearing on crop diversification (Whitmee et al. 2015). Food insecurity affects human health and exacerbates environmental degradation hence the need for sustainable food production practices that produce food to meet rising demands while protecting the environment. Available evidence suggests that food security achievement in SSA is a major challenge (Poppy et al. 2014). This is due to the region's reliance on rainwater and lack of irrigation technologies for crop production (Poppy et al. 2014).

With a projected increase in population, the need to adopt environmentally friendly farming practices is eminent. An increase in research towards food security has been observed to increase over the years. With food security depending on the environment for its attainment, scientists and researchers have shifted their research towards the development of sustainable farming activities to ensure food security globally (Kleijn et al. 2019). This resulted in the discovery and promotion of various sustainable farming practices including AF, organic agriculture, conservation farming and sustainable woodland systems all which centre on the principles of EI. With low adoption of the practices that guarantee food security in the region, there is a need for evidence-based research on the economic benefits of EI by expressing the benefits of ecosystem services in terms of costs and benefits which are of value to the farmer (IFAD 2013).

There is a need for concrete evidence of high production resulting in high profits using EI than conventional farming methods to enhance adoption. The current switch by consumers from conventionally grown to organically grown crops could be used to motivate the adoption of sustainable farming practices. There is still need for elaborate research on the social benefits of EI, various systems that could be adopted by small-scale farmers especially in rural areas for food security.

17.7 Policy and Legal Aspects of Ecological Intensification

Different countries in the SSA have attempted to develop policies to promote EI. The choice, implementation and legal aspects of these policies are unclear and need to be properly rectified. However, it is vital that these policy designs consider the multidimensionality, the costs and benefits, synergies and trade-offs that strike a balance among the diverse benefits. Garibaldi et al. (2019), for instance, proposed ten science-based policy targets and their multidimensionality towards EI for increased crop yield and sustainable environmental services. However, in view of the wide preferences, a diversity of ecosystem services is essential to produce an environment that optimises benefits for all. Therefore, unlike conventional policies (Garibaldi et al. 2017), policies on EI should account for legitimacy and relevance to the needs of the users (Garibaldi et al. 2019). For example, what should be the extent of landscape alteration or transition in the quest for EI? What targets promote habitat diversity and biodiversity? What trade-offs are possible without compromising provision of environmental services and goods?

In general, policies on EI should promote green infrastructure that ensures connectivity across agroecosystems and natural habitat landscapes. On this front, several attempts have been made in SSA and a number of bottlenecks encountered. Despite the high potential for development in wildlife ranching in many African countries, the industry is performing poorly. This is due to challenges including poaching, bureaucracy, lack of government support and unclear policies on game ranching (Lindsey et al. 2013b). Adugna and Jebessa Debella (2019) emphasise on the need to have an effective wildlife policy in SSA countries in order to enhance economic contribution of the sector. Some of the policy considerations include strengthening community-based conservation, practicing of wildlife-friendly farming, regular monitoring of wildlife populations and creating national and regional atlases of wildlife for managing and conserving wildlife resources efficiently.

17.8 Conclusion

EI offers a practical pathway towards enhancing food security and environmental conservation in the SSA. It intensifies food production while sustaining the environmental services on which agriculture and natural facets depend. In addition, the natural ecosystem can also be optimised to provide different products and services. In SSA, the rapid population growth, increasing demand for food, scarcities in natural resources and climate change are associated with dwindling farm yields and high rates of hunger and malnutrition.

In this chapter, the various forms of EI have been discussed from the agroecosystems perspective. The rationale behind this approach is that many agricultural and ecological systems are way below their productive potential, given the accelerated socioeconomic growth and development. To deal with complex and diverse, local specific problems essential for development, poverty alleviation and sustainability, sustainable EI and environmental management are required. In the natural ecosystems, woodland products have been harvested for domestic and commercial purposes. AF in many regions of the SSA continues to be more promising in sustaining animal and human food supply, woody and non-woody forest products and maintenance of biodiversity. The high biological diversities have encouraged various economic facets from environmental exploitation. Game ranching in the SSA has promoted tourism and related activities causing socioeconomic transformation in the areas. Another form of EI, livestock-wildlife ranching has incorporated the niche concept in spatial and temporal resource utilisation between livestock and wildlife.

However, since policies on EI are inadequate, there is need to develop robust policies and strengthen rules and regulations governing EI and resources utilisation to promote food production and environmental stewardship for environmental sustainability in SSA. From the discussion, there is clear indication of weak or non-existent legal frameworks to govern the existing and emerging components of agro EI in the region. The 'grey areas' on matters relating to incentives, trade-offs and compensation must be harmonised to minimise cross-sectoral conflicts during resource allocation and utilisation.

17.9 Future Perspective

Population in SSA is projected to increase (Zuberi and Thomas 2012) with rapid increase in urban areas where land is restricted to small holdings. The increased demand for food due to population will increase pressure on the scarce natural resources (land), thereby increasing environmental degradation if conventional methods of farming are continued (Zuberi and Thomas 2012). Population increase in SSA will exacerbate the food security and environmental degradation challenge. These challenges can be addressed by switching to environmentally friendly farming methods that aim to enhance environmental conservation while increasing yields such as EI practices (Bommarco et al. 2013). Besides EI role in sustainable food supply, it plays a major role in combating climate change and sustainable supply of ecosystem services for enhanced human well-being. Notwithstanding the benefits of EI towards food and environmental security in SSA, its adoption and implementation on large scales is lagging.

Available evidence suggests that the low adoption of EI by farmers is due to its focus on ecosystem services (processes) by scientists which are not important to farmers other than profits (Kleijn et al. 2019). The different perceptions by farmers and scientists on the benefits of EI hamper its adoption. Therefore, a change in farmer's perception of the practice will enhance its adoption. This could be achieved by providing evidence of economic benefits of EI to farmers, hence motivating its adoption.

Another challenge hampering the adoption of EI is inadequate policies that support the adoption of environmentally friendly farming practices. By putting policies that reduce the use of artificial farming inputs, promote soil fertility maintenance, support efficient water use in agriculture and enhance biodiversity conservation would ultimately enhance the adoption of EI. Additionally, public awareness of the effects of conventional farming could directly or indirectly enhance the adoption of EI. The adverse effects of climate change coupled with low yields experienced by farmers provide an opportunity for EI to be recognised as a valuable agricultural practice that would increase food production and security and reduce environmental degradation in SSA.

EI role in mitigating climate change could be used to promote its adoption and implementation in the region. Small-scale farmers could be encouraged to adopt EI for C trading as a way to increase income and to combat the effects of climate change on farmers.

References

Adugna C, Jebessa Debella H (2019) Historical wildlife policy development and major wildlife threats under the recent three regimes of Ethiopia. World J Zool 14(1):9–22. https://doi.org/10.5829/idosi.wjz.2019.09.22

Afam I, Jideani O, Netshiheni RK (2017) Selected edible insects and their products in traditional medicine, food and pharmaceutical industries in Africa: utilisation and prospects (additional information is available at the end of the chapter). https://doi.org/10.5772/intechopen.68330

Affholder F, Jourdain D, Quang DD, Tuong TP, Morize M, Ricome A (2010) Constraints to farmers' adoption of direct-seeding mulch-based cropping systems: a farm scale modelling approach applied to the mountainous slopes of Vietnam. Agric Syst 103:51–62. https://doi.org/10.1016/j.agsy.2009.09.001

Agera S (2011) Role of beekeeping in the conservation of forests. Global J Agric Sci 10(1):27–32

Agropolis (2013) DMC-an ecological intensification engineering tool. www.agropolis.org/agronomy/research.php?id=10c. Accessed 04 Mar 2020

Akinnifesi F, Kwesiga F, Mhango J, Chilanga T, Mkonda A, Kadu C, Kadzere I, Mithofer D, Saka J, Sileshi G (2006) Towards the development of Miombo fruit trees as commercial tree crops in southern Africa. Forests Trees Livelihoods 16(1):103–121

Akinnifesi FK, Sileshi G, Ajayi OC, Chirwa PW, Kwesiga FR, Harawa R (2008) Contributions of agroforestry research and development to livelihood of smallholder farmers in southern Africa: 2. Fruit, medicinal, fuelwood and fodder tree systems. Agric J 3(1):76–88

Balinga M, Mapunzu P, Moussa J, N'gasse G (2004) Contribution of forest insects to food security. The example of caterpillars in Central Africa. FAO, Rome

Banerjee A, Jhariya MK, Yadav DK, Raj A (2020) Environmental and sustainable development through forestry and other resources. Apple Academic Press Inc., CRC Press - a Taylor and Francis Group, US & Canada. ISBN: 9781771888110, p 400. https://doi.org/10.1201/9780429276026

Bommarco R, Kleijn D, Potts SG (2013) Ecological intensification: harnessing ecosystem services for food security. Trends Ecol Evol 28(4):230–238

Bond I (2013) Private land contribution to conservation in South Africa. In: Parks in transition. Routledge, pp 46–79

Bos D, Grootenhuis JG, Prins HH (2000) Financial feasibility of game cropping in Machakos District, Kenya. In: Wildlife conservation by sustainable use. Springer, pp 277–294

Bothma JDP (2002) Game ranch management. Van Schaik Publishers

Bradbear N (2009) Bees and their role in forest livelihoods: a guide to the services provided by bees and the sustainable harvesting, processing and marketing of their products. Non-wood Forest Products (19), http://www.fao.org/3/a-i0842e.pdf

Brigham T, Chihongo A, Chidumayo E (1996) Trade in woodland products from Miombo region. FAO, Rome

Brussaard L, Caron P, Campbell B, Lipper L, Mainka S, Rabbinge R, Babin D, Pulleman M (2010) Reconciling biodiversity conservation and food security: scientific challenges for a new agriculture. Curr Opin Environ Sustain 2(1–2):34–42. https://doi.org/10.1016/j.cosust.2010.03.007

Bucagu C, Vanlauwe B, Van Wijk M, Giller KE (2013) Assessing farmers' interest in agroforestry in two contrasting agro-ecological zones of Rwanda. Agrofor Syst 87(1):141–158

Campbell B, Brigham T (1993) Non-wood forest products in Zimbabwe. In: FAO/Commonwealth Science Council Regional Expert Consultation Meeting on Non-Wood Forest Products, Arusha, 17–22 October 1993

Carsan S, Stroebel A, Dawson I, Kindt R, Mbow C, Mowo J, Jamnadass R (2014) Can agroforestry option values improve the functioning of drivers of agricultural intensification in Africa? Curr Opin Environ Sustain 6:35–40

Cassman KG (2005) Ecological intensification of agriculture and implications for improved water and nutrient management. In: Imas P, Price R (eds) Fertigation: optimizing the utilization of water and nutrients. International symposium on fertigation, Beijing, 20–24 September 2005. International Potash Institute, Horgen, pp 23–34

Chidumayo EN (1997) Miombo ecology and management: an introduction. Intermediate technology. FAO, Rome

Chirwa T, Mafongoya P, Chintu R (2003) Mixed planted-fallows using coppicing and non-coppicing tree species for degraded Acrisols in eastern Zambia. Agrofor Syst 59 (3):243–251

Chirwa P, Syampungani S, Geldenhuys C (2008) The ecology and management of the Miombo woodlands for sustainable livelihoods in southern Africa: the case for non-timber forest products. Southern Forests J For Sci 70(3):237–245

CIRAD (Centre de coopération internationale en recherche agronomique pour le développement) (2008) CIRAD Strategic Vision 2008–2012. www.cirad.fr/en/content/download/970/31343/version/6/file/CIRAD_Strategie_GB_web.pdf. Accessed 3 Oct 2013

Connolly-Boutin L, Smit B (2016) Climate change, food security, and livelihoods in sub-Saharan Africa. Reg Environ Chang 16(2):385–399. https://doi.org/10.1007/s10113-015-0761-x

Daramola TM (2012) Opportunities for climate change mitigation through afforestation and reforestation of degraded lands in the sub-Saharan Africa. Carbon Sink Development Initiatives, Nigeria. Lane F5 Ponle Ojo Close Oremeta Ologuneru Ibadan, Nigeria https://pdfs.semanticscholar.org/d8d2/7a5f3804dcc6fa45a5e10f0c128b9e6a4de8.pdf

DeFoliart GR (1992) Insects as human food: gene DeFoliart discusses some nutritional and economic aspects. Crop Prot 11(5):395–399

Egger K (1986) L'intensification écologique. Conservation (LAE) et amélioration des sols tropicaux par les systèmes agro-sylvopastoraux. In: Département Systèmes Agraires du CIRAD (ed) Aménagement hydro-agricoles et systèmes de production. Collection Documents Systèmes Agraires 6, pp 129–135

El Bilali H (2019) Research on agro-food sustainability transitions: where are food security and nutrition? Food Secur 11(3):559–577. https://doi.org/10.1007/s12571-019-00922-1

Eloff T (2000) The game industry: delicately poised. SA Game Hunt 6(6):21–23

Ericksen PJ (2008) Conceptualizing food systems for global environmental change research. Glob Environ Chang 18(1):234–245

FAO (2008) Land degradation on the rise. http://www.fao.org/newsroom/en/new/2008/1000874/index.html. Accessed 09 Mar 2020

FAO (2009) Organic agriculture.glossary onorganic agriculture. FAO, Rome, pp 1–173s

Fox T, Du Plessis P (2000) Hunting in southern Africa. Afr Indigo 3:40–47

Gaba S, Bretagnolle F, Rigaud T, Philippot L (2014) Managing biotic interactions for ecological intensification of agroecosystems. Front Ecol Evol 2:29

Gama B, Otsyina R, Nyadzi G, Banzi F, Shirima D, Mumba M (2004) Improved fallows for soil fertility improvement at Tabora in western Tanzania: a synthesis. World Agroforestry Centre, Nairobi

Gamula GE, Hui L, Peng W (2013) An overview of the energy sector in Malawi. Energy Power Eng 5:8–17. https://doi.org/10.4236/epe.2013.51002

Garibaldi LA, Gemmill-Herren B, D'Annolfo R, Graeub BE, Cunningham SA, Breeze TD (2017) Farming approaches for greater biodiversity, livelihoods, and food security. Trends Ecol Evol 32(1):68–80

Garibaldi LA, Pérez-Méndez N, Garratt MP, Gemmill-Herren B, Miguez FE, Dicks LV (2019) Policies for ecological intensification of crop production. Trends Ecol Evol 34(4):282–286
Geldenhuys CJ (2010) Managing forest complexity through application of disturbance–recovery knowledge in development of silvicultural systems and ecological rehabilitation in natural forest systems in Africa. J For Res 15(1):3–13
Genschick S, Kaminski AM, Kefi AS, Cole SM (2017) Aquaculture in Zambia: an overview and evaluation of the sector's responsiveness to the needs of the poor. Penang, Malaysia: CGIAR research program on fish Agri-food systems and Lusaka, Zambia: Department of Fisheries. Working paper: FISH-2017-08
Girvetz E, Ramirez-Villegas J, Claessens L, Lamanna C, Navarro-Racines C, Nowak A, Thornton P, Rosenstock TS (2019) Future climate projections in Africa: where are we headed? In: Rosenstock TS, Nowak A, Girvetz E (eds) The climate-smart agriculture papers: investigating the business of a productive, resilient and low emission future. Springer Open
GIZ (2012) Green economy in sub-Saharan Africa-Lessons from Benin, Ethiopia, Ghana, Namibia and Nigeria. www.greengrowthknowledge.org/resource/green-economysub-saharan-africa-lessons-benin-ethiopia-ghana-namibia-and-nigeria. Accessed 09 Mar 2020
Griffin K (1999) Alternative strategies for economic development. Springer
Gumbo D, Dumas-Johansen M, Muir G, Boerstler F, Zuzhang X (2018) Sustainable management of Miombo woodlands: food security, nutrition and wood energy. FAO, Rome
Handavu F, Chirwa PW, Syampungani S (2019) Socio-economic factors influencing land-use and land-cover changes in the Miombo woodlands of the Copperbelt province in Zambia. For Policy Econ 100:75–94
Harrison N, Smith M, Chatré B (2015) Benefits of a Green economy transformation in sub-Saharan Africa. Deutsche Gesellschaft für Internationale Zusammenarbeit (GIZ) GmbH
Harsch E (1992) Drought devastates southern Africa. Drought Network News 4(2):17–19
Hartmann DL, Klein Tank AMG, Rusticucci M, Alexander L, Brönnimann S, Charabi Y, Dentener F, Dlugokencky E, Easterling D, Kaplan A, Soden B, Thorne P, Wild M, Zhai PM (2013) Observations: atmosphere and surface supplementary material. In: Stocker TF, Qin D, Plattner G-K, Tignor M, Allen SK, Boschung J, Nauels A, Xia Y, Bex V, Midgley PM (eds) Climate change 2013: the physical science basis. Contribution of Working Group I to the Fifth Assessment Report of the Intergovernmental Panel on Climate Change. www.climatechange2013.org and www.ipcc.ch
Haule C, Kanyama-Phiri G, Nyirenda M, Mafongoya P, Kwapata M (2003) Maize and biomass production, and soil fertility improvement under Sesbania sesban improved fallows in Kasungu district, central Malawi. Malawi J Agric Sci 2(1):21–33
IFAD (2013) Smallholders, food security, and the environment. United Nations Environment Programme
IPCC (2012) IPCC 2012: summary for policymakers. In: Field CB, Barros V, Stocker TF et al (eds) Managing the risks of extreme events and disasters to advance climate change adaptation, a special report of Working Groups I and II of the Intergovernmental Panel on Climate Change. Cambridge University Press, Cambridge
James R, Washington R (2013) Changes in African temperature and precipitation associated with degrees of global warming. Clim Chang 117(4):859–872
Jhariya MK (2017) Vegetation ecology and carbon sequestration potential of shrubs in tropics of Chhattisgarh, India. Environ Monit Assess 189(10):518. https://doi.org/10.1007/s10661-017-6246-2
Jhariya MK, Singh L (2020) Herbaceous diversity and biomass under different fire regimes in a seasonally dry forest ecosystem. Environ Dev Sustain 22:1–19. https://doi.org/10.1007/s10668-020-00892-x
Jhariya MK, Bargali SS, Raj A (2015) Possibilities and perspectives of agroforestry in Chhattisgarh. In: Zlatic M (ed) Precious forests-precious earth. InTech, Croatia, pp 237–257, ISBN: 978–953–51-2175-6, 286 pages. https://doi.org/10.5772/60841

Jhariya MK, Banerjee A, Meena RS, Yadav DK (2019a) Sustainable agriculture, forest and environmental management. Springer Nature Singapore Pte Ltd., Singapore. eISBN: 978-981-13-6830-1, hardcover ISBN: 978-981-13-6829-5, p 606. https://doi.org/10.1007/978-981-13-6830-1

Jhariya MK, Yadav DK, Banerjee A (2019b) Agroforestry and climate change: issues and challenges. Apple Academic Press Inc., CRC Press - a Taylor and Francis Group, US & Canada. ISBN: 978-1-77188-790-8 (hardcover), 978-0-42957-274-8 (E-book), p 335. https://doi.org/10.1201/9780429057274

Jideani AI, Netshiheni RK (2017) Selected edible insects and their products in traditional medicine, food and pharmaceutical industries in Africa: utilisation and prospects. Future Foods, 55

Jindal R, Swallow B, Kerr J (2008) Forestry-based carbon sequestration projects in Africa: potential benefits and challenges. In: Natural Resources Forum, 2008, vol 2. Wiley Online Library, pp 116–130

Kamwi JM, Chirwa PW, Manda SO, Graz PF, Kätsch C (2015) Livelihoods, land use and land cover change in the Zambezi Region, Namibia. Popul Environ 37(2):207–230

Kandji S, Verchot L, Mackensen J (2006) Climate change and variability in Southern Africa. Impacts and adaptation in the agricultural sector. ICRAF, UNEP

Khan N, Jhariya MK, Yadav DK, Banerjee A (2020a) Herbaceous dynamics and CO_2 mitigation in an urban setup – a case study from Chhattisgarh, India. Environ Sci Pollut Res 27 (3):2881–2897. https://doi.org/10.1007/s11356-019-07182-8

Khan N, Jhariya MK, Yadav DK, Banerjee A (2020b) Structure, diversity and ecological function of shrub species in an urban setup of Sarguja, Chhattisgarh, India. Environ Sci Pollut Res 27 (5):5418–5432. https://doi.org/10.1007/s11356-019-07172-w

Kirby KR, Potvin C (2007) Variation in carbon storage among tree species: implications for the management of a small-scale carbon sink project. For Ecol Manag 246(2–3):208–221

Kleijn D, Bommarco R, Fijen TP, Garibaldi LA, Potts SG, van der Putten WH (2019) Ecological intensification: bridging the gap between science and practice. Trends Ecol Evol 34(2):154–166

Kumar B (2006) Carbon sequestration potential of tropical homegardens. In: Tropical homegardens. Springer, pp 185–204

Kumar S, Meena RS, Jhariya MK (2020) Resources use efficiency in agriculture. Springer Nature Singapore Pte Ltd., Singapore. eBook ISBN: 978-981-15-6953-1, hardcover: 978-981-15-6952-4, p 760. https://doi.org/10.1007/978-981-15-6953-1

Leakey R (1996) Definition of agroforestry revisited. Agrofor Today 8:5–5

Leakey R, Tchoundjeu Z, Schreckenberg K, Simons T, Shackleton S, Mander M, Wynberg R, Shackleton C, Sullivan C (2006) Trees and markets for agroforestry tree products: targeting poverty reduction and enhanced livelihoods. World agroforestry into the future, pp 11–22

Lindsey PA, Barnes J, Nyirenda V, Pumfrett B, Tambling CJ, Taylor WA, Rolfes MS (2013a) The Zambian wildlife ranching industry: scale, associated benefits, and limitations affecting its development. PLoS One 8(12):e81761. https://doi.org/10.1371/journal.pone.0081761

Lindsey PA, Havemann CP, Lines RM, Price A, Retief T, Rhebergen T, Van der Waal C, Romañach S (2013b) Benefits of wildlife-based land uses on private lands in Namibia and limitations affecting their development. Oryx 47(1):41–53

Liniger H, Mekdaschi R (2019) Sustainable rangeland management in sub-Saharan Africa-guidelines to good practice (conference edition). TerrAfrica; World Bank; world overview of conservation approaches

Makumba W, Akinnifesi FK, Janssen B, Oenema O (2007) Long-term impact of a Gliricidia-maize intercropping system on carbon sequestration in southern Malawi. Agric Ecosyst Environ 118 (1–4):237–243

Malaisse (1997) Se nourir en foret claire africaine: approche écologique et nutritionnelle. Les Presses Agronomiques de Gembloux, Gembloux

Mbata K, Chidumayo E, Lwatula C (2002) Traditional regulation of edible caterpillar exploitation in the Kopa area of Mpika district in northern Zambia. J Insect Conserv 6(2):115–130

Mbow C, Skole D, Dieng M, Justice C, Kwesha D, Mane L, El Gamri M, Von Vordzogbe V, Virji H (2012) Challenges and prospects for REDD+ in Africa: desk review of REDD+ implementation in Africa. Global Land Project Reports 5

McCracken ME, Woodcock BA, Lobley M, Pywell RF, Saratsi E, Swetnam RD, Mortimer SR, Harris SJ, Winter M, Hinsley S (2015) Social and ecological drivers of success in agri-environment schemes: the roles of farmers and environmental context. J Appl Ecol 52 (3):696–705

Meena RS, Lal R (2018) Legumes for soil health and sustainable management. Springer Singapore, Singapore, p 541. ISBN 978-981-13-0253-4 (eBook), ISBN: 978-981-13-0252-7(hardcover). https://doi.org/10.1007/978-981-13-0253-4_10

Meena RS, Kumar V, Yadav GS, Mitran T (2018) Response and interaction of *Bradyrhizobium japonicum* and Arbuscular mycorrhizal fungi in the soybean rhizosphere: a review. Plant Growth Regul 84:207–223

Meena RS, Lal R, Yadav GS (2020a) Long term impacts of topsoil depth and amendments on soil physical and hydrological properties of an Alfisol in Central Ohio, USA. Geoderma 363:1141164

Meena RS, Lal R, Yadav GS (2020b) Long-term impact of topsoil depth and amendments on carbon and nitrogen budgets in the surface layer of an Alfisol in Central Ohio. Catena 194:104752

Meena RS, Kumar S, Datta R, Lal R, Vijaykumar V, Brtnicky M, Sharma MP, Yadav GS, Jhariya MK, Jangir CK, Pathan SI, Dokulilova T, Pecina V, Marfo TD (2020c) Impact of agrochemicals on soil microbiota and management: a review. Land (MDPI) 9(2):34. https://doi.org/10.3390/land9020034

Montagnini F, Nair P (2004) Carbon sequestration: an underexploited environmental benefit of agroforestry systems. In: New vistas in agroforestry. Springer, pp 281–295

Morris B (2004) Insects and human life. Berg, Oxford

Murthy IK, Gupta M, Tomar S, Munsi M, Tiwari R, Hegde G, Ravindranath N (2013) Carbon sequestration potential of agroforestry systems in India. J Earth Sci Clim Chang 4(1):1–7

Mutunga C, Mwangi S (2006) Inventory for ecosystem service payment in Kenya. Forest Trends, Washington DC

Niang I, Ruppel OC, Abdrabo MA et al (2014) Africa. In: Barros VR et al (eds) Food security and food production systems. Climate Change 2014: impacts, adaptation, and vulnerability. Part B: regional aspects. Contribution of adaptation and vulnerability, Working Group II Contribution to the IPCC 5th Assessment Report of the Intergovernmental Panel on Climate Change. Cambridge University Press, Cambridge, pp 1199–1265

NOAA (2018) Climate at a glance: global time series. NOAA National Centers for Environmental Information. http://www.ncdc.noaa.gov/cag/

Ntiamoa-Baidu Y (1997) Wildlife and food security in Africa, vol 33. Food & Agriculture Org, Rome

Otsuka K, Place F (2014) Changes in land tenure and agricultural intensification in sub-Saharan Africa. WIDER Working Paper

Painkra GP, Bhagat PK, Jhariya MK, Yadav DK (2016) Beekeeping for poverty alleviation and livelihood security in Chhattisgarh, India. In: Narain S, Rawat SK (eds) Innovative technology for sustainable agriculture development. Biotech Books, New Delhi, pp 429–453, ISBN: 978–81–7622-375-1

Parish F, Sirin A, Charman D, Joosten H, Minayeva T, Silvius M, Stringer L (eds) (2008) Assessment on peatlands, biodiversity and climate change: main report. Global Environment Centre/Wetlands International, Kuala Lumpur/Wageningen

Poppy G, Jepson P, Pickett J, Birkett M (2014) Achieving food and environmental security: new approaches to close the gap. Philos Trans R Soc Lond Ser B Biol Sci 369:20120272. https://doi.org/10.1098/rstb.2012.0272

Pretty J, Bharucha ZP (2014) Sustainable intensification in agricultural systems. Ann Bot 114:1571–1596

Pywell RF, Heard MS, Bradbury RB, Hinsley S, Nowakowski M, Walker KJ, Bullock JM (2012) Wildlife-friendly farming benefits rare birds, bees and plants. Biol Lett 8(5):772–775
Raj A, Jhariya MK, Yadav DK, Banerjee A, Meena RS (2019a) Agroforestry: a holistic approach for agricultural sustainability. In: Jhariya MK, Banerjee A, Meena RS, Yadav DK (eds) Sustainable agriculture, forest and environmental management. Springer Nature Singapore Pte Ltd., Singapore, pp 101–131, eISBN: 978-981-13-6830-1, hardcover ISBN: 978-981-13-6829-5. https://doi.org/10.1007/978-981-13-6830-1, p 606
Raj A, Jhariya MK, Banerjee A, Yadav DK, Meena RS (2019b) Soil for sustainable environment and ecosystems management. In: Jhariya MK, Banerjee A, Meena RS, Yadav DK (eds) Sustainable agriculture, forest and environmental management. Springer Nature Singapore Pte Ltd., Singapore, pp 189–221. eISBN: 978-981-13-6830-1, hardcover ISBN: 978-981-13-6829-5. https://doi.org/10.1007/978-981-13-6830-1, p 606
Raj A, Jhariya MK, Yadav DK, Banerjee A (2020) Climate change and agroforestry systems: adaptation and mitigation strategies. Apple Academic Press Inc., CRC Press - a Tayler and Francis Group, US & Canada. ISBN: 9781771888226, p 383. https://doi.org/10.1201/9780429286759
Rosenzweiga C, Elliott J, Deryng D, Ruane AC, Müllere C, Arneth A, Boote KJ, Folberth C, Glotter M, Khabarov N, Neumann K, Piontek F, Pugh TAM, Schmid E, Stehfest E, Yang H, Jones JW (2014) Assessing agricultural risks of climate change in the 21st century in a global gridded crop model intercomparison. PNAS 111(9):3268–3273
Sage C (2013) The interconnected challenges for food security from a food regimes perspective: energy, climate and malconsumption. J Rural Stud 29:71–80
Sasson A (2012) Food security for Africa: an urgent global challenge. Agric Food Secur 1(1):2
Schabel HG (2006) Forest entomology in East Africa: forest insects of Tanzania. Springer Science & Business Media
Scholes RJ, Montanarella L, Brainich E, Brainich E, Barger N, ten Brink B, Cantele M, Erasmus B, Fisher J, Gardner T, Holland TG, Kohler F, Kotiaho S, von Maltitz G, Nangendo G, Pandit R, Parrotta J, Potts MD, Prince S, Sankaran M, Willemen L (eds) (2018) IPBES (2018): Summary for policymakers of the assessment report on land degradation and restoration of the Intergovernmental Science - Policy Platform on Biodiversity and Ecosystem Services. Intergovernmental Science-Policy Platform on Biodiversity and Ecosystem Services, Bonn
Scurrah-Ehrhart C, Blomley T (2006) Amani butterfly forest-based enterprise, Tanga, Tanzania. Rights and Resources, unpublished paper
Serdeczny O, Adams S, Baarsch F, Coumou D, Robinson A, Hare W, Schaeffer M, Perrette M, Reinhard J (2016) Climate change impacts in sub-Saharan Africa: from physical changes to their social repercussions. Reg Environ Chang. https://doi.org/10.1007/s10113-015-0910-2
Shackleton C, Shackleton SE, Gambiza J, Nel E, Rowntree K, Urquhart P, Fabricius C, Ainslie A (2010) Livelihoods and vulnerability in the arid and semi-arid lands of southern Africa: exploring the links between ecosystem services and poverty alleviation. Nova Publishers
Sileshi G, Akinnifesi FK, Ajayi OC, Place F (2008) Meta-analysis of maize yield response to woody and herbaceous legumes in sub-Saharan Africa. Plant Soil 307(1–2):1–19
Stack J, Dorward A, Gondo T, Frost P, Taylor F, Kurebgaseka N (2003) Mopane worm utilisation and rural livelihoods in southern Africa. In: International conference on rural livelihoods, forests and biodiversity, Bonn, pp 19–20
Stott PA, Gillett NP, Hegerl GC, Karoly D, Stone D, Zhang X, Zwiers FW (2010) Detection and attribution of climate change: a regional perspective. Wiley Interdiscip Rev Clim Chang 1 (2):192–211
Syampungani S (2009) Vegetation change analysis and ecological recovery of the Copperbelt Miombo woodland of Zambia. University of Stellenbosch, Stellenbosch
Syampungani S, Chirwa P (2011) Miombo woodland productivity: the potential contribution to carbon sequestration and payment for environmental services in east and southern Africa. Ecology, Management and Conservation, Woodlands, pp 187–201

Syampungani S, Chirwa P, Akinnifesi F, Ajayi O (2010) The potential of using agroforestry as a win-win solution to climate change mitigation and adaptation and meeting food security challenges in southern Africa. Agric J 5(2):80–88

Syampungani S, Geldenhuys CJ, Chirwa PW (2016) Regeneration dynamics of Miombo woodland in response to different anthropogenic disturbances: forest characterisation for sustainable management. Agrofor Syst 90(4):563–576

Syampungani S, Tigabu M, Matakala N, Handavu F, Oden PC (2017) Coppicing ability of dry Miombo woodland species harvested for traditional charcoal production in Zambia: a win–win strategy for sustaining rural livelihoods and recovering a woodland ecosystem. J For Res 28 (3):549–556

Taylor A, Lindsey P, Davies-Mostert H, Goodman P (2016) An assessment of the economic, social and conservation value of the wildlife ranching industry and its potential to support the green economy in South Africa. The Endangered Wildlife Trust, Johannesburg, pp 96–109

The Economist (2011) Africa's impressive growth. www.economist.com/blogs/dailychart/2011/01/daily_chart. Accessed 06 Mar 2020

The World Bank Brief (2014) Sustainable aquaculture. https://www.worldbank.org/en/topic/environment/brief/sustainable-aquaculture. Accessed 06 Mar 2020

Thresher RE (1980) Reef fish: behavior and ecology on the reef and in the aquarium. Palmetto Publishing Company

Tilman D, Balzer C, Hill J, Befort BL (2011) Global food demand and the sustainable intensification of agriculture. Proc Natl Acad Sci 108(50):20260–20264

Tittonell P, Klerkx L, Baudron F, Félix GF, Ruggia A, van Apeldoorn D, Dogliotti S, Mapfumo P, Rossing WA (2016) Ecological intensification: local innovation to address global challenges. In: Sustainable agriculture reviews. Springer, pp 1–34

UNEP (2011) Towards a green economy-pathways to sustainable development and poverty eradication. www.unep.org/greeneconomy/Portals/88/documents/ger/ger_final_dec_2011/Green%20EconomyReport_Final_Dec2011.pdf. Accessed 06 Mar 2020

Van der Merwe P, Saayman M (2003) Determining the economic value of game farm tourism. Koedoe 46(2):103–112

van Huis A (2003) Insects as food in sub-Saharan Africa. Insect Sci Appl 23:163–185

van Huis A, Vantomme P (2014) Conference report: insects to feed the world. Food Chain 4:184–193

van Huis H, van Gurp H, Dicke M (2013a) Potential of insects as food and feed in assuring food security. Annu Rev Entomol 58:563–583

van Huis A, van Itterbeeck J, Klunder H, Mertens E, Halloran A, Muir G, Vantomme P (2013b) Edible insects: future prospects for food and feed security. FAO Forestry Paper 171. FAO, Rome

Wagner S, Huth F, Mohren F, Herrmann I (2013) Silvicultural systems and multiple service forestry (Chapter 1.5). In: Integrative approaches as an opportunity for the conservation of forest biodiversity, p 64

Wezel A, Soboksa G, McClelland S, Delespesse F, Boissau A (2015) The blurred boundaries of ecological, sustainable, and agroecological intensification: a review. Agron Sustain Dev 35 (4):1283–1295

White A, Scherr S, Khare A (2004) For services rendered: the current status and future potential of markets for the ecosystem services provided by tropical forests. International Tropical Timber Organization (ITTO)

Whitmee S, Haines A, Beyrer C, Boltz F, Capon AG, de Souza Dias BF, Ezeh A, Frumkin H, Gong P, Head P (2015) Safeguarding human health in the Anthropocene epoch: report of the Rockefeller Foundation–Lancet Commission on planetary health. Lancet 386 (10007):1973–2028

World Bank (2009) Living Standard Measurement Study (LSMS) surveys. LSMS – Integrated Surveys on Agriculture (LSMS–ISA). www.worldbank.org/lsms-isa. Accessed 04 Mar 2020

Yhoung-Aree J, Viwatpanich K, Paoletti M (2005) Edible insects in the Laos PDR, Myanmar, Thailand, and Vietnam. In: Ecological implications of Minilivestock, pp 415–440
Young A (2002) Effects of trees on soils–special supplement on agroforestry. The Natural Farmer
Zitzmann G (1999) Multiple use and livelihood strategies in Mopane woodland: the case of Ditladi, north-east district, Botswana. Insitut für Internationale Forst-und Holzwirtschaft Technische Universitat
Zuberi T, Thomas KJ (2012) Demographic projections, the environment and food security in sub-Saharan Africa. Regional bureau for Africa, United Nations Working Group 22

Eco-Intensified Breeding Strategies for Improving Climate Resilience in Goats

18

V. Sejian, M. V. Silpa, S. S. Chauhan, M. Bagath, C. Devaraj, G. Krishnan, M. R. Reshma Nair, J. P. Anisha, A. Manimaran, S. Koenig, R. Bhatta, and F. R. Dunshea

Abstract

Goats are projected as an ideal climate adapted animal because of the various advantages associated with this species such as higher thermo-tolerance, drought tolerance, ability to survive on limited pastures, and high disease resistance. With

V. Sejian (✉) · M. Bagath · C. Devaraj · G. Krishnan · R. Bhatta
ICAR-National Institute of Animal Nutrition and Physiology, Adugodi, Hosur Road, Bangalore, Karnataka, India

M. V. Silpa
ICAR-National Institute of Animal Nutrition and Physiology, Adugodi, Hosur Road, Bangalore, Karnataka, India

Institute of Animal Breeding and Genetics, Justus-Liebig-Universität Gießen, Gießen, Germany

S. S. Chauhan · F. R. Dunshea
School of Agriculture and Food, Faculty of Veterinary and Agricultural Sciences, The University of Melbourne, Melbourne, VIC, Australia
e-mail: ss.chauhan@unimelb.edu.au; fdunshea@unimelb.edu.au

M. R. R. Nair
ICAR-National Institute of Animal Nutrition and Physiology, Adugodi, Hosur Road, Bangalore, Karnataka, India

Academy of Climate Change Education and Research, Kerala Agricultural University, Vellanikkara, Thrissur, Kerala, India

J. P. Anisha
College of Veterinary and Animal Sciences, Kerala Veterinary and Animal Sciences University, Mannuthy, Thrissur, Kerala, India

A. Manimaran
Southern Reginal Station, National Dairy Research Institute, Adugodi, Bangalore, India

S. Koenig
Institute of Animal Breeding and Genetics, Justus-Liebig-Universität Gießen, Gießen, Germany
e-mail: Sven.Koenig@agrar.uni-giessen.de

M. K. Jhariya et al. (eds.), *Ecological Intensification of Natural Resources for Sustainable Agriculture*, https://doi.org/10.1007/978-981-33-4203-3_18

the projected climate associated adversities such as high temperature, shrinking grazing lands, less feed, fodder and water resources, and emerging new diseases, goat farming seems to be the more profitable enterprise. With its exceptional adaptability to the adversities of climate change, goat farming seems to be the better option available for poor and marginal farmers across the globe to ensure their livelihood and food security. The main challenges associated with goat production are the improvement of the productive potential of the indigenous breeds and the conservation of indigenous germplasm. The phenotypic traits coat color, respiration rate, rectal temperature, skin temperature, thyroid hormones, and other genotypic markers such as heat shock proteins and thyroid hormone receptors are considered reliable markers of metabolic adaptation of goat during heat. Study involving genome-wide selection signatures and genomic inbreeding has identified MTOR, MAPK3, SLC27, NR2F6, and DRD2 genes for thermal adaptation in goat. Further research efforts can help in identifying agro-ecological zone-specific goat breeds to be disseminated to the local farmers for obtaining optimum economic return. In addition, refinements in existing breeding programs to develop eco-intensified breeding strategies by incorporating traits governing production, adaptation, and low methane emission could revolutionize goat sector reducing the impact on the ecosystem.

Keywords

Breeding · Biomarkers · Climate change · Eco-intensification · Goat · Heat stress

Abbreviations

ACACA	Acetyl-coenzyme A carboxylase alpha
CBBP	Community-based breeding programs
CSN3	Kappa casein
EI	Ecological intensification
ELFI	Electrochemical lateral flow immunosensor
ELISA	Enzyme linked immunosorbent assay
eNOS	endothelial type III NOS
GEBV	Genomic estimated breeding value
GH	Growth hormone
GHG	Greenhouse gas
GS	Genomic selection
GWAS	Genome-wide association study
Hb	Hemoglobin
HS	Heat stress
HSP	Heat shock proteins
HSPBAP1	Heat shock 27 kDa associated protein 1
MAS	Marker assisted selection
ME1	malic enzyme 1

Ne-ELISA	Nanoelectronic-enzyme linked immunosorbent assays
NEFA	Non-esterified fatty acids
NOS	Nitric oxide synthase
PBMC	Peripheral blood mononuclear cell
PCV	Packed cell volume
PPAR	Peroxisome proliferator-activated receptor
RT-qPCR	Quantitative reverse transcription polymerase chain reaction
SDS-PAGE	Sodium dodecyl sulfate–polyacrylamide gel electrophoresis
SNP	Single nucleotide polymorphism
SPGE	Screen-printed gold electrodes
T3	Triiodothyronine
T4	Thyroxine
TLR	Toll like receptors

18.1 Introduction

Climate change is no longer a fictional concept; it is a true event, the consequences of which humans must face. The over-exploitation of natural resources by humans has altered the ecological balance thereby resulting in rapidly changing climatic scenarios (Raj et al. 2020; Banerjee et al. 2020; Khan et al. 2020a, 2020b). Agriculture is one of the most influential sectors which can ensure human survival and the livestock sector plays a crucial role in it. The livestock sector is predicted to ensure the food security of the alarmingly rising human population (Thornton et al. 2007; Meena and Lal 2018). However, this sector also faces severe threat due to climate change and therefore ameliorative and mitigation measures must be enforced so as to sustain the livestock farming (Sejian 2013; Meena et al. 2018). Goats (*Capra hircus*) are the earliest domesticated animals from the ancestors of wild bezoar goat (*Capra aegagrus*) near Middle East and Western Asia and its versatility made rapid spread to the other parts of the world along with human migration (Zeder 2005; Tresset and Vigne 2011). At present, there are more than 1000 breeds of goats all over the world including commercial and indigenous breeds (Kim et al. 2019). Among all the domestic livestock species, goats are the ideal climate adaptive species as they adapt well to all the extreme climatic conditions. In addition to this, they support the livelihood of many small and marginal farmers across the globe. The livestock sector holds a unique characteristic of contributing to climate change and at the same time being affected by it. Hence, utmost care should be taken while enforcing adaptation strategies against climate change in livestock production such that it does not cause any ecological imbalances.

Over the past several decades, livestock breeders have been practicing several breeding strategies to boost livestock production. Undoubtedly most of them have succeeded in meeting set targets while few of such attempts failed. The biotechnological advancement in recent years has also resulted in the development of improved breeding programs wherein the genetic information of the animal is incorporated along with its phenotypic and pedigree information. These improved the breeding programs such as marker assisted selection (MAS) and genomic selection (GS), which yields quicker, accurate, and higher selection gains when compared to the conventional breeding programs (Rashamol et al. 2019). Although these methodologies have remarkably improved production, they have been adopted well in only a few livestock species like dairy cattle, beef cattle, pigs, and chicken. Limited studies have been reported in goats with extremely minimal research on adaptive traits in goats. With the rising issue of climate change, researchers from various regions across the globe are now focusing on adaptation and low methane emission traits. Moreover, these traits should also be considered while framing the breeding programs for specific region so as to develop agro-ecological zone-specific climate resilient goat breeds.

Animal breeding and conservation are often said to be two sides of a coin, in that improvement in one is to the detriment of the other. While, animal breeding is undoubtedly beneficial, intensive crossbreeding breeding program has altered the natural livestock diversity leading to the dilution of native indigenous breeds in many parts of the world. Indigenous goat breeds possess excellent adaptive traits to thrive during climatic extremes and therefore breeding strategies should focus on exploiting these traits. However, such strategies should also focus on eco-intensified breeding approaches to improve goat farming and ensure long-term sustainability. The aim of ecological intensification (EI), as stated by Cassman (1999), is "further intensification of production systems that satisfy the anticipated increase in food demand while meeting acceptable standards of environmental quality." Achieving eco-intensification would ensure increased production with decreased production losses and damage to the ecology (Kumar et al. 2020; Jhariya et al. 2019a, 2019b). The EI primarily targets the efficiency in improvement of the key resources such as the indigenous goats, production norms, and standards or their substitution, via ecological technological solutions and innovations (Dubeuf et al. 2018). This concept thereby would look at a holistic approach to amalgamate the benefits of other suitable farming systems and options to maintain efficient production and reduce losses. Therefore, in the current erratically varying climatic condition, eco-intensified goat breeding approaches would be the ideal strategy.

18.2 Climate Change and Goat Farming Scenario across the Globe

Goat enterprise is gaining significance from economic point of view globally over the last decades with the current goat population of over 860 billion (Gupta et al. 2013). A vast proportion of economically weaker section of the farmers around the globe relies on goat farming to ensure their livelihood (Meena et al. 2020a, b). Small ruminants, specifically goats, possess distinctive biological features like rapid growth rate, shorter gestation period, higher prolificacy, improved feed conversion efficiency, relatively higher disease resistance capacity, and better marketability, when compared to other ruminants and livestock thereby making them relevant for subsistent agricultural systems (Sejian et al. 2019).

Several local breeds of goats that are well adapted to climate, feed, fodder, and enzootic diseases have evolved in different agro-ecological niches through the process of natural selection. They are well known for maintaining their production potential despite harsh environmental conditions. For example, Osmanabadi goat breeds which originated from semi-arid regions in India have better adaptability by altering their physiological responses to heat, nutritional, and combination of both these stressors (Shilja et al. 2016). Similarly, the Salem black goat is found to be more superior in adapting to heat stress conditions than Malabari and Osmanabadi breeds in Southern India making it better suited for climate resilient farming (Pragna et al. 2018b).

Goat enterprise plays a significant role in sustaining livestock production in the changing climate (Darcan and Silanikove 2018). Goats are the most resilient among the domesticated ruminant species. The advantageous characteristics of goats that allow them to thrive in harsh environments include their adaptability to broad environmental conditions, thermo-tolerance, drought tolerance and their ability to prosper on a wide range of pastures, forages, and browse. They have better morphological, physiological, and metabolic features providing them the ability to thrive in various agro-ecological zones. Goats can live in mountains and areas with soils poor in nutrition. Their anatomical advantage of upper lip and browsing habit enable them to flourish with limited feedstuffs of low quality. Goats reared in tropical region dig up the soil in search of immersed parts of plants for consuming during feed scarcity period in dry seasons. Goat also contributed towards protecting the environment by controlling the weed, diversity maintenance, and mitigating the adverse impact of changing climate through their ingestive behavior and ability to select their preferred diet (Miller and Lu 2019).

Goat farming is characterized by its low initial investment and high turnover rate. Climate change is associated with increased incidences of drought, flood, and diseases and goats which are resistant to these are preferred under changing climatic scenarios. Goat adapts themselves during conditions of drought through their ability to conserve water by desiccating their feces, concentrating their urine, and reducing the evaporative water loss and ability to utilize the rumen as a water reservoir (Silanikove 2000). The low body mass and metabolic rate of goats help them to minimize their maintenance and water requirements and hence allows them to thrive in adverse climatic conditions with limited feed and water availability (Pragna et al. 2018a). Heat stressed Osmanabadi goat breed had higher plasma concentration of aldosterone hormone which controls electrolyte balance in animals. Because of their browsing diet, these goats require less water and moreover the rumen plays the crucial role of water reservoir during periods of dehydration. The plasma cortisol concentration was also higher in these breeds exposed to multiple stresses highlighting the immense potential of this species to counter cumulative impacts of different environmental stresses (Shilja et al. 2017). Goats are efficient converters of low-quality feeds into economically valuable products such as meat and milk and thrive well on low-quality pastures and in addition to producing less methane per unit of feed consumed than other ruminant species. There is an increasing demand for goat milk and products globally because of their higher digestibility and lipid metabolism when compared to cow milk, which thereby enables health and nutritional benefits. Additionally certain consumers also prefer goat milk over cow milk due to their taste. Apart from producing milk, goats also offer additional benefits to producers like skin, meat, manure, and fiber in some instances. The low cost of investment, easy maintenance, and ability to thrive in adverse climatic conditions and moreover the high profit returns make goats the ideal animal under changing climatic conditions.

18.3 Goat as Ideal Climate Animal Model

Agriculture is considered as a vital sector, providing food, fiber, and energy to meet and satisfy the demands of an increasing human population. Among all the allied sectors of agriculture, livestock plays a significant role in ensuring optimum economic return for economically weaker sections of the farming community. However, climate change acts as a potential threat to curb animal agriculture with a direct influence on their growth, milk production, meat production, reproduction, and health (Sejian 2013). Besides the direct impacts, climate change also hampers the livestock production through non-availability of feed, fodder, and water resources as well as through sudden disease outbreaks (Sejian et al. 2015).

High ambient temperature along with the effect of humidity impairs the availability of forages by affect both the quality and quantity, especially in tropical countries where indigenous animals are predominantly distributed. The reasons for the potential depletion of pasture resources are due to the occurrence of extreme events, pandemic disease outbreaks, and elevated CO_2 concentration (IPCC 2000). This can significantly hamper the nutritional requirements of livestock.

Besides being susceptible to the reduced forage availability, animals are exposed to the heavy risk of drought. Livestock which are extensively reared in tropical regions are extremely vulnerable to the adverse effects of drought. The indicator for water use, the water footprint, associated with the growth of forages, milking, servicing water and milk production indicates the quantum of water requirement in each component of animal agriculture (Ibidhi and Salem 2020). As per the earlier estimates at the beginning of this century, global livestock water demand is expected to increase 71% by 2025, with the greatest increase in consumption occurring in developing countries (Bruinsma 2003; Rosegrant et al. 2002). As the dry period progresses, animals are forced to mobilize their available body fat reserve to cope with the nutritional deficiency, eventually resulting in reduction of livestock production. It is therefore important that the scientific fraternities should put their efforts to find the most climate resilient animal cutting across the livestock species so that it can be disseminated to the poor and marginal farmers to maintain their livelihood in challenging climatic conditions.

Goats are opportunistic feeders and thus the depletion of pasture lands may hardly impact their feed base. In addition, selective feeding behavior of goat helps them to consume even the poor quality forages and have the ability to convert them into high-quality products (Dossa et al. 2015). Moreover, goats exhibit a bipedal stance which enables them to get access to tree leaves and this is considered advantageous as compared to other livestock species. Further, goat has the better feed conversion efficiency than other ruminant species (Silanikove 2000). In addition, goats do not require specialized shelter structures and they could ideally survive in any location with minimum protection from the weather variables. Therefore, goats are projected to be the future animal model for ensuring food security.

18.4 Goat as Most Significant Livelihood for Small and Marginal Farmers

Goats are often considered as the alternate to cow especially to the farmers with poor economic resources ensuring them a sustainable economic return. They believe to survive with human over several thousands of years and are known to be the animals of choice for resource poor and marginal farmers (Aziz 2010). Apart from being an enterprise to ensure a livelihood, the rural farmers rear goats in an attempt to maintain their good old traditional life (Singh et al. 2013).

In the context of the anticipated increase in human population, goats can play a substantial part in catering for the nutritional demands of future generations through the production of milk and meat (Kumar et al. 2010). The demand for goat meat and milk has been rising exponentially and surpassed other livestock species for their purported health benefits and therapeutic values and thus the goat products are becoming popular (Raut and Kurpatwar 2020). In recent years, goat enterprises have turned out to be of more commercial value as a result of the marketing preference of goat products all over the world (Kumar et al. 2006). Thus, having the potential scope to ensure the nutritional security, goat production was considered an important source of economy for the poor farmers.

Apart from their role in bringing nutritional security, goats are reared for their socio-economic as well as management advantages when compared to other livestock species (Shivakumar et al. 2020). In perspectives of less initial investment and low maintenance cost, the often landless rural farmers, who cannot afford the resource demands of large ruminants, depend on goats for their livelihood. In contrast to the large ruminants, goats can be kept with limited housing facilities which are usually made with the residues obtained from crop harvest (Kumar and Upadhyay 2009). Also, the nutritional requirements of goats are less than large ruminants (Kumar et al. 2010). They can survive and produce efficiently on poor quality forages which thereby reduce their production expenses. Additionally, the labor requirement for goat rearing is less and can often be met effectively by family members. Because of the smaller size and ease in handling these animals, women, and children are involved in management of goats, which decreases the overall production costs in goat rearing (Paul et al. 2020).

Moreover, in the context of ecological footprint, goat rearing can be considered as a viable form of animal agriculture and can be integrated with other livestock species, agricultural crops, and trees (Herrero et al. 2009). They are the ideal animal for mixed farming for their versatile nature to adapt to intensive, semi-intensive, and extensive systems of rearing (Kumar et al. 2010). It was identified that most of the rural farmers rely on goats for risk aversion and economic security to cope with climate change associated crop failures.

18.5 Challenges Associated with Goat Production from Climate Change Perspectives

The climate change is expected to have far reaching adverse consequences especially in developing part of world particularly due to the lack of advanced infrastructure to tackle the situation (Darcan and Silanikove 2018; Naqvi and Sejian 2011). Sustaining livestock production to meet the increased demands of growing human population is a great challenge in the twenty-first century (Osei-Amponsah et al. 2019). Although goat can be an important source of economy to the rural poor they are also are vulnerable to the adverse impacts of climate change (Feleke et al. 2016).

Out of the 351 goat breeds found across the world, India possess a vast diversity of 23 goat breeds. These breeds acquired their adaptive potential to specific agro-ecological conditions through natural selection and specific breeding programs over the years (Hoffmann 2013). The loss of natural habitat and introduction of exotic breeds is a threat to many of these indigenous breeds and some are on the verge of extinction. Shrinkage of grazing land, changing agricultural patterns, decrease in natural browse and cross breeding with other breeds are the reasons for the declining populations of Jamnapari, Beetal, Barbari, Surti, Jakhrana, and Changthangi breeds of goats. Several other indigenous breeds in southern, eastern, and Himalayan regions are also considered as threatened breeds (Mandal et al. 2014). This warrants appropriate research efforts to conserve these indigenous germplasms which could help in reversing the rural poverty through their extreme climate resilient capacity.

Climate change has adverse effect on goat farming as it has a great impact on the production and quality of fodder and the availability of water as well as large pasture lands (Rojas-Downing et al. 2017). Apart from the increased incidences of natural calamities such as flood and drought, threats to animal health and productivity are also associated with climate change. The farmers usually sell their animals before or during the crisis eventually resulting in an economic loss (Thornton 2010). Substantial benefits are possible with proper adaptation technologies to cope with situations of changing climate and rearing livestock are tipped to be one such important strategy to ensure economic return to farmers (Nardone et al. 2010).

Building farmer's capacity through new technologies, management practices, knowledge, and education to adapt immediate environmental change is one of the coping mechanisms to combat climate change (Abdul-Razak and Kruse 2017). Technological developments such as new forage varieties, efficient weather forecasting, and effective water management strategies also have to be adopted. Also, supplemental irrigation and diversification of farm production can help to meet the challenges associated with environmental variability and economic risks (Giridhar and Samireddypalle 2015).

18.6 Climate Resilient Traits in Goats

Heat stress, consequence of climate change is detrimentally affecting the growth, production, reproduction, and health of ruminants (Joy et al. 2020). However, small ruminants, especially goats, have unique adaptive traits which make them tolerant and able to survive in extreme harsh environmental condition via changes in behavioral, morphological, physiological, and genetic bases (Bernabucci et al. 2010; Berihulay et al. 2019). Therefore, goats are more resilient to climate change than other ruminants (Silanikove 2000). This resilience is influenced by species ecology, phenotypic, and genetic diversity. These traits are categorized broadly into phenotypic and genotypic traits. Table 18.1 describes both the phenotypic and genotypic climate resilient traits in goats.

Table 18.1 Climate resilient traits in goats

Climate resilient traits		
Phenotypic traits		References
Morphological	Coat thickness, coat color and length, long legs and ears, body size, shape, sweat gland capacity	Naandam and Assan (2014)
Behavioral	Shade seeking, eating, drinking	Garner et al. (2017), Sejian et al. (2019)
Physiological	Respiration rate, heart rate, skin temperature, rectal temperature, sweating rate	Marai et al. (2007), Okoruwa (2014), Shilja et al. (2016)
Haemato-biochemical	Hemoglobin, PCV, NEFA, cortisol, T3, T4	McManus et al. (2009), Okoruwa (2014), Shilja et al. (2017), Aleena et al. (2018)
Genotypic traits		
Thermo-tolerant genes	HSP110, HSP100, HSSSSP90, HSP70, HSP60, HSP 40, HSP10, and small HSPs. HSPA1 A, HSPA6, HSPA8, HSPA1 L, and HSPA2 FGF2, GNAI3, PLCB1, GRID2 MTOR, and MAPK3	Banerjee et al. (2014), Angel et al. (2018), Lv et al. (2014), Kemper et al. (2014), Elbeltagy et al. (2016)
Genes associated with growth pathways	GH, GH receptor, insulin like growth factor-1, thyroid hormone receptor, leptin and its receptor, BMP2, BMP4, GJA3, GJB2	Banerjee et al. (2014), Angel et al. (2018), Onzima et al. (2018)
Genes associated with immunity	TLR1, TLR2, TLR3, TLR5, TLR6, TLR8, and TLR10 GRIA1, IL2, IL7, IL21, IL1R1, IL10RB, and IL23A	Contreras-Jodar et al. (2018), Kim et al. (2016), Sophia et al. (2016, 2017), Vandana et al. (2019)

18.6.1 Phenotypic Traits

Identification of climate resilient phenotypic markers would aid in the selection of superior goat breeds, which can adapt to harsh environmental conditions and produce effectively (Hoekstra 2006). In general, animals respond to varying environment by altering their phenotypic characteristics like behavior, physiology, biochemistry, and metabolism which govern the adaptive potential of an animal (Sejian et al. 2019).

Goats possess diverse phenotypic characters which enable them to thrive in harsh environment. The coat characteristics are considered to be important traits aiding in assessing the climate resilience in farm animals (Naandam and Assan 2014). Generally, the skin temperatures of animals with dark coat color are higher as compared to light colored coat. Thermal insulation increases by increasing the coat depth which increases the air captured between the hair fibers (Helal et al. 2010). The long hair serves as an insulator from the heat, providing an air buffer zone between the outer environment and the animal's body. Skin provides the next line of protection against solar radiation. Dark color skin is preferred since it will absorb any UV light penetrate the coat thereby protecting the underlying tissues (Naandam and Assan 2014).

The behavior of the animal is an important phenotypic trait that could be used as an indicator to assess the severity of climatic stress (Garner et al. 2017). Shade seeking, reduced feed intake, and increased water intake and increased frequency of drinking are considered as typical behavior of goats under heat stress condition (Sejian et al. 2019). Decline in feed intake was also observed in three indigenous goat breeds of South India, Osmanabadi, Malabari, and Salem Black on exposure to heat stress (Shilja et al. 2016; Aleena et al. 2018). Such variations in feed intake could be attributed to their adaptive mechanism to reduce metabolic heat load.

Physiological variables commonly evaluated to study the adaptability of small ruminants to hot environment are respiratory rate, rectal temperature, heart rate, and skin temperature (Marai et al. 2007; Shilja et al. 2016). During hot environmental conditions animals increase their respiratory rate so as to enhance evaporative heat loss and maintain homeothermy. High skin temperature favors heat dissipation in animals whereas increased rectal temperature during heat stress is indicative of the animal's inability to maintain homeothermy (Marai et al. 2007; Renaudeau et al. 2012). However, rectal temperature of goats was found to be higher during heat stress exposure (Marai et al. 2007; Okoruwa 2014) which has been considered as natural mechanism to protect from heat stress damage. Increased heart rate or pulse rate indicates increased blood flow from core to the periphery resulting in higher heat loss through sensitive and insensitive mechanism (Marai et al. 2007; Shilja et al. 2016).

Blood biochemical variables are important tools used to evaluate heat stress in animals. The prime biochemical indicators for heat stress in goats are hemoglobin (Hb), packed cell volume (PCV), and blood cortisol (Okoruwa 2014). Heat stressed animals exhibit severe signs of panting which aids to meet the increased oxygen demand. This process, however, leads to water loss from the animal thereby leading to dehydration. The consequence of this mechanism is reflected on the blood biochemical variables as increased Hb and PCV (McManus et al. 2009). Higher levels of cortisol are beneficial to the animal since cortisol is considered as a principal stress reliever in ruminant species which helps them to cope up with heat stress condition (Shilja et al. 2017).

Metabolic hormones such as triiodothyronine (T3) and thyroxine (T4) are important for metabolic adaptation and growth performance of animals (Aleena et al. 2018). Reduced plasma concentrations of T3 and T4 was observed in Osmanabadi, Malabari, and Salem Black goat breeds on exposure to heat stress which was proposed to be an effort to limit the basal metabolism thereby reducing the metabolic heat production. Non-esterified fatty acid (NEFA) is another important metabolic regulator reported to have reduced levels heat stress. Lower NEFA levels are indicative of increased lipid utilization by liver. Moreover, this scenario concurs with reduced glucose levels which further complicate the adaptive ability in heat stressed animals (Aleena et al. 2018). Therefore, T3, T4, and NEFA are proposed to serve as reliable markers of metabolic adaptation of goat during heat stress.

18.6.2 Genotypic Traits

The genetic basis of thermo-tolerance at the physiological and cellular levels in ruminant species have been well documented (Osei-Amponsah et al. 2019). Goats adapt to different ecological conditions and maintain their productive performance effectively in harsh environment. Numerous candidate genes have been identified in goats that are highly associated with thermal tolerance (Banerjee et al. 2014; Angel et al. 2018). Genetic diversity within a population offers flexibility to acclimate to the changing climate and it is crucial for their survival over the period. The genetic basis of thermal adaptation is more complex and mediated by network of several genes. The role of several genes to determine an individual's ability to adapt to environmental stress has been established in goats (Lv et al. 2014; Kemper et al. 2014).

There are several genes associated with heat tolerance and investigation of these candidate genes in farm animals, especially in goats are very much needed for future genetic selection in tropical conditions (Hassan et al. 2019; Sejian et al. 2019). Therefore, it is very much important to select a resistant goat breed which is particularly to harsh environments and drought. Important traits in such genetic selection schemes include skin and coat type, sweat gland capacity, ability to maintain production and reproduction rate, diseases resistance, metabolic heat production, drought tolerance, anatomic and morphologic structure (Darcan et al. 2009). Indigenous goat breeds possess high resistance to diseases and ecto-parasites and are well adapted to harsh environment. Thus, this quality could give them additional shield against vector borne diseases which increase due to varying climatic conditions (Silanikove 2000).

Heat shock proteins (HSP) are commonly expressed group of proteins that protect cells against heat stress. The HSP family consists of HSP110, HSP100, HSSSSP90,

HSP70, HSP60, HSP 40, HSP10, and small HSPs (Feder and Hofmann 1999). Among the HSPs, HSP70 encoded by the HSP70 gene is an important molecular chaperone of primary importance to all mammalian cells. It plays a significant role in protecting cells, tissues, and organs during heat stress by promoting the folding and assembly of nascent polypeptides, refolding of the misfolded or aggregated proteins, preventing protein aggregation and degradation of misfolded proteins (Turturici et al. 2011; Hassan et al. 2019). Nucleotide variations in 5′-UTR and the promoter region of HSP70 are associated with various productive and reproductive traits in animals. These highly polymorphic regions in the HSP70 genes associated with thermo-tolerance and performance traits make them potential candidate for marker assisted selection to develop climate resilient animals with superior thermo-tolerance with enhanced productive potential (Hassan et al. 2019). Therefore, thermo-tolerance related HSPs could be used as potential biomarkers in advanced breeding programs to develop thermo-tolerant breeds (Archana et al. 2017).

Heat stress decreases the productive and reproductive efficiency of small ruminants. Growth traits in farm animals are quantitative variables which are regulated by several genes. Genes associated with growth pathways which are downregulated in goats during heat stress are genes encoding growth hormone (GH), GH receptor, insulin like growth factor-1, thyroid hormone receptor, leptin and its receptor (Angel et al. 2018). Downregulation of these genes could be attributed to the thermal tolerance capacity of goats. Heat stress hampers the function of immune cells which would possibly make the immune system of heat stressed goats less capable of defending against diseases (Contreras-Jodar et al. 2018). Different toll like receptors (TLR1, TLR2, TLR3, TLR5, TLR6, TLR8, and TLR10) are highly expressed in the spleen of heat stressed Osmanabadi goats (Sophia et al. 2017). These TLR expressions in spleen indicate that it could be used as potential markers during heat stress in goats. In another study, TLR 8 and TLR10 gene expression increased significantly in liver tissues of Osmanabadi goats (Sophia et al. 2016). In a similar study, Vandana et al. (2019) reported that TLR2 gene highly expressed in mesenteric lymph node while other TLRs (TLR1, TLR4, TLR5, and TLR6) were downregulated during heat stress. Likewise, TLR expression differs in various tissues and this effect is breed specific to heat stress.

18.7 Significance of Identifying Agro-Ecological Zone-Specific Goat Breeds

The ecosystem benefits human and plays a great role in the agricultural and livestock production systems. The ecosystem supports the living creatures by providing food, water, genetic resources through regulation of climate and precipitation pattern along with maintenance of cultural inheritance (De Groot et al. 2010). Over time, the geographical expansion of goats to a broad range of environmental conditions covering hot to cold climates, humid to dry climates, and tropical rainforests to hypoxic high-altitude regions (Guo et al. 2019). The long-term selection for morphological characters has resulted in the evolution of many goat breeds with diverse phenotypic traits such as coat color, horn shape, and hair type specific to agro-climatic conditions of their habitat (Amills et al. 2017). The natural selection plays an important role in the selection of an individual that are highly adapted to new environmental conditions (Naderi et al. 2008). The selection processes resulted in different breeds emerging that are specific for milk, meat and fiber or dual-purpose breeds in different regions of the world (Brito et al. 2017). Therefore, goats have the potential to adapt to varying divergent environments with the development of specific characteristics (Kim et al. 2019). Benjelloun et al. (2015) reported that the indigenous goat breeds of desert areas developed a specific characteristic of intermittent gasping to regulate body temperature. On the other hand, the high-altitude indigenous goats have acquired a unique feature of oxygen-sensing capacity to adapt to hypoxic conditions (Wang et al. 2016). Further, the goat breeds adapted to the tropical environments have enhanced immune competence to resist infections and parasites (Onzima et al. 2018).

Goats are significant component of livestock production system with high adaptability to unfavorable climatic conditions which can survive on available fodder and pasture grasses (Fonseka et al. 2018). Goats play a vital role in providing the livelihood of rural populations particularly those in the hills and mountains, tropical and semi-arid regions which adapt to various farming systems, climatic conditions, and terrains and produce high-quality products form low-quality resources (Morales et al. 2019). In general, goats reared under smallholder farming systems are indigenous breeds of populations of ecotypes according to the geographical region (Mdladla et al. 2018). The native breeds are sturdy and perform better under specific environmental conditions than the high producing exotic breeds (Lozano-Jaramillo et al. 2019). The indigenous goat breeds have evolved by natural selection processes and acquiring specific qualities that make them more adaptable to their respective environment (Guo et al. 2018). Hence, native breeds are locally well adapted to specific agro-ecological environment with disease resistance capacity (Mirkena et al. 2010). These characteristics impart to them the potential for sustainable production under the varying agro-climatic conditions in their natural habitat (Morales et al. 2019). Goats possess better feed conversion efficiency with high dry matter and fiber digestibility that facilitates survival in arid regions with poor vegetation. In developed countries, most goats are of high-yielding breeds meant for dairy production while in developing countries particularly in tropical regions they are used for meat and fiber production (Amills et al. 2017). The genetic resources of agro-ecological zone-specific goats are essential to ensure to the livelihoods of rural communities under a changing climatic scenario (Monau et al. 2020). The agro-ecological zone-specific breeds have great variation in adaptability, productivity, feed utilization, and disease resistance and therefore it is vital to conserve the unique characteristics of these indigenous breeds.

In a series of studies conducted in indigenous Osmanabadi, Malabari, and Salem Black goats, attempts were made to test their ability to survive in multiple locations (Pragna et al. 2018b; Afsal et al. 2019; Savitha et al. 2019; Rashamol et al. 2019). Based on several variables such as growth, adaptation, meat production, and immune system related variables it was established that Salem Black breed possessed the better ability to survive in multiple locations (Pragna et al. 2018c; Archana et al. 2018; Vandana et al. 2019; Aleena et al. 2018). This breed was able to maintain their growth and meat production apart from adapting to the heat stress challenges. Therefore, it was concluded that Salem Black breed could be propagated to different agro-ecological zones in Southern India. This may help the resource poor and marginal farmers to ensure their livelihood security. Such research efforts pertaining to identifying the best breed among the various indigenous breeds is the current need to sustain livestock production in the changing climate scenario.

18.8 Time for Revisiting the Breeding Programs in Goats

Breed substitution for indigenous breeds with exotic breed and cross breeding caused dismal performance in the tropical countries (Kosgey et al. 2006). Disadvantages in breeding programs include the larger superior animals which are the first to be sold in the market as they fetch higher prices leading to negative selection (Ojango et al. 2015). Similarly, during the breeding process the genetic selection in male is done using few males, while no selection is done with females. Thus, the risk of higher inbreeding prevails in small flocks with a small population size (Gatenby 1986). Hence, during breed selection within the indigenous small ruminant's biodiversity needs to be maintained. Genetic diversity is a prerequisite for environmental stability and also for food security and only genetically conserved genotypes will have the ability to sustain food production now and, in the future (Drucker and Scarpa 2003). Breeding programs involving crossbreeding, upgrading, and breed replacement targeted only the production traits without considering the threats imposed on indigenous genetic resources especially the indigenous well adapted genotype. Whereas in dairy breeds, the exotic breeds adaptation to harsh weather and production was not researched in depth in extreme weather conditions (Rewe et al. 2002). Selection within the breed ranges from 0.5 to 3% per year, but such change is small, cumulative, and permanent (Smith 1984). However, the recent breeding programs must look into those breeds that are most suitable during drought conditions, periods of feed and water scarcity, disease resistance without compromising the productive quality. For example, Community-Based Breeding Programs (CBBP) was launched in Ethiopia concentrating the smallholder farmers targeting the genetic improvement within their livestock (Gizaw et al. 2011). The same program was launched in Liberia to improve the productivity and profitability of indigenous breeds without compromising their resilience and genetic integrity, and without high cost interventions (Karnuah et al. 2018). Breeding strategy changes can produce better breeds with heat tolerant and diseases resistance with improved growth, development, and reproduction (Henry et al. 2012). Hence additional research is needed to have better producing breeds without compromising the local adaption traits. In future policies that improve the adaptive capability of the breed will be crucial. Similarly, Colditz and Hine emphasized on the breeding of animals with improved resilience capability with respect to their respective stressors (Colditz and Hine 2016). One must look for the lower GHG producing animals without much alteration in the production traits wise choosing the animal for selection also. For a successful breeding program one has to look into the futuristic approaches of selecting adapted breeds with heat resistance, higher feed conversion efficiency, better drought and water tolerability, disease resistance, adapt to rainfall fluctuations, lower methane emission for the next 30–50 years.

18.9 Eco-Intensification Practices for Breeding in Goats

Eco-intensification primarily focuses on intensifying the production system by improving yield while at the same time taking into consideration not to harm the environment. This is an important system especially in the current varying climate change scenario wherein improving, or at least maintaining, the quality of the environment could enable to reduce the adversities of climate change (Meena et al. 2020c; Raj et al. 2019a, 2019b). However, this system has been studied more in broad acre agriculture as compared to livestock (Gomes et al. 2014). Studies should also focus on evaluating the ecological intensification in goats as they are ideal species for future farming practices wherein drastic decline in cultivable land is predicted.

Eco-intensification of agriculture collates the principles of integrated livestock farming, conservation, pest management, and biodiversity preservation. Similar strategies may also be followed in goat farming especially while programming their breeding policies. Conventional breeding strategies have been focusing on improving the productive traits in goat. Though this has boosted the production system, it has severely hampered the native goat breeds which were neglected. Therefore, it is time to revisit the existing breeding policies and take up efforts to upscale the indigenous goat germplasm. These indigenous goat breeds are known for their hardy nature and possess excellent adaptive traits which must be uplifted. Amalgamation of production and adaptation traits could ensure optimal outputs in the current climate change scenario. Moreover, from the eco-intensification point of view, it would be suggested to look beyond production traits and consider adaptation traits in the breeding programs. In addition to this indigenous goat breeds having the resilience to perform well to different agro-ecological zones should be promoted (Sejian et al. 2019). This would therefore ensure maximum output to the poor marginal small-scale farmers in terms of economic returns.

Another important trait that is also gaining more importance especially from the environmental sustainability aspect is methane emission. The relative contribution of livestock to total anthropogenic GHG emissions is 9–11%, in which a major proportion accounts from enteric methane emission. Though goats emit relatively low enteric methane when compared to other ruminant species, further reduction in these levels can ensure improved feed consumption efficiency. Therefore, in a true sense, an ideal eco-intensified goat breeding approach should amalgamate adaptation, low methane emission, and productivity traits. This would ensure a long-term solution to the threat on goat farming due to climate change.

18.10 Prospects of Eco-Intensified Breeding Strategies

The eco-intensified breeding strategies could enable the livestock industry to meet the desired farm goal. Implementing such breeding strategies ensures optimum production in livestock. The eco-intensified breeding strategies increase the productive potential by ensuring desired output per unit of feed consumed. This will help the farm managers to get optimum return for the input cost they are investing. Implementing such eco-intensified breeding strategies could increase the productive response of the individuals through genetically driven feed conversion and digestibility efficiency of the animals which will impart them the potential to convert the low or medium quality feed resources into desired products. This may ensure the appropriate economic return for the farm households. Further, such eco-intensified breeding strategies also could be cost effective as the input cost can be minimized to a greater extent through increased productive efficiency of the farm animals. This cost effectiveness could be attributed to the fact that these animals would consume less feed to bring per unit of output by acquiring sound genetic merit through such breeding programs. In addition, the eco-intensified breeding strategies also can ensure proper protection for the ecosystem as it ensures producing minimal GHGs from the enterprise.

18.11 Breeding Strategies for Improving Climate Resilience in Goats

Livestock breeders across the globe have worked extensively to develop appropriate breeding strategies to improve the production. Until a few years back, selection strategies primarily focused on phenotypic traits and genetic potential of an animal was evaluated based on the performance of pedigree and phenotype records. Though these strategies caused a revolution all over world by increasing the production potential of animals, they have their own pros and cons. The existing conventional selection strategies are time consuming and could be effectively applied only for certain traits having good heritability. Most of the economically important traits have medium to low heritability, this limits their application in conventional breeding programs as their selection accuracy and gain are often low.

Decades ago, climatologists had predicted environmental conditions to worsen in the near future and had also sent the warning to all stakeholders across the globe. Unfortunately, this warning was not taken up seriously by the sectors, including the livestock sector. The livestock sector being both a contributor and being affecting by climate change should be on notice. It took scientists years to prove the impact of climate change on livestock production and now the stage has been set to take this further to effectively counter the adversities. Researchers have reported several ameliorative and mitigation strategies to counter climate change, particularly heat stress. Among them the most promising and long lasting approach would be to improve the climate resilience capacity of livestock via improved breeding strategies. Breeding strategies to develop climate resilient goats will not alter the

Table 18.2 Breeding strategies for developing climate resilient goats

Breeding strategies	Significance	References
Marker assisted selection	Identification of biomarkers	Hayes and Goddard (2010)
Genome-wide association study	Incorporation of desired biomarkers to predict breeding value	
Genomic selection	Screens the whole genome to identify all markers and their association with the climate resilient traits	Hayes and Goddard (2010); Shumbusho et al. (2013); Nguyen et al. (2016)
	Find out genomic estimated breeding value for adaptive traits	

basis of the existing breeding programs rather it will only include variables and traits pertaining to climate change adaptation. The target should be the selection within breed to enhance the production performance of locally adapted breed and propagation of those animals as breeding stock to produce goats.

Similar to dairy cattle breeding, goat breeding has also followed the conventional breeding methods to improve productivity mainly by incorporating exotic germplasm into the local flocks. Though this has significantly increased their productive ability, it has also lead to a serious dilution of the existing herd in addition to the loss of the valuable indigenous goats. These indigenous goats have the vital adaptability trait which is the need of the hour. Thus in the current climate change scenario, breeders are urged to conserve the local germplasm, revive them and also their superior adaptation traits into the goat breeding program. Inclusion of adaptive traits into conventional breeding programs has a lot of constraints due to its low heritability and extremely low selection gain. Hence researchers are now targeting marker assisted selection (MAS) and genomic selection (GS) to develop climate resilient animals.

The biotechnological advancement in recent years have led to the development of various tools which aid in the identification of potential biomarkers for heat stress in goats (Aleena et al. 2018; Sejian et al. 2019). The MAS predicts the breeding values after incorporating the marker information for the trait. This broadly involves two steps, firstly identification of significant markers by genome-wide association study (GWAS) followed by incorporation of significant markers to predict the breeding value (Hayes and Goddard 2010). Although this methodology is now being practiced in other livestock species, there are comparatively fewer reports on goats. MAS have aided researchers to identify several potential biomarkers for heat stress in livestock. Thus, there is a need to use this methodology to explore the adaptation related genetic markers in goats. Table 18.2 describes the different breeding strategies for developing climate resilient goats.

Genomic selection (GS) can be simplified as an advanced approach after MAS. While MAS selects only those markers which are significant in GWAS, GS screens the entire genome of the population to identify all markers, their association with the trait and finally derive a genomic estimated breeding value (GEBV). This methodology is predicted to set a new milestone in the field of animal genetics and breeding. The GS has two main advantages over MAS, the first one being the inclusion of all markers from the SNP panel irrespective of its significance. This ensures to track all the genetic variance for a trait (Hayes and Goddard 2010). The second benefit of selecting all markers in GS is that it avoids any biasness of marker effects. Though GS has not been widely practiced, it is still expected to revolutionize breeding programs in livestock. In goats, GS was proved to increase the annual genetic gain for production traits like meat and dairy traits (Shumbusho et al. 2013). When compared to other selection approaches GS can accelerate genetic gain by reducing the generation interval and also the GEBV for adaptive traits are more accurate than the progeny only breeding value estimates (Nguyen et al. 2016). Therefore, GS is a promising approach to develop climate resilient goat breeds specific for each agro-ecological zone.

While considering heat stress adaptation traits, another trait of similar importance which also must be looked at is low methane emission trait. As mentioned earlier, livestock are one of the contributors to climate change mainly via enteric methane emission. Though goats have a relatively lower enteric methane emission when compared to other livestock species, reducing the methane emission levels would indirectly increase the productivity of the animal. Therefore, improved breeding strategies should necessary include both adaptation and low methane emission traits in addition to production traits. This not only would ensure optimum production by the animals but would also reduce the carbon and water footprint thereby ensuring eco-intensified goat farming. Figure 18.1 describes the eco-intensified breeding strategies for improving climate resilience in goats.

18.12 Advances in Genomics and Proteomics Tools for Identifying Thermo-Tolerance in Goats

Identification of few or individual genes of interest are the most common to identify the thermo-tolerant goats. However, thermo-tolerance identification can also be done with the help of different omics approaches using genomic, transcriptomic, proteomic, and metabolomics. However, literature shows that relative expression of genes using RT-qPCR (Quantitative reverse transcription PCR) is the most commonly used method to identify thermo-tolerant breed (Banerjee et al. 2014; Rout et al. 2016; Aleena et al. 2018; Nagayach et al. 2017; Madhusoodan et al. 2020). RT-qPCR was the prominently used biotechnological tool to study the thermo-tolerance in goats based on the target genes. The PBMC mRNA expression of HSPA1 A, HSPA6, and HSPA8 and HSPA1 L and HSPA2 was studied in Gaddi, Barbari, Sirohi, and Chegu Indian breed of goats using RT-qPCR (Banerjee et al. 2014). Similarly RT-qPCR was used to study liver HSP70 mRNA expression to identify the best thermo-tolerant breed of goats among Barbari, Jamunapari, Jakhrana, and Sirohi during HS (Rout et al. 2016). Likewise the same technique

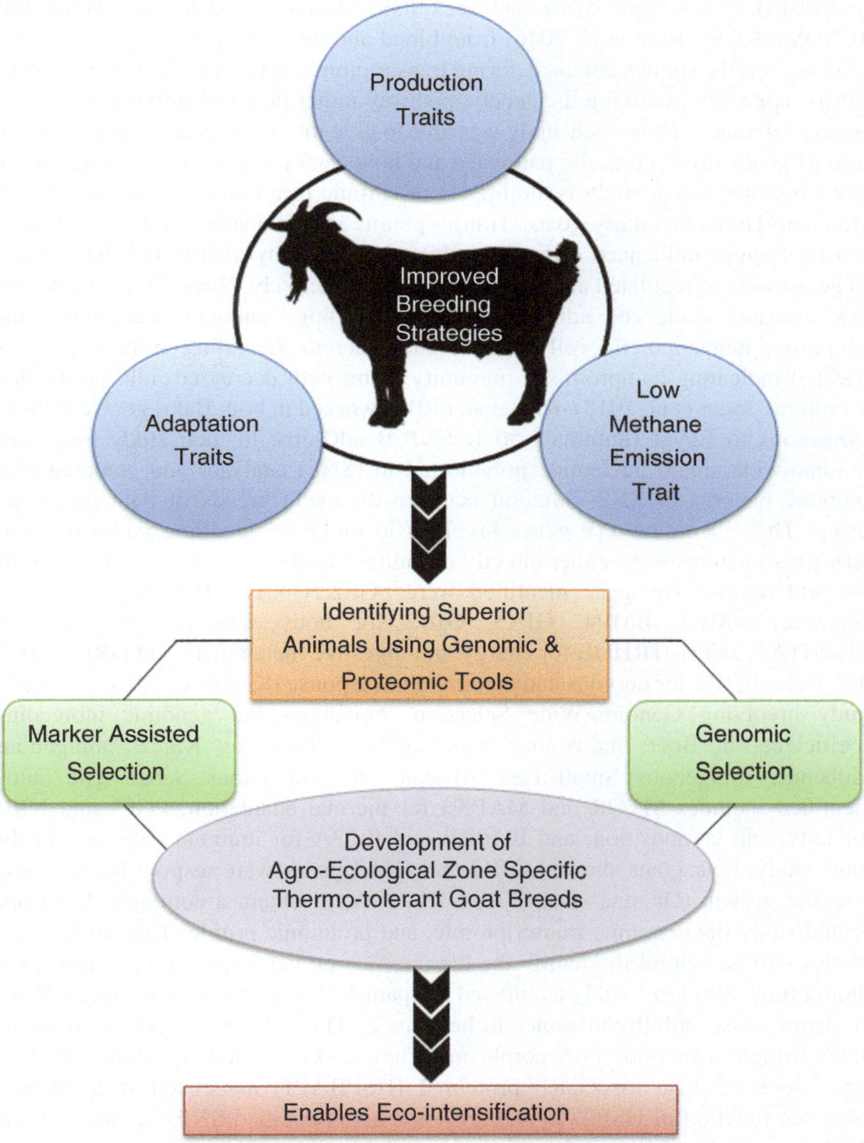

Fig. 18.1 Eco-intensified breeding strategies for improving climate resilience in goats

was used to study the relative gene expression of thermo-tolerant genes like HSP60 (Dangi et al. 2015), HSP70 (Dangi et al. 2012; Shilja et al. 2016), goats (Chaidanya et al. 2017; Aleena et al. 2018; Archana et al. 2018), HSP90 (Rout et al. 2016; Dangi et al. 2015), type II Nitric oxide synthase (NOS) (Rout et al. 2016), endothelial type III NOS (eNOS) (Rout et al. 2016) from blood and different organs.

However, the studies obtained during transcriptomic approach were able to give a holistic approach involving the genetic pathway rather than the individual genes of thermo-tolerance. Thus, such study was able to give more clarification and details of various genes involved in the pathways and how each pathway are affected during heat stress. One such study is the impact of chronic heat stress on the multiparous Murciano-Granadina dairy goats. Transcriptomic studies using microarray showed that heat stress influenced different pathways. The study also found that 55 and 88 genes were upregulated and downregulated, respectively. Nucleotide metabolism was activated while cell adhesion, PPAR signalling, leukocyte transendothelial migration, hematopoietic cell lineage, and calcium signalling were negatively affected indicating compromised immunity along with decreased milk production (Contreras-Jodar et al. 2018). Kim et al. (2016) worked in both Bakri goat and sheep indigenous to Egypt Illumina's 50 K SNP BeadChips. In their study they used genome-wide single nucleotide polymorphism (SNP) analyses and compared the genomic patterns of SNP variation between divergent breeds of both goats and sheep. They identified 119 genes involved in multiple signalling pathways. The selected signatures were either directly or indirectly related to adaption traits in the hot arid region. The gene identified were FGF2, GNAI3, PLCB1 for thermo-tolerance; BMP2, BMP4, GJA3, GJB2 for body size and development; ALDH1A3, MYH, TRHDE for energy and digestive metabolism and GRIA1, IL2, IL7, IL21, IL1R1 for nervous and autoimmune response (Kim et al. 2016). Similarly study involving Genome-Wide Selection Signatures and genomic inbreeding coefficients in Boer, indigenous Karamojong, indigenous Kigezi, indigenous Mubende, indigenous Small East African, and indigenous Sebei. The study identified includes MTOR and MAPK3 for thermal adaptation, FGF9 and IGF1 for fatty acid composition, and IL10RB and IL23A for immune response. In the same study, boar goats showed GJB2 and GJA3 genes were responsible for body size and growth (Onzima et al. 2018). However, to obtain a complete detail one should study the genomic, transcriptomic, and proteomic profile. Thus such omics studies will be helpful to identify the thermo-tolerant genotypes and phenotypes at shorter time. Similarly, study conducted in Spanish Florida dairy goats using GWAS to identify susceptibility/tolerance to heat stress. They identified several important SNPs (Single nucleotide polymorphism). They also identified significant SNPs at heat shock 27 kDa associated protein 1 (HSPBAP1) associated with thermo-tolerance function of HSP27 at cell level and also identified SNPs in kappa casein (CSN3), acetyl-coenzyme A carboxylase alpha (ACACA), and malic enzyme 1 (ME1) which are involved in milk production (Zidi et al. 2014). Another study using GWAS was conducted using 50 K Illumina SNP Beadchips from 394 and 366 Egyptian desert sheep and goats, and 895 and 464 non-desert sheep and goats, respectively. The study identified several candidate regions exhibiting selection signatures, genomic selection, and thermo-tolerant genes in desert stress conditions. Important thermo-tolerant genes identified were GRID2 neurotransmitter receptor as genes associated with HS, the SLC27, NR2F6, and DRD2 associated with heat generation and detection of temperature stimulus or homeostatic processes (Elbeltagy et al. 2016).

18.13 Policy, Planning, and Legal Perspectives

The government should work towards the goal of developing policy and legal framework for breeding goats for thermo-tolerance. Implementation of the refined policy and legal framework needs to be improved. The complex relationship between goat farming and country planning needs to be modeled to facilitate an approach that allows changes to production systems. All these innovations emphasize the need to articulate scientific and expert knowledge. Implementation of the new rural development policy will force the government to continue to increase its support to the goat sector through direct payments and rural development programs. This will create an even better environment for the development of goat production in the country. The government must also plan properly to improve the goat breeding policy. This can be achieved by the following the simple framework of: (i) developing animal registration and data management systems for goat breeding; (ii) goat farmers should be educated about animal health and breeding strategies; (iii) male and female goats of high genetic capacity should be provided to farmers; and (iv) goat farmers should be supported in adopting modern enterprise systems. Apart from government agencies, involvement of private sector, non-government organizations, local co-operatives, self-help groups, and self-sustainable community-based breeding programs can strengthen market linkages in goat sector. Strengthening the existing infrastructure along with technical input and skilled manpower is very crucial for achieving the breeding objectives in the goat sector. Further to promote goat farming, efforts are needed to develop legal framework pertaining to (i) facilitating establishment of a goat cryoconservation bank; (ii) allowing goats for grazing in forest areas by drawing a clear cut grazing plan by the Ministry of Forestry; and (iii) restricting the use of hormonal treatments. All these measures by the government may help to refine the existing breeding policies in goat sector.

18.14 Future Proteomic Approaches in Identification of Thermo-Tolerant Breeds

Future proteomic approaches provide easy and faster identification methods than nucleic acid based methods in identification of thermo-tolerant breed. Compared to nucleic acid approaches proteomic approaches were used less frequent. Absolute quantification of thermo-tolerant proteins was estimated using ELISA. Quantification of HSP70, HSP90 in goats subjected to heat stress was studied by few researchers (Nagayach et al. 2017; Rout et al. 2016, 2018). The HSP70 was characterized using SDS-PAGE, immunodetection, and quantification by optic densitometry (Meza-Herrera et al. 2006).

Identification of thermo-tolerant breed using proteomic approach is neglected in goats and sheep than larger ruminants. However, just extending the below mentioned technology to all the ruminants species will be a boon at the farm level as primary screening method. Lateral flow is usually a qualitative one, however, if it is a quantitative one then one can easily find the concentration of protein or its range in the field itself and to implement mitigation measures at farm level without involvement of lab. One such test that can be extrapolated for HSP70 and marker proteins, where instead of coating single gold bound monoclonal antibody to the strip, ascending concentration of antibody is coated from lower concentration to higher concentration at logarithmically spaced sample concentrations from the sample side to the opposite end. This graded ladder bar format could be used for quantification of sample concentrations, while retaining the same chemistry of lateral flow assay using gold nano particles (Gasperino et al. 2018). Lateral flow assay test to identify the presence of HSPs especially for the presence of HSP70 antibody can be a very easy and rapid test under field condition. However, Electrochemical Lateral Flow Immunosensor (ELFI), a combination of screen-printed gold electrodes (SPGE) with lateral flow test strip is a hybrid for both detection and quantification of dengue NS1 protein (Sinawang et al. 2016). Such technology can be extended for detection and quantification of different thermo-tolerant proteins in goats. Such tests can be very powerful primary tool to screen and identify the thermo-tolerance capacity in a very large group. Similarly, advancement in the ELISA methods will also be helpful in the identification of the protein in a physiological solution at lower concentration. For example, Nanowire field effect transistors sense the charges of the molecule, thus enabling it to find not only small molecules but also DNA and proteins too. During ELISA's enzymatic conversion of bound substrate mediates pH changes proportionally to the bound ligand concentration. Field effect transistors configured as pH sensors can detect protein in physiologically buffered solution. These nanoelectronic-enzyme linked immunosorbent assays (Ne-ELISA) are promising tools of future that can find the concentration even at picro grams/ml levels (Stern et al. 2010). Thus Ne-ELISA can be extrapolated to detect HSPs and other thermo-tolerant protein markers.

18.15 Concluding Remarks

The goat undoubtedly needs to be the priority species in the changing climate scenario. The low input cost and better output in terms of secured economy for the poor and marginal farmers makes goat rearing a vital livelihood option for the economically weaker sections of population. The various advantages associated with goat rearing over other species and its natural adaptive mechanisms makes them suitable for rearing in any agro-ecological zones of the world. Although there are several advances associated with goats still the species has been a neglected one as compared to cattle and sheep. This warrants more research efforts in this species. The organized research efforts particularly from the conservation of indigenous breeds in this species will be an important step forward for the researchers. A systematic approach in designing appropriate research methodologies involving intensive selection can improve the productive performance of this neglected species and this could help the farming community a lot. The greater progress that has been made in the field of biotechnology may revolutionize breeding programs in goat sector and this may pave way for either uplifting the genetic merits of the indigenous germplasm which may help to develop more thermo-tolerant goat breeds. More studies also should focus on evaluating the ecological intensification in goats as they are ideal species for future farming practices especially from the predicted loss of climate change associated feed and fodder resources. With the advantage of producing less methane per unit of feed consumed by goats, introducing such traits which governs the enteric methane emission into the existing breeding programs could revolutionize goat production in addition to maintaining the stability of the environment. As a way forward scientific fraternities must ensure that such breeding policies involving intensive selection within indigenous breeds are developed and put into use in near future to get the best out this once neglected species.

References

Abdul-Razak M, Kruse S (2017) The adaptive capacity of smallholder farmers to climate change in the northern region of Ghana. Climate Risk Mgt 17:104–122

Afsal A, Bagath M, Sejian V, Krishnan G, Beena V, Bhatta R (2019) Effect of heat stress on HSP70 gene expression pattern in different vital organs of Malabari goats. Biol Rhythm Res:1–15. https://doi.org/10.1080/09291016.2019.1600270

Aleena J, Sejian V, Bagath M, Krishnan G, Beena V, Bhatta R (2018) Resilience of three indigenous goat breeds to heat stress based on phenotypic traits and PBMC HSP70 expression. Int J Biometeorol 62(11):1995–2005

Amills M, Capote J, Tosser-Klopp G (2017) Goat domestication and breeding: a jigsaw of historical, biological and molecular data with missing pieces. Anim Genet 48:631–644

Angel SP, Bagath M, Sejian V, Krishnan G, Bhatta R (2018) Expression patterns of candidate genes reflecting the growth performance of goats subjected to heat stress. Mol Biol Rep 45 (6):2847–2856

Archana PR, Aleena J, Pragna P, Vidya MK, Niyas APA, Bagath M, Krishnan G, Manimaran A, Beena V, Kurien EK, Sejian V (2017) Role of heat shock proteins in livestock adaptation to heat stress. J Dairy Veterinary Animal Res 5(1):1–8

Archana PR, Sejian V, Ruban W, Bagath M, Krishnan G, Aleena Manjunathareddy GB, Beena V, Bhatta R (2018) Comparative assessment of heat stress induced changes in carcass traits, plasma leptin profile and skeletal muscle myostatin and HSP70 gene expression patterns between indigenous Osmanabadi and Salem black goat breeds. Meat Sci 141:66–80

Aziz MA (2010) Present status of the world goat populations and their productivity. World 861 (10782):1

Banerjee D, Upadhyay RC, Chaudhary UB, Kumar R, Singh S, Polley S, Mukherjee A, Das TK, De S (2014) Seasonal variation in expression pattern of genes under HSP70. Cell Stress Chaperones 19(3):401–408

Banerjee A, Jhariya MK, Yadav DK, Raj A (2020) Environmental and sustainable development through forestry and other resources. CRC Press, Boca Raton, FL, p 400. https://doi.org/10.1201/9780429276026

Benjelloun B, Alberto FJ, Streeter I, Boyer F, Coissac E, Stucki S (2015) Characterizing neutral genomic diversity and selection signatures in indigenous populations of Moroccan goats (*Capra hircus*) using WGS data. Front Genet 6:107. https://doi.org/10.3389/fgene.2015.00107

Berihulay H, Abied A, He X, Jiang L, Ma Y (2019) Adaptation mechanisms of small ruminants to environmental heat stress. Animals 9(3):75

Bernabucci U, Lacetera N, Baumgard LH, Rhoads RP, Ronchi B, Nardone A (2010) Metabolic and hormonal acclimation to heat stress in domesticated ruminants. Animal 4(7):1167–1183

Brito LF, Kijas JW, Ventura RV, Sargolzaei M, Porto-Neto LR, Cánovas A, Feng Z, Jafarikia M, Schenkel FS (2017) Genetic diversity and signatures of selection in various goat breeds revealed by genome-wide SNP markers. BMC Genomics 18:229. https://doi.org/10.1186/s12864-017-3610-0

Bruinsma J (2003) In: Bruinsma J (ed) World agriculture: towards 2015/2030 an FAO perspective. FAO, Earthscan Publications, London, UK

Cassman KG (1999) Ecological intensification of cereal production systems: yield potential, soil quality, and precision agriculture. Proc Natl Acad Sci U S A 96:5952–5959

Chaidanya K, Soren NK, Sejian V, Bagath M, Kurien EK, Manjunathareddy GB, Varma G, Bhatta R (2017) Impact of heat stress, nutritional stress and combined (heat and nutritional) stresses on rumen associated fermentation characteristics, histopathology and HSP70 gene expression in goats. J Anim Behav Biometeorol 5:36–48

Colditz IG, Hine BC (2016) Resilience in farm animals: biology, management, breeding and implications for animal welfare. Anim Prod Sci 56:1961–1983. https://doi.org/10.1071/an15297

Contreras-Jodar A, Salama AA, Hamzaoui S, Vailati-Riboni M, Caja G, Loor JJ (2018) Effects of chronic heat stress on lactational performance and the transcriptomic profile of blood cells in lactating dairy goats. J Dairy Res 85(4):423–430

Dangi SS, Gupta M, Maurya D, Yadav VP, Panda RP, Singh G, Mohan NH, Bhure SK, Das BC, Bag S (2012) Expression profile of HSP genes during different seasons in goats (Capra hircus). Trop Anim Health Prod 44(8):1905–1912

Dangi SS, Gupta M, Dangi SK, Chouhan VS, Maurya VP, Kumar P, Singh G, Sarkar M (2015) Expression of HSPs: an adaptive mechanism during long-term heat stress in goats (Capra hircus). Int J Biometeorol 59(8):1095–1106

Darcan KN, Silanikove N (2018) The advantages of goats for future adaptation to climate change: a conceptual overview. Small Ruminant Res 163:34–38

Darcan NK, Karakök SG, Daşkıran I (2009) Strategy of adapting Turkish animal production to global warming, 1. National symposium of drought and desertificiation, 14–16 May 2009, Konya-Turkey

De Groot RS, Alkemade R, Braat L, Hein L, Willemen L (2010) Challenges in integrating the concept of ecosystem services and values in landscape planning, management and decision making. Ecol Complex 7:260–272

Dossa LH, Sangaré M, Buerkert A, Schlecht E (2015) Production objectives and breeding practices of urban goat and sheep keepers in West Africa: regional analysis and implications for the development of supportive breeding programs. Springerplus 4(1):281

Drucker AG, Scarpa R (2003) Introduction and overview to the special issue on animal genetic resources. Ecol Econ 45(3):315–317

Dubeuf JP, Morales FDAR, Guerrero YM (2018) Evolution of goat production systems in the Mediterranean basin: between ecological intensification and ecologically intensive production systems. Small Ruminant Res 163:2–9

Elbeltagy AR, Kim ES, Rischkowsky B, Aboul-naga AM, Mwacharo JM, Rothschild MF (2016) Genome-wide analysis of small ruminant tolerance to grazing stress under Arid Desert. Animal Ind Rep 662(1):67

Feder ME, Hofmann GE (1999) Heat-shock proteins, molecular chaperones, and the stress response: evolutionary and ecological physiology. Annu Rev Physiol 61(1):243–282

Feleke FB, Berhe M, Gebru G, Hoag D (2016) Determinants of adaptation choices to climate change by sheep and goat farmers in northern Ethiopia: the case of southern and Central Tigray, Ethiopia. Springerplus 5:1692

Fonseka WSR, Mahusoon MM, Narmhikaa K (2018) The rearing system of goats in Mahaoya veterinary range in Ampara District, Sri Lanka. Int Res J Biological Sci 7(12):26–31

Garner JB, Douglas M, Williams SRO, Wales WJ, Marett LC, DiGiacomo K, Leury BJ, Hayes BJ (2017) Responses of dairy cows to short-term heat stress in controlled-climate chambers. Animal Prod Sci 57(7):1233–1241

Gasperino DJ, Leon D, Lutz B, Cate DM, Nichols KP, Bell D, Weigl BH (2018) Threshold-based quantification in a multiline lateral flow assay via computationally designed capture efficiency. Analytical Chem 90(11):6643–6650

Gatenby RM (1986) Sheep production in the tropics and sub-tropics. Longman Press, New York, p 351

Giridhar K, Samireddypalle A (2015) Impact of climate change on forage availability for livestock. In: Climate change impact on livestock: adaptation and mitigation. Springer, Cham, pp 97–112

Gizaw S, Kassahun A, Alemu Y (2011) A practical guide on village-based sheep and goat cooperative breeding schemes. Technical Bulletin No. 42. Addis Ababa, Ethiopia: ESGPIP

Gomes L, Pailleux JP, Dedieu B, Claudete R, Cournut S (2014) An approach for assessing the ecological intensification of livestock systems. International Farming Systems Association, Berlin, Germany

Guo J, Tao H, Li PLL, Zhong T, Wang L, Ma J, Chen X, Song T, Zhang H (2018) Whole-genome sequencing reveals selection signatures associated with important traits in six goat breeds. Sci Rep 8:10405. https://doi.org/10.1038/s41598-018-28719-w

Guo J, Zhong J, Li L, Zhong T, Wang L, Song T, Zhang H (2019) Comparative genome analyses reveal the unique genetic composition and selection signals underlying the phenotypic characteristics of three Chinese domestic goat breeds. Genet Sel Evol 51:70. https://doi.org/10.1186/s12711-019-0512-4

Gupta M, Kumar S, Dangi SS, Jangir BL (2013) Physiological, biochemical and molecular responses to thermal stress in goats. Int J Livest Res 3(2):27–38

Hassan FU, Nawaz A, Rehman MS, Ali MA, Dilshad SM, Yang C (2019) Prospects of HSP70 as a genetic marker for thermo-tolerance and immuno-modulation in animals under climate change scenario. Anim Nutr 5:340–350

Hayes B, Goddard M (2010) Genome-wide association and genomic selection in animal breeding. Genome 53(11):876–883

Helal A, Hashem ALS, Abdel-Fattah MS, El-Shaer HM (2010) Effect of heat stress on coat characteristics and physiological responses of Balady and Damascus goats in Sinai, Egypt. American-Eurasian J Agric Environ Sci 7(1):60–69

Henry B, Charmley E, Eckard R, Gaughan JB, Hegarty R (2012) Livestock production in a changing climate: adaptation and mitigation research in Australia. Crop Pasture Sci 63 (3):191–202

Herrero M, Thornton PK, Gerber P, Reid RS (2009) Livestock, livelihoods and the environment: understanding the trade-offs. Curr Opin Environ Sustain 1(2):111–120

Hoekstra HE (2006) Genetics, development and evolution of adaptive pigmentation in vertebrates. Heredity 97(3):222–234

Hoffmann I (2013) Adaptation to climate change – exploring the potential of locally adapted breeds. Animal 7(s2):346–362

Ibidhi R, Salem HB (2020) Water footprint and economic water productivity assessment of eight dairy cattle farms based on field measurement. Animal 14(1):180–189

IPCC (Intergovernmental Panel on Climate Change) (2000) In: Watson RT, Noble IR, Bolin B, Ravindranath NH, Verardo DJ, Dokken DJ (eds) Special report on land use, land use change and forestry. Cambridge University Press, Cambridge, UK

Jhariya MK, Banerjee A, Meena RS, Yadav DK (2019a) Sustainable agriculture, forest and environmental management. Springer, Singapore. https://doi.org/10.1007/978-981-13-6830-1

Jhariya MK, Yadav DK, Banerjee A (2019b) Agroforestry and climate change: issues and challenges. CRC Press, Boca Raton, FL, p 335. https://doi.org/10.1201/9780429057274

Joy A, Dunshea FR, Leury BJ, Clarke IJ, DiGiacomo K, Chauhan SS (2020) Resilience of small ruminants to climate change and increased environmental temperature: a review. Animals 10:867

Karnuah AB, Dunga G, Rewe T (2018) Community based breeding program for improved goat production in Liberia. MOJ Curr Res Rev 1:216–221

Kemper KE, Saxton SJ, Bolormaa S, Hayes BJ, Goddard ME (2014) Selection for complex traits leaves little or no classic signatures of selection. BMC Genom 15(1):246

Khan N, Jhariya MK, Yadav DK, Banerjee A (2020a) Herbaceous dynamics and CO_2 mitigation in an urban setup- a case study from Chhattisgarh, India. Environ Sci Pollut Res 27(3):2881–2897. https://doi.org/10.1007/s11356-019-07182-8

Khan N, Jhariya MK, Yadav DK, Banerjee A (2020b) Structure, diversity and ecological function of shrub species in an urban setup of Sarguja, Chhattisgarh, India. Environ Sci Pollut Res 27 (5):5418–5432. https://doi.org/10.1007/s11356-019-07172-w

Kim E, Elbeltagy A, Aboul-Naga A, Rischkowsky B, Sayre B, Mwacharo JM, Rothschild MF (2016) Multiple genomic signatures of selection in goats and sheep indigenous to a hot arid environment. Heredity 116:255–264. https://doi.org/10.1038/hdy.2015.94

Kim JY, Jeong S, Kim KH, Lim WJ, Lee HY, Kim N (2019) Discovery of genomic characteristics and selection signatures in Korean indigenous goats through comparison of 10 goat breeds. Front Genet 10:699. https://doi.org/10.3389/fgene.2019.00699

Kosgey IS, Baker RL, Udo HMJ, Van Arendonk JA (2006) Successes and failures of small ruminant breeding programmes in the tropics: a review. Small Ruminant Res 61(1):13–28

Kumar S, Upadhyay AD (2009) Goat Farmers' coping strategy for sustainable livelihood security in arid Rajasthan: an empirical Analysis1. Agric Econ Res Rev 22(2):281–290

Kumar S, Vaid RK, Sagar RL (2006) Contribution of goats to livelihood security of small ruminant farmers in semiarid region. Indian J Small Rumin 12(1):61–66

Kumar S, Rao CA, Kareemulla K, Venkateswarlu B (2010) Role of goats in livelihood security of rural poor in the less favoured environments. Indian J Agric Econ 65(4):1–22

Kumar S, Meena RS, Jhariya MK (2020) Resources use efficiency in agriculture. Springer, Singapore, p 760. https://doi.org/10.1007/978-981-15-6953-1

Lozano-Jaramillo M, Bastiaansen JWM, Dessie T, Komen H (2019) Use of geographic information system tools to predict animal breed suitability for different agro-ecological zones. Animal 13 (7):1536–1543

Lv FH, Agha S, Kantanen J, Colli L, Stucki S, Kijas JW, Joost S, Li MH, Ajmone Marsan P (2014) Adaptations to climate-mediated selective pressures in sheep. Mol Biol Evol 31(12):3324–3343

Madhusoodan AP, Bagath M, Sejian V, Krishnan G, Rashamol VP, Savitha ST, Awachat VB, Bhatta R (2020) Summer season induced changes in quantitative expression patterns of different heat shock response genes in Salem black goats. Trop Anim Health Prod 52:2725–2730

Mandal A, Karunakaran M, Rout PK, Roy R (2014) Conservation of threatened goat breeds in India. Anim Genet Resour 55:47–55

Marai IFM, El-Darawany AA, Fadiel A, Abdel-Hafez MAM (2007) Physiological traits as affected by heat stress in sheep—a review. Small Ruminant Res 71(1–3):1–12

McManus C, Paludo GR, Louvandini H, Gugel R, Sasaki LCB, Paiva SR (2009) Heat tolerance in Brazilian sheep: physiological and blood parameters. Trop Anim Health Prod 41(1):95–101

Mdladla K, Dzomba EF, Muchadeyi FC (2018) Landscape genomics and pathway analysis to understand genetic adaptation of south African indigenous goat populations. Heredity 120:369–378

Meena RS, Lal R (2018) Legumes for soil health and sustainable management. Springer, Singapore, p 541. https://doi.org/10.1007/978-981-13-0253-4_10

Meena RS, Kumar V, Yadav GS, Mitran T (2018) Response and interaction of *Bradyrhizobium japonicum* and Arbuscular mycorrhizal fungi in the soybean rhizosphere: A review. Plant Growth Regul 84:207–223

Meena RS, Lal R, Yadav GS (2020a) Long term impacts of topsoil depth and amendments on soil physical and hydrological properties of an Alfisol in Central Ohio, USA. Geoderma 363:1141164

Meena RS, Lal R, Yadav GS (2020b) Long-term impact of topsoil depth and amendments on carbon and nitrogen budgets in the surface layer of an Alfisol in Central Ohio. Catena 194:104752
Meena RS, Kumar S, Datta R, Lal R, Vijaykumar V, Brtnicky M, Sharma MP, Yadav GS, Jhariya MK, Jangir CK, Pathan SI, Dokulilova T, Pecina V, Marfo TD (2020c) Impact of agrochemicals on soil microbiota and management: a review. Land 9(2):34. https://doi.org/10.3390/land9020034
Meza-Herrera CA, Martinez L, Arechiga C, Bañuelos R, Rincon RM, Urrutia J, Salinas H, Mellado M (2006) Circannual identification and quantification of constitutive heat shock proteins (HSP 70) in goats. J Appl Anim Res 29:9–12
Miller BA, Lu CD (2019) Current status of global dairy goat production: an overview. Asian-Australasian J Anim Sci 32(8):1219
Mirkena T, Duguma G, Haile A, Tibbo M, Okeyo A, Wurzinger M, Sölkner J (2010) Genetics of adaptation in domestic farm animals: a review. Livest Sci 132:1–12
Monau P, Raphaka K, Zvinorova-Chimboza P, Gondwe T (2020) Sustainable utilization of indigenous goats in southern Africa. Diversity 12:20. https://doi.org/10.3390/d12010020
Morales FAR, Genís JMC, Guerrero YM (2019) Current status, challenges and the way forward for dairy goat production in Europe. Asian-Australas J Anim Sci 32(8):1256–1265
Naandam J, Assan IK (2014) Effect of coat color, ecotype, location and sex on hair density of west African dwarf (WAD) goats in northern Ghana. Sky J Agric Res 3(2):25–30
Naderi S, Rezaei HR, Pompanon F, Blum MG, Negrini R, Naghash HR, Balkız O, Mashkour M, Gaggiotti OE, Ajmone-Marsan P (2008) The goat domestication process inferred from large-scale mitochondrial DNA analysis of wild and domestic individuals. Proc Natl Acad Sci 105 (46):17659–17664
Nagayach R, Gupta UD, Prakash A (2017) Expression profiling of hsp70 gene during different seasons in goats (Capra hircus) under sub-tropical humid climatic conditions. Small Ruminant Res 147:41–47
Naqvi SMK, Sejian V (2011) Global climate change: role of livestock. Asian J Agric Sci 3:19–25
Nardone A, Ronchi B, Lacetera N, Raniere MS, Bernabucci U (2010) Effects of climate change on animal production and sustainability of livestock systems. Livest Sci 130:57–69
Nguyen TT, Bowman PJ, Haile-Mariam M, Pryce JE, Hayes BJ (2016) Genomic selection for tolerance to heat stress in Australian dairy cattle. J Dairy Sci 99(4):2849–2862
Ojango JMK, Audho J, Oyieng E, Recha J, Muigai A (2015) Sustainable small ruminant breeding program for climate-smart villages in Kenya: Baseline household survey report. CCAFS Working Paper no. 127. Copenhagen, Denmark: CGIAR Research Program on Climate Change, Agriculture and Food Security (CCAFS)
Okoruwa MI (2014) Effect of heat stress on thermoregulatory, live bodyweight and physiological responses of dwarf goats in southern Nigeria. Eur Sci J 10(27):255–264
Onzima RB, Upadhyay MR, Doekes HP, Brito LF, Bosse M, Kanis E (2018) Genome-wide characterization of selection signatures and runs of homozygosity in ugandan goat breeds. Front Genet 9:318. https://doi.org/10.3389/fgene.2018.00318
Osei-Amponsah R, Chauhan SS, Leury BJ, Cheng L, Cullen B, Clarke IJ, Dunshea FR (2019) Genetic selection for Thermotolerance in ruminants. Animals 9:948–948
Paul AK, Sultana MJ, Nasrin F (2020) Management practices, reproductive performance and constraints of goat rearing in Bangladesh. Indian J Small Ruminants 26(1):138–141
Pragna P, Chauhan SS, Sejian V, Leury BJ, Dunshea FR (2018a) Climate change and goat production: enteric methane emission and its mitigation. Animals 8(12):235
Pragna P, Sejian V, Bagath M, Krishnan G, Archana PR, Soren NM, Beena V, Bhatta R (2018b) Comparative assessment of growth performance of three different indigenous goat breeds exposed to summer heat stress. J Animal Physiol Animal Nutrition 102(4):825–836
Pragna P, Sejian V, Soren NM, Bagath M, Krishnan G, Beena V, Devi PI, Bhatta R (2018c) Summer season induced rhythmic alterations in metabolic activities to adapt to heat stress in

three indigenous (Osmanabadi, Malabari and Salem black) goat breeds. Biol Rhythm Res 49 (4):551–565

Raj A, Jhariya MK, Yadav DK, Banerjee A, Meena RS (2019a) Agroforestry: a holistic approach for agricultural sustainability. In: Jhariya MK, Banerjee A, Meena RS, Yadav DK (eds) Sustainable agriculture, forest and environmental management. Springer, Singapore, pp 101–131. https://doi.org/10.1007/978-981-13-6830-1

Raj A, Jhariya MK, Banerjee A, Yadav DK, Meena RS (2019b) Soil for sustainable environment and ecosystems management. In: Jhariya MK, Banerjee A, Meena RS, Yadav DK (eds) Sustainable agriculture, forest and environmental management. Springer, Singapore, pp 189–221. https://doi.org/10.1007/978-981-13-6830-1

Raj A, Jhariya MK, Yadav DK, Banerjee A (2020) Climate change and agroforestry systems: adaptation and mitigation strategies. CRC Press, Boca Raton, FL, p 383. https://doi.org/10.1201/9780429286759

Rashamol VP, Sejian V, Bagath M, Krishnan G, Beena V, Bhatta R (2019) Effect of heat stress on the quantitative expression patterns of different cytokine genes in Malabari goats. Int J Biometeorol 63(8):1005–1013

Raut MS, Kurpatwar LC (2020) Commercial goat farming in India: an emerging Agri-business opportunity. Stud Indian Place Names 40(38):1034–1039

Renaudeau D, Collin A, Yahav S, De Basilio V, Gourdine JL, Collier RJ (2012) Adaptation to hot climate and strategies to alleviate heat stress in livestock production. Animal 6(5):707–728

Rewe TO, Ogore PB, Kahi AK (2002) Integrated goat projects in Kenya: impact on genetic improvement. In: Proceedings of the seventh world congress on genetics applied to livestock production, Vol. 33, pp. 385–387

Rojas-Downing MM, Nejadhashemi AP, Harrigan T, Woznicki SA (2017) Climate change and livestock: impacts, adaptation, and mitigation. Climate Risk Mgt 16:145–163

Rosegrant MW, Cai X, Cline SA (2002) World water and food to 2025: dealing with scarcity. IFPRI, Washington, DC

Rout PK, Kaushik R, Ramachandran N (2016) Differential expression pattern of heat shock protein 70 gene in tissues and heat stress phenotypes in goats during peak heat stress period. Cell Stress Chaperones 21(4):645–651

Rout PK, Kaushik R, Ramachandran N, Jindal SK (2018) Identification of heat stress-susceptible and-tolerant phenotypes in goats in semiarid tropics. Anim Prod Sci 58(7):1349–1357

Savitha ST, Girish Kumar V, Amitha JP, Sejian V, Bagath M, Krishnan G, Devaraj C, Bhatta R (2019) Comparative assessment of thermo-tolerance between indigenous Osmanabadi and Salem black goat breeds based on expression patterns of different intracellular toll-like receptor genes during exposure to summer heat stress. Biol Rhythm Res 1:1–9

Sejian V (2013) Climate change: impact on production and reproduction, adaptation mechanisms and mitigation strategies in small ruminants: a review. Indian J Small Rumin 19(1):1–21

Sejian V, Bagath M, Parthipan S, Manjunathareddy BG, Selvaraju S, Archana SS, Soren NM, Rao SBN, Giridhar K, Ravindra JP, Bhatta R (2015) Effect of different diet level on the physiological adaptability, biochemical and endocrine responses and relative hepatic HSP 70 and HSP 90 genes expression during summer season in Osmanabadi kids. J Agric Sci Tech A 5:755–769

Sejian V, Bagath M, Krishnan G, Rashamol VP, Pragna P, Devaraj C, Bhatta R (2019) Genes for resilience to heat stress in small ruminants: a review. Small Ruminant Res 173:42–53

Shilja S, Sejian V, Bagath M, Mech A, David CG, Kurien EK, Varma G, Bhatta R (2016) Adaptive capability as indicated by behavioral and physiological responses, plasma HSP70 level, and PBMC HSP70 mRNA expression in Osmanabadi goats subjected to combined (heat and nutritional) stressors. Int J Biometeorol 60(9):1311–1323

Shilja S, Sejian V, Bagath M, Manjunathareddy GB, Kurien EK, Varma G, Bhatta R (2017) Summer season related heat and nutritional stresses on the adaptive capability of goats based on blood biochemical response and hepatic HSP70 gene expression. Biol Rhythm Res 48 (1):65–83

Shivakumar C, Reddy BS, Patil SS (2020) Socio-economic characteristics and composition of sheep and goat farming under extensive system of rearing. Agric Sci Digest 40:1
Shumbusho F, Raoul J, Astruc JM, Palhiere I, Elsen JM (2013) Potential benefits of genomic selection on genetic gain of small ruminant breeding programs. J Animal Sci 91(8):3644–3657
Silanikove N (2000) The physiological basis of adaptation in goats to harsh environments. Small Ruminant Res 35(3):181–193
Sinawang PD, Rai V, Ionescu RE, Marks RS (2016) Electrochemical lateral flow immunosensor for detection and quantification of dengue NS1 protein. Biosens Bioelectron 77:400–408
Singh MK, Dixit AK, Roy AK, Singh SK (2013) Goat rearing: a pathway for sustainable livelihood security in Bundelkhand region. Agric Econ Res Rev 26(conf):79–88
Smith C (1984) Rates of genetic change in farm livestock. Res Dev Agric 1:79–85
Sophia I, Sejian V, Bagath M, Bhatta R (2016) Quantitative expression of hepatic toll-like receptors 1–10 mRNA in Osmanabadi goats during different climatic stresses. Small Ruminant Res 141:11–16
Sophia I, Sejian V, Bagath M, Bhatta R (2017) Influence of different environmental stresses on various spleen toll like receptor genes expression in Osmanabadi goats. Sciences 10(1):9–16
Stern E, Vacic A, Li C, Ishikawa FN, Zhou C, Reed MA, Fahmy TM (2010) A nanoelectronic enzyme-linked immunosorbent assay for detection of proteins in physiological solutions. Small 6(2):232–238
Thornton PK (2010) Livestock production: recent trends, future prospects. Philos Trans Royal Soc London Series B: Biol Sci 365(1554):2853–2867
Thornton PK, Herrero MT, Freeman HA, Okeyo Mwai A, Rege JEO, Jones PG, McDermott JJ (2007) Vulnerability, climate change and livestock-opportunities and challenges for the poor. J Semi-Arid Trop Agric Res 4(1):1–23
Tresset A, Vigne JD (2011) Last hunter-gatherers and first farmers of Europe. C R Biol 334 (3):182–189
Turturici G, Sconzo G, Geraci F (2011) Hsp70 and its molecular role in nervous system diseases. Biochem Res Int 2011:618127
Vandana GD, Bagath M, Sejian V, Krishnan G, Beena V, Bhatta R (2019) Summer season induced heat stress impact on the expression patterns of different toll-like receptor genes in Malabari goats. Biol Rhythm Res 50(3):466–482
Wang X, Liu J, Zhou G, Guo J, Yan H, Niu Y (2016) Whole-genome sequencing of eight goat populations for the detection of selection signatures underlying production and adaptive traits. Sci Rep 6:38932. https://doi.org/10.1038/srep38932
Zeder MA (2005) A view from the Zagros: new perspectives on livestock domestication in the Fertile Crescent. In: Vigne JD, Peters J, Helmer D (eds) The first steps of animal domestication: new archaeozoological approaches. Oxbow Books, Oxford, FL, pp 125–146
Zidi A, Abo-Shady H, Molina A, Menéndez-Buxadera A, Sánchez-Rodríguez M, Díaz C, Carabaño MJ, Serradilla JM (2014) Genome wide association for heat stress tolerance/susceptibility in Florida dairy goats. In 10th World Congress on Genetics Applied to Livestock Production, pp 17–22

The manufacturer's authorised representative in the EU is Springer Nature Customer Service Centre GmbH, Europaplatz 3, 69115 Heidelberg, Germany. If you have any concerns regarding our products, please contact ProductSafety@springernature.com

Printed and bound by CPI Group (UK) Ltd, Croydon, CR0 4YY
30/09/2026
02225297-0020